AF879586

The Study of
Plant Disease Epidemics

Laurence V. Madden
Ohio State University, Wooster
U.S.A.

Gareth Hughes
University of Edinburgh
Scotland

Frank van den Bosch
Rothamsted Research
Harpenden, England

APS PRESS
The American Phytopathological Society
St. Paul, Minnesota U.S.A.

Cover photograph: PHOTODISC®, Volume 19, Agriculture CD-Rom

This book has been formatted directly from computer-generated copy submitted in final form to APS PRESS by the authors. No editing or proofreading has been done by the Press.

Reference in this publication to a trademark, proprietary product, or company name is intended for explicit description only and does not imply approval or recommendation to the exclusion of others that may be suitable.

Library of Congress Control Number: 2006940915
ISBN: 978-0-89054-354-2

©2007 by The American Phytopathological Society
Second printing 2008
Third printing 2011

All rights reserved.
No portion of this book may be reproduced in any form, including photocopy, microfilm, information storage and retrieval system, computer database, or software, or by any means, including electronic or mechanical, without written permission from the publisher.

Printed in the United States of America on acid-free paper

The American Phytopathological Society
3340 Pilot Knob Road
St. Paul, Minnesota 55121, U.S.A.

To the Memory of Lee Campbell, whose contributions to epidemiology were interrupted far too soon

Preface

In every phenomenon the beginning remains always the most notable moment.

Thomas Carlyle

Jürgen Kranz (1990b) wrote that "Epidemiology provides plant pathologists with refined and deepened knowledge of the behavior of diseases in the field for improved disease control or management." This book is about refining and deepening our understanding of diseases in the field through the study of the temporal and spatial dynamics of plant diseases in populations, that is, through the study of epidemics. Over the last 40+ years there have been considerable developments in our approaches to the study of plant disease epidemics, with a noteworthy increase in complexity in the analytical methodology applied. The casual reader of the epidemiology literature may feel overwhelmed by the seemingly endless collection of statistical and mathematical techniques utilized over the years to describe, compare, understand, or predict epidemics. We would argue strongly that statistical and mathematical approaches *are* essential both to study epidemics and to interpret epidemiological data for the purpose of developing and evaluating control strategies and tactics. However, without exposure to an organized exposition of the subject, plant pathologists may have great difficulty in evaluating the epidemiological literature, understanding why various methods are being applied, and interpreting the significance of published results. This book is meant to provide the needed exposition of the subject.

Epidemiology—the study of epidemics—encompasses portions of many disciplines, including, as noted above, mathematics and statistics, as well as crop protection and production, ecology, genetics, aerobiology, meteorology, systems analysis, and other fields. No book could come close to covering all of these subjects in any depth, including this one. What we have tried to do is introduce readers to the essential principles and concepts of plant disease epidemiology and show how to utilize and interpret epidemiological data. We consider models to be essential tools for describing and understanding epidemics, so we routinely utilize mathematical and statistical models throughout the text to characterize epidemics.

In most of the chapters, we have used data collected by others, or by us, to demonstrate concepts or to show how to perform necessary calculations. In some cases, we simply "read off" points in published graphs, or copied values from tables in papers, but in many cases we obtained data directly from the scientists who performed the original studies. We are extremely grateful to our many colleagues who have been gracious enough to share their data with us, including: E. Adipala, L. Amorim, A. Bergamin Filho, A. Culbreath, F. Ferrandino, T. Gottwald, Q. Guo, M. Jeger, R. Jiménez-Díaz, M. Kelly, R. Kolnaar, B. Landa del Castillo, P. Lucas, L. Maffia, C. Mundt, J. Navas-Cortés, M. Nita, H. Ngugi, F. Nutter, Jr., J. Pataky, K. Sackett, H. Scherm, A. Schoeny, W. Sserubombwe, E. Twengstrom, X. Xu, and J. Yuen. We also thank M. Shaw and T. Gottwald for allowing us to use software they wrote.

We are indebted to our colleagues at our home institutions and throughout the world who have given us useful comments on portions of the manuscript and positive feedback and encouragement on the preparation of this book. L.V.M. would also like to thank Lee Wilson for his invaluable assistance in preparation of the manuscript and Mizuho Nita for assistance with several figures. L.V.M. also thanks his wife, Susan, and daughter, Zoe, for their continuous support throughout the time this book was being written. G.H. thanks his parents, William and Norma Hughes, and children, Lilian and Warren Hughes, for all their support and understanding. F.vdB. thanks Weeraphan for her support, enthusiasm, and the change in perspective. F.vdB. also thanks J.C. Zadoks, J.A.J. Metz and O. Diekmann for introducing him to the exciting worlds of plant disease epidemiology and theoretical biology.

Laurence V. Madden
Gareth Hughes
Frank van den Bosch
2006

Table of Contents

The Study of
Plant Disease Epidemics

1

Introduction

Chemical industry and plant breeders have forged fine tactical weapons; but only epidemiology sets the strategy.

J. E. Vanderplank

1.1 Plant Disease Epidemics

Plants are susceptible to many pathogens, such as fungi, oomycetes, bacteria, viruses, viroids, and nematodes (Agrios, 2005). The book by Farr et al. (1989) lists over 13,000 fungal and oomycete pathogens of plants in the United States alone, with more than 75,000 listed diseases (because some pathogens can infect many species of plants). A given crop or forest tree species may be affected by over 200 distinct pathogens world-wide; however, in a given region, there are generally 5–20 serious diseases to control on an annual basis (Madden and Wheelis, 2003).

The consequences of plant diseases can be high, especially in terms of reductions in yield and costs of control (James et al., 1991; Teng and Ohima, 1983). Tremendous advances have been made over the last 100 years in the management of plant diseases, especially in the development of disease-resistant cultivars, development of many chemical and biological-control compounds, as well as the use of many cultural methods (Fry, 1982; Strange, 1993). Nevertheless, many plant diseases remain difficult to control using affordable methods, both in the developed and developing world.

Although infection of a single high-value plant may sometimes be of importance, such as a mature elm tree on a city street infected by *Ophiostoma ulmi* (the cause of Dutch elm disease), infection of an individual plant is generally not of any significance. Rather, plant diseases usually matter when many plants or plant parts (such as fruits or leaves) in a field, forest, or region are infected. Unlike the situation with several highly contagious animal diseases, very few plant diseases have been eradicated (Madden and Wheelis, 2003; NRC, 2003); thus, some inoculum of the pathogen can usually be found in a given area for the important diseases of a given crop. Therefore, the typical goal of disease management is to maintain the intensity of disease to a *low* (but not necessarily zero) level, or as stated by Merrill (1977), "preventing an intolerable build-up of disease within a population." Just what constitutes intolerable will depend on the value of the crop and the impact of the disease.

The issue can also be addressed in terms of the pathogen population and not just the population of diseased plants or plant parts (such as leaves). Vanderplank wrote in the first paragraph of the first chapter of his 1963 book: "In order to control rust in wheat fields (caused by *Puccinia graminis* f.sp. *tritici*) one must stop, or at least retard, the growth of millions or billions of rust pustules on thousands or millions of wheat plants. The fate of a single wheat plant and the growth of a single pustule are relatively trivial details in a large picture." It can be reasonably argued that one must study plant diseases in populations in order to understand diseases at this scale, and also to predict their outbreaks, quantify their impact on yield, and, ultimately, to better control them (Jones, 1998b). This book deals with quantifying diseases in populations.

1.2 Some Concepts

1.2.1 Epidemics

Epidemiology can be considered, quite simply, the study of epidemics. There are several possible definitions of an epidemic, and the one we prefer is

> *EPIDEMIC: Change in disease intensity in a host population over time and space.*

We generally do not overly emphasize the nuances of definitions in this book, but it is helpful here for instructional purposes to consider the components of this definition of an epidemic. First of all, the use of "change" emphasizes that epidemics are *dynamic* processes, and many aspects of their study involve characterizing *rates* of change. The use of "disease" emphasizes that epidemics specifically involve diseases, and not just pathogens or hosts. "Intensity" is a general term that characterizes the magnitude of disease or the infection, and will be defined and explained in the next chapter. The use of "host population" emphasizes that an epidemic is primarily a *population phenomenon*, involving in this case, a population of plants susceptible to one or more species of pathogen. Finally, "time" and "space" are indications of the two physical dimensions over which the dynamic process of an epidemic occurs. Specifically, disease intensity varies from time to time and from location to location during epidemics.

There are variations of this definition. Kranz (1990b), for instance, succinctly states that "An epidemic is the progress of disease in time and space." Often, "space" is omitted from the definition. As will be clear from Chapters 7 to 9, diseases do not exist at the same intensity everywhere at any given time. Thus, the spatial dimension would then simply be implicit in any definition when space is not mentioned. Commonly, "increase" is used instead of "change". An epidemic could then be considered an increase in disease intensity in a host population over time (with an implied spatial component). This can be, in fact, a very useful way of looking at an epidemic, because it is often of interest (both theoretically and practically) to know what factors (or combination of factors) lead to a rapid increase in disease intensity and what factors (or combination of factors) prevent an increase in disease intensity or slow the rate of increase to a tolerable level. This approach underlies much of Chapters 4 and 5.

However, we prefer to maintain the term of "change" in the general definition instead of "increase" to better reflect the full range of situations that can be studied. For instance, consider a foliar disease of a perennial plant. Although the intensity of disease may, in general, increase during the growing season, there will eventually be a decrease as defoliation occurs at the end of the season. Moreover, there may be a decline in the proportion of plant tissue affected by disease if the host is producing new foliage faster than infections are occurring. In addition, for the spatial component of epidemic, the concept of monotonic increase (or decrease) in disease is not always easy to apply, as will be seen in Chapters 7 and 9 and elsewhere.

Sometimes our concept of an epidemic becomes more restrictive in this book, depending on the specific topic being addressed. For instance, in Chapter 5 we formally address the conditions necessary for disease to *increase* in time for diseases that spread from plant to plant, leaf to leaf, root to root, and so on.

It should be emphasized that the everyday concept of an epidemic, as well as the technical definition used by some non-epidemiologists, is different from the concept and definition espoused here. For instance, the Merriam-Webster definitions of epidemic include: "affecting or tending to affect a disproportionately large number of individuals within a population," or "excessively prevalent." A Dictionary of Plant Pathology (Holliday, 1989) defines an epidemic as "a widespread increase, usually limited in time, in the incidence of an infectious disease." In contrast, the operational definition of epidemic of this book makes no mention of large numbers, excessive prevalence, or widespread incidence. The reason can be seen with an example.

Consider late blight of potato, caused by the oomycete *Phytophthora infestans*, which may be the most infectious disease known of any crop based on rates of increase (Vanderplank, 1963). Assume there is a potato field, of a susceptible cultivar, with 4 million (4×10^6) plants, and that the epidemic starts with one lesion on a single plant. The density of lesions per field would be $1/(4 \times 10^6)$. The source of the lesion could be spores (sporangia) produced on an infected tuber in or near the field of interest. According to van der Zaag (1956), one late blight lesion covers 0.1% of the foliage, on average (i.e., 0.001 = 1/1000 of a plant with visible disease symptoms). The relation between lesion number and foliage coverage would likely depend on the plant cultivar and pathogen strain (i.e., some cultivars will have higher leaf area than others, and some strains may result in larger lesions than others), but we can accept this equivalence for the exercise. The proportion of foliage in the field covered by lesions is our measure here of disease intensity, which we denote as y. In the next chapter, we will provide a detailed discussion of different types of disease intensity measurements.

We refer to time (in days here) as t, and indicate the start of the epidemic as $t = 0$. With one lesion per field, intensity for the field is $y = 0.001/(4 \times 10^6) = 2.5 \times 10^{-10}$. If the number of lesions doubles every 2 days, there will be an average of one lesion per plant (or 4×10^6 total lesions per field) after about 44 days. Although doubling of disease intensity this often may seem exaggerated, it is a typical value for late blight epidemics with susceptible cultivars grown under favorable environmental conditions (Fry, 1978). So, $y = 0.001$ at $t = 44$ days. If disease intensity continues to increase in the same manner, except for the slowing effect of loss of disease-free foliage (new spores are more and more likely to land on lesions rather than on disease-free foliage), $y \approx 0.5$ at $t = 64$ (i.e., half the foliage with lesions), $y \approx 0.90$ at $t = 70$, $y \approx 0.99$ at $t = 77$, and $y \approx 0.999$ at $t = 83$ days.

A plot of y versus t, known as a *disease progress curve* or epidemic growth curve, exhibits an S-shaped pattern in this example (Fig. 1.1), typical for diseases that spread from plant to plant or leaf to leaf (see Chapter 4). Note that at one lesion per plant ($y = 0.001$), the curve is barely above the zero axis. Interestingly, one lesion per plant may be about the level of intensity at which a person (e.g., grower, field scout) would first detect the disease in the field under many circumstances. That is, with an average of one lesion per plant, some plants will have multiple lesions, and many plants will have no lesions. So, unless a person is observing many plants, there is a good chance that lesions will not be found until disease intensity reaches 0.001 or so. (See Chapter 9 for a presentation on patterns of disease and Chapter 10 for a presentation on sampling for disease.)

There is a 1000-fold increase in disease intensity from day 44 to day 84. With the common layperson definition of an epidemic, 100% coverage of foliage with lesions would certainly be considered an epidemic, but 0.1% (i.e., $y = 0.001$) or less foliar coverage would probably not be considered an epidemic, given that a casual

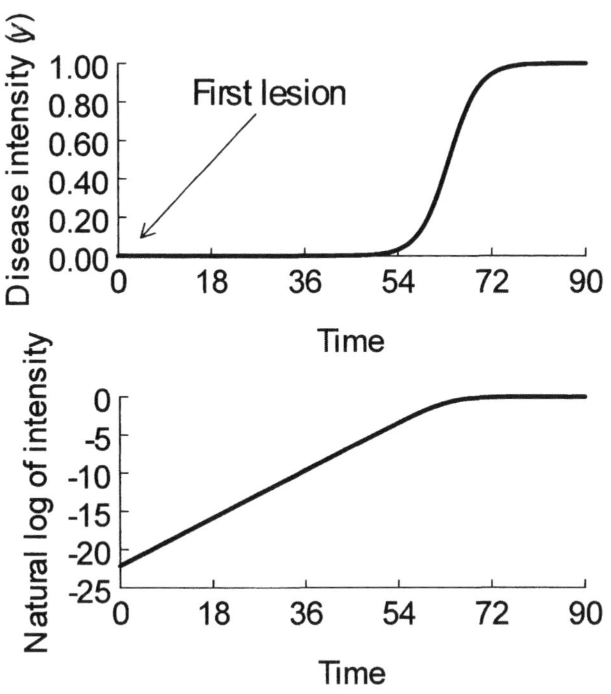

FIG. 1.1. Hypothetical epidemic growth curve (i.e., a disease progress curve) for potato late blight that started with a single lesion in a field of 4 million plants. Upper frame: proportion of foliage with lesions (a measure of disease intensity). Lower frame: natural logarithm (log) of the proportion with lesions.

observer of the field may not even notice any disease at all in the latter case. However, between days 0 and 44, there was a 4-million-fold increase in disease intensity (from one lesion to 4 million lesions per field, or from $y = 2.5 \times 10^{-10}$ to $y = 0.001$). Clearly, there was a much greater *increase* in disease during the first 40 or so days than in the subsequent 40 days. From a population biology perspective, there is no difference in the increase in disease during the first part of the time span than in the later part, and as will be shown in Chapter 4, a single model with a constant rate parameter for disease increase can represent the entire epidemic process from the formation of the first lesion until all foliage is diseased. In fact, the curve in Fig. 1.1 was drawn based on a model known to accurately represent disease increase for potato late blight (Fry, 1978).

There is no practical way to draw a line separating epidemic from non-epidemic phases in this example; rather, there is a continuum of disease increase over the given time period. In fact, when one plots the log of y versus t, a straight line is produced over most of the time period (Fig. 1.1, lower frame). Only late in the epidemic does the log line flatten, because of the limitation to disease increase due to depletion of disease-free foliage available for infection. Other transformations can produce a straight line over the entire time period for many epidemics, a subject addressed later. For practical necessity, plant pathologists mostly study epidemics where y increases from about 0.001 to 1.0, but this increase is really just an extension of the increase in disease that is often not observed because of sampling limitations.

Thus, the magnitude of disease intensity is not an indication of whether an epidemic has occurred or not. Depending on the initial level of disease and the rate of increase, the disease intensity can be either low or high at a given time during an epidemic. Of course, this does not mean that all levels of disease intensity are equally important, either in affecting crop yield or in the need to impose controls. Chapter 11 covers some topics related to predicting whether or not a control is needed based on sampling for disease intensity, and Chapter 12 covers the relationship between crop yield and disease epidemics.

There is sometimes a need to refer specifically to situations with widespread occurrence of high disease intensity. The term *pandemic* can be used to label epidemics that occur over large areas with high levels of disease. The late blight epidemics in Western Europe in 1845 and 1846 and the southern corn leaf blight epidemics in the United States in 1970 are example pandemics that were also of high consequence (Horsfall and Cowling, 1978a).

1.2.2 Epidemiology

As stated above, *epidemiology is the study of epidemics.* For those who prefer a more detailed definition, Diekmann and Heesterbeek (2000) define epidemiology as "the study of the spread of diseases, in space and time, with the objective to trace factors that are responsible for, or contribute to, their occurrence." More specifically to plant diseases, Kranz (1990b) refers to epidemiology as "the science of populations of pathogens in populations of host plants, and the diseases resulting therefrom under the influence of the environment and human interferences." Because of the emphasis on populations, it is sometimes useful to think of epidemiology as the ecology of disease.

Clearly, this science involves the classic disease triangle of host, pathogen, and environment (Stevens, 1960), often taught in beginning plant pathology courses, but at the population and not the individual level. For pedagogical purposes, the triangle is sometimes expanded to a square, pyramid, or other multi-dimensional structure to either emphasize the effects of time and/or space on disease, or to reflect the influence of humans on diseases (Francl, 2001). To be complete, the environment can be partitioned, when relevant, into abiotic (e.g., weather, soil moisture) and biotic (e.g., insect vectors) components.

Populations comprise individuals, so the study of epidemics often involves characterizing or summarizing the disease status of individuals in order to understand the dynamics of disease in populations. For example, the sporulation of fungal pathogens on individual plants is one of the determinants of the rate of disease intensity increase in a population.

1.2.3 Epidemic versus epiphytotic

Some authors, such as Unger (1833), Whetzel (1929), and Ryan and Birch (1978), have argued in favor of the word "epiphytotic" instead of epidemic to describe plant diseases in populations. The argument centers on the derivation of the word epidemic from the ancient Greek adjective, which is most often translated as *what is among the people* or *upon the people*. However, as pointed out by Millar (1978), the Greek adjective can also be correctly translated as *upon the population*, with the population consisting of humans, other animals, or plants.

The term epidemic apparently was first used by Hippocrates (ca. 400 BC), and the word has received increasing use over time. As pointed out by Zadoks and Koster (1976), there is also early precedence for the use of epidemic over epiphytotic to describe plant diseases in populations. Duhamel de Monceau (1728), von Martius (1842), Kühn (1858), and Ward (1901), among others, used the term epidemic in their extensive writings on plant diseases in populations. Perhaps an even stronger argument for the use of epidemic is based on population biology and dynamics. In particular, many of the population-dynamic processes are the same for diseases of humans, other animals and plants, and some of the useful analytical approaches for characterizing and modeling plant disease epidemics originated in medical epidemiology (as will become evident in later chapters) or were first motivated by medical problems.

Thus, although it is not wrong to use epiphytotic for diseases of plants, there is no reason to do so. Incidentally, for those who prefer to use epiphytotic, the unusual term *epiphytotiology* should be used to describe the science of plant diseases in populations. This term has had virtually no use in plant pathology. When it is necessary to refer specifically to the study of plant disease epidemics, *botanical epidemiology* is a very useful label. We use botanical epidemiology, plant disease epidemiology, and plant epidemiology as synonyms in this book.

1.3 Some Historical Developments

1.3.1 Up to 1963

There are several excellent articles on the history of plant disease epidemiology. Zadoks and Koster (1976) present an especially thorough review of the early history of the study of plant diseases in populations, with some interesting translations of passages of some early texts. Cowling and Horsfall (1978), Zadoks (2001), and Waggoner and Aylor (2000) all provide valuable information. We present here only a very limited history of some developments in the field.

Hippocrates (Jones, 1972) wrote 2,400 years ago about epidemics of humans, such as mumps and malaria, and indicated the importance of the environment on disease outbreaks. Theophrastus (372–287 BC) and Pliny (23–79 AD) commented on plant diseases and also emphasized the role of the environment in epidemic occurrences. In a publication far ahead of its time, but ignored for centuries, Duhamel de Monceau (1728) wrote about epidemics of saffron crocus, caused by the soil-borne fungus *Rhizoctonia violacea*, and provided control methods based on his knowledge of the epidemics. Moreover, he even gave evidence that the disease was caused by a "biological entity." Zadoks and Koster (1976) believe that this is the first documentation of a plant epidemiological study.

The role of the environment (especially weather) in disease outbreaks in crops, especially for rust of wheat, was addressed by various authors in the 18th and 19th centuries, including Fontana and Tozzetti (Zadoks and Koster, 1976). In possibly the first textbook of plant pathology, Kühn (1858) wrote clearly about the importance of epidemics of disease in crops, and compared them to epidemics of disease in human and other animal populations. At this stage, epidemics were still mostly considered to be the widespread occurrence of high disease intensity. A more ecological approach to understanding epidemics was taken by Ward in his influential 1901 textbook *Diseases in Plants*. He states "When we come to enquire into what circumstances bring about those severe and apparently sudden attacks on our crops, orchards, gardens, and forests ... we soon discover the existence of a series of complex problems of intertwined relationships between one organism and another, and between both and the non-living environment which fully justify the caution already given against concluding that any cause of disease can be a single agent working alone."

Historically, the emphasis generally was on describing or understanding the epidemics with high disease intensity (e.g., the "severe and apparently sudden attacks" in Ward, 1901). A broader view began to emerge by the middle of the 20th century, where the diversity of possible epidemics (e.g., those with high and low disease intensity) was appreciated. The classic treatise by Gäumann, *Pflanzliche Infektionslehre* (Principles of Plant Infection) in 1946, with an English translation 1950, is especially relevant in this regard. Gäumann states "... an epidemic is delimited in two ways; it has a beginning and an end, both in time and space", and explicitly writes about the specific course followed by each epidemic. Moreover, he developed a clear listing of the conditions necessary for an epidemic, such as the presence of a susceptible host, high reproductive capacity and efficient dispersal ability of the pathogen, and optimal weather conditions, and presented the initial ideas to compare epidemics, at least in a qualitative sense.

The period from about 1945 to 1963 was of fundamental importance in the transition of botanical epidemiology from a mostly qualitative to a mostly quantitative discipline. Large (1952) showed the value of disease progress curves, such as the one in Fig. 1.1, for understanding and comparing epidemics, as well as for determining yield losses due to disease. P. H. Gregory performed elegant research on the dispersal of pathogen spores, and utilized detailed physical models to describe the results. His book on *The Microbiology of the Atmosphere* developed an integrated theory of spore movement and had a great influence on many plant pathologists who were concerned with the spread of disease (Gregory, 1961). The second edition of his book (Gregory, 1973) should still be required reading for all those interested in studying microbes in the atmosphere. Focusing just on books (for now), an important event was the publication of the last volume of a three-volume treatise on plant pathology. The treatise editors, J. G. Horsfall and A. E. Dimond, took the bold and visionary step of dedicating the entire third volume to plant diseases in populations (Horsfall and Dimond, 1960). Finally, the modern concepts of epidemics had started to take hold, and the quantitative aspects of the discipline were on display. Populations were emphasized in chapters that dealt with, among other things, disease forecasting or prediction, physics of spore dispersal, inoculum density, and mathematical modeling. Of special note is the chapter entitled "Analysis of Epidemics", by a little-known potato breeder from South Africa, J. E. Vanderplank (who used to spell his last name as van der Plank). Vanderplank (1960) had a radical idea (at least for most plant pathologists at the time) that epidemics could actually be modeled, and that analysis based on models was essential for understanding the population process of disease progress and for determining control strategies.

Writing that chapter for the 1960 book inspired Vanderplank to write several other books on epidemiology, host resistance, and plant-pathogen interactions (see discussion in Cowling and Horsfall, 1978). Of great significance was his first book, *Plant Diseases: Epidemics and Control*, published in 1963. The book expanded on the ideas in the 1960 article and developed a theory of plant disease epidemics as dynamic population processes that was firmly rooted on empirical data found in the literature (Vanderplank, 1963). In particular, he developed a thorough rationale for the characterization of epidemics using rates of change that could be expressed in the form of nonlinear mathematical models (see Chapter 3 and later chapters of this book). Moreover, the linkage between epidemiology and control was firmly established, with a great deal of emphasis on disease resistance, and chemical control, as well as inoculum reduction. Here, the concepts of vertical and horizontal resistance were developed, which were later expanded upon in his 1968 book (Vanderplank, 1968), updated in 1984 (Vanderplank, 1984), and argued about for decades.

In many ways, Vanderplank's 1963 book was the birth of modern botanical epidemiology. Rarely has a single publication had such a large effect on a discipline. Within a few years, epidemiology was considered to be as important as mycology, virology, and so on, in plant pathology departments (Campbell and Madden, 1990; Cowling and Horsfall, 1978; Zadoks, 1974). Of course, the mathematical treatment was too much for many pathologists to comprehend at the time, but the basic concepts underlying the mathematics were accepted by many. By the 1970s, graduate education included more statistics, mathematics, and computer programming than in the past, and many quantitative epidemiologists entered the profession and began to make important contributions (Campbell, 1998). As Cowling and Horsfall wrote in 1978, Vanderplank is "the man who 'pulled the bung on the epidemiological barrel'. At present new knowledge is spilling out rapidly in ever-widening pools of understanding. With his stimulation, quantitative thinking has spread like an epidemic across the field of plant pathology and it appears that the field will never again be the same."

Vanderplank was certainly not the first to use mathematics or statistics in plant pathology, nor the first to quantify disease progress curves. But it was his elegant exposition of the topic, and innovative integration of biology and mathematics, that convinced many others to change directions. Before him, several researchers calculated disease progress curves (see Campbell, 1998, for brief review), and some, such as Fracker (1936) and Large (1952), used some statistical methods to quantify rates of increase in time. For the spatial component of epidemics, various authors, especially Gregory and Read (1949) and Gregory (1961), characterized gradients of disease in relation to inoculum sources. Vanderplank (1960) considered some spatial aspects of epidemics in 1960, and even made an early contribution to the quantification of spatial patterns of disease in 1946 (Vanderplank, 1946). A significant major early study on spatial patterns of disease was by Bald (1937), which was the basis for some detailed statistical modeling by Cochran (1936).

There had been many important developments in the study of human disease epidemics, including some very elaborate and complicated modeling by Kermack and McKendrick (1927) which still serves as the foundation for theoretical research in quantitative epidemiology (see Chapter 5). Even earlier, the physician Ross (1911) developed a model for malaria epidemics that incorporated dynamics of the vector as well as the infected human population in the form of differential equations. Ross was also quite prophetic in stating:

> *As a matter of fact all epidemiology, concerned as it is with variation of disease from time to time or from place to place, must be considered mathematically, however many variables are implicated, if it is to be considered scientifically at all. ... And the mathematical method of treatment is really nothing but the application of careful reasoning to the problems at hand.*

Outside of epidemiology, there is also a long history of important quantitative work on the population growth of organisms, especially humans and other animals (see Renshaw, 1991). Good examples include the development of the logistic model for representing the increase in numbers of a population (Pearl and Reed, 1920; Verhulst, 1838).

Although Vanderplank was well aware of the early use of the logistic model in ecology, there is no apparent evidence that he was directly influenced by the advances in medical epidemiology, such as the work of Ross and of Kermack and McKendrick in the first part of the 20th century. However, later work showed the strong linkage between the early medical models and those utilized and advocated by Vanderplank and others in botanical epidemiology (see, for example, Jeger, 1986b; Jeger and van den Bosch, 1994a; Segarra et al., 2001).

1.3.2 After 1963

As mentioned above, studying the dynamics of disease in populations, and using the obtained information in evaluating disease controls, certainly became fashionable in the years after Vanderplank's article in 1960 and book in 1963. Of special importance are the long-term contributions by three individuals, Jan C. Zadoks of Wageningen University in the Netherlands; Jürgen Kranz of Giessen University, Germany, and Paul E. Waggoner of the Connecticut Agricultural Experiment Station, USA, as well as the early contributions of the Department of Plant Pathology at the Pennsylvania State University in the United States. The three listed individuals and their colleagues made pioneering contributions over more than three decades to the quantitative aspects of botanical epidemiology. Through their roles as university professors, both Zadoks and Kranz trained numerous students in botanical epidemiology, and directed research projects in many developing countries on the temporal and spatial dynamics of disease and prediction of disease outbreaks. Both scientists produced important books on the subject (Kranz, 1974a, 1990a, 2003; Zadoks and Schein, 1979) that influenced many aspiring epidemiologists over the years.

The Department of Plant Pathology at Penn State was established in 1963, right around the publication of Vanderplank's book. Many of the faculty members of this fledgling department fully recognized the value of epidemiology in the study and control of plant diseases, and built a new curriculum that focused heavily on this population-based discipline. Many of the faculty members and their graduate students made major contributions to quantitative epidemiology in the 1960s and 1970s, and some of the graduates from this department went on to train many of the next generation of epidemiologists in the United States (and elsewhere). Because Waggoner was based at a research station, he was less involved in graduate education, but his research contributions were also highly influential, especially the development of models of epidemics (see, for instance, reviews in Waggoner, 1974, 1981).

With the ubiquitous nature of computers today, readers may be surprised to know that the modeling and analysis performed by Vanderplank (1963), and other early researchers, was done without the use of computers. There was, however, a perceived limit in the number of details of the host–pathogen–environment system that could be incorporated into a model that was being evaluated with just "pen and paper." Waggoner and colleagues made a breakthrough by developing so called computer simulation models that represented the components of disease cycles (e.g., infection, sporulation) in relation to the weather and simulated the increase in disease intensity in populations (Waggoner and Horsfall, 1969; Waggoner et al., 1972). This led to a major activity in developing computer simulation models for many diseases, utilizing the principles of systems analysis (Hau, 1988; Teng, 1981).

The computer-simulation-model results showed that plant disease epidemics can be treated in a quantitative manner, and that collections of equations can be utilized to describe disease development, and to understand and compare epidemics. The systems-analysis emphasis of botanical epidemiology, however, generally declined towards the end of the 20th century as researchers realized that elaborate computer simulation was not necessarily needed to answer many fundamental questions about epidemics. In fact, as summarized by Jeger (1986a) and Jeger and van den Bosch (1994b), a few properly constructed differential equations may be more useful to answer general questions about thresholds for epidemics and final level of disease in epidemics. Waggoner (1990) shows several ways of modeling epidemics with and without the use of computer simulation. Of course, computers remain indispensable tools for quickly determining solutions to model equations over the time and space dimensions, for graphing results, and for statistically fitting models to data and comparing the fit of models for multiple epidemics. We do not take a formal systems-analysis approach to the quantification of epidemics in this book, but, as will be discussed later (e.g., Chapters 5 and 8), there are situations where simulation modeling is clearly of benefit in epidemiology.

1.3.3 Some conferences and books, starting in 1963

A scientific discipline can be judged, in part, by the professional conferences and publications dedicated to the subject. For botanical epidemiology, 1963 was a key year because of the publication of *Plant Diseases: Epidemics and Control*, and *also* because it was the year of the first international meeting dedicated to the topic. The meeting, which was sponsored as a NATO Advanced Study Institute, was held in Pau, France, and was organized by R. D. Schein, J. M. Hirst, with assistance by A. J. P. Oort and J. C. Zadoks. All sessions were informal, and there were no proceedings (Zadoks, 2001). Forty participants from 14 countries contributed. The next dedicated meeting was held

TABLE 1.1. Some text and reference books in botanical epidemiology published since 1970.

Year	*Title*	*Authors (a) or Editors (e)*
1974	*Epidemics of Plant Diseases: Mathematical Analysis and Modeling*	J. Kranz (e)
1978	*Plant Disease: An Advanced Treatise, Vol. 1: How Disease Develops in Populations*	J. G. Horsfall and E. B. Cowling (e)
1978	*Plant Disease Epidemiology*	P. R. Scott and A. Bainbridge (e)
1979	*Epidemiology and Plant Disease Management*	J. C. Zadoks and R. D. Schein (a)
1980	*Comparative Epidemiology: A Tool for Better Disease Management*	J. Palti and J. Kranz (e)
1983	*Plant Virus Epidemiology: The Spread of Insect-borne Viruses*	R. T. Plumb and J. M. Thresh (e)
1985	*Advances in Plant Pathology, Vol. 3: Mathematical Modelling of Crop Disease*	C. A. Gilligan (e)
1986	*Plant Virus Epidemiology: Monitoring, Modelling, and Predicting Outbreaks*	G. D. McLean, R. G. Garrett, and W. G. Ruesink (e)
1986	*Plant Disease Epidemiology, Vol. 1: Population Dynamics and Management*	K. J. Leonard and W. E. Fry (e)
1986	[*Handbook of Plant Disease Epidemiology*], in Chinese	Zeng Shimai (a)
1987	*Crop Loss Assessment and Pest Management*	P. S. Teng (e)
1988	*Experimental Techniques in Plant Disease Epidemiology*	J. Kranz and J. Rotem (e)
1989	*Plant Disease Epidemiology, Vol. 2: Genetics, Resistance, and Management*	K. J. Leonard and W. E. Fry (e)
1989	*The Spatial Components of Plant Disease Epidemics*	M. J. Jeger (e)
1990	*Epidemics of Plant Diseases: Mathematical Analysis and Modeling, 2nd edition.*	J. Kranz (e)
1990	*Introduction to Plant Disease Epidemiology*	C. L. Campbell and L. V. Madden (a)
1991	*L'Épidémiologie en Pathologie Végétale*	F. Rapilly (a)
1994	*Epidemiology and Management of Root Diseases*	C. L. Campbell and D. M. Benson (e)
1996	*Doenças de Planta Tropicais: Epidemiologia e Controle Econômico*	A. Bergamin Filho and L. Amorim (a)
1996	*Epidemiologie der Pflanzenkrankheiten*	J. Kranz (a)
1997	*Exercises in Plant Disease Epidemiology*	L. J. Francl and D. A. Neher (e)
1998	*The Epidemiology of Plant Diseases*	D. Gareth Jones (e)
2003	*Comparative Epidemiology of Plant Diseases*	J. Kranz (a)
2006	*The Epidemiology of Plant Diseases, 2nd edition*	B. M. Cooke, D. Gareth Jones, and B. Kayle (e)

at Wageningen, the Netherlands, in 1971, also as a NATO Advanced Study Institute session. This meeting was organized by Zadoks, Schein, Hirst, and H. D. Frinking. There were 74 participants from 24 countries.

The next international workshop was held at the Pennsylvania State University in 1979. The fourth workshop was held in 1983 at North Carolina State University, and the fifth international workshop was held in Jerusalem in 1986. The next three international workshops were held in Giessen, Germany, in 1990; Papendal, the Netherlands, in 1994; and Ouro Preto, Brazil, in 2001. The ninth international workshop was held in Landerneau, France, in 2005. These meetings generally have been fairly informal, with considerable discussion of various topics of interest. Attendance was limited to about 100 in order to facilitate discussion. Until recently, there was no official structure linking these workshops, except for the common subject matter. Then, the decision was made during the workshop in 1994 to place these meetings under the umbrella of the Epidemiology Committee of the International Society of Plant Pathology (ISPP). Although the meetings have continued to be operated locally, the selection of the venue, date, and organizer is now done by the ISPP committee.

Of course, epidemiology has been, and continues to be, the subject of numerous other symposia, colloquia, and workshops, often organized as parts of larger general conferences, such as national and international congresses of plant pathology. Other meetings focus on particular crops or diseases. However, the international workshops in epidemiology provide some continuity for the discipline and serve as an avenue for the epidemiologists to exchange ideas.

In addition to the many journal articles published, many of which are cited throughout this book, several valuable text and reference books have been written on botanical epidemiology since 1963. Some of the books are listed in Table 1.1 in chronological order. Zadoks (2001) lists additional books of relevance. Some of these books focused on groups of pathogens (e.g., viruses, soilborne pathogens), and others focused on themes, such as comparative epidemiology or epidemic analysis and modeling. In terms of the latter, the multi-author book edited by Kranz in 1974 was especially important historically because it synthesized many of the mathematical and statistical approaches to the study of epidemics that were introduced in the decade after Vanderplank's first book. This work further set the stage for many of the quantitative advances in the 1970s and 1980s in epidemiology.

The first textbook of botanical epidemiology was written by Zadoks and Schein and published in 1979. The authors of the book did an excellent job of conveying much of the excitement of the field, showing the need for mathematics in understanding epidemics, and clearly demonstrating how management of disease is linked to epidemiology. The book dealt with diseases caused by fungi. The textbook of Campbell and Madden (1990) expanded the coverage to diseases caused by all pathogen groups and gave detailed instructions on the quantification, statistical analysis, and modeling of plant disease epidemics. The "laboratory manual" edited by Francl and Neher (1997) expanded on the Campbell and Madden (1990) textbook and gave more details on many forms of epidemic quantification and analysis. Other textbooks have been written in Chinese, French, and Portuguese (see Table 1.1).

1.3.4 Final thoughts on the review of historical developments

The above sections addressed only *some* of the major developments in botanical epidemiology, with an emphasis on the progression toward quantification in the field. The tremendous growth in the discipline within plant pathology during the 1960s, 1970s, and 1980s has been eclipsed somewhat in recent decades by the growth in the larger discipline of molecular biology. However, until molecular biology (or more traditional breeding) results in durable resistance to all plant pathogens on all crops, coupled with the acceptance of the new cultivars by growers and the public, there will be plant disease epidemics, and many of these will have a substantial impact on growers, consumers, and society as a whole. Thus, botanical epidemiology continues to be of utmost importance in understanding when and how epidemics occur, and in giving a sound theoretical and practical basis for disease management decisions.

In addition to the individuals mentioned in the previous sections, *many* other scientists have made important contributions to the field of botanical epidemiology, both recently and in the past. The problem is that we cannot adequately explain the historical importance of many scientific contributions until the contributions are first explained in some detail. Thus, comments on historical developments in botanical epidemiology are presented, where appropriate, throughout this book. It is not our goal to provide a thorough review of the history of the discipline, but we feel it is of value for the reader to gain some understanding of major early contributions in epidemiology.

1.4 Prelude to the Rest of the Book

This book is concerned with the study of epidemics. It builds on the contributions of the scientists mentioned above, and many others, and attempts to provide an integrated approach to the quantification of disease development in time and space in host populations. It could be equally stated that this book is about relationships between variables (see Fig. 1.2). For instance, the disease progress curves of Fig. 1.1 and Fig. 1.2A show the relationship between disease intensity and time. A considerable amount of attention is given in Chapters 4, 5, and 6 (and elsewhere) to characterizing the manner in which some disease variable changes over time. Concurrent with the increase (or decrease) in disease over time is the spread of disease, through pathogen dispersal, in space. This is usually expressed as some measure of disease versus distance from an inoculum source, as shown in Fig. 1.2B. Characterization of this relationship is discussed in Chapters 7 and 8. The simultaneous dynamics of disease over time and space are also considered in these chapters.

Often, there are many inoculum sources (e.g., infected weeds, overwintering spores in the soil, infected seeds), possibly at unknown locations, at the start of an epidemic, so there is no unambiguous way to relate disease intensity to distance from the inoculum. This has led to the development of alternate ways of characterizing the spatial component of epidemics, often borrowed from ecology (Pielou, 1977), which are mostly based on statistical methods (Madden, 1989; Madden and Hughes, 1995, 2002). One important approach involves, in part, determining the frequency distribution—and related measures of variability—of diseased individuals per sampling unit (such as diseased leaves per plant or diseased plants per crop row). This is exemplified in Fig. 1.2D for a situation with 10 individuals in each of 194 sampling units. Utilization of a frequency dataset such as this to determine spatial heterogeneity (and, as will be shown, small-scale aggregation) of disease will be explained in Chapter 9. In Chapters 10 and 11, we show how information on disease heterogeneity, or information on the relationship between heterogeneity and magnitude of some measure of disease intensity, is utilized to develop sampling plans. These sampling plans can be for either estimating disease intensity with a desired level of precision (Chapter 10) or for making a decision on whether or not to invoke a control measure, such as a fungicide application (Chapter 11).

When the physical locations of sampling units are known, very extensive and multifaceted quantification of the spatial pattern of disease is possible. Then, instead of simply determining the frequency of diseased individuals,

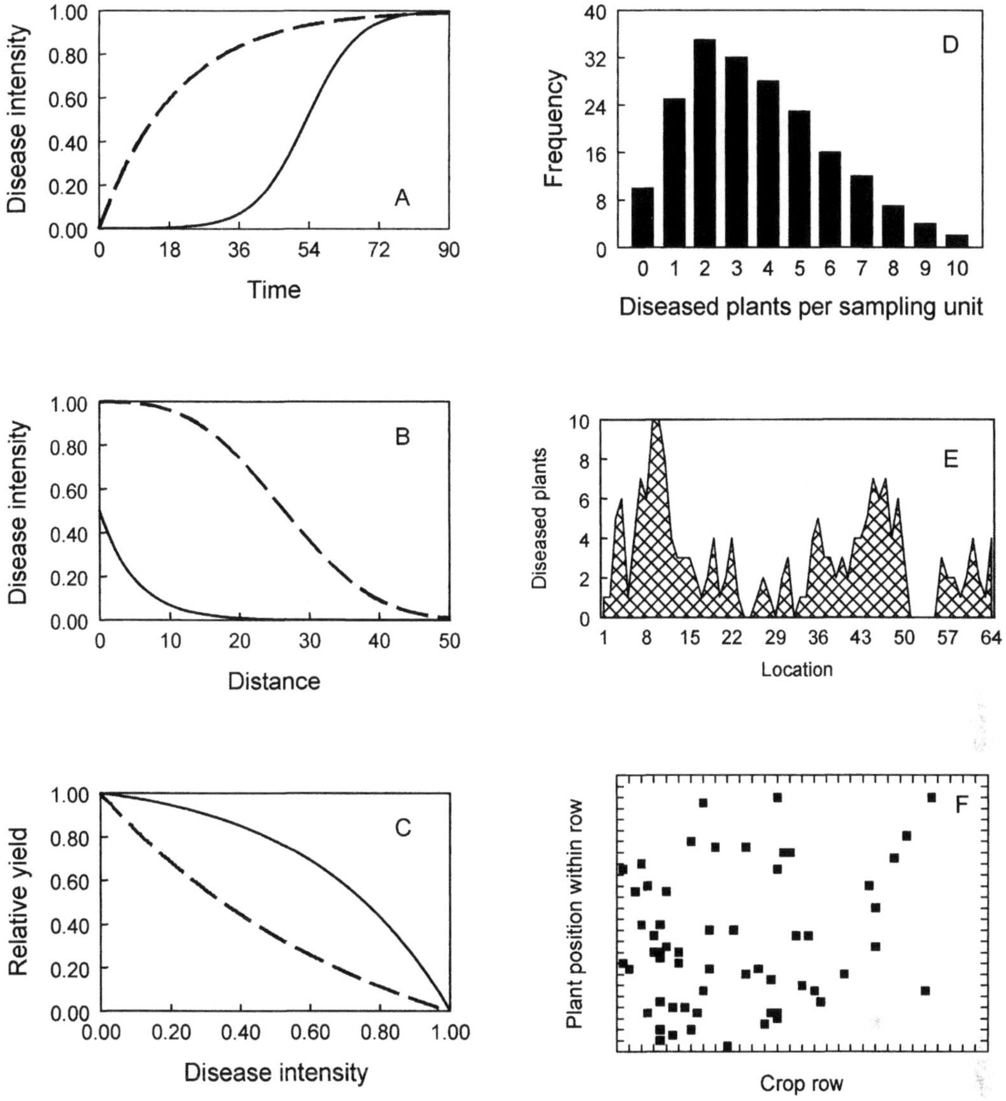

FIG. 1.2. Example relationships of great interest in botanical epidemiology. (**A**) Disease intensity versus time during an epidemic. (**B**) Disease intensity versus distance from an inoculum source. (**C**) Relative yield of a crop versus disease intensity at a particular time during epidemics. (**D**) Frequency of diseased plants per sampling unit (out of 10 possible plants diseased). (**E**) One-dimensional map of diseased plants; that is, number of diseased plants (out of 10) diseased at each of 64 spatially referenced locations in a transect of a field. (**F**) Two-dimensional map of diseased plants in a field. Each solid square represents a diseased plant (with a blank space representing the disease-free plants).

one can show the relationship between disease and location, as demonstrated in Fig. 1.2E for a situation where the number of diseased plants (out of 10) was determined at each of 64 locations in a transect through a field. Two-dimensional maps can also be prepared, as well. An example is shown in Fig. 1.2F, where each sampling unit is a plant (and not a group of plants), so that each disease observation is binary (diseased or not). The relationship between disease and spatial location—or relationship between disease status of neighboring plants—can be characterized with a wide collection of interrelated statistical methods to elucidate small-to-large scale patterns of disease and, hopefully, to gain an understanding of the dispersal of the pathogen causing the measured disease. The statistical methods are discussed in Chapter 9.

A frequent consequence of plant disease epidemics in crops is a reduction in the yield of harvested product. Indeed, this has probably been, historically, the most important driving force for innovation and quantification in botanical epidemiology. Crop losses are often characterized based on the relationship between yield (or yield relative to a disease-free control, or the reduction in yield relative to a control) and a measure of disease intensity, as exemplified in Fig. 1.2C. This topic is addressed in some detail in Chapter 12.

The graphs in Fig. 1.2 demonstrate just some of the most common relationships quantified by botanical epidemiologists. Other relationships are shown throughout the book.

To study disease progress in time and space, or to relate yield to the epidemic, one must first measure disease. Thus, before addressing key relationships of interest in epidemiology, we discuss in Chapter 2 ways of measuring disease of plants and quantifying the accuracy and reliability of the obtained measurements. Moreover, characterization of relationships, such as those in Fig. 1.2, requires the use of mathematical and statistical models. To help the reader in this regard, we introduce the concepts of modeling in Chapter 3, and show some of the ways of fitting models to data and comparing model-fitting results for different epidemics. Many other aspects of modeling and data analysis are introduced throughout the book, as appropriate, for quantifying aspects of epidemics, sampling, and making disease-management decisions based on disease or other interrelated indicator variables.

There are many facets to the study of epidemics, and no single book can cover all aspects of the topic. Throughout the book we direct readers to journal articles, book chapters, and books that deal with topics that we do not fully address. Two particular topics not covered need to be mentioned here. In order to keep the book to a reasonable size, we decided to mostly deal with epidemics caused by a single pathogen species in a single host species. In crops and forests, of course, diseases caused by several pathogens can generally be found, although one disease may dominate in a growing season in a given year. The study of multiple diseases in populations simultaneously requires expansions of the approaches discussed in this book (e.g., Madden et al., 1987b; Ngugi et al., 2001; Turechek and Madden, 2000), and we provide some relevant citations in various chapters for this general topic.

Ever since Vanderplank's first book in 1963, there has been a strong link between the disciplines of botanical epidemiology and genetics—both in terms of host resistance and pathogen population genetics and evolution (Leonard and Fry, 1989; Robinson, 1976; Vanderplank, 1968, 1975, 1984). Although we acknowledge the great importance of taking an epidemiological approach to studying host-pathogen genetics, and in taking a genetics approach to studying epidemics, we do not give explicit coverage of genetics in this book. The book hopefully does provide enough on the theory and practice of epidemiology to allow interested readers to study epidemics and their control in terms of host and pathogen genetics.

1.5 Possible Course Outlines

This book can be used as either a reference or a textbook. As a textbook, there is sufficient material probably for two courses. However, depending on the objectives of the instructor, the book can be used for a single course by selecting different tracks. We list just a few here.

A purely temporal-analysis course could be based on Chapters 4–6, with the earlier chapters used (as needed) to explain disease variables and types of modeling. Chapter 12 could also be covered in this case to show the consequences of epidemics. A course in theoretical epidemiology could be based on Chapters 5, 6, and 8, with background information in Chapters 3, 4, and 7. A data-analysis, sampling, and decision-making course could be based on Chapters 2–4, 7, and 9–12. A primarily spatial-analysis course could be based on Chapters 7–10, with earlier chapters used to "fill in the blanks." Finally, a standard (classical) course in epidemiology would best be based on parts of *all* of the chapters (with simple avoidance of the more advanced sections or subsections).

1.6 Suggested Readings

At the end of each chapter we list some useful articles, book chapters, or books that deal in more detail with some of the major themes of the chapter. We do not necessarily agree with everything in the suggested readings, but think the citations provide valuable material to assist the reader in understanding one or more topics. For a general background on botanical epidemiology, including historical developments in the field, the following articles are useful.

Cowling, E. B., and Horsfall, J. G. 1978. Prologue: How disease develops in populations. In: *Plant Disease: An Advanced Treatise, Vol. 1: How Disease Develops in Populations* (J. G. Horsfall and E. B. Cowling, editors). Academic Press, NY, pp. 1–15.

Gilligan, C. A. 1985. Introduction. In: *Advances in Plant Pathology, Vol. 3: Mathematical Modelling of Crop Disease* (C. A. Gilligan, editor). Academic Press, London, pp. 1–10.

Jones, D. G. 1998. An introduction to plant disease epidemiology. In: *Plant Disease Epidemiology* (D. G. Jones, editor). Kluwer, London, pp. 3–13.

Zadoks, J. C. 2001. Plant disease epidemiology in the twentieth century: A picture by means of selected controversies. *Plant Dis.* 85: 808–816.

2

Measuring Plant Diseases

As always in biology, the problem was to know what you were looking at.

Richard Preston (*The Hot Zone*)

2.1 Introduction

An epidemic consists of a population of infected individuals in a (generally larger) host population, and the change in diseased individuals over time and space. Before one can understand or compare epidemics, one must first monitor them. We use the Merriam-Webster dictionary and define *monitor* as: "to watch, keep track of, or check usually for a special purpose." For the purpose of epidemic analysis, this involves sampling for disease, (Chapter 10), and measurement of disease in the sample. Measurement is the primary subject of this chapter.

Because plant disease involves the interaction of a plant and a pathogen as influenced by the environment, it is natural to consider measurements of plants (e.g., crops, forest trees), pathogens (e.g., spores), and the physical environment (e.g., ambient air temperature) when quantifying epidemics. Depending on the objectives of the study, an investigator may need to quantify variables explicitly representing one or more components of the disease triangle. Because of space limitations, with a few exceptions, we do not explicitly discuss measures of the host, pathogen, or environment. Interested readers should refer to Chapters 3–5 in Campbell and Madden (1990), and also Rotem (1988), Seem (1988), and Sutton et al. (1988). What we do cover here is measurement of disease. As stated by Kranz (1988), "without quantification of disease, no studies in epidemiology, no assessment of crop losses, and no plant disease surveys and their applications would be possible." Campbell and Neher (1994) summarized this nicely by calling disease measurement "the cornerstone of epidemic analysis."

Different approaches to the measurement of disease may be appropriate for different pathosystems. For instance, measuring a systemic disease caused by a virus may be very different from measuring the area of lesions caused by a fungus on leaves. There are, however, general principles and approaches that apply for all types of measurements, and these are discussed in this chapter. We start with an overview of concepts, and give relevant definitions and explanations of disease, measurement, and some statistical terms. Then, we present a review of common disease measurement methods, and discuss ways to determine the so-called reliability and accuracy of the disease measurements.

2.2 Plant Disease Intensity

2.2.1 Concepts

Disease assessment and disease estimation are terms commonly used to describe the measurement of plant disease. The assessed data may be quantitative, qualitative, or a combination of both. A plant or plant unit (e.g., leaf, branch) can be classified as diseased or not diseased, based on either symptoms or some other type of assay. Typically, a sample of plants is assessed in this manner, and the number diseased out of the total (N) is recorded. The term *disease incidence* is used for the proportion of plants (or plant units) diseased or the number diseased out of N. Incidence is a discrete variable (see section 2.3).

There can be various scales to incidence, depending on what plant units are assessed for diseased individuals (Hughes et al., 1997; Turechek and Madden, 2001). For instance, if one is assessing individual leaves for disease, then *leaf* disease incidence is defined as the proportion or number of diseased leaves. If one is assessing individual plants, then *plant* disease incidence is defined as the proportion or number of diseased plants. A single diseased plant can have one or many diseased leaves. This concept can be extended to even higher scales. For instance, one can determine the proportion of fields with one or more diseased plants as a measure of field disease incidence in a region. However, the specialized term *disease prevalence* often is used for the proportion or number of fields with diseased plants (Nutter, 2002a).

Besides determining whether or not a plant is diseased, some aspect of the "degree of infection" can be ascertained. For instance, with foliar diseases caused by fungi, one can assess the area of lesions on plants. Although the measurements can be in absolute units (for instance, square centimeters of lesions), it is much more common to assess the proportion or percentage of host area affected by disease. The term *disease severity* is used for (relative or absolute) area of plant tissue affected by disease. With a sample of plants, one averages the individual values to obtain a mean disease severity. Severity can refer to either the diseased area of an individual plant or the mean diseased area of a sample. As used here, severity is a continuous variable (see section 2.3.2). In practice, however, severity may also be assessed by assigning a disease severity category or class value to each observed plant; then, severity is a discrete variable, although possibly with many distinct values.

Generally, mean disease severity in a plot, field, etc., is the average of the severity of all assessed plants (or plant units), whether the plant is diseased or not. In other words, a severity value of zero is used for all observations with no disease, and the calculated mean is a measure of disease severity for the sample of interest. One could also determine the mean severity for just the diseased plants; that is, one uses the disease-free observations in determining the mean. This is really a "conditional severity," because it is a measure of disease severity that is conditional on the individuals being infected. Conditional severity is easily obtained by dividing the mean severity of *all* assessed plants (or plant units) by the proportion of those plants (or plant units) diseased.

When a plant is diseased, instead of assessing the area affected, one can count the number of lesions or other units of infection for certain types of disease. In agreement with McRoberts et al. (2003), we refer to the number of lesions per plant or per unit area (say, lesions per square centimeter) simply as a *disease count*. The count values can be averaged over the observations in the sample to obtain a mean disease count. A count is a discrete variable, because only distinct integer values can be obtained for an individual. However, fractional values can be achieved for means.

Often the term severity is also used for a disease count (Seem, 1984). This is reasonable for some purposes, especially when lesions are all of a similar size. Then, the area diseased is simply the number of lesions multiplied by the average area of an individual lesion. When one is modeling and analyzing disease increase for entire plots or fields, then a mean count of lesions per plant (leaf, etc.) may be a quite reasonable substitute for severity (or severity may be a reasonable substitute for a count). However, the statistical properties of counts and severity are different, and one cannot interchange the variables for all purposes. For instance, some useful methods to determine the spatial pattern of disease depend on the use of discrete variables and are not valid for continuous variables such as severity (Chapter 9).

A disease count may also be referred to as *disease density* because the counts are expressed relative to a leaf, root, stem, and so on (McRoberts et al., 2003). With common usage of the term, density refers to a quantity per unit area or volume. Thus, it should be pointed out that incidence, severity, and counts can all be expressed as densities. Examples include: the proportion of diseased plants per square meter of crop or the mean proportion of leaf area with disease symptoms per square meter of crop.

Several (but certainly not all) of the principles to be discussed in later chapters correspond to any form of measured disease in a host population. Thus, for ease of communication, it is often desirable to use a general term for plant disease that includes all of the measurements just discussed. We use the term *disease intensity* to encompass disease incidence, severity, and counts. As mentioned above, sometimes we lump severity and counts together as forms of severity (e.g., some data sets in Chapter 4), and sometimes we keep them separate, especially for statistical purposes (e.g., Chapter 9).

As with most definitions in plant pathology, these definitions of disease are not universally used. For instance, one might find intensity being used for what we call severity. Some sub-disciplines have their own terminology. For instance, researchers working on Fusarium head blight of wheat in the United States use the term "severity" for a disease variable equivalent to conditional severity (the mean severity of just the diseased plants), and the term "index" for a disease variable equivalent to mean severity for a sample (Groth et al., 1999). Moreover, disease terms may be used differently in other scientific disciplines. In medicine, as an example, "prevalence" is used for what plant pathologists generally call incidence. The conclusion here is that one should be careful when reading the literature to find out how the authors are using the disease terms.

2.2.2 Severity versus incidence: some considerations

In conducting epidemiological studies, one must decide, among other things, which measures of disease intensity to obtain, how often during the epidemic to obtain disease data, and how much sampling to do at each time. The second point involves the characterization of disease progress curves, the subject of Chapters 4 and 5, and only a few comments on frequency of data collection are presented here. The third point is not addressed until Chapter 10, after many statistical details are introduced and explained. The first point is addressed fully in this chapter.

Whether to assess disease incidence, severity or counts depends on the pathosystem being studied, the overall magnitude of disease intensity, and the objectives of the investigator. With many foliar diseases of plants caused by fungi, the concept of severity as a continuous variable is relatively easy to visualize and put into practice. That is, an investigator can use one of several techniques (discussed below) to assess the proportion of the total leaf surface covered by lesions (pustules, etc.). With many systemic diseases, however, such as those caused by viruses and soil-borne pathogens, the concept of leaf (or plant) area covered by lesions is not a useful one. Rather, severity involves harder-to-define aspects of the degree of infection, such as extent of wilting, yellowing of leaves, senescence, and so on. Severity of disease can also be a difficult variable to use for many root diseases because direct assessment would involve destructive sampling—digging up the plants to observe the roots. For these situations, depending on the study objectives, incidence may be preferable for quantifying disease because the proportion of plants with symptoms has a direct and objective interpretation.

For many applied and basic studies in epidemiology, crop yield in relation to disease intensity is of major interest (Chapter 11). Initially, it may appear that disease severity is a more useful measure of intensity than incidence for yield–loss purposes (at least for certain types of diseases). For instance, if there was one early blight lesion on every potato plant in a field, incidence would be 1 (100%) but mean severity would be very low (<0.001). Under these conditions, it is unlikely that yield would be different from that obtained in a disease-free field. However, if half of the leaf surface of plants were covered by lesions in half of the field, and there were no lesions at all on the plants in the other half of the field, incidence would be 0.5 and mean severity for the field would be $0.5 \times 0.5 = 0.25$. Based on our knowledge of crop loss (see Chapter 11), one can expect a substantial yield reduction in this case compared with a disease-free field, even though incidence is half that in the first case. For this and probably several other reasons, it is very common to assess severity in basic and applied epidemiological studies. Even in situations with difficult-to-measure severity (systemic diseases, root diseases), efforts are often made to quantify severity in the best way possible (e.g., Harveson and Rush, 2002; Krause et al., 2001; Murphy et al., 2000).

The extreme combinations of incidence and severity of the previous paragraph, although certainly possible, are not the norm. There is often a relationship between incidence and severity which means that one can predict one from the other (see McRoberts et al., 2003). Consider the epidemic curves shown in Fig. 2.1 These types of curves were introduced in Chapter 1 and will be considered in detail starting in Chapter 4. For now, simply accept that the curves of intensity over time are realistic for many diseases. Severity was generated for two treatments (1 and 2) based on a classic epidemic model, and incidence was predicted based on a known relationship between severity and incidence (McRoberts et al., 2003; see section 9.8). The same relation between severity and incidence was used for both treatments.

Although there are clear differences between the two treatments in Fig. 2.1, one would not see this if only severity was considered and disease assessments were made just over the first 30 days or so of the epidemic. In this time span, severity barely is above 0 on the vertical axis. Although a logarithmic scale would show differences in

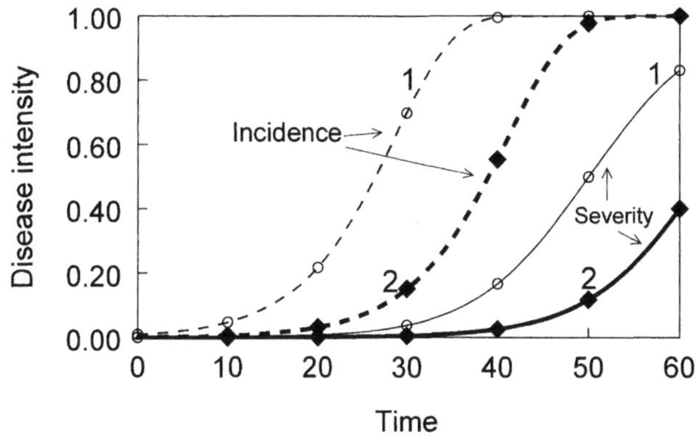

FIG. 2.1. Theoretical disease progress curves for two epidemics (e.g., corresponding to two treatments). Both incidence and severity of disease are shown. The relationship between severity and incidence is identical for both epidemics (see section 9.7). Epidemic 1: thin lines and open symbols. Epidemic 2: thick lines and closed symbols.

severity for this theoretical exercise, given the variability associated with sampled disease values, it is unlikely that one could find significant differences of the two treatments in terms of disease severity at individual times before day 30. On the other hand, disease incidence is much higher over this time span, and clear differences between the treatments are apparent at individual times (even on this linear scale). Later in the epidemics (>40 days), severities were well separated for the two treatments, but incidence was very close to 1 (100%). An investigator making the first disease assessment at day 40 or so, would not find incidence to be of much value in discriminating between the two groups. When one has data over the entire course of the epidemic, one can quantify changes of incidence and severity, and possibly characterize a relationship between them.

2.3 Measurement Levels and Random Variables

We have been using terms like "discrete," and "continuous" so far without really explaining them. Since the properties of variables determine how one analyzes the data, we think it is very useful to take a slight diversion from disease assessment material and discuss some terminology from statistics and measurement science. There are different systems for classifying (labeling) the information obtained from planned experiments or observations of natural phenomena (Schabenberger and Pearce, 2002; Sheskin 1997). Sometimes these systems are merged, which can be a little confusing. We discuss two systems here.

2.3.1 Measurement level

One classification system is based on the level of information obtained with a measurement procedure.

"Measurement" in this sense is considered very broadly and could be visual observations or values obtained with electronic or physical devices. There are four levels in this system (Fig. 2.2, left-hand side).

Nominal level. A label is given to each measured individual with respect to the property of interest. The property is qualitative in nature. For example, a plant is labeled diseased or not based on some criterion. Diseased could be coded as "D", "+", "1", or other convenient label. Disease free (healthy) could be coded as "H", "−", "0", or other label. Thus, disease incidence at the individual level (where assessment occurs) is a nominal measurement. Many measured properties of plants and pathogens are also nominal in form. Examples include plant species, cultivar, and pathogen species.

Ordinal level. An ordered value is given to each measured individual with respect to the property of interest. The values used are interpretable only in terms of their arrangement in a given order, not to the magnitude of the differences of the values used. For example, the severity of a plant is measured by assigning it to one of four categories: (0) no disease, (1) slight disease, (2) moderate disease, and (3) very severe disease. These numerical values may be called scores or ratings. Assuming no error in assessing disease, a measurement of "2" does reflect higher severity than a measurement of "1", but the numerical difference between 1 and 2 has no interpretable meaning. Likewise, the increase in actual severity between scores of 1 and 2 is not necessarily the same as the increase in actual severity between scores 3 and 4. One could just as easily uses values of −20, 0, 50, and 100, or A, B, C, and D for the four levels

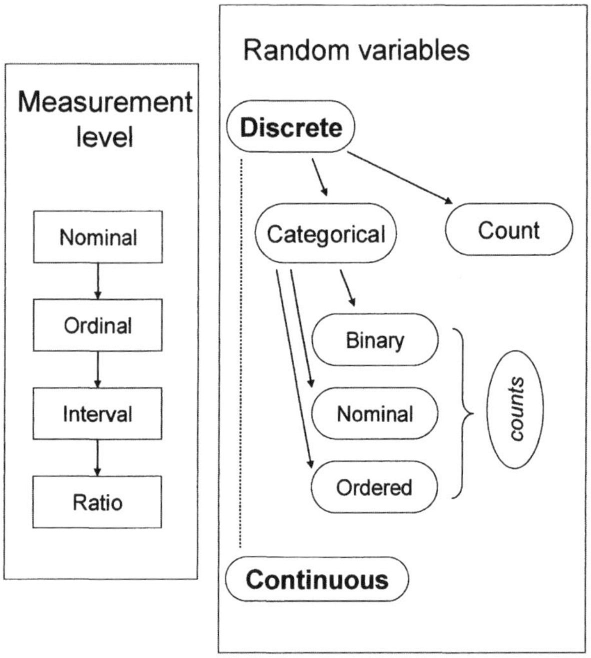

FIG. 2.2. Classification of measurement scales and random variables, based on concepts in Schabenberger and Pierce (2002) and Sheskin (1997).

of severity. No difference is inferred among individuals with the same ordinal measurement rating, but individuals with different ratings can be ordered. This type of measurement is fairly common in plant pathology, especially for dealing with the severity of root diseases and systemic diseases (Shah and Madden, 2004).

Interval level. The measurement of each individual not only can be ordered, but also the differences in measurement values have direct interpretation. One can say how much larger (or smaller) one observation is, compared with another. Ambient air temperature in Celsius is an example of interval level of measurement. A defining property is that equal differences between measurements correspond to equal differences in the property being measured. For instance, the difference between, say, 20 and 25 is the same as (i.e., has the same meaning as) the difference between 25 and 30, since the numbers are not arbitrary values.

Ratio level. Like an interval level of measurement, the ratio-level measurements and their differences have direct interpretation. However, there is also a "zero point" or fixed origin. Plant height, yield, area of a single lesion, or disease severity for a plant are examples of ratio level measurements. A defining feature of this scale is that the ratio of two values has direct interpretation. For two severities that were 20 and 40%, then the second is twice as large as the first.

These four levels here are presented in increasing order of information content. That is, assuming that the measured values are correct, a ratio- or interval-level measurement is more informative than an ordinal- or nominal-scale measurement (Gibbons, 1976). This does not mean, however, that it is necessarily better to use ratio- or interval-scale measurements. This is discussed later in the chapter.

2.3.2 Random variables

Generally there will be variability when measurements of properties are obtained from individuals (animate or inanimate). Two plant pathologists may obtain different visual estimates of disease severity for the same plants, or ELISA absorbances may vary for multiple runs of the assay. For this reason, something measured is commonly referred to as a variable or *random variable*. The term random variable can be used more generally for any characteristic that differs from observation to observation (individual to individual), or from time to time, or location to location, but here we focus on its use in terms of measurements.

A variable can be classified as to whether it is continuous or discrete (Fig. 2.2). For a *continuous variable*, the number of possible values is not countable and the "distance" between values is directly interpretable. A continuous variable in this system is directly analogous to interval- and ratio-level variables in section 2.3.1. For example, consider the yield of two plots, which are recorded as 10.1 and 10.3 kg. The difference of 0.2 kg is well-defined in a physical sense. Moreover, an indefinitely large number of weights are possible between these two. Although the accuracy of the measuring and recording device may not allow it to be determined, or the person taking notes may round all values to one decimal place, there could be plots with yields of 10.11, 10.101, 10.100001, and so on. When disease severity represents the actual area of leaves or roots with symptoms, or the proportions, then severity is a continuous variable.

For a *discrete variable*, the number of possible values is countable. That is, unlike the situation for continuous variables, there is a finite number of possible values. If the discrete variable consists of counts of something, it is known as a *count variable*. Numbers of spores in volumetric spore samplers or numbers of overwintering spores per volume of soil are examples of count variables. In theory, counts may range from 0 upwards, without upper limit, although in practice there must be a finite upper bound. As long as the upper bound is very high relative to the actual counts, the simplifying assumption of an infinite upper bound is reasonable. A disease count (such as lesions per leaf) obviously is a discrete random variable.

If the values taken on by a discrete variable indicate membership of some group, it is known as a *categorical variable* (Fig. 2.2). There may as few as two categories, in which case it is known as a *binary variable*. Disease incidence at the individual (single leaf, plant, etc.) scale is an example of a binary variable. If there are more than two categories, but there is no order to the categories, then it is known as a *nominal variable*. An example would be the species of *Phytophthora* found in a soil sample. A binary variable is a special case of a nominal variable. It is clear that there is an equivalence of the binary and nominal categorical random variables here and a nominal-level measurement of the previous section.

If the category labels reflect an order, then the variable in question is known as an *ordered categorical variable*. Examples could be categories of plant health, or ordinal scores of disease severity (Shah and Madden, 2004). The scores, however, are just labels, and it is generally inappropriate to consider an ordered categorical variable to be the same as a continuous variable. An ordered categorical variable obviously corresponds to an ordinal level of measurement (Fig. 2.2).

There is a natural link between discrete count variables and discrete categorical variables. Consider disease incidence. At the scale of measurement (the individual plant or leaf), the disease variable is binary. However, one can total the number of diseased (and disease-free) plants to obtain a count for a plot, field, or other unit of interest. Schabenberger and Pierce (2002) call this a *count with a natural denominator*. These counts can be converted into proportions. In contrast, counts without a natural denominator cannot be converted into proportions. One can also obtain counts for nominal and ordered categorical variables; here, the total for each category is obtained, just as the total was determined for the disease and disease-free categories.

The relationship between count variables with and without a natural denominator can be even further elaborated. For instance, consider lesions on leaves. One can think of each lesion as occupying a finite site on a leaf, with a very large number of sites that could be occupied (Daamen, 1986; Hughes et al., 2004). The actual count of lesions is equivalent, conceptually, to the count of sites that are occupied. Of course, it is not possible to actually count these sites (until they are occupied), or know the actual total number of sites.

2.3.3 Plant disease variables

There are at least two reasons to be concerned about the level of measurement or the type of random variable. First, as described above, the measurement level is an indication of how much information is conveyed about the property of interest. Second, the type of variable dictates what types of statistical and mathematical analyses are appropriate. Many of the common statistical methods used by plant pathologists, such as analysis of variance and regression analysis, were developed and are most appropriate for continuous random variables. For counts without a natural denominator and for counts of binary variables (i.e., counts with an upper bound), the discrete variable may be reasonably approximated as a continuous variable if there is a high number of counts and a wide range of values (Schabenberger and Pierce, 2002). There are, however, many statistical methods explicitly developed for discrete variables (Agresti, 1990; Collett, 2003) which can be extremely effective in characterizing the variables of interest. For instance, several aspects of spatial pattern analysis of plant diseases (Chapter 9) depend on statistical properties of count data (with and without a natural denominator). Some alternatives to classical regression analysis and analysis of variance are found in Chapters 4 and 10.

In assessments of plant disease, for the most part there are five types of random variables that are commonly encountered:

- Continuous [e.g., severity on a 0–1 scale (0–100%), either for individual plants (leaves, etc.), or for the mean of several individuals].
- Ordered categorical [e.g., ordered rating classes of disease severity on individual plants (or leaves, etc.)].
- Binary [e.g., disease incidence, consisting of disease status of individual plants (leaves, etc.)].
- Counts with a natural denominator [disease incidence, consisting of number of diseased plants (leaves, etc.) out of a total number assessed; often expressed as a proportion of the total].
- Counts with no natural denominator [e.g., disease count, such as numbers of lesions per plant (or per leaf, etc.)].

With a few exceptions (see section 2.5.3), we use Y and y for the disease intensity random variable throughout this book, upper case Y for disease in absolute units (e.g., area of lesions per plant, number of diseased plants), and lower case y for relative units (e.g., proportion of plant leaf area with lesions, proportion of diseased plants).

2.4 Assessing Disease Intensity

2.4.1 Incidence, counts, and severity: some general comments

Assessing disease incidence is relatively straightforward, at least in principle. One simply categorizes each individual plant unit (e.g., whole plant, leaf, root, etc.) as being diseased or not, and adds up the total number observed and number in the disease category. The categorization can be based on visual symptoms (the most common approach in epidemiology) or other technique, such as use of biochemical or molecular assays (Fox, 1998). Of course, there can be errors in assessing for disease, in particular, declaring a plant diseased when it is not or declaring a plant disease free when it is diseased.

Assessing disease counts is also straightforward in principle. For density of lesions, for instance, one simply counts the number of lesions on plants, leaves, etc. In practice, this can be more difficult than it appears, since it may not be easy to see all the lesions, or they overlap or grow together, or they exhibit misleading signs and symptoms. Counting lesions on roots is an especially difficult process (Campbell and Neher, 1994). As an alternative one can move back in the infection cycle and count the density of spores in the air, on plant surfaces, or in the soil (Benson, 1994; Fitt and McCartney, 1986).

Assessing disease severity, either as a proportion of leaf or root area affected, or as an actual area affected, is a challenging endeavor, although it may appear deceptively simple. Much has been written on the subject (e.g., Cooke, 1998, 2006; Gaunt, 1987; James, 1974; Large, 1966; Nutter and Schultz, 1995; Nutter et al., 2006) and the subject continues to be of interest (Nita et al., 2003; Nutter, 2002a; Vereijssen et al., 2003). The general approaches mostly involve visual assessment, based on symptoms, and electronic assessment, based on reflected electromagnetic radiation (e.g., light, radiation in the visible part of the spectrum). Because of their widespread use, these methods are addressed in some detail in the following subsections. With very intensive data collection and measurement efforts, one can also use disease count values to obtain directly a measure of severity. For instance, one measures the size of every observed lesion, and uses the number of lesions and their average area to calculate total area affected. Of course, it is rare that investigators can afford such intensive measurement efforts, except for a very a limited number of plant units.

2.4.2 Visual assessment of disease severity

Plant pathologists have been attempting to obtain quantitative information—i.e., interval or ratio-level measurements (Fig. 2.2)—on disease severity through visual assessment for over a century (Cobb, 1892). Efforts to use ordinal scale measurements go back nearly as far (e.g., McKinney, 1923). The basic idea is for an individual person (called a *rater* here) to look at a *specimen* (individual leaf, root section, entire plant, etc.) and estimate the area of the specimen affected by the disease. Generally, the visual measurement is of the *relative* area, that is, the proportion or percentage of the specimen that is affected. The assumption is that raters can, indeed, make reasonable estimates of relative area by visual inspection. The validity of this assumption is discussed in section 2.5.

To facilitate assessment of disease severity, various aids have been developed to assist raters, such as text descriptions of specific severity values and diagrams of severity percentages (Campbell and Madden, 1990). These aids form a convenient basis for classifying the assessment methods. We consider four visual assessment methods.

- Direct estimation
- Direct estimation with use of disease diagrams
- Use of disease scales
- Use of ordinal rating scales.

Other classifications could be used, of course.

2.4.2.1 Direct estimation. With direct estimation, the rater observes a specimen and assigns it a disease severity value from 0 to 100% (0 to 1) based on perceived area affected. If the rater feels it is justified, fractions of a percentage can be recorded. One does not use any other aid to make the assignment.

2.4.2.2 Direct estimation with use of disease diagrams. A disease diagram is a pictorial or graphic representation of selected levels (classes) of disease severity. Sometimes they are called *standard area diagrams*. Although Cobb (1892) was the first to use disease diagrams, James (1971) is probably most responsible for their widespread use. Some examples are shown in Figs. 2.3 and 2.4. They consist of drawings of plant parts of interest (e.g., leaves, fruit, stem) covered, in part, by pictorial representations of typical lesions. Usually, the plant surface is in white and the lesion area is in black, although color photographs could be used. It is important that the lesion area in the drawing has the same shape as typical lesion areas on plants. For practical purposes, only a selected (small) number of severity levels are shown.

The rater uses a set of diagrams as a guide. In particular, the rater compares an actual diseased specimen to the set of diagrams, and decides if the severity of the specimen is close to one of the pictorial representations. In most cases, interpolation or extrapolation is needed to

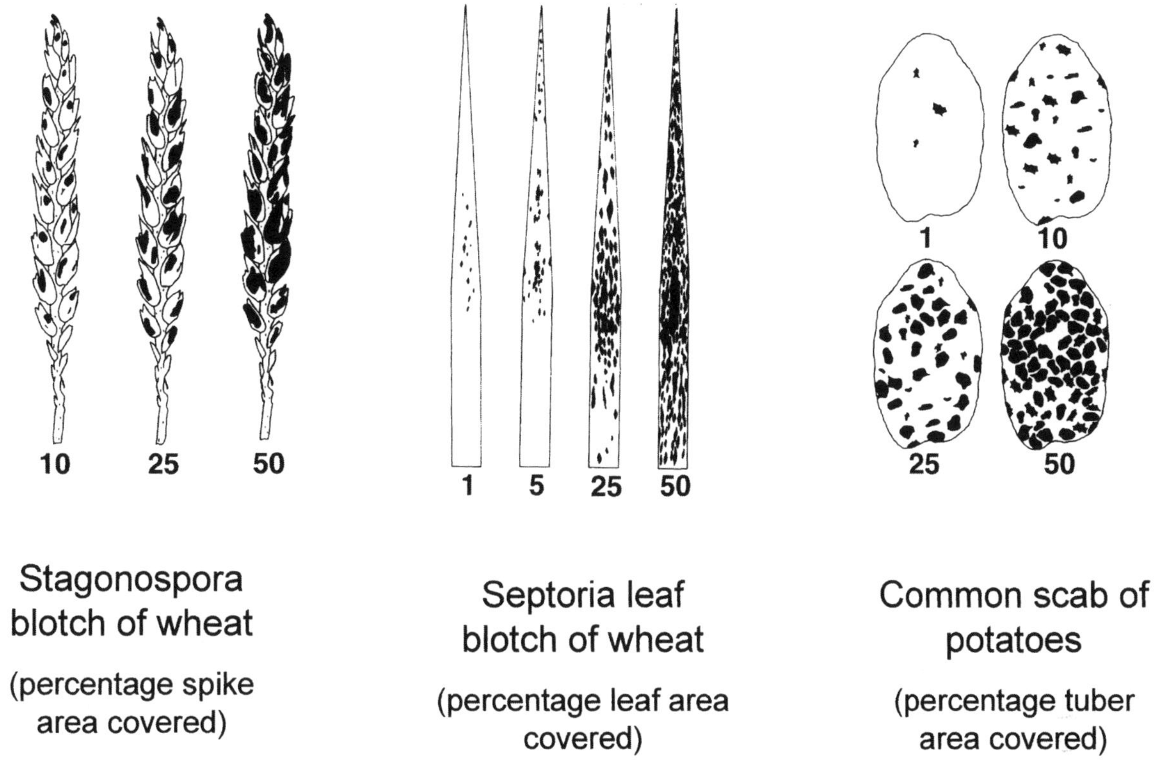

FIG. 2.3. Example disease diagrams, taken from James (1971). Selected severities as percentages are shown.

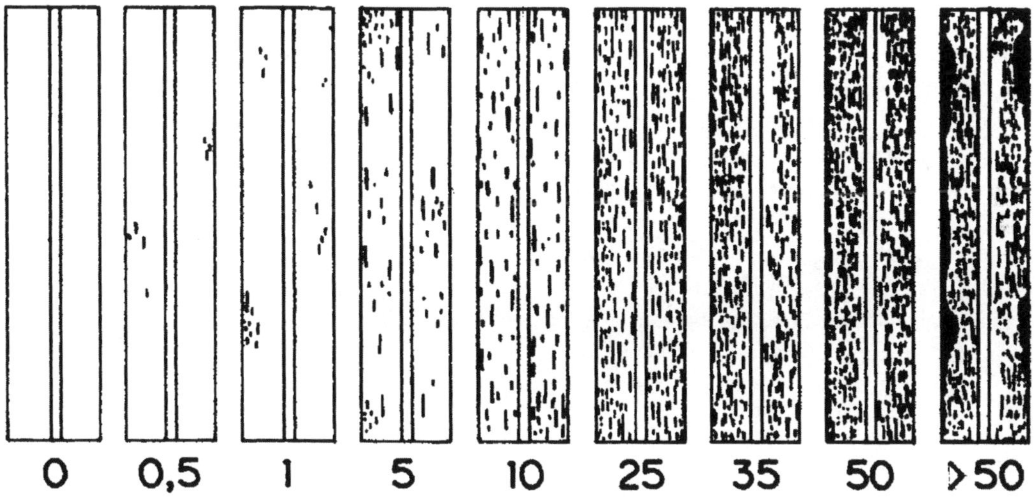

FIG. 2.4. Disease diagram for sugarcane rust (courtesy of A. Bergamin Filho and Lilian Amorim), which is similar to the Cobb diagram for rust of small grains (Cobb, 1892). Each number represents disease severity (as a percentage).

assign an appropriate severity percentage to the specimen. Obviously, the severities shown on the diagrams could influence how the rater interpolates between values. Although a rater may initially compare every actual specimen to the diagram, ultimately the rater "learns" the different severities of the diagrams, and just makes direct estimates of severity without reference to the diagrams (as in section 2.4.2.1).

One problem with standard area diagrams is that most show only one example of a particular severity percentage. In reality, of course, there are many ways that, for example, 25% of the surface could be covered by lesions. There may be many small lesions or a few large ones. The lesions may be aggregated at the base (or tip), or dispersed over the entire surface. Some diagrams address this issue by showing multiple versions of a particular percentage (e.g., Fig. 2.4).

2.4.2.3 Estimation with use of disease scales. A disease scale is a partition of the continuous severity values from 0 to 100% into a finite number of classes. Basically, a disease scale is a written and numerical representation of the severity classes to be distinguished (Cole, 1981). A simple example is:

Class	*Severity range*
0	0
1	0^+ up to 25%
2	25^+ up to 50%
3	50^+ up to 75%
4	75^+ up to 100%

Here, the superscript + means a value just slightly above the indicated value (e.g., for Class 1, any low level of severity above 0 would be assigned a 1). To use a disease scale, a rater observes a specimen (leaf, root, entire plant, etc.) and assigns it a class value based on the description of the class. The written description of the classes can be as simple as a listing of the range of percentages in the classes (as in the above example), or a detailed explanation of what diseased plants (leaves, etc.) look like when in each class.

One reason for using a scale is for convenience and speed of rating. For instance, it may take a rater a relatively long time to decide whether the severity of a specimen is 28, 29, or 30%, but the rater could quickly decide that the specimen belongs to Class 2 (as an example), and move on to the next specimen. A second reason is based on an assumption that a rater cannot easily distinguish severity values within a class. That is, there is an assumption that trying to discriminate between 28% and 32%, for example, is not worth the effort, so one should use a measurement scale that reflects measurement abilities of raters. This idea underlies the Horsfall-Barratt (H-B) scale (Horsfall and Cowling, 1978b). This scale was originally proposed in 1945 (but only as an abstract), and has since become one of the most common scales, if not the most common scale, for disease assessment in plant pathology. It consists of 12 classes of disease as listed here.

Class	*Severity*	*Midpoint*
0	0	0
1	0^+–3	1.5
2	3^+–6	4.5
3	6^+–12	9
4	12^+–25	18.5
5	25^+–50	37.5
6	50^+–75	62.5
7	75^+–88	81.5
8	88^+–94	91
9	94^+–97	95.5
10	97^+–100	98.5
11	100	100

The most striking feature of the H-B scale is the unequal severity intervals (widths) corresponding to the different classes of disease, and the pattern to the interval widths. Close to zero, the intervals are small (e.g., 0^+–3 for Class 1, with 0^+ a nonzero severity slightly above 0), and the interval width generally increases with increasing severity up to 50%, then declines again in a symmetric fashion. Variations of the scale in use include smaller and larger number of classes (e.g., Green et al., 1998).

The H-B scale is based on two assumptions. The first is that there is a logarithmic relationship between the intensity of a stimulus (i.e., reflected light from a diseased specimen) and sensation (i.e., estimated area diseased) (Horsfall and Cowling, 1978b). This is a consequence of the Weber and Fechner laws of psychophysics, which are sometimes combined into the so-called Weber-Fechner law (Bringmann, et al., 1997; Murray and Ross, 1988; Titchener, 1928). If the Weber-Fechner law held, a rater would be able to discriminate between small levels of severity (e.g., between 2% and 5%) when severity was low, but only be able to discriminate between large levels (e.g., between 20% and 40%) at higher severity.

The second assumption behind the H-B scale is that when observing an object consisting of two colors or forms, a rater focuses on the one that is smaller in size. In terms of disease assessment, this means that the human eye perceives diseased area below 50% and healthy area above 50% (Horsfall and Cowling, 1978b). This second assumption explains why the severity intervals decrease above 50% in the H-B scale, all the way down to an interval width of 0 at 100% severity. If only the first assumption held, the interval widths would continue to increase above 50%.

The problem is that there is little, if any, experimental evidence that the assumptions behind the H-B scale hold for assessing disease severity (Forbes and Korva, 1994; Hebert, 1982; Nita et al., 2003; Nutter and Schultz, 1995; Parker et al., 1995; Sherwood et al., 1983). There is even question of the general appropriateness of the Weber-Fechner law(s) for stimuli such as light and color (Murray and Ross, 1988). In fact, there is much stronger empirical evidence that a linear scale (e.g., equal interval width with increasing disease) is more appropriate for severity assessment (e.g., Nita et al., 2003; Nutter and Schultz, 1995). For instance, a straight line relationship between estimated disease and actual disease severity is often found, although a nonlinear relation is expected on the basis of the H-B scale. (The concepts of linearity and nonlinearity are explained in Chapter 3). One also expects that the greatest error in estimation is around 50% if the H-B assumptions hold. However, in studies, there is either no consistent relationship between magnitude of the error and actual severity, or if there is relationship, the highest error occurs at less than 50% severity (e.g., Forbes and Jeger, 1987; Nita et al., 2003; Parker et al., 1995).

Detailed evaluation of visual assessment methods has been done for only a few plant diseases. Certainly, it cannot be claimed that the H-B assumptions are unreasonable for all plant diseases. However, it is a mistake for plant pathologists to use the H-B scale, thinking that it is somehow justified by measurement/assessment theory. This does not mean that a scale such as the H-B one has no value. As discussed above, one reason to use a disease scale is for convenience and ease of use. The H-B scale may sometimes meet this criterion. Specifically, we have found the H-B scale or similar ones to be very useful for rapidly assigning a severity class to each specimen in the field and greenhouse. However, as a general rule, equal-increment scales may be preferable to the H-B scale (Nutter, 2002a; Nita et al., 2003; Nutter et al., 2006).

Specimens considered for disease assessment can be individual leaflets, leaves, branches, roots, stems, or even whole plants. The "specimen" assessed can even be an entire field. For example, the British Mycological Society (1947) issued a scale to determine the percentage of foliage affected by late blight in entire fields based on symptoms and patterns of severity in the field. For instance, 5% mean severity corresponds to ~50 lesions per plant or up to 10% of the leaflets with symptoms. This type of disease scale is known sometimes as a *field key*.

2.4.2.4 Estimation with use of ordinal rating scales. As discussed above, it can be very difficult to measure the severity of certain types of diseases. This is especially true for systemic diseases, those caused by viruses, and many root diseases. Even the concept of severity becomes a more difficult one. A popular approach to circumvent such difficulties is to use an ordinal rating scale. That is, one observes a specimen and assigns it to one of a fixed number of labeled severity rating classes. Such a scale is different from the disease scales in the previous section because there is less information in the description of the ordinal rating classes. Here, the specific ratings used do not have any particular physical interpretation.

A recent example of a rating scale is one used for Rhizoctonia damping-off of plants (Krause et al., 2001). There are five possible severity ratings for a specimen:

Rating	*Description*
1	Symptomless
2	Small root or stem lesions
3	Large root or stem lesions
4	Post-emergence damping-off
5	Pre-emergence damping-off

Higher ratings are indicative of increasing severity of the disease. However, only the order of the ratings has any direct interpretation—a 4 is worse than a 3, a 5 worse than a 4, and so on. The difference between ratings provides no quantitative information about disease severity. Even though the difference between successive ratings is 1 in all cases (in this example), the increase in disease between a rating of 2 and rating of 3 is not necessarily equal to the increase from a 3 to a 4. One could just as easily have used 0, 10, 30, 60, and 100 for the five ratings, or even A, B, C, D, and E.

Another example of an ordinal rating scale is found in Murphy et al. (2000) for assessing tomato plants infected by tomato mottle virus. The six classes are:

Rating	*Description*
0	Symptomless
1	Mild mottling on young leaves
2	Obvious mottling on leaves from at least one of the main stems
3	Obvious mottling on leaves over most of the plant
4	Obvious mottling on leaves and leaf distortion over the entire plant
5	Obvious mottling on leaves, leaf distortion, and severe stunting

While the descriptions of the rating classes vary in ease of interpretation, the scale does nicely summarize the sequence of symptoms that can develop during plant infection. It would be a difficult task to derive an alternative scale that could capture the meaning of some of these properties of disease severity in a more quantitative manner.

The obvious advantage of an ordinal rating scale is ease of use, making it possible to assess specimens rapidly for disease. It is probably (relatively) easy to train raters to use such a scale, compared with a quantitative one. In addition, it may not always be possible to develop quantitative scales. The major disadvantage of an ordinal rating scale is that the actual rating values used are arbitrary. This can have a major impact on data analysis (Gibbons, 1976; Shah and Madden, 2004), as described at the end of the next sub-section.

2.4.2.5 Random variables for severity of disease. Directly estimated severity over the 0–100% range is a continuous variable, whether or not a disease diagram is used as an aid. It may appear that disease scales (section 2.4.2.3; Fig. 2.2) that involve a finite number of severity classes result in a discrete variable (ordered categorical). In one sense this is true since a countable number of possible severities are obtained. However, the classes used are representing a continuous (0–100%) scale of measurement, and the assigned numbers are used just for convenience not because of any inherent meaning in terms of biology or epidemiology. In a sense, the severity obtained with a disease scale (such as the 12-level H-B one) can be (and usually is) considered a continuous random variable with just a great deal of rounding, so that only a fixed number of possible severities are recorded.

The question arises whether one can use analysis of variance and regression analysis for severity data obtained with a finite-class disease scale, since these statistical procedures were developed for continuous variables.

Snedecor and Cochran (1989) stated that these types of analyses may be appropriate for ordered categorical data *if* the classes used in the scale represent equal gradations on an underlying continuous scale. Certainly, this criterion is not met for the 12 class values of the H-B scale. For example, a "2" represents severities from 3 to 6% (a range of 4), but a "5" represents severities from 25 to 50% (a range of 25%). Thus, one should *not* use analysis of variance on the actual severity class values. Rather, one should convert each class value to a severity on a percentage scale. The most common approach is to use the mid-point of the severity range for each class (see the table). Then, because the severity data are on a continuous scale, analysis of variance may be performed. Obviously, it may be that the mid-point of a class range is not very close to the actual severity, especially with severities around 50%. This is another disadvantage of using a scale such as the H-B one, especially if the rater is better able to measure severity than assumed for the H-B scale.

Severity of disease obtained with an ordinal disease rating scale (see section 2.4.2.4) is clearly an ordered categorical variable (see Fig. 2.2), corresponding to an ordinal measurement level. Based on the conclusion of Snedecor and Cochran (1989), one should generally not use analysis of variance, because there is no way to know if the numerical classes chosen have any relation to some unknown variable characterizing degree of infection on a continuous scale. In most cases, it is not possible to identify such variables. To analyze data of this type, one should use either: (1) non-parametric statistical methods as discussed by Sprent and Smeeton (2001) and Shah and Madden (2004); or (2) parametric generalized linear models for multiple-category data, as discussed by Schabenberger and Pierce (2002). These two general statistical methods could also be used to analyze directly the ratings obtained from disease scales with unequal class widths (such as the H-B scale), without the need for conversion of ratings to percentages based on class mid-points.

2.4.3 Remote-sensing and electronic assessment of disease severity

2.4.3.1 Spectral signature. The electromagnetic radiation emerging from a plant or plant canopy is determined by many interactions—reflections, transmission, and absorptions—between the incident radiation (joules per unit area) and the plant tissue (Nilsson, 1995; West et al., 2003). The nature of the interaction varies with the wavelength of the radiation. In broad terms, one can divide the electromagnetic spectrum into broad regions, the most important of which for studies of plant disease are:

- visible (VIS; wavelength ~400–700 nm),
- near-infrared (NIR; wavelength ~700–1200 nm),
- shortwave infrared (SWIR; wavelength ~1200–2400 nm), and
- thermal (TIR; wavelength ~10 μm).

The VIS region can be divided by wavelength into the violet-blue (~380–450 nm), green (~550 nm), and red (~650–700 nm) regions. The reflected radiation divided by the incident radiation for each wavelength and multiplied by 100 is the *spectral signature* of the plant or plant canopy. Because of the division of reflected by incident radiation, the spectral signature is unitless, and ranges from 0 to 100% at any wavelength (although the realized range may be much less than this). Through this signature, many aspects of crop growth and physiology can be characterized, including infection by plant diseases (Nilsson, 1995; West et al., 2003).

In broad terms, a healthy plant (or plant canopy) is characterized by low reflectance (as a percentage of incident radiation) in the VIS and SWIR regions, and high reflectance in the NIR region. Within the VIS region, there is relatively high reflectance in the green region, but low reflectance in the other visible regions of the spectrum. This is why plants appear green. For a plant disease with necrotic or chlorotic lesions on leaves, there is an overall increase in reflectance in the VIS region. More exactly, because of the reduction in chlorophyll, there is a reduction in reflectance in the green part of the VIS region, and an increase in reflectance in the red and blue parts. Moreover, reduction in total leaf area or biomass of plants, due to senescence, reduced growth, and defoliation, results in a decrease in reflectance in the NIR region.

Many other changes in the spectral signature can occur due to infection (see Nilsson, 1995; West et al., 2003). Some of these changes may be diagnostic for particular pathogens or types of pathogens, and may be exhibited before visible symptoms are obvious. Some of the changes, however, are evidence of general stress, and may not be diagnostic for infection, *per se*.

2.4.3.2 Multispectral radiometry. Reflected electromagnetic radiation has been found to be of value in measuring degree of infection of some plant diseases (Nutter and Schultz, 1995; Newton et al., 2004). A multispectral radiometer is used to measure reflected and incident radiation, in order to determine a spectral signature of the plant specimens (e.g., individual leaves, whole plants, sections of a field). Changes in percent reflectance at certain wavelengths or wavebands (i.e., fixed ranges of wavelength, such as 50-nm wide regions), or changes in reflectance in combinations of wavelengths, are associated with changes in the proportion of plant tissue with lesions (or other forms of disease severity). This is generally due to a reduction in chlorophyll content of cells. The methodology is known as *remote sensing*, which is a technique for measuring the characteristic manner in which a substance emits, absorbs, transmits, or reflects electromagnetic radiation using a sensing device that is situated at some distance from the surface of the substance of interest (Campbell and Madden, 1990). As used here, emphasis primarily is on reflectance. In one sense, of course, visual assessment of disease severity is a form of remote sensing since the human eye (the sensor) is recording a portion of the electromagnetic spectrum (i.e., visible light) reflected from the plant, and the eye is not in contact with the specimens of interest. As commonly used, however, the term remote sensing is used for photographic and electronic recording of the reflected radiation.

One can classify remote sensing methods, *inter alia*, based on the distance of the sensor from the plant or crop. At the lowest level, measurements are taken within ~2 m of the crop. At an intermediate level, sensors are ~75–1500 m above the canopy, usually in airplanes or hot-air balloons. At the highest level, sensors are housed in satellites orbiting the earth. For the purposes of this chapter, we are concerned with the lowest level for data acquisition, although all levels can provide useful information.

For many foliar diseases, reflectance (as a percentage of incident radiation) at the low end of the NIR region (~800 nm) has been found to be highly correlated with disease severity (as a percentage of leaf area covered by lesions). Increasing disease severity is associated with decreasing reflectance around 800 nm (Nutter, 2002a; Nutter et al., 1990). By conducting calibration studies, where reflectance and visual (or other) measures of severity are obtained for a particular pathosystem, one can develop a relationship between the two. A typical example of such a relationship is given in Fig. 2.5. On the basis of such a relationship, one can then use percentage reflectance (in the appropriate wavebands) to estimate severity (or degree of infection) for other specimens.

Sometimes, there is also a positive correlation between severity and reflectance in the red portion of the VIS region (~600 nm). Moreover, indices based on differences and ratios of reflectance in the VIS and NIR regions have also been found to be related to severity (Nilsson and Johnsson, 1996). To characterize other aspects of disease on the plant canopy, a wide range of wavebands can be used (West et al., 2003).

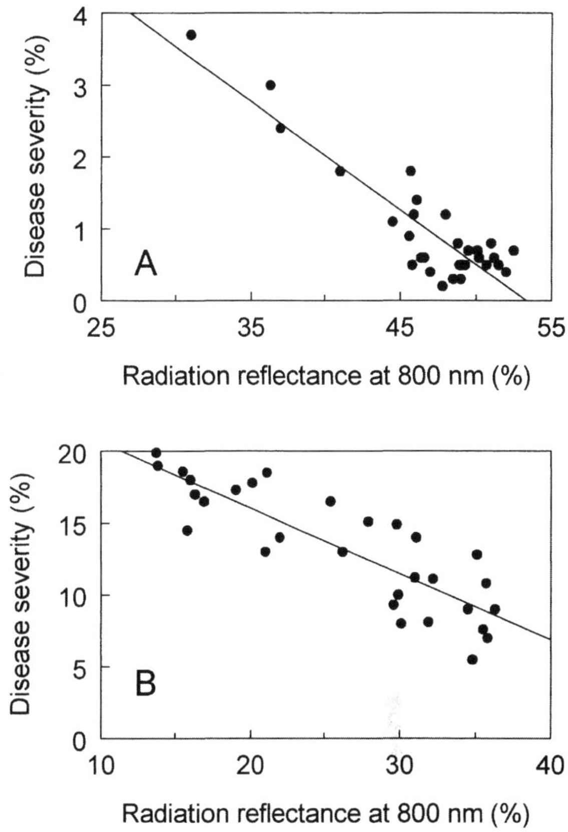

FIG. 2.5. Example relationship between severity of late leaf spot of peanut and reflected radiation at 800 nm wavelength as a percentage of the incident radiation (Nutter et al., 1990). Results for 11 September (**A**) and 7 October (**B**).

2.4.3.3 Image analysis. One specialized type of remote sensing of particular value in assessing disease severity is image analysis (Nilsson, 1995; Lamari, *ASSESS*, APS Press; Lindow and Webb, 1983). A plant specimen is recorded by some device, such as a video or still camera, and the recorded image is digitized into a large number of pixels, generally several thousand. With digital cameras, the digital image is directly determined. Then, the reflected light—usually just the VIS is used—is recorded for each pixel, and the pixels are classified based on the light reflectance. When the healthy and the infected plant tissue can easily be distinguished on the basis of color(s), it is relatively straightforward to place all pixels into one of two categories, representing healthy or infected areas of plant tissue on the image. Dividing the number of pixels representative of disease by the total number of pixels encompassing the plant specimen produces a measurement of disease severity.

There are at least two broad uses for image analysis in disease assessment. The first is for direct estimation of severity. That is, one uses this approach instead of visual assessment (or other method). This might entail taking photographs of specimens in the field and then determining severity in the laboratory using an image-analysis system. With portable equipment, the severity values could be determined right in the field. One could determine mean severities and other statistics from a sample of specimens. This approach might also entail destructive sampling of leaves or roots, where the specimens are actually brought into the laboratory for analysis.

The second use is in training. That is, visual assessments of severity are made as discussed in section 2.4.2, and the image analysis is used only to develop better visual assessment approaches. As discussed in section 2.5.5, one can determine the accuracy and reliability of raters' estimates of disease severity, and, in principle, raters can learn to improve their visual assessments. Image analysis of selected specimens can provide accurate standards for training purposes. This approach can also be used to prepare standard area diagrams for certain classes of severity, based on some real specimens, that raters can refer to when estimating severity (e.g., Falloon et al., 1995).

Most disease assessments in epidemiological studies rely on visual assessments. This is partly because of the costs and time involved in using image analysis. However, with advances in digital technology, inexpensive computers and cameras, and availability of software for processing and manipulating the data, the application of image analysis will continue to become more and more common.

2.4.4 Indirect measurement of severity

For certain diseases, direct visual or remote-sensing-based assessment of severity can be problematical. For example, without destructive sampling, one cannot "see" root disease severity. Thus, indirect approaches are sometimes required. With root diseases, one might assess plant senescence and wilting as indicators of diseased roots. Cooke (1998, 2006) should be consulted for more examples of indirect estimation.

Another approach to determine disease severity is based on disease incidence. At the individual plant-unit level, knowing that a single plant is diseased tells you nothing about the severity of disease. However, as discussed briefly in 2.2.2 (see Fig. 2.1), for populations of plant units (e.g., plants, leaves), there is often a relationship between mean severity and incidence (McRoberts et al., 2003; Seem, 1984). Based on the premise that assessing incidence is easier than assessing severity, one can use a relationship between severity and incidence to predict mean severity from assessments of disease incidence. This is discussed further in section 9.8.

2.5 Reliability, Accuracy, Agreement

2.5.1 General concepts

Ideally, a disease assessment scheme should be both reliable and accurate (Campbell and Madden, 1990; Nutter and Schultz, 1995). These terms can have more than one meaning in sampling and data analysis, survivor analysis, and measurement science, so it is important to or explain the terms. The definition of *reliability* of relevance here is: "the extent to which the same measurements of individuals [e.g., diseased specimens] obtained under different conditions yield similar results" (Everitt, 1999). There are two kinds of reliability of importance in assessing disease. Intra-rater reliability is the agreement between measurements of some property of specimens taken at repeated times by the same raters. For instance, intra-rater reliability is the extent to which a particular rater obtains the same estimates of disease severity for the same specimens at two or more different assessment times. This is also known as intra-rater variability, intra-rater repeatability, or simply, repeatability (Nutter, 2002a).

Inter-rater reliability is the agreement between measurements of some property of specimens among different raters. For instance, inter-rater reliability is the extent to which two or more raters obtain the same estimates of disease severity of the same specimens (at a single assessment time). This is also known as reproducibility or inter-assessor variability. High reliability (low variability) does not necessarily mean that measurements are close to the true values.

Accuracy is the degree of closeness of measured values to some recognized standard, true, or actual values (Everitt, 1999; Nutter, 2002a). For instance, accurate visual measures of disease severity are close to the true severity values. The concept of accuracy is also of importance in estimating parameters in models (including the mean), a subject we address in later chapters. For now, we look at accuracy strictly in terms of measurements. One obvious issue in disease assessment is the lack of knowledge of the true value for a given specimen. Thus, in some ways, accuracy can never be fully assessed. In many disciplines, where there are multiple ways of measuring a property of interest, one of the measurements is considered a "gold standard", and other measurements are compared with this (Lin, 1989; Shoukri, 2004). Gold standards may change as advances in science and technology result in new ways of measuring quantities of interest. For the purposes of disease assessment, especially for diseases with necrotic or chlorotic lesions, the image-analysis-based percentage of leaf area covered by lesions may be considered a gold standard, especially compared with visual estimation of severity.

With the assessment of disease on a single specimen, one can determine the difference in the measurement made at two different times, or by two different raters, or between the visual assessment and another assessment (e.g., based on spectral reflectance of radiation). These are all *components* of reliability. The difference between the measurement and the gold-standard value involves accuracy. However, a single number does not tell one anything about the reliability or accuracy of measurements of the population or sample of interest—the specimens in a field, plot, greenhouse, or laboratory. One must obtain measurements of disease for a sample of specimens from a larger population to have some quantitative information on measurement accuracy (Shoukri, 2004; Shoukri and Pause, 1999). In the following subsections, some relevant statistics based on measurements of disease are discussed. Also, in section 2.5.3, a general statistical framework is established for agreement studies. Once the statistical framework is established, it is more straightforward to consider formally the concept of agreement in relation to disease measurements.

Despite the importance of having reliable and accurate measurements of disease intensity, there have been relatively few studies in which these properties have been addressed. Cooke (2006), Nutter and Schultz (1995), Nutter (2002a), Nutter et al. (2006), and Nita et al. (2003) should be consulted for reviews of the subject.

2.5.2 Reliability

Fig. 2.6 is based on a small portion of a much larger data set in which reliability and accuracy were determined for

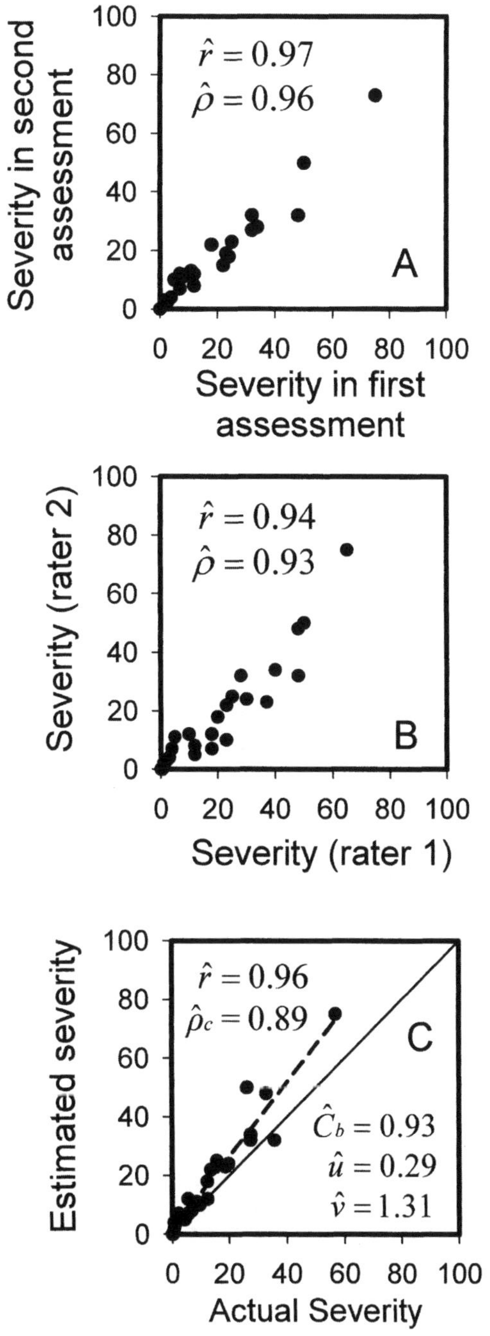

FIG. 2.6. A, Estimated severity of Phomopsis leaf blight of strawberry for the same rater at two different assessment times (intra-rater reliability). B, Estimated severity between two different raters at the same assessment time (inter-rater reliability). C, Estimated severity in relation to assumed actual (true) severity, as determined by the actual size of the leaflets and each lesion (Nita et al., 2003). Correlations (*r*) are shown for each case; moreover, the concordance correlation coefficient and related statistics are shown in C (see text for details).

visual assessments of the severity of Phomopsis leaf blight of strawberry (Nita et al., 2003). The top frame shows disease severity estimates for 25 leaflets, made by the same rater at two separate assessments. All the leaflets were assessed once, and then the rater assessed them again later (in a different order) without looking at the notes from the first occasion. Thus, these data provide evidence of the degree of intra-rater reliability. The middle frame shows disease severity estimates for the same 25 leaflets, made by two different raters at a single assessment. Thus, these data provide evidence of the degree of inter-rater reliability. Rater 1 in this graph is the one represented in the top frame of Fig. 2.6. It should be noted the strawberry leaflets were randomly chosen by a person who did not make the disease assessments.

These graphs show a strong relationship between the two variables (either the measurements made at two times, or by two raters). This type of graph suggests that the (Pearson product-moment) correlation coefficient, r, would provide a reasonable characterization of reliability. For continuous data, r varies from –1 to 1, with 0 indicating no relationship between a pair of variables (Shoukri and Pause, 1999). A value $r = +1$ means that there is a perfect straight-line relationship (with a positive slope) between the variables. Increasing variability of the data around a straight line is reflected by decreasing r. It is common in statistics for the term *precision* to be used for the degree of variability (Everitt, 1999), with greater variability being equated with less precision. In this sense, then, r is a measure of precision, suggesting that precision and reliability are synonymous. This is not quite true in the context of studying measurement accuracy and reliability (the current subject), although the synonym *is* appropriate for sampling studies (when measurement accuracy is not considered). For the top two graphs in Fig. 2.6, the estimated correlation coefficient was above 0.9, indicative of high reliability (high precision, low variability).

The requirement for $r = 1$ for continuous random variables is a perfect straight-line relationship between two variables. However, $r = 1$ does not mean that the two variables are equal. For example, it might happen that one of the variables is always twice the value of the other. Or, one of the variables is 10 points higher than the other over the entire range of measured values. Neither of these situations occurs with the example in Fig. 2.6.

One parameter that *can* be used to characterize both the variability and general disagreement (or agreement) between two variables is the *intra-class correlation* coefficient (ρ). The intra-class correlation, also known as the intra-cluster correlation, is calculated for a wide range of analyses for continuous and discrete data (Donner, 1986; Ridout et al., 1999). For instance, ρ is used in a very different manner for characterizing sampled disease incidence data in Chapter 9 (for problems that have nothing to do with measurement reliability). One way to calculate ρ is through analysis of variance using a random effects linear model (Donner, 1986; Shoukri and Pause, 1999). ρ is then defined as the ratio of inter-specimen variability to total variability.

It is generally accepted that ρ is a better indicator of agreement than r (Lin, 1989; Lin and Chinchilli, 1997; Shoukri and Pause, 1999). For a pair of raters or assessment times, and a continuous variable, ρ generally will be less than r. One additional advantage of the use of ρ is that it can be determined for any number of raters (or times), not just for pairs. For the example, estimated ρ values were almost as high as estimated r values, showing high agreement. However, because disease severity estimated by one rater is not necessarily more correct than that estimated by another rater, one may not need to know the agreement (or disagreement) in an absolute sense. In these circumstances, knowing the strength of the relationship (association), as quantified by r, may be sufficient if one simply wants to know how severity measurements depend on rater or time of assessment.

2.5.3 Accuracy

The bottom frame of Fig. 2.6 shows Phomposis leaf blight severity estimates on 25 randomly selected leaflets, made by rater 1, and the corresponding severity values determined by a combined photographic and weight method (Nita et al., 2003; Sherwood et al., 1983). The latter are taken to be the actual (i.e., true) severity values. To understand accuracy properly, some statistical results are needed. Here, these are presented in terms of characterizing accuracy, but the theory also has direct bearing on reliability.

Consider two random variables, W and U. For our purposes, we assume that W is the actual severity and U is the observed severity. Obviously, if measured and actual severity were equal, $U = W$. However, even with the best possible measurement approaches, one generally cannot expect such a result. Thus, one needs a statistical protocol to characterize the degree of agreement (or disagreement) between U and W. There are various approaches for this (Lin et al., 2002; Shoukri, 2004), and we present only a few of the possibilities.

The mean squared deviation (MSD) between U and W can be written as

$$\text{MSD} = E(U - W)^2$$

where $E(U - W)^2$ is the expected value of the squared difference. This expectation can be expanded using the rules of statistics and expressed as:

$$\text{MSD} = (\mu_U - \mu_W)^2 + (\sigma_U^2 + \sigma_W^2 - 2\sigma_{UW}) \qquad (2.1)$$

in which μ_U and μ_W are expected values (means), σ_U^2 and σ_W^2 are variances, and σ_{UW} is the covariance of the two variables. Each component is directly and easily estimated from a data set comprising corresponding values of U and W. The first term of equation 2.1 [$(\mu_U - \mu_W)^2$] is a measure of the squared *bias* of the measured variable because it characterizes the difference between the mean observed severity and the actual mean. The second term characterizes the total variability of the differences. The σ_W^2 value is strictly a property of the population of interest (plant specimens), and is unrelated to the measurement of severity. The other variance term (σ_U^2) and the covariance are functions of the ability to measure the actual severity; that is, σ_U^2 and σ_{UW} are indicators of the extent to which the observed severity values reflect actual severity. σ_U^2 could be different from σ_W^2 if, for example, the range of measured severities is less than (or greater than) the range of the actual severities. σ_{UW} equals a positive number if observed and actual severity increase together, and equals zero if observations are unrelated to actual values.

Perfect agreement occurs when: the means for the observed and actual severities are equal, the variance of the observed and actual severities is the same ($\sigma_U^2 = \sigma_W^2$), and the covariance equals the variance ($\sigma_{UW} = \sigma_U^2 = \sigma_W^2$). Then, MSD = 0. Obviously, one desires a small MSD for any measurement approach, but it can be difficult to interpret the MSD intuitively (other than when MSE = 0), since there is no natural upper bound. There are, fortunately, several elaborations for this measurement of agreement that are of intuitive value. We discuss in a little detail one that has become very popular in some disciplines. Lin (1989) developed a so-called *concordance correlation coefficient* (ρ_c) as a measurement of agreement that is a standardized version of MSD. In particular, to obtain an index in the form of a correlation coefficient, MSD is divided by the largest possible mean square of a difference for *uncorrelated* variables [$= (\mu_U - \mu_W)^2 + (\sigma_U^2 + \sigma_W^2)$], and the resulting value is subtracted from 1. After a little algebra, one obtains:

$$\rho_c = \frac{2\sigma_{UW}}{(\mu_U - \mu_W)^2 + \sigma_U^2 + \sigma_W^2} \cdot \qquad (2.2a)$$

The estimate is obtained by using means, variances, and covariances estimated from the sample of observed and actual severity values:

$$\hat{\rho}_c = \frac{2s_{UW}}{(\bar{U} - \bar{W})^2 + s_U^2 + s_W^2} \cdot \qquad (2.2b)$$

The desirable property of equation 2.2a is that values of ρ_c range from –1 to 1 when applied to continuous variables. ρ_c equals 1 when there is perfect agreement, and 0 when a plot of U versus W is a random scatter, which one can think of as total lack of agreement (i.e., where the observed value is independent of the actual). A negative ρ_c occurs when observed values decrease as the actual values increase, that is, when a plot of U versus W displays a negative trend. Less than perfect agreement could arise because of bias or variability in the observed data. It can be shown that equation 2.2a can be written as the product of two terms:

$$\rho_c = r \cdot C_b \qquad (2.3)$$

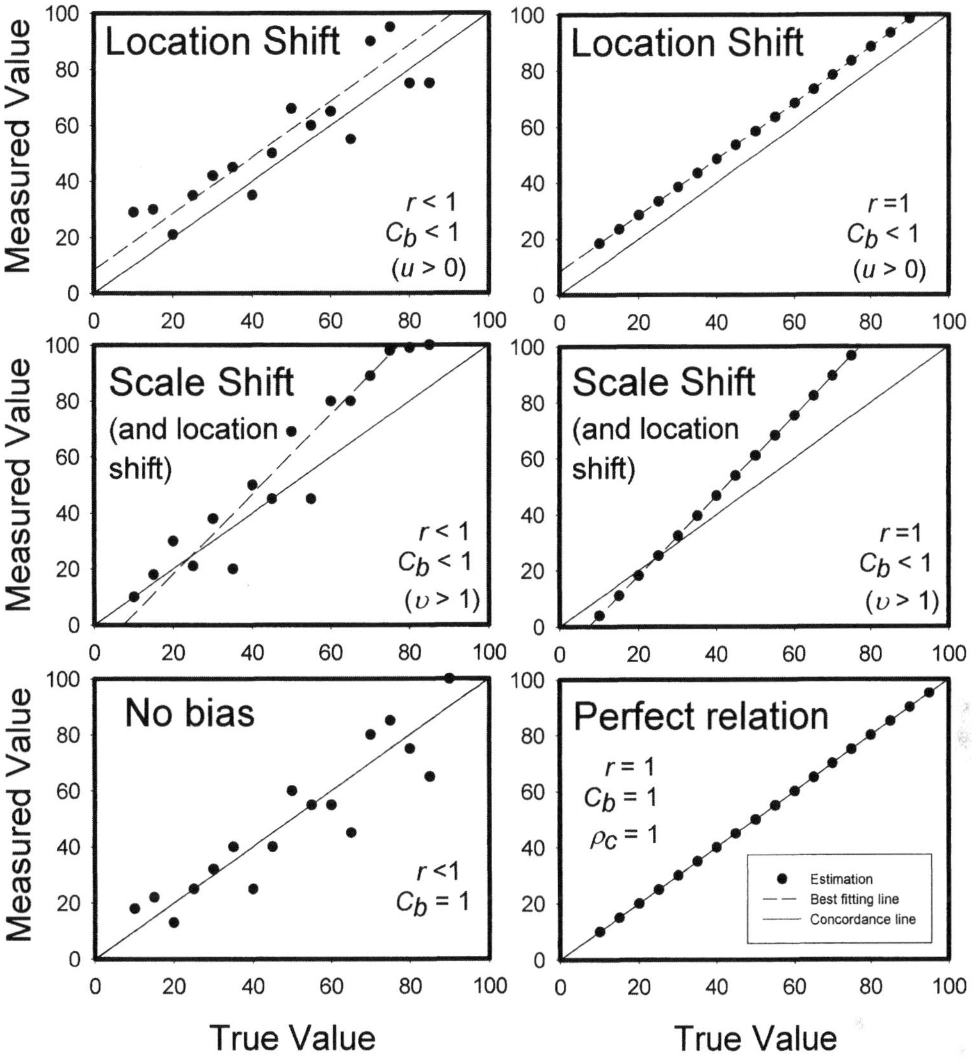

FIG. 2.7. Conceptual graphs demonstrating aspects of agreement between measured and true values, or between values obtained from two different measuring methods. Left-hand frames: a certain level of imprecision is manifested (correlation, r, less than 1). Right-hand frames: no imprecision ($r = 1$). Upper frames: systematic bias of measurements at all true values. Middle frames: bias in measurements that varies with true values. Bottom frames: no bias.

in which r is the usual Pearson product-moment correlation coefficient between U and W and C_b is a coefficient equal to

$$C_b = 2/(\upsilon + 1/\upsilon + u^2),$$

where $\upsilon = \sigma_U/\sigma_W$, and $u = (\mu_U - \mu_W)/\sqrt{\sigma_U \cdot \sigma_W}$. The correlation coefficient, r, is an indication of variability about a straight line [the best fitting line from least squares (see Chapter 3)], and thus is a scaled measure of precision, as mentioned in the previous section. The scaling guarantees that r must range from –1 to 1. The best fitting line is not necessarily the line of perfect agreement, however, as seen in the right hand side of Fig. 2.7. All these graphs for the relationship between measured and true values have $r = 1$, meaning there is an exact straight line between the two. However, only the lower right graph is representative of perfect agreement.

The C_b coefficient is an indication of the difference between the best fitting line and the perfect agreement (concordance) line, which is a straight line with slope of 1 and intercept of 0:

$$U = 0 + 1 \cdot W = W.$$

C_b equals 1 when the best fitting line is identical to the concordance line. $C_b < 1$ when there is a difference in the two lines. The four upper frames of Fig. 2.7 are for situations where $C_b < 1$. There can be two kinds of disagreement between observed and actual values. The first case is when there are larger (or smaller) observed values than the actual values, on average. This is known as a "location shift", and corresponds to u being larger than 0 (i.e., when the means for the observed and actual values are not the same) (top frames of Fig. 2.7). A location shift represents the classic case of bias; the situation where the slope of the best fitting line is 1 but the intercept is not 0. The

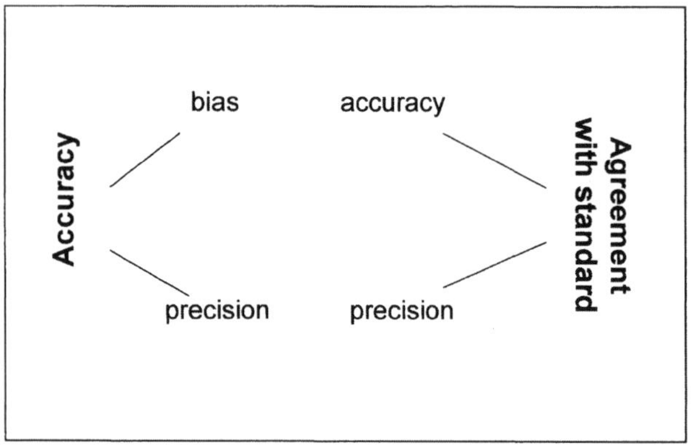

FIG. 2.8. Two versions of the conceptualizations of agreement, accuracy, bias, and precision.

second case in which $C_b < 1$ is when the slope of the best fitting line differs from that of the concordance line. This is known as a "scale shift" and corresponds to ν being different from 1 (i.e., when the variance for the observed values is different from the variance for the actual values). A scale shift represents the case where the slope of the best fitting line is not equal to 1; this means that the bias depends on the magnitude of the actual value. It is possible, of course, for there to be both a location and scale shift (see the middle frames of Fig. 2.7). Since C_b involves more than just the difference between the means, it is probably best to consider it a *generalized* bias parameter.

Lin (1989) has called C_b a measure of "accuracy" because it involves both the standard bias metric (difference between means of observed and expected values) and scale differences. Then, one can think of degree of agreement (as measured by ρ_c) to be the product of (scaled) precision and accuracy (Fig. 2.8). This conceptualization is very common in measurement science. However, using the general definition of accuracy as closeness of estimates to the true value(s) (Binns et al., 2000), then agreement and accuracy can be considered equivalent concepts when one of the variables is a set of data regarded as being the true values. With this conceptualization, accuracy (agreement with a standard) is a product of (scaled) precision and generalized bias (Fig. 2.8). This latter conceptualization is based on the premise that perfect accuracy can only occur if there is no variability and no bias. Depending on one's terminology preferences, perfect accuracy or perfect agreement with a standard is characterized by MSD $= 0$ and $\rho_c = 1$. The reader is reminded here that we use the term reliability when referring to agreement between measurements that do not necessarily involve true or standard values.

The concordance correlation coefficient has received considerable statistical attention in recent years for the study of measurement agreement (Barnhart et al., 2002; Carrasco and Jover, 2003; King and Chinchilli, 2001; Lin et al., 2002). It turns out that ρ_c is a generalized type of intra-class correlation (Carrasco and Jover, 2003; Lin et al., 2002); however, unlike the intra-class correlation coefficient, the standard formulation for ρ_c (eq. 2.3) allows for straight-forward and intuitive partitioning into components of bias and precision. In particular, through equation 2.3, one can determine the agreement of observed values to actual values, and for cases with $\rho_c < 1$, determine whether the disagreement is due to imprecision ($r < 1$), generalized bias ($C_b < 1$), or both. The situation with $r < 1$ and $C_b = 1$ would occur when the disease measurements are equal to the true values, on average, but there is (considerable) uncertainty about the closeness of any given measurement to the true value (lower-left frame of Fig. 2.7). On the other hand, the situation with $r = 1$ and $C_b < 1$ would occur when there are systematic differences between measurements and true values, with either a consistent difference ($u > 0$) or a difference that depends on the actual values ($\nu > 1$) (see upper right frames of Fig. 2.7). The latter case is the most troubling, because the error depends on the generally unknown true value. In most situations, both r and C_b are expected to be less than 1.

Returning to the example in the bottom frame of Fig. 2.6, one can see that the precision was very high for this rater, with a correlation coefficient of 0.96, but the overall agreement of measured severity with actual severity was not quite as high (0.89), although still high. In addition to the slight imprecision (variation about a best fitting line), there was some generalized bias ($C_b < 1$), with evidence of both a location shift ($u > 0$; shift in line height) and scale shift ($\nu > 1$; slope different from 1). It is important to note that visual disease assessment efforts are not all as precise or accurate as in this example (see Nutter and Schultz, 1995; Nutter et al., 1993). Even some other raters in the study by Nita et al. (2003) had poorer agreement with the assumed actual severities than the rater used in Fig. 2.6. Moreover, use of the H-B scale produced lower agreement than simple direct estimation of percentage severity in this study. This is further evidence against the principles underlying the H-B scale.

Standard errors can be calculated for estimates of r, C_b and ρ_c, or for transformations of the estimated coefficients (Lin, 1989). The formulae are quite complicated and are not repeated here. They can be calculated using a specialized program in SAS described by Lin et al. (2002). Confidence intervals can also be calculated. In practice, one can test hypotheses about achieved agreement by specifying (before the study is done) the minimum acceptable values of estimated r, C_b, and ρ_c. If the lower limits of the confidence intervals for these statistics are below the pre-selected values, then one concludes that acceptable agreement (accuracy) has not been achieved.

In some situations, one can use the concordance correlation coefficient (or MSD) to evaluate inter-rater reliability. For example, it may be felt that one rater is especially proficient at visually assessing severity. This person could be considered the gold-standard rater, and measurements by other raters can be compared to this rater's measurements. This approach can also be used to compare different approaches to measuring disease, such as spectral reflectance measurements and visual estimates (see Fig. 2.5). Shoukri (2004) and Shoukri and Pause (1999) discuss many other aspects of studying agreement. It should be noted that Barnhart et al. (2002) have recently extended the approach of Lin (1989) by showing how to simultaneously determine agreement of more than two raters at a time, using a generalized type of concordance correlation coefficient.

The concordance coefficient of correlation has not yet been used much in plant pathology. It is much more common to use regression analysis of U on W (see Chapter 3). In particular, one could write:

$$U = \beta_0 + \beta_1 W$$

where β_1 is the slope and β_0 is the intercept (value of U when $W = 0$). As mentioned above, a perfect agreement is represented by a best fitting line with slope of 1 and an intercept of 0. Regression analysis is a very useful method to describe and characterize relationships, and we use variations of regression methods throughout the book. However, regression analysis is *not* a good approach for testing for agreement (Lin, 1989; Shoukri and Pause, 1999). In particular, testing for the equivalence of the slope and intercepts to certain constants can lead to misleading results in agreement studies. When there is high variability (low precision), one can too easily accept that the slope and intercept are equal to the hypothetical values, leading to the false impression of agreement. Conversely, with very low variability (high precision), it is too easy to conclude that the slope and intercept are not equal to the hypothetical values, even when the best fitting line is very close to the concordance line. This would lead to the false impression of poor agreement. Thus, it is much better to use the concordance correlation coefficient [or MSD or related statistic (Lin et al., 2002)], and not regression analysis, to test for agreement.

As an aside, we point out that concordance correlation methodology can be used for several other purposes. For instance, the problems of using regression analysis for testing agreement also arise in model validation. That is, the degree of precision of model predictions (say, for crop loss equations) can unduly influence the testing of bias (Madden and Nutter, 1995). Thus, the agreement between model predictions and observations can often be better evaluated using concordance correlation methodology than regression analysis.

2.5.4 Ordinal and binary data

The above discussion was primarily focused on disease severity as a continuous variable (i.e., an interval- or ratio-level of measurement). For ordered categorical data (see Fig. 2.2), such as that obtained with an ordinal disease rating scale (see section 2.4.2.3), one can determine the degree of agreement of measurements using Cohen's (1968) so-called *weighted kappa* statistic. The statistic and its standard error are calculated by some standard statistical programs such as the FREQ procedure of SAS. It turns out that Cohen's weighted kappa is directly analogous to Lin's concordance correlation coefficient for continuous data (Shoukri and Pause, 1999). We do not give further details here. We do think, however, that such evaluations of ordinal disease assessments are warranted.

For binary data, such as disease incidence, measurements (visual or otherwise) of disease can take one of only two values (e.g., diseased or disease free). True incidence can take only one of these two values, as well. It is common in epidemiology to assume (perhaps implicitly) that estimates of incidence are more precise and less biased than estimates of severity (McRoberts et al., 2003). Thus, it is not common to consider the accuracy or reliability of measured disease incidence data. However, with increasing use of biochemical and molecular tests for disease (Fox, 1998), the question arises whether there is agreement between visual and other measurements of disease incidence. For instance, Nutter (2002a) showed that the percentage of plants visibly diseased by a virus was less than the percentage testing positive by ELISA. As will become apparent in Chapter 5, some disagreement may be temporary, because the biochemical test may detect infection before symptoms are visible (i.e, infections are in the "latent" state). Of greater interest are the disagreements that are not temporary, due, perhaps, to the inability of the rater to see the symptoms, or to the similar appearance of symptoms caused by more than one pathogen.

With binary data, there are many measures of agreement or association (Agresti, 1990; Gibbons, 1976; Schabenberger and Pierce, 2002; Shoukri and Pause, 1999). Most can easily be calculated using commercial software programs. In medicine, Cohen's kappa (for binary data) (Cohen, 1960) is the most popular index for determining agreement. It takes on a value of 0 for no agreement and 1 for perfect agreement. Kappa is a standardized version of the observed proportion of values in agreement minus the expected proportion in agreement by chance alone. A standard error of the estimated kappa can be calculated.

The calculation of Cohen's kappa is demonstrated in Fig. 2.9 for data taken from example 3.9 in Shoukri and Pause (1999). The data are from a medical study, but they are useful for demonstrating degree of agreement of two raters for any binary variable. One could consider one of the raters as providing gold-standard measurements. Estimated incidence of disease was close for the two raters (40 versus 37%), indicating little bias. A test for bias in this case can be done with McNemar χ^2-test (Shoukri and Pause, 1999); the test was not significant, indicating that one cannot reject the hypothesis of equal disease incidences by the two raters. Overall agreement

Rater 2 \ Rater 1	1	0	
1	a	b	a+b
0	c	d	c+d
	a+c	b+d	

Rater 2 \ Rater 1	1	0	
1	32	7	39
0	10	56	66
	42	63	

$$\text{Rater 1 incidence}: \frac{a+c}{a+b+c+d} = \frac{42}{105} = 0.40$$

$$\text{Rater 2 incidence}: \frac{a+b}{a+b+c+d} = \frac{39}{105} = 0.37$$

$$\text{Agreement (A)}: \frac{a+d}{a+b+c+d} = \frac{88}{105} = 0.838$$

Chance agreement (A_C):

$$\frac{(a+c)(a+b)+(b+d)(c+d)}{(a+b+c+d)^2} = \frac{42 \cdot 39 + 63 \cdot 66}{105^2} = 0.526$$

$$\text{Cohen's kappa}: \frac{A - A_C}{1 - A_C} = \frac{0.838 - 0.526}{1 - 0.526} = 0.658$$

FIG. 2.9. Demonstration of the calculation of Cohen's kappa for disease incidence. Disease indicated by a "1" and disease-free (healthy) by a "0".

consists of the number of values where both raters indicated diseased and both raters indicated disease free, divided by the total number of observations. For these data, agreement was 83.8%. However, there would be 52.6% agreement in this 2 × 2 table simply by random chance. Combining these two agreement values, one obtains 0.658 for Cohen's kappa (see details in Fig. 2.9). A 95% confidence interval for kappa, obtained with the FREQ procedure in SAS, was 0.51–0.81. Whether or not this is acceptable degree of agreement would depend on the objectives of the investigator. For instance, Landis and Koch (1977) would classify this data set as exhibiting "substantial" agreement because kappa was between 0.6 and 0.8.

2.5.5 Improving disease measurements

As mentioned above, a valuable reason for evaluation of agreement—reliability and accuracy—is to improve disease intensity, especially severity, measurements. One can determine whether particular methods or raters are better than others, and the types of disagreements that occur. For instance, there is some general evidence of (slight) overestimation of severity based on visual symptoms, and number of lesions can affect the estimate (at a given severity value) (Nita et al., 2003; Sherwood et al., 1983; Nutter, 2002a). If assessments are unsatisfactory in some sense for raters, it may be possible to improve assessments through training (Nutter and Schultz, 1995). Computer programs have been written that generate plant-specimen drawings (e.g., leaves) with varying number of lesions, of different sizes and possibly shapes, and occupying different positions on the specimen (Nutter, 2002a). The programs can provide either immediate or delayed feedback on actual severity percentages, in order to educate the user on what different percentage areas actually look like. These programs can be very useful for training raters. The generated images were originally in black and white, just like the traditional standard area diagrams (Figs. 2.3 and 2.4). Newer programs use color images to mimic more closely actual plant specimens.

Training can be of value in improving assessments of disease. One way to document any change in bias, precision, or both is through the concordance coefficient (ρ_c). In particular, a pre- and post-training determination of accuracy could be performed, and the results compared. Some statistical methods have been developed for formally comparing the reliability or accuracy of measurement devices (of which raters are a particular example) (Shoukri, 2004).

2.6 Attributes and Properties of the Crop

Since this is a book about disease in populations, we have spent the majority of the chapter on disease measurements. However, since this is a book specifically about plant disease, we should not ignore the plant part of the system. Of course, some aspects of the crop are considered *implicitly* in any disease assessment procedure when one is visually observing symptoms on plants or electronically recording reflected electromagnetic radiation from plant surfaces. Depending on the objectives of the study, a large number of plant or crop properties could be explicitly measured or determined (see Chapter 3 in Campbell and Madden, 1990). In other situations, minimal crop information may be required. In this section, we briefly mention a few crop properties that can be important.

2.6.1 Some useful static and dynamic properties

Many of the crop attributes of value are obvious, such as plant species, cultivar or variety, growing location, time of planting and harvesting, and planting pattern (e.g., in rows, or broadcast). Planting density (seeds per unit area, plants per unit length of row, distance between rows) and area of planting (field or plot size) are important because they influence disease dynamics and crop yield.

The crop characteristics mentioned in the previous paragraph are mostly static in nature. There are also some important dynamic variables, that is, those that change over the growing season or other time period of interest. One is the *growth* or *phenological stage* of the crop at different times. Growth stages are key times in the life cycle of plants that have meaning in relation to crop physiology and/or yield (Campbell and Madden, 1990). Examples are given in Tables 2.1 and 2.2 for

TABLE 2.1. Some stages of wheat development based on the Feekes and Zadoks scales (Zadoks et al., 1974).

Feekes scale	*Zadoks scale*	*Description*	*Feekes scale*	*Zadoks scale*	*Description*
		Germination			**Inflorescence emergence**
	00	Dry Seed	10.1	50	First spikelet of inflorescence visible
	01	Start of imbibition	10.2	53	¼ of inflorescence emerged
	07	Coleoptile emerged from seed	10.3	55	½ of inflorescence emerged
			10.4	57	¾ of inflorescence emerged
		Seedling and main stem growth	10.5	59	Emergence of inflorescence complete
1	10	First leaf through coleoptile			
	11	First leaf unfolded			**Anthesis**
	19	Nine or more leaves unfolded	10.5.1	60	Beginning of anthesis
				65	Anthesis half complete
		Tillering		69	Anthesis complete
2	21	Main stem and 1 tiller			
	25	Main stem and 5 tillers			**Milk Development**
3	26	Main stem and 6 tillers	10.5.4	71	Kernel watery ripe
	27	Main stem and 7 tillers		73	Early milk
			11.1	75	Medium milk
		Stem elongation		77	Late milk
4–5	30	Pseudostem erection			
6	31	1st node detectable			**Dough Development**
7	32	2nd node detectable		83	Early dough
	33	3rd node detectable	11.2	85	Soft dough
8	37	Flag leaf just visible		87	Hard dough
9	39	Flag leaf ligule/collar just visible			
					Ripening
			11.3	91	Kernel hard (difficult to divide by thumbnail)
		Booting			
10	45	Boots just swollen	11.4	92	Kernel hard (can no longer be dented by thumbnail)

TABLE 2.2. Growth stages of peanut (Boote, 1982).

Stage	*Abbreviated title of stage*	*Description*
	Vegetative Stages	
VE	Emergence	Cotyledons near soil surface, with the seedling showing; some part of the plant visible
VO		Cotyledons flat and open at or below the soil surface
V−1 to V−(*N*)	First tetrafoliolate to *N*th tetrafoliolate	One to *N* developed nodes on the main axis (a node is counted when its tetrafoliolate is unfolded and its leaflets are flat)
	Reproductive Stages	
R1	Beginning bloom	One open flower at any node on the plant
R2	Beginning peg	One elongated peg (gynophore)
R3	Beginning pod	One peg in the soil with turned, swollen ovary at least twice the width of the peg
R4	Full pod	One fully expanded pod, to dimensions characteristic of the cultivar
R5	Beginning seed	One fully expanded pod in which seed cotyledon growth is visible when the fruit is cut in cross section with a razor blade (past the liquid endosperm phase)
R6	Full seed	One pod with cavity apparently filled by the seeds when fresh
R7	Beginning maturity	One pod showing visible natural coloration or blotching of the inner pericarp or testa
R8	Harvest maturity	Two-thirds to three-fourths of all developed pods showing coloration of the pericarp or testa (the fraction is cultivar-dependent, being lower for Virginia types)
R9	Over-matured pod	One undamaged pod showing orange-tan coloration of the testa or natural peg deterioration

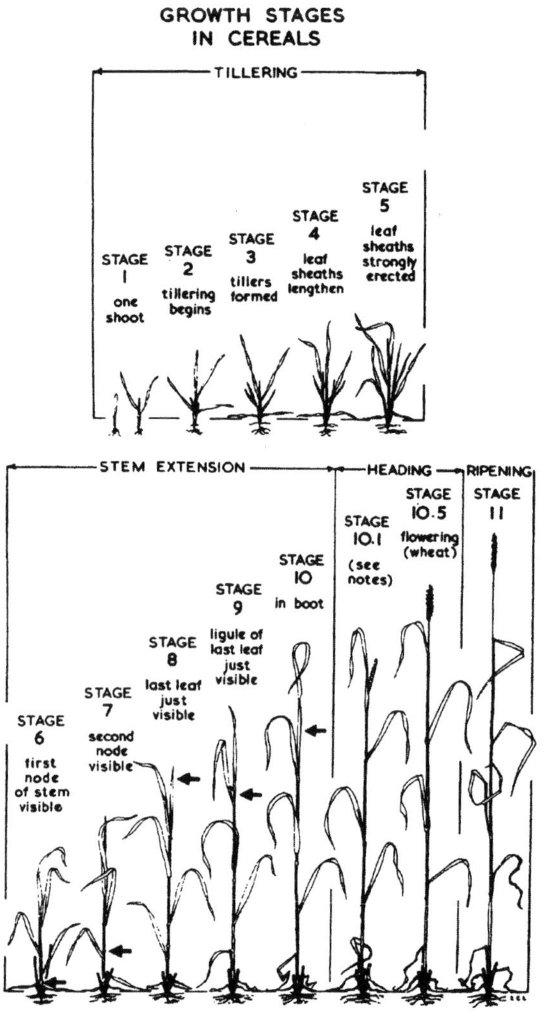

FIG. 2.10. Growth (phenological) stages of grains (e.g., wheat) based on the Feekes scale (James, 1971; Large, 1954).

wheat and peanuts, respectively. Often, drawings, pictures, or diagrams are used to represent specific growth stages. An example is shown in Fig. 2.10, which is known as the Feekes scale for growth stages (James, 1971). Events such as emergence of the top (flag) leaf and beginning of flowering, for example, are clearly identified. The numbering system, which is really an ordinal level of measurement (ordered categorical variable), consists of a mixture of integers, fractions, and "double fractions" (e.g., "10.5.4"). An alternative ordinal scale for growth stages, consisting of just integers (called a "decimal code") (Zadoks et al., 1974) is widely- used because it allows easy data entry into databases and spreadsheets (see Table 2.1).

For crops with clearly identifiable growth stages, it is desirable to record the stage at which each assessment of disease intensity is made. This makes it easier to compare results from different regions, or under different environments, or simply from different experiments. Moreover, since the yield of some crops is correlated with disease intensity (or other stresses) at certain stages (see Chapter 12), it is of practical advantage to provide growth stage observations with the measurements of disease.

2.6.2 Leaf area index

For studies on crop physiology, crop growth, and yield production, it is common to determine the total leaf area at various times because this is related to yield (Goudriaan and van Laar, 1994; Monteith, 1977; Watson, 1947). Rather than determining leaf area per plant, it is most useful to determine total leaf area per unit area of ground, called *leaf area index* (LAI). Since LAI is a ratio of two values, both with units of area (e.g., m^2), LAI is a dimensionless value. The value of LAI changes greatly during a growing season (often from <1 to >5), but characteristic values, and changes in LAI values over time, are known for various crops. The importance of LAI will become more apparent in Chapter 12 in terms of crop losses in relation to disease intensity. For crops in which other plant organs in addition to leaves (e.g., stems, pods) contribute significantly to photosynthesis, the terms green area index (GAI) or photosynthetic area index (PAI) are often used.

Direct determination of LAI generally requires the measurement of leaf area for a sample of leaves in a field or plot. This is most readily done now using an electronic and portable image analysis device, such as that sold by LI-COR (1990). There are also several indirect techniques of great value. Remote sensing methods, in general, can be used to estimate LAI based on the relationship between radiation penetration and canopy structure (Goel and Norman, 1990). For instance, the proportion of the sky visible through the canopy at various angles can be used to estimate LAI.

Another valuable indirect method is to use easy-to-obtain measurements of the crop to estimate LAI. For example, the mean width or length of leaves (perhaps at a particular height or position on the plant), or the mean height of plants can be used. This requires a calibration study, where a direct measurement of leaf area is obtained, in addition to an indirect measure, and a regression equation is developed (see Chapter 3). The objective is to be able to predict LAI from relatively easy-to-obtain measurements of the crop. Visual estimation with the aid of diagrams or keys can be used, as well (Campbell and Madden, 1990). This would be advantageous when no simple indirect method has been devised, or it is impractical to measure individual leaves, or electronic equipment is not available. Previously discussed methods for evaluating the reliability and/or accuracy of measurements apply to leaf area data, as well.

Determining root area or volume is of importance for many diseases, especially those caused by soil-borne pathogens. Root "size" is especially difficult to measure, and generally requires destructive sampling. English and Mitchell (1994) and Chapter 3 in Campbell and Madden (1990) should be consulted for details.

2.7 Conclusions and Prelude to Following Chapters

Plant pathologists typically measure the incidence, lesion counts, or severity of disease. Depending on the objectives of the study, all of these variables can be of value, as will be made apparent in the following chapters. The measurement of disease severity has attracted the most attention in epidemiology, probably because it is perceived that accurate and reliable measurements can be difficult to obtain. Often, aids such as disease diagrams and disease scales (such as the Horsfall-Barratt scale) are used. Although one can usually consider severity as a continuous variable, corresponding to an interval- or ratio-level of measurement (see Fig. 2.2), many researchers use an ordinal-scale measurement of severity. This may be for convenience, or because of difficulty in visually estimating severity on a continuous scale.

In principle, continuous variables provide more information than ordinal variables of the property of interest. However, if the measurement of the continuous variable is not reliable or accurate, then the perceived information content is illusory. More studies are needed to determine the reliability and accuracy of measured disease intensity for a range of disease types. These may show, among other things, whether a particular scale or diagram is of value, and what type of scale (if any) to use. Although scales in which the classes correspond to unequal ranges of percentages are commonly used, the limited evidence available does not support the nonlinear aspects of this type of scale. With the advances in remote sensing and image analysis, alternatives to visual estimation of severity are becoming more common. These approaches have much to offer epidemiology, although it is unlikely that visual estimation of disease intensity will disappear anytime in the near future.

In terms of any measurement, one can consider (at least conceptually) the actual value of the property (W) and the value obtained as the measurement (U). In all the later chapters, we use Y (or y) for disease intensity (whether incidence or severity) and do not, in general, distinguish between actual and measured values. This is partly because we often do not have enough information to distinguish between these. Readers should keep in mind, however, that the interpretation of results from studies of epidemics can be affected by the accuracy and reliability of the observed disease intensity data collected over time and space.

The analysis discussed in this chapter was strictly within the context of measurement of plant-unit (e.g., leaves, roots) specimens. In other words, for practical purposes there was no direct interest in using the estimated mean or variance to make statements about properties of a larger population, other than in terms of measurement reliability and accuracy. In later chapters, most of the focus is on disease intensity collected by one (or one type of) measurement device, and used explicitly for the purpose of making statements about the mean, variance, or other attributes of a larger population.

Some statistical methods were used in this chapter to deal with accuracy, precision, reliability, and agreement, in general. Many more statistical methods are used in the following chapters. Although it is not possible to teach statistics in this book on plant disease epidemics, knowing something about modeling is of value to readers who wish to understand statistical and mathematical analyses. Thus, an introduction to modeling and data analysis is presented in the next chapter.

2.8 Suggested Readings

Cooke, B. M. 2006. Disease assessment and yield loss. In: *The Epidemiology of Plant Diseases*, 2nd Ed. (B. M. Cooke, D. G. Jones, and B. Kayle, editor). Springer, The Netherlands, pp. 43–80.

Gaunt, R. E. 1987. Measurement of disease and pathogens. In: *Crop Loss Assessment and Pest Management* (P. S. Teng, editor). APS Press, St. Paul, MN., pp. 6–18.

Nutter, F. W., Jr. 2002. Disease assessment. In: *Encyclopedia of Plant Pathology* (O. C. Maloy, O. C. and T. D. Murray, editors) Wiley, New York, pp. 312–323.

3

Introduction to Modeling in Epidemiology

Do not worry about your difficulties in mathematics,
I assure you that mine are greater.

Albert Einstein

3.1 Introduction

Plant disease epidemics involve changes in disease intensity over time and space in a host population. If the entire host population of interest is very small, it may be possible to actually measure or observe every individual (e.g., plant) for disease intensity (and other variables of interest). However, even having such a complete collection of data for disease intensity does not automatically lead to any insight into epidemic processes or help in communicating results to others. In general, one uses models (as defined below) to summarize the essential features of the observations or measurements of interest. With models, the many individual observations are reduced to a few model terms, making it much easier to visualize and ultimately to understand the phenomena being studied, as well as to communicate that understanding to others.

In fact, the populations that epidemiologists are interested in are seldom very small. For instance, even when epidemics are studied in small experimental plots, where every plant (or leaf, fruit, etc.) is actually observed, the host population of interest is generally much larger, such as all plants of the same crop cultivar in commercial fields, grown under conditions similar to those studied in the small experimental plots. In other words, researchers usually want to apply results obtained with small data sets to larger populations that are not observed. Thus, models are needed here to allow inferences to be made about these large populations based on more limited direct information.

Epidemiology is a science that deals with relationships at the population scale. The fundamental relationship is between disease intensity (y or Y) and time (t), which in graphic form is known as the disease progress curve (see Fig. 1.1). Other relationships include disease intensity versus distance from an inoculum source (the disease gradient), and yield loss versus mean disease intensity for a field. Through various types of models, one can capture the essential features of these relationships in fairly simple and efficient form, facilitating understanding and comparison of epidemics.

There are many types of models, and even ways of labeling models. In this chapter, we first give a general overview of modeling, and then introduce some of the most commonly used models in plant disease epidemiology. This leads directly to the topic of fitting models to data. We briefly review some of the essential features of model fitting, emphasizing so-called least squares regression methods. This chapter is intended primarily for readers who do not have a great deal of experience in mathematical and statistical modeling of biological systems. Even those who have substantial experience with analysis of variance and linear regression analysis should benefit from reading about nonlinear models, since this material is usually not taught in introductory courses. By studying this chapter, readers will become better versed in the language of modeling and model fitting, and hopefully appreciate the quantification of epidemics presented in subsequent chapters. After studying the material presented here, readers should be able to fit certain types of models to data and interpret results obtained in model fitting.

3.2 Models

3.2.1 Definition and general classification

There are many ways of defining and classifying models, depending on one's perspective and the type of research one is conducting. A useful definition of a *model is a simplification of reality* (Edminster, 1978). Another valuable, if longer, definition is an abstraction of a real phenomenon or process that emphasizes those aspects relevant to the objectives of the user (Schabenberger and Pierce, 2002). The premise here is that any real phenomenon or process is very complicated and we may never have a complete understanding or description of reality. Even if a complete understanding was actually achieved, we would have no way of recognizing that we have reached this full understanding. The "reality" here could be anything from the multiplication and movement of viruses in plants to the spatio-temporal population dynamics of multiple diseases in large commercial fields.

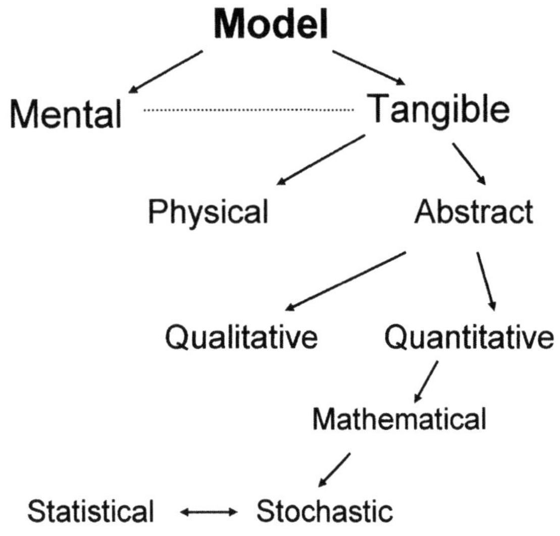

FIG. 3.1. Schematic showing a classification system of models, based on Edminster (1978) and Campbell and Madden (1990).

Even if we could have a complete description of a phenomenon or process, the description would be so complicated that simplifications would be needed to meet the objectives of the investigator in relation to understanding and/or managing plant disease epidemics.

Based on the concept that models are simplifications of reality, Box (1979), a famous statistician and heavy user of models, made the often-repeated and fanciful declaration that: "All models are wrong, but some are useful". That is, with the supposition that reality is so complicated that no model can capture all aspects of the reality of interest, there will always be some inaccuracy or incompleteness of any model. However, this does not mean that a model is necessarily wrong within the constraints of its intended use. Models are developed for many uses or objectives. The most common ones are *description, understanding, prediction, comparison, and communication*. A given model may be used for one or several of these objectives, although it is unlikely that a single model would be used for all of them.

It is important to note that the definitions above for model do not involve equations, symbols, functions or formulae. As anyone who has read some of the primary literature in epidemiology knows, models for epidemics generally *do* involve equations. However, these are special kinds of models. To gain a better appreciation of models and modeling, it is helpful to consider a classification system mostly based on Edminster (1978), which is summarized in Fig. 3.1. Models can be first classified into *mental* and *tangible* ones. A mental model is a conceptual image of a phenomenon or process (Campbell and Madden, 1990). We all use mental models to reduce the complexity of everyday reality to some manageable form. Of course, a mental model of any particular phenomenon or process depends on a person's previous experience and education. For instance, a practicing epidemiologist has a different (and, it is to be hoped, more detailed) conceptual image of an epidemic than does an agronomist or a schoolteacher.

A *tangible model* is a mental model that has been given some explicit form. By themselves, mental models are not very useful for communicating with others, so it is helpful to be able to express conceptual images in some tangible way. Examples could be word descriptions, flow charts, or equations. Tangible models can be of two types, *physical* and *abstract* (Fig. 3.1). A physical model represents some aspects of the form or function of reality in a suitably simplified form, generally using another physical medium. A classic example would be a plastic scale-model of an automobile. A well-known example from biology is the physical model on which the original proposal of a double-helix structure for the DNA molecule was based.

In contrast to a physical model, an abstract model is a tangible model based on the use of symbols and rules to represent aspects of the form and function of the particular reality of interest. Equations, flow charts and schematic drawings (say, of the structure of a chemical compound) may be used to capture aspects of the phenomenon or process in question. Abstract models can be considered either as either *qualitative* or *quantitative*. A qualitative model, as used, here, represents some essential aspects of reality in an abstract way that is of concern to the user, without the use of equations, mathematics or statistics. A disease cycle ["infection chain", see Gäumann (1946)] is a classic example of a qualitative model; it shows key components of the chain of events during and between epidemics without giving quantitative information on disease development. The disease triangle is another good example of a qualitative model (Francl, 2001).

A quantitative model uses mathematical notation, symbols, and rules, to represent some aspects of reality of interest to the user (Fig. 3.1). For our purposes, a quantitative model is a mathematical model, which is often simply called a *function*. The variable or variables in the model, however, can be either discrete or continuous (see Chapter 2). For instance, even if one is recording disease as a binary variable (a plant either is or is not diseased), one can still summarize the data in a such way as to be able to obtain an estimate of the probability of a plant being diseased, which is a continuous variable ranging from 0 to 1. In this book, we use quantitative or mathematical model in a very general way, and most of the models presented in this and other chapters can be considered mathematical models.

We realize there are more ways to classify models, and we present some of these below, but the system in Fig. 3.1 should serve to give a framework for the general concept of modeling. In the next section, some special kinds of mathematical models are explained.

3.2.2 Quantitative (mathematical) models—some general concepts

It may be easiest to discuss models by using a simple hypothetical example. This is done by linking an equation (a mathematical model) to its common depiction, the graph. Consider the upper frame of Fig. 3.2, which shows an example relationship between two variables, Y and X. We call Y the response or *dependent variable* and X the independent or *predictor variable*. Note that we continue to use Y or y for disease intensity for the most part in this book (with exceptions for a few special cases); upper case Y for intensity in absolute units (such as number of diseased leaves) and lower case y for proportions (such as the number of diseased leaves expressed as a proportion of the total number of leaves observed) (see Chapter 2). The dependent variable we are discussing in this chapter could be something other than disease intensity, but it is convenient to think of disease intensity as the response variable (for now). The original concept for the model terminology is the situation where Y is actually dependent on X, or that Y responds to X in a systematic way, or that Y is predicted from X. However, the relationship is not necessarily a cause-and-effect type, even though the names of the terms typically are still used. An example of a pair of response and predictor variables would be when Y represents spores per lesion and X represents lesion area. A quantitative model for this relationship is given by:

$$Y = \beta_0 + \beta_1 X \tag{3.1}$$

in which β_0 and β_1 are constants. In other words, the upper frame of Fig. 3.2 is the depiction of the function written as equation 3.1, when β_1 is greater than 0. When $X = 0$, $Y = \beta_0 + \beta_1 0 = \beta_0$; thus, β_0 is the value of Y when $X = 0$. The β_1 constant is the change in Y with unit change in X (if X increases from, say, 6 to 7, Y increases by β_1). Using terminology from statistics, when the constants are not known *a priori*, but are (ultimately) estimated from data, they are known as *parameters*. The term parameter may be used differently outside of statistics, but we use the statistical definition.

Equation 3.1 is not the only way to express the relationship in Fig. 3.2. For instance, taking the first derivative of Y with respect to $X(dY/dX)$ for this model results in:

$$\frac{dY}{dX} = \beta_1$$

In other words, the first derivative is a constant, equal to the β_1 constant of equation 3.1. Note that here and below, we do not teach how one calculates derivatives.

If Y does not truly depend on X, but is just associated with it, one could just as easily write the equation as X in relation to Y. Rearrangement of equation 3.1 produces:

$$X = \frac{Y - \beta_0}{\beta_1} = -\frac{\beta_0}{\beta_1} + \frac{1}{\beta_1} Y \tag{3.2}$$

Equation 3.1 is known as a *deterministic model*. That is, in the model Y is completely determined from X and the

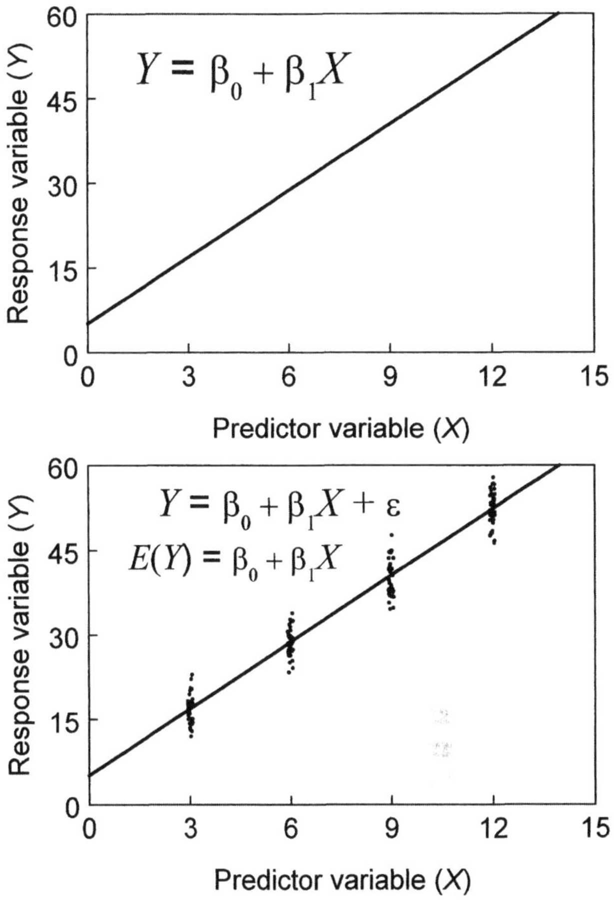

FIG. 3.2. Response variable, Y, versus predictor variable X. Upper frame: deterministic relationship (equation 3.1). Lower frame: deterministic plus stochastic relationship (points generated using equation 3.3a, with ε as a normal distribution with mean 0 and constant variance). The points are slightly scattered horizontally at X values of 3, 6, 9, and 12 in order to see the values better.

two constants (β values). If X equals 10, for instance, Y must equal $\beta_0 + \beta_1(10)$. Put another way, there is no part of the model that allows for any value of Y other than $\beta_0 + \beta_1 X$. Since models are simplifications of reality, one cannot expect the value of a response variable ever to be *completely* specified by a function such as equation 3.1, or indeed any other deterministic function of a predictor variable. The model can, however, be easily expanded in an efficient manner to encompass deviations in Y from that specified by $\beta_0 + \beta_1 X$. Consider the lower frame of Fig. 3.2. The line is the same as in the upper frame; now, however, there are points shown at four different values of X. In fact, there are 40 data points at each of these four chosen X levels, and the average (central value) of these values falls on the line. To represent each point on this graph, a random variable (ε) is added $\beta_0 + \beta_1 X$ to obtain the actual Y. The specific values of ε vary from observation to observation, and are not known *a priori*. The new model is written as:

$$Y = \beta_0 + \beta_1 X + \varepsilon \tag{3.3a}$$

One can think of ε as representing the combined influences of all factors or variables that affect Y, other than that represented by the deterministic part of the model ($\beta_0 + \beta_1X$). In other words, it is a term for the unexplained variation in Y. In typical formulations of models of this type (see section 3.3 below), it is standard to assume that the random variable ε has an expected value (mean) of 0 and a constant variance given by σ^2. Means of populations are known as expected values, and we can use the notation $E(\bullet)$ for the expectation. Thus, we write $E(\varepsilon) = 0$. It is further generally assumed that the ε values are independent (i.e., they do not affect each other). Finally, it is often assumed that the εs have a normal distribution. (See the next section for an introduction to the normal distribution function.) There are ways, fortunately, of dealing with violations of the assumptions of constant variance, independence, and normality, and some of these issues are addressed in this or later chapters, where relevant. The assumption of normality is not necessarily that important (Neter et al., 1983) and it is generally satisfactory to assume that ε has any continuous distribution for many models.

Recall that a random variable is an observation or measurement of a variable phenomenon (Chapter 2). Because Y is a function of a random variable (ε) in equation 3.3a, it also is a random variable. As indicated above, the β values are constants. X is a variable that could be either a collection of fixed values (e.g., of time, or distance), or an actual random variable. We take the usual approach (Schabenberger and Pierce, 2002) and assume the former (this simplifies the statistics considerably). Equation 3.3a is referred to as a *stochastic model* because it involves random variation, as represented by the ε term (in addition to the deterministic part represented by $\beta_0 + \beta_1X$). A stochastic model is a special form of a mathematical model (see Fig. 3.1). The deterministic component of a model could be very complicated, a simple function of X (as exemplified in equation 3.3a), a single parameter (β_0; i.e., $\beta_1 = 0$ in equation 3.3a), or even be 0 (i.e., $\beta_0 = \beta_1 = 0$ in equation 3.3a). Viewed from a slightly different perspective, ε is the difference between an observed value of the response value and the value specified by the deterministic component of the stochastic model (e.g., equation 3.3a), namely $\varepsilon = Y - (\beta_0 + \beta_1X)$. Often, ε is called the *error term* in the model. Use of "error" here does not imply a mistake, but is simply an indication of the discrepancy between the calculation from the deterministic part of the model and the observation. One could also refer to equation 3.3a as a *statistical model*. Schabenberger and Pierce (2002) prefer to label a stochastic model as statistical only if one or more of the constants or parameters are unknown and are actually estimated from observed data. In this section, since we have not addressed the issue of parameter estimation at all, we tend to refer to models such as equation 3.3a as stochastic.

Using statistical theory, we can define the expected (mean) value of the random variable Y by determining the expected values of all terms on the right-hand side of the equation. This can be written as:

$$E(Y) = E(\beta_0 + \beta_1X) + E(\varepsilon)$$

Expectations of constants are just the constants themselves. For the case when X is a collection of fixed values (e.g., time during the growing season), $E(\beta_0 + \beta_1X) = E(\beta_0) + E(\beta_1X) = \beta_0 + \beta_1X$. Thus, one can write equation 3.3a as:

$$E(Y) = \beta_0 + \beta_1X \qquad (3.3b)$$

(since $E(\varepsilon) = 0$). Equations 3.3a and 3.3b are equivalent ways of expressing the same relationship. Using this formulation, one can see that the error (ε, as defined above) is just $Y - E(Y)$, the difference between the observed value and its expectation at the given X. From a slightly different perspective, one can think of a stochastic (or statistical) model as being correct on average (as long as the deterministic component has been correctly specified).

For teaching purposes, sometimes it is useful to write equation 3.3a as a pseudo-equation. Then, the lower frame of Fig. 3.2 can be expressed as:

Response = (systematic or deterministic part)
+ (random part)

or

Response = (structure) + (error)

The structural (i.e., deterministic) part of the equation 3.3a is very simple (representing a straight line), but much more complicated deterministic functions are common in epidemiology. As indicated by equation 3.3b, the structural part is describing the mean (expected) response at each level of X, and the random part of the model is describing the deviation of the actual Y from the mean Y (for the specific level of X).

Consider the depiction of equation 3.1 for the situation with $\beta_0 = 5$ and $\beta_1 = 4$. When $X = 9$, the straight-line equation gives $5 + 4 \times 9 = 41$ for Y. Obviously, most points on the graph at $X = 9$ do not equal 41. For instance, the largest value of Y at $X = 9$ is 47. Here, ε equals $47 - 41 = 6$. In other words, one can write $47 = 5 + 4 \times 9 + 6$ for this specific value of Y. Since each value of Y can be different (potentially) even when X is the same, ε is different (potentially) for every observation. To keep track of the individual values, it is sometimes useful to add a subscript (j) to the terms of equation 3.3a that can change from observation to observation, giving:

$$Y_j = \beta_0 + \beta_1X_j + \varepsilon_j \qquad (3.3c)$$

With N observations in total, j ranges from 1 to N (i.e., $j = 1, 2, \ldots, N$).

3.2.3 Probability distributions

There is a probability distribution associated with every random variable, although it may not be known. For discrete data (e.g., number of spores on a lesion), the probability distribution is a function (equation) that specifies the probability of obtaining each possible value of the variable (e.g., probability of 0, 1, 2, and so on, spores). For discrete data, the probability distribution may be called the *probability mass function*. Properties of some discrete probability distributions useful in epidemiology are used extensively in Chapters 9 and 10, and little further is said about them here.

Because, by definition, the number of values taken on by a continuous variable is not countable (see Chapter 2), the meaning of a probability distribution for a continuous variable is a little more complicated. In this situation the distribution is a function that is used to specify the probability of a continuous variable falling within a particular *range* of values (Everitt, 1999). For continuous data, the probability distribution may be called the *probability density function*. In the realm of data analysis, it is common to assume that the variable of interest has a normal distribution, or at least that that the normal distribution provides a reasonable approximation to the unknown true distribution. The probability density function of the normal distribution can be written as:

$$f(Y_i) = \frac{1}{\sqrt{(2\pi)}\sigma}\exp\left[-\frac{(Y_i-\mu)^2}{2\sigma^2}\right] \tag{3.4}$$

in which μ and σ are constants (parameters), and exp[•] means that e (the base of the natural log system; $e = 2.718\ldots$) is raised to a power. In particular, μ represents the expectation (mean) of Y [$\mu = E(Y)$], and σ^2 is the variance of Y (i.e., σ is the standard deviation). To represent the parameters explicitly, the left-hand side of equation 3.4 could be written as $f(Y_i; \mu, \sigma^2)$, where $f(\bullet)$ indicates a function (equation).

Equation 3.4 is actually a stochastic model for the distribution of random variable Y. In fact, any statistical probability distribution can be considered a type of stochastic model because it describes aspects of variability through the calculated probabilities. In addition to their importance in fitting models to data, such distributions are of use in simulating biological and other processes (Renshaw, 1991).

3.2.4 Is the model linear?

An extremely important property of a model is its linearity (or lack thereof, in which case we are interested in its nonlinearity). The concept can be difficult to appreciate at first, especially for those who are not especially well versed in mathematics or statistics. We can apply the concept both to purely deterministic models (such as equation 3.1) and to stochastic models (such as equation 3.3a). For ease of presentation, we deal only with deterministic models here.

One can ask whether a model is "linear in the parameters" and/or "linear in the variables"? A model is linear in the variables if it does not contain powers, other transformations, or products of two or more variables on the right-hand side of the model equation. Equation 3.1 is linear in this regard. However, a model $Y = \beta_0 + \beta_1 X^3$ is not linear in the variables with regard to X. Of course, one can define a new variable as the cube of the original (i.e., let $X^* = X^3$); then the model $Y = \beta_0 + \beta_1 X^*$ is linear in terms of X^*. The following model, with two variables on the right-hand side (e.g., X_1 for temperature and X_2 for relative humidity), is linear in the variables:

$$Y = \beta_0 + \beta_1 X_1 + \beta_2 X_2$$

However, the following two expressions are examples of models nonlinear in the variables (in terms of the original two variables X_1 and X_2):

$$Y = \beta_0 + \beta_1 X_1 + \beta_3 X_2 \cdot \log(X_2)$$
$$Y = \beta_0 + \beta_1 (X_1)^2 + \beta_2 (X_2)^{1/2}$$

For a one-variable model, the derivative of Y with respect to X does not involve X when the model is linear in the variables. For example, as stated above, dY/dX is a constant for equation 3.1, not a function of X ($dY/dX = \beta_1$).

In working with models, especially in parameter estimation (see section 3.4), it is of relatively little concern whether or not a model is linear in the variables. Of much greater importance is whether or not a model is *linear in the parameters*. A model that is linear in this sense can be written as a summation of terms, where each term is a product of a parameter (i.e., a constant that is estimated from data) and a variable. Equation 3.1 is linear in the parameters. Using β with subscripts for parameters, some other models linear in the parameters are:

$$Y = \beta_1 X^{1/2} \tag{3.5a}$$
$$Y = \beta_0 + \beta_1 \ln(X + 1) \tag{3.5b}$$
$$Y = \beta_0 + \beta_1 X + \beta_2 X^2 + \beta_3 X^3 \tag{3.5c}$$
$$Y = \beta_0 + \beta_1 X_1 + \beta_2 X_2 \tag{3.5d}$$

where ln(•) is the natural logarithm (base e) function. These models can be written generally as:

$$Y = \sum_{k=0}^{K} \beta_k X_k \tag{3.5e}$$

where the subscript represents the different variables in the model (e.g., temperature and relative humidity) or different functions (transformations) of variables [e.g., X, $\ln(X + 1)$, X^3, etc.], and the βs represent different parameters. With this formulation, there are $K + 1$ parameters in the model (e.g., in equation 3.1, there are two parameters, so $K = 1$). For the mathematically inclined, linearity is defined in terms of the *partial* derivative of Y with respect to each parameter (e.g., $\partial Y/\partial \beta_1$), not the derivative(s) with respect to the predictor variable(s). For a linear model, the partial derivative is *not* a function of the parameter of interest. For instance, for equation 3.1, it can be shown that $\partial Y/\partial \beta_1 = X$, so the partial derivative of Y with respect to β_1 is not a function of β_1 and linearity with respect to the β_1 parameter is established.

Some of the examples above, including equation 3.1, contain a single parameter that does not multiply a variable (β_0). These models still meet the linear-in-the-parameters definition of equation 3.5e because a stand-alone additive term of β_0 is equivalent to $\beta_0 X_0$, where X_0 equals 1 for all cases ($X_0 \equiv 1$).

Models that are nonlinear in the parameters cannot be written as a summation of terms, as in equation 3.5. In particular, one or more of the parameters appear as exponents, or as functions that involve variables, or are multiplied or divided by other parameters. Using Greek symbols for parameters, a few examples include:

$$Y = \alpha \exp(\beta X) \tag{3.6a}$$

$$Y = \alpha + \ln(X - \delta) \tag{3.6b}$$

$$Y = 1 - \exp\left(-\left(\frac{X - \delta}{\beta}\right)^{\gamma}\right) \tag{3.6c}$$

$$Y = \beta_0 + \frac{\beta_1}{\beta_2} X \tag{3.6d}$$

Consider equation 3.6b. It looks little like linear equation 3.5b. However, one parameter appears inside the log function in equation 3.6b. In terms of partial derivatives, $\partial Y/\partial \delta = -1/(X - \delta)$; that is, the derivative with respect to δ is a function of both X and the parameter, making the model equation nonlinear with respect to the δ parameter.

Now consider equation 3.6d. In terms of the three parameters listed, it is nonlinear, since β_1 is divided by β_2. However, a constant divided by a constant is still a constant. Thus, one can think of β_1/β_2 as being equal to a (*new*) parameter labeled γ, in this case, equation 3.6d is linear in terms of parameters β_0 and γ.

The concept of linearity (or nonlinearity) is particularly important when it comes to the estimation of parameters from data. Parameter estimation is much easier for linear models, compared with nonlinear ones. This issue is dealt with on several occasions, starting in section 3.4.

Some models that are nonlinear in the parameters can be transformed algebraically into a linear form. Consider equation 3.6a, which is a classic model for exponential growth (see Chapter 4) of a population. Taking natural logs of both sides results in:

$$\begin{aligned} \ln(Y) &= \ln[\alpha \exp(\beta X)] \\ &= \ln(\alpha) + \ln[\exp(\beta X)] = \ln(\alpha) + \beta X \end{aligned}$$

This is a linear model for $\ln(Y)$ as the dependent variable ($=Y^*$) and X as the predictor variable. In fact, a straight-line model is produced. The new intercept parameter is $\ln(\alpha)$, which we call α^*, and the slope is β. We can write the new equation as:

$$Y^* = \alpha^* + \beta X$$

This model is not linear in terms of the original α parameter, but in terms of its natural logarithm. Models that can be transformed into linear form are called *intrinsically linear*.

Unfortunately, many nonlinear models are not intrinsically linear. Consider equation 3.6c, which is known as the Weibull function and describes and increase in Y from 0 at $X = \delta$ to 1 at large X. Use of logs and some algebra results in the transformed version:

$$\ln(-\ln(1 - Y)) = -\gamma \ln(\beta) + \gamma \ln(X - \delta)$$

which is an equation for a straight line relationship between $Y^* = \ln(-\ln(1-Y))$ and $X^* = \ln(X - \delta)$. The slope is γ and the new intercept is $-\gamma\ln(\beta)$. However, this model is still nonlinear because the parameter δ appears as part of a log function involving variable X. Further algebra cannot separate out the δ term to produce a linear model. If δ was known (e.g., $\delta = 0$), and so was not a parameter to be estimated, one can consider this new model as being linear. A different algebraic manipulation of equation 3.6c produces another straight-line equation:

$$(-\ln(1 - Y))^{1/\gamma} = -(\delta/\beta) + (1/\beta)X.$$

This model is also nonlinear, because γ appears as the exponent of a variable, and on the left-hand side of the model equation. If γ was known (e.g., $\gamma = 1$) and so was not a parameter to be estimated, then this model is linear, with slope $1/\beta$, and intercept $-\delta/\beta$. Finding a linearized form of a nonlinear model can be tricky, and may involve several algebraic steps. We demonstrate the approach with another nonlinear model in 3.5.2.

3.3 Methods of Model Development

Models for biological or other processes can be developed in numerous ways. The methodology of model development provides another useful classification system for models. We first present two approaches for model development in an idealized sense, and then show that in practice many models are developed using a hybrid of both approaches.

An *empirical model* is developed to describe an observed process, phenomenon, or relationship between variables, using accepted statistical principles, and does not use previously developed theory or concepts to establish the relationship between the response variable and predictor variables. In other words, empirical models have the potential to provide a good fit to observed data, but in so doing do not take into account the biological mechanisms (molecular, biochemical, population, etc.) that generate the relationship characterized by the data. For instance, given only the data points in the lower frame of Fig. 3.2, and no information on the mechanism(s) that generated the observed relationship between variables Y and X, it is natural to consider a straight-line equation (such as equation 3.1) as an empirical model to represent the relationship. Empirical models may also be referred to as correlative models or descriptive models. These models are especially useful in exploratory data analysis for determining whether there is a relationship between two (or more) variables and for comparing the effects of experimentally imposed treatments (e.g., fungicide applications) on biological processes (e.g., epidemic development). Empirical modeling is the method of model development taught in many introductory data analysis courses.

The steps involved in empirical model development can be summarized as follows:

- Obtain data; that is, collect the observations or measurements of disease intensity and other variables of interest from a planned experiment or from a survey;
- Describe the phenomenon or process in question, generally using simple models (often, but not exclusively, of the linear type);
- Evaluate the fit of the model (see section 3.4.2);
- Use other empirical models if the fit does not appear to be acceptable (see section 3.4.5);
- Once a model is selected (on the basis of an acceptable fit), use it to achieve the previously-stated objectives, for example, predict future Y, compare treatments, determine if a variable significantly affects the response, and so on.

In contrast to empirical model development, a *mechanistic model* is developed by starting with a theory, hypothesis, or concept of how a (biological, in our case) phenomenon or process occurs. Data are only considered *after* a mechanistic model is developed. The steps involved in (quantitative) mechanistic model development can be summarized as follows:

- Represent the theory by specifying a quantitative model in mathematical notation;
- Obtain relevant data, either from new experiments or surveys, or from previous studies;
- Evaluate the fit of the specified model to the data;
- If the model is consistent with data (i.e., provides a satisfactory fit), then use the model to achieve previously-stated objectives;
- If the model is not consistent with the data (i.e., it provides an unsatisfactory fit), then reject the initial theory (as represented by the quantitative model) and consider how or why the initial theory does not (fully) apply to the data.

This may lead to specification of a new mechanistic model for investigation. Mechanistic models may also be referred to as explanatory, theoretical, biological, or physical models.

It may be argued (at least by some) that mechanistic models are superior to empirical models (Diekmann and Heesterbeek, 2000; Waggoner, 1990) for various reasons. For instance, there can be inadequacies with observed data used to develop an empirical model. As discussed in the previous chapter, measurements of disease severity are not (and probably can never be) completely accurate (reliable and unbiased). But even if the individual observations were fully accurate, only a fraction of a larger host population typically is observed in characterizing epidemics. Because of this, estimates of the typical response variable in epidemiological models, mean intensity, can be imprecise, depending (in part) on the spatial pattern or heterogeneity of disease intensity and number of samples used (see Chapters 9 and 10). Moreover, empirical models are often unsatisfactory in extrapolation, that is, in predicting Y at values of X outside the range of the data used in developing the model.

If we think that we understand a phenomenon, we should be able to specify a model for it, at least in a simplified way, before seeing the data. Population genetics and epidemiology are two areas of biology (not just plant pathology) where mechanistic models are commonly developed and used (Diekmann and Heesterbeek, 2000; Daley and Gani, 1999). For population dynamic processes, it turns out that almost all reasonable quantitative models used to characterize the mechanisms of biological processes are nonlinear in the parameters. This is of concern when a mechanistic model is specified but some (at least) of the parameter values are to be estimated empirically from data. Although mechanistic models appear at various places throughout this book, Chapters 5, and 8 have a special emphasis on the use of these types of models in understanding epidemic processes.

In practice, model development may involve both mechanistic and empirical approaches, and a model, once developed, may be considered neither fully empirical nor fully mechanistic. For instance, the theory used to develop a mechanistic model may itself have arisen from a consideration of previously observed data and empirical models fitted to those data. Mechanistic models are ultimately tested against data, using principles of empirical modeling for the appraisal of their validity. Thus, in epidemiology, a considerable amount of modeling, and related data analysis, is based on a combination of the methods of model development outlined above. For instance, in the next chapter, several alternative mechanistic models for disease progress curves are presented in some detail, and it is shown how to use modeling results to both understand epidemics (including the evaluation of control strategies) and to compare epidemics occurring under different conditions. The different models shown can all be developed mechanistically, based on some of the known or accepted properties of population dynamics for diseases and other organisms, independent of observed data points. Using wording of Cousens (1985) for a different situation, such models are constructed "based on biologically sound premises", whether or not they account for all the relevant features of the reality being represented. However, until there are data for comparison, it may not be clear which of the alternatives is most appropriate, among those that are candidates as mechanistic models for the process under study. This leads to an empirical approach, where a few candidate models are evaluated for their fit to data points. The candidate models are *not* chosen from the collection of all possible models that will provide a satisfactory fit to the observed data, but rather from the limited number of previously developed models that are based on sound biological premises. This general approach to model development can be called *semi-empirical model* building (Cousens, 1985), and the approach is followed often in botanical epidemiology.

To conclude this section, we point out that intense argument about the general value of empirical and mechanistic modeling is not really productive. Models are developed and used for many purposes, and different objectives may require different types of models. The relevant issues are whether or not the type of modeling used is reasonable for the objectives of the investigator, and (when model fitting is being done) whether or not the method of model fitting is consistent with the type of random variable being analyzed, the nature of the process being represented, and the manner in which the data are obtained. Throughout this book, readers will find a wide range of models, and a corresponding range of approaches to model fitting. Because plant pathologists study epidemics for many reasons, many modeling approaches are used, and there is probably no textbook in mathematics or statistics that covers the entire range of models (and associated analyses) used here. Thus, in the various chapters and sections of chapters, we direct the reader to books and articles to gain more background on the topics being presented.

3.4 Fitting of Linear Models to Data

"Data! data! data!" he cried impatiently. "I can't make bricks without clay".
Sherlock Holmes ("*The Copper Beeches*" by Arthur Conan-Doyle)

3.4.1 Introduction

Although some research in epidemiology only deals with data indirectly, most quantitative research involves observed data *and* models that may describe the data. Here, and elsewhere, we use data to refer to a collection of observations or measurements. Usually, models are fitted to data, whether the models were developed strictly for descriptive purposes, or were developed based on elaborate theoretical considerations of the biological processes involved in epidemics. In this section, we briefly discuss some standard methods for fitting models to data. Other methods, somewhat less common than those mentioned here, are discussed elsewhere in the book with regard to specific applications. This presentation is intended for those readers with little familiarity with statistical model fitting. By studying the following sections, it will hopefully be easier to follow the presentation in other parts of the book and to fit linear models to data. To conduct research in epidemiology, however, individuals will need to learn considerably more about model fitting than what we present here. Some excellent textbooks, not specifically orientated towards plant disease epidemiology, include: Collett (2003), Neter et al. (1983), Draper and Smith (1998), and Schabenberger and Pierce (2002). The last book is especially helpful, since it deals exclusively with agricultural (i.e., plant and soil) topics.

3.4.2 Least squares regression—general concepts

Fitting models to data is equivalent to estimating the parameters in a model such as equation 3.1. The two most common approaches to estimate parameters of models are the method of *least squares* (or least squares regression) and the method of *maximum likelihood*. The two approaches can give the same results under *some* circumstances. In this section, the emphasis is on least squares regression, and we consider models that are linear in the parameters. For such linear models, the method is often referred to as linear least squares, ordinary linear least squares, or ordinary least squares regression.

We present the concept of least squares regression through examples, one taken from the plant pathology literature and one small hypothetical one. Jeger and Lyda (1986) studied the relationship between incidence of Phymatotrichum root rot of cotton (caused by *Phymatotrichum omnivorum*) and ambient weather conditions in Texas. Their goal was to develop a predictive model for this disease based on easily obtained environmental variables. Incidence of disease was determined yearly in field plots and empirical models were developed to predict incidence based on temperature and precipitation over periods of time during the growing season. Fig. 3.3A is a display of the relationship between final incidence in each year and accumulated precipitation (in centimeters) during the growing season (up to 31 August),

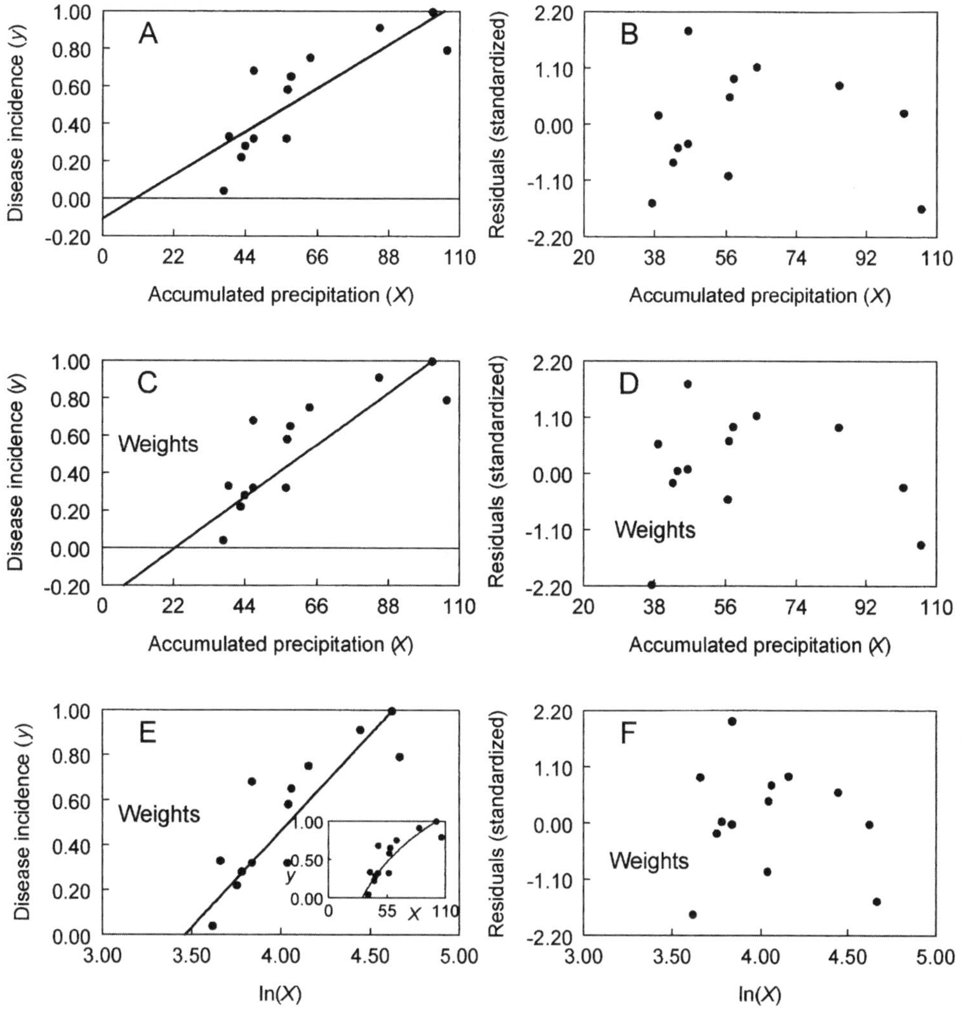

FIG. 3.3. Incidence of Phymatotrichum root rot of cotton at the end of the growing season (y) and cumulative precipitation (X) in centimeters up to 31 August during the growing season. (**A**) fit of equation 3.7c based on ordinary least squares. (**B**) standardized residuals from A. (**C**) fit of equation 3.7c based on weighted least squares. (**D**) standardized residuals from C. (**E**) fit of equation 3.17 based on weighted least squares. (**F**) standardized residuals from E. Inset in E: same results as in E, but shown on original X scale. Data taken from Jeger and Lyda (1986). Note: the time period for cumulative precipitation used here is somewhat different from that used in the model developed by Jeger and Lyda.

together with a best-fitting line that is explained below. The data are given in the Appendix of this chapter.

Disease incidence is a discrete variable, specifically a count with a natural denominator [i.e., number diseased (Y) out of a total of N observed] (see Chapter 2). As usual, we use lower case y to indicate incidence as a proportion ($y = Y/N$), so we present the models here with y on the left-hand side. There are specific model-fitting methods for discrete response variables (when the actual count and N are known), some of which are covered in other chapters (see also Collett, 2003); however, if the number of observations made to determine incidence is fairly large (>30), then y can be approximated as a continuous random variable and represented with a continuous probability distribution (Neter et al., 1983). The assumption of a continuous random variable is very often made in model fitting, and we do so here.

We consider disease incidence to be a deterministic function of accumulated precipitation (X) and a stochastic (error) term with a normal distribution. We can use a general expression, $g(X)$, to represent the deterministic function. Using the convention of a j subscript to indicate the specific observation, we can write a model for incidence as:

$$y_j = g(X_j) + \varepsilon_j \tag{3.7a}$$

We assume that the random variable ε has a mean of 0 and a constant variance of σ^2, and that the ε values are independent. We consider alternatives to the assumption of constant variance below. Taking expectations of both sides of equation 3.7a, one obtains:

$$E(y_j) = g(X_j) \tag{3.7b}$$

In other words, the deterministic function of $X[g(X_i)]$ gives the expected (average, mean) disease incidence at X_j, $E(y_j)$. It is equivalent, then, to use either equation 3.7a or 3.7b for describing the relationship; the latter shows how the mean varies with X, and the former shows how the individual observations vary with X and with the random error.

The specific form of $g(X)$ could be based on mechanistic modeling (where a specific equation is developed before seeing the data points) or empirical modeling (where the data fully guides the choice of equation). We take the empirical approach here, for the most part. The data in Fig. 3.3A clearly do not follow a straight line, but it is informative to consider, as a first step, the fit of equation 3.7a when $\beta_0 + \beta_1 X$ is used for $g(X)$. The model then is written as:

$$y_j = \beta_0 + \beta_1 X_j + \varepsilon_j \tag{3.7c}$$

or

$$E(y_j) = \beta_0 + \beta_1 X_j \tag{3.7d}$$

or

$$y_j = E(y_j) + \varepsilon_j \tag{3.7e}$$

These are all equivalent ways of writing the linear model. Equations 3.7c and 3.7d explicitly show the functional relationship between the response and predictor variables (a simple linear model in this case). In contrast, equation 3.7a and 3.7e are more general, emphasizing the stochastic nature of the response. Equation 3.7e demonstrates that an individual y value is composed of a mean and a random term, but does give any information on how $E(y_j)$ is related to X.

Fitting a model like equation 3.7a to data by least squares requires estimation of the parameters of the deterministic part of the model so that $g(X_j)$ comes "as close as possible" to the collection of data points. This means the same thing as finding parameter values so that the ε_js are "as small as possible", given the data points and the choice for $g(X)$. To distinguish estimates of parameters from the unknown parameters themselves, a "hat" is placed over the symbols when estimates are indicated (e.g., $\hat{\beta}_1, \hat{\beta}_0$). When one uses estimates of the parameters in the model, one calculates the so-called predicted or fitted value of y at each X. This is written as:

$$\hat{y}_j = \hat{\beta}_0 + \hat{\beta}_1 X_j \tag{3.7f}$$

It should be noted that $\hat{y}$ is not just a predicted individual value of y at a given X; it is also an estimate of the expected (mean) y at that given X (Neter et al., 1983). We do not provide the proof of this; however, a comparison of equations 3.7d and 3.7f may help the reader see that the right-hand side of the equations without an error term specify expected responses (either actual or predicted). Continuing, the difference between the observed y and the fitted y is the estimated error term for the model, usually known as the *residual*. This is written as:

$$\hat{\varepsilon}_j = y_j - \hat{y}_j = y_j - (\hat{\beta}_0 + \hat{\beta}_1 X_j) \tag{3.8}$$

Although the model parameters are unknown constants, their estimates are random variables. If one could repeat the same study, generating data several times, and fit the same model to the data from each repetition, different estimates of the parameters would (in all probability) be obtained each time.

The principle of least squares in model fitting is to consider the *sum of squared differences* between y_j and $g(X_j)$. One can define:

$$S = \sum_j \left(y_j - g(X_j)\right)^2 = \sum_j \left(y_j - E(y_j)\right)^2 = \sum_j \varepsilon_j^2 \tag{3.9a}$$

as the sum of squares based on any parameter values. For the case where equation 3.7c is the chosen model, S is given by:

$$S = \sum_j \left(y_j - (\beta_0 + \beta_1 X_j)\right)^2 \tag{3.9b}$$

With least squares, one finds the parameter values that give the minimum sum of squares (the "least squares"). In practice, one *could* try a large number of possible values of the parameters and determine the achieved S value for each, and choose the parameter values that give the smallest obtained S. This is equivalent (for equation 3.7c) to drawing a large number of possible lines through the data points in Fig. 3.3A and using the line that gives the minimum observed S. The method can be visualized by the simple hypothetical example with four points in Fig. 3.4. Three possible straight lines are drawn, and the residual ($\hat{\varepsilon}_j$; equation 3.8) for each of these is shown (as the thin vertical line) for the observation with the smallest X. The residuals for the other data points would be calculated in the same way for the different lines. To calculate S for a given line, the residuals for all the different X values are squared and totaled. In this case it turns out that the minimum sum of squares (for the three possibilities considered in Fig. 3.4) corresponds to the solid line.

Fortunately, because of the properties of linear models (Neter et al., 1983), trial and error consideration of parameter values is totally unnecessary. As shown in most elementary statistical textbooks, one can determine the minimum value of the sum of squared differences for any linear model fitted to a data set (equation 3.9a or 3.9b) through formulae derived from calculus. The formulae are not repeated here, but are incorporated into standard statistical analysis computer programs. When the least squares estimates of the parameters are inserted into equation 3.9a or 3.9b, we call S the sum of squares for *error* (SSE) or the residual sum of squares (RSS or SSR), and write it as:

$$\text{SSE} = \sum_j \left(y_j - \hat{y}_j\right)^2 = \sum_j \hat{\varepsilon}_j^2 \tag{3.10a}$$

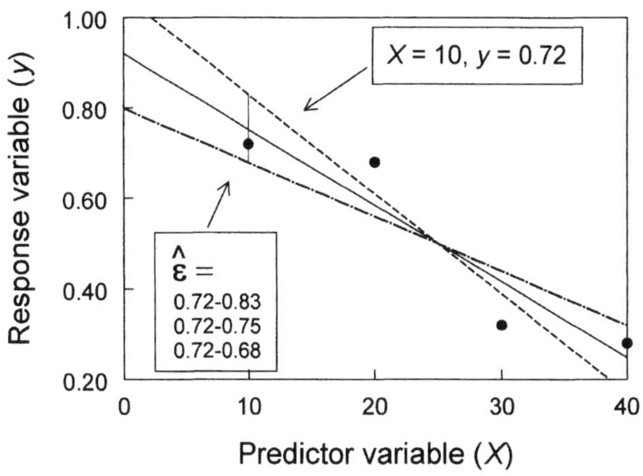

FIG. 3.4. Hypothetical example of ordinary least squares for model fitting. Some possible lines through data points are shown, and the solid line represents the least squares fit (equation 3.7f). For the first observation (lowest X), residuals (estimated errors) are shown for the three example lines.

For the case of the straight-line model, one can write:

$$SSE = \sum_i \left(y_i - (\hat{\beta}_0 + \hat{\beta}_1 X_i)\right)^2 \quad (3.10b)$$

Returning to the root rot example, the straight line in Fig. 3.3A represents the ordinary least squares fit of equation 3.7c to the data. The fitted model can be written as:

$$\hat{y}_i = -0.108 + 0.0105X_i$$

In other words, $\hat{\beta}_0 = -0.108$ and $\hat{\beta}_1 = 0.0105$. Then, for instance, with 60 mm of precipitation, the model provides a predicted incidence of $-0.108 + 0.0105 \times 60 = 0.52$. This is the estimated mean incidence for all observations with $X = 60$. These interpretations are based on the assumption that the correct model (i.e., the correct deterministic component) was chosen and all the statistical assumptions were met. These issues are addressed below.

Using the values of the parameter estimates in equation 3.10b, it is found that SSE = 0.33. It is pretty clear from Fig. 3.3A that a straight line does not provide a good fit of the data; in fact, the example was chosen for this reason. Before showing alternative models, we discuss the meaning of the results presented for this model, and show how to evaluate the model fit.

As indicated above for the general situation, the slope parameter (β_1) in a linear model such as equation 3.1 represents the change in y with change in X (dy/dX). Thus, $\hat{\beta}_1$ is an estimate of this derivative. The intercept parameter (β_0) represents the expected value of y when X is zero, and $\hat{\beta}_0$ is an estimate of this. In this case, the negative value for the estimate of the intercept may seem to be nonsensical, since disease incidence (by definition) cannot be below 0. However, there are no data points at $X < 30$, so there is considerable extrapolation to get to $X = 0$. Moreover, when the intercept is less than 0, this means that y is predicted to be 0 at some X greater than 0. This can be seen in Fig. 3.3A where the best fitting line crosses the thin horizontal line (which corresponds to $y = 0$). The point where the fitted and horizontal lines cross is found by placing 0 on the left-hand side of equation 3.7f, and solving for X (the outcome of which we call $\hat{X}_0$):

$$\hat{X}_0 = \left(\frac{-\hat{\beta}_0}{\hat{\beta}_1}\right) = 10.3$$

This can be thought of as an estimated threshold value of X required for disease to occur. Note that this value of X is considerably below the smallest X value used, so one would need to be cautious in using this estimated threshold (and especially so here, since the simple straight-line model is not satisfactory). Evidence for a threshold would be stronger if there were observed data points with $X > 0$ for which $y = 0$.

If the fit of a model to data were perfect (so that $y = \hat{y}$ at every X), SSE = 0 and $\hat{\varepsilon}_i = 0$ for all observations. Values of SSE greater than 0, therefore, indicate some deviation from a perfect fit ($|\hat{\varepsilon}_i| > 0$ for some or all observations). The magnitude of SSE (when it is not equal to zero) depends on the number of data points and the scale (magnitude) of the y values. For instance, an SSE value of 0.1 would be considered relatively small if the response variable was the number of rust spores per leaf, but relatively large if the response variable was the proportion of diseased leaves. Thus, it is helpful to have a unitless measure of goodness of fit (or lack of fit). This measure is easily constructed by considering another sum of squares, this time for the difference between y and overall mean $y(\bar{y})$. We call this the total sum of squares (SST), which is written as:

$$SST = \sum_i (y_i - \bar{y})^2 \quad (3.11)$$

Conceptually, one can think of SST as the error sum of squares when one fits to data a straight-line model (equation 3.7c) in which the slope parameter (β_1) is set to zero. This model can be written as:

$$y_i = \beta_0 + 0 \cdot X_i + \varepsilon_i = \beta_0 + \varepsilon_i \quad (3.12)$$

Graphically, equation 3.12 specifies a horizontal line, where the height of the line is β_0. For linear models, SSE is always less than or equal to SST. The ratio of SSE to SST is the proportion of variability that is unexplained by the model (i.e., the one that includes X). Thus, the proportion of *explained* variability is determined by subtracting SSE/SST from 1. One defines the coefficient of determination (R^2) as:

$$R^2 = 1 - \frac{SSE}{SST} \quad (3.13)$$

which, for linear models, ranges from 0 to 1. Sometimes R^2 is multiplied by 100 so that it is presented as a percentage. R^2 does not depend of the scale of the response variable or of the number of data points. For the Phymatotrichum root rot example, SST equals 1.043, so that $R^2 = 1 - (0.331/1.043) = 0.683$.

3.4.3 Distributional results

It is often assumed that ε (and hence y) has a normal distribution. If the dependent variable is normally distributed, then the estimated parameters of a linear model and the predicted y based on the estimated parameters ($\hat{y}_i$) also have normal distributions. Even when the distribution of y is not normal, the estimated parameters may be approximately normally distributed at typical sample sizes (Schabenberger and Pierce, 2002). We assume here that the assumption of normality is reasonable for the data in the example.

The estimated variance of the y (or ε) values is called the mean-square error (MSE) or residual mean square:

$$\hat{\sigma}^2 = s^2 = \text{MSE} = \frac{\text{SSE}}{N-2}$$

where N is the number of data points. The "2" in the denominator arises because in this case there are two parameters estimated in the model; for linear models generally, the denominator is the number of data points minus the number of estimated parameters. The estimated variance of $\hat{y}(s^2_{\hat{y}})$, as well as of the estimated parameters, $\hat{\beta}_0(s^2_{\hat{\beta}_0})$, and $\hat{\beta}_1(s^2_{\hat{\beta}_1})$, are all functions of MSE, and increase with increasing MSE. We do not show the formulae here since these are routinely calculated using standard statistical software for regression analysis.

It is often desirable to determine a confidence interval estimate for a parameter. To do so, one needs the (estimated) standard error of the estimated parameter, which is just the square root of its (estimated) variance. We use different convenient forms of notation for representing a standard error, both here and other places in the book. For instance, the standard error of the estimated slope can be written as:

$$s(\hat{\beta}_1) = s_{\hat{\beta}_1} = \sqrt{s^2_{\hat{\beta}_1}}$$

Although this formula is for the estimated standard error of the estimated slope, we often omit the first "estimated" for ease of expression. For the example, $s(\hat{\beta}_1) = 0.00217$. (More generally in some other chapters, we use SE for the standard error a random variable). A confidence interval is calculated as:

$$\hat{\beta}_1 \pm t_{\alpha/2,df}\, s(\hat{\beta}_1) \tag{3.14}$$

in which $t_{\alpha/2,df}$ is the upper $100\alpha/2$ percentile of the Student t distribution with df degrees of freedom. For a linear model with two parameters, $df = N - 2$, the same as the denominator in the formula for MSE. For the example data, $df = (13 - 2) = 11$. More generally, $df = N -$ (number of estimated parameters).

For a 95% confidence interval, one specifies $\alpha = 0.05$, so that one uses the t-value for an upper percentile of 0.025 (which may be given as the lower $1 - 0.025 = 0.975$ percentile in some references). The value of $t_{0.025,11}$ equals 2.2 here. Thus, a 95% confidence interval for the slope is

$$0.0105 \pm 2.2 \times 0.00217$$

which is the interval [0.0057, 0.0153]. One general way of interpreting this interval is that there is a probability $1 - \alpha$ that the calculated interval contains the true slope β_1. A more thorough explanation is that if the experiment resulting in the estimated β_1 was repeated a very large number of times, and the slope was estimated for each repetition, $(1 - \alpha) \times 100\%$ (e.g., 95%) of the calculated intervals would contain the true parameter value. In terms of hypothesis testing, for this example, any slope values from 0.0057 to 0.0153 cannot be rejected as an hypothesized slope of the model. If X really had no effect on y, then the true slope would be 0 (for a linear model). Since the calculated confidence interval for the estimated slope $\hat{\beta}_1$ does not include 0, there is evidence that X does affect y.

The confidence interval estimate for the intercept is calculated in the same way. Here, $s(\hat{\beta}_0) = 0.139$. The reader can confirm that the 95% confidence interval is [−0.414, 0.198]. Since this interval includes 0, there is evidence that the intercept is not significantly different from 0 (at $\alpha = 0.05$).

The concept of hypothesis testing mentioned above can be expressed much more formally. For instance, one can test the null hypothesis (H_0) that there is no relation between y and X, compared with the alternative hypothesis (H_a) that there is a relationship. For the simple linear model in equation 3.7c, it turns out that the null (H_0) and alternative (H_a) hypotheses are identical to:

$$H_0\colon \beta_1 = 0$$

$$H_a\colon \beta_1 \neq 0.$$

A t-test can be done to test this hypothesis. Specifically, one calculates:

$$t = \frac{\hat{\beta}_1}{s(\hat{\beta}_1)}$$

and rejects H_0 if the absolute value of t is larger than the $\alpha/2$ percentile of the tabulated Student's t distribution (with $df = N - 2$). The t statistic is routinely displayed by statistical computer programs used for the kind of data analysis described here. The t test will result in rejection of H_0 only if the confidence interval estimate for β_1 does not include 0.

A short comment on significance levels is warranted here to conclude this section. A significance level is a probability, and we use both P and α for probability in this and several other chapters, as well as the $Pr(\bullet)$ function. In general, P is used to represent the realized or *achieved* significance value when a statistical test is performed (i.e., the calculated probability of rejecting the null hypothesis when the null hypothesis is true), and it can take any value between 0 and 1. In contrast, α is an investigator-chosen specific value of the significance level. Any time that P is less than the chosen α, one rejects the null hypothesis (which is equivalent to rejecting the null hypothesis when the calculated test statistic is larger than the theoretical value calculated (or looked up in a table) with $P = \alpha$).

3.4.4 Model evaluation

The determination of estimated variances or standard errors, and their subsequent use for making inferences about parameters, are conditional on the use of an appropriate (i.e., reasonable) model for the data. Evaluation of the appropriateness of a model is often simply called model evaluation, model checking, or model criticism. Most evaluations focus on the residuals, i.e., the differences between the observed responses and those predicted by the model ($\hat{\varepsilon}_i = y_i - \hat{y}_i$). Often, the residual is represented by e ($\hat{\varepsilon}_i = e_i$).

If a model is appropriate, a plot of the residuals versus the predictor variable will be a random scatter. Like the unknown error term (ε), the estimated error [i.e., residual (($\hat{\varepsilon}$))] is a random variable. However, even if ε has a constant variance, the residual variance is not (quite) constant (Neter et al., 1983). The variance of the residual is proportional to MSE; in fact, MSE can be used as an approximate estimate of the residual variance, although statistical computer programs used for data analysis directly calculate the actual value for each residual. A standardized (or "studentized") residual ($\check{\varepsilon}$) is determined by dividing $\hat{\varepsilon}$ by the estimated standard deviation of the residuals (i.e., the square root of the estimated variance of the residuals). This standardization ensures that most of the $\hat{\varepsilon}$ values fall in the range from about -2 to $+2$; therefore, it is easy to identify unusually large or small values that might warrant some investigation of data-collection or data-handling procedures.

A plot of ($\check{\varepsilon}$) versus X is shown in Fig. 3.3B for the root rot example. Although there are no extremely large or small residuals, there is a strong pattern visible in the plot, with $\hat{\varepsilon}$ values tending to first increase and then decrease with increasing X. Thus, a straight-line model (equation 3.7c) is clearly not appropriate for the data. This conclusion can be just as easily reached by a plot of y and $\hat{y}$ versus X (Fig. 3.3A) in this example, because the departure from a straight line is so pronounced. However, the residual plot can be diagnostic for an inappropriate model choice when the evidence from the plot of y versus X is ambiguous. Moreover, the residual plot can easily be used for models with multiple predictor variables or when the model is not a straight line. For these situations, viewing the residuals is an essential part of model evaluation.

In addition to the evidence provided by studying the residuals against the use of equation 3.7c as an appropriate model for the data in question, there is another potentially important problem here. The standard (default) assumption is that there is a constant variance of y (or ε) for all data points. However, the response variable here is a proportion. As discussed in Chapters 9 and 10, the variance of a proportion is not expected to be constant, but is a function of the proportion. In particular, $\sigma^2 = \xi y(1-y)$, where ξ is a constant [i.e., the variance is proportional to $y(1-y)$]. For count data with a natural denominator (i.e., disease incidence; see Chapter 2), ξ is a function of the number of individuals observed in determining the proportion diseased and, for certain sampling approaches, the degree of heterogeneity of incidence from sampling unit to sampling unit. For the purpose of fitting a model for y as a function of X using least squares methodology, one only needs to know that the variance is proportional to $y(1-y)$, and not the actual value of the variance (that is, one does not need to know ξ).

The precision of a measured variable (see section 2.5) is inversely proportional to its variance. Therefore, it is often assumed that more weight should be given to observations with smaller variances (i.e., to more precise observations). One can attach a weight (w_i) to each data point, defined by:

$$w_i = \frac{1}{y_i(1-y_i)} \tag{3.15}$$

The sum of squares is now calculated by multiplying each squared difference between y and $\hat{y}$ by the corresponding weight given by equation 3.15. The error sum of squares of equation 3.10a is thus generalized to:

$$\text{SSE} = \sum_i w_i\left(y_i - \hat{y}_i\right)^2 = \sum_i w_i \hat{\varepsilon}_i^2 \tag{3.16a}$$

For the case of the straight-line model (equation 3.7c), as with the root rot example, one can write:

$$\text{SSE} = \sum_i w_i\left(y_i - (\hat{\beta}_0 + \hat{\beta}_1 X_i)\right)^2 \tag{3.16b}$$

In other words, parameter estimates are calculated that give the minimum possible value for equation 3.16a or

3.16b (for a given data set and model choice). Because the weights vary among observations, the estimates of slope and intercept that minimize equation 3.16b are not the same as the estimates that minimize equation 3.10b.

Use of equation 3.16b for fitting a model to data is known as *weighted least squares* or weighted linear least squares (when models such as equation 3.7c are used), or weighted regression analysis. Note that $y(1 - y)$ is at its largest value when $y = 0.5$, and declines towards 0 as y approaches either 0 or 1. Thus, the weights calculated from equation 3.15 are smallest in value at $y = 0.5$, and increase as y approaches 0 and 1. Thus, more weight is given in parameter estimation to y values near the extremes, and less to y values near the midpoint of the range of possible incidence values. The residual plot can be very useful for identifying situations where the magnitude of the variance is related to the response variable. In such cases, there will be regions of the residual plot where the variation in residuals (or standardized residuals) is noticeably larger than in other regions on the plot. Identification of regions where the variation is relatively large may require a considerable number of data points (certainly more than are available here).

Fig. 3.3B did not provide any real evidence of unequal variances, and so the need to use weighted least squares. However, we do so for demonstration purposes. Typically, in purely empirical models, one first attempts to find a reasonable model for the relation between y and X, and then determines whether weights are needed. However, we use weights here at the outset because of the known theoretical relationship between the variance and y for incidence data. Fig. 3.3C shows the fit of equation 3.7c to the root rot data based on weighted least squares. Model parameters were estimated as: $\hat{\beta}_0 = -0.283$ [$s(\hat{\beta}_0) = 0.105$], and $\hat{\beta}_1 = 0.0126$ [$s(\hat{\beta}_1) = 0.0012$]. In comparison to ordinary (i.e., unweighted) least squares, the slope was slightly higher and the intercept lower (Fig. 3.3C versus Fig. 3.3A); however, the difference between the two lines was minor. The standardized residual plot (Fig. 3.3D) was very similar to that obtained without use of weights; this is not surprising since the fitted line was not very different for the original (i.e., unweighted) one.

For the weighted root rot analysis, SSE = 2.50, SST = 29.51, $R^2 = 0.915$. Despite the larger R^2, the fit is not necessarily better. This is because in assessing a model fit based on weighted least squares, the summary statistics are also calculated using the weights. For instance, calculation of SSE (equation 3.16a) and the corresponding equation for calculating SST (not shown) involve the chosen weights. Thus, the magnitudes of these values, and those derived from them (e.g., R^2, MSE) cannot be directly and quantitatively compared to the values from ordinary (i.e., unweighted) least squares.

For alternative models in the next section, we continue to use weighted least squares, even though in this case the weights may not be necessary.

3.4.5 Model adjustments

Clearly, equation 3.7c is not satisfactory for representing the root rot data. If the model was chosen based on mechanistic grounds, one might stop here and simply conclude that either the theory is wrong or that it does not apply here for various reasons. However, as is often (but certainly not always) the case in epidemiological studies of disease incidence and composite environmental variables (such as cumulative hours of rain, or cumulative precipitation over time), there is no *a priori* theoretical model for the relationship under investigation here. It is a standard, and good, practice to consider alternatives to straight-line models for describing the data from field observations. The approach we explain here is based on a continuous response variable (or where a continuous distribution for the response variable is a reasonable approximation) and a single predictor variable. For examples with more than one predictor variable, readers will need to consult textbooks dealing in more detail with regression analysis in more detail than is possible here (e.g., Draper and Smith, 1998; Neter et al., 1983).

The plot of y versus X, and resulting residual (or standardized residual) plot, give guidance on modifying the empirical model. When a plot of y versus X is concave (i.e., decreasing slope of y versus X, so that there is "bending" of the curve towards higher X values), certain transformations of X are generally useful for obtaining a more satisfactory fit than is possible with a straight-line model for y versus X. In particular, with data of the form in Fig. 3.3A, instead of modeling y as a linear function of X, one often considers models for y as a function of $\sqrt{X}$ ($=X^{1/2}$), $\ln(X)$, or $1/X$. Both the logarithm and the reciprocal of zero are undefined, so for empirical modeling a small constant is added to X before either of these transformations are applied [e.g., $\ln(X + 1)$, $1/(X + 1)$]. For demonstration purposes, we consider only the (natural) log transformation, and fit the following model to the data using weighted least squares:

$$y_j = \beta_0 + \beta_1 \ln(X_j) + \varepsilon_j \qquad (3.17)$$

Equation 3.17 is linear in the parameters. One can think of $\ln(X)$ as a new predictor variable (X^*). The fit of this model to the data is shown in Fig. 3.3E, based on weighted least squares. Although the standardized residual plot (Fig. 3.3F) is still not ideal, this model does provide a reasonable fit to the data, and there was just a little curvature of y versus $\ln(X)$. Estimated parameters were $\hat{\beta}_0 = -2.97$ [$s(\hat{\beta}_0) = 0.285$], and $\hat{\beta}_1 = 0.858$ [$s(\hat{\beta}_1) = 0.0644$]. This can be written in equation form as:

$$\hat{y}_j = -2.97 + 0.858\ln(X_j).$$

Other statistics were: SSE = 1.72, SST = 29.51, $R^2 = 0.942$. Note that both the response variable and the weights are the same here as used when fitting simple equation 3.7c to the data using weighted least squares (as above). Thus, one can directly compare the goodness-of-fit results. SSE was reduced by 0.78 (2.50 − 1.72), indicating an improvement in fit. There was also an increase in R^2.

Because the predictor variable is different between equations 3.7c and 3.17, the parameters (or their estimates) have different meanings. The slope represents the change in y with change in ln(X) [i.e., $\beta_1 = dy/d\ln(X)$], not with change in X. In fact, if X^* is any transformation of X, then dy/dX cannot be constant if dy/dX^* is constant. This can be seen by the inset graph in Fig. 3.3E. Both y and predicted y ($\hat{y}$) from equation 3.17 are shown versus X, not versus ln(X).

The intercept parameter in equation 3.17 represents expected y when ln(X) equals 0, not when X equals 0. Because $\ln(1) = 0$, $\hat{\beta}_0$ is the estimate of expected y when $X = 1$. The negative value indicates that y is predicted to be 0 at some value of ln(X) greater than 0. This issue is especially relevant for the analysis of plant disease gradients with models that use log of distance as the predictor variable (see Chapter 7). A little algebra shows that the value of predicted X when y is 0 is estimated as:

$$\hat{X}_0 = \exp\left(\frac{-\hat{\beta}_0}{\hat{\beta}_1}\right) = 31.9$$

Note that this estimate of X_0 is much closer to the lowest observed X in the data set, where the observed incidence was near 0, than for the straight line model (for which $\hat{X}_0 = 10.3$, above). This is further evidence that equation 3.17 is a more appropriate model for describing these data.

With the empirical modeling approach, one would not necessarily just accept this "ln(X)" model without trying some alternative transformations of X, such as $1/X$, for the predictor. One can think of equation 3.17 as the (new) starting model, and determine whether other descriptive (generally, linear) models can be found that provide a better fit. We do not pursue this here. As discussed in section 3.5 (below), there are still some problems with the use of equation 3.17 (or similar ones) for representing the relationship described by the observed data.

We return briefly here to the general topic of finding an appropriate linear model when the straight-line model (equation 3.7c) between y and X is not satisfactory. If the plot of y versus X is convex to the X-axis (i.e., increasing slope of y versus X, so that there is "bending" of the curve upwards), which is not the case for the root rot observations, transformations of y (rather than transformations of X) can be considered. Standard transformations in this case include $\sqrt{y}$, ln(y), and $1/y$. When $y = 0$ is found in the data, a small constant is added to y before either of the transformations ln(y) and $1/y$ are applied, to avoid problems of undefined values of the transformed variable. An example of such a model would be:

$$\ln(y_j) = \beta_0 + \beta_1 X_j + \varepsilon_j \qquad (3.18)$$

Some forms of curvature in $y : X$ graphs are not readily described using linear models with just a transformation of X or y. In these cases, transformations of both the predictor and response variables may be very beneficial. For instance, a common model in many disciplines involves the log transformation of X and y. These so-called log–log models appear elsewhere in this book for different situations.

A word of caution is needed for those comparing the fit of various models that involve different transformations of y. When the models being compared involve only differences on the right-hand side of the equal sign [e.g., X, ln(X), etc.], direct comparison of statistics such as SSE can made in evaluating fits (when using a consistent model fitting method, with the same weights for different models). This is because the y values are the same for each model, and model evaluations are based on y, $\hat{y}$, and statistics based on these values (e.g., SSE). However, if one model involves y as the response variable and another involves ln(y), for example, then one cannot *directly* use SSE (and similar statistics) to compare fits, at least in a quantitative sense. This is because the left-hand-sides of the equations are different for the different models, so there is no common scale when calculating SSE. For instance, suppose that equation 3.7c *is* appropriate for a data set, but that one also fitted equation 3.18 to the same data. The SSE may well be lower for the fit of (inappropriate) equation 3.18 because the log transformation reduces the scale of the observed and predicted values. The scale changes for SST (equation 3.11) as well as for SSE when the response variable represents different transformations of y, so R^2 (equation 3.13) still reflects the relative fit of the data by the model, and provides some basis for comparing fits of different models. However, there is still not a common left-hand-side to the models, which means that one should be cautious in over-interpreting small "improvements" in R^2 when comparing models. This is relevant in modeling of disease progress curves (see section 4.6.2), where theoretical and empirical considerations lead to linear models of disease intensity that involve different transformations of y as dependent variables.

3.4.6 Other considerations

In empirical modeling of data, finding an appropriate equation to represent the data is the primary objective. Although there are useful formal tests to choose among some models and to determine the goodness of fit of a given model (Draper and Smith, 1998), simple residual plots and plots of the responses versus the predictor are often most beneficial for choosing a reasonable model.

More sophisticated and elaborate assessments of the residuals can also be performed (Neter et al., 1983). Once a reasonable model is chosen, the statistical assumptions underlying the model fitting (parameter estimation) are evaluated.

The main statistical assumption considered so far involves the variance of y (or ε). This is taken to be constant under the standard ("default") assumptions for least squares analysis, although weighted least squares can be used when constancy of the variance is not expected. Often, with empirical modeling, one chooses a reasonable model first [i.e., the form of $g(X)$ in equation 3.7a], and then addresses the underlying statistical assumptions utilized in model fitting. This is different from the approach taken above where the decision to use weights was made before finding an appropriate model. Our approach here was used for demonstration purposes and to emphasize that some random variables have a variance that is a function of the variable itself.

The other two standard statistical assumptions of relevance are that the ε values are independent and normally distributed. Least squares regression is robust to *reasonable* departures from normality under many circumstances (Neter et al., 1983). There are formal statistical tests as well as powerful graphical methods to evaluate departures from normality of the residuals, but we do not elaborate these further here.

A more serious assumption is independence of the errors. When not independent, estimates standard errors of the parameter estimates using ordinary least squares can be considerably smaller than the correct values (Neter et al., 1983). This affects calculations of confidence intervals and tests of the equality of parameters for different treatments. The issue is especially relevant for models of disease progress curves, and approaches for assessing this assumption, and making adjustments when the assumption is violated, are discussed in section 4.6.2.

3.5 Fitting of Nonlinear Models to Data

3.5.1 General considerations

As already mentioned above, many theoretical considerations in epidemiology, especially for population dynamics, lead to the development of models that are nonlinear in the parameters (e.g., equation 3.6a). Moreover, to account for some essential features of observed data, nonlinear empirical models may also be quite useful. Here, we briefly introduce some of the methodology for fitting nonlinear models to data, a very complicated and difficult subject. Other textbooks and articles must be read to gain a fuller understanding of the subject (e.g., Madden and Campbell, 1990; Ratkowsky, 1983, 1990; Schabenberger and Pierce, 2002). We use the Phymatotrichum root rot data set (Fig. 3.3A) as the main example to demonstrate the procedure.

Although equation 3.17 provides a fairly good description of the data, there are still some problems that, depending on the objectives of the investigator, may or may not be of concern. First, the fitted model predicts negative values for y at values of X below 31.9. As discussed above, it is easy to consider this value of X as a threshold, and just specify that predicted y equals 0 for all values of X below the threshold. However, such a conclusion would be more justified if there were actual (observed) values of X between 0 and 32 with corresponding observed y values equal to 0. On the other hand, it is entirely possibly that y is nonzero (but small) at X values $0 < X < 32$. Second, the predictions from the model at large X may become nonsensical. Using the estimated parameters for equation 3.17, one can determine the estimated X ($\hat{X}_1$) where one predicts 100% incidence (1.0). By replacing the left-hand-side by 1, and rearranging, one obtains:

$$\hat{X}_1 = \exp\left(\frac{1-\hat{\beta}_0}{\hat{\beta}_1}\right) = 102.2.$$

Any value of X larger than 102.2 (or, more generally, $\hat{X}_1$) results in $\hat{y} > 1$, an undesirable result since this has sssssno physical meaning. Both of these problems may be resolved, either by using certain transformations of y in linear models or by using a nonlinear model for the data. Here, we describe the use of a nonlinear model, and then apply the model to derive a linear (linearized) model for the data.

3.5.2 Nonlinear least squares

Equations 3.7a and 3.7b provide a very general expression for a response variable in relation to a predictor variable. In the previous section on model fitting, we considered forms of $g(X)$ that were strictly linear in the parameters. Here we abandon this restriction. There are many specific forms of $g(X)$ that could be used when y takes values only between 0 and 1 (see Chapter 5 in Schabenberger and Pierce, 2002), but we only consider one of these here. One useful two-parameter function can be written, without subscripts for observation number, as:

$$g(X) = \frac{1}{1+(X/\lambda)^{-\beta}} \tag{3.19}$$

where λ and β are parameters (both greater than 0). Representing this as a statistical model, with explicit consideration of unexplained variability, and adding a subscript for each individual observation, one obtains:

$$y_i = \frac{1}{1+(X_i/\lambda)^{-\beta}} + \varepsilon_i \tag{3.20a}$$

We assume that ε has a mean of 0 and a constant variance (σ^2), and that the ε_j values are independent. By determining the expected value for both sides of the equation, one can write equation 3.20a equivalently as:

$$E(y_j) = \frac{1}{1+(X_j/\lambda)^{-\beta}} \quad (3.20b)$$

In other words, the average (expected) value of disease incidence at a given value of X is represented by the right-hand side of 3.20b. This model, or variations of it, have been found to be very useful for representing nutritional uptake by plants, plant response to herbicide dose, and other relationships (Brain and Cousens, 1989; Morgan et al., 1975; Schabenberger et al., 1999).

One appealing feature of equation 3.20a or 3.20b is that the parameter λ has a very direct interpretation: λ is the value of X for which $E(y) = 1/2$. That is, λ represents the X value that results in expected y being exactly halfway between its lowest and highest possible values (0 and 1, respectively, in this model formulation). This can be easily seen by substituting λ for X in equation 3.20b:

$$\frac{1}{1+(\lambda/\lambda)^{-\beta}} = \frac{1}{1+1^{-\beta}} = \frac{1}{2}$$

In other words, the interpretation does not depend on the other parameter (β) because 1 raised to any power is still equal to 1.

Fitting nonlinear models to data via least squares is an iterative process because there is no exact (analytic) solution to the equation for SSE when a model is nonlinear (equations 3.10a or 3.16a). The methodology by means of which the numerically smallest SSE is found is often called *optimization*. Interested readers should consult textbooks or reviews to learn more about the methodology (e.g., Madden and Campbell, 1990; Schabenberger and Pierce, 2002). Some key points to keep in mind are: that the model fitting does not always work (i.e., an acceptable numerical solution, based on the rules for the chosen optimization method, may not be achieved); and that even when the computer program indicates that model fitting was "successful", the resulting estimates of parameters may not be the definitive least squares estimates (i.e., the method is iterative, and the definitive parameter estimates may not be found).

The investigator must supply initial guesses of the parameter values as starting points in the iterative search for the numerical solution (the parameter values that give the smallest calculated SSE). For some models and data sets, success depends heavily on the initial guesses, and investigators generally should try a few different initial guesses to see if the same numerical solution is obtained each time. If the initial guesses are too far away from the true values, parameter values near the true values may not be evaluated during the parameter

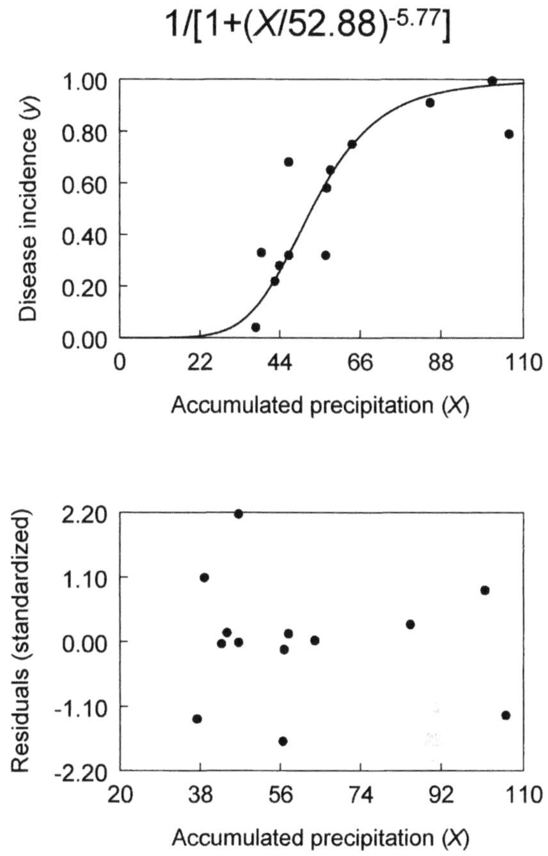

FIG. 3.5. Upper frame: Incidence of Phymatotrichum root rot of cotton at the end of the growing season (y) and cumulative precipitation (X) in centimeters up to 31 August during the growing season, together with the fitted values from equation 3.20a. Lower frame: standardized residuals for the fitted model.

estimation process. In such cases, what is reported as the smallest SSE is actually not the smallest SSE for the data set and chosen model—a better initial guess would have provided a smaller SSE. Despite these cautions, nonlinear model fitting is fairly routine in statistics, because either empirical evidence or theoretical arguments lead to nonlinear models for many biological phenomena and processes.

We fitted equation 3.20a to the root rot data by weighted least squares, using the NLIN procedure of SAS (Fig. 3.5). The default options were utilized, which means that the so-called Gauss–Newton optimization method was used. The weights were the same as used with equations 3.7c and 3.17 (for linear models). In this case, different attempted initial guess of the parameters gave the same results, so we feel confident in the estimates. The parameter estimates (and their estimated standard errors) were: $\hat{\lambda} = 52.88$ [$s(\hat{\lambda}) = 2.34$], and $\hat{\beta} = 5.77$ [$s(\hat{\beta}) = 1.07$]. The estimated value of λ means that y is predicted to be 50% at ~53 mm of rainfall.

The model is useful for representing responses that increase monotonically between 0 and 1 with increasing X. In Fig. 3.6, example responses for three different values of λ (at a single β) and three different values of

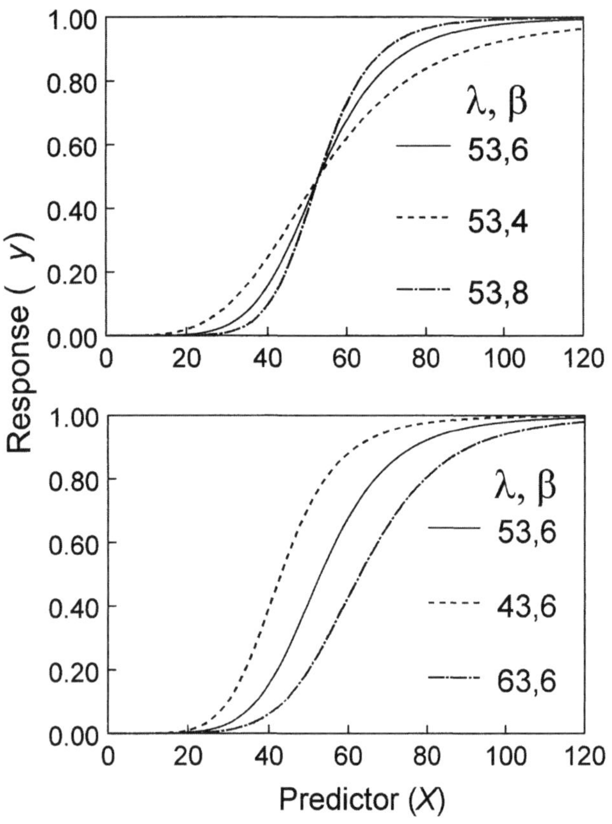

FIG. 3.6. Values of y versus X based on equation 3.21, with values of β and λ shown on the graphs.

β (at a single λ) are shown. With increasing λ, the curve is simply shifted to the right (lower frame of Fig. 3.6). With increasing β, the steepness of the curve increases (primarily at X values near the chosen λ value).

The predicted values of y (based on the estimated parameters) are shown, together with the observed y values, in the top frame of Fig. 3.5. SSE (see equation 3.16a) was equal to 1.55, which was smaller numerically than the SSE found when linear equation 3.17 [with ln(X) as the predictor variable] was fitted to the data. In general, the prediction curve was consistent with the scatter plot of y versus X. The standardized residual plot (Fig. 3.5, lower frame) displayed a random scatter, indicating that the model choice was reasonable. Although the predictions are very close to 0 at small X, they are never quite 0 for this model. For instance, at $X = 10$, one obtains: $\hat{y} = 6.7 \times 10^{-5}$. Moreover, $\hat{y}$ does not exceed 1 no matter how large the value of X is. Consider $X = 200$ (almost double the largest observed X); predicted incidence is calculated to be $\hat{y} = 0.99954$.

The R^2 summary statistic (equation 3.13), so commonly used with linear models, is somewhat controversial with nonlinear models. In fact, some authors indicate that it should never be calculated (Ratkowsky, 1990), and statistical computing programs often do not display it. One problem is that with nonlinear models, SSE could actually be greater than SST, meaning that R^2 could be less than 0! Of course, this would only happen if there was no relation at all between y and X, or the model choice or the parameter estimates were very far from being appropriate. Thus, with nonlinear models, R^2 is no longer interpreted in an exact sense as the proportion of explained variability. We agree with Schabenberger and Pierce (2002) who state that $1-(\text{SSE/SST})$ is, nevertheless, a useful measure of the goodness of fit of a model to data. It has the desirable property of having an upper limit of 1, which is only achieved when observed and predicted y values are identical at all X values. Since both SSE and SST are printed by most model fitting programs, it is easy for the investigator to calculate the value of R^2. For nonlinear-estimation programs that do not display SST, it can be easily determined by multiplying the variance of y (ignoring the predictors) by $N-1$. It could be argued that the left-hand side of equation 3.13 should be called a pseudo-R^2 when calculated for nonlinear model fits, but for simplicity of presentation, we often just refer to it as R^2. For the example in Fig. 3.5, $R^2 = 1-(1.55/29.51) = 0.947$. Because both SSE and SST were determined based on the weights, the relevant comparisons are with linear models (with y as the response variable) fitted with weighted least squares.

Another relevant issue for fitting of nonlinear models to data can be brought up here. If the errors (εs) are normally and independently distributed with a linear model (such as the straight-line equation 3.7c), the least-squares estimated parameters are, among other things: (1) normally distributed (very useful for determining confidence intervals); (2) unbiased when there is one predictor variable (meaning that the expected values of estimated parameters are equal to the true but unknown parameter values); and (3) the estimates that have the smallest variances of all possible estimates of the true parameters (meaning that they have narrower confidence intervals than other possible parameter estimates) (Ratkowsky, 1990; Schabenberger and Pierce, 2002). Even if the εs are not normally distributed, the parameter estimates approximately have these same desirable properties at moderate sample sizes.

In contrast, with nonlinear models, statistical theory shows that the least-squares estimated parameters do not have any of these properties at finite sample sizes, whether the εs are normally distributed or not. These properties are achieved at infinite sample sizes, but this does not help researchers working with finite and typically small sample sizes. It turns out that these properties *can* be approximately achieved at small sample sizes, but this actually depends on the specific form of the nonlinear model and the actual X and y values used in model fitting (Ratkowsky, 1990). In this book, we generally accept that the estimated parameters have these properties as an approximation.

Finally, the residuals obtained from nonlinear model fitting have more complicated statistical properties than those obtained when fitting linear models. In particular, the residuals (or standardized residuals) may have a mean that is not zero, even though the assumed unknown error has an expected value of 0. Furthermore, there may be a small negative correlation of the residuals with $\hat{y}$ (Seber and Wild, 1989), meaning that there may be a (slight) trend of residuals with X even when an appropriate model is chosen. This depends on the form of the nonlinear model used. Thus, examination of residual plots is a little more difficult than with linear models. However, with these caveats, we still find the examination of residual plots (see Fig. 3.5, lower frame) to be helpful for many of the nonlinear models of value for epidemiologists.

3.5.3 Linearized models

3.5.3.1 From nonlinear to linear. As discussed in section 3.2.4, some nonlinear mathematical models are intrinsically linear. Thus, although the inherent biological process may lead to a nonlinear model as a fundamental representation of the process (e.g., population growth, as discussed starting in Chapter 4), sometimes one can use linear model fitting methods instead of the more difficult nonlinear model fitting methods. Consider the model suggested above for the Phymatotrichum root rot data (equation 3.20b). Temporarily, assume that the nonlinear function in equation 3.19 provides a perfect description of the data, so that $\varepsilon = 0$ for all observations. Dropping the subscript denoting individual observations, we can write the model as:

$$y = \frac{1}{1+(X/\lambda)^{-\beta}} \qquad (3.21)$$

Equation 3.21 represents a deterministic model. When there is no variability (as assumed here), the expected and actual y values are the same [i.e., $y = E(y)$].

This model can be expressed in linear form. For this one example, we give some of the intermediate algebraic manipulations to help the reader see the steps needed to obtain a linear model. First, one takes the reciprocal of each side of equation 3.21, and then subtracts 1 from each side. The left-hand side is then $(1/y)-1$, which is the same as $(1-y)/y$, and the right-hand side is $(X/\lambda)^{-\beta}$. Taking natural logs of both sides gives:

$$\ln\left(\frac{1-y}{y}\right) = -\beta\ln\left(\frac{X}{\lambda}\right)$$

which is still not linear because λ appears within the log function. However, because $\ln(A/B) = \ln(A) - \ln(B)$, by definition, the right-hand side can be written as: $-\beta[\ln(X)-\ln(\lambda)]$. Multiplying through by $-\beta$, one then obtains:

$$\ln\left(\frac{1-y}{y}\right) = -\beta\ln(X) + \beta\ln(\lambda) \qquad (3.22a)$$

It is customary (but not required) to express the function of y on the left-hand side in a slightly different way, as $\ln[y/(1-y)]$ instead of $\ln[(1-y)/y]$. This change is easily done by first noting that $\ln(A/B) = -\ln(B/A)$. So, multiplying both sides of equation 3.22a by -1, one obtains:

$$\ln\left(\frac{y}{1-y}\right) = -\beta\ln(\lambda) + \beta\ln(X) \qquad (3.22b)$$

The function of y on the left-hand side of the model represented by equation 3.22b, known as the *logit* of y, is now the new response variable ($y^* = \ln[y/(1-y)]$), and $\ln(X)$ is the new predictor variable [$X^* = \ln(X)$]. This is a straight-line model with slope equal to $\beta(=\beta_1)$ and intercept equal to $-\beta\ln(\lambda)$ $(=\beta_0^*)$. Equation 3.22b is not linear in terms of λ, but in terms of $-\beta$ times the natural log of λ (recall that products, ratios, and other functions of constants not involving variables are still constants). We can write equation 3.22b as:

$$y^* = \beta_0^* + \beta_1 X^* \qquad (3.22c)$$

in order to emphasize the linear relationship between the logit of y and the natural log of X. Because the linearization of equation 3.21 results in the logit of y as a linear function of the natural log of X, the model (in either nonlinear or linear form) is often called the *log-logistic model.*

The deterministic equation 3.22b can be expanded into a stochastic model in order to account for unexplained variation in y^* at each value of X^*. This is done by adding an error term (a random variable with mean 0 and constant variance σ^2). Using a j subscript to denote an individual observation, as we did in the previous sections, one obtains from equation 3.22b:

$$\ln\left(\frac{y}{1-y}\right)_j = -\beta\ln(\lambda) + \beta\ln(X)_j + \varepsilon_j \qquad (3.23a)$$

which can also be written as:

$$\ln\left(\frac{y}{1-y}\right)_j = \beta_0^* + \beta_1\ln(X)_j + \varepsilon_j \qquad (3.23b)$$

in which $\beta_0^* = -\beta\ln(\lambda)$ and $\beta_1 = \beta$. In fact, this is the model formulation used by Jeger and Lyda (1986) in their development of a forecasting system. One could also directly choose to model the relationship between disease incidence and X using this linear model without first considering an intermediate nonlinear model. Taking expectations of both sides, one obtains:

$$E\left(\ln\left(\frac{y}{1-y}\right)_j\right) = \beta_0^* + \beta_1\ln(X)_j \qquad (3.23c)$$

which indicates that the mean value of the logit (not the mean y) is a linear function of $\ln(X)$. Note, the left-hand side of equation 3.23c is the mean or expected value of the logit $[E(y^*)]$, *not* the mean of y, at each value of $\ln(X)$. Thus, ε equals $y^*-E(y^*)$ here.

3.5.3.2. Model fitting. Fitting equation 3.23b to the logit transformation of the root rot incidence values can be done directly with ordinary least squares, using the logit-transformed y values as the response variable and $\ln(X)$ as the predictor variable. Using ordinary least squares, one obtains: $\hat{\beta}_0^* = -19.67$ [$s(\hat{\beta}_0^*) = 3.891$], and $\hat{\beta}_1 = 4.95$ [$s(\hat{\beta}_1) = 0.960$]. SSE = 14.73, SST = 50.31, and $R^2 = 0.707$. Note that $\hat{\beta}_1$ is a direct estimate of β of nonlinear equation 3.20a (i.e., there is a direct correspondence). However, this $\hat{\beta}_1$ has no relation to the slope of the linear models above with y as the response variable. One has to do a little algebraic manipulation to determine the estimated λ based on $\hat{\beta}_0^*$:

$$\hat{\lambda} = \exp\left(\frac{\hat{\beta}_0^*}{-\hat{\beta}}\right) = \exp\left(\frac{-19.67}{-4.95}\right) = 53.2$$

For the regression analysis involving y as the response variable, we made the point that the variance of incidence is not expected to be constant but to be proportional to $y(1-y)$. If this is true (and we were assuming so above), then the variance of the logit will be neither constant nor proportional to $y(1-y)$ (Schabenberger and Pierce, 2002). There are rules to determine (or approximate) the variance of a function of a random variable that can be applied to this problem. Details are given in the next chapter (specifically, section 4.6.2) for disease progress curve models, and we do not pursue this any further here.

The parameter estimates are not the same as those obtained with nonlinear least squares. In general, unless there is a perfect fit (SSE = 0), the estimated parameters will vary between different estimation methods (ordinary versus weighted least squares) and different ways of expressing the model (linear or nonlinear versions). As stated before, one cannot directly compare the SSE (and related statistics) for unweighted (i.e., ordinary) and weighted least squares results. Moreover, because the response variable differed between equation 3.20a (y) and equation 3.23b (logit), one should be cautious in comparing goodness of fit of the models.

The interested reader should plot graphs of the logit(y) versus the $\ln(X)$, and the standardized residuals versus $\ln(X)$, to see if a linear model seems appropriate.

3.5.3.3 Where is the error additive? Linear equation 3.22b and nonlinear equation 3.21 are two equivalent ways of expressing a *deterministic* relationship between y and X. It is tempting to think of linear equation 3.23a and nonlinear equation 3.20a as two equivalent ways of expressing a *statistical* (or stochastic) relationship between y and X. However, the latter interpretation would be incorrect. Although one can algebraically rearrange equation 3.21 to obtain equation 3.22b, there is no way to rearrange the nonlinear equation with the additive error term (equation 3.20a) to obtain a linear model with a function of y (and no parameters) on the left-hand side and an additive error term on the right-hand side.

One *can* obtain the statistical linear equation 3.23b from a different version of a statistical nonlinear model for y. That is, instead of equation 3.20a, one can write:

$$y_i = \frac{1}{1+(X_i/\lambda)^{-\beta}\varepsilon_i'} \qquad (3.24)$$

in which ε' is an error term (a random variable). However, this error term has a mean of 1, and takes on only positive values. When observed y is fully specified by the deterministic part of the model, then $\varepsilon' = 1$ (not 0). When y is differs from that given by the deterministic part of the model, then ε' will be either smaller or larger than 1. Algebraic manipulation of equation 3.24 results in:

$$\ln\left(\frac{y}{1-y}\right)_i = -\beta\ln(\lambda)+\beta\ln(X)_i - \ln(\varepsilon_i') \qquad (3.25)$$

where $-\ln(\varepsilon'_i)$ is the "new" error term ($=\varepsilon$). Note that when ε' is 1, $\ln(\varepsilon') = 0$, the desired property for an additive error. Linear statistical equation 3.25 is the same as linear statistical equation 3.23a [with $\varepsilon = -\ln(\varepsilon')$]. In fact, if ε' has a log–normal distribution with expectation of 1, then ε has a normal distribution with expectation of 0.

It can be seen that the error (unexplained variation) can be either additive in y (equation 3.20a) or in y^* (equation 3.25 or 3.23a), but not both. If the error is additive on a transformed scale, then it cannot be additive in the original scale. Conversely, if the error is additive in the original scale, then it cannot be additive in the transformed scale (i.e., the scale that gives a linear model). For the latter situation, use of equation 3.25 to estimate parameters is undesirable, since the errors would not be added to the deterministic part of the linearized model.

When a deterministic model (such as equation 3.21) is appropriate, it is not necessarily clear on general theoretical grounds whether the error (discrepancy between observed y and that determined by deterministic part of the model) is additive in a nonlinear model or a linear one. In practice, investigators simply assume the error is additive in the form of the model they find convenient or practical to work with (whether or not this is theoretically justified). For instance, when using a linearized version of a model, it is assumed that the discrepancy between each y^* and $E(y^*)$ can be expressed as the difference of the two [$\varepsilon = y^* - E(y^*)$]. When it is possible to model a nonlinear relationship by both nonlinear and linear equations, and number of data points is large, it is possible to examine the residuals for both models to determine which appear to be closer to random when plotted against the predictor variable. However, the results of this comparison may be affected by some of the complex distributional properties of residuals for some nonlinear models (Seber and Wild, 1989) mentioned above.

One more way to help understand the linkage between nonlinear and linear formulations of a model is to consider the expected (mean) value of y for a nonlinear model. As mentioned in section 3.5.2, equations 3.20a and 3.20b are two equivalent statistical ways of showing the relationship between disease incidence and X, one directly for y and one for $E(y)$. Although one cannot transform equation 3.20a into a linear equation, one *can* transform equation 3.20b for the expectation [E(y)] into a linear equation. Using the same algebraic manipulations, one obtains:

$$\ln\left(\frac{E(y_j)}{1-E(y_j)}\right) = \beta_0^* + \beta\ln(X)_j \qquad (3.26)$$

in which $\beta_0^* = -\beta\ln(\lambda)$. This equation may look somewhat like equation 3.23c for expectations, but there is a fundamental difference. Equation 3.23c is a model for the *expectation of the logit of* y, and is the proper formulation for regression analysis, but equation 3.26 is a model for the *logit of the expectation of* y. Equation 3.26 is not a formulation that can be used for linear regression analysis. Note, however, that equation 3.26 does serve as the basis for another type of model fitting, known as generalized linear modeling (Collett, 2003).

3.5.3.4 Nonlinear or linearized statistical models? To conclude these sub-sections, we reiterate that some deterministic models can be expressed in linear form, after one or many algebraic steps. However, it may not be possible to linearize a statistical nonlinear model (with an additive error term) even when the deterministic component of the model (i.e., without the error term) can be converted into linear form. Whether or not a statistical nonlinear model can be linearized depends on the nature of the error term, that is, the discrepancy between the observed response variable and that specified by the deterministic component in the model. In other words, it depends on whether ε is additive to the deterministic component, or appears as a more complicated function on the right-hand side of the nonlinear model.

Researchers often find it desirable to work with linear models, even when the biological process may be directly described by a nonlinear model. In fact, we place a great deal of emphasis in this book on the use of linear models (for those nonlinear models that can be linearized). These are especially easy to fit to data, and to use in comparing epidemics. The estimated parameters and the $\hat{y}$ values have well understood and desirable statistical properties when the model is appropriate. However, if the error is truly additive in the nonlinear model, then linear least squares will only result in rough estimates of the actual parameters. Although various theoretical arguments as well as empirical evidence often leads to the choice of deterministic nonlinear models for biological processes, considerably less attention is paid to determining the form of the error term in the models. In fact, some detailed statistical research shows that the so-called error structure can be quite complicated for many population-dynamic processes (see Chapters 6 and 7 in Seber and Wild (1989) for details). Thus, all the statistical models discussed may be approximations for the true error structure. Since all models are simplifications, this is not necessarily a problem. What matters is that one is using a model that is reasonable for the data and for the specified objectives of the study in question. Whether one is fitting linear or nonlinear models to data, residuals should be checked to see if the statistical model appears reasonable. If residuals appear reasonable, this is evidence (although definitely not proof) that the chosen statistical model (including the way the error term enters the model) provides a satisfactory description to the data.

3.6 Applications

Two useful applications of nonlinear modeling are given here to help the reader become more familiar with model fitting procedures. We will assume that the statistical model for the response comprises a nonlinear deterministic component and an additive error term (ε, which we assume is independently and normally distributed with mean 0 and constant variance, σ^2). That is to say, the response is given by $g(X_j) + \varepsilon_j$, where $g(\bullet)$ is nonlinear in the parameters (see equation 3.7c for a linear version). For simplicity of presentation, we generally show the following models only in deterministic form (i.e., with an implied additive error and without a subscript for the specific observation). This approach is consistent with many other parts of the book.

3.6.1 Disease intensity in relation to inoculum density

Epidemiologists often inoculate plants with different densities of inoculum (e.g., conidia per milliliter of water), or expose plants to different inoculum densities in the soil, and determine resulting disease intensity at each density. The relationship can be used to characterize the susceptibility of the host genotype and aggressiveness of the pathogen strain under a given set of environmental conditions (Baker, 1978; Gilligan, 1985). The relation is important for determining the rate of disease increase over time, as will be discussed in the next two chapters.

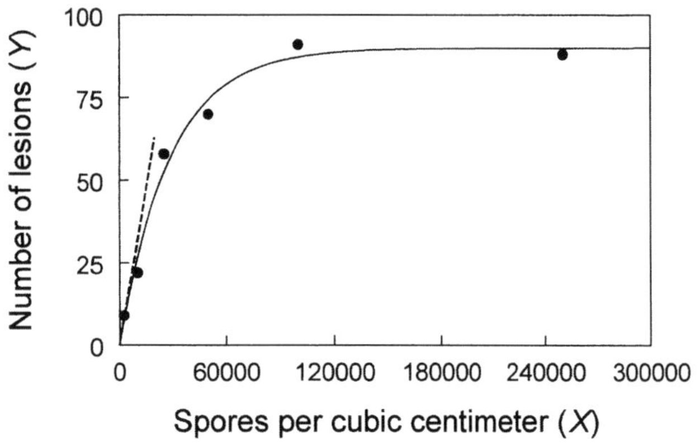

FIG. 3.7. Number of lesions on apples caused by *Venturia inaequalis* in relation to inoculum density (spores/ml). Data read off of Fig. 1 ("16C treatment") in Hartman et al. (1999). Curve represents the predicted Y values based on the fit of equation 3.27 to the data. Straight line represents an approximation to the nonlinear model at small X.

We consider an example taken from Hartman et al. (1999), where they determined the number of apple scab lesions per plant [a type of disease density (see Chapter 2); Y] in relation to inoculum dose of *Venturia inaequalis* (spores/ml of water in the spore suspension; X). Because disease is measured in numbers of lesions, rather than as a proportion, we use upper case Y for intensity. The observed relationship between Y and X in Fig. 3.7 is typical for disease intensity in relation to inoculum density (Vanderplank, 1975). There is an initial sharp increase in Y with X, and the increase becomes less and less with increasing X, and eventually very little or no increase is seen at very high values of X. An upper limit to Y is expected because there is a limit to the possible number of infections with a finite area of plant tissue (whether or not that upper limit is obvious from the observations). One useful model to describe this relationship is:

$$Y = \kappa(1 - \exp(-\beta X)) \tag{3.27}$$

in which β and κ are parameters. Here, κ is the *asymptote*, the maximum possible Y (under the conditions being studied) at very large X. Mathematically, Y never quite reaches κ (which is why it is labeled an asymptote), but in practice, it is convenient to think of κ as an actual achievable maximum (at very large X). β is a type of rate parameter, reflecting the steepness of the $Y:X$ curve (but, of course, the steepness changes with increasing X). Equation 3.27, or variations of it, are used in many disciplines; it is sometimes known as the Mitscherlich or von Bertalanffy model, in honor of earlier users of this equation (Ratkowsky, 1990). The model is also known as the *negative exponential* or restricted exponential.

At $X = 0$, $Y = 0$ in equation 3.27. This is a desirable property for the relationship being modeled, since no disease is expected when there is no inoculum. There are more general versions of the equation, which include a third parameter to allow for a possibly nonzero value of Y at $X = 0$ (for other types of relations where such a result is expected theoretically), and one formulation is seen in the next chapter (see section 4.4.2). This model is nonlinear in the two parameters, and cannot be linearized unless κ is known (so that equation 3.27 is just a one-parameter model). An attempt to linearize this model results in an unknown parameter appearing on the left-hand side of a straight-line equation.

The statistical version of equation 3.27 was fitted to the data using the default least squares algorithm of the NLIN procedure of SAS. No weights were used, although one could argue that the variance of a count will increase with the mean (see Chapter 9). The parameter estimates were: $\hat{\beta} = 5.8 \times 10^{-5}[s(\hat{\beta}) = 6.5 \times 10^{-6}]$, and $\hat{\kappa} = 88.3$ $[s(\hat{\kappa}) = 2.94]$. SSE = 67.6, and $R^2 = 0.988$. As can be seen in Fig. 3.7, the predicted Y values are very close to the observed ones.

There is an interesting property of equation 3.27 when the exponent value is small. It can be shown that $\exp(-\beta X)$ is approximately equal to $1-\beta X$ when βX is small, in which case $Y \approx \kappa\beta X$. In other words, there is (to a reasonable approximation) a straight-line relation between disease intensity and inoculum dose at low spore density, with an intercept of 0 (no spores means no disease) and slope of $\kappa\beta$. For the example, the slope for the example data set is approximated as $88.3 \times 5.8 \times 10^{-5} \approx 0.0051$. The straight line with this slope is shown in Fig. 3.7, which overlaps the fitted values for the more general equation 3.27 at low X. This straight-line relationship is of epidemiological relevance because inoculum density is likely to be much lower in actual epidemics than in controlled-inoculation studies.

If one knew the milliliters of spore suspension per plant, one could convert spores/ml (X; spores/cm^3) to spores/plant (X'). If one uses X' instead of X in equation 3.27, then $\kappa\beta$ represents infection efficiency ("lesions/spore") of the pathogen for the given host genotype under the studied environmental conditions. Infection efficiency is a key parameter in determining the rate of disease increase in epidemics, a topic dealt with in Chapter 5.

3.6.2 The cumulative response

Plants infected by fungi and oomycetes produce spores for a certain amount of time. Typically, the rate of production of spores per unit time (e.g., spores/day) is low at first, followed by an increased spore production rate for a period of time. Eventually, no new spores are produced. One way (but not the only way) of depicting this process is by plotting the cumulative number of spores produced over time. For example, consider the data in Fig. 3.8, which is based on some of the data collected by Kolnaar and van den Bosch (2001). The upper frame shows the cumulative average number of spores produced per rust plant (caused by *Puccinia lagenophorae*). The lower frame shows the average spores produced per day.

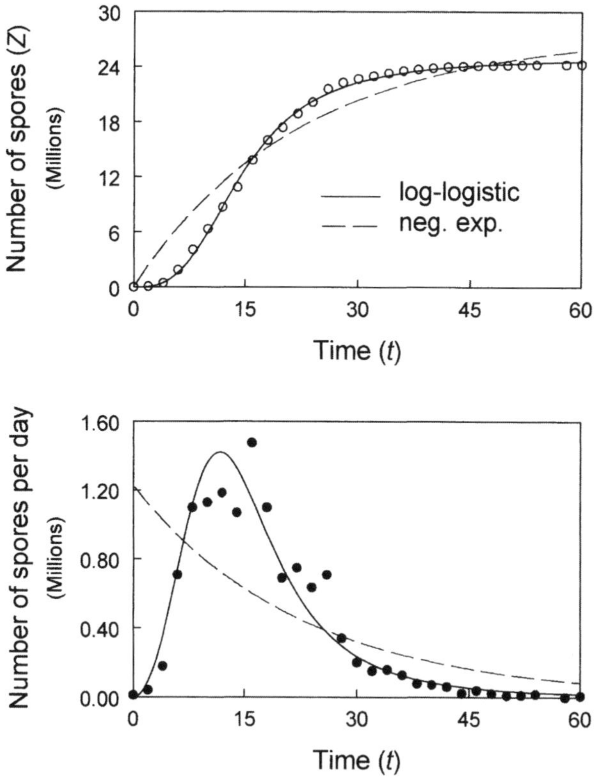

FIG. 3.8. Upper frame: Cumulative number of spores per plant produced by *Puccinia lagenophorae* on inoculated *Senecio vulgaris* plants over time (Z), based on an example in Kolnaar and van den Bosch (2001). Data are from the "16 C pre-inoculation, 16 C incubation" treatment. Each point is from a mean of seven plants. Time 0 is the day of first observed spore. Curves represent the fit of equation 3.27 (negative exponential) and 3.28 (log-logistic). Lower frame: mean number of new spores produced per day (daily increments; dZ/dt) corresponding to the cumulative values. Curves are based on the first derivatives of the two equations.

One can model either the cumulative numbers or the daily increments in spore numbers, but here we restrict ourselves to the cumulative response. For comparative purposes, we first consider two previously described nonlinear models, the negative exponential (equation 3.27) and the log-logistic (equation 3.21), after which two alternatives are presented. Because the response variable is not a version of disease intensity here, we use the generic Z symbol for the dependent variable; that is, Z replaces y or Y in the previously given equations. Moreover, because the predictor variable is time, we denote it t instead of X. As usual, we assume the error is additive, and is independently and normally distributed with mean 0 and variance σ^2. Because Z represents cumulative number of spores, then the *new* spores per unit time (really, the instantaneous rate of change of Z with t) is represented as dZ/dt. One can think of dZ/dt loosely as the daily spore production since the analyzed data are expressed as daily counts (see below).

Equation 3.21 was used previously when the maximum possible response (the asymptote) was 1, which is not satisfactory here. So, the model was expanded to include a κ parameter for the asymptote by multiplying equation 3.21 by κ. The model can be written as:

$$Z = \frac{\kappa}{1 + (t/\lambda)^{-\beta}} \tag{3.28}$$

With unknown κ, this nonlinear model cannot be linearized.

The negative exponential model (equation 3.27, using Z and t) provided a relatively poor fit to the data, as seen by the dashed line in Fig. 3.8. Fitted values were consistently above the observed Z values at very small and very long times, and below the observed Z values at intermediate times. The (pseudo-) R^2 value was $1 - (1.35 \times 10^{14}/2.32 \times 10^{15}) = 0.942$. It should be noted that for increment data expressed as cumulative values, as here, there is a *forced* increasing trend of Z with t, which results in small SSE and large R^2 even for model fits that are not acceptable. We do not show residual plots, but the pattern should be obvious from the discrepancies between observed and fitted values noted above.

The expanded log-logistic model (equation 3.28) provided a much better fit to the data. The SSE was 2.8×10^{12}, which may seem like a large number; however, the SST was also very large (2.32×10^{15}). The (pseudo-) R^2 was 0.9988. Normally, it is sufficient to express R^2 with three decimal places, but we show four here to emphasize the very good fits for some models. Estimated parameters were: $\hat{\beta} = 2.99$ [$s(\hat{\beta}) = 0.071$], $\hat{\lambda} = 14.72$ [$s(\hat{\lambda}) = 0.119$], and $\hat{\kappa} = 24.9 \times 10^6$ [$s(\hat{\kappa}) = 1.24 \times 10^5$]. Based on the estimated κ, ~25 million spores were produced per plant over the "lifetime" of the lesions (under the conditions studied). Moreover, based on estimated λ, it took ~15 days for a lesion (on average) to produce half of its spores (i.e., $\hat{\lambda}$ is the estimate of the time when $Z = \kappa/2$ in this model formulation).

The lower frame in Fig. 3.8 shows the daily increments in spore numbers and the predictions from the two models. The spores were collected every 2 days in the original study, so that the original data represented new spores produced per 2-day interval. Thus, we divided each spore-count by 2 to obtain average spore production per day over time. The predicted increments were determined by inserting the estimated parameters into the equation for the first derivative of Z with respect to $t(dZ/dt)$ for the two models. The instantaneous rate of change ("daily increment" here) is predicted to be:

$$\hat{\kappa}\hat{\beta}\exp(-\hat{\beta}t)$$

for the negative exponential model (equation 3.27), and:

$$\frac{\hat{\kappa}}{\left(1 + (t/\hat{\lambda})^{-\hat{\beta}}\right)^2} \left(\frac{t}{\hat{\lambda}}\right)^{-\hat{\beta}} \frac{\hat{\beta}}{t}$$

for the log-logistic model (equation 3.28). With this view the data, one can readily see the poor fit of the negative exponential model to the data, because the curve described by the model is nothing like that of the observed points.

If an investigator simply wished to obtain a good fit to the data, and possibly to compare the estimated parameters corresponding to different conditions, the log-logistic would be a satisfactory choice of model, based on model-fitting results shown here. However, this model may not be satisfactory for different conditions, or for different imposed treatments (e.g., for different temperature regimes). In fact, the main reason for adopting the log-logistic model here was simply that it can describe a relationship where Z increases monotonically with t up to an asymptote κ. One can also consider other models that may have more of a theoretical justification. The next two models are presented to demonstrate (at least in part) the rich diversity of models that can be used to describe and understand biological processes.

To motivate the next model, we temporarily stop thinking of the data in Fig. 3.8 as a relationship between two variables, Z and t, and the resulting population of Z values at each time t. Instead, we think of a *single* population of κ spores produced by lesions on a plant, and consider a *single* random variable, T, which is the time until each individual spore is produced. The random variable T cannot be any smaller than 0, but has no upper limit. Since the normal probability distribution (equation 3.4) has no lower bound, it is unlikely that T can adequately be represented by this distribution. In such circumstances, a potentially useful model for T is the two-parameter gamma distribution, a statistical probability distribution for a continuous random variable that takes on only positive values. The probability of a spore being formed at or before a given time t [i.e., $\Pr(T \leq t)$] is specified by integrating the gamma distribution (or probability density) function from 0 up to t. This is written as:

$$G(t;\eta,\lambda)=\Pr(T \leq t)=\int_0^t \frac{x^{\eta-1}\, e^{-x/\lambda}}{\lambda^{\eta}\Gamma(\eta)}dx$$

in which λ and η are parameters, $\Gamma(\bullet)$ is the gamma function (sometimes known as the generalized factorial), and x here is just an index for any value of time between 0 and t. Then the cumulative number of spores produced per plant at a given time t is modeled as:

$$Z = \kappa \cdot G(t;\, \eta,\, \lambda) \qquad (3.29)$$

Thus, the Z-axis in Fig. 3.8 (the ordinate) can be viewed as the cumulative probability of a spore being formed at or before t, multiplied by the total number of spores that are produced. It is interesting to note that there is no analytical solution to the integration of the gamma distribution; that is, one cannot write out an equation for $G(\bullet)$ without including the integral symbol. Fortunately, the cumulative gamma distribution is calculated (really, approximated with very high accuracy) by many statistical programs, such as SAS, simply by specifying the parameter values in a program-defined function. Then one can use the function $G(\bullet)$ just as one would use the negative exponential or any other specified equation. In other words, $G(\bullet)$ is simply utilized as a flexible function to represent a monotonically increasing response. One reason to use it here is to demonstrate explicitly that many useful models for biological processes may not be expressible as linear or nonlinear equations without the use of the integral symbol. In fact, this situation is quite common for theoretical models of disease dynamics (see Chapter 5). In these circumstances, we say that the model does not have an analytical solution.

There are specialized statistical methods in the lifetime and survival analysis literature for estimating parameters of (cumulative) distribution functions for time of occurrence of events (Scherm, 2004; Lawless, 1982). However, these are difficult to apply when population sizes are in the millions. Moreover, the relationship illustrated in Fig. 3.8 is not quite the same as that typically analyzed in lifetime studies. This is because the number of spores produced is really an estimate based on sampling, and represents an average across multiple plants and lesions per plant (Kolnaar and van den Bosch, 2001). Thus, we return to modeling a response variable as a function of a predictor variable, but now use $\kappa \cdot G(t;\, \eta, \lambda)$ as the nonlinear function of time (plus the assumed additive error term, ε). Using the NLIN procedure in SAS, which permits the use of the gamma distribution function in addition to standard functions for nonlinear models, the following parameter estimates for equation 3.29 were obtained: $\hat{\eta} = 3.60$ [$s(\hat{\eta}) = 0.070$], $\hat{\lambda} = 4.48$ [$s(\hat{\lambda}) = 0.091$], and $\hat{\kappa} = 24.2 \times 10^6 = [s(\hat{\kappa}) = 4.5 \times 10^4]$ spores per plant. A useful property of the gamma distribution is that expected (mean) time of spore formation is given by $\eta \cdot \lambda$. The estimated mean here is $3.60 \times 4.48 = 16.1$ days. This is close to the value determined by Kolnaar and van den Bosch using a different parameter estimation method.

The (pseudo-) R^2 for the analysis based on equation 3.29 was $1 - (7.03 \times 10^{11}/2.32 \times 10^{15}) = 0.9997$, slightly larger than that obtained with the log-logistic model. The top frame of Fig. 3.9 shows the fit of equation 3.29 to the data. For comparison purposes, the fit of the log-logistic model (equation 3.28) is repeated from Fig. 3.8. Clearly the cumulative gamma function provided an excellent fit to the data. When shown in terms of daily increments of spores (obtained as the first derivative of equation 3.29 with respect to time), one can better see the differences in fit for the cumulative gamma and log-logistic models (bottom frame of Fig. 3.9). The predicted daily increment (i.e., estimate of dZ/dt) was determined by multiplying the estimated κ by the

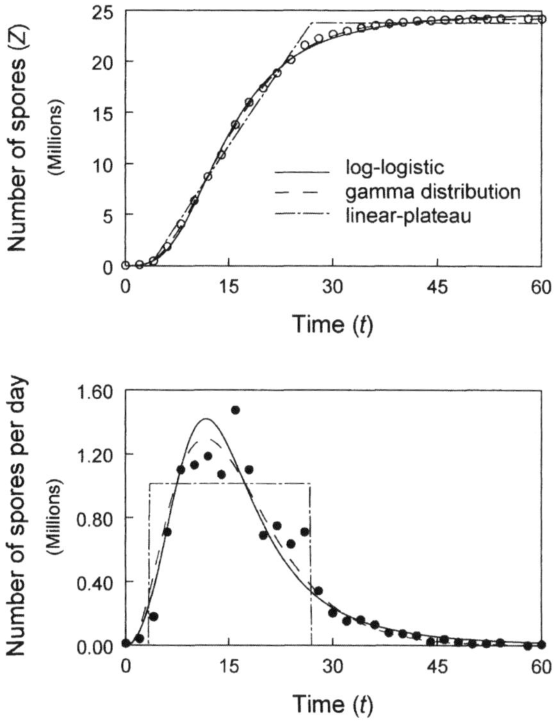

FIG. 3.9. Upper frame: Cumulative number of spores per plant produced by *Puccinia lagenophorae* on inoculated *Senecio vulgaris* plants over time (Z). See Fig. 3.8 for more details. Curves represent the fit of equation 3.28 (log-logistic), equation 3.29 (cumulative γ distribution function), and equation 3.30a, b (segmented equations). Time 0 is the day of first observed spore. Lower frame: mean number of new spores produced per day (daily increments; dZ/dt) corresponding to the cumulative values. Curves are based on the first derivatives of the equations.

gamma distribution (density) function (not cumulative function) using estimated parameters:

$$\hat{\kappa}\frac{t^{\hat{\eta}-1}e^{-t/\hat{\lambda}}}{\hat{\lambda}^{\hat{\eta}}\Gamma(\hat{\eta})}.$$

We now consider an additional approach for representing these data. Between times $t \sim 3$ and $t \sim 30$ days, there is *roughly* a straight-line relation between Z and t. Before $t = 3$ days and after $t = 30$ days (or so), there appears to be no or little change in Z with change in t; it may be reasonable to describe these regions using an equation with a slope of 0. One can, in fact, model the overall relation with a mixture of equations for the different segments of the time axis. We can consider Z to be a constant (β_0) at all times between 0 and α_0 (another constant), and then a linear increase in Z with time between times α_0 and α_1 (another constant). Finally, we can consider that Z is a constant for all times above α_1. This can be written as:

$$\begin{aligned} Z &= \beta_0, && \text{if } t \leq \alpha_0 \\ Z &= \beta_0 + \beta_1(t - \alpha_0), && \text{if } \alpha_0 < t \leq \alpha_1 \qquad (3.30\text{a}) \\ Z &= \beta_0 + \beta_1(\alpha_1 - \alpha_0), && \text{if } t > \alpha \end{aligned}$$

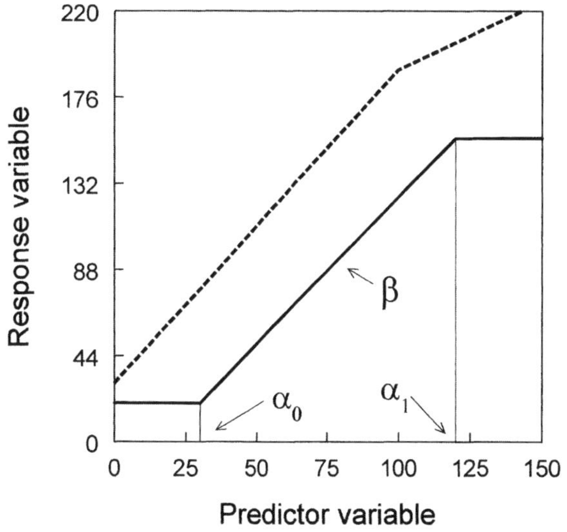

FIG. 3.10. Theoretical relationships between a response and predictor variable based on a segmented model. Solid line: three-piece model (see equation 3.30a, b), consisting of two pieces with zero slope. Dashed line: two-piece model, consisting of two nonzero slopes. The joint points (values of the predictor variable where the slope abruptly changes) are identified with α symbols.

The middle equation represents the straight line increase in Z with t, with slope of β_1 and intercept of $\beta_0 - \beta_1\alpha_0$. The third equation is just a function of constants, so it is also a constant at all times. Essentially, this third equation indicates that all times greater than α_1 are identical to α_1 in terms of predicting Z. Note that $\beta_0 + \beta_1(\alpha_1 - \alpha_0)$ is analogous to κ in the previously discussed models.

Equation 3.30a is a type of segmented or "broken stick" model, in which the transition between segments is not smooth (Schabenberger and Pierce, 2002). In other words, there is an abrupt change in slope at each of the *join points* (as specified by the α values). The model is also called a linear-plateau model. A hypothetical graph for equation 3.30a is shown in Fig. 3.10 (solid line). For comparison, a graph of a simpler segmented model, with one join point rather than two, is also shown (dashed line). Equation 3.30a has four parameters (the unknown constants α_0, α_1, β_0, and β_1). If one does not estimate the join points, but uses assumed known values, then one can use linear least squares to estimate parameters. For instance, if one could assume that there is a linear increase between times of 4 and 30; then model fitting would be done just using the data between times of 4 and 30. However, here we assume that the join points are unknown and must be estimated, together with the other model parameters.

Equation 3.30a can be written as a single equation by using an indicator function, $\zeta(\bullet)$, that takes on two possible values, 0 and 1. For instance, $\zeta(t \leq \alpha_0)$ equals 1 if t is any value at or below α_0, and equals 0 otherwise. The segmented model can then be written as:

$$\begin{aligned} Z = {} & \beta_0 \cdot \zeta(t \leq \alpha_0) \\ & + [\beta_0 + \beta_1 \cdot (t - \alpha_0)] \cdot \zeta(\alpha_0 < t \leq \alpha_1) \quad (3.30\text{b}) \\ & + [\beta_0 + \beta_1 \cdot (\alpha_1 - \alpha_0)] \cdot \zeta(t > \alpha_1) \end{aligned}$$

This model is nonlinear in the parameters. Although we do not show the details, equation 3.30b is equivalent to a scaled version of the cumulative uniform (rectangular) distribution function with parameters α_0 and α_1. In Chapter 12 it is shown that the uniform distribution serves as a good tool to model crop loss data in relation to disease intensity (see section 12.4.2.2).

We fitted equation 3.30b to the cumulative rust spore data using the NLIN procedure of SAS, because this program allows the user to specify indicator functions. However, we specified β_0 as 0 instead of as an unknown constant; this means we assume no sporulation before α_0. There actually are some spores produced at all times shown in the graph, but here we make this assumption about β_0 for the purpose of demonstration. Thus, the model as fitted had three parameters (β_1, α_0, and α_1), the same number in the cumulative gamma (equation 3.29) and log-logistic (equation 3.28) models.

The fitted values are shown in Fig. 3.9 for equation 3.30a. There was a reasonable fit to the data, with a (pseudo-) R^2 of $1 - (1.23 \times 10^{13}/2.32 \times 10^{15}) = 0.995$. Estimated parameters were: $\hat{\beta}_1 = 1.01 \times 10^6 [s(\hat{\beta}_1) = 2.7 \times 10^4]$, $\hat{\alpha}_0 = 3.5\ [s(\hat{\alpha}_0) = 0.36]$, and $\hat{\alpha}_1 = 26.9 [s(\hat{\alpha}_1) = 0.40]$. This means that the model predicts no sporulation (because $\beta_0 = 0$) until $t = 3.5$ days, then a linear increase in cumulative number of spores until $t = 26.9$ days. After this time, there is no further increase in spores. The maximum cumulative amount of sporulation is predicted as:

$$\hat{Z} = 1.01 \times 10^6 \times (26.9 - 3.5) = 23.6 \times 10^6 \text{ spores/plant.}$$

This prediction is close to $\hat{\kappa}$ in the previous models.

The nature of the relationship represented by equation 3.30a is that there is no change in Z with increasing t outside of the time period between α_0 and α_1 (i.e., $dZ/dt = 0$), and a *constant* change in Z with change in t between these two times ($dZ/dt = \beta_1$). From Fig. 3.9, one can see that the segmented model underpredicts daily spore production for $t < \alpha_0$ and $t > \alpha_1$. For $\alpha_0 < t < \alpha_1$ the model underpredicts daily spore production in the period with the highest observed increments, but overpredicts in the periods before and after the period with the highest observed increments. Clearly, this model is less desirable than the other two (i.e., those based on equations 3.28 and 3.29), judged in terms of goodness-of-fit for this data set. However, the segmented model may be acceptable in some circumstances, depending on the objectives of the user.

The four models demonstrated here for the cumulative response are examples of possible ways of describing this relationship. For those interested, Ratkowsky (1990) should be read for a presentation of a very large number of useful nonlinear models for responses that increase with the predictor variable. We evaluated the models presented in this section primarily in terms of fit to the data, and did not discuss the reasonableness of the standard statistical assumptions about the implied additive error (ε). For cumulative responses, the error term may have a non-constant variance and individual errors may not be independent (Schabenberger and Pierce, 2002). This is addressed in Chapter 4. There was no substantial evidence for these departures from standard assumptions with the current data.

All of the models presented can be expanded for more complicated scenarios. For instance, instead of modeling spore production over time from the start of sporulation, one can model spore production from the time of inoculation. This will require incorporation of a parameter in all the potential models that is equivalent to α_0 in the segmented model (equation 3.30b) that represents the time from inoculation until the first spore is produced.

3.7 Maximum Likelihood

Least squares regression is the most common method for fitting linear and nonlinear models to continuous data, or to data that—as an approximation—are taken to be continuous. However, there are other methods of model fitting, some of which depend explicitly on the statistical distribution of the response variable, and some which are distribution free (Sprent and Smeeton, 2001; Schabenberger and Pierce, 2002). One alternative to the method of least squares is the method of maximum likelihood. The likelihood is the *joint* probability or probability density of the observed data as a function of the parameters in the model. This depends explicitly on the probability distribution or probability density function (e.g., equation 3.4) of the response variable. Here, we briefly demonstrate the method for normally distributed data and a simple linear model for Y as a function of X (e.g., equation 3.7c).

The normal probability density function for a single observation is given in equation 3.4. This function is easily generalized for relationships between variables, after noting that μ is the expected value of Y, $E(Y)$, which is expressed as a function of predictor variables. With equation 3.7d, for example, $E(Y_j)$ is $\beta_0 + \beta_1 X_j$ for the jth observation. One can then write the density for the two-variable situation as:

$$f(Y_j) = \frac{1}{\sqrt{(2\pi)}\sigma} \exp\left[-\frac{(Y_j - (\beta_0 + \beta_1 X_j))^2}{2\sigma^2}\right] \quad (3.31a)$$

The product of the $f(Y_j)$ values for all N observations is the likelihood, $L = \Pi_j f(Y_j)$. For the normal distribution, and the linear two-parameter model, this is written as:

$$L = \frac{1}{(2\pi\sigma^2)^{N/2}} \exp\left[-\frac{\Sigma(Y_j - (\beta_0 + \beta_1 X_j))^2}{2\sigma^2}\right] \quad (3.31b)$$

The principle of maximum likelihood estimation (MLE) is to find the parameter values that result in the largest value of L, given the observed data. These values of the parameters, the maximum likelihood estimates, are the most likely ones to have produced the observations in the data set. Note that in equation 3.31b, the term outside the bracket does not depend on the observations. Because e is raised to a negative power, the largest L is obtained when the summation in the exponent is at its smallest possible value. However, this expression $[\Sigma_j(Y_j - (\beta_0 + \beta_1 X_j))^2]$ is identical to the sum of squared differences minimized with ordinary least squares (see equation 3.9b). Thus, for normal distributions and a constant variance, least squares and maximum likelihood result in the same parameter estimates!

The equivalence of least squares and maximum-likelihood parameter estimates does not extend beyond the normal distribution case, nor to situations with multiple error terms (say, with complicated experimental designs such as split plots). If the response variable has (or is assumed to have) a binomial distribution, for instance, least squares and maximum likelihood approaches will provide different estimated of model parameters. Generally, the maximum likelihood estimates will be considered more appropriate, because they are based on the actual distribution of the random variable of interest. However, assuming an inappropriate distribution for the random variable of interest could lead to misleading results, because the expression from which the maximum L is obtained is of the wrong form.

The disadvantage of MLE is that a unique analytical solution may not exist for some distributions, even for a linear model. That is, parameter estimation would entail an iterative search for the parameter values that give the largest achieved L for a data set (much as is done with least squares model fitting for nonlinear models). Conversely, there are some advantages of MLE in terms of hypothesis testing, inference, and the distributional properties of estimated parameters (e.g., Schabenberger and Pierce, 2002).

In addition to the direct link between least squares and MLE for normal distributions, there is an additional link for some other situations. In particular, for certain distributions in the so-called *exponential family* (which include the Poisson, binomial, and gamma), the iterative maximum likelihood estimation can be conducted by the utilization of an iterative least squares methodology. That is, iterative and weighted least squares estimation is performed on a transformation of $E(Y_j)$, not on the expectation of the transformation of Y_j. Models of this type are known as *generalized linear models* (GLMs) (see equation 3.26 for an example). Some practical applications of GLMs are demonstrated in Chapter 4, and (in rather more detail) in Chapter 10.

3.8 Discussion and Prelude to Later Chapters

Models are indispensable tools for investigators in probably all fields of study. For plant disease epidemiologists, mathematical models are used to describe, understand, predict, and compare epidemics or components (e.g., sporulation) of epidemics. Deterministic mathematical models are especially popular and useful to characterize the dynamics of disease intensity in time and space, and numerous versions of these models are used throughout the book. There is an inherent randomness to biological processes, and sometimes—depending on the objectives of the study—explicitly stochastic models are needed (Renshaw, 1991; Gibson et al., 1999). Based on the principles of population dynamics, as well as on observations of relations between variables, model that are nonlinear in the parameters are often needed to represent processes or phenomena of interest. Fortunately, some nonlinear models can be re-expressed in linear form using algebraic manipulation, facilitating both model fitting and making various types of calculations with the models (as will be demonstrated in the next chapter).

Even when one takes a deterministic approach to modeling of epidemics, or their components, ultimately parameters are estimated by fitting models to observed data. The two main methods of model fitting are least squares and maximum likelihood, with the former being most commonly used when the response variable is continuous, or assumed to be so. When one fits models to data, randomness (i.e., stochasticity) is incorporated as part of the model, typically expressed as an additive error term—a random variable with expectation of 0, representing the combined effects of all the variables and factors affecting the response that are not considered in the deterministic component. Essentially, this means that one is using a statistical model at this stage, whether or not one is considering variability in the original modeling formulation. The magnitude of the estimated variance of the error term (σ^2), or of related metrics (e.g., SSE), serve as a basis, among other things, for: determination of the precision of estimated parameters; testing hypotheses about the relationship between the response and predictor variables; and comparison of estimated parameters.

In this book, we show how to use models—and the results of model fitting—to describe and understand dynamics of disease development over time (Chapters 4–6) and over space (Chapters 7–9), and to relate yield loss of crops to disease intensity (Chapter 12). We apply statistical models, and results from previous chapters, to a range of problems in epidemiology, including sampling, in Chapters 10 and 11. The principles of modeling are further applied in Chapter 11 to evaluate controls and management decision making. The current chapter focused on model fitting methods that are most applicable to continuous random variables, or variables that are approximately continuous, and where a relatively simple function (linear or nonlinear) can be used to relate a response variable to a predictor variable. Some of the more complicated scenarios that can arise are considered in the context of epidemiological problems presented in other chapters.

Modeling and analysis in plant disease epidemiology have been directly influenced by many other disciplines, including statistics, biomathematics, medical and theoretical epidemiology and population dynamics, geostatistics, ecology, and crop physiology. The symbols used in equations thus vary considerably within botanical epidemlology (e.g. between crop loss assessment and spatial pattern analysis), and there is no practical way to use a notation throughout the book where each symbol has one and only one interpretation. As mentioned in this and the previous chapters, we use Y and y in this book for disease intensity, but most other symbols have multiple interpretations depending on the topic. Even the r symbol, generally understood to be a rate parameter for disease increase (as readers will see in the next two chapters), also can be used as a calculated correlation coefficient. Moreover, the symbol s can represent an estimated standard deviation and also the spatial dimension (distance from an inoculum source). We have strived to be consistent regarding notation within a specific topic in this book (temporal analysis, spatial analysis, etc.). Where a symbol has multiple meanings, the context should make the specific meaning clear.

It should be noted that there are several applications of modeling in botanical epidemiology that we do not cover in much detail, beyond what was covered briefly in this chapter. In particular, a major aspect of epidemiology is determining the relationship between observed disease intensity and the abiotic environment (e.g., climatic and weather variables, edaphic soil conditions), biotic environment (e.g., biocontrol agents, competitors of the pathogens), and anthropogenic factors (e.g., cropping history, crop production systems). We touched on this subject with the Phymatotrichum root rot example (Fig. 3.5) for the situation with a single predictor variable. Typically, there are many interrelated potential predictor variables, and a major part of the modeling is determining which variables should be included in the model, coupled with an evaluation of whether or not transformations of the variables are needed. Some examples of this approach (out of the many) can be found in: De Wolf et al. (2003), Mila et al. (2004), Chtioui et al. (1999), Coakley et al. (1988), Reynolds et al. (1988b), and Wheeler et al. (1994). Although model choice (and interpretation) should always be guided by knowledge of the epidemiology of the disease, the modeling approach is more statistical than it is epidemiological. That is, the analysis is based more on determining statistically significant relationships, rather than building models incorporating temporal, spatial, or spatio-temporal features of the epidemics. This is a very valuable exercise in epidemiology, which can lead to some useful predictive models of disease to be used in disease-management decision making (see Chapter 11; also see Chapter 15 in Campbell and Madden, 1990). Alternatives to traditional statistical models, such as artificial neural networks, have also found good use for these purposes (Francl, 2004). Interested readers who wish to pursue this kind of modeling in an epidemiological context will have to study textbooks, or part of textbooks, such as Neter et al. (1983), Schabenberger and Pierce (2002), Mead et al. (1993), and Collett (2003), in conjunction with this textbook. Two excellent book chapters on regression methods written for plant pathologists or ecologists are Butt and Royle (1990) and Philippi (1993).

To conclude, it must be emphasized that in addition to the mathematical and statistical methodology of modeling, there are philosophical issues of concern. Apart from the argument over empirical and mechanistic modeling, there are general issues that transcend which model-fitting protocols to use and other such technical details. For example, as will be seen in Chapter 5, one mechanistic approach for capturing relevant details of reality is to use a relatively small number of differential equations that result in a large number of epidemic outcomes (Jeger, 1986a). Known sometimes as the "analytical" approach, it is very useful in epidemiology and other areas of population dynamics for deriving general principles of disease dynamics. An alternative mechanistic approach, based on a more reductionist philosophy, is to partition the epidemic into many small parts or components (e.g., spores in air, spores on leaves, infection rate as a function of individual leaf age and position, spore production rate, and so on), with the assumption that all components require description, develop models for each component, and assemble the models into a "systems" type or "simulation" model of epidemics (see chapter 12 in Campbell and Madden, 1990; Miyai et al., 1986; Waggoner, 1990). This simulation approach was very popular in the 1970s and early 1980s, but is less commonly pursued today. As with all models and model types, the value of simulation models is conditional on the objectives of the user. There are some excellent simulation models that have been of tremendous value in representing the complexities of plant disease epidemics and evaluating management approaches (Bruhn and Fry, 1981; Gilligan, 1994; Gilligan et al., 1994; Reynolds and Arneson, 1997; Xu and Ridout, 1998). Here, we generally prefer to take the analytical approach to modeling plant disease epidemics, and to use simulation when, for various reasons, analytical models are inadequate.

Finally, we point out that both empirical and mechanistic models, or their hybrids, are based on the premise that models are simpler than reality, or in the words of Buckland et al. (1997), "that truth is high (effectively infinite) dimensional." An obvious question is: how simple (or complex) should a model be? Quoting again from Buckland et al.:

> *The more information that is gathered, the greater is the model complexity that the data can support. If data are sparse, they can support only a simple model with few parameters. In our view, model selection is the process of identifying the best approximating model,*

accepting that the data can never support, and can never identify the true model.

We would add that model selection should be based on our understanding of the underlying process or phenomenon of investigation (determined from previous investigations), and the intended use of the model, in addition to the sparseness of the data used to develop the model. Many mathematicians, statisticians, and philosophers have argued about how best to represent the complexity of reality in an efficient and useful manner and to characterize the resulting goodness (or "badness") of fit. Interested readers should consult Forster and Sober (1994), Zucchini (2000), and Bozdogan (2000) for a more thorough, although statistically challenging, discussion of these issues.

3.9 Suggested Readings

Hau, B. 1990. Mathematics and statistics for analysis in epidemiology. In: *Epidemics of Plant Diseases: Mathematical Analysis and Modeling*, 2nd Edn. (J. Kranz, editor). Springer, Berlin, pp. 12–52.

Quinn, G. P., and Keough, M. J. 2002. *Experimental Design and Data Analysis for Biologists*. Cambridge University Press, Cambridge, UK (Chapters 2–6).

Schabenberger, O., and Pierce, F. J. 2002. *Contemporary Statistical Models for the Plant and Soil Sciences*. CRC Press, Boca Raton, FL (Chapters 4 and 5).

Appendix

Data for incidence Phymatotrichum root rot in cotton (y) and cumulative precipitation (X) (Jeger and Lyda, 1986), and cumulative number of rust spores per plant (Z) and time (t) (Kolnaar and van den Bosch, 2001) used in example analyses of Chapter 3.

Root rot		*Cumulative spores*			
X	Y	t	Z ($\times 10^4$)	t	Z ($\times 10^4$)
42.7	0.22	0	2	30	2266
43.9	0.28	2	11	32	2297
56.6	0.32	4	46	34	2329
57.9	0.65	6	188	36	2355
37.3	0.04	8	408	38	2371
85.1	0.91	10	634	40	2385
63.8	0.75	12	871	42	2397
56.9	0.58	14	1085	44	2402
46.5	0.32	16	1380	46	2409
101.6	0.995	18	1600	48	2414
46.5	0.68	20	1738	50	2416
106.1	0.79	22	1888	52	2418
38.9	0.33	24	2015	54	2422
		26	2157	58	2422
		28	2226	60	2424
				62	2425

4

Temporal Analysis I: Quantifying and Comparing Epidemics

And time waits for no one, and it won't wait for me.
M. Jagger, K. Richards (*The Rolling Stones*)

4.1 Introduction

As discussed in the first chapter, a plant disease epidemic can be viewed as a change in disease intensity in a host population over time and space. In many situations, the spatial component is not explicitly considered, and the focus is on the change in intensity over time. As we show in this and the next chapter, a tremendous amount of epidemiological insight can be gained by characterizing disease progress over time. In fact, this characterization is essential to understand how plant diseases develop in populations and how disease control measures affect epidemics.

A graphical plot of disease intensity (y) versus time (t) is known as a *disease progress curve* (DPC). For many purposes, this is the primary depiction of an epidemic, serving to summarize the interactions of host, pathogen, and the biological and physical environment on disease development. For convenience of expression, 'epidemic' and 'disease progress curve' often are used interchangeably in this book, although, in fact, the latter is merely a graphical summary of the former biological process. Although researchers have published disease progress curves throughout the 20th century, it was Vanderplank (1960, 1963) who made the most compelling case for the preparation and analysis of disease progress curves (see section 1.3.1). Vanderplank's concepts, originally proposed four decades ago, will be brought up several times in this and later chapters. Although considerable advances have been made in recent years, it should be pointed out here that the ideas and approaches of Vanderplank still have a large influence on plant pathologists who are trying to understand epidemics.

Researchers may have many (possibly overlapping) goals in characterizing a plant disease epidemic. At one level, a researcher may only be interested in determining if two or more epidemics are different (in some sense) from each other. In this situation, the population dynamics of disease are not of direct interest; rather, one is interested in which 'treatments' have higher (or lower) intensities of disease or rates of disease change. At another level, a researcher may be interested in quantifying how the environment, host, and pathogen affect disease dynamics for the explicit purpose of understanding the complex interactions that influence disease development. This understanding can be used, among other things: to predict disease intensity at a particular time; to quantify the effects of control strategies on epidemics; or ultimately, to develop a theoretical basis for determining whether or not an epidemic will occur, and in the event of occurrence, to identify factors affecting the magnitude of the rate of disease increase and the final intensity of disease.

A central feature of this book is that models (of various degrees of complexity) are used to quantify epidemiological processes of interest. This use of modeling is certainly reflected in the approach taken in this and the following chapters to characterize the typical types of disease progress curves that can be found in plant pathology. Several population-dynamics models that are of direct value in summarizing the essential features of epidemics are introduced and described (more-or-less in order of increasing complexity). By choosing an appropriate model, an entire epidemic can be characterized with a small number of parameters. This makes it relatively easy to compare epidemics and evaluate the effects of biological and environmental factors, as well as possible disease controls, on epidemics. None of these models will provide a perfect fit to observed epidemics. Thus, unexplained variability, or *error* (in a statistical sense, see section 3.4), must be considered, and the models viewed from a statistical as well as a mathematical perspective. Fitting these models to data, and comparing epidemics based on estimated parameters, can be done using various statistical methods (Campbell and Madden, 1990; Madden, 1986), the choice depending on the design of the experiment or survey that produced the data and the assumptions made about the unexplained variability.

When the investigator is only interested in determining if various treatments affect intensity of disease over time, comparisons of disease progress curves can be done without assuming a particular population-dynamics model. Two general approaches are then possible. First, one uses a form of repeated-measures analysis of variance (ANOVA) to test for treatment (or multiple factor) effects. Here, disease assessment time is considered an explicit factor that affects y, and the appropriate error term for this 'repeated measure' is determined (Wolfinger, 1996; Zeger and Liang, 1992). Second, one can calculate the so-called area under the disease progress curve (AUDPC) for each replicated epidemic, and then compare epidemics based on ANOVA or t-test results, depending on the size of the study.

A considerable amount of theoretical research has been conducted on disease dynamics over the last decade or so. Several of these advances are covered in Chapter 5, partly because the research depends on the use of more complicated models than used in this chapter. The current chapter, however, sets the stage for more elaborate material by showing some alternative ways of representing disease progress curves.

4.2 General Concepts

4.2.1 Notation and introduction to models

Epidemics are dynamic processes, and dynamic processes are typically represented by rates of change. If we use the symbol y to represent disease intensity (either incidence or severity) and t to represent time (days, weeks, years, etc.), then $y(t)$ is disease intensity at time t, and $dy(t)/dt$ denotes the absolute rate of change in y at t. Loosely speaking, the absolute rate is a representation of how much y changes over a very small change in time, relative to the small change in time (at any time point during the epidemic). As shown in the subsequent sections, it is not expected that the absolute rate of change $dy(t)/dt$ will be constant for any epidemic; rather, $dy(t)/dt$ generally changes with time in an often predictable manner, given the type of disease being considered. For convenience of presentation, it is standard with population dynamics models to write the absolute rate of change for a variable y over time as dy/dt [i.e., the "(t)" is omitted], because the dependence of the rate on t is implicit. In some more complicated models or analyses, returning to the more detailed version of the notation is useful, however.

Several things should be noted about this formulation for disease dynamics. First, lower-case y is used to express disease intensity on a proportion scale $(0-1)$. Thus, y can represent the proportion of plants diseased or the proportion of leaf area covered by lesions. For incidence data, y is an estimate of the expected probability of a plant (or leaf, root, etc.) being diseased, p. The symbol Y is used to represent disease intensity on an absolute-unit scale (e.g., square centimeters of lesions per plant, number of diseased plants), although most of the discussion of disease on an absolute scale takes places in later chapters. Sometimes, Y specifically refers to *density* of disease intensity (e.g., number of diseased leaves per m^2); this is important for certain formulations of models of disease dynamics (see Chapter 5).

For general modeling purposes, it is assumed that y (or Y) changes continuously over time (Madden and Hughes, 2002). Models based on this assumption are called *continuous-time models*. Obviously, this is an oversimplification, as an example will demonstrate. Spores of many fungal diseases are only produced, and infections of plants only occur, when certain environmental conditions are met. In many cases, free moisture is required for sporulation, infection, or both. Thus, y does not increase every day, minute, and second, but only during certain periods of time. However, it turns out that the continuous-time model formulation is not problematic for theoretical and empirical studies of epidemics, unless one is interested in what is happening over very small increments of time. Actual observations of disease intensity seldom occur more frequently than every 5–7 days during epidemics of annual crop diseases, and perhaps only every year or two for epidemics of forest tree diseases. Thus, a range of actual absolute rate values are 'averaged out' between the well-separated times at which observations of disease are recorded, and as a result, discrete-time models—in general, more complicated than continuous-time models—are not required. There are some situations, however, in which discrete-time models can be used to develop an intuitive understanding of the population mechanisms underlying epidemics, as described in section 5.2.1 of the next chapter. Chapter 8 uses some discrete-time models as well.

The population-dynamics models considered in this chapter can all be considered deterministic (see Chapter 3), at least in their original derivation. That is, for a given set of parameter values, a given model will always predict the same absolute rate of change, dy/dt, at any given time t. Put another way, variability is not explicitly considered in the models. This is obviously a simplification, because in reality there is variability in biological processes. For instance, the lesions resulting from infection of a particular host by a particular pathogen do not always produce the same number of spores 10 days after infection, and infection of a leaf by a deposited spore may or may not occur on any given day. Stochastic models can be used to deal with this kind of variability, but the complexity of the modeling increases dramatically (Gibson et al., 1999; Renshaw, 1991). Stochastic models probably are of greatest value when the population size of interest is small (e.g., a small host population), or one is interested in the disease status of the individual plant (e.g., characterizing the probability of a plant at location s being infected). The latter is relevant when characterizing spatial patterns of disease (discussed in Chapter 9). When one is interested in modeling the mean disease intensity for a fairly large population of plants, the deterministic approach is reasonable.

We do discuss and use one specific model formulation that is stochastic. This is in relation to fitting models to observed data. Because the models shown here (and elsewhere) never fully describe the observed disease intensity values, it is typically assumed that the statistical model for an epidemic involves a deterministic component and a random error (unexplained variability) component (Madden and Campbell, 1990; Neter et al., 1983), the latter encompassing all the factors that can determine $y(t)$ that are not explicitly considered in the deterministic part of the model. In this chapter we expand the presentation of Chapter 3 (section 3.4) on statistical models to show how to fit population-dynamic models to data and compare two or more epidemics.

Relative to the alternatives, continuous-time deterministic models are the easy to use, to fit to data, and to manipulate, and this probably explains their popularity in the general fields of population dynamics and medical, animal, and plant disease epidemiology (Anderson and May, 1991; Diekmann and Heesterbeek, 2000; Gilligan et al., 1997). For most applications, researchers give up very little (if anything) in using these models over the (much) more complicated stochastic or discrete-time models. In fact, continuous-time deterministic models can represent a rich diversity of epidemic types and outcomes, and form the basis for a general set of concepts of disease progression. It should be pointed out that some strong arguments have been made in favor of stochastic models for the general characterization of population dynamics (Daley and Gani, 1999; Renshaw, 1991; Shaw, 1995). Nevertheless, few plant pathologists have found the need to use stochastic models as their first choice in characterizing disease progress curves.

4.2.2 Disease progress curves

Integration of the absolute rate of change leads to the determination of y [i.e., $y(t)$] as a function of t. A graph of y versus t for a particular model is a theoretical disease progress curve—based on the assumed model and parameter values. A plot of observed y versus t is an observed disease progress curve. Fig. 4.1 exhibits some observed disease progress curves corresponding to a range of pathogens, host crops, and biological/environmental conditions. Multiple curves on the same frame are for a given pathosystem with a range of different experimental treatments, cultivars, or other conditions. The examples were chosen to demonstrate the diversity of curves that can be obtained, and to stress that there is a great deal of similarity of many of the curves. The time scale varies from days to years, and intensity of disease can be given as incidence or severity. Observed intensity could also be based on a (scaled) disease rating scale (see Fusarium wilt of chickpea). Chapter 2 discusses disease-assessment schemes in some detail.

Many, but not all, of the disease progress curves have a so-called sigmoid ('slanted *S*') shape. There are biological reasons for this, as will be made clear below. The potato late blight data were originally published in 1954(!), and were used by Vanderplank (1963) in his classic book (also see the hypothetical DPC in Fig. 1.1). Other curves were published after 2000, thus showing the continuing value of disease progress curves in epidemiology. Most of the curves increase monotonically towards a maximum value of disease intensity which may be 1 (100%) or lower value. The Dutch elm disease progress curve rises to a maximum and then declines somewhat. This is due to the dynamics of the host population (i.e., the planting of new disease-free elms), a subject addressed in Chapter 6. Although the host crop is not necessarily static in the other examples, it is with the Dutch elm disease pathosystem that the impact of host dynamics is most obvious in Fig. 4.1.

4.3 How Does an Epidemic Occur?

4.3.1 Contact of inoculum with the crop host

Plant disease epidemics occur through the *'contact'* between inoculum of the pathogen and the disease-free host. The units used to quantify 'host' here depend on the disease and crop. Examples include plants, leaves, stems, roots, and leaf area. For convenience of expression, we often refer to plants or just individuals, but the concepts hold for other units. Inoculum is being used fairly loosely here, on purpose. Agrios (2005) defines inoculum as "the pathogen or its parts that can cause infections". Often the term *infectious unit* is used for a unit of inoculum produced by a diseased individual (see section 5.1.1). For many fungal pathogens, inoculum consists of spores. However, the inoculum could be comprised of mycelia. Bacterial cells constitute the inoculum for bacterial diseases. For plant viruses, the concept of inoculum becomes more nebulous. Obviously virus particles (virions) comprise the inoculum that can cause infections. However, for most plant viruses, infections occur (naturally) only through virus transmission by arthropod (insect, mite), nematode, or fungal vectors (Thresh, 1983; 1985). Thus, the 'inoculum' in the epidemiological sense is comprised of the vectors that have acquired the pathogen and are capable of transmitting the pathogen to another individual. This concept also holds for other prokaryotes (e.g., spiroplasmas, phytoplasmas) that are transmitted by vectors. For a detailed understanding of plant disease epidemics in relation to the biology of the pathogen and host, chapters in the following books are beneficial: Agrios (2005), Campbell and Benson (1994), Fry (1982), Jones (1998a), McLean et al. (1986), and Zadoks and Schein (1979).

4.3.2 Epidemic classification

In principle, plant disease epidemics can be classified into two basic types, depending on the source of the inoculum that comes in contact with the host over the course of disease development. In the first type, inoculum that can cause infections is produced *during* the epidemic by the pathogen in or on individuals that had

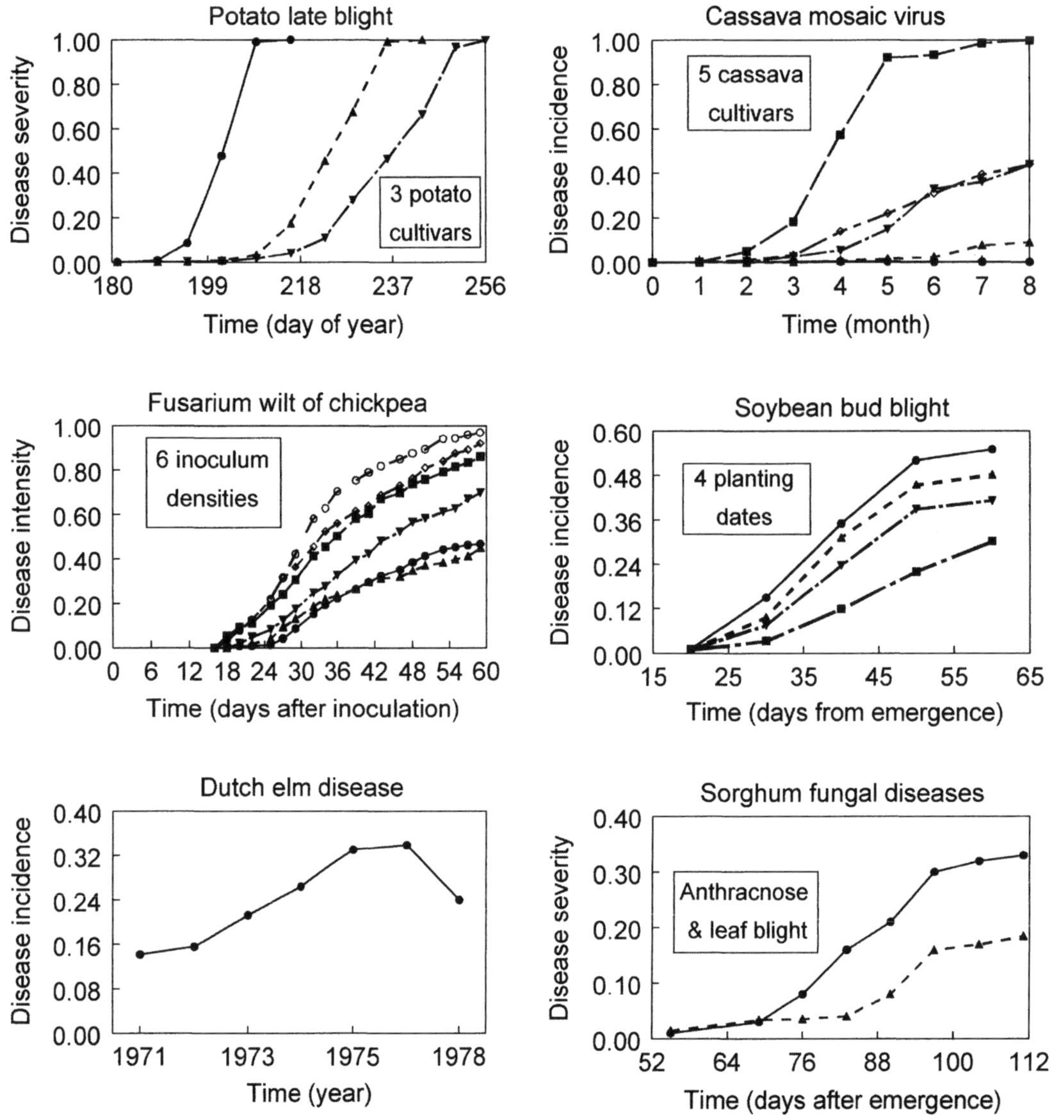

FIG. 4.1. Example disease progress curves from several pathosystems. Severity of *late blight* of potatoes, caused by *Phytophthora infestans*, corresponding to three potato treatments. Data originally collected in 1954 and reported in Vanderplank (1963). Incidence of plants with *cassava mosaic* virus symptoms in Africa (Sserubombwe et al., 2001). Intensity of *Fusarium wilt* of chickpea for a controlled study in which plants were grown in soil with different inoculum densities of *Fusarium oxysporum* f.sp. *ciceris* (Landa et al., 2001). Intensity determined from an ordinal disease-rating scale and converted to a 0−1 scale. Incidence of *bud blight* of soybeans, caused by tobacco streak virus (Almeida et al., 1994). Incidence of elms with *Dutch elm disease*, caused by *Ophiostoma* spp. (Swinton and Gilligan, 1996). Data compiled for the total number of elms in the UK that were healthy, weakly infected or strongly infected. Severity of sorghum with two diseases in the same field, *anthracnose* (caused by *Colletotrichum sublineolum*; solid line) and *leaf blight* (caused by *Exserohilum turcicum*; broken line) (Ngugi et al., 2001).

been previously infected during the current epidemic. Potato late blight is a good example of this. Lesions on infected leaves produce sporangia, some of these sporangia are dispersed to other parts of the same leaves or other leaves or other plants, and some of these cause infections. These new infections result in new lesions, new sporangia, and then additional infections. Epidemics of this type are called polycyclic, and diseases that cause them are called *polycyclic diseases*. This is because there are (typically) multiple cycles of infection (e.g., lesion → spore → lesion → ...) during the epidemic. There may be many cycles or just a few in an epidemic, depending on how long it takes to complete an infection cycle. Many of the foliar diseases caused by fungi, oomycetes, and bacteria are polycyclic. These diseases are also called *compound interest* diseases (Vanderplank, 1963), because in the early part of an epidemic, a model that describes disease increase is identical to the model that describes the compound-interest accumulation of money. A convenient way of thinking of a polycyclic epidemic is one in which there is *spread from plant to plant* (or from individual to individual, etc.) during the epidemic. The term *transmission* may be used broadly to represent the process of transfer of infectious units from individual to individual (e.g., plant to plant), including the production of inoculum, dispersal, and subsequent infection of contacted individuals.

In the second type of epidemic, inoculum that can cause infections is *not* produced during the epidemic by the pathogen in or on individuals that had been infected during the current epidemic. Rather, the inoculum that infects the host individuals during the current epidemic may be in the soil, on weed plants, in infected crop plants in another field, or elsewhere. During the current epidemic, infected plants may produce inoculum, but the inoculum is not available (in some sense) to cause additional infections in the current epidemic. Many soil-borne pathogens cause epidemics of this type. For example, with Verticillium wilt of potato, caused by *Verticillium dahliae*, roots of plants become infected by the germinating microsclerotia (the survival spores of the fungus) in the soil. The pathogen moves through the vascular tissue of the plant and ultimately produces new microsclerotia within the infected plant. This inoculum is only released into the soil when the plant decays at the end of the season. Thus, healthy plants do not contact the newly produced inoculum during current epidemic. In other cases, inoculum is only produced at the end of a growing season, such as with maize smut (caused by *Ustilago maydis*). Several plant virus diseases are also of this type (Thresh, 1983), possibly including soybean bud break (Almeida et al., 1994), one of the examples in Fig. 4.1. A typical situation would be when the arthropod vector is capable of acquiring the virus from infected plants of one species and then transmitting the virus to plants of a second species, but the vector is not capable of acquiring the virus from infected plants of the second species. Then, there can be no spread from plant to plant within a crop comprising plants of the second species.

Epidemics of this latter type are called monocyclic, and diseases that cause them are called *monocyclic diseases*. This is because there is just one cycle of infection in an epidemic, although the infections can be spread out over a very long period of time. Vanderplank (1963) called these diseases *simple interest*, because one mathematical model that describes disease increase is the same as a model that describes the simple-interest accumulation of money. A convenient way of thinking of a monocyclic epidemic is one in which there is *no spread from plant to plant* (or from individual to individual, etc.) during the epidemic. One might be tempted to think of simple-interest diseases as causing less serious epidemics because infections do not lead to further infections. Indeed, in medical and veterinary epidemiology, only polycyclic epidemics would generally be considered as 'real' epidemics. However, simple-interest diseases of crops are important. With either type of disease it is possible for 100% of the plants to become infected, depending on factors which will be considered below.

In the discussion of polycyclic diseases, we ignored the initial infections that start an epidemic. Previously infected plants do not cause this initial infection, since there are no previously infected plants at the start of the epidemic. This initial level of disease, often called the *primary infection*, is a result of the contact between disease-free host individuals and inoculum produced elsewhere or in a previous year (at the same location). Any infection that results ultimately from the primary infection is called a *secondary infection*. A secondary infection could be a late blight lesion formed from a spore produced in the first lesion in the field, or from spore that was produced in a lesion that resulted from another secondary infection. It should be noted that the primary infection process that starts a polycyclic epidemic is analogous to the process that occurs throughout a monocyclic epidemic. In this sense, one can think of monocyclic epidemics as consisting of only primary infections.

4.3.3 Nuances of classification of epidemics

Classification of an epidemic as being monocyclic or polycyclic can depend on several factors. In particular, time scale and host-plant scale both affect the classification. For instance, although Verticillium wilt of potato is monocyclic in a single growing season, over the course of multiple growing seasons the disease can be considered polycyclic. This is because the inoculum produced in one season infects plants the next season; thus, inoculum originating in previously infected plants (from the previous year) can infect other plants in the current year. Sometimes the term *polyetic* epidemic or polyetic disease is used for disease development over multiple growing seasons. Often, this book focuses on what happens in a single growing season, because this is a natural unit of time to consider for disease dynamics, at least in the context of diseases of annual crops. For diseases of annual crops, the natural time scale for an epidemic, then, is a fraction of a year. For diseases of perennials, however, the natural time scale may cover a number of growing seasons, and an epidemic can take place over many years.

If the host individual of direct interest, and the basis for disease assessment, is the plant, then the statements above about soil-borne pathogens causing monocyclic diseases are reasonable for many situations. However, if the unit of interest is the individual root, or a site on a root, then many soil-borne pathogens cause polycyclic diseases (of roots) (Gilligan et al., 1994). For instance, epidemics of take-all disease of wheat, caused by *Gaeumannomyces graminis* var. *tritici*, occur by root contact with inoculum in the soil. In addition, there are secondary infections of other parts of the same roots or different roots of the same plant, resulting from mycelial growth. In fact, with this disease, there also could be some spread from plant to plant within a season by mycelial growth.

4.4 Models

Five population-dynamics models that can be of value for representing, comparing, and understanding plant disease epidemics are presented in some detail. These can be also called disease-progress or *growth-curve models*. The latter term arises because the models used here in a population-dynamics context can also be used to represent the growth in size (mass, length, etc.) of individual organisms over time (Richards, 1969). The first three of the models presented below are especially important, because they serve as the bases for the most detailed empirical and theoretical research and application in epidemiology.

4.4.1 Exponential model

The first model considered is called the exponential, and has been used to represent population growth for more than two centuries. The equation for the absolute rate of disease increase is written as:

$$\frac{dy}{dt} = r_E y \tag{4.1}$$

in which r_E is a rate parameter (constant). All of the models discussed here have a rate parameter, and we use a subscript to identify the model. This subscript is needed because the scale (magnitude) of the parameter for a set of disease observations depends on the model. It is often convenient to refer to the rate parameter in general, rather than for a specific model; we use r with an asterisk subscript ($r*$) for this purpose.

The absolute rate of increase divided by $y[(dy/dt) \cdot (1/y)]$ is called the *relative rate of increase*. Rearrangement of equation 4.1 shows that the relative rate for the exponential model is equal to the parameter r_E. Thus, exponential growth is characterized by a constant relative rate of increase. If y is representing the proportion of diseased individuals (e.g., plants), and time is expressed in days, then dy/dt represents the new diseased individuals (on a proportion basis) per day. The parameter r_E, in this case, would be interpreted as the new diseased individuals per diseased individual per day. The "diseased individual" terms cancel out, and the units simple become days^{-1} (or more generally, time^{-1}).

The absolute rate of disease increase is plotted versus time for two values r_E in Fig. 4.2. A property of this model is that dy/dt increases continuously, without limit, over time. The higher r_E value results in higher dy/dt. The exponential model may be useful for representing the early part of polycyclic epidemics. This is a logical initial model to consider for such circumstances because the absolute rate of disease increase is directly proportional to the level of intensity at a given time. Given that polycyclic epidemics are characterized by the production of inoculum by diseased individuals during the epidemic, the absolute rate needs to be related to the level of disease intensity.

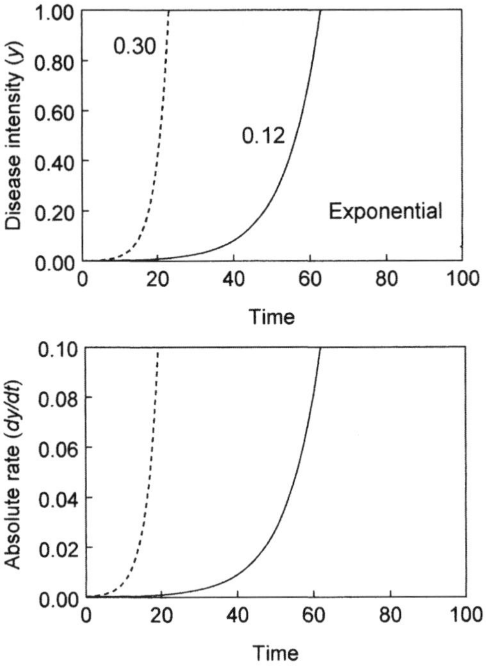

FIG. 4.2. Absolute rate of disease increase over time (dy/dt; bottom frame) and disease intensity (y; top frame) over time for the exponential model for two different rate parameters (indicated on the figure; units of time^{-1}). Initial disease intensity (y_0) was 0.001.

Assuming that the exponential model is representing a polycyclic epidemic, then r_E summarizes—in some complex sense—all the biological and physical factors that can determine the increase in disease, other than the intensity of disease itself (which is explicitly incorporated into the model). Inoculum production per diseased individual, probability of the inoculum contacting and infecting a disease-free host plant individual, time for a new infected individual to start producing spores, the length of time that an infected individual produces inoculum, virus titer in infected plants, probability of a vector acquiring and then transmitting a virus to a plant, and other factors, all affect the magnitude of r_E. Although the inter-relationships between the variables that contribute to the value of r_E may be complex, the exponential model provides a simple way of summarizing their overall effect.

Although absolute rates are useful for understanding dynamic processes, dy/dt is not directly measured with actual epidemics. Rather, disease intensity (i.e., y) at selected times are observed, as shown in Fig. 4.1. To obtain y as a function of t, one integrates equation 4.1. Integration leads to:

$$y = y_0 e^{r_E t} \tag{4.2}$$

in which y_0 is a (new) parameter, and e is the base of the natural log system (2.718...). Mathematically, y_0 is the constant of integration, that is, the additional parameter (constant) that is obtained when analytically integrating an equation. It also has a biological interpretation. If $t = 0$ represents the start of the epidemic, then y_0 is the initial disease intensity (i.e., using equation 4.2, y_0 is the value of y when t is equal to zero). Reverting back to the more thorough notation, which is generally not needed here, $y_0 = y(0)$ and equation 4.2 could be written as $y(t) = y(0)\exp(r_E t)$. Note that exp(•) is an alternate notation for e raised to a power. It is important for the reader to realize that the description of the epidemic given above started with one parameter in the model for the absolute rate of disease increase (equation 4.1), but that after integration the resulting model for the disease progress curve (equation 4.2) has two parameters.

Equation 4.2 can be used to model any variable that 'grows' exponentially. In fact, the equation even represents the compound-interest accumulation of capital when the interest is added to the principal continuously; here, y_0 represents the initial investment and r_E represents the compound-interest rate. A plot of equation 4.2 shows that y increases continuously over time, with no upper limit (Fig. 4.2 upper frame). That is, for any positive value of r_E, y will increase indefinitely over time. Thus, this model is not realistic for any epidemic in the long-term, but it can be useful—often as an approximation of more complicated models—before intensity of disease is too large. Just what constitutes 'too large' depends on the extent of the approximation one is willing to accept: for intensities up to about 0.05 the simplicity of the exponential model is often worth the trade-off involved in accepting an approximate calculation.

Equation 4.2 is nonlinear in the parameters (see sections 3.2.4 and 3.5). However, it can be linearized by taking logarithms of both sides. This results in the following equation:

$$\ln(y) = \ln(y_0) + r_E t \tag{4.3}$$

This linear model describes a straight line relation between $\ln(y)$ and t, with slope of r_E, and intercept $\ln(y_0)$. Each of the models discussed in this section can be expressed in linear form. However, the transformation of y is different in each case. We use the symbol y^* to represent the transformation of y that produces a linear model. For the exponential, $y^* = \ln(y)$. One advantage of equation 4.3 is that parameters can be estimated with linear regression. Another advantage is in ease of manipulation in calculations.

Some calculations. In section 4.4.6, a considerable amount of detail is given on performing calculations with the growth models. However, we take a slight diversion here from the presentation on models and show some example calculations in order to preview the application of disease progress models in an epidemiological context.

Assuming that one is describing a tree disease over multiple years, that y_0 is 0.005, and that r_E equals 0.1/year (see Fig. 4.3), one can calculate, for example,

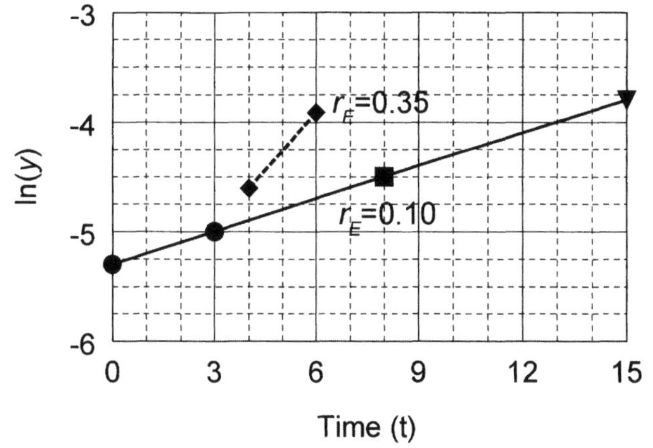

FIG. 4.3. Examples of exponential disease progress curve on a linearized scale; that is, $\ln(y)$ versus t. Disease intensity corresponding to the points along the solid line (for times 0, 3, 8, and 15) which equal the backtransformation of the logs of disease intensity shown on the graph, are: $y(0) = 0.005$, $y(3) = 0.0067$, $y(8) = 0.0111$, $y(15) = 0.022$. Disease intensity corresponding to times 4 and 6 along the broken line are: $y(4) = 0.01$ and $y(6) = 0.02$.

disease intensity at year 3. One way of doing the calculation is based on equation 4.3:

$$\ln(y) = \ln(0.005) + (0.1) \cdot (3) = -5.30 + 0.3 = -5.0$$

In this calculation, $\ln(y)$ is determined at year 3, *not* y. One needs to 'backtransform' this value. For the exponential model, the backtransformation is

$$y = e^{\ln(y)} = \exp[\ln(y)] \tag{4.4}$$

So, for the example, $y = \exp(-5.0) = 0.0067$. With this model, it is probably just as convenient to use the nonlinear version of the model directly for the calculation:

$$y = 0.005\exp[(0.1) \cdot (3)] = 0.0067.$$

However, for the more complicated models discussed subsequently, calculations such as this one are more straightforward with the linear, rather than the nonlinear version of the model.

A conceptual advantage of a straight-line model is that any point can be treated as the starting point for some calculations. Using the same example, assume that one knows that y at $t = 8$ years into the epidemic is 0.0111, r_E is 0.1/year, but that one does not know y at the start of the epidemic. If one wanted to predict y at year 15, one could first extrapolate backwards to first find y_0. Rearrangement of equation 4.3 gives:

$$\ln(y_0) = \ln(y) - r_E t = \ln(0.0111) - (0.1) \cdot (8)$$
$$= -4.501 - 0.8 = 5.30$$

Backtransformation gives a y_0 of $\exp(-5.30) = 0.005$, the original starting disease level. Then, the calculation of $\ln(y)$ at year 15 could be done with equation 4.3 as $-5.30 + (15) \cdot (0.10) = -3.8$. Backtransformation gives $\exp(-3.8) = 0.022$ for $y(15)$. The determination of y_0 is *not* needed however, by simply considering $t = 8$ as the first time. The same prediction for y at $t = 15$ years is obtained by using $t = 7$ in equation 4.3, $\ln(0.0111) + (7) \cdot (0.10) = -3.8$, and then backtransforming the resulting $\ln(y)$ value.

This is one place where a version of the alternate notation is useful. If one is interested in the change in disease between two specific times, labeled t_1 and t_2, one can write equation 4.3 in the following equivalent form:

$$\ln[y(t_2)] = \ln[y(t_1)] + r_E(t_2 - t_1) \qquad (4.5)$$

Here, the starting time is t_1, the later time is t_2, and the time period is $t_2 - t_1$. Using this formulation, $\ln(y)$ at $t = 15$ can be calculated as:

$$\ln(0.011) + (0.1) \cdot (15 - 8) = -4.501 + 0.7 = -3.80.$$

Backtransformation, once again, gives $y = 0.022$ for disease intensity at $t = 15$ years.

Through rearrangement of equation 4.5, one can easily determine the rate parameter (with the assumed exponential model) as:

$$r_E = \frac{\ln[y(t_2)] - \ln[y(t_1)]}{t_2 - t_1} \qquad (4.6)$$

If, as a different example (Fig. 4.3), $y = 0.01$ at $t_1 = 4$ years and $y = 0.02$ at $t_2 = 6$ years, then r_E is calculated as:

$$[\ln(0.02) - \ln(0.01)]/(6 - 4) = [-3.912 - (-4.6050)]/2 = 0.346/\text{year}.$$

More sophisticated approaches are available for parameter estimation, as discussed later.

In the examples used here, it was either assumed that r_E equaled 0.1/year, or that it was calculated as 0.346/year (Fig. 4.3). It is natural to ask: what is a large value for this rate parameter? Although researchers can answer this question with experience, there is also a direct way of gaining a general understanding of large and small rates. In particular, one can determine how long it takes disease intensity to double with a particular value of the rate parameter. Using symbols, one can determine the time to increase from one specific value of intensity, $y(t_1)$, to $y(t_2) = 2y(t_1)$. Then the time period in equation 4.5, $(t_2 - t_1)$, is the time to double, which can be written simply as t_D. Rearrangement of equation 4.5 with the symbol substitutions given here results in:

$$t_D = \frac{\ln[2y(t_1)] - \ln[y(t_1)]}{r_E} \qquad (4.7a)$$

Based on the rules of algebra, the numerator can be written as: $\ln[2y(t_1)/y(t_1)] = \ln(2)$. In other words, the disease-intensity terms cancel out, and the doubling time is given very simply as:

$$t_D = \ln(2)/r_E = 0.693/r_E \qquad (4.7b)$$

This relation can be easily derived directly from equation 4.2, by writing it as $2y = y\exp(r_E\, t_D)$, and algebraically rearranging the formula so that t_D appears alone on the left-hand side. With this model, for a given value of r_E, it takes the same amount of time for disease to increase from 0.01 to 0.02 as it does to increase from 0.05 to 0.10. Note that the doubling time is inversely related to the rate parameter, meaning that a small rate constant corresponds to a large doubling time. When $r_E = 0.1$/year, then disease intensity doubles every $0.693/0.1 = 6.93$ years; when $r_E = 0.35$/year, disease doubles every 2 years (or every 2 days if $r_E = 0.35$/day).

4.4.2 Monomolecular model

The so-called monomolecular model has been used to describe numerous phenomena, including certain chemical reactions (from where the name derives) and the growth of plants and animals. It can also be called the negative exponential or restricted exponential model. With this model the absolute rate of disease increase is given by:

$$\frac{dy}{dt} = r_M(1 - y) \qquad (4.8)$$

in which r_M is a rate parameter with units of time^{-1}. For this model, dy/dt is not directly proportional to y. Instead, the absolute rate is directly proportional to $1 - y$. Because disease intensity is represented as a proportion, $1 - y$ is equal to disease-free individuals (e.g., leaves, plants, etc.), or plant area, when we assume (as is normally done) that maximum possible y is 1. It is common to refer to the disease-free individuals (or area) as healthy individuals (or area). This does not mean that the plants are healthy in an absolute physiological sense, but simply that they are deemed not infected (for example, by visual inspection) by the pathogen of interest.

With a constant and positive r_M, dy/dt for this model *decreases* towards 0 as y increases, because $1 - y$ becomes smaller with increasing y. This decline can be seen in Fig. 4.4 for two values of the rate parameter. The mathematical reason for the decline can be seen by writing equation 4.8 in an expanded form:

$$\frac{dy}{dt} = r_M - r_M y \qquad (4.9)$$

The absolute rate is highest when $y = 0$ (with $dy/dt = r_M$), and then declines to 0 when $y = 1$. In epidemiological

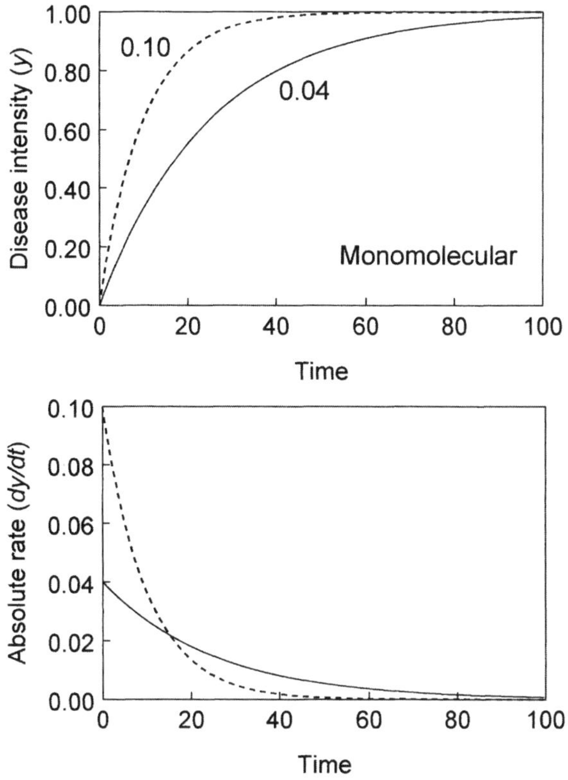

FIG. 4.4. Absolute rate of disease increase over time (*dy/dt*; bottom frame) and disease intensity (*y*; top frame) over time for the monomolecular model for two different rate parameters (indicated on the figure; units of time^{-1}). Initial disease intensity (y_0) was 0.001.

terms, the absolute rate declines because healthy individuals are fewer (or healthy plant area is smaller) as disease intensity increases. Once an individual plant (or an individual unit of plant tissue) is infected it cannot be infected again. Thus, as y increases, inoculum is more likely to come in contact with already infected individuals and so less likely to cause a new infection.

Because dy/dt in equation 4.8 is directly proportional $1 - y$, and not to y, the monomolecular is an appropriate model to consider for monocyclic epidemics. Infected individuals during a monocyclic epidemic do not produce inoculum that can infect other individuals or plant area (during the *current* epidemic). Thus, disease intensity does not contribute to new disease, and dy/dt should not be proportional to y for this type of disease.

If equation 4.8 is used to represent monocyclic epidemics, then a straightforward interpretation of the rate parameter is possible. Dividing both sides of equation 4.8 by $1 - y$, r_M [$=(dy/dt)\cdot(1/(1-y))$) is seen to be equal to the increase in diseased individuals per healthy individual (both on a proportion scale) per unit of time. Moreover, r_M is the product of the amount of inoculum (x') available for infection and the rate (i.e., per unit of time) at which this inoculum causes infection (φ); that is, $r_M = \varphi x'$. The fuller definition for φ is the number of new diseased individuals per healthy individual (on a proportion scale) per unit of inoculum per unit of time. This simplifies to "per unit of inoculum per unit of time" because disease and healthy individuals are represented as proportions (essentially, unitless). The φ term can be considered a type of efficiency. By multiplying φ by x', we see that the units simplify to just time^{-1}.

The concept behind r_M is embedded in disease-inoculum curves that are often determined for plant diseases, in which plants are exposed to increasing densities of inoculum (in the soil or on the leaf surface) and disease intensity is determined after so much time has elapsed (Baker, 1978). Although these curves are almost always concave over the full range of inoculum densities, the curves are actually straight lines at low densities with slope equal to φ or a scaled value of φ (Vanderplank, 1975). See sections 3.6.1 and Fig. 3.7 for an example data set. Both an increase in inoculum and an increase in probability of inoculum causing infections (say, through a more favorable physical and biotic environment) result in higher r_M. There is some similarity of $\varphi x'$ of the monomolecular model and y_0 of the exponential (and other polycyclic-disease models). However, there is a difference: y_0 in equation 4.3 is an instantaneous value, not a change over time.

This concept of r_M as the product of inoculum and inoculum efficiency can be expressed another way. Instead of thinking of inoculum density as being fixed in time, one can consider a continuing influx of inoculum into a field. Then, inoculum has units of "per unit time" and written here as x_0. This influx could be in the form of spores blown in per hour from outside sources (e.g., from weeds) or viruliferous insect vectors arriving per day from other fields. Then, the efficiency of infection term (φ) is unitless, giving 1/time units to the product φx_0. As long as the product of inoculum and its efficiency is a constant (or approximated reasonably well by a constant), with 1/time units, the following mathematical derivation is appropriate. When the product of inoculum and efficiency varies over time then more complicated mathematical models are needed, as discussed in section 6.3.

Analytical integration of equation 4.8 results in the following expression for y:

$$y = 1 - (1 - y_0)e^{-r_M t} \tag{4.10}$$

in which y_0 is a constant of integration (equal to the value of y at $t = 0$). A graphical plot of y versus t is concave to the time axis (Fig. 4.4); y increases rapidly at low t, and steadily approaches an upper limit of 1 with ever-decreasing slope. The shape of this curve is fundamentally different from the $y:t$ curve for the exponential model (see Fig. 4.2). At small times, the monomolecular curve is nearly a straight line. This can be seen by considering the second version of the differential equation (equation 4.9) for monocyclic diseases. At small y ($y \approx 0$), $dy/dt \approx r_M$.

Integration then leads to $y = y_0 + r_M t$, an equation for a straight line with slope r_M on an untransformed scale. This is the equation for the simple-interest accumulation of money at rate r_M. Disease intensity must be very low for the approximation to hold, and we will not explore this situation any further here.

It should be noted that one can use equation 4.10 for an increase in disease intensity over time even when $y_0 = 0$. The equation then simplifies to: $y = 1 - \exp(-r_M t)$. On the other hand, if $y_0 = 0$ for the exponential model, $y = 0$ at all times, and hence there is no epidemic. This difference between the two models can be explained by consideration of the underlying epidemiology. For polycyclic diseases, epidemics occur by the multiplication of effective inoculum, and the compounding of disease, through the epidemic; hence, disease at any time is based on disease at the previous times (back to time 0). With monocyclic diseases, however, the increase in disease (during the current epidemic) is not due to multiplication of inoculum, or compounding of disease; rather, the increase in disease is due exclusively to the plant contact with inoculum produced in previous epidemics or in epidemics at other locations (and restrained by the available disease-free host area). Thus, there need not be any disease at a specific time (e.g., $t = 0$) for an epidemic to occur.

Equation 4.10 can be linearized to:

$$\ln\left(\frac{1}{1-y}\right) = \ln\left(\frac{1}{1-y_0}\right) + r_M t \qquad (4.11)$$

in which $\ln[1/(1 - y)]$ is known as the multiple-infection transformation. This transformation can be obtained in a different context based on the Poisson probability distribution (see sections 5.2.1 and 9.7.2). It should be noted that $\ln[1/(1 - y)]$ is equivalent to $-\ln(1 - y)$. Equation 4.11 is an expression for a straight line with intercept of $\ln[1/(1 - y_0)]$ and slope of r_M. At low y, $\ln[1/(1 - y)]$ is roughly equal to y, but the transformation deviates substantially from y as y increases. For instance, transformed values of a few disease intensities are given in Table 4.1. Given that the solution to equation 4.8 is $y_0 + r_M t$ at small y, it should not be surprising that y and $\ln[1/(1 - y)]$ are similar at small y. The linearized version of the model can also be written as:

$$\ln\left(\frac{1}{1-y(t_2)}\right) = \ln\left(\frac{1}{1-y(t_1)}\right) + r_M(t_2 - t_1) \qquad (4.12)$$

The linear form of the monomolecular model (either equation 4.11 or 4.12) can be used to predict the multiple-infection transformation of y at a given t. One can then obtain the corresponding predicted value of y by backtransformation:

$$y = 1 - e^{-\ln[1/(1-y)]} \qquad (4.13)$$

For example, suppose that $y_0 = 0.01$ and $r_M = 0.05$/day. One can predict y at $t = 6$ days by first calculating:

$$\ln[1/(1 - 0.01)] + (0.05) \cdot (6) = 0.010 + 0.30 = 0.31.$$

To obtain a prediction of y, one backtransforms 0.31 with equation 4.13 to obtain $y = 0.267$. Further discussion of calculations involving the monomolecular model is delayed until after some other models have been presented.

4.4.3 Logistic model

We now consider a model that is more generally appropriate than the exponential model for representing polycyclic epidemics over the full range of disease intensity values. The model incorporates features of both the exponential and monomolecular models, as already discussed, and thus is presented here, following the discussion of the two simpler models. The model is known as the logistic and has a long history in population biology,

TABLE 4.1. Transformation of disease intensity (y) for some different disease progress models.

y	*ln(y)* *equation 4.3*	*ln[1/(1 – y)]* *equation 4.11*	*ln[y/(1 – y)]* *equation 4.16*	*–ln[–ln(y)]* *equation 4.21*
0.005	−5.298	0.0050	−5.293	-1.67
0.01	−4.605	0.0101	−4.595	-1.53
0.05	−2.996	0.0513	−2.944	-1.10
0.10	−2.303	0.1054	−2.197	-0.83
0.25	−1.386	0.2877	−1.099	-0.33
0.50	−0.693	0.6932	0.0	0.37
0.75	−0.288	1.386	1.099	1.25
0.90	−0.105	2.303	2.197	2.25
0.95	−0.051	2.996	2.944	2.97
0.99	−0.010	4.605	4.595	4.60
0.995	−0.005	5.298	5.293	5.30

going back at least to 1838 (Madden and Campbell, 1990). It has been widely used in plant pathology over the last 40 years or so, on the basis of compelling arguments presented by Vanderplank (1963). The logistic model also serves a 'stepping stone' to the more sophisticated models presented in the next chapter.

The differential equation form of the logistic model is given by:

$$\frac{dy}{dt} = r_L y(1-y) \qquad (4.14)$$

in which r_L is a rate parameter. For this model, the absolute rate of disease increase is proportional to both diseased and healthy plant individuals (both on a proportion basis in this formulation). As such, the logistic model can be considered a combination of the exponential, where dy/dt is proportional to y, and the monomolecular, where dy/dt is proportional to $1 - y$. We assume here, as with the previous model, that maximum possible value of y is 1. By itself, the y term in equation 4.14 accounts for the fact that the absolute rate increases with increasing disease, since increasing disease means more inoculum produced for further infections. By itself, the $1 - y$ term accounts for the fact that the absolute rate decreases with increasing y because there is a decreasing number of healthy individuals (or a decreasing amount of healthy plant tissue) for this inoculum to infect as y increases. This combination of properties of the exponential and monomolecular models provides a basis for a model description of polycyclic epidemics.

With the logistic model and positive r_L, dy/dt starts low, increases to a maximum, and then declines towards 0 (Fig. 4.5). Early in the epidemic (specifically, at values of y close to 0), the rate curve resembles the exponential model (Fig. 4.2); mathematically, this is because $1 - y \approx 1$ when y is near 0. Late in the epidemic (specifically, as y approaches the maximum value of 1), the rate curve resembles the monomolecular model (Fig. 4.4); mathematically, this because $y \approx 1$ when $1 - y$ is near 0. This pattern in the absolute rate can also be seen by rewriting equation 4.14 in an alternative format, $dy/dt = r_L y - r_L y^2$, which makes it clear that the right-hand side is a quadratic function of y. With a positive r_L, the quadratic function increases to a maximum and then declines. The maximum can be obtained by differentiating equation 4.14 with respect to y, setting it equal to 0, and solving for y. The derivative of equation 4.14 is $r_L - 2r_L y$. Equating this to zero and rearranging shows that the maximum value of dy/dt occurs at $y = 1/2$, and is not dependent on r_L. In other words, for a logistic-type epidemic, the maximum absolute rate always occurs at a disease intensity that is exactly 50% of the maximum. The value of dy/dt at the maximum is obtained by substituting 1/2 for y in equation 4.14. This results in $dy_{max}/dt = r_L/4$.

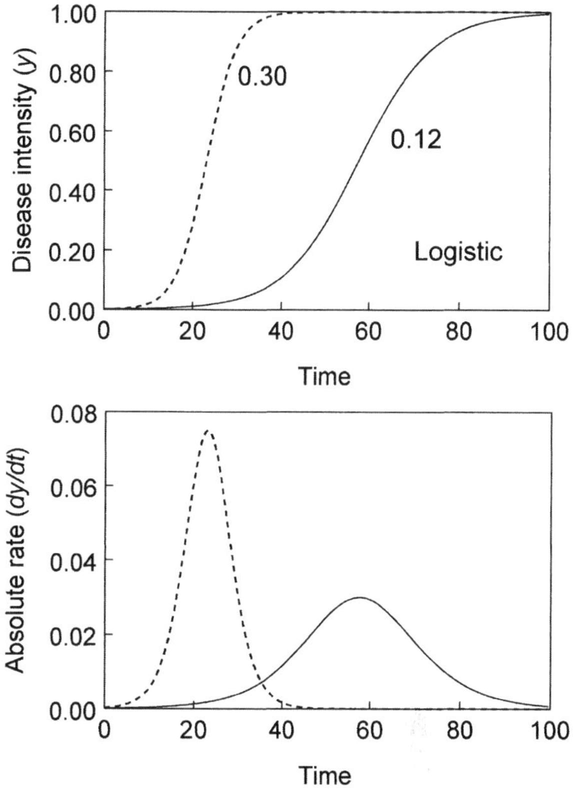

FIG. 4.5. Absolute rate of disease increase over time (dy/dt; bottom frame) and disease intensity (y; top frame) over time for the logistic model for two different rate parameters (indicated on the figure; units of time^{-1}). Initial disease intensity (y_0) was 0.001.

At the beginning of this chapter we emphasized that an epidemic occurs because of the contact between inoculum and healthy individuals. The healthy individuals are incorporated in the logistic model with the $1 - y$ term. The reader may now notice that there appears to be an essential shortcoming of the exponential model (as discussed above) for polycyclic diseases—it does not have an explicit healthy-individual term. In fact, the exponential can be viewed as a special case of the logistic in which healthy individuals (on a proportion scale) are fixed at the maximum possible value (=1) at all times. Obviously, this is not realistic in general, but can be useful near the start of an epidemic, as already mentioned.

Inclusion of the y term in the model recognizes that the inoculum in a polycyclic epidemic comes from the infected individuals during the epidemic. The production of inoculum from diseased individuals, the movement of inoculum to other individuals, and the infection of healthy individuals by dispersed inoculum are all incorporated into the rate parameter r_L. This parameter also accounts (in a rather complicated manner) for the time between infection and the production of inoculum by infected individuals and the time period over which infected individuals can produce inoculum. These latter properties are addressed more fully in the next chapter.

More formally, r_L in equation 4.14 is equal to $(dy/dt) \cdot (1/y(1-y))$. That is, the logistic rate parameter is the increase in diseased individuals per diseased individual per healthy individual (all as proportions) per time. The diseased individuals cancel out in the definition. Furthermore, since "healthy individuals" is represented as a proportion, the only unit remaining in the definition is time. Thus, r_L has units of time^{-1}. The parameter is sometimes known as the secondary infection rate because it summarizes the secondary infections occurring throughout polycyclic epidemics.

Integration of equation 4.14 results in the following expression for y as a function of t:

$$y = \frac{1}{1+\left(\frac{1-y_0}{y_0}\right)\exp(-r_L t)} \qquad (4.15)$$

As above, y_0 is a constant of integration and represents disease intensity at $t = 0$. A graph of equation 4.15 produces a 'classic' sigmoid (*S*-shaped) disease progress curve (Fig. 4.5), in which y increases slowly at first, rises rapidly in the middle range of intensities, and then slowly approaches the maximum. One can see that the maximum dy/dt occurs at $y = 1/2$ by plotting both the rate and disease on the same graph (Fig. 4.6). The time at which the maximum rate occurs can now be found by substituting 1/2 for y in equation 4.15 and solving for t. We call this time t'. Mathematically, this is known as the inflection point—dy/dt is increasing at times less than t', and decreasing at times greater than t'. A little algebra results in the following expression: $t' = -\ln[y_0/(1-y_0)]/r_L$. That is, although the value of disease intensity where maximum dy/dt is reached does not depend on r_L, r_L (and y_0) determine how long it takes to reach the maximum. In general, higher values of r_L mean shorter times to reach any particular y value. For the two curves in Fig. 4.5, with $y_0 = 0.001$, one can calculate t' values of 69 and 23 days.

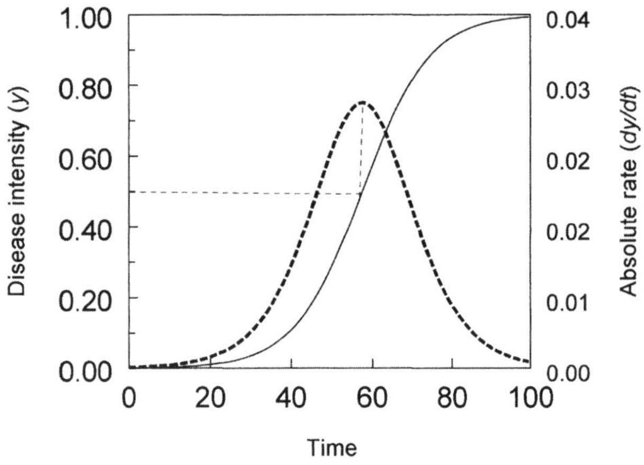

FIG. 4.6. Absolute rate of disease increase over time (dy/dt; right-hand axis) and disease intensity (y; left-hand axis) over time for the logistic model for $r_L = 0.12$/day. Initial disease intensity (y_0) was 0.001.

It should be noted that there are many equivalent ways of writing the nonlinear solution to the differential equations presented for disease progression. Because of the widespread application of the logistic model in epidemiology, it is helpful to see some of the alternatives. For example, we can write:

$$y = \frac{1}{1+\exp\left(-\left(\ln\left[y_0/(1-y_0)\right]+r_L t\right)\right)} \qquad (4.15a)$$

$$y = \left(1+Be^{-r_L t}\right)^{-1} \qquad (4.15b)$$

and

$$y = \left(1+e^{-r_L(t-t')}\right)^{-1} \qquad (4.15c)$$

in which $B = (1-y_0)/y_0$, and, as above, $t' = -\ln[y_0/(1-y_0)]/r_L$. As with the exponential model, the logistic model is undefined when initial disease is 0. Biologically, this makes sense since disease increase during the current epidemic for a polycyclic disease is dependent on diseased individuals that are already present.

The logistic model of equation 4.15 can be linearized to:

$$\ln\left(\frac{y}{1-y}\right) = \ln\left(\frac{y_0}{1-y_0}\right) + r_L t \qquad (4.16)$$

which is an equation for a straight line with intercept of $\ln[y_0/(1-y_0)]$ and slope of r_L. The term $\ln[y/(1-y)]$ is known as the *logit transformation*. Logits are similar to natural logs at small values of y (say, <0.05), but then deviate substantially as y increases (see Table 4.1). At large y, logits are similar to the multiple infection transformation values of the monomolecular model (equation 4.11). The logit of y is never similar to y. It should be noted that logits are negative at $y < 0.5$, and positive at $y > 0.5$. The linearized logistic model can also be written as:

$$\ln\left(\frac{y(t_2)}{1-y(t_2)}\right) = \ln\left(\frac{y(t_1)}{1-y(t_1)}\right) + r_L(t_2 - t_1) \qquad (4.17)$$

for transformed disease intensity at two times. The rationale for an alternative expression for a linearized model was given in the discussion of the exponential model. Specifically, when one is dealing with a straight line, y at any time (e.g., t_1) can be considered the starting point for performing calculations.

As a simple example using equation 4.16 (or 4.17), assume that $y_0 = 0.01$ and $r_L = 0.10$/day. Then disease intensity at day 60 can be calculated by first determining the logit of disease intensity at $t = 60$

days: ln(0.01/0.99) + (0.10) · (60) = −4.595 + 6 = 1.405. To obtain the disease intensity at this time, back-transformation is accomplished using:

$$y = \frac{1}{1+e^{-\ln[y/(1-y)]}} \tag{4.18}$$

For the example here, $y = 1/[1 + \exp(-1.405)] = 1/[1 + 0.2454] = 0.803$.

4.4.4 Some other population dynamics models

We made the point that the logistic model is often appropriate for quantifying polycyclic epidemics. However, it should be noted that there are other nonlinear models that have been used for representing such epidemics. The more common ones are discussed briefly in this section.

4.4.4.1 Gompertz model. A characteristic feature of the logistic model is that the absolute rate curve is symmetrical about $y = 1/2$. This can be quite restrictive, because there is no biological reason why the maximum observed dy/dt for an epidemic should not be at either a higher or a lower value of y. One commonly-used alternative to the logistic is the Gompertz model, which is named after the 19th Century scientist who proposed its use for animal growth (Madden and Campbell, 1990). The Gompertz model is written as:

$$\frac{dy}{dt} = r_G y[\ln(1) - \ln(y)] \tag{4.19}$$

in which r_G is a rate parameter. This is similar to the logistic model, except that the "$1 - y$" term (i.e., the *correction* for healthy plant individuals) of equation 4.14 is replaced by "[ln(1) − ln(y)]". The effect of this is to produce a skewed (asymmetrical) dy/dt curve, which rises relatively rapidly (depending on the rate parameter) to a maximum at $y = 1/e$ (≈ 0.367), and then declines slowly towards zero (Fig. 4.7).

Integration of equation 4.19 results in:

$$y = \exp(-Be^{-r_G t}) \tag{4.20}$$

in which B is a constant of integration equal to $-\ln(y_0)$. A plot of y versus t is S-shaped (Fig. 4.7), but is not quite of the same form as the logistic graph. With the logistic, the disease progress curve above $y = 1/2$ is just a "flipped-around-and-up" version of the disease progress curve below $y = 1/2$; such symmetry is absent with the Gompertz curve (a consequence of the asymmetrical dy/dt curve). A linear form of equation 4.20 is:

$$-\ln(-\ln(y)) = -\ln(-\ln(y_0)) + r_G t \tag{4.21}$$

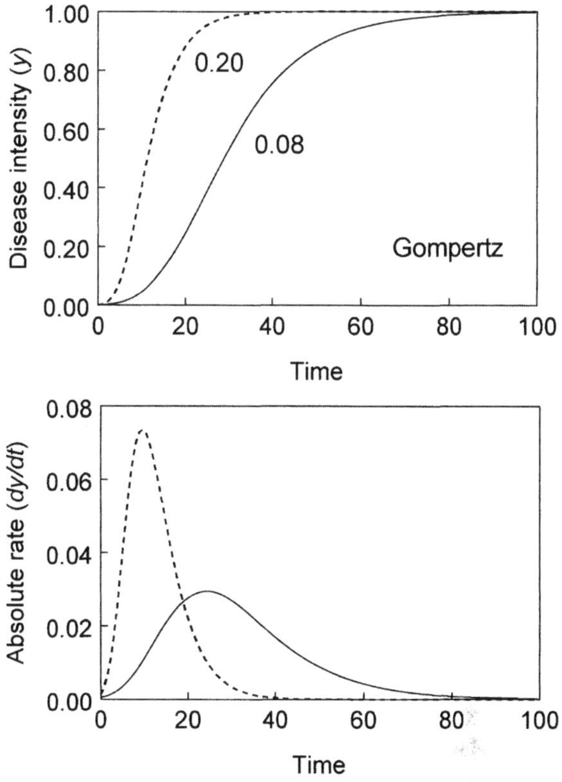

FIG. 4.7. Absolute rate of disease increase over time (dy/dt; bottom frame) and disease intensity (y; top frame) over time for the Gompertz model for two different rate parameters (indicated on the figure; units of time^{-1}). Initial disease intensity (y_0) was 0.001.

Note that there is a minus sign in front of the log. Back transformation is accomplished using:

$$y = \exp\left(-e^{-[-\ln(-\ln(y))]}\right) = \exp\left(-e^{\ln(-\ln(y))}\right).$$

The Gompertz model has often been found to provide a better fit to observed disease progress curves than the logistic model (Berger, 1981).

4.4.4.2 Richards model. There are numerous other ways of modifying the logistic model to account for asymmetry of the dy/dt curve. Most of these involve modifying the simple "$1 - y$" term, into a more complicated expression. In the Richards model, the modification involves use of an additional unit-less parameter (η). This model can be written as:

$$\frac{dy}{dt} = \frac{r_R y\left(1 - y^{\eta-1}\right)}{\eta - 1} \tag{4.22}$$

in which r_R is a rate parameter. Actually, the last term in the numerator of equation 4.22 can be written as: $1^{(\eta-1)} - y^{(\eta-1)}$, but 1 raised to a power is equal to 1, so the simpler version is used.

The parameter η can take on any real value between 0 and infinity. However, at specific values of η, the Richards model reduces to some of the models already discussed. The reader can perform the necessary algebra to see that when $\eta = 0$, equation 4.22 reduces to the monomolecular model (equation 4.8), and that when $\eta = 2$, equation 4.22 reduces to the logistic model (equation 4.14). The Richards model is undefined when $\eta = 1$, but when η approaches 1 in the limit (i.e., gets very close to 1 without actually reaching it), the Richards model reduces to the Gompertz (equation 4.19), in that the two models are indistinguishable. When η approaches infinity in the limit, the Richards model approaches the exponential; that is, dy/dt increases continuously (without an inflection point) until $y = 1$ is reached. Because different shapes are obtained depending on the exponent in equation 4.22, η typically is called a *shape parameter.* When $0 < \eta < 2$, the dy/dt curve rises rapidly to a maximum at a disease incidence less than 1/2; when $\eta > 2$, the dy/dt curve rises slowly to a maximum, the maximum occurring at a value of y greater than 1/2, and then the absolute rate declines quickly. An example of the latter is given in Fig. 4.8. Because the Richards model can accommodate a wide variety of shapes, and includes simpler models as special cases, it is sometimes referred to as a *general or flexible disease progress model.*

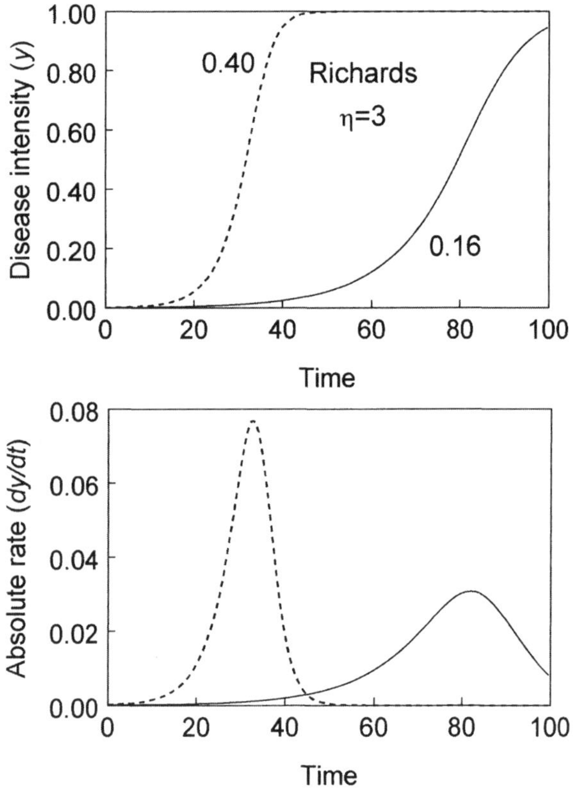

FIG. 4.8. Absolute rate of disease increase over time (dy/dt; bottom frame) and disease intensity (y; top frame) over time for the Richards model with $\eta = 3$ for two different rate parameters (indicated on the figure; units of time^{-1}). Initial disease intensity (y_0) was 0.001.

Integration of equation 4.22 leads to two possible solutions, depending on whether η is less than or greater than 1:

$$y = \left(1 - Be^{-r_R t}\right)^{1/(1-\eta)} \tag{4.23a}$$

if $\eta < 1$, and

$$y = \left(1 + Be^{-r_R t}\right)^{1/(1-\eta)} \tag{4.23b}$$

if $\eta > 1$. B is the constant of integration, and its specific form depends on the shape parameter:

$$B = 1 - \left(y_0^{1-\eta}\right), \quad \text{if } \eta < 1,$$

and

$$B = \left(y_0^{1-\eta}\right) - 1, \quad \text{if } \eta > 1.$$

S-shaped disease progress curves can be obtained for many values of η, especially for η values greater than 1/2 or so. The example disease progress curve in Fig. 4.8 is for a shape parameter of $\eta = 3$.

Equations 4.23a and 4.23b can be linearized to obtain the following equations:

$$\ln\left(\frac{1}{1 - y^{1-\eta}}\right) = -\ln(B) + r_R t \tag{4.24a}$$

if $\eta < 1$, and

$$\ln\left(\frac{1}{y^{1-\eta} - 1}\right) = -\ln(B) + r_R t \tag{4.24b}$$

if $\eta > 1$. The reader can confirm that substitution of $\eta = 0$ into the left-hand side of equation 4.24a results in the multiple infection transformation of the monomolecular model [$\ln(1/(1 - y))$; see equation 4.11] and substitution of $\eta = 2$ into the left-hand side of equation 4.24b results in the logit transformation [$\ln(y/(1 - y))$; see equation 4.16]. We emphasize here that the Gompertz model is a special case that is determined in the limit (i.e., $\eta \to 1$). Back transformation depends on whether the shape parameter is greater than or less than 1:

$$y = \left(1 - e^{-\ln[1/(1-y^{1-\eta})]}\right)^{1/(1-\eta)}, \quad \text{if } \eta < 1$$

and

$$y = \left(1 + e^{-\ln[1/(y^{1-\eta}-1)]}\right)^{1/(1-\eta)} \quad \text{if } \eta > 1.$$

Calculations with the Richards model can be done as with the other models. The maximum dy/dt occurs at

$y' = \eta^{1/(1-\eta)}$. This is obtained, as discussed for the logistic model, by taking the derivative of the dy/dt equation, setting this to zero, and solving for y. Increasing values of η produce larger values of y where the maximum rate occurs. The reader can try using very large values of η to see that y' approaches 1 as η increases indefinitely. The actual value of the maximum rate can be obtained by substituting this disease value into equation 4.22. Although one cannot use $\eta = 1$ directly for the Gompertz model, a value of $\eta = 1.01$ can be used as an approximation. This results in an estimated maximum dy/dt occurring at $1.01^{(1/-0.01)} = 0.3697$, fairly close to the true (i.e., directly from the Gompertz) value of 1/e (=0.3678...).

4.4.5 Model comparisons

Several models have been presented in this chapter that can be useful for representing disease progress curves. The models have several properties in common. In all cases, there is only one variable being analyzed, disease intensity (y). All the models are expressed in terms of the absolute rate being proportional to: (1) a rate parameter (r_*, in general); (2) a function of diseased individuals [$f(y)$]; and (3) a function of disease-free individuals or area [$f(1-y)$]. These functions can be the variables themselves [e.g., $f(y) = y$], 1 [e.g., $f(y) = 1$], or a transformation of the variable [e.g., $f(1-y) = 1 - y^{\eta-1}$]. Here is a summary of the models discussed:

Model	*f(y)*	*f(1 − y)*
Exponential	y	1
Monomolecular	1	$1 - y$
Logistic	y	$1 - y$
Gompertz	y	$\ln(1) - \ln(y)$
Richards	y	$1 - y^{\eta-1}$

There are other models that are of the same general form as the ones shown here that can provide a range of shapes for the $dy/dt : t$ curve. These include the Weibull, the log-logistic, and the Turner generic model. Some of these are discussed in some detail in Madden and Campbell (1990), Birch (1999) and Ratkowsky (1990). Although one or more of these may give a good fit to a given data set, for our purposes here, these other models do not provide a great deal of additional information for epidemiological interpretation of disease progress curves. There *are* more general models of epidemics that are of tremendous value in interpretation, but these are discussed later.

Integration of the differential equation versions of the models discussed above leads in each case to a nonlinear model (in terms of the parameters), with one additional parameter (a constant of integration) that can be equated to a transformation of disease intensity at $t = 0$. Although the nonlinear solutions can be of direct use, it is often convenient to express the models in linear form. All of the models presented here can be linearized in terms of some of the parameters, and written in a form so that there is a straight-line relationship between a transformation of y (y^*) and t:

$$y^* = y_0^* + r_* t \qquad (4.25a)$$

or

$$y^*(t_2) = y^*(t_1) + r_*(t_2 - t_1) \qquad (4.25b)$$

Here, y_0^* is a transformation of initial disease intensity, y^* or $y^*(\bullet)$ is a transformation of y at any given time, and r_* is the slope of the $y^* : t$ line. Besides aiding in interpretation and in making calculations (see next section), the straight-line form is useful for parameter estimation with most of the models. However, y^* for the Richards model contains a parameter (η); thus, one cannot use equation 4.24 to estimate η. For the linear model, one must assume that a particular value of η is correct and then use the appropriate transformation to achieve linearity and perform the desired calculations. In this sense, one can think of choosing a η of the Richards model as being equivalent to choosing a particular (inflexible) model to represent an epidemic. For instance, using the logistic model (equation 4.16) is the same as using the Richards model with $\eta = 2$ (equation 4.24b).

Even though we will not directly use the Richards model very much in this book, the model is of interest at the introductory stage in that it nicely shows that many of the models used for representing disease progress curves are, in fact, part of a continuum that covers a range of curves of various shapes. It also serves as a basis for making formal comparisons between models, especially in terms of the rate parameter. As shown in equation 4.25, r_* is a slope of the $y^* : t$ straight line. However, the transformation of y required to achieve linearity depends on the specific model adopted (or, equivalently, on the choice of η). Because of this, one cannot directly compare rate parameters from two different models, such as the logistic and monomolecular. For instance, if one knew (or determined) that $r_L = 0.1$/day for one epidemic and that $r_M = 0.033$/day for another, one could not conclude that disease increases faster for the first epidemic. The dynamic process of disease increase is qualitatively different for these two epidemics—dy/dt is high at low y in a monomolecular epidemic, whereas dy/dt is high around $y = 1/2$ in a logistic epidemic. In one sense, then, if two epidemics are described by different models, there is no single number that can be used to capture fully the differences in the pattern of disease increase between the two epidemics.

All is not lost when wanting to compare epidemics described by different models. Richards (1959) proposed a summary parameter for comparing rates between different models. The parameter is known as the weighted mean absolute rate (ϖ), which can be written as

$$\varpi = \frac{r*}{2\eta + 2} \tag{4.26}$$

This composite parameter is equal to the mean weighted value of dy/dt over the entire epidemic, with weights equal to dy/dt. It is a measure of the average height of the dy/dt curve. To calculate ϖ for one of the inflexible models, one uses the implied value of the shape parameter for the model. With the logistic model, $\varpi = r_L/[2 \times 2 + 2] = r_L/6$; with the monomolecular model, $\varpi = r_M/[2 \times 0 + 2] = r_M/2$. For the example in the previous paragraph, one can determine that the two apparently different rates correspond to virtually the same weighted mean rate: $\varpi = 0.1/6 = 0.0167$/day for the logistic epidemic and $\varpi = 0.033/2 = 0.0165$/day for the monomolecular epidemic. If one is using the Richards model directly, and estimates parameters from observed data (see section 4.6), then ϖ would be calculated using the estimate of η and r_R.

A final note on ϖ and η should be made, concerning the exponential model. It was previously stated that at very large η values (i.e., as $\eta \to \infty$ in the limit), the exponential model is approximated by the Richards model over the full range of y values ($y \leq 1$). This would also require the use of a large $r*$ to obtain a specified ϖ (e.g., for $\varpi = 0.0165$/day and $\eta = 100$, $r* = 3.33$/day; see equation 4.26). On the other hand, it was also stated that the exponential is a good approximation of the logistic model when y is small (i.e., $r_L = r_E$ at $y < 0.05$). The latter situation corresponds to using a η value of 2 in the Richards model to obtain the logistic equation; the exponential model and corresponding r_E ($=r_L$) that approximates the logistic at low y would quickly breakdown at higher y values, producing nonsensical values of disease intensity (e.g., >1). The exponential-type epidemic obtained by using a very large η in the Richards model (and corresponding $r*$) would not necessarily approximate very well the logistic model at low y. In general, we use the exponential model as an approximation for the logistic (at low y). This means that we are assuming that $\eta = 2$, and $r_L \approx r_E$ at $y \leq 0.05$.

4.4.6 Calculations with the models

One can use the linear formulation of the disease progress models (equation 4.25a or b) to predict disease intensity at any time, to determine the value of the rate parameter, and to evaluate the effects of changing initial disease intensity and the rate parameter on disease development. Of course, one can directly use the nonlinear versions of all the models for these purposes, but many users find it more convenient to work with the straight line versions.

Consider a situation with $y_0 = 0.01$ and $r* = 0.15$/day. It is helpful in performing calculations to realize that equation 4.25a consists of four terms: 1) a transformation of initial disease intensity (y_0^*); 2) a rate of increase or slope term ($r*$); 3) a time period beyond the starting time (t); and 4) a transformation of disease intensity at a selected time period (y^*). In the problems we address here, three of the four terms are known, and one solves for the fourth. The calculations shown here can be done with any of the specific models, but we primarily use the logistic (equation 4.16 or 4.17) as the basis of the examples, because of its widespread application in epidemiology (some similar calculations were done previously for the simple exponential model; section 4.4.1). Thus, for the logistic model, using the values given above, we have $y_0^* = \ln[y_0/(1 - y_0)] = \ln(0.01/0.99) = -4.595$, and $r* = r_L = 0.15$/day.

What is disease intensity at t = 20 days and t = 40 days? The transformation of disease intensity at 20 days is obtained from:

$$-4.595 + (0.15) \cdot (20) = -4.595 + 3 = -1.595.$$

Disease intensity at 20 days [$y(20)$] is obtained by backtransformation. For the logistic model, one uses equation 4.18:

$$1/[1 + \exp(1.595)] = 1/5.928 = 0.169.$$

Note that the e is raised to the power of minus the predicted logit; thus, -1.595 becomes $+1.595$ in the backtransformation. Care is needed with this particular aspect of the calculations for several of the models, because y^* may be negative or positive, depending on the value of y. This calculation is demonstrated graphically in Fig. 4.9. To calculate y at 40 days, one can continue to use $t = 0$ as the starting point, and calculate the logit as:

$$-4.595 + (0.15) \cdot (40) = 1.405.$$

Backtransformation produces: $1/[1 + \exp(-1.405)] = 0.803$. Alternatively, one could treat $t = 20$ days as the

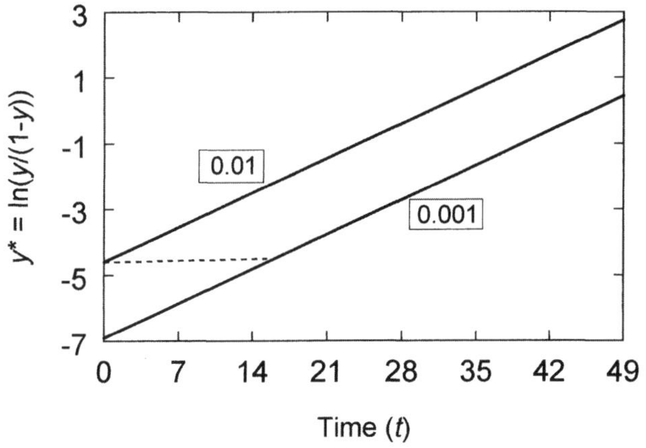

FIG. 4.9. Example logit-transformed disease intensity versus time for epidemics with $r_L = 0.15$/day, and y_0 equal either 0.01 or 0.001 (identified in boxes in the figure).

starting point (because one advantage of a straight line is that any point can be considered a new starting point), and calculate the logit for 20 days later:

$$-1.595 + (0.15) \cdot (20) = 1.405.$$

The general steps here using equation 4.25a apply to any model (assuming that the right transformation was used to obtain y_0^*. However, the appropriate backtransformation is model dependent. One has to be careful to use the backtransformation for the assumed/selected model to properly obtain y from y^*.

What was initial disease intensity? Suppose that at t = 35 days, disease intensity is 0.16 [i.e., $y(35) = 0.16$], and one "knows" that $r_L = 0.15$/day. Initial disease can be determined by rearranging equation 4.25a to:

$$y_0^* = y^* - r_* t.$$

For the example, $y^*(35) = \ln(0.16/0.84) = -1.657$, and transformed initial disease is calculated as:

$$-1.657 - (0.15) \cdot (35) = -6.907.$$

Backtransformation shows that y_0^* is

$$1/[1 + \exp(6.907)] = 0.001.$$

This can also be seen in Fig. 4.9.

How long does it take disease intensity to reach a certain level? Returning to the first set of conditions in this section ($y_0 = 0.01$, $r_* = 0.15$/day), one can determine how long it takes to reach some specific intensity. This is done by rearranging equation 4.25a to solve for t:

$$t = (y^* - y_0^*)/r_*.$$

For example, the time taken to reach $y = 0.75$ is determined by first calculating the logit of 0.75 [$=\ln(0.75/0.25) = 1.099$], and then finding t:

$$t = [1.099 - (-4.595)]/0.15 = 37.96 \approx 38 \text{ days}.$$

The reader should see that the time taken to reach a given disease level is inversely related to the rate parameter, meaning that small changes in the rate can have a big effect on time taken. For instance, if r_L was reduced to 0.12/day, the time taken to reach $y = 0.75$ would be 47.4 days, almost 10 days longer than when $r_L = 0.15$/day. Note also that the time taken to reach a certain y is inversely dependent on initial disease intensity—the smaller the initial intensity, the longer the time taken.

How long does it take disease intensity to double? A special case of this "time" concept is the doubling time (t_D), already discussed for the exponential model (see equations 4.7a,b). Here, it may be intuitive to use the alternative version of the linear model (equation 4.25b). One can then think of doubling time as $t_D = t_2 - t_1$ when intensity at the second time is double disease at the first time, that is, $y(t_2) = 2y(t_1)$. This is equivalent to simply writing equation 4.25a as

$$(2y)^* = y^* + r_* t_D$$

in which y^* is now the transformed starting disease intensity and $(2y)^*$ is the transformation of the doubled diseased intensity. Note that although $2y$ is double the magnitude of y, in general, $(2y)^*$ is not double the magnitude of y^*. Using this notation, one can write doubling time as:

$$t_D = [(2y)^* - y^*]/r_*.$$

Using the logistic example with $r_L = 0.15$/day, the time for y to increase from 0.01 to 0.02 is calculated as:

$$[\ln(0.02/0.98) - \ln(0.01/0.99)]/0.15 = [-3.892 - (-4.595)]/0.15 = 4.69 \text{ days}.$$

If one used the exponential-based equation (equation 4.7b), doubling time would be calculated as $t_D = 0.693/0.15 = 4.62$ days, which is almost the same as the logistic-based calculation here. This is because the exponential and logistic models are almost the same at low y.

The time for y to increase from 0.10 to 0.20 is:

$$[\ln(0.2/0.8) - \ln(0.1/0.9)]/0.15 = [-1.386 - (-2.197)]/0.15 = 5.41 \text{ days}.$$

One can see from these two examples with the logistic model that t_D depends on the starting value of y in the calculations. The reader can confirm that t_D increases as y increases for all of the models discussed, except for the exponential. Obviously, doubling time can only be calculated when the starting intensity is less than 1/2. At moderate-to-high values of y (e.g., $y > 0.05$ or > 0.10), it is important to take the specific disease values into account in making calculations (and not use the calculations based on the exponential model, where the specific disease values are not utilized).

What is r_* between two times? One can easily determine the value of r_* between two times by rearranging equation 4.25b to:

$$r_* = \frac{y^*(t_2) - y^*(t_1)}{t_2 - t_1} \qquad (4.27)$$

For instance, if disease intensity at day 16 was 0.10, and at day 25 was 0.30, for a logistic epidemic we can write

$t_1 = 16$, $t_2 = 25$, $y(t_1) = 0.10$, $y^*(t_1) = \ln(0.10/0.90) = -2.197$, $y(t_2) = 0.30$, $y^*(t_2) = \ln(0.30/0.70) = -0.847$. Using equation 4.27 we obtain:

$$[-0.847 - (-2.197)]/[25-16] = 1.35/9 = 0.15.$$

Thus, the estimate of r_L equals 0.15/day. This corresponds to the top line in Fig. 4.9 that was used in several of the exercises here. The reader can confirm on their own that if the monomolecular model is assumed, the estimated r_M equals 0.021/day, based on the same disease intensity and time values.

4.5 Control

4.5.1 Control strategies for polycyclic diseases

Vanderplank (1963) elegantly explained that disease control for polycyclic diseases involves a reduction in either the initial disease intensity (y_0) or the rate of disease increase (as summarized by the parameter $r*$), or in both. Initial disease intensity is reduced by any control that lowers either the amount of inoculum available for infection of the host or the probability of the inoculum causing infection. Examples include crop rotation, soil fumigation, removal of weed hosts, elimination of seed-borne pathogens in the planting material, addition of certain biocontrol agents, and so on (Fry, 1982; Zadoks and Schein, 1979). The term *sanitation* is sometimes used to refer to any control that reduces inoculum and, hence, initial disease intensity. Although others might prefer to use sanitation in a more specific sense, we use the term here for any control method that results in reduction of y_0.

The rate parameter $r*$ is reduced by any control that lowers either the number of infections or the increase in lesion area throughout the entire time span of the epidemic subsequent to $t = 0$. For polycyclic diseases, reducing inoculum production per diseased individual per unit time, probability of the produced inoculum of causing new infections, growth of lesions, and time that a diseased individual can produce inoculum will all result in a reduction in $r*$. Furthermore, increasing the time between the start of the infection process and the production of inoculum can lower $r*$. Environmental manipulation that results in less favorable conditions for disease development, such as use of some cultural practices, can also reduce $r*$.

Pesticides (fungicides) and plant disease resistance can affect y_0, $r*$, or both. Protectant fungicides typically reduce the probability of infection of deposited spores, either at the beginning of the epidemic or throughout the epidemic, depending on when the chemical is applied. Systemic fungicides may also reduce the probability of infection or reduce spore production, at the start or throughout the epidemic. Cultivars with so-called vertical resistance (major gene resistance) reduce y_0, possibly to 0, if no individuals in the pathogen population can overcome the plant resistance gene(s). If a fraction (<1) of the pathogen population can overcome the resistance gene(s), then y_0 will be reduced, but not to zero, and secondary infections will continue to occur throughout the epidemic as if no resistance genes were present. Cultivars with so-called horizontal resistance (minor gene resistance, rate-reducing resistance, partial resistance, field resistance) do not prevent infection from occurring. Instead, probability of a spore causing infection is reduced (but not to zero), or the time interval between infection and spore production by an infected individual is increased, or the number of spores produced per diseased individual per unit time is reduced, or the period over which a diseased individual produces spores is reduced. These changes mean there are fewer new infected individuals per unit time and $r*$ is reduced. Of course, a cultivar may have both vertical and horizontal resistance. There is a considerable body of literature on vertical and horizontal resistance, and a great deal of controversy has erupted over the years regarding the genetic basis of these two forms of resistance, but we do not consider the specifics here. Interested readers should consult Fry (1982) and Zadoks and Schein (1979).

Of particular relevance in this chapter is the relative impact of reducing $r*$ or y_0 on disease development. Although epidemics are nonlinear processes, there are some fairly straightforward ways of seeing the effects of changes in $r*$ or y_0 on plant disease epidemics, using the types of calculations done in section 4.4.6. More general information on control strategies can be found in Fry (1982), Strange (1993), and Arneson (2001). We discuss the situation first for polycyclic diseases, and then deal with the special circumstances of monocyclic diseases. Subsequent chapters will deal with many other aspects of plant disease management.

4.5.2 Calculations for polycyclic diseases

The linearized version of the epidemic model shown in equation 4.25a provides a convenient way of evaluating the impact of reductions in y_0 on disease development. Although one could simply determine predicted y at any given t with a new y_0, a useful alternative is to use the kind of calculation presented previously, relating to the time taken to reach a certain value of y. We take the latter approach here. Assume that without control, $r* = 0.15$/day and $y_0 = 0.01$ (top line in Fig. 4.9). If y_0 is reduced by a factor of 10 to $y_0 = 0.001$, one obtains the lower line in Fig. 4.9 (if $r*$ remains unchanged). We refer to the new (lower) initial disease as y_{0S}, where the "S" subscript indicates "sanitation". One can ask the question: is it useful to reduce initial disease 10-fold if no other controls are used during the epidemic (i.e., if $r*$ is not changed)?

One can determine the effect of this sanitation by calculating the time required for disease intensity to increase from 0.001 (initial disease after sanitation) to 0.01 (initial disease if sanitation had not occurred). Generally, this is the time it takes for disease intensity "to get back to" the value it *would* have had at the start of the epidemic if sanitation had not been used. By using the straight-line equation 4.25a, we can write:

$$y_0^* = y_{0S}^* + r^* t_S$$

in which y_{0S}^* is the transformed disease intensity after sanitation (starting disease value in the equation), y_0^* is the transformed disease intensity before sanitation (now, the "final" disease value), and t_S is the time for disease to increase from y_{0S}^* to y_0^*, the *sanitation time*. One can think of t_S as the time savings from sanitation. In Fig. 4.9, this equation represents the broken horizontal line from $t = 0$ to the point where it intercepts the lower line. Rearrangement leads to:

$$t_S = \frac{y_0^* - y_{0S}^*}{r_*} \quad (4.28)$$

If the logistic model is appropriate, then $r_* = r_L = 0.15$/day in the example. The logistic version of equation 4.28 is written as:

$$t_S = \frac{\ln\left(\frac{y_0}{1-y_0}\right) - \ln\left(\frac{y_{0S}}{1-y_{0S}}\right)}{r_L} \quad (4.29)$$

The calculated time is given by:

$$[\ln(0.01/0.99) - \ln(0.001/0.999)]/0.15 = [-4.595 - (-6.907)]/0.15 = 15.41 \text{ days.}$$

Obviously, based on equation 4.28, the greater the reduction in initial inoculum (and, hence, initial disease intensity), the longer the time it takes disease to reach the original intensity (as if initial disease intensity had not been lowered). The value of t_S is strongly influenced by the rate parameter; the higher is r_*, the lower is the value of t_S. In other words, a big reduction in initial disease intensity may have little effect on t_S if r_* is high. This can be seen most easily with the exponential version of equation 4.28. To see this, we must first give background on the exponential model. Here, $r_* = r_E$, $y_0^* = \ln(y_0)$, and $y_{0S}^* = \ln(y_{0S})$. Because $\ln(A) - \ln(B) = \ln(A/B)$ for any two positive A and B, one can write:

$$t_S = \frac{\ln(y_0 / y_{0S})}{r_E} \quad (4.30)$$

for the exponential model. With the example values used here, one finds that $t_S = \ln(10)/0.15 = 15.35$ days. This is almost identical to that obtained with equation 4.29 for the logistic model, which is expected because the exponential and logistic models are very similar at small y.

The use of equation 4.30 is specific to the exponential model. However, this is a reasonable equation to use for the early part of a polycyclic epidemic when y is small and most newly produced inoculum does not contact previously infected individuals. A key feature of this equation is that the specific values of y_0 and y_{0S} do not matter in determining the effect of sanitation; all that matters is the ratio, which is known as the *sanitation ratio* (SR). Use of sanitation ratio when referring to initial disease intensities is warranted because it is assumed that y_0 is proportional to inoculum (especially at low y). That is, if x and x_S represent inoculum before and after sanitation, respectively, then SR $= y_0/y_{0S} = x/x_S$. The percent reduction in y_0 is related to the SR using: $100 \times [1-(\text{SR})^{-1}]$. For instance, an SR of 10 means that $100 \times (1-0.1) = 90\%$ of the inoculum was removed or eliminated.

The specific values of disease intensity are not relevant for calculating t_S in equation 4.30, *if* the exponential model holds. For instance, a situation with $y_0 = 0.05$ and $y_{0S} = 0.01$ gives the same SR as a situation with $y_0 = 0.01$ and $y_{0S} = 0.002$ (in both cases, SR = 5). This makes it easy to evaluate, in a general sense, the impact of reducing initial disease intensity at different values of the rate parameter r_E. Fig. 4.10 shows the timesavings for SRs of 5, 10, 20, 100, and 1000 for r_E values up to 0.5/day. At $r_E = 0.4$/day, SRs of 20 or less have only minor effects on timesavings. Even an SR of 100 provides only ~11 days of savings at this rate. Although 0.4/day is high, it is also typical of diseases such as potato late blight under favorable environmental conditions. At lower rates, the effect of sanitation is much greater. At $r_E \leq 0.1$/day, and especially at $r_E \leq 0.05$/day, reducing initial disease intensity can have a big impact on the epidemic. At $r_E = 0.05$/day, a SR of 20 (95% reduction in y_0) gives a t_S value of ~60 days. Even an 80% reduction in y_0 (i.e., SR = 5) gives a t_S of ~32 days when $r_E = 0.05$/day.

The general conclusion that can be obtained from assessment of equation 4.30 and Fig. 4.10 is that reducing

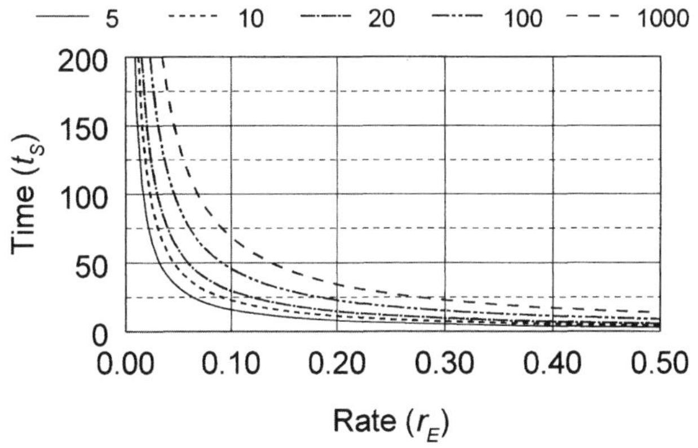

FIG. 4.10. Time savings (t_S) due to reducing initial disease intensity from y_0 down to y_{0S}, for the exponential model at different values of the rate parameter r_E. SR = y_0/y_{0S}.

initial disease intensity for a polycyclic disease is most beneficial when the rate parameter is low. Put another way, unless the rate is relatively low, the reduction in y_0 must be large (enormous) to have much of an effect on the epidemic. Thus, in general, control of polycyclic diseases cannot be based on reducing y_0 alone; one must reduce the rate to obtain a substantial change in the epidemic. Fry (1982) and Vanderplank (1963) discuss this at greater length.

One can, in fact, directly evaluate the effect of changes in both y_0 and r_* using the linear version of the epidemic model (equation 4.25a). We demonstrate this graphically in Fig. 4.11 for a logistic epidemic, showing both logits (upper frame) and backtransformed disease intensities (lower frame). The solid and dashed lines represent identical conditions to those in Fig. 4.9, although the time period is extended to 98 days (solid line: $y_0 = 0.01$ and $r_L = 0.15$/day; dashed line: $y_0 = 0.001$ and $r_L = 0.15$/day). The other two lines represent the same two initial disease values, but with an r_L value of 0.075/day. In this example, the outcome of reducing the rate parameter by a factor of 2 is that, there is actually less disease after about 30 days than when initial disease is reduced 10-fold. Reducing the rate and initial disease simultaneously has an even greater impact. One could quantify the effects of

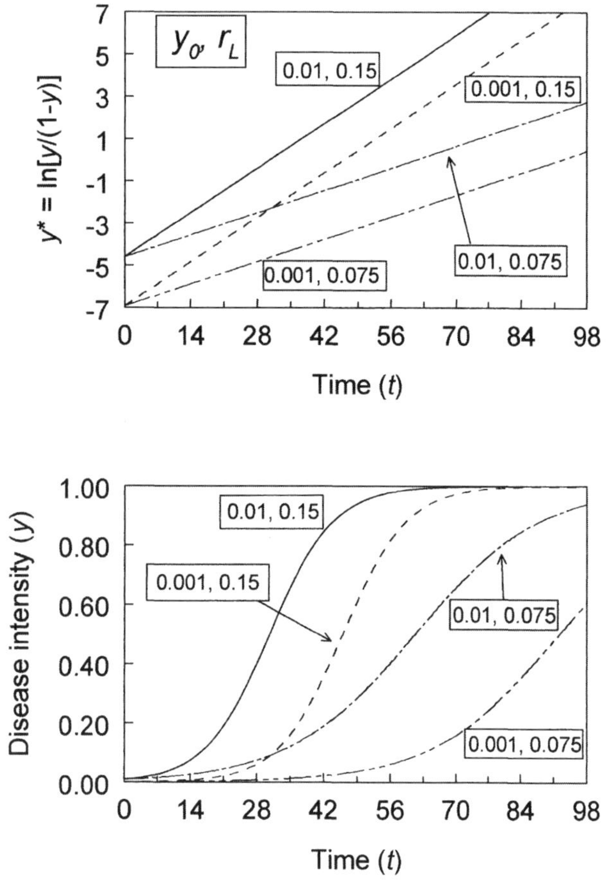

FIG. 4.11. Example logit-transformed disease intensity versus time for epidemics with $r_L = 0.15$/day or 0.075/day, and y_0 equal either 0.01 or 0.001 (identified in boxes in the figure).

changing y_0 and r_* by determining disease intensity at some selected time (e.g., at $t = 70$). The reader can confirm that $y^*(70)$ equals 5.90, 3.59, 0.65, and −1.62 for the four epidemics of Fig. 4.11. These correspond to $y(70)$ values of 0.997, 0.98, 0.67, and 0.16. Thus, cutting the rate in half produces a larger reduction in y ($0.997 \rightarrow 0.67$) than does a 10-fold reduction in initial disease intensity ($0.997 \rightarrow 0.98$).

Another way of evaluating the reduction in r_L and/or y_0 is to determine the time to reach some specified disease intensity larger than y_0. This is analogous to the calculation of t_S for the reduction in initial disease intensity. However, if r_L is reduced but y_0 is unchanged, one does not calculate the time to reach y_0 from y_{0S}, because there is no y_{0S}. We show this with the examples of Fig. 4.11. The time to reach $y = 0.50$ (or $\ln[y/(1-y)] = 0$), labeled as $t_{0.5}$, for a logistic epidemic can be calculated by rearranging equation 4.25a to:

$$t_{0.5} = [0 - (-4.595)]/0.15 = 30.6 \text{ days}$$
$$(y_0 = 0.01,\ r_L = 0.15);$$
$$t_{0.5} = [0 - (-6.907)]/0.15 = 46.0 \text{ days}$$
$$(y_0 = 0.001,\ r_L = 0.15);$$
$$t_{0.5} = [0 - (-4.595)]/0.075 = 61.3 \text{ days}$$
$$(y_0 = 0.01,\ r_L = 0.075);$$
$$t_{0.5} = [0 - (-6.907)]/0.075 = 92.1 \text{ days}$$
$$(y_0 = 0.001,\ r_L = 0.075).$$

It takes over 61.3−30.6 = 30.7 days longer to reach 50% intensity when r_L is reduced to half of its original value; in contrast, it takes only 46.0−30.6 = 15.4 days longer when y_0 is reduced to a tenth of its original value. When both initial y and the rate are reduced to the values used in this example, it takes 92.1−30.6 = 61.5 days longer to reach $y = 0.50$. This calculation confirms the fact that a relatively small change in r_L is can produce the same effect as a large change in y_0. The reader can determine these results graphically (and approximately), by drawing a horizontal line in the top frame of Fig. 4.11 at a logit of 0, and finding where this line crosses each of the four disease progress lines. As another exercise, the reader can determine how much y_0 must be reduced to give the same result as halving r_L. That is, with $r_L = 0.15$/day, what value of y_0 (or y_0^*) is needed so that $t_{0.5} = 61.3$ days? Remember: there are four terms in equation 4.25b, one knows three of them, and solves for the fourth.

Because the above calculations were based on use of a disease intensity considerably above 0.05, the exponential model could not be used as an approximation. Specific results will depend on the initial disease, the rate parameter, and the selected disease intensity that must be reached. As always, the specific result will also depend on which model was used. For comparison, the reader can try these calculations with, for example, the Gompertz model.

One final item can be noted about the disease progress curves of Fig. 11 shown in untransformed and transformed versions. When observing y versus t, it is very difficult to see the impact of changes in y_0 and r_L at low y. In fact, at small t, the curves appear to virtually overlap. In fact, there are major differences in the epidemics at small t that can clearly be seen when logits are graphed.

4.5.3 Control for monocyclic diseases

In reading section 4.5.2, one might get the impression that one should evaluate disease controls for monocyclic diseases using the same calculations as for polycyclic diseases. This would be a mistake. Of course, one can use the linear form of monomolecular model (equation 4.11 or 4.12) to determine disease intensity at a particular time, calculate r_M between two times, or determine how long it takes disease intensity to reach a certain value, for specified values of y_0 and r_M. However, one should *not* consider the effects of reducing y_0 independent of r_M or of reducing r_M independent of y_0. The reason for this is related to the epidemiological interpretation of the model parameters.

As discussed previously, r_M is a product of amount of inoculum and probability per unit time of a unit of inoculum causing an infection. Thus, the slope of the y^* line for the monomolecular model reflects the amount of inoculum. Any reduction in inoculum or the effectiveness of the inoculum (e.g., as a result of sanitation) does not just change y_0 (y_0^*), it also changes the rate parameter r_M. Moreover, y_0 (or y_0^*) is just an instantaneous value of disease intensity (or transformed disease intensity) at $t = 0$. Unlike the models for polycyclic diseases, y_0 can even equal 0. Therefore, one does not necessarily calculate a time for disease to increase from y_{0S} to y_0 (t_S), because y_0 could be 0.

Disease control can be quantified in terms of the change in disease intensity at a particular time. We show this for day 60 of an epidemic using equation 4.11, with $y_0^* = 0$. If $r_M = 0.02$/day, then y^* at $t = 60$ [i.e., $y^*(60)$] is given by:

$$y^* = 0 + (0.02)(60) = 1.2.$$

Backtransformation (equation 4.13) gives: $y = 1 - \exp(-1.2) = 0.70$. Now, if 90% of the inoculum is removed or eliminated (say, by crop rotation), so that SR = 10, what is y at $t = 60$? This could be calculated by dividing r_M by SR to obtain a new rate r_{MS} (=r_M/SR), and using equation 4.13 to obtain disease intensity at day 60:

$$y^* = 0 + (0.02/10)(60) = 0.12.$$

Backtransformation gives $y = 1 - \exp(-0.12) = 0.11$. Note that the ratio of $\ln[1/(1-y)]$ at a particular time before and after sanitation (=1.2/0.12 in example) is directly related to the SR, but the ratio of the corresponding untransformed y values (=0.70/0.12) is not (although when disease intensity is low (e.g., <0.1), the ratio of y values will be fairly close to the SR). We emphasize that the specific results for y depend on the specific time during the monocyclic epidemic (e.g., $t = 60$). For simplicity, one would not even have to use equation 4.11 to evaluate sanitation. One could simply divide an assumed y^* (at a particular time) without sanitation by SR, and then backtransform to find the corresponding value of y with sanitation.

Other calculations can be done as well. For an example, we use the same original r_M as above (=0.02/day), but specify that $y_0 = 0.01$ (or, $y_0^* = 0.01$ in this case). The reader can determine that y^* at $t = 60$ is $0.01 + 1.2 = 1.21$, resulting, through backtransformation, in a y value of 0.702. One can ask: with a SR of 10, how long will it take disease to increase to 0.702? To calculate this, one reduces y_0 *and* r_M by a factor of 10, and writes equation 4.11 as:

$$1.21 = \ln(1/[1-(y_0/10)]) + (0.02/10)(t_{0.702})$$

in which $t_{0.702}$ is the time taken to reach $y = 0.702$. Solving this equation to find $t_{0.702}$ gives: $t_{0.702} \approx 604$ days. Thus, for an annual crop, such a high level of disease intensity would never be reached with a SR of 10.

4.5.4 Summary of disease control strategies

The discussion on control can be summarized as follows.

1. Monocyclic diseases are controlled by reducing the amount of inoculum or the efficacy of inoculum.
2. Effective (or efficient) control of polycyclic diseases depends on initial disease intensity (a function of inoculum and its efficacy) and the rate parameter.
   a. At low $r*$, reducing y_0 can be valuable;
   b. At high $r*$, reducing y_0 is not effective unless SR is very large;
   c. $r*$ must be kept at a relatively low value to obtain effective control, unless the time period of the epidemic is very short.

Although control of polycyclic diseases emphasizes reduction in $r*$, it may be necessary to reduce y_0 as well, for example in cases where there is a high value of y_0 that requires reduction even in the absence of further disease progress. For example, initial disease intensity of apple scab from overwintering inoculum ('primary scab') can be very high, with 50% of the leaves infected (MacHardy, 1996). Even if there was no further increase in disease intensity (perhaps, by excellent disease control with fungicides), the initial disease intensity would be sufficient to cause economic losses.

As shown above with theoretical examples, reducing $r*$ a small amount can have a big effect on a polycyclic epidemic. This is fundamental to Vanderplank's thesis on disease control. The question remains: how difficult is it to reduce r_* by, say, 20%? This can be ascertained empirically and theoretically. The empirical method involves determining r_* for observed epidemics corresponding to a range of disease control treatments. One could then evaluate the changes in r_* with different treatments. The theoretical method involves formally relating r_* to specific components of the epidemic, such as the production of inoculum per diseased individual per unit time. This requires a more detailed characterization of plant disease epidemics, especially for polycyclic diseases, which is covered in the next chapter.

Most of the discussion on control here has focused on diseases of annual crops. With perennials (e.g., forest trees), the perspective can be different. For instance, even with a low r_*, ultimately disease intensity approaches high values if the models presented so far are appropriate. For instance, with r_L = 0.10/year (=0.00027/day), 50% of the trees would be infected (y = 0.5) in 69 years if y_0 = 0.001. For a long-lived tree species, 69 years is not an unrealistic time frame to consider. To obtain a better perspective on disease epidemics in perennials, and the control of diseases of perennials, the dynamics of the host—such as planting, harvesting, replanting; or change in size (biomass or numbers) over time—often need to be considered. Host dynamics will also be discussed at length in Chapter 6.

4.6 Model Fitting

For the most part, plant disease epidemics have been considered theoretically so far in this chapter. We introduced various differential-equation models for the dynamics of disease development, and presented the biological rationale for the terms in the model. We discussed some of the properties of the models and showed how to make various calculations using the model terms. We further demonstrated that the models produce predicted disease progress curves that are at least qualitatively similar to observed curves (see Fig. 4.1).

It is uncommon for epidemic model parameters to be known *a priori* for a particular disease-host-environment situation. Even when reasonable values can be given for r_L, as an example, based on the known susceptibility of the host genotype and aggressiveness of the pathogen strain, the favorableness of the environment can dramatically influence the actual rate in a given year and location (Kim et al., 2005). Thus, one usually needs to estimate model parameters. Parameter estimation is done through fitting a model to observed data. Of course, model fitting involves use of a statistical method, such as ordinary least squares or maximum likelihood (see Chapter 3) (Neter et al., 1983; Schabenberger and Pierce, 2002). Model fitting can be viewed much more broadly, however. First, a decision must be made on which model to fit. With a mechanistic approach, the decision will be made ahead of time regarding the model to use (e.g., logistic); with an empirical approach, the decision will be made based on the observed data. A hybrid approach is often taken when analyzing disease progress curves. Graphs of the observed data are used to suggest models to use *within* the set of models developed for epidemics. That is, the investigator would choose among the models presented so far, or those to be presented later, based on the observed data and knowledge of the biological system. Many of the typical linear models used by statisticians to provide good fits to data that do not fall on straight lines (e.g., polynomials) would *not* be considered because they have no biological basis. We go through this hybrid approach in the following sections.

Once the parameters are estimated, then the general model fitting protocol involves an appraisal of the goodness of fit. As discussed in Chapter 3, the investigator can use various statistics and graphical methods to determine if a chosen model is appropriate. If it is decided that a model is not appropriate for describing the data, then this appraisal can suggest alternative models, which can then be fitted to data (depending on the objectives of the investigator).

4.6.1 Choosing a model

We demonstrate the protocol for model fitting with data for an epidemic of tobacco etch disease of peppers, caused by tobacco etch potyvirus (from Fig. 9 in Nutter, 1997). The proportion of diseased plants (i.e., disease incidence) was determined eight times during the epidemic, and the data are given in the first two columns of Table 4.2. A plot of y versus t shows a classic *S*-shaped curve (Fig. 4.12). Based on this graph, the logistic model should certainly be considered as a reasonable way of representing the data. However, the Gompertz and the Richards model (with a range of η values >1/2 also produce *S*-shaped curves, so this graph alone cannot indicate the most appropriate model. It is clear from the graphical plot of the data that the disease progress curve is not at all consistent with the monomolecular or exponential models (see Figs. 4.2 and 4.4).

A plot of the absolute rate of disease increase (dy/dt) versus time can be very informative for discriminating among the models. However, dy/dt is not observed directly, and must be estimated. The simplest way of approximating or estimating dy/dt ($\Delta y/\Delta t$) is to use the difference in y between each successive pair of times:

$$\frac{\Delta y}{\Delta t} = \frac{y_2 - y_1}{t_2 - t_1} \quad (4.31)$$

where the number subscripts here refer to two successive times. Essentially, this equation provides an estimate of the *average* rate between times 1 and 2. To plot these estimated rates, one immediately faces the issue of what value of time to use for each $\Delta y/\Delta t$ value, since

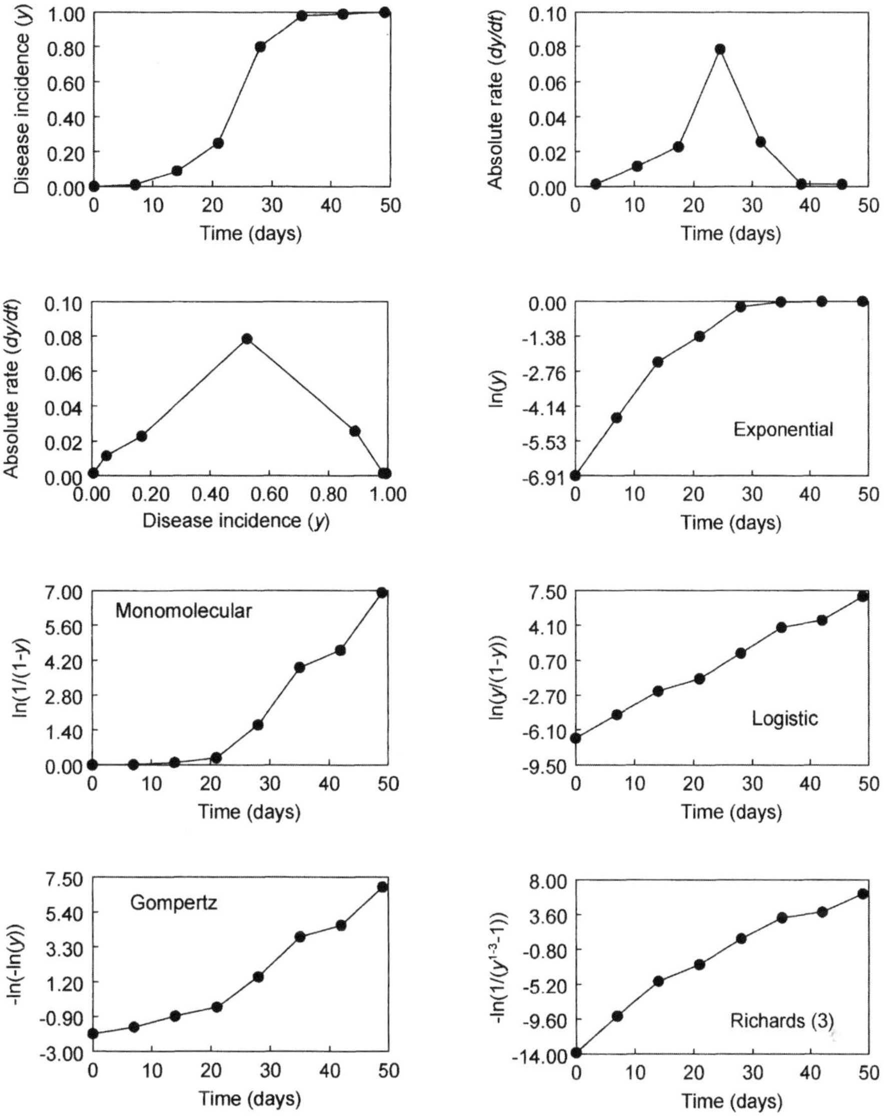

FIG. 4.12. Example graphical analysis of an epidemic of tobacco etch disease of pepper (from Nutter, 1997). Frames show incidence (y) versus time (t), estimated absolute rate of disease increase (dy/dt) versus time, dy/dt versus y, and transformations of y for the exponential, monomolecular, logistic, Gompertz, and Richards model ($\eta = 3$).

TABLE 4.2. Results of fitting the linearized logistic model (equation 4.32a) to the tobacco etch disease data of Fig. 4.12 using ordinary least squares regression and the predicted (fitted) values based on nonlinear least squares regression (last column only).

t (time)	*y (observed incidence)*	*y* (observed logit)*	*ŷ* (predicted logit)*	*Residual (based on logits)*	*ŷ (back-transformation)*	*ŷ (nonlinear fit)*
0	0.001	−6.907	−6.590	−0.918	0.0014	0.00034
7	0.010	−4.565	−4.640	0.117	0.0096	0.0035
14	0.09	−2.314	−2.691	0.920	0.064	0.035
21	0.25	−1.099	−0.741	−0.847	0.323	0.269
28	0.80	1.386	1.208	0.423	0.770	0.790
35	0.98	3.892	3.157	1.791	0.959	0.975
42	0.99	4.595	5.107	−1.327	0.994	0.997
49	0.999	6.907	7.056	−0.432	0.999	0.9998

equation 4.31 is not giving a rate at each time, but a rate of disease progress *between* times. A convenient approach is to use the mid-point between times 1 and 2. For instance, for the estimated rate between $t_1 = 0$ and $t_2 = 7$ days, one would use $t = 3.5$ days. Following the same approach, to relate the estimated rate to level of disease, one would use the mid-point (average) between y_1 and y_2. With the tobacco etch example, between

days 7 and 14, $\Delta y/\Delta t = (0.09 - 0.01)/(14 - 7) = 0.011$/day. This rate can be paired with $t = 10.5$ [$=(7 + 14)/2$] days or $y = 0.05$ [$=(0.09 + 0.01)/2$].

Plotting the estimated dy/dt versus t for the tobacco etch epidemic produces a fairly symmetrical curve, with a maximum around 25 days (Fig. 4.12 top right). A plot of the estimated dy/dt versus y is also fairly symmetric, with a maximum rate near $y = 1/2$. Both graphs strongly suggest a logistic model for the epidemic. It should be noted that the rate curves depend heavily on interpolation. Unlike the curve of y versus t, which is monotonically increasing (for the situations we are considering here), dy/dt can increase to a maximum somewhere between $y = 0$ and $y = 1$. There can be large gaps in time or disease where no estimated dy/dt information is available, or major interpolation is needed, depending on how often disease is measured over time. Consider the graph of estimated dy/dt versus y: there is no estimated dy/dt value between intensities of about 0.2 and 0.5. Thus, actual dy/dt might be increasing to a maximum at $y = 0.37$ (the Gompertz model), but there is no information on this available.

A further limitation of using dy/dt graphically is that it is only useful when epidemics go close to completion. If disease intensity is low throughout the epidemic, because of a low r_*, the inflection point (if any) may never be reached within the time period of disease assessment. For instance, if only the first four data points of the tobacco etch epidemic were available, then the dy/dt curve would have increased monotonically with t or y, highly suggestive of an exponential model. Because the exponential model is a good approximation of the logistic when y is low, this would not be an unreasonable conclusion. Although an exponential model would be a satisfactory equation to use for characterizing this epidemic (at low y), and evaluating control strategies, one would make inaccurate predictions of future y values if the exponential model were used over the full range of times. The main point we emphasize here is that an absolute rate graph (as well as the disease progress curve) can suggest appropriate models, and definitely eliminate other models, but these graphs will not necessarily indicate the ideal model to represent the data.

If one of the models presented so far in this chapter is appropriate, then there will be a straight line when the appropriate linearizing transformation of y (y^*) is plotted versus t. In Fig. 4.12, y^* is plotted for the exponential, monomolecular, logistic, Gompertz, and Richards ($\eta = 3$) models (see equations 4.3, 4.11, 4.16, 4.21, and 4.24b). As expected, based on the graphs of y versus t, dy/dt versus t and dy/dt versus y, neither the exponential nor the monomolecular transformation produce anything close to a straight line, indicating that these models are not appropriate. There is also a fair amount of curvature with the Gompertz-based transformation. The logistic-based (logit) transformation does result in a fairly straight line, consistent with the $y : t$ curve, and the $dy/dt : t$ and $dy/dt : y$ curves (Fig. 4.12). The logistic model corresponds to $\eta = 2$ of the Richards model. When y^* for the Richards model is based on $\eta = 3$ (bottom right frame of Fig. 4.12), curvature is returned to the $y^*: t$ graph. In fact, if one looks at the y^* graphs in order of increasing η [monomolecular(0) $\rightarrow$ Gompertz(1) $\rightarrow$ logistic(2) $\rightarrow$ Richards(3)], one can see a pattern to the changing line shapes, with concave lines to the time axis for $\eta < 2$, little if any curvature for $\eta = 2$, and convex line for $\eta > 2$. For the example epidemic, it should also be noted that there is a fairly straight line with the exponential-based transformation for the first three times; as stated in the previous paragraph, this is expected here because the exponential is a good approximation of the logistic at small y. If the epidemic was truly of the monomolecular type, then the $y^*: t$ graph for this model would be a straight line, and there would be increasing degree of curvature with increasing (implied) η of the model.

The conclusion based on the graphs in Fig. 4.12 is that the logistic is the most appropriate model, among those considered, for the example epidemic. Readers should note that such straightforward conclusions often do not happen. This is typically because a smaller number of assessment times is used, because there is more variability than seen here, and because y has not leveled off by the last assessment time.

4.6.2 Estimating parameters and assessing model fit—linear least squares

We have already dealt with the issue of parameter estimation a little, by showing how to determine r_* between two times (equation 4.27). One must assume a particular disease progress model to use this approach, because there is no information available from the calculation to know if the assumed model is reasonable or not. Of course, the graphs of Fig. 4.12 can be used to determine if the assumed model is reasonable when there are multiple assessment times. However, equation 4.27 does not lend itself to calculations based on data from multiple assessment times (such as in Figs. 4.1 and 4.12), because it works only on pairs of times. Moreover, there is no way to determine the uncertainty associated with the estimate of r_* using equation 4.27. For these reasons, most investigators do not use equation 4.27 to estimate the rate parameter in detailed studies; however, the equation is quite useful to quickly obtain an estimate of r_*, especially when one only has access to published disease progress curves and not the raw data.

We can estimate y_0 (or a transformation of y_0) and r_* using linear or nonlinear parameter-estimation methods such as least squares or maximum likelihood (see sections 3.4 and 3.5; Neter et al., 1983). This requires the mathematical model of the epidemic to be written as a statistical model, taking account of the fact that fitted disease progress curves do not exactly follow observed data. We mostly work with the linearized versions of the epidemic models here. Equation 4.25a can be written as the following statistical model:

$$y_j^* = y_0^* + r_* t_j + \varepsilon_j \tag{4.32a}$$

in which the j subscript indicates j-th data point or time ($j = 1,..., n_j$) and ε is the 'error', a random variable which is the difference between the appropriate transformation of observed disease intensity and the same transformation of disease intensity as determined from the model, with intercept y_0^* and slope r_*. For the data in Fig. 4.12, $n_j = 8$. The part of equation 4.32a before the ε term is known as the deterministic component (corresponding to the simple mathematical model of an epidemic), and the ε part is known as the stochastic (random) component (see Chapter 3 for background). It is typically assumed (at least in the initial analysis) that ε is normally and independently distributed with mean 0 and constant variance σ^2, that is, $\varepsilon \sim$ NID $(0, \sigma^2)$ (where ~ means 'is distributed as'). As one specific example, the logistic version of equation 4.32a can be written as:

$$\ln\left(\frac{y}{1-y}\right)_j = \ln\left(\frac{y_0}{1-y_0}\right) + r_L t_j + \varepsilon_j \tag{4.32b}$$

Although the evidence is fairly strong that the logistic model is the most appropriate model for representing the disease progress data in Fig. 4.12, we show graphic results for fitting the linearized logistic (equation 4.16) and Gompertz (equation 4.21) models to the data (Fig. 4.13), and provide some statistics for the other models. High coefficients of determination (R^2) were obtained by fitting all of the considered models to the data. R^2 values were 0.81, 0.86, 0.99, 0.94, and 0.96 for the exponential, monomolecular, logistic, Gompertz, and Richards ($\eta = 3$) models, respectively. High values are expected, in general, because of the cumulative nature of disease progress curves—when host 'size' is not

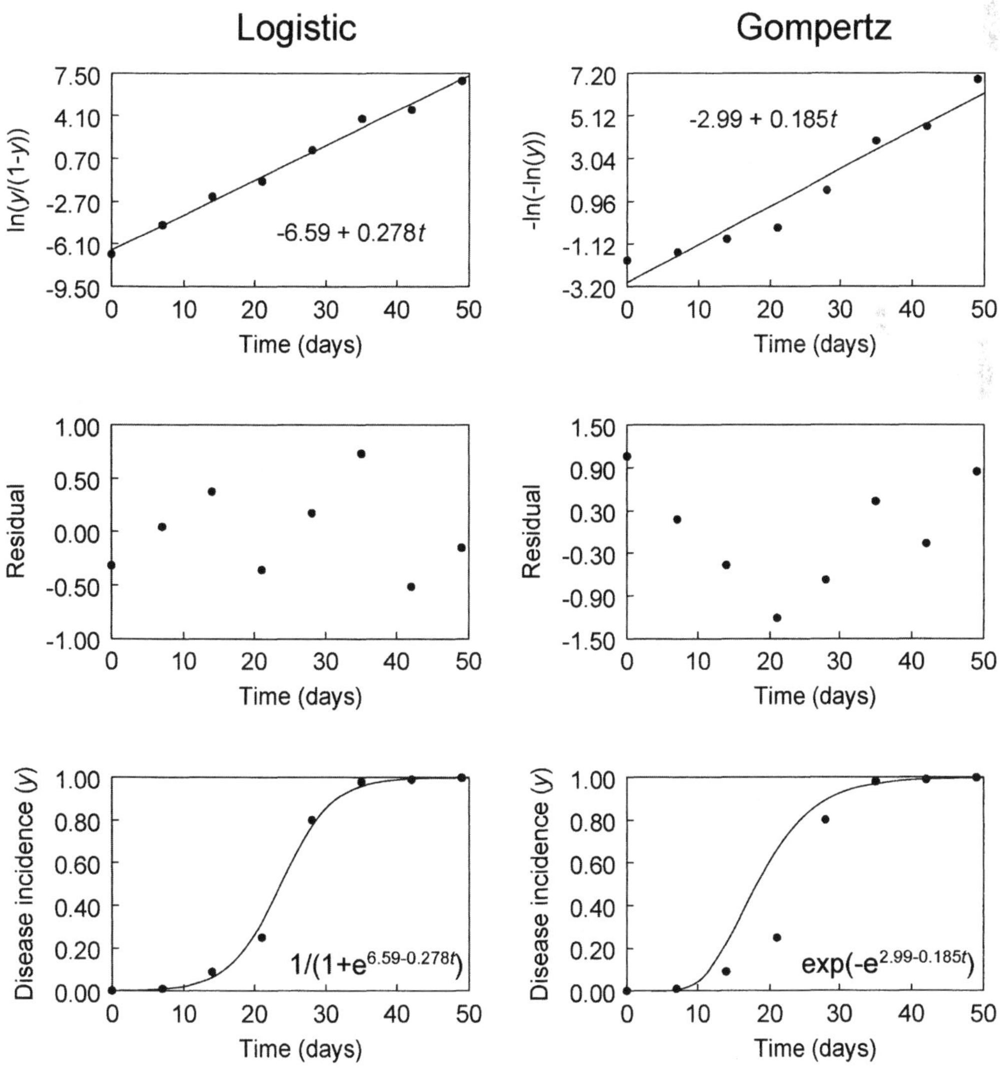

FIG. 4.13. Fit of the linear versions of the logistic and Gompertz models to the example epidemic of tobacco etch disease of pepper. Frames show the fit of the models to the transformed data, the residual plots, and the plots of backtransformed disease incidence (i.e., solving for y using the predicted transformation of y).

changing, or not changing much, y increases between times. Almost any model that specifies increasing y over time will have a high R^2. Likewise, all of the models considered resulted in good fits when assessed in terms of a significance probability based on the F-test ($P < 0.01$ in each case). However, as indicated in Chapter 3, high values of R^2 and statistically significant F-tests are not sufficient indicators of an appropriate model. For the example data set, the logistic provided a better fit compared to the Gompertz model (Fig. 4.13). The residuals ($\hat{\varepsilon}$) for the logistic model appeared to have a random scatter about the fitted line, a strong indicator of an appropriate model being used. In contrast, the residuals for the Gompertz model exhibited a strong pattern, reflecting the overprediction of y^* values in the middle portion of the disease progress curve and underprediction of y^* values at times near the beginning and the end of the observed data.

The fitted logistic model can be written as:

$$\hat{y}^* = -6.59(0.293) + 0.278(0.0100)t$$

in which the numbers in parentheses are the estimated standard errors of the parameter estimates. Note that the estimated intercept (−6.59) is the logit of y_0. To calculate the estimate of y_0 using this value, one utilizes the backtransformation (equation 4.18):

$$\hat{y}_0 = \frac{1}{1 + e^{6.59}} = 0.0014.$$

The same backtransformation (which is specific to the logistic model) can be used to determine the predicted disease intensity from the predicted logit of disease. There is a subtle point to make here. The predicted values, based on ordinary least squares regression, are estimates of the expected (i.e., mean) values of y^* at each t (for the assumed distribution of y^* values at each t; see Chapter 3). Because the backtransformation function (e.g., equation 4.18) is a nonlinear expression, the value of $\hat{y}$ calculated from $\hat{y}^*$ is *not* the expected y at that time (Neter et al., 1983). When the exponential model [$y^* = \ln(y)$] is fitted to data, the backtransformation [$\hat{y} = \exp(\hat{y}^*)$]) actually calculates the *median* y of the assumed distribution of y values at each t. For more complicated models, it is sufficient to know that the backtransformation is producing some measure of central tendency of the distribution of y at each t, although not necessarily the mean.

One can use the estimated parameters (from ordinary least squares regression) to calculate predicted y values at the full range of ts by using the nonlinear versions of the models (e.g., equations 4.15 and 4.20). This is shown in the bottom frames of Fig. 4.13. The curves also could be generated by predicting y^* at t values between 0 and 50 and using the backtransformation to obtain y. It is easy to see here that the Gompertz model results in substantial overprediction of y between times of 15 and 30, and that the logistic model provides an overall good fit with these data. This is a good place to note that the R^2 values given above were for the agreement between observed and predicted y^* values, not for the agreement between observed and predicted y values. Furthermore, y^* was different for each model. A type of R^2 (R^{*2}) can be determined for the goodness of fit of the observed y values when comparing results for models with different transformations of the dependent variable. The easiest way of calculating R^{*2} is to backtransform the $\hat{y}^*$ (using the appropriate function) to obtain $\hat{y}$, calculate the correlation coefficient between $\hat{y}$ and y, and then square this correlation. The reader can use the results in Table 4.2 to confirm that the R^{*2} equals 0.996 for the fit of the logistic model. With a little more work, the reader can fit the Gompertz model to the data, obtain the predicted y^* and then y values, and determine that R^{*2} equals 0.903. The R^{*2} values for the exponential, monomolecular, and Richards models were 0.428, 0.449, and 0.972, respectively. One can more easily see the poor fit of the exponential and monomolecular models reflected in R^{*2} than in R^2.

The overall conclusion from the assessment of the example disease progress curve is that the logistic model is the most appropriate. Further discussion primarily centers on an evaluation of the appropriateness of the standard model assumptions for the particular data set. For the most part, evaluation focuses on the residuals. In addition to assessing the residuals for randomness (one indicator of an appropriate model), they can be checked for normality, independence over time, and constant variance. Normality is not overly critical to model fitting; however, a so-called normal probability plot (Neter et al., 1983) indicated that normality of the residuals was a reasonable assumption in this case.

Independence is critical in model fitting (Madden, 1986). Because of the cumulative nature of disease progress curves, discussed previously, there is a possibility that the errors, and hence their estimates (residuals) will be positively correlated over time (i.e., autocorrelated or serially correlated). If this occurs, the value of ε_j will be affected by ε_{j-1}, and so on. A so-called first-order autoregressive model [AR(1)] is often considered as a possible description of the serial correlation of the errors for many situations (Schabenberger and Pierce, 2002). In this setting, the correlation between errors one time period apart (i.e., between ε_j and ε_{j-1}) is given by $\rho^1 = \rho$, and correlation between errors two time periods apart (i.e., between ε_j and ε_{j-2}) is given by ρ^2, and so on. A statistical model for this is:

$$\varepsilon_j = \rho\varepsilon_{j-1} + \zeta_j$$

in which ζ is a random variable with mean 0 that *is* independent of ε and independent over time. The parameter

ρ is known as the first-order autocorrelation coefficient ($-1 \leq \rho \leq 1$). The general formula for the correlation between errors at times j and j' is given by $\text{Corr}(\varepsilon_j, \varepsilon_{j'}) = \rho^{|j-j'|}$, where $|j - j'|$ is the number of time periods separating j and j'. Because of this serial correlation, one can also say there is a nonzero covarariance in ε between times. This is discussed in detail by Madden (1986). What this means in practice when $\rho > 0$, is that large positive residuals will, on average, be followed by large positive residuals, and that large negative residuals will, on average, be followed by large negative residuals. The consequence of positive autocorrelation among the residuals (i.e., of the assumption of independence being violated) is that standard errors of the estimated parameters (especially the slope) will, without correction, be artificially small. This can lead, among other things, to falsely declaring significant differences between slopes that are not truly different.

One way of checking for this is to analyze the residuals (estimates of ε) with a time series procedure that determines first-order temporal autocorrelations. This analysis of the logistic model residuals (see Table 4.2) indicated that the first-order autocorrelation was -0.296, which was not significantly different from 0 ($P > 0.05$). Although the general concern is of a positive autocorrelation, a negative autocorrelation of residuals can occur when there is a very good fit of a model to the data, and the temporal sequence of y_j^* values alternate above and below the predictions ($\hat{y}_j^*$). Thus, there is no reason to correct for the autocorrelation with this particular data set. There are specialized statistical procedures [e.g., PROC AUTOREG in SAS (part of the ETS system of SAS for analyzing data collected over time)] that can directly adjust for the first-order autocorrelation in fitting models to data. This procedure can also automatically determine the first-order autocorrelation coefficient (ρ) using different methods. There are also specialized tests for autocorrelation which are not covered here. Madden (1986) gives a detailed description of the ways to correct for serial correlation without using specialized programs. If one corrected for the autocorrelation of residuals with the example (even though this is not actually required in this case), one obtains an estimate of $r*$ of 0.279/day and standard error of the estimated rate parameter of 0.0087, nearly identical to the ordinary least squares results because of the lack of significant autocorrelation in this case. When the serial correlation is more negative than found here, standard errors of the estimated slope will be too large when no corrections are made; corrections for large negative serial correlations would lead to smaller standard errors.

Although first-order autoregressive serial correlation of residuals is the most common assumption (when independence is unlikely), there are many other possible so-called structures to the correlations (see Wolfinger, 1996). In other words, the appropriate equation for $\text{Corr}(\varepsilon_j, \varepsilon_{j'})$ may be simpler, or more likely, much more complicated than $\rho^{|j-j'|}$. Some possibilities are discussed in section 4.7.2.3. The complexity is an outcome of the cumulative process of epidemics.

The residual plot also showed no evidence for nonconstant variance for the error term (unexplained variability). With nonconstant variance, there will be some parts of the residual plot with a high scatter of points and other parts with a low scatter. However, with only one observed disease value at each time, it is difficult to determine nonconstancy using a residual plot, because such a condition can be confused with nonrandomness. Based on statistical properties of disease intensity, one would actually expect the error term not to be constant. For instance, if disease incidence is being considered, then the random variable y is an estimate of the probability of an individual (e.g., plant) being diseased, p. For binary data such as incidence, the variance of y [$\text{Var}(y)$] is proportional to $y(1 - y)$. Further details of this are given in Chapter 9.

Because one is using a transformation of y (i.e., y^*) with the (ordinary least squares) linear regression analysis, one needs to know the variance of y^*, or at least know if the variance of y^* is a function of y. Large-sample statistical theory shows that the variance of a function of a random variable can be determined based on the variance of the variable and the mathematical form of the function. Specifically, if we write $g(y)$ for y^*, to clearly indicate that the transformation is a function of y [e.g., $y^* = g(y) = \ln(y)$ for the exponential model], then the variance of $g(y)$ is given by:

$$\text{Var}[g(y)] = [g'(y)]^2 \text{Var}(y) \tag{4.33}$$

in which $g'(y)$ is the first derivative of $g(y)$ with respect to y. For instance, with the logistic model, $g(y) = \ln[y/(1 - y)]$, and $g'(y) = [y(1 - y)]^{-1}$. Assuming that $\text{Var}(y) = y(1 - y)$ (i.e., ignoring a proportionality constant), one can show that the variance of the logit transformation of y is $[y(1 - y)]^{-1}$. For the models considered so far in this chapter, the variances of y^* are:

Model	*Var[g(y)]*
Exponential	$\left(\frac{1-y}{y}\right)$
Monomolecular	$\left(\frac{y}{1-y}\right)$
Logistic	$\left(\frac{1}{y(1-y)}\right)$
Gompertz	$\left(\frac{1-y}{y\ln(y)^2}\right)$
Richards	$\left(\frac{y^{1-2\eta}(\eta-1)^2(1-y)}{(y^{1-\eta}-1)^2}\right)$

Note that the variance for the Richards model is the same whether η is greater than or less than 1. Strictly speaking, the above formulae hold only asymptotically (i.e, at very large sample sizes). At small sample sizes, the variance formulae for functions of y should be considered approximations.

If disease is measured as severity (a continuous variable) rather than incidence, there is no *a priori* relationship between y and the variance of y based on statistical theory. However, it is reasonable to assume that Var(y) is still proportional to $y(1 - y)$ because the maximum possible variation in severity is at $y = 1/2$, and the range of possible severity values declines at y approaches 0 and 1. Thus, one could still use these functions for Var[$g(y)$] with severity data, at least as an approximation. The important point is that the variance of y^* is expected to vary with y in a nonlinear fashion. To account for the nonconstant variance, one can use *weighted* least squares, with weights (w) equal to $w = 1/\text{Var}[g(y)]$. Using this method for fitting the logistic model to the example data resulted in the following fit:

$$\hat{y}^* = -6.62(0.487) + 0.282(0.0204)t$$

with $R^2 = 0.970$. It should be noted that the R^2 value will always be lower for weighted least squares compared with unweighted least squares, even if weighted least squares is the appropriate approach to take (Neter et al., 1983). The interested reader can confirm that the residual plot exhibited a random pattern for this fit. Estimated initial disease ($\hat{y}_0$) is calculated from the intercept to be 0.0013. With this example, the estimates of the parameters were very similar to the estimates obtained from the unweighted analysis; this is primarily because the fit is so good here for the logistic model. When there is greater variation (lower R^2), there will be greater differences between the parameter estimates obtained from the two approaches.

When there is just one observation at each time, there is no simple test for unequal variances; rather, one uses residual plots as an indication of this heterogeneity. If the residual plot does not reveal unequal variability, then it is debatable whether or not weighting is needed. It is a good idea to use weighted least squares for a variable that is expected to have unequal variances *unless* there is evidence that the unweighted approach gives satisfactory results.

4.6.3 Estimating parameters—nonlinear least squares

Except for the Richards model with unknown shape parameter, the models considered so far in this chapter can be written in linear form to estimate all parameters, even though y is actually a nonlinear function of t (see section 3.5). Many models to be considered later cannot be linearized, so model fitting (or parameter estimation) must involve nonlinear methods. Even when a model can be linearized, fitting can still be done using nonlinear least squares. This is demonstrated with the logistic model fitted to the example tobacco etch data. We use equation 4.15a as the mathematical representation of the epidemic.

The statistical version of equation 4.15a can be written as:

$$y_i = \frac{1}{1 + \exp\left(-\left(\ln[y_0/(1 - y_0)] + r_L t_i\right)\right)} + \varepsilon_i \qquad (4.34)$$

where ε_j is the error for the j-th time, which is assumed (here) to be normally and independently distributed with mean 0 and constant variance. Fitting this model to the data using nonlinear least squares (e.g., utilizing PROC MODEL of the ETS system of SAS, or the more common PROC NLIN of the STAT system of SAS), resulted in the following parameter estimates (with standard errors in parentheses): $\hat{y}_0 = 0.00034$ (0.00021), $\hat{r}_L = 0.332$ (0.0249). The R^2 value is 0.997. Note that the estimate of initial disease incidence is lower than with linear least squares, and that the estimate of the rate parameter is higher. Predicted (fitted) values of y are not quite the same as with linear least squares (see Table 4.2). In particular, fitted values are lower at times 14 and 21 days.

The fit of the nonlinear model can be evaluated in the same way as is done for linear models. In particular, residuals are evaluated for random scatter, normality, independence (over time), and constant variance. We do not show these results here, but point out that the diagnostics indicated that the logistic model is appropriate and that the statistical assumptions are reasonable. Correcting for serial correlation of the residuals (i.e., lack of independence) is more difficult for nonlinear models, but options in PROC MODEL can be used to adjust for the correlation. The issue of constant variance should be addressed. As discussed above, it is reasonable to assume that Var(y) is proportional to $y(1 - y)$. Thus, weighted nonlinear least squares, with weights of $1/[y(1 - y)]$, should be considered. Unlike the case for the linearized versions of the epidemic models (see section 4.4), the same weight function is used for all the nonlinear versions of the models. This is because the dependent (response) variable, y, is the same for all models. Using weighted nonlinear least squares, the following estimates were obtained: $\hat{y}_0 = 0.0006$ (0.00035), $\hat{r}_L =$ 0.314 (0.0247). For this example data set, there was no empirical evidence that weighting was needed because the fitted values were so close to the observed.

The reader may ask why the parameter estimates obtained with linear and nonlinear least squares are different. The answer involves the difference between equations 4.32a (with the logit used for y^*) and 4.34. This was addressed in Chapter 3, but is discussed here specifically in terms of disease progress curves. Without the error terms added, nonlinear equation 4.15a and linear equation 4.16 are equivalent ways of representing a logistic epidemic. However, equations 4.32a and 4.34 are not equivalent. For instance, one *cannot* rearrange nonlinear equation 4.34 to obtain linear equation 4.32a. To obtain linear equation 4.32a (with logit as the dependent variable), the nonlinear version would have to be of the following form:

$$y_j = \frac{1}{1+\exp\left(-\left(\ln\left[y_0/(1-y_0)\right]+r_L t_j+\varepsilon_j\right)\right)} \qquad (4.35)$$

Note that the error term is not additive, as in $y_j = f(t_j) + \varepsilon_j$, but is related to y in a more complicated (nonlinear) manner. Put another way, if one fits a disease progress model using linear least squares, one is implicitly assuming that the error is not additive for the original untransformed disease data. Conversely, if one fits a disease progress model using nonlinear least squares, one is implicitly assuming that the error is *not* additive on the transformed (i.e., logit) scale. Note that the ε of equation 4.34 is not equal to the ε of equation 4.35, even though they both represent the unexplained variability.

There is no straightforward way of knowing which model form for the error is appropriate before analyzing a disease progress curve. In fact, because an epidemic is determined by the integration of dy over time, the actual error form may be more complicated than represented by either of the above formulations (Schabenberger and Pierce, 2002). For instance, to incorporate all the factors that can influence an epidemic that are not incorporated into the deterministic part of a model, it could be argued that an error term be added directly to the model for dy/dt. With the logistic model, we could write this as: $dy/dt = r_L y(1-y) + \zeta$, in which ζ is an error term (Marcus, 1991; Sandland and McGilchrist, 1979). Models of this type can be called stochastic differential equations. [Note that the term 'stochastic epidemic model' often refers to other types of models that explicitly deal with probabilities of plant individuals making transitions, for example, transition from healthy to infected per unit time (Renshaw, 1991). This is different from the model discussed here]. For reasons we will not go into in this chapter, ζ can actually be a function of (i.e., proportional to) y, so that $\zeta = \varepsilon \cdot y$ [in which $\varepsilon \sim \mathrm{NID}(0, \sigma^2)$]. y at a given time is given by:

$$y = \int [ry(1-y)+\zeta]d\hat{t} \qquad (4.36)$$

where $\hat{t}$ indicates a time between 0 and t. In general, there are not analytical solutions to stochastic differential equations of this type, except under special circumstances. The point here is that if equation 4.36 is correct (in some sense), there is no reason to believe that numerical solutions will produce results that will be totally consistent with either equation 4.32a (or 4.35) or 4.34. Both of equations 4.32a and 4.34 are approximations to a more complicated stochastic process. Since all models are simplifications of reality, this is not necessarily a problem. The important issue is whether or not the epidemic model of interest (e.g., logistic) in one of its standard statistical forms (equation 4.32a or 4.34) adequately describes the disease progress curve. In our experience, both the nonlinear and linear versions can provide very good descriptions. With more complicated models, the researcher typically must use nonlinear modeling methods.

4.6.4 Parameter estimation—generalized linear models for disease incidence

As discussed in Chapter 9, the number of diseased individuals per plot has a discrete statistical distribution (binomial or related), not a continuous normal distribution. The expected value of the binomial distribution can be written as p, which is the probability of a plant individual being diseased. An estimate of p is y. Thus, the standard methods of using linear or nonlinear least squares may appear to be inappropriate, since these methods assume a normal distribution. However, it is known that the binomial distribution is well approximated by the normal if the sample size is large enough (Snedecor and Cochran, 1989). So, if the number of observations used to determine y is large (e.g., >50), the normality assumption is quite reasonable. If number of observations is low, results could be strongly affected by use of a normal-based parameter estimation method.

As an alternative to least squares, one can express the epidemic models of this chapter in the form of generalized linear models (GLMs). A GLM is not the same as the linear model such as equation 4.32a, although the distinction may not be obvious at first. To see the difference, we consider an alternative way of writing equation 4.32a. By starting with this equation, one can write the expected (mean) value of y^* at any time, $E(y_j^*)$, as:

$$E(y_j^*) = E(y_0^* + r_* t_j + \varepsilon_j) = E(y_0^*) + E(r_* t_j) + E(\varepsilon_j)$$

Because y_0^* and r_* are constants (not random variables), $E(y_0^*) = y_0^*$ and $E(r_* t_j) = r_* t_j$. Moreover, $E(\varepsilon_j) = 0$ (with $\mathrm{Var}(\varepsilon) = \sigma^2$), by definition. Thus, one can write equation 4.32a as:

$$E(y_j^*) = y_0^* + r_* t_j$$

In a statistical sense, this indicates that the mean *transformed* disease intensity (e.g., logit) for the population of logits at time t_j is given by $y_0^* + r_* t_j$, and the variance of this population of logits is given by σ^2. For the logistic model, we would write:

$$E\left(\ln\left(\frac{y}{1-y}\right)_j\right) = a + r_L t_j \qquad (4.37a)$$

with a representing the logit transformation at $t = 0$ ($a = y_0^*$).

The GLM version of the logistic model would be written as:

$$\ln\left(\frac{p}{1-p}\right)_j = a + r_L t_j \qquad (4.37b)$$

with the assumption that y has a binomial distribution with mean of p [i.e., $E(y) = p$]. In other words, the logit on the left-hand side of equation 4.37b could be written as:

$$\ln\left(\frac{E(y)}{1-E(y)}\right)_j .$$

Thus, in a linear model for a logistic epidemic, the *expected value of the logit* is a straight-line function of t, but in a generalized linear model, the *logit of the expected value* of disease incidence is a straight-line function of t. This was discussed more in section 3.5.3.3 of the previous chapter. The transformation of the expected value is known as the *link function* in the terminology of generalized linear models. In order to estimate parameters of a GLM, maximum likelihood or other related method based on the binomial distribution is needed, rather than least squares. PROC LOGISTIC or GENMOD in SAS, or the GLIM program of NAG can be used for the analysis. For this analysis, one must specify both the number of diseased individuals (Y) and total number of individuals (M) at each time used to calculate y ($=Y/M$), not just the proportion diseased. One does not calculate the logit before the model fitting; rather, one specifies the selected link, and the program does the calculations. The assumed variance of a binomial variable, which is not a constant but a function of y and M, is incorporated in the model fitting (Collett, 2003).

Using $M = 500$, equation 4.37b was fitted to the tobacco etch epidemic example using PROC LOGISTIC. Parameter estimates were: $\hat{a} = -6.856$ (0.250) and $\hat{r}_L = 0.290$ (0.0102). These parameter estimates are close to those obtained with linear ordinary (normal-distribution-based) least squares regression, either weighted or unweighted. This is mostly because of the large value for M used and the good fit obtained with the logistic model. With a small M, both the estimates and the standard errors can be considerably different, especially if there is more variation around the predictions than found for this epidemic. Our general recommendation is to use GLMs when M is small (especially if $M < 30$). Even when total number of individuals is large for a field, sometimes one is interested in characterizing disease progress at individual sites (with small M) within the field. Then GLMs should be used for these data.

Because the default link for binomial data is the logit, fitting the GLM version of the logistic model to epidemic data in the form of disease incidence is usually quite straightforward. However, fitting the GLM version of models other than the logistic to epidemic data is more difficult than for the logistic model. Some of the link functions for the other models are not standard ones, so one must use several options with PROC GENMOD, and one must know a fair amount of GLM theory to be able to correctly specify these procedure options (Collett, 2003).

4.7 Comparing Disease Progress Curves

Although sometimes the goal of a researcher is to fit a model to a single disease progress curve to characterize an epidemic, it is often the case that what is required is to fit a model (or models) to several curves representing different epidemics. Using the estimated parameters, the epidemics can then be compared in terms of initial disease intensities and rate parameters. The different disease progress curves may originate from planned experiments, where treatments (e.g., fungicides) are imposed on experimental plots, or from observations of natural epidemics in commercial fields or in forests.

In the models discussed so far, epidemics can be characterized by two parameters (y_0 and r_*). For more detailed or complicated descriptions of epidemics, additional parameters are required. Epidemic comparisons can involve a consideration of several parameters, and involve some elaborate multivariate statistical procedures that we will not discuss here (e.g., see Mora-Aguilera et al., 1996; Sanogo and Yang, 2004). The term 'Comparative Epidemiology' is used for the general approach (Campbell, 1998; Kranz, 1980; 2003). Comparative epidemiology seeks to define "what is different, similar, or identical among diseases" and to "derive models and principles of more general application from the multitude of singular events in plant diseases and their epidemics" (Kranz, 1974b). The models used so far in this chapter, and those to be presented later, have the general applicability needed to compare epidemics using the principles of population dynamics. The new book by Kranz (2003) should be required reading for those interested in the subject.

In this section, we show how to compare epidemics through the estimated parameters of disease progress models fitted to the data. We show how to do this in the general sense, when no information is available on the experimental design, or when the data resulted from a survey of natural epidemics. Then we show how to compare epidemics by fully utilizing information from the experimental design used to obtain the data. We use covariance and repeated measures concepts throughout this section.

4.7.1 Simple comparison of epidemics

In section 4.6, we showed that the logistic model provided a good description of the tobacco etch epidemic in

```
data a;
      input group t y @@;
      *--String multiple observations per line;
      wt=y*(1-y); logit=log(y/(1-y));
datalines;
 1  0   0.080   2  0   0.001   3  0   0.001
 1  7   0.130   2  7   0.010   3  7   0.001
 1 14   0.780   2 14   0.090   3 14   0.010
 1 21   0.920   2 21   0.250   3 21   0.050
 1 28   0.990   2 28   0.800   3 28   0.180
 1 35   0.995   2 35   0.980   3 35   0.340
 1 42   0.999   2 42   0.990   3 42   0.480
 1 49   0.9991  2 49   0.999   3 49   0.740
 ;
title 'TEV on peppers, NUTTER data -- three DPCs';
proc sort data=a out=a; by group;
proc reg data=a;
      by group;
      model logit = t;
proc mixed data=a;
      class group;
      model logit = group t group*t / solution;
      estimate 'group 1 INT' int 1 group 1 0 0;
      estimate 'group 2 INT' int 1 group 0 1 0;
      estimate 'group 3 INT' int 1 group 0 0 1;
      estimate 'group 1 SLOPE' t 1 group*t 1 0 0;
      estimate 'group 2 SLOPE' t 1 group*t 0 1 0;
      estimate 'group 3 SLOPE' t 1 group*t 0 0 1;
      estimate 'INT, one vs two'  group 1 -1 0;
      estimate 'INT, one vs three'  group 1 0 -1;
      estimate 'INT, two vs three'  group 0 1 -1;
      estimate 'SLOPE, one vs two'  group*t 1 -1 ;
      estimate 'SLOPE, one vs three'  group*t 1 0 -1;
      estimate 'SLOPE, two vs three'  group*t 0 1 -1;
proc mixed data=a; *--Do not use F tests here, just parameters;
      title2 'Direct estimates of three intercepts and slopes';
      class group ;
      model logit = group group*t / noint solution;
run;
*-------------------------------------------------------------;
data b; set a; if (group ne 2) then delete;
w=1/(y*(1-y)); M=500; bigY=y*M;
proc autoreg data=b;
      title2 'Autoregressive least squares, 1st order';
      model logit = t / nlag=1;
proc nlin data=b;
      title2 'Nonlinear least squares, weights';
      *--Give initial estimates of parameters. Define weights-->;
      parameters y0 .001 rL .2;   _weight_ = w;
      model y = 1/(1 + exp(-(log(y0/(1-y0)) + rL*t)));
run;
proc genmod data=b;
      title2 'Generalized linear model fit, assume M=500';
      model bigY/M = t / dist=binomial link=logit;
run;
```

FIG. **4.14.** Input program file of SAS to read in incidence data for three treatments of pepper with tobacco etch disease, transform the data, and fit the logistic model to the data. Data input format is: treatment, time, and y. The "@@" on the input record indicates that a single record can contain multiple observations for the three variables: treatment, time, and y. Data are sorted by treatment, and then the REG procedure is used to fit the linear version of the logistic model to the data. The "by" statement in REG indicates that separate analyses are to be done for each treatment. Then the MIXED procedure is used to perform a covariance analysis, with treatment (cult) as a factor (class variable) and time (t) as a continuous variable.

peppers (Fig. 4.13). The data in Fig. 11 of Nutter (1997) actually consists of three disease progress curves, corresponding to three treatments (which we can just call groups). The data for all three treatments are given in Fig. 4.14 in the form of a program file for SAS, and y is plotted versus t in the top frame of Fig. 4.15. The data used above in the example correspond to treatment "2". In this book we do not, in general, give specific instruction on the use of statistical software, because for many analyses there are different programs that can be used for the same purpose, and the format is different for each. However, for the covariance analysis to be presented below (and for some other analyses in this and subsequent chapters), we feel it is difficult to follow the steps without seeing example input and output from a statistical program. Thus, we show how to do some of the analysis with SAS code. The last part of the SAS program in Fig. 4.14 also shows how to do some of the

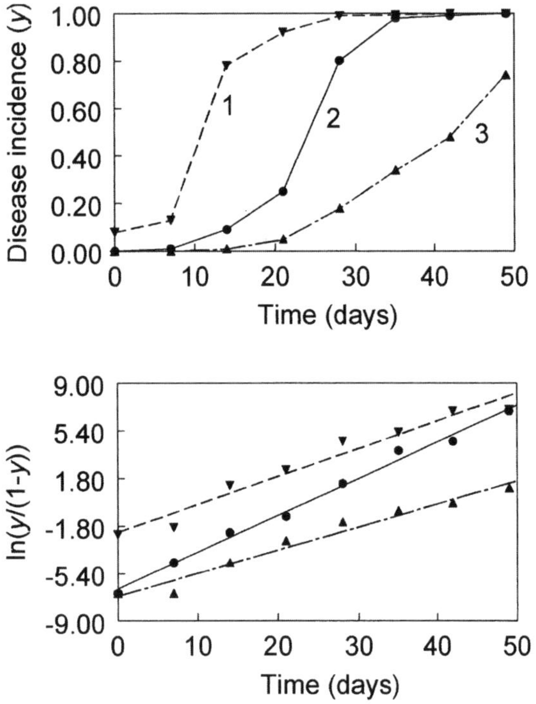

FIG. 4.15. Disease progress curves for three tobacco etch epidemics in peppers (see Nutter, 1997). Disease incidence is shown in the upper frame and logits in the lower frame. Predicted logits are based on fitting equation 4.32b to the data without weights.

linear and nonlinear modeling of the single disease progress curve presented above.

We assume that the logistic model is appropriate for all three epidemics and use only the linear version of the model (equation 4.32b or 4.37a). The interested reader can confirm that the logistic model is reasonable for all three disease progress curves. It should be noted that because y for the third treatment does not come close to 1.0 by the last assessment time, one must be careful in use of the *dy/dt* graphs in making model selections. Based on use of the REG procedure, the following model fits were obtained for the treatments (G_1, G_2, G_3):

$$G_1{:}\ \hat{y}^* = -2.29(0.516) + 0.211(0.0176)t, \quad R^2 = 0.960$$
$$G_2{:}\ \hat{y}^* = -6.59(0.293) + 0.278(0.0100)t, \quad R^2 = 0.992$$
$$G_3{:}\ \hat{y}^* = -7.11(0.390) + 0.175(0.0133)t, \quad R^2 = 0.966.$$

The observed and predicted logits are shown in Fig. 4.15. Backtransformation of the estimated intercepts gave 0.092, 0.0014, and 0.0008 for y_0. All results are based on unweighted least squares regression analysis. To use weighted least squares, one would simply add a statement to the REG procedure as "weights wt", in which wt is the weight variable for the logit transformation (created in the data step in Fig. 4.14). Note that we use Arial font to indicate exact SAS statements.

Just based on the estimated parameters, it appears that there is a much higher initial disease incidence for treatment 1 than the others, and that the rate parameter for treatment 3 was lower than for treatment 2 and perhaps lower than for treatment 1. These observations, and less obvious ones, can be evaluated with *t*-tests of the estimated parameters. For instance, one can test the hypothesis that r_L for the first and second treatments are equal. In statistical notation, one would write this in terms of a null (H_0) and alternative (H_a) hypothesis:

$$H_0{:}\ r_{L1} = r_{L2}, \quad \text{or} \quad H_0{:}\ r_{L1} - r_{L2} = 0$$
$$H_a{:}\ r_{L1} \neq r_{L2}, \quad \text{or} \quad H_a{:}\ r_{L1} - r_{L2} \neq 0$$

in which the number subscripts refer to the treatments. A *t*-test is used to test the null hypothesis; this is done by calculating the *t*-statistic:

$$t = \frac{\hat{r}_{L1} - \hat{r}_{L2} - 0}{s(\hat{r}_{L1} - \hat{r}_{L2})}$$

in which the denominator is the standard error of the difference between the two estimated slopes. This can be estimated (for independent data) as:

$$s(\hat{r}_{L1} - \hat{r}_{L2}) = \left(s^2(\hat{r}_{L1}) + s^2(\hat{r}_{L2})\right)^{1/2}$$

in which $s^2(\hat{r}_{L1})$ and $s^2(\hat{r}_{L2})$ are the squares of the standard errors of the estimated slopes. For the comparison of estimated slopes for treatments 1 and 2, the t-statistic is given by:

$$t = \frac{0.211 - 0.278}{\left(0.0176^2 + 0.0100^2\right)^{1/2}} = \frac{-0.067}{0.02024} = -3.3.$$

The degrees of freedom for the *t*-test is equal to the sum of the error degrees of freedom for the two disease progress curves. There were 6 error degrees of freedom for each epidemic (because in each case there were 8 data points and two parameters were estimated), so degrees of freedom (*df*) for the *t*-statistic here is 6 + 6 = 12. Since the alternative hypothesis is not directional (i.e., we did not specify that one of the slopes was larger, just that they were not equal), large absolute values of t indicate statistical significance. With $df = 12$, any value of t greater than 2.179 or less than −2.179 indicate significance at a pre-selected probability (α) of 0.05. The achieved significance level (i.e., the value of P exactly corresponding to $|t| = 3.3$) is $P = 0.0016$. Thus, there is clear evidence that these two slopes are different.

The interested reader can do the *t*-test for the equality of initial disease incidence for treatments 1 and 2. Since a transformation of y_0 is estimated, the *t*-test must be based on the transformed values. The reader should be able to show that the achieved t-statistic is 7.2, which corresponds to $P < 0.001$. One can continue this to compare the estimated slopes and intercepts between treatments 2 and 3, and 1 and 3. This general method works for any number of disease progress curves, but can become quite tedious with many epidemics. Moreover, there is a danger of performing these unrestricted multiple comparisons of parameter estimates (Neter et al., 1983) because overall error rate can be (much) higher than 5% (when, for instance, a 0.05 significance level is used for each comparison). However, there is a more direct and efficient approach to determine if the slopes and/or intercepts vary among epidemics. This is done with an *analysis of covariance*. The use of 'covariance analysis' here should not be confused with the covariance (=correlation × variance) that can exist between two random variables, or the same variable (such as ε) at two times. The approach we take here continues to use normal-distribution-based ordinary least squares, but expands the model to explicitly account for treatment.

A linear model for the logit of disease intensity in the i-th treatment at the j-th time, y^*_{ij}, can be written as:

$$y^*_{ij} = y^*_{0i} + r*_i t_{ij} + \varepsilon_{ij} \qquad (4.38a)$$

in which t_{ij} is j-th time for the i-th treatment, and y^*_{0i} and $r*_i$ are the intercepts and slopes for the i-th treatment. Equation 4.38a indicates that there are three intercepts and slopes, because i = 1, 2, 3 (e.g., $r*_1$, $r*_2$, $r*_3$). Analysis based on models (such as equation 4.38a) that include continuous variables (time) and class or factor variables (treatment) is traditionally called covariance analysis. Models of this type can be fitted using many statistical programs. However, some covariance programs or procedures only allow a single slope parameter, not slopes that depend on the factors of interest (e.g., treatment). PROC GLM and MIXED in SAS do allow for the more general situation encountered here. We use MIXED below because some later analysis (section 4.7.2) can be done most readily with MIXED. This procedure actually uses restricted maximum likelihood, which is a special form of maximum likelihood. But as discussed in Chapter 3, for a normal distribution, this estimation method is equivalent to least squares. Before working with equation 4.38a, we present an even more general model for covariance analysis. One can write y^*_{ij} as:

$$y^*_{ij} = a + y^*_{0i} + \iota\, t_{ij} + r*_i t_{ij} + \varepsilon_{ij} \qquad (4.38b)$$

in which a is constant (an overall intercept term), y^*_{0i} is now the effect of the i-th treatment on the intercept, ι is a slope parameter, and $r*_i$ is now the effect of the i-th treatment on the slope. This model can be written in a more mnemonic form without subscripts as:

YSTAR = CONSTANT + TREATMENT + TIME
+ (TREATMENT × TIME) + ERROR

in which YSTAR is the transformed y, TREATMENT is a class (factor) variable, TIME is a continuous variable with an *implied* slope term, and the product of these two terms means that there is a treatment-dependent slope. In common statistical packages, when a listed variable is continuous (time), then a slope parameter is estimated for the continuous variable. When a class variable is listed, then a separate parameter is estimated for each level of the variable (and the means for these levels are determined from the parameters). When a continuous variable appears as a product with a class (factor) variable, then a separate parameter is estimated for each level of the class variable, corresponding to the effect of each class level (e.g., treatment) on the slope.

With the more cumbersome equation 4.38b (relative to equation 4.38a), the intercept for treatment 1 is $a + y^*_{01}$, and the slope is $\iota + r*_1$. Other intercepts and slopes would be determined in the same way (e.g., slope for treatment 3: $\iota + r*_3$). The main reason for using the cumbersome version is for hypothesis testing. In particular, if one fits equation 4.38b to data, an F-test of y^*_0 indicates whether there is a treatment effect on initial disease intensity (the intercept) and an F-test of $r*$ (really, a test of the interaction of time and treatment) indicates whether there is a treatment effect on the rate parameter. For instance, the F-test for interaction of treatment and time is evaluating the following hypothesis:

H_0: $r*_i = 0$, for all i
H_a: at least one of the $r*_i$ values is not 0.

The 'model' statement in the MIXED procedure shown in Fig. 4.14 fits equation 4.38b to the data. As with many statistical program, the intercept parameter (a), as well as the error (ε) term, are assumed and not explicitly written. This is consistent with the REG procedure, where intercept and error terms were not written in the model statement. The y^*_{0i} term is specified by group (in essence, a main-effect factor [or class] variable), t is specified by t in the SAS program (the covariable or continuous variable), and $r*_i t_{ij}$ (the interaction of the covariable and the factor) is specified by group*t (where '*' is used for multiplication in SAS). Because of prior evidence (shown for treatment 2), we did not adjust for nonconstant variance or serial correlation of the residuals. We do make adjustments of this kind later, for another data set (section 4.7.2).

Output of the MIXED procedure (Fig. 4.16) shows that large and significant F values are achieved for treatment, time, and their interaction (Type 3 Tests of Fixed Effects). Thus, the overall tests indicate that there are treatment effects on both the intercept and slope of the disease progress curves, on a linear scale. The output under Solution for Fixed Effects contains the estimated parameters of equation 4.38b. Using the notation of this equation, one can write the estimates (with standard errors in parentheses) as:

$$\hat{a} = -7.11(0.410), \quad \hat{y}^*_{01} = 4.82(0.580),$$
$$\hat{y}^*_{02} = 0.52(0.580), \quad \hat{y}^*_{03} = 0(-),$$
$$\hat{\iota} = 0.175(0.014), \quad \hat{r}*_1 = 0.036(0.0198),$$
$$\hat{r}*_2 = 0.103(0.0198), \quad \hat{r}*_3 = 0(-)$$

This format for the parameters is not directly useful without some manipulation, but as indicated above, the formulation allows for the *F*-tests of interest. Note that the effects of the third treatment on the intercept and slope are 0 (by definition). This is a consequence of using more parameters than the number of treatments to represent either the intercept or slope. Consider the slope as an example. There are four parameters for specifying the slope in the model, ι, $r*_1$, $r*_2$, and $r*_3$, but only three treatments. In a statistical sense, the model is overparameterized. SAS deals with this by assigning the parameter for the last level (treatment 3 in this case) a value of 0. Estimates of the actual slopes for the three treatments are obtained by:

$$G_1: \hat{\iota} + \hat{r}*_1 = 0.211$$
$$G_2: \hat{\iota} + \hat{r}*_2 = 0.278$$
$$G_3: \hat{\iota} + \hat{r}*_3 = \hat{\iota} = 0.175$$

It should be noted that these are the slopes given above for the individual regression analyses with each treatment. Similar calculations would be made for the intercept (i.e., $\hat{a} + \hat{y}^*_{0i}$):

$$G_1: \hat{a} + \hat{y}^*_{01} = -2.29$$
$$G_2: \hat{a} + \hat{y}^*_{02} = -6.59$$
$$G_3: \hat{a} + \hat{y}^*_{03} = \hat{a} = -7.11$$

The first six **estimate** lines following the **model** statement in Fig. 4.14 use SAS conventions to calculate the intercepts and slopes for each treatment using the method given in the previous paragraph. These estimates and their standard errors are given in the last part of the printed output in Fig. 4.16. Several things can be noted. The *t*-values in the output test the equality of the parameter estimate to 0. Thus significance means that the slope or intercept is different from 0. The *df* (**DF**) are higher than for the individual regressions. This is because all 3 × 8 = 24 observations are used in a single analysis instead of 8 observations for a single treatment. The increased *df* is one practical outcome of doing the full covariance analysis compared with the individual regressions for each group. The standard errors in the output for the three slopes and three intercepts are not the same as shown above for the separate regressions. This is

Type 3 Tests of Fixed Effects

Effect	Num DF	Den DF	F Value	Pr > F
group	2	18	41.71	<.0001
t	1	18	753.16	<.0001
t*group	2	18	14.00	0.0002

Solution for Fixed Effects

Effect	group	Estimate	Standard Error	DF	t Value	Pr > |t|
Intercept		-7.1136	0.4099	18	-17.35	<.0001
group	1	4.8244	0.5797	18	8.32	<.0001
group	2	0.5240	0.5797	18	0.90	0.3779
group	3	0	.	.	.	.
t		0.1752	0.01400	18	12.52	<.0001
t*group	1	0.03645	0.01980	18	1.84	0.0821
t*group	2	0.1033	0.01980	18	5.22	<.0001
t*group	3	0	.	.	.	.

Estimates

Label	Estimate	Standard Error	DF	t Value	Pr > |t|
group 1 INT	-2.2892	0.4099	18	-5.58	<.0001
group 2 INT	-6.5896	0.4099	18	-16.08	<.0001
group 3 INT	-7.1136	0.4099	18	-17.35	<.0001
group 1 SLOPE	0.2117	0.01400	18	15.12	<.0001
group 2 SLOPE	0.2785	0.01400	18	19.90	<.0001
group 3 SLOPE	0.1752	0.01400	18	12.52	<.0001
INT, one vs two	4.3004	0.5797	18	7.42	<.0001
INT, one vs three	4.8244	0.5797	18	8.32	<.0001
INT, two vs three	0.5240	0.5797	18	0.90	0.3779
SLOPE, one vs two	-0.06682	0.01980	18	-3.38	0.0034
SLOPE, one vs three	0.03645	0.01980	18	1.84	0.0821
SLOPE, two vs three	0.1033	0.01980	18	5.22	<.0001

FIG. 4.16. Abridged output of the MIXED procedure of SAS based on the program file in Fig. 4.14. Results are explained in the text.

because a single error variance (estimated as the residual variance) is assumed, based on the input program (value not shown in the partial output given here). Separate error variances could be determined for each treatment by adding the following line after the model statement:

```
repeated/group =group;
```

In the convention of MIXED, the repeated statement refers to the (residual) error variation. In other words, one uses a repeated statement when one wants to specify properties of ε that are different from the default. With this option, the same standard errors of estimated parameters are obtained as calculated with individual regressions (results not shown).

The estimate statements (Fig. 4.14) can be used for many valuable manipulations of the parameter estimates. The last six estimate statements calculate the differences between the intercepts and between the slopes for the pairs of treatments. In the jargon of statistical models, these are called contrasts. Consider the third estimate statement from the end of the first mixed procedure (SLOPE, one vs two):

```
estimate 'SLOPE, one vs two' group*t 1 – 1 0;
```

which estimates: $(\hat{\iota} + \hat{r}_{*1}) - (\hat{\iota} + \hat{r}_{*2}) = \hat{r}_{*1} - \hat{r}_{*2} = -0.067$. The standard error of the difference between the slopes, and the corresponding t-statistic, are given; the t value is for testing the equality of the difference to 0. The stated value (-3.38) is close to the t-statistic determined above for the comparison of two separate regression slopes; the difference in the t-statistic arises as a result of a slightly different standard error estimate (because we are using a single error variance).

The main conclusion from the covariance analysis with this example is that the initial disease intensity (on a logit scale) and the rate parameter are both affected by treatment (F-tests). The estimates of parameter differences indicated that initial disease is different for each pairwise comparison of treatments, except for treatments 2 and 3 (P = 0.378; see Fig. 4.16). The slopes are different for each pairwise comparison of treatments, except possibly for treatments 1 and 3 (P = 0.082).

One can avoid some of the awkwardness of using equation 4.38b and fit equation 4.38a directly to the data. This is done in the last MIXED procedure in Fig. 4.14. Since MIXED (and most programs of this type) implicitly assume that an intercept term is included (a), one must explicitly state that there is not an intercept (see the noint option in the model statement). The model statement specifies group and group*t for the treatment-dependent intercepts and slopes, respectively, for equation 4.38a. Note that no individual time (t) terms are given for the model in this form. Then the solution gives the three intercepts and slopes directly, without any extra manipulation, because there are exactly three intercepts and three slopes for the three treatments (results not shown). Note, however, that the F-tests for treatment and the interaction *do not* have the same interpretation. Specifically, the F-test for treatment (F = 196.94) is testing the null hypothesis that the intercepts are all equal to zero (*not* that they are equal among the treatments). In general, the latter MIXED version of the SAS analysis (or similar from another statistical package) should only be used to easily get the slopes and intercepts (and their standard errors, and contrasts of the parameters), and *not* to test for treatment (or any grouping) factor.

Performing the same type of covariance analysis for the nonlinear version of the logistic (or other) disease progress model is much more tedious, and we do not discuss this here (see Schabenberger and Pierce, 2002, for some innovative new approaches). Most researchers find it more convenient to fit a nonlinear model to the epidemic data for each group separately, as done above with the linear version of the logistic model, before the formal covariance analysis, and then to compare parameter estimates with t-tests. A more general approach to this problem with nonlinear models is described in Gilligan (1990).

4.7.2 Epidemics in designed experiments

Researchers often collect disease intensity data over time in designed experiments. For example, a randomized complete block experiment may be used, in which an area is divided into four blocks and then six plots are established within each block, corresponding to six different treatments. Observations of disease intensity are recorded in each plot at, say, weekly intervals. Although more complicated designs can be considered, we limit the discussion here to this randomized complete block. Because data are collected over time in the experimental unit (the plot), this is often called a *repeated measures* or a randomized complete block repeated measures design. The data in section 4.7.1 were also from a repeated measures design (Nutter, 1997). However, in that case we calculated means across replications at each time, and therefore did not really utilize the experimental design to the full.

4.7.2.1 Choosing a disease progress model. Fig. 4.17 is an example of the type of disease progress data that one might obtain in a designed experiment. The data are in Fig. 4.18 in the form of a SAS program. Almeida et al. (1994) conducted an experiment to determine the effects of planting date and cultivar on epidemics of soybean bud blight, caused by tobacco streak virus (TSV). The experimental design consisted of four blocks, with two cultivars and four planting dates, and the study was done in two different years. We only consider here the results for one of the cultivars ('IAC-4') from one year.

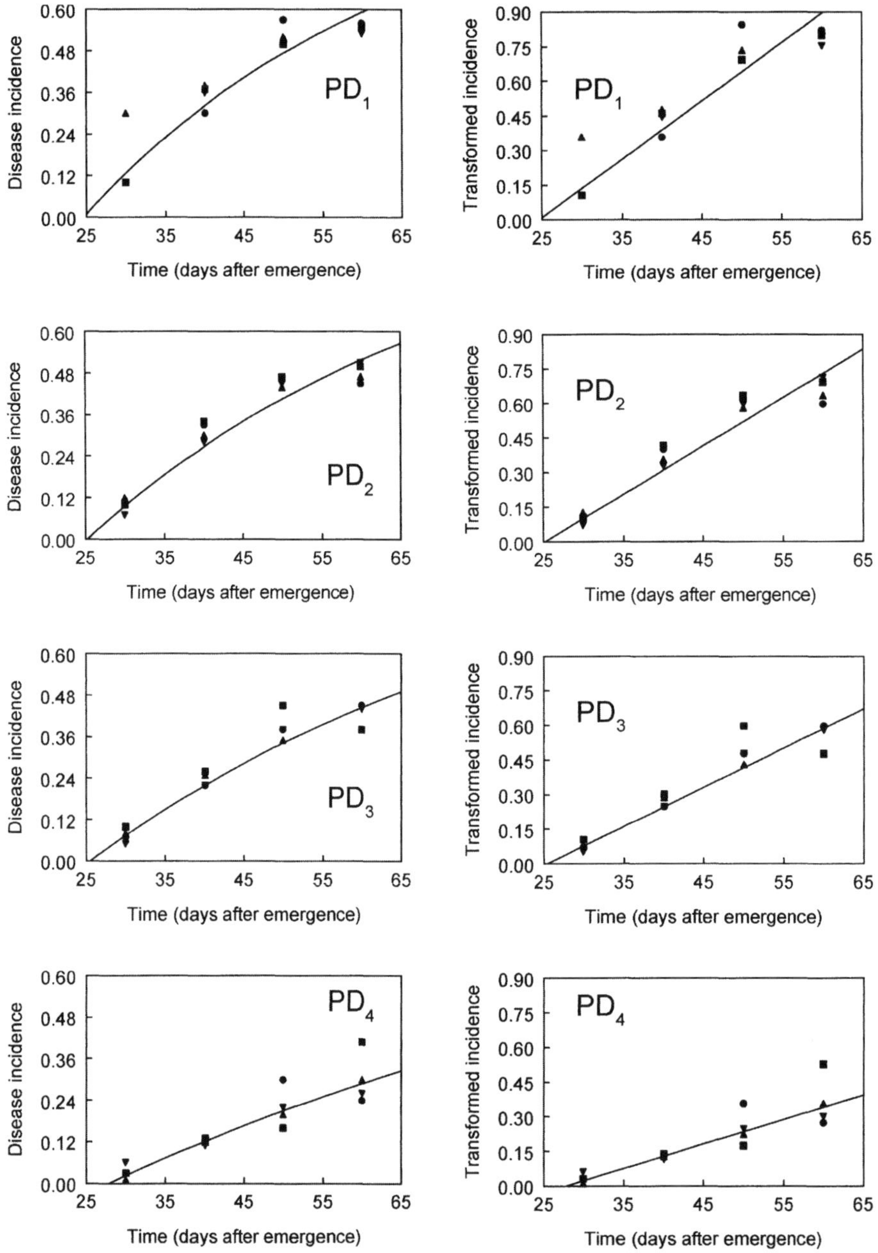

FIG. 4.17. The incidence (left-hand column) and monomolecular-transformed incidence (right-hand column) of soybeans with bud blight, caused by tobacco streak virus, in Brazil for four planting dates (PD) (Almeida et al., 1994). Data for each of the four blocks are shown as separate symbols. The straight lines in the right-hand column of frames are the fits of the linear version of the monomolecular model to the transformed data. The curves in the left-hand column of frames are the backtransformations of the predicted transformed values.

The proportion of plants with symptoms was determined at five assessment times in each plot. For some cultivars and planting dates, no diseased plants were found at the first assessment date, so there is uncertainty when the first infections occurred in some plots. Thus, we analyzed data from the last four assessment times for all planting dates and blocks. The first question the investigator might ask is: which disease progress model is reasonable to use for these data? Because the data from each block are being considered separately, one has, in fact, four disease progress curves for each group (planting date). Thus, there are 16 disease progress curves in this relatively small experiment (in terms of number of experimental factors [treatments]). One should generally follow the methods discussed above to choose a model for the data over time. For the graphical appraisal of y versus t, dy/dt versus t and dy/dt versus y, one considers the epidemic for each planting date and block separately (i.e., for each plot). One can also determine the mean incidence for each planting date at each time (across the four blocks), and assess these data for an appropriate model. On the other hand, the graphical appraisal of y^* versus t (where several transformations are tried) can be done for each block separately or for the combined data for all the blocks (i.e., using four incidence values at each time for each planting date).

```
data a;
input PD t block y @@; *--String multiple observations per line;
   ystar=-log(1-y); *--Note: -ln(1-y) = ln[1/(1-y)];
   wt = (1-y)/y ;
   title 'Analysis of ALV year 7, variety 1';
datalines;
1  30 1  0.10  2  30 1  0.10  3  30 1  0.07  4  30 1  0.03
1  30 2  0.30  2  30 2  0.12  3  30 2  0.08  4  30 2  0.01
1  30 3  0.10  2  30 3  0.07  3  30 3  0.05  4  30 3  0.06
1  30 4  0.10  2  30 4  0.10  3  30 4  0.10  4  30 4  0.03
1  40 1  0.30  2  40 1  0.33  3  40 1  0.22  4  40 1  0.12
1  40 2  0.38  2  40 2  0.30  3  40 2  0.25  4  40 2  0.12
1  40 3  0.36  2  40 3  0.28  3  40 3  0.22  4  40 3  0.11
1  40 4  0.37  2  40 4  0.34  3  40 4  0.26  4  40 4  0.13
1  50 1  0.57  2  50 1  0.46  3  50 1  0.38  4  50 1  0.30
1  50 2  0.52  2  50 2  0.44  3  50 2  0.35  4  50 2  0.20
1  50 3  0.50  2  50 3  0.45  3  50 3  0.38  4  50 3  0.22
1  50 4  0.50  2  50 4  0.47  3  50 4  0.45  4  50 4  0.16
1  60 1  0.56  2  60 1  0.45  3  60 1  0.45  4  60 1  0.24
1  60 2  0.56  2  60 2  0.47  3  60 2  0.38  4  60 2  0.30
1  60 3  0.53  2  60 3  0.51  3  60 3  0.44  4  60 3  0.26
1  60 4  0.55  2  60 4  0.50  3  60 4  0.38  4  60 4  0.41
;
proc sort;
   by PD ;

proc reg data=a;
   title2 'Individual linear regressions';
   by PD;
   model ystar = t;
   weight wt;

proc mixed data=a covtest ;
   title2 'Repeated measures covariance, autoregressive error';
   class  PD block ;
   model ystar = PD t PD*t /  solution ddfm=betwithin;
   random block;
   weight wt;
   repeated  / subject=PD*block type=ar(1) r rcorr;

run;
```

FIG. 4.18. Input program file of SAS to read in incidence data for four planting dates (PD) of soybean bud blight, transform the data, and fit the monomolecular model to the data with weights based on the inverse of the theoretical variance. Data input format is: planting date, time, block, and y. The "@@" on the input record indicates that a single record can contain multiple observations. Data are sorted by planting date, and then the REG procedure is used to fit the linear version of the logistic model to the data. The "by" statement in REG indicates that separate analyses are to be done for each planting date. Then the MIXED procedure is used to perform a linear mixed model (covariance) analysis with a autoregressive structure for the error.

Residual-plot assessment, based on regression analysis, can be based on individual blocks or the combined data.

When there is a large number of data sets, it is not reasonable for the investigator to carry out the complete set of model assessments described in this chapter for each disease progress curve. Furthermore, it is very unlikely that a single disease progress model will be found to be the best model for every epidemic. However, it is important that a model is used that is reasonable for the epidemics, at least for most of them. It is also important that a population-dynamics model is used that is consistent with the biological properties of the disease. Because one ultimately needs to perform regression (or similar) analyses, researchers often find it satisfactory to simply use the residual plots for different transformations to find a reasonable model. For soybean bud blight, Almeida et al. (1994) make the case that the disease is monocyclic under the conditions in the study. Thus, the monomolecular model (equations 4.10–4.11) should be considered. Graphical appraisal of the disease progress curves showed that maximum estimated dy/dt (i.e., $\Delta y/\Delta t$) occurred very early in the epidemics (i.e., at low y), so that a Richards-type model with a shape parameter $0 \leq \eta < 1$ is a possibility. However, because y does not approach a maximum in several cases, it is difficult to make a decision on the appropriate model to use on the basis of the graphical plots of dy/dt versus y. The monomolecular model was found to be adequate for describing many of the disease progress curves, and the Gompertz model was also found to be adequate for several curves (including those not shown here). Although using the Richards model with η equal 1/4 or 1/2 would result in a good description of the data, these shape-parameter values are not commonly used by epidemiologists. Assuming the disease to be of the monocyclic type, we have chosen to use the monomolecular model here.

A final note on model selection for these epidemics is warranted. In many ways, the data in Fig. 4.17 are 'messier' than the tobacco etch data in Fig. 4.15. However, the soybean bud blight data are fairly typical example of epidemiological data in plant pathology. Often, the number of disease assessment times is not sufficient to observe the 'entire' epidemic. Plus, no single model is found to be completely satisfactory for all the data sets. Nevertheless, a reasonable model can often be chosen and used to characterize and compare the epidemics.

4.7.2.2 Fitting one or more disease progress models. For the purpose of analysis, we consider only linear models here, and fit equation 4.32a with $y^* = \ln[1/(1-y)] = -\ln(1-y)$ (i.e, the monomolecular transformation) to the data. We use weights equal to the inverse of the theoretical variance of y^* when y has a binomial distribution. That is, $w = (1-y)/y$. The SAS program is shown in Fig. 4.18. The fitted equations for the four planting dates ($PD_1 - PD_4$) based on using PROC REG are:

$$PD_1:\ \hat{y}^* = -0.62(0.079) + 0.025(0.0022)t, \quad R^2 = 0.908$$
$$PD_2:\ \hat{y}^* = -0.53(0.049) + 0.021(0.0014)t, \quad R^2 = 0.946$$
$$PD_3:\ \hat{y}^* = -0.43(0.043) + 0.017(0.0012)t, \quad R^2 = 0.939$$
$$PD_4:\ \hat{y}^* = -0.29(0.032) + 0.011(0.0009)t, \quad R^2 = 0.904$$

The fitted lines are given on the right-hand graphs in Fig. 4.17; backtransformation is done to obtain the predicted y values on the left-hand graphs. The residual plots indicate that the monomolecular model is reasonable for these data. All of these model fits are based on using the data from the four individual blocks at each time. Thus, a high R^2 indicates that the incidence values for a given block are consistent with the chosen disease progress model *and* that there is relatively little variation in incidence among the blocks at each time.

To recap, graphs of the data and of the residual plots after regression analyses indicate that the monomolecular model was a reasonable choice for the epidemics. We used weighted regression to adjust for the lack of constant variance. With this example, using weights only makes a small difference in the results, but we continue with the weights below. The reader can determine that the estimates of the rate parameters are higher, and the intercepts are lower, if no weighting is used with these data. The standard errors also are different. With this example data set, there is strong evidence that the rate decreases with increasing (i.e., later) planting date. Based on the calculations in previous sections of this chapter, readers should be able to predict when a certain disease incidence is reached (e.g., 0.80) based on the estimated parameters. If the epidemics are truly of the monocyclic type, then the rate ($r* = r_M$ here) is the product of inoculum amount (x') and the per unit probability per time of the inoculum causing an infection (φ). Later planting dates may be associated with lower numbers of vectors or lower numbers of vectors that are viruliferous (both components of x'), or with environmental conditions that result in a lower probability of a thrip transmitting the virus (a component of φ).

The estimated intercepts shown here are all negative. If one used the backtransformation (equation 4.13), one would obtain negative estimates of y_0. For example, for the first planting date, one would obtain $\hat{y}_0 = 1 - \exp[-(-0.62)] = -0.86$. This is epidemiologically meaningless, of course. However, the correct interpretation of a negative intercept is based on the proper use of the monomolecular model. We remind the reader that initial disease intensity for the monomolecular model can be 0, because, for this model, disease increase is not based on the how much disease is initially present. The first nonzero y is at some point along the time axis, but not necessarily at exactly $t = 0$, since the time of the first assessment is chosen by the observer. In the analysis, we used time after plant emergence as t. Because $y^* = 0$ when $y = 0$ for this model, one can set 0 on the left-hand side of equation 4.11, and rearrange to solve for the estimated "starting time" (t_0). In y^* notation, this is written as: $t_0 = -y_0^*/r_M = \ln(1 - y_0)/r_M$. With the first planting date, one obtains 0.62/0.025 = 24.8 days (about half way between the first and second assessment dates). So, when the estimate of y_0^* is negative for the monomolecular model, this simply means that the estimated start of the epidemic is at $t > 0$.

One could use t-tests to determine which slopes and intercepts are different from each other, or one could use covariance analysis to more formally test for effects of planting date on the intercepts and slopes (rates). However, the analysis shown so far does not take into account the potential for serial correlation of the error term, as indicated by nonzero correlation of the residuals. It is more difficult to test for serial correlation with replicated data at each time. In particular, procedures such as AUTOREG in SAS are designed for a *single* series of observations (residuals in this case) over time. The serial correlation correction methods in Madden (1986) for regression analysis also are all directly applicable to a single series, not for the situation with multiple observations at each time.

Statistical methods that can be used to determine serial correlation for the type of data in Fig. 4.17 are based on advances in theory and application of linear mixed models for data analysis (Littell et al., 1996, 2006; Piepho et al., 2004; Wolfinger, 1996). To understand some of the concepts behind this analysis, one needs to consider the four values of y^* at each time in Fig. 4.17 for one of the planting dates. They might appear to be four independent values of incidence, with no relation between times in the four replicate values.

In reality, the data from the four blocks actually comprise four (transformed) disease progress curves. One can think of these as four profiles of observations over time for each planting date. Each profile comes from a single experimental unit, which in this example is a plot in a single block with one specific planting date. So, one could have connected the points for each of the four blocks in the figures, but this is often not done because the investigator or reader typically is more interested in expected or mean results for a given treatment at a given time. One could determine the fit of a selected model to the data in each profile, determine the variation within each profile (i.e., residual variance or mean square error about the line), and determine the serial correlation of the errors. All of this, and more, can be done simultaneously with a mixed model (Garrett et al., 2004; Wolfinger, 1996).

The linear model in equation 4.38b for the covariance analysis of tobacco etch in pepper over time is expanded to incorporate the specific features of the experimental design and correct for serial correlation (if it exists). One way of writing the linear mixed model is:

$$y^*_{ijk} = a + A_i + B_k + \iota\, t_{ijk} + r*_i t_{ijk} + \varepsilon_{ijk} \qquad (4.39)$$

in which y^*_{ijk} is the transformed disease intensity (multiple infection transformation of incidence in this case; see equation 4.11) in treatment i (planting date here; $i = 1, \ldots 4$), block k ($k = 1, \ldots, 4$ here), and time j ($j = 1, \ldots 4$), a is a constant, B_k is the effect of the k-th block on transformed disease, A_i is the overall effect of the i-th planting date (or, in general, the i-th level of the experimental factor A) on transformed disease, ι is a slope (rate) parameter, $r*_i$ is the effect of the i-th planting date (i-th level of factor A) on the slope, and ε_{ijk} is the (random) error term (the difference between observed y^* and that determined by all the other terms in the model). Unlike the simpler regression or analysis of variance situations, it is assumed that the ε_{ijk} within a plot are not independent. In fact, the correlations of the error values within a plot (combination of planting date and block) between times could follow some complicated structure (Littell et al., 2006; Wolfinger, 1996), including first order autoregressive (as shown for the TEV).

Using this model is a form of covariance analysis for this repeated measures design. The term A_i could be equivalently written using y^*_{0i}; then it is obvious that this term is representing the effect of planting date on transformed initial disease incidence. The estimated transformed initial incidence is given by $a + A_i$, and the estimated slope by $\iota + r*_i$. Note that there is an i subscript on both A and $r*$; this signifies that both terms refer to planting date.

Block is often considered a so-called random effect factor (but not always; see Piepho et al., 2004), with an assumed normal distribution (with a mean 0 and variance σ^2_B), and treatment (planting date here) and time are considered so-called fixed effect factors (Littell et al., 1996; Schabenberger and Pierce, 2002). It is assumed that block and error are independent. We do not go into the concepts of random and fixed effects in any detail here, but simply state that with fixed effects, one is interested in how the specific levels of the factor studied (e.g., the four specific planting dates) affect the response variable (e.g., transformed diseased intensity). With random effects, the levels of the factor studied are only considered a sample from a much larger population, and one is not interested in how specific levels of the factor affect the response (e.g., one is not specifically interested in how block 1 affects transformed disease). Rather, one is interested in the variation (variance) of the effects of the factor on the response.

As with the simpler covariance analysis for tobacco etch, readers may find it useful to see equation 4.39 written in mnemonics, without subscripts:

YSTAR = CONSTANT + PLANTING DATE + BLOCK
+ TIME + (PLANTING DATE)
× TIME + ERROR

As noted above, the constant term and the error term are implied when writing models in many statistical programs. A slope parameter is indicated when there is a continuous variable in the model (time); also, separate slope-effect parameters are specified for each level of the factor if an interaction is specified with a continuous variable ((planting date) × time). In Fig. 4.18, the MIXED procedure has the code to fit equation 4.39 to the soybean bud blight incidence data. Because there are fixed- (planting date, time) and random-effects (block) terms in equation 4.39 (or in the pseudo-equation), it is considered a *linear mixed model*. Analysis based on this model can be called mixed-model analysis. (As an aside, we point out that even if the random block-effect term was not in the model, it would still be a linear mixed model if the error [ε] values within a plot were correlated, although we do not go into the theory here). Furthermore, because the model contains both continuous and factor (planting date) variables, the analysis can be termed a covariance analysis. Finally, because the data are collected over time, and time is a variable in the model, the analysis can also be called a repeated measures analysis. The most general label is a linear mixed model, and the covariance and repeated measures labels are special features of this particular linear mixed model.

The model statement in PROC MIXED specifies transformed disease incidence as a function of planting date, time, and the interaction of planting date and time (Fig. 4.18). The convention of MIXED is to specify all random-effect terms (except for the error) of the model (equation 4.39) using random statements (and *not* place these in the model statement). The weight statement specifies the weight variable (created in the data step). The possible serial correlation of the errors, or any special features of the errors, is specified with the repeated statement. When one is accepting the default assumptions about ε, as in the previous section for the tobacco etch epidemics, the repeated statement is not needed. The full statement here is:

```
repeated/subject = block * PD type = AR(1);
```

in which the statements after the "/" are important options. The subject statement defines the experimental unit (plot) where data are recorded over time; here, block and planting date together give a unique label for each plot ($4 \times 4 = 16$ unique plots). The type statement defines the type of serial correlation being hypothesized in the model. For this model fit, we adopted a first-order autoregressive correlation for the residuals, which is a widely-used specification. Finally, note that in the model statement, an option of ddfm = betwithin was specified. This is instructing MIXED to calculate the degrees of freedom for the repeated measures design with the so-called "Between-Within" method. This is the default for repeated measures models when there are no random effects, so we explicitly specify it here since there is a random effect in the model. In fact, there has been considerable recent advances in the proper determination of degrees of freedom for mixed models with random effects or repeated measures. Many of the 'conventional' rules about determining degrees of freedom are now considered inadequate. It has been argued that the so-called 'Kenward-Roger' method be used in general (this is obtained with the ddfm = KR option in MIXED) (Schaalje et al., 2002).

The linear mixed model analysis for the soybean bud blight in Fig. 4.17 indicated that the effects of planting date ($F = 9.8$; $df = 3, 12$), time ($F = 720.1$; $df = 1, 44$), and interaction ($F = 26.7$; $df = 3, 44$) were significant (all at $P < 0.001$). Thus, planting date affected the intercept and slope (rate). The estimate of the block variance (σ_B^2) was 0 in this example, and the estimate of the first-order autocorrelation (-0.076) was not significantly different from 0 on the basis of a standard normal test (see Littell et al., 1996). (As an aside, because the block variance was 0, the denominator *df* for the planting date factor test are larger than the conventional number). Because of the small level of 1st order autoregressive correlation (if any) and the lack of effect of block on y^*, estimates of the intercepts and slopes, determined from the model parameters, were very similar to those obtained with standard weighted regression with no block effects and no adjustments for serial correlation (see above). Just one example is given here. For planting date 3, the estimated slope is

$$\hat{\iota} + \hat{r}_{*3} = 0.0106 + 0.0065 = 0.0171.$$

As with the tobacco etch epidemics in pepper, a simpler way of obtaining the slopes and intercepts for each planting date is to take out the t term and specify the noint option; thus one would replace the model statement with:

```
model ystar = PD PD *t / noint solution
        ddfm = betwithin;
```

and rerun the program.

Using this approach, the results for each planting date are:

$$\begin{aligned}
PD_1&: \hat{y}^* = -0.63(0.064) + 0.025(0.0017)t,\\
PD_2&: \hat{y}^* = -0.53(0.056) + 0.021(0.0015)t,\\
PD_3&: \hat{y}^* = -0.43(0.048) + 0.017(0.0013)t,\\
PD_4&: \hat{y}^* = -0.30(0.031) + 0.011(0.0009)t,
\end{aligned}$$

Because the block variance is 0 here, one could obtain identical results as obtained with individual regressions (above) by specifying a separate error variance for each planting date, and ignoring serial correlation. This can be done by replacing the repeated statement with:

```
repeated/group = PD;
```

By not specifying a type in the statement, an assumption of independence is made. In general, however, if the block variance is not zero and there is serial correlation, one would not expect the model parameters or their standard errors to be the same as those obtained with linear regressions that do not take into account random effects or the repeated nature of the time variable.

Before moving on to further analysis of this data set, we point out that PROC MIXED can be used instead of PROC AUTOREG to adjust for serial autocorrelation with single disease progress curves (with no replications), such as the TEV example in Fig. 4.13. The statistical method of estimating ρ is different (at least with default options for the two procedures), however, and estimates could be different for the two procedures for small data sets.

4.7.2.3 Comparing models with different error (residual) variance-covariance structures. Because the estimate of ρ was very close to 0, one might believe that that there is no serial correlation. This would not necessarily be correct. The estimate of ρ in the MIXED procedure is based explicitly on fitting a power-law model between error autocorrelation and temporal lag (difference in time) *within* each combination of planting date and block, $\rho^{|j-j'|}$. This means that the correlation declines towards 0 at increasing lags. Put another way, ρ is not just calculated from the residuals and assumed to be of the autoregressive form, but the autoregressive form is part of the mixed model. If there is correlation of the errors that is not of this form, then the estimate of ρ could be (very) inaccurate. There are many other possible autocorrelation structures (Wolfinger, 1996) that can now be considered and compared, thanks to advances with linear mixed models. One simple correlation model we consider here is referred to as 'compound symmetry' (CS). In this case, the serial correlation in error between *any* two times is constant, so that $\text{corr}(\varepsilon_j, \varepsilon_{j'}) = \rho$, no matter how far apart the times are. This structure for serial correlation can be incorporated into the program of PROC MIXED by changing the repeated statement in Fig. 4.18 to:

```
repeated/subject = block * PD type = CS;
```

In this formulation, the parameter estimated is the covariance between times ($\sigma_{jj'}$) *within* a plot (planting date and block combination), equivalent to the *between*-plot (but within a block) variance when the covariance is nonnegative. The estimate of this correlation is given by the estimate of ($\sigma_{jj'}$) divided by the estimate of the total variance (here, total variance is residual variance plus the covariance). Fitting this model to the soybean bud blight data gives an estimate of ($\sigma_{jj'}$) equal to -0.00136. The correlation estimate is equal to $-0.00136/(0.0131 - 0.00136) = -0.116$. Thus, the average serial correlation in this model is small and negative. The parameter estimates and their standard errors, as well as *F*-tests and other statistics, are similar to those obtained with the first-order autorgressive error structure. (As an aside, it should be noted that within-plot covariance can also depend on the variance between blocks, but since this is estimated as 0 here, then there is no block effect on results).

The relationship, if any, in the errors over time (within plots), as well as the variability of errors within each plot (within plots) can be totally specified with a variance–covariance matrix of the errors, with the serial correlations determined from the variances and covariances. For simplicity of expression, we refer to the lack of independence of the errors as the error structure (which is really the variance–covariance structure). The fitting of a model with two different error structures (first-order autoregressive and compound symmetry) exemplifies a common and valuable practice in fitting linear mixed models to data. That is, where appropriate (e.g., in repeated measures designs), several possible error structures are evaluated for the fit of the model (Littell et al., 1996). This relatively new concept—that is possible because of advances in statistical algorithms—is analogous to fitting different disease progress models to a single data set consisting of one *y* value at each time. With mixed models, one can decide which model provides the best fit by using the log-likelihood ($\ln L$) obtained from the fit of each model. If the error structure in one model is a special case of a more general structure, then the difference in $\ln L$ values can be used in a likelihood ratio test to determine the model fit [see Hughes and Madden (1995, 1998) for a general discussion of this approach for another type of problem]. However, very often the variance–covariance error structures in the models are not special cases of each other. This is the case with the compound symmetry and autoregressive errors (Wolfinger, 1996). One valuable approach for comparison purposes is to calculate a version of Akaike's Information Criterion (AIC), which is written as:

$$\text{AIC} = -2\ln L + 2q,$$

in which q is the 'effective' number of random-effects parameters in the model (i.e., number of nonzero variance or covariance parameters). The model with the smallest value of AIC is considered to provide the best fit. For the model considered here, q could be 3, because the variance parameters are σ_B^2 (block variance), σ^2 (residual variance), and either $\sigma_{jj'}$ (compound symmetry) or ρ (first order autoregressive). However, because the block variance was estimated as 0, $q = 2$. PROC MIXED automatically calculates AIC for any model. For the two model forms considered here, AIC was -103.4 for first-order autoregressive and -104.3 for compound symmetry. Thus, the latter error structure was (somewhat) more appropriate.

There are numerous error variability structures that can be considered (see SAS manual for MIXED). One very common one (additional to those already used here) is known as unstructured. In other words, for the transformed within-plot disease values, a separate variance is determined for each time and a separate covariance is calculated for each pair of times. With four times, there are $4 \times (4 + 1)/2 = 10$ within-plot variance–covariance parameters. This is a substantially larger number of random-effects parameters than for the other models, and it may not always be possible to fit such a model to data (i.e., the maximum likelihood procedure may fail to converge). One can fit the unstructured model to the soybean bud blight data by replacing the model statement in Fig. 4.17 with:

```
repeated/subject = block * PD type = UN;
```

With this error structure, an AIC of -113.2 was obtained, considerably lower than for the other models. Although not shown here, the covariance (and hence serial correlation) between times did not decline with increasing time lags or remain constant, indicating a much more complicated structure than specified by the previous two models. The *F*-tests were substantially different from those calculated with an assumed first-order autoregressive error structure. In particular, the test statistics for effects of planting date ($F = 19.7$; $df = 3, 12$), time ($F = 1585.4$; $df = 1, 44$), and interaction ($F = 74.6$; $df = 3, 44$) were all much larger than previously found. The reader should note that the test statistics for the intercept, slope, and interaction will not necessarily be larger when an unstructured error variance/covariance is assumed, although they were larger in this example. The important point is that choosing an appropriate error structure will result in more accurate tests of the effects of experimental factors (e.g., planting date) on the intercepts and slopes, and more realistic estimates of the parameters and their standard errors.

With the unstructured case, the following model fits were obtained for each planting date:

$$\begin{aligned} PD_1:&\quad \hat{y}^* = -0.71(0.061) + 0.028(0.0013)\,t,\\ PD_2:&\quad \hat{y}^* = -0.61(0.053) + 0.023(0.0011)\,t,\\ PD_3:&\quad \hat{y}^* = -0.48(0.046) + 0.018(0.0009)\,t,\\ PD_4:&\quad \hat{y}^* = -0.30(0.026) + 0.011(0.0005)\,t. \end{aligned}$$

The estimates of the intercepts and slopes were close to those obtained with the first-order autoregressive model form (see above). Moreover, the standard errors were all smaller, especially for the slopes. Thus, tests of the equality of pairs of intercepts or slopes could be affected. Note that because we now explicitly allow for unequal variances, we could certainly drop the weighting, although we do not do so here (by using consistent weighting for all modeling, it is easier to compare AIC values).

The reader should realize that often AIC does not decline when a model with unstructured error is fitted to data compared with a simpler error structure. Specifically, although $\ln L$ will increase with the more complicated model (by definition), the increase in q can more than offset this larger $\ln L$ (or smaller $-2\ln L$), resulting in a larger AIC. There are many error structures to choose from for repeated measures designs that allow for more complicated error structure but require fewer parameters than the unstructured choice (Littell et al., 2006; Wolfinger, 1996). We will not discuss these further here. Readers should also note that for the autoregressive error structure (but not the others described above), a more complete model would include a u_{ik} random-effects term in equation 4.39 (representing overall plot variability). This is specified in the MIXED procedure with `random block block*PD;` in addition to the other statements. This made no difference in the example analysis.

4.7.2.4 Summary of model fitting and comparisons. When data are from a designed experiment, with multiple blocks (and possibly other random-effect factors), one should strongly consider using linear mixed models for analyzing the data. Compared with the use of nonmixed models, this will lead to more accurate tests for significance and more accurate estimates of standard errors of estimated parameters (Schabenberger and Pierce, 2002). For characterizing disease progress, it is first necessary to find an appropriate population dynamics model. It may be difficult to find a single model that is satisfactory for all observed disease progress curves from an experiment, and some compromise over model selection may be necessary to find a generally reasonable model. Model selection may become overly tedious if one works with the data individually for each block of each level of each experimental factor, but residual plots from regression analysis may give enough information to select an appropriate model. The monomolecular and logistic models are simple enough to be easily applied in the analysis of disease progress curves, yet sufficiently detailed to embody some key epidemiological concepts, and thus usually deserve serious consideration.

A model such as equation 4.39 can then be used to determine the effects of the experimental factor on the intercept (transformed disease intensity at time 0) and slope (rate of increase on a transformed scale). This is done by testing for, and then estimating, the main effects of the experimental factor (a class variable) and time (a continuous variable), as well as their interaction, on y^*. Weighting can be used to correct for nonconstant variances. With modern linear mixed statistical programs (such as PROC MIXED in SAS), one can simultaneously estimate the variances-covariances (correlations) of the errors within plots over time, and incorporate other parameters for representing this correlation (Garrett et al., 2004; Schaalje et al., 2002). It is natural to consider a first-order autocorrelation error structure, because this is consistent with the assumed error structure in many other regression algorithms that characterize data over time. However, more complicated structures, especially those that allow for unequal variances, may be more appropriate. More research is required in this area.

4.7.2.5 General repeated measures analysis. Sometimes an investigator collects data over time but does not want to characterize the epidemics with parameters for transformed initial disease intensity and rate of increase on the transformed scale. This might be because disease was only assessed at a small number of times (e.g., three), making it difficult to choose a reasonable disease progress model. Or, the disease progress curves for the different levels of the experimental factor(s) (e.g., different treatments) were very different in shape, and no single model seemed satisfactory. Or, the disease progress curves were not typical of those resulting from the (relatively) simple models that are usually used for describing plant disease epidemics. An example of this would be a disease progress curve that first increases and then decreases over time. Then, general repeated measures analysis is an appropriate method of analysis. We call this 'general' here to distinguish it from more specialized analysis for repeated measures designs, such as the covariance analysis in the previous sub-section with time as a continuous variable.

For simplicity of presentation, we consider only a randomized complete block design with data collected over time in each plot, and continue to use the soybean bud blight example. In this case, an area is divided into blocks, and a single plot is randomly selected for each of the levels of the experimental factor within each block. Then, data are collected over time. The analysis here is, in many ways, conceptually the same as the covariance analysis presented in the previous subsection. The only difference is that time is now a so-called class variable (with discrete possible levels, corresponding to each assessment time) rather than a continuous variable. One now views the soybean experiment as a factorial, with fixed-effects factors of planting date and time, plus the random-effects factor of block. Time, however, is not the same kind of factor as an experimental treatment factor because of the possible correlations of errors over time within plots (see Brown and Prescott, 2001).

The linear mixed model for the soybean bud blight experiment can be written as:

$$y^*_{ijk} = a + A_i + B_k + T_j + AT_{ij} + \varepsilon_{ijk} \qquad (4.40)$$

in which y^*_{ijk} is the transformed disease intensity (multiple infection transformation of incidence in this case; see equation 4.11) in treatment i [planting date here; $i = 1, \ldots n_i$ (=4 here)], block k ($k = 1, \ldots, 4$ here), and time j [$j = 1, \ldots n_j$ (=4 here)], a is a constant, B_k is the effect of the k-th block on transformed disease, A_i is the overall effect of the i-th planting date (or, in general, the i-th level of the experimental factor A), T_j is the effect of the j-th time, AT_{ij} is the interaction effect of the experimental factor (e.g., planting date) and time, and ε_{ik} is the error term (difference between observed y^* and that determined by all the other terms in the model). It is assumed that the ε values are correlated within plots. All terms are for fixed effects except for B and ε. Compared to equation 4.39, T_j and AT_{ij} replace the terms for the overall slope and the effects of the experimental factor on the slope. Often, one uses a transformation of disease intensity that stabilizes the variances rather than a transformation based on a disease progress model. For binary data, such as plant disease incidence, the angular transformation is often appropriate. This transformation is written as $y^* = \arcsin(\sqrt{y})$. Readers should note that even with this transformation, the variance may still be nonconstant when y is very close to 0 and 1. Moreover, when incidence data are aggregated (see Chapter 9; Madden and Hughes, 1995), the angular transformation also may not produce a constant variance at all y values (Hughes and Madden, 1992). To be consistent with the previous sub-section, we use the monomolecular-based transformation for y, and use weights to adjust for the nonconstant variance.

The code with PROC MIXED of SAS is *almost* the same as for the earlier analysis:

```
class PD t block;
model ystar = PD t PD*t/ddfm = betwithin;
random block;
weight wt;
repeated/subject = block*PD type = un;
lsmeans PD t PD*t;
```

Note that t (time) is now defined as a class (factor) variable. Also, the solution option is not specified in the model statement; this could be kept, but the solution is not relevant here. Means for the main effects and interaction are specified with lsmeans ('least squares means'—estimated expected values based on the model parameters, not necessarily the same as the arithmetic means). For this example, we use the unstructured variability specification (un) for the errors based on the results in the section 4.7.2.3. We could have used ar(1), cs, or any of several other options (Kowalchuk et al., 2004; Littell et al., 2006; Wolfinger, 1996), and chosen the model with the lowest AIC.

Since the emphasis in this chapter is in characterizing disease progress curves, we considered different error structures when time was a continuous variable; then, we simply accepted this structure (un here) for the general repeated measures. It may actually be more advantageous to start with the general repeated measures (time as a class factor) and find the error structure with the lowest AIC; then use this structure for any covariance (continuous time) analysis (Wolfinger, 1996). Interested readers can try this with the data set.

The analysis indicated that planting date ($F = 360.7$; $df = 3, 13$), time ($F = 517.5$; $df = 3, 35$), and the interaction ($F = 25.1$; $df = 9, 35$) all had highly significant effects on transformed disease incidence. It should be noted that the numerator df (first of the pair of df values) for the F-test of time is 3, one less than the number of assessment times. (With continuous time, the numerator df was only 1). The variance for block was estimated as slightly larger than 0 (0.00018). Results from this type of analysis are presented graphically or in tabular form as means for each factor level at each time. These are shown in Table 4.3 (readers

TABLE 4.3. Means (and standard errors) of transformed disease incidence of soybean bud blight (see Fig. 4.17) for four planting dates based on fitting a linear mixed model (equation 4.40) to the data.

Planting date	*t = 30 days (j = 1)*	*t = 40 days (j = 2)*	*t = 50 days (j = 3)*	*t = 60 days (j = 4)*	*Mean*
1	0.12 (0.02)	0.42 (0.02)	0.73 (0.05)	0.81 (0.05)	0.53 (0.01)
2	0.10 (0.02)	0.37 (0.02)	0.61 (0.04)	0.66 (0.05)	0.43 (0.01)
3	0.08 (0.02)	0.27 (0.02)	0.49 (0.04)	0.53 (0.04)	0.34 (0.01)
4	0.02 (0.01)	0.12 (0.01)	0.24 (0.02)	0.35 (0.03)	0.18 (0.008)
Mean	0.08 (0.01)	0.30 (0.01)	0.52 (0.02)	0.59 (0.02)	

Standard errors vary with the planting date and time because of the weighting in the mixed model analysis and also because an unstructured error variability was specified.

should now know how to convert the presented estimates of y^* back to estimates of y). Because of the interaction, one compares means among the planting dates at each time separately, or compares means among times at each planting date separately. It is typical in this kind of analysis for there to be no significant difference among means for factor levels at the early assessment times, because disease intensity starts low for all treatments in most studies. Then, at later times there are more substantial differences among the treatment means. This is a reason for a significant interaction. This scenario was found with the current example. When there is no interaction, one compares the main-effect means across all times or all planting dates, depending on which factors were significant. Note that in cases where disease intensity approaches 1 towards the final assessment times (this is not the case here) there may be no significance differences among treatment means.

Fitting equation 4.40 to data requires computer software that allows flexibility in defining the error structure. Before the development of programs such as MIXED in SAS, the common statistical model for a general repeated measures was of the form:

$$y^*_{ijk} = a + A_i + B_k + AB_{ik} + T_j + AT_{ij} + \varepsilon_{ijk} \quad (4.41)$$

in which AB_{ik} is the interaction of block and the experimental factor (a random-effects term), which is also considered to be the proper error term for testing the effects of planting date (A) on y^* (Madden, 1986). Equation 4.41 is, essentially, an analysis of variance (ANOVA) model. Here, at least initially, it *is* assumed that the εs are independently distributed over time (with a constant variance). In this model, there are three random-effects terms, corresponding to B, AB, and ε, and it is assumed that they are independent. Equation 4.41 is also appropriate for a split plot (split block) design, in which T corresponds to the 'sub-plot' factor that is randomized within the 'whole plots' (A). It can be shown that use of equation 4.41 is equivalent to using equation 4.40 with compound symmetry (cs) for the error structure (Littell et al., 1996) as long as the within-plot covariance is not negative. This is a very restrictive assumption for repeated measures, because under compound symmetry the correlation of the error between any two disease intensities [within a plot (the primary experimental unit)] does not depend on the time period between the two disease values. Because of this, specialized statistical methods were developed and incorporated into some ANOVA procedures (e.g., PROC GLM in SAS) to evaluate whether or not compound symmetry was reasonable based on the data collected over time; when there was evidence that compound symmetry was not appropriate (e.g., when an autoregressive error structure was reasonable), some *ad hoc* adjustments were calculated to improve the accuracy of F-tests. Equation 4.41 is still routinely used by many, even though it is quite limiting in terms of the error structure. In general, use of equation 4.40 is preferable if one has access to statistical software that allows flexible specification of ε in a mixed model framework.

From an epidemiological perspective, characterizing epidemics in terms of estimated rate and initial disease parameters (equation 4.39) is more intuitive than characterizing epidemics in terms of a general time factor and the interaction of time and experimental factors (equation 4.40). However, as mentioned at the start of this subsection, there are some circumstances where the latter approach is of value, especially when it is not realistic to characterize the observed epidemics with one of the standard disease progress models or when there are a small number of assessment times. When there are many assessment times, repeated measures analysis is possible but can be quite difficult to interpret, and presentation of the results is more cumbersome. For instance, means of two treatments could be significantly different at times 2, 3, 7, and 9, but not at other times. There is no simple way to relate this result to the dynamics of disease development. Furthermore, when there are many experimental factors, it also can be extremely difficult to interpret the F-tests for main effects and two-way, three-way, and higher-order interactions. As discussed next, there are alternatives to the general repeated measures analysis that do not require the use of one of the disease progress models of section 4.4.

4.7.2.6 Area under the disease progress curve. A simpler empirical approach than using repeated measures analysis is to represent the profile of disease observations in each plot [i.e., for each combination of block and level of the experimental factor (e.g., planting date)] with a summary variable. A very common and useful summary variable is the *area under the disease progress curve (AUDPC)* (Jeger, 2004). It should be noted that the concept of using area under a curve is common in many disciplines where data are collected over time (see Brunner et al., 2002). As with general repeated measures analysis, analysis of AUDPC may be beneficial when the observed disease progress curves are not typical of the ones that can be described with the common population dynamics models presented in this chapter, or when the disease progress curves do not have fairly consistent shapes in the different plots. Unlike the situation with general repeated measures, however, use of AUDPC can also be beneficial when there are many assessment times, since the observations are summarized by a single number.

There is more than one way to estimate area under a curve. A common approach with observed disease progress curves is to use the so-called mid-point or trapezoidal method. The concept is to break up a disease progress curve (or, in fact, any curve of interest) into a series of trapezoids, calculate the area of each, and then add up the areas. If l is the length of a trapezoid and the short and tall heights are, respectively, h_1 and h_2, then the

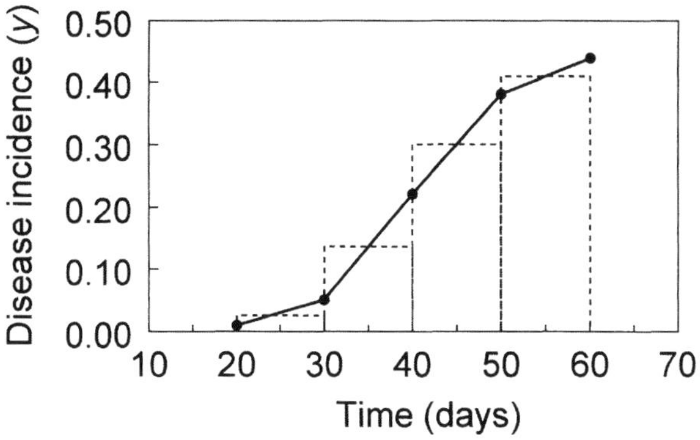

FIG. 4.19. Incidence of soybean bud blight in a single plot, corresponding to planting date 3 and block 3 in Fig. 4.18. Broken lines are the rectangles used for calculating area under the curve.

area is $[(h_1 + h_2)/2] \cdot l$. The area of a trapezoid is equivalent to the area of a rectangle $(h \cdot l)$ with length l and height $h = (h_1 + h_2)/2$. The AUDPC calculation is demonstrated in Fig. 4.19 for soybean bud blight progress in one plot. The observed (t, y) values are: (20, 0.01), (30, 0.05), (40, 0.22), (50, 0.38), and (60, 0.44). Note that because there was a nonzero incidence at $t = 20$ days in this plot, results from five times are being used (the first time was not used in the above modeling because sometimes there were no visible infections at $t = 20$). With five times, there are four areas that are calculated (i.e., one less than the number of times), for time periods of 30−20, 40−30, 50−40, and 60−50. Thus, the length of each rectangle in this example is $l = 10$ (days). Now consider the area under the curve between the last two times. We use the midpoint of the last two y values, $(0.38 + 0.44)/2 = 0.41$, as the height (h) of the last rectangle in Fig. 4.19. The area of this rectangle (or the corresponding trapezoid) is $0.41 \times 10 = 4.1$. The total area under the curve simply involves adding up all the areas.

For a single disease progress curve, the mathematical formula for calculating the AUDPC is:

$$\text{AUDPC} = \sum_{j=1}^{n_j-1} \left(\frac{y_j + y_{j+1}}{2} \right) \left(t_{j+1} - t_j \right) \qquad (4.42)$$

where j is the order index for the times (and n_j is the number of times). The total area for the example is calculated using this equation as:

$$[((0.01 + 0.05)/2) \cdot 10] + [((0.05 + 0.22)/2) \cdot 10] + [(0.22 + 0.38)/2) \cdot 10] + [((0.38 + 0.44)/2) \cdot 10] = 8.75.$$

As an exercise, the reader can confirm that the area under the tobacco etch disease progress curve in Table 2 (and Fig. 4.12) is 25.3. Units for AUDPC are intensity · days (or whatever time units are being used). If disease intensity is a proportion (as is typical), then the units are time.

Equation 4.42 is based on the assumption that there is a straight-line change in y between times. Of course, the evidence is strong that y rarely changes in a straight-line fashion over time, based on all the realistic disease progress models that represent epidemics and our theoretical understanding of epidemics. However, with short intervals between disease assessments, the straight-line assumption is reasonable. Thus, the accuracy of equation 4.42 increases with increasing number of assessment times (over a fixed total time period). Researchers often use equation 4.42 for both small and large numbers of assessment times. With a small number of times, the calculated AUDPC may not be an accurate estimate of the true (but unknown) area, but it may still be a useful summary variable for comparing epidemics.

The linear mixed model for analyzing AUDPC is simpler than the ones given above. This may be one of the reasons why many investigators calculate and compares AUDPCs. Because there is only a single number for each experimental unit (e.g., a plot), there is no issue with serial correlations within experimental units. A model for the soybean bud blight experiment is:

$$\text{AUDPC}_{ik} = a + B_k + A_i + \varepsilon_{ik} \qquad (4.43)$$

in which AUDPC_{ik} is the area for planting date i and block k. Here it is assumed that ε is normally and independently distributed with constant variance. The assumption of constant variance can be easily relaxed to accommodate a variance proportional to AUDPC; then weighting would be used. Fitting this model to the data, it was found that planting date had a highly significant effect on AUDPC ($F = 83.1$; $df = 3, 9$). Mean areas (and standard errors) for planting dates 1 through 4 were: 13.06 (0.36), 11.1 (0.36), 9.2 (0.36), and 5.3 (0.36). These were all significantly different from each other.

The areas in the example are directly related to the magnitude of the estimated r_M parameters for these four planting dates (see above). This makes sense because the higher the rate, the greater y will be at any time (for a single y_0), and thus the greater the AUDPC. Although it cannot be seen in this example because y_0 is essentially 0 (because of negative intercept on the linear scale; see section 4.7.2.2), increasing y_0 also leads to increasing AUDPC. In fact, there is an exact relationship between AUDPC and the parameters of the disease progress models presented in this chapter. We discuss this in general and then give specific results for the logistic and monomolecular models.

One can mathematically write the area as AUDPC = $\int y(t)dt$. By substituting the logistic (equation 4.15) or some other population growth models for $y(t)$, it is possible to perform the integration analytically and express AUDPC as a function of total time of the epidemic and the epidemic parameters y_0 and r_*. Some algebraic manipulation of the integration solution can produce an equation written terms of y at the first and last assessment times (Jeger and Viljanen-Rollinson, 2001). For the logistic model, AUDPC is equal to:

$$\text{AUDPC} = t_F - t_0 + \left(\frac{\ln(y_0 / y_F)}{r_L} \right) \qquad (4.44)$$

in which t_0 and t_F are the initial and final times, y_F is the final disease intensity, and y_0 and r_L are the logistic parameters as defined previously (see equation 4.15). Thus, if disease increase is logistic, one can determine AUDPC using estimates of the logistic parameters, the final observed disease intensity, and the time period over which disease was assessed. Even y_F can be estimated by using the estimates of y_0 and r_L to determine disease intensity at t_F [i.e., $y_F = y(t_F)$]. Although it appears in equation 4.44 that AUDPC would decrease with increasing r_L (because the rate is in the denominator), this is not true. The reason is the last term in equation 4.44 is a negative number (because $y_0/y_F < 1$), meaning that an increase in r_L leads to an increase in area. Moreover, it should be noted that y_F will also change with changing rate for a given final assessment time. Using equation 4.44 for the tobacco etch epidemic, one finds that AUDPC is predicted to be:

$$(49 - 0) + [\ln(0.0014/0.999)/0.278] = 49 - 23.6 = 25.4.$$

This is almost identical to the value obtained with the observed ys and equation 4.42. The strong agreement is because the logistic model provided a very good fit in this example.

If one assesses disease intensity for long enough (i.e., t_F is a large number), final disease intensity ultimately will reach 1 if the logistic model is correct. Then, the last term in equation 4.44 becomes a constant, $\ln(y_0)/r_L$. Once this happens, AUDPC increases in a direct linear manner with increasing t_F. For instance, if t_F was increased by 7 days to 56 (and not 49) with the tobacco etch epidemic in Table 4.2, one would calculate AUDPC as $56 - 23.6 = 32.4$, which is simply an increase of 7 units over the value calculated above. This time dependence is a good reason to scale the calculated AUDPC values (AUDPCS or AUDPC*)—either estimated from model parameters or from the mid-point rule—by dividing by the time duration $(t_F - t_0)$. Then, one can write:

$$\text{AUDPC}^* = \frac{\text{AUDPC}}{(t_F - t_0)}.$$

In general, one should use the scaled version of the area whenever one is comparing epidemics over different time durations.

For the monomolecular model, the area under the curve is given by:

$$\text{AUDPC} = t_F - t_0 + \left(\frac{y_0 - y_F}{r_M} \right) \qquad (4.45a)$$

This model has the same qualitative properties as equation 4.44. Quantitatively, there are some differences in that the area is based on the difference of initial and final disease intensity and not on the logarithms of the ratio. There is also one special feature that must be considered for monocyclic epidemics. As shown above for the soybean bud blight example, *estimates* of y_0 can be negative, which really means that the predicted start of the epidemic is at a time after $t = 0$ (i.e., $t_0 > 0$) and that $y_0 = 0$. Then, equation 4.45a reduces to:

$$\text{AUDPC} = t_F - t_0 - \left(\frac{y_F}{r_M} \right) \qquad (4.45b)$$

For the example data in Fig. 4.19, and using $\hat{r}_M = 0.0156$/day, and $\hat{y}_0^* = -0.351$ (obtained in a separate regression analysis for just this single disease progress curve, based on the *five* assessment times), one obtains $\hat{t}_0 = 22.5$ days. Using equation 4.45b, AUDPC $= (60 - 22.5) - (0.44/0.0156) = 9.3$. This is fairly close to the AUDPC of 8.75 calculated above directly from the data (without curve fitting). The reason for the discrepancy is that the monomolecular model did not provide a perfect fit to the data.

AUDPC is a very convenient summary of a plant disease epidemic that incorporates initial intensity, the rate parameter, and the duration of the epidemic (which determines final disease intensity). The essential point that can be derived from equations 4.44 and 4.45a is that one does not need to empirically determine AUDPC from a profile of y values over time (equation 4.42) because one can calculate the area from the parameters of a population-dynamic model of the epidemic. In fact, with only a small number of assessment times, it may be more accurate to fit the logistic model to the data and use the estimated parameters to determine AUDPC than to directly determine AUDPC as the summation of areas (equation 4.42) (Jeger and Viljanen-Rollinson, 2001). Of course, equations 4.44 and 4.45a are based on the assumption that disease progress is adequately described by the chosen model. If there is no credible evidence that one of the standard models fits the observed disease progress data, or that the disease progress data varies substantially in shape for different factor levels, then the empirical determination of AUDPC is justified.

4.7.2.7 Some other approaches. At the risk of being excessive, some other approaches for analyzing epidemics from designed experiments will be briefly considered here. One can revert to treating time as continuous rather than as a class (factor) variable, and once again look at fitting disease progress models (e.g., logistic) to the data. It is possible to fit a model separately to the data from *each* plot and each block, and then to use a linear mixed model to determine if the factor levels affected results. For instance, one can determine $r*$ for each of the 16 (4 blocks × 4 planting dates) disease progress curves in Fig. 4.17, and then use equation 4.43

(with r_{ik}^* substituted for $AUDPC_{ik}$) to statistically determine the effects of planting date on the rate. One could then perform the analysis on the transformed initial disease values. Fortunately, there is a more efficient way of carrying out this analysis that does not individually fit a model to separate disease progress curves followed by ANOVA or linear mixed model analysis. The approach is based on so-called *random coefficients models*, which has a strong connection to Bayesian analysis (Littell et al., 2006; Wolfinger, 1996).

With this approach, one is implicitly assuming that the y_0^* and $r*$ values are not constants (to be estimated) but a random sample from a population of possible intercept and slope values. Using this concept, the random coefficients model for the soybean bud blight data set can be written as:

$$y_{ijk}^* = a + A_i + u_{ik} + B_k + \iota\, t_{ijk} + r*_i t_{ijk} + v_{ik} t_{ijk} + \varepsilon_{ijk} \quad (4.46)$$

in which u_{ik} and v_{ik} are random variables, with variances of σ_u^2 and σ_v^2, respectively, and covariance σ_{uv}. The mean of both u and v is 0. It is usually assumed that here that the error (ε) is independently distributed over time, and that ε is independent of the other random terms. However, a consequence of the additional variance terms is that the y^* values within plots are correlated, with possibly complicated structure. One can fit this model to the soybean data using PROC MIXED of SAS. Thus, one does not have to first calculate slope and intercept values for each epidemic, store the results, and analyze the effects of planting date (or other factor) on them in a separate step. Instead, one can do all the calculations with one step. Interested readers should read Littell et al. (1996) or Littell et al. (2006) to learn how to fit equation 4.46 to data.

One can compare the fit of equation 4.46 to others such as equation 4.39 through the AIC values. Thus, equation 4.46 can compete with others as being the most appropriate for representing the data. For the soybean example, AIC = −110.7, smaller than found with outoregressive or compound symmetry models, but considerably larger than found for the unstructured errors. Although the random coefficients model is very flexible, it also is known to produce conservative (i.e., excessively large) estimates of the standard errors of model parameters (e.g., slopes). Thus, in using this model, one is likely to find an artificially small number of significant differences. Although the random-coefficients approach makes some intuitive sense, it is probably desirable to use equation 4.39 (covariance and repeated measures analysis, with appropriate error structure), because of its higher power.

One can also use nonlinear models to determine the effects of factors in designed experiments (see Chapter 8 in Schabenberger and Pierce, 2002). However, the methodology is quite complicated, and it is not possible to fully address many of the features that can be handled with linear models, such as multiple sources of variation. With more complicated population-dynamic models, researchers do not have the option of using linear versions for analysis. In these cases, direct use of nonlinear models is required. Comparisons then are typically based on fitting the model individually to the data for each treatment (e.g., each planting date) and comparing epidemics with *t*-tests of pairs of estimated parameters. One can fit the nonlinear model to data in each plot (each combination of treatment and block), and then fit a linear mixed model to the estimated parameters (such as the slopes). As mentioned in the previous paragraph, this is analogous to performing a random coefficients analysis (Wolfinger and Lin, 1997). Although the approach is conservative (in a statistical sense), there are fewer alternatives for analysis of this type within the nonlinear framework than are available for linear models.

All of the analysis in this section is based on an assumed normal distribution for the dependent or response variable (y^* or y). As discussed in section 4.6.4, the normal distribution is an approximation for disease incidence because of the binary nature of the variable. The normality approximation of the binomial distribution is reasonable when assessing relatively large numbers of plant individuals per plot (the basic experimental unit), but will be unreasonable when incidence is based on small numbers of individuals. Generalized linear models (GLMs) and generalized linear mixed models (GLMMs) can and probably should be used instead of equation 4.39 for analyzing disease progress curves obtained in designed experiments (Littell et al., 1996; Madden et al., 2002). We do not give any detail here because there has been little consideration of these non-normal models for characterizing epidemics in the general sense. Detailed information on the use of GLMs and GLMMs for non-temporal disease data has been published (Hughes and Madden, 1995; Madden et al., 2002).

4.8 Models with Maximum Disease Intensity as a Parameter

4.8.1 General concepts

In most of the disease progress models discussed so far, two parameters were used to characterize an epidemic, y_0 (initial disease or constant of integration) and $r*$ (rate). Through these two parameters, two or more epidemics can be compared and a considerable understanding of disease dynamics is achieved. However, not all epidemics can be described by models with two parameters. Through the Richards model, we have already showed that an additional parameter is needed to account for disease progress curves that do not have the same basic shape. In the next chapter, we develop a general theory of plant disease epidemics that involves additional parameters and equations that have direct biological interpretation. Here, we deal with one specific generalization of the models presented in section 4.4.

The versions of the logistic, monomolecular, and other models for epidemics presented in previous sections are based on the assumption that maximum disease intensity is 1 (100%). This can be seen by the "$1 - y$" or similar [e.g., "$\ln(1) - \ln(y)$"] term for disease-free (healthy) individuals or area in the models. The "1" is the maximum of disease intensity on proportion scale. However, it is well known that maximum disease intensity can be less than 1 (Gilligan, 1990; Kranz, 2003; Madden and Campbell, 1990). Some of the disease progress curves in Fig. 4.1 clearly show the behavior of intensity leveling off before 100% intensity is reached. With some rust diseases, only a certain percentage (around 37%) can be visibly diseased. If this maximum was, in fact, fixed, so that it applied to all epidemics of a given disease, then one could simply define y as the observed severity divided by the known maximum severity. However, diseases do not always have a fixed or repeatable maximum severity. This maximum (denoted here K) then becomes a parameter in a disease progress model. There may or may not be a mechanistic interpretation for $K < 1$. As presented in Chapter 5, theoretical population-dynamics theory leads to a maximum y less than 100% for polycyclic diseases. However, it is also possible that unfavorable environment, lack of insect vectors (for pathogens obligately transmitted by vectors), or other unspecified physical or biological factors may cause dy/dt to decline to 0 before y reaches 1.

The approach discussed here is applicable to all of the population-dynamic models discussed in this chapter (see also Madden and Campbell, 1990), and is valid whether or not there is a clear reason for K being less than 1. For brevity we focus on logistic disease increase. In Table 4.4 we give model forms for other cases. With the symbol K representing the maximum possible disease intensity for a given epidemic ($K \leq 1$), then $K - y$ is a measure of how far disease intensity is from it maximum (on a proportion scale). Although $1 - y$ represents the disease-free plant individuals (e.g., healthy leaf area), $K - y$ represents the disease-free plant area or individuals that *could* become diseased. For this situation, the logistic model can be written in various ways, such as:

$$\frac{dy}{dt} = r_L y\left(\frac{K-y}{K}\right) = r_L y(1 - y/K) = r_L(y/K)(K-y) \quad (4.47)$$

At very small y, the term in parentheses in the first two versions of equation 4.47 is almost 1; then the logistic model reduces to the exponential ($r_L y$), as was shown previously for the case of maximum disease intensity equal to 1 (i.e., implicitly, $K = 1$). Furthermore, when y increases and comes close to K, y/K approaches 1, and the logistic model reduces to $r_L(K - y)$. This latter expression is the general (K) form of the monomolecular model. Integrating equation 4.47 results in:

$$y = \frac{K}{1 + \exp\left[-\left(\ln[y_0/(K - y_0)] + r_L t\right)\right]} \quad (4.48)$$

At large values of t (depending on the other parameters), y gets very close to K. However, it never quite reaches this value. Thus, K is known as an asymptote. Numerically, y can be so close to K (say, to five or six decimal places) that, through rounding, y appears to equal K.

TABLE 4.4. Versions of four disease progress models when maximum disease intensity (K) is a parameter.

Model	*dy/dt* =	*y* =	*y** =
Monomolecular	$r_M(K-y)$	$K\left(1 - Be^{-r_M t}\right)$	$\ln\left(\frac{K}{K-y}\right)$
Logistic	$r_L y\left(1-(y/K)\right)$	$K\left(1 + Be^{-r_L t}\right)^{-1}$	$\ln\left(\frac{y}{K-y}\right)$
Gompertz	$r_G y(\ln(K) - \ln(y))$	$K\exp\left(-Be^{-r_G t}\right)$	$-\ln\left(-\ln(y/K)\right)$
Richards ($\eta < 1$)	$\frac{r_R y\left(K^{\eta-1} - y^{\eta-1}\right)}{(\eta-1)K^{\eta-1}}$	$K\left(1 - Be^{-r_R t}\right)^{1/(1-\eta)}$	$\ln\left(\frac{1}{1-(y/K)^{1-\eta}}\right)$
Richards ($\eta > 1$)		$K\left(1 + Be^{-r_R t}\right)^{1/(1-\eta)}$	$\ln\left(\frac{1}{(y/K)^{1-\eta} - 1}\right)$

The definition of B depends on the model: $(K - y_0)/K$ (monomolecular), $(K - y_0)/y_0$ (logistic), $-\ln(y_0/K)$ (Gompertz), $1-(y_0/K)^{1-\eta}$ (Richards, $\eta < 1$), and $(y_0/K)^{1-\eta} - 1$ (Richards, $\eta > 1$).

With some algebraic manipulation, a straight-line equation can be obtained from equation 4.48 as:

$$\ln\left(\frac{y}{K-y}\right) = \ln\left(\frac{y_0}{K-y_0}\right) + r_L t \qquad (4.49)$$

It should be noted that in the logit transformation, K replaces 1 for this general situation. The transformation can also be written as

$$\ln\left(\frac{y/K}{1-(y/K)}\right)$$

which should help make it clear that the transformation involves expressing y as a fraction of its maximum. The relevant equations for the other models are given in Table 4.4.

The situation with a maximum disease intensity less than 1 is demonstrated in Fig. 4.20. The absolute rate increases and decreases symmetrically around a maximum; disease intensity increases in a sigmoid manner, and levels off as it approaches K. This brings up one major comment: the $dy/dt : t$ and $y : t$ curves have the

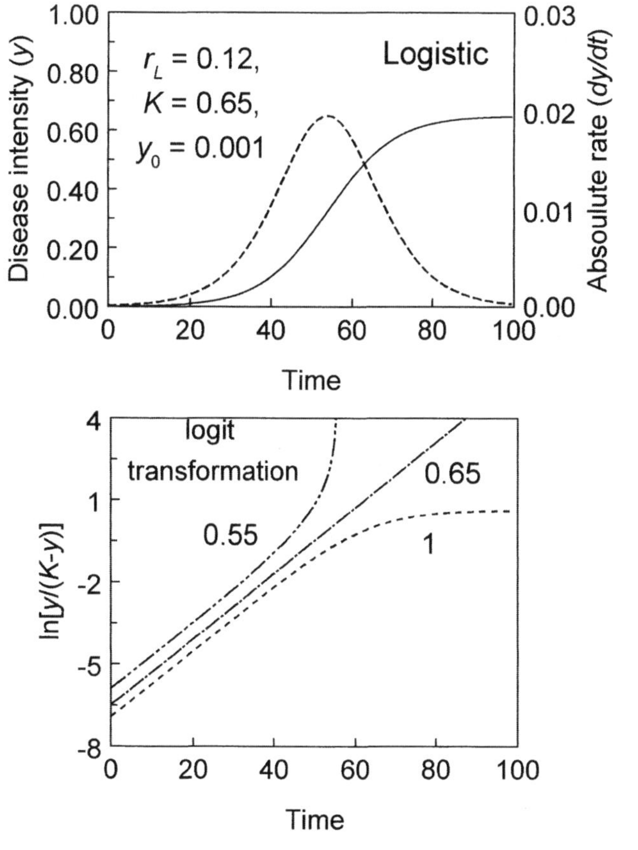

FIG. 4.20. Theoretical epidemic corresponding to a logistic model with maximim y (K) less than 1 (see equations 4.47–4.49). Upper frame: y versus t (solid line) and dy/dt versus t (dashed line). Lower frame: logit [$\ln(y/(K-y))$] versus t, for three values of K (indicated on the graph).

same general shape with $K < 1$ as they do with $K = 1$. This applies to all the population-dynamic models presented in section 4.4. In Fig. 4.20, the maximum dy/dt does not occur at $y = 1/2$, but at $y = 0.325$. This value of y equals $K/2$. Stated another way, the maximum rate for the logistic model occurs when $y/K = 1/2$. For the Gompertz model, the maximum rate occurs when $y/K = 1/e \approx 0.37$, or when $y = K/e$. For the general Richards model, the maximum rate occurs at: $y = K\eta^{1/(1-\eta)}$.

The lower frame in Fig. 4.20 shows the logit transformation of y for three K values, when the true $K = 0.65$. If one used $K = 1$ in the logit transformation (i.e., one used the 'standard' equation 4.16), the line is originally straight but then curves towards the right as y approaches 0.65. If one used $K = 0.55$ in the logit transformation, the obtained line is also straight originally, and then curves upward as y nears K. In this case, the logit is undefined at $y > 0.55$. If one used $K = 0.65$ (the correct value), then the line is straight over the entire range. An important point here is that the choice of K has little effect on the logit until y is close to K. This is especially true when a value of K larger than the true one is used. If one did a regression analysis on the data from $t = 0$ to $t = 60$ (in this example), one would obtain roughly the same r_L estimates using a wide range of K values between 0.65 and 1.0. This is analogous to the situation where one can use the exponential model as a good approximation of the logistic model when y is low. The intercept is also affected by the choice of K (Fig. 4.20). This is because the intercept is now a function of initial disease and K (equation 4.49). It is helpful that K does not have much of an effect on the rate calculation when y is far from K, because it is very difficult to determine K precisely if the largest y is not close to K.

The epidemics depicted in Figs. 4.6 and 4.20 both correspond to a rate of $r_L = 0.12$, athough the Ks are different (1 versus 0.65). Both y and dy/dt are higher at a given t for the epidemic in Fig. 4.6 ($K = 1$) than for the epidemic depicted in Fig. 4.20 ($K = 0.65$), even though they both started with the same initial disease intensity ($y_0 = 0.001$). This is because r_L is a measure of how fast y approaches K and not how fast it approaches 1 (unless $K = 1$). Thus, equality of the two r_L values is misleading whenever the Ks are different. Fortunately, one can adjust for difference in K values in comparing epidemics. This is done by calculating the more general form of the weighted mean absolute rate (equation 4.26):

$$\varpi = r{*}K/(2\eta + 2) \qquad (4.50)$$

In fact, the expression for ϖ in equation 4.26 was specifically for $K = 1$. When all the curve shapes are the same, so that one is using a single disease progress model for the epidemics of interest (such as the logistic here), one can simply multiply $r{*}$ and K to obtain a scaled version of the rate for each epidemic. Using this argument, one can see that $r_L K$ is higher for the epidemic in Fig. 4.6 than for the one in Fig. 4.20 (0.12 versus 0.078/day).

Besides the modification of the population growth models for an asymptote less than 1, other generalizations are possible. Gilligan (1990) and Schabenberger and Pierce (2002) show how to incorporate a *lower* asymptote into nonlinear models (where K would then be considered an upper asymptote). We do not pursue this here.

4.8.2 Choosing a model

One uses the same general procedures, presented in section 4.6, to choose a model when there is evidence that K is less than 1. In particular, one plots y versus t, estimated dy/dt versus t, and various transformations. This is demonstrated in Fig. 4.21 for leaf blight of sorghum (Ngugi et al., 2001). The disease progress curve is sigmoid shaped and shows strong evidence of a leveling off at a value of y considerably less than 1. It should be noted that if the last two disease assessments were not made, there would be no evidence of a leveling off, and one would be justified in assuming that $K = 1$. With some disease progress curves, there is no increase in y over the last few times. Then one could use this final y as an empirical estimate of K, divide all y values by this K, and proceed as discussed in 4.6. However, there is still a slight increase in y over the last few times in the example, and therefore one must assume that K is probably (somewhat) larger than the highest observed y.

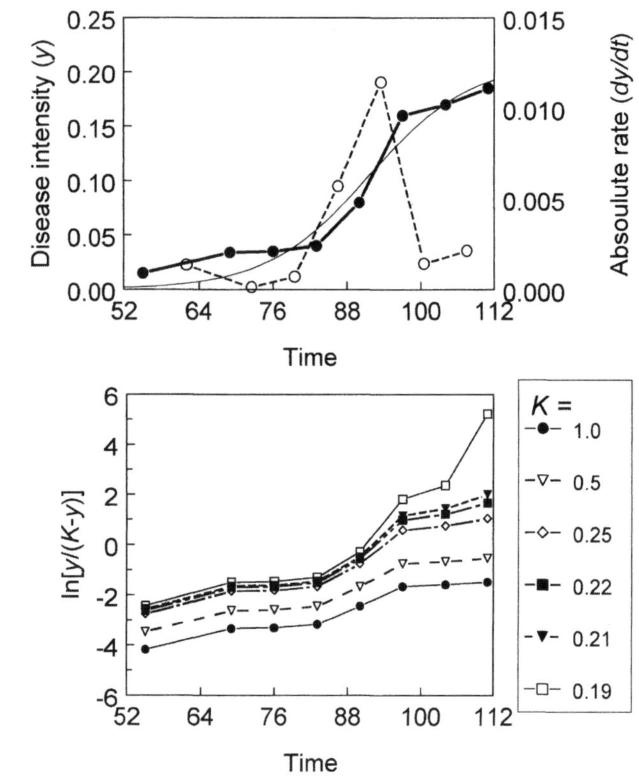

FIG. 4.21. Example disease progress curve of leaf blight of sorghum (y versus time), caused by *Exserohilum turcicum*, estimated absolute rate of disease increase (dy/dt) versus time, and logit transformation of y (see equation 4.49) versus time, for several values of the maximum y (=K). Data from Fig. 3 in Ngugi et al. (2001).

The plot of the absolute rates shows that the largest values of dy/dt in the example is around $t = 92$ days, corresponding to $y \approx 0.12$. This brings up the one tricky issue in working with disease progress curves that level off at $y < 1$. If one ignored the actual disease progress curve, the dy/dt curve would suggest that a model such as the Richards with $\eta \approx 1/4$ would be appropriate because the absolute rate peaks at a low y. However, except for the logistic model, theoretical dy/dt curves are asymmetrical; thus, it should be possible to distinguish between, for instance, a logistic model with $K < 1$ and another model with $K = 1$. The problem is that it may be difficult to observe the asymmetry unless there are many disease assessment times. By looking at the disease progress curve, one can clearly see that y is not approaching 1 in this epidemic. If one used $K = 0.25$ as a rough guess of the maximum, one can see that the highest dy/dt is around $y = 0.12/0.25 = 0.48$. This is consistent with a logistic model (with K of about 0.25). Thus, whenever K is clearly less than 1 (because of leveling off of the disease progress curve at $y < 1$), one should adjust all the calculations for the value of K.

One can plot the various transformations of y versus t to find the one that gives the straightest line. However, the transformation also depends on choice of K. Fortunately, as shown in Fig. 4.20, choice of K mostly affects the curvature of the $y^* : t$ line when y is close to K. If $y < K/2$, the $y^* : t$ line will be fairly straight for the appropriate model, even if the guess of K is far from the true value. Plots of y^* versus t (all for $K = 1$) for the first four times produced no evidence that the logistic model was inappropriate, so we assume here that the logistic model is correct for this polycyclic disease. The lower frame in Fig. 4.21 shows the logits for several values of K. When one uses $K = 1$, the logits level off at large t values. The leveling off is less pronounced as smaller values of K are used. Moreover, if too small of a K is used (e.g., $K = 0.19$ here), the transformed values curve upwards at large t. Based on this graph, the appropriate value of K is between 0.19 and 0.21.

4.8.3 Parameter estimation

As with the situation when $K = 1$, various methods can be used to fit models to data and estimate y_0, r_*, and K. Although equation 4.49 produces a straight line when r_L is fixed, it is not linear in terms of all the parameters (see Chapter 3). In particular, K appears on the left-hand side of the equation, and thus cannot be estimated directly with ordinary least squares regression or other fitting procedures for linear models. One *ad hoc* approach for using linear methodology follows from the graphical appraisal of logits in Fig. 4.21. One calculates logits for a range of K values, from 1 down to a number slightly above the largest observed y, and finds the K that produces the straightest $y^*:t$ line and most desirable residual plot. Then, the estimates of the intercept and slope are *conditional* on the chosen value of K. There is no easily obtained measure of precision of this value of K, but this approach is useful when there are not many observations in a disease progress curve. Following the *ad hoc* method of finding a K that produces the straightest $y^*:t$ line (see Fig. 4.21), a value of $K = 0.20$ was chosen for subsequent analysis of this disease progress curve. Using this K, the rate estimate was found to be $\hat{r}_L = 0.10$ (0.012), and the R^2 value was 0.908.

Nonlinear least squares, with an assumed normal distribution for the errors, typically is the most useful method for estimating all three parameters. If one was using the general form of the Richards model (see Table 4.4), then one could directly estimate all four parameters. Usually one should only use nonlinear least squares (or other nonlinear method) when there are many disease values (e.g., at least twice as many disease values as parameters). To estimate the logistic-model parameters for the leaf blight data (Fig. 4.21), we used equation 4.48 in its statistical form. That is, we add an error term (ε) that was assumed to be normally distributed with mean 0 and constant variance. The reader should go back to equation 4.34 to see how this is done for the simpler logistic model with $K = 1$. Using PROC MODEL (or NLIN) in SAS, we obtained the following parameter estimates: $\hat{y}_0 = 0.0025$ (0.0029), $\hat{r}_L = 0.121$ (0.038), and $\hat{K} = 0.209$ (0.034). The R^2 value is 0.953. Readers should be reminded here that parameter estimates from nonlinear modeling may be biased, and the various statistics calculated are only valid at large sample sizes (Bates and Watts, 1988; Ratkowsky, 1990). How large is "large" depends on the data and the choice of model (Ratkowsky, 1990). The predicted values from this model fit are given on the top frame of Fig. 4.21 as the smooth curve. The scaled version of the rate ($r_L K$) is estimated as $0.121 \times 0.209 = 0.025$/day. The residual plot seemed acceptable for this model fit, and the estimated serial correlation of the residuals was only 0.07, indicating no evidence for lack of independence of the errors. Weighted nonlinear least squares can, and typically should, be performed because of the dependence of the variance on y. As an exercise, the reader can determine the parameter estimates for the data set using binomial-based weights.

As with the simpler situation in section 4.6, there are multiple approaches for comparing epidemics. We do not go into these approaches here. With an unknown K that must be estimated, it is more difficult to use the general approach of linear covariance analysis (equation 4.38b) and linear mixed models (equation 4.39) for comparing epidemics. This is because K is on the left-hand side of the equations (part of y^*) and cannot be simultaneously estimated, for example, with the parameters for the effects of treatment on the intercept and slope. One approach is to find a reasonable single value of K (<1) based on a detailed assessment of select disease progress curves. One would use this K in the y^* transformation, and then carry out the analysis with this new transformed disease intensity. As shown above, if most of the observed intensity values are not close to the maximum (i.e., no clear leveling off of y at different maxima), then the results will not be greatly affected by the exact choice of K. If there are large apparent differences in K among the epidemics, and observed intensity values often are near the maximum (leveling-off point), then one will typically needs to use nonlinear methods.

Comparisons of epidemics can be done with t-tests of pairs of estimated parameters, in which the parameters are based on fitting the chosen model to each disease progress curve separately. This was done in section 4.7.1 for a simpler ($K = 1$) situation. For planned experiments with blocks or replications, one can estimate the parameters for each epidemic, and then use linear mixed models to relate the estimated parameters (as the response variable on the left-hand side) to the treatment and block levels. This is analogous to a random coefficients model for linear models. Or, comparisons can be made with more general and sophisticated methods based on changes in residual sums of squares or deviances with the subtraction of terms for treatment effects on K, y_0, and r_L (see Schabenberger and Pierce, 2002; Gilligan, 1990). However, it is not generally possible (yet) to account for all the features of complicated experimental designs with standard nonlinear modeling programs, although progress is being made with nonlinear mixed models (Wolfinger and Lin, 1997).

4.9 Time-Varying Rate Term

In this chapter, we have generally considered models with a single rate parameter, r_*. However, there is no reason why r_* must be constant, even as an approximation. One scenario that could result in a systematically changing r_* is a systematically changing environment, where conditions for infections become progressively more favorable. One can write a time-varying r_* as $r_*(t)$.

The population growth models can all be generalized to accommodate such a phenomenon (Hau et al., 1993; Jeger, 1987). For instance, the logistic model (with $K = 1$) would be written as $dy/dt = r_L(t)y(1 - y)$ and the monomolecular model as $dy/dt = r_M(t)(1 - y)$. Whether or not these models can be integrated analytically will depend on actual form of $r_*(t)$. A very simply example is:

$$r_*(t) = \zeta_0 + \zeta_1 t \qquad (4.51)$$

where ζ_0 and ζ_1 are parameters. A positive ζ_1 means that the r_* rate term increases during the epidemic, and a negative means that the r_* rate term decreases during the epidemic. When $\zeta_1 = 0$, the rate term is a constant

($r*(t) = \zeta_0 = r*$, the original situation). Using equation 4.51, the general solution for the epidemic models (such as exponential, monomolecular, logistic, and Gompertz) in linear form is given as:

$$y^* = y_0^* + \zeta_0 t + (\zeta_1/2)t^2 \quad (4.52)$$

Note that a linear increase in $r*$ (e.g., r_L) during the epidemic results in a quadratic change in y^* over time (and that the coefficient for t^2 is $\zeta_1/2$).

Examples of the use equation 4.51 for the rate term (with both positive and negative ζ_1, as well as with $\zeta_1 = 0$) are shown in Fig. 4.22, for both the logistic and monomolecular models. One can see that a time-varying $r*$ complicates the possible curves. For instance, the $dy/dt : t$ and $y : t$ curves of the monomolecular model can appear more like Gompertz curves if r_M increases over time. In contrast, the $dy/dt : t$ and $y : t$ curves of the logistic model can appear more Gompertz like if r_L decreases over time. The clearest evidence of a time-varying rate term is the plot of y^* versus t (lower panels Fig. 4.22), where curvature results when ζ_1 is not 0.

We indicated previously that curvature to a logit:t plot can mean that the logistic model is not the most appropriate for describing a data set (see Fig. 4.12), or that the assumed value of K is not reasonable (see Fig. 4.20). Similar comments can be made regarding curvature to a plot of the multiple-infection transformation versus t for assumed monomolecular-type disease dynamics. Now, the graphs in Fig. 4.22 show that nonconstant $r*$ can also lead to curvature of a $y^* : t$ graph. An improved fit can be achieved (when one has the objective of finding a very good description of observed data) by choosing a different function for y in the dy/dt equation [e.g., $y \cdot (-\ln(y))$ instead of $y(1 - y)$; see section 4.4.4], or selecting a different K in the $1 - y$ portion of the dy/dt equation (if the

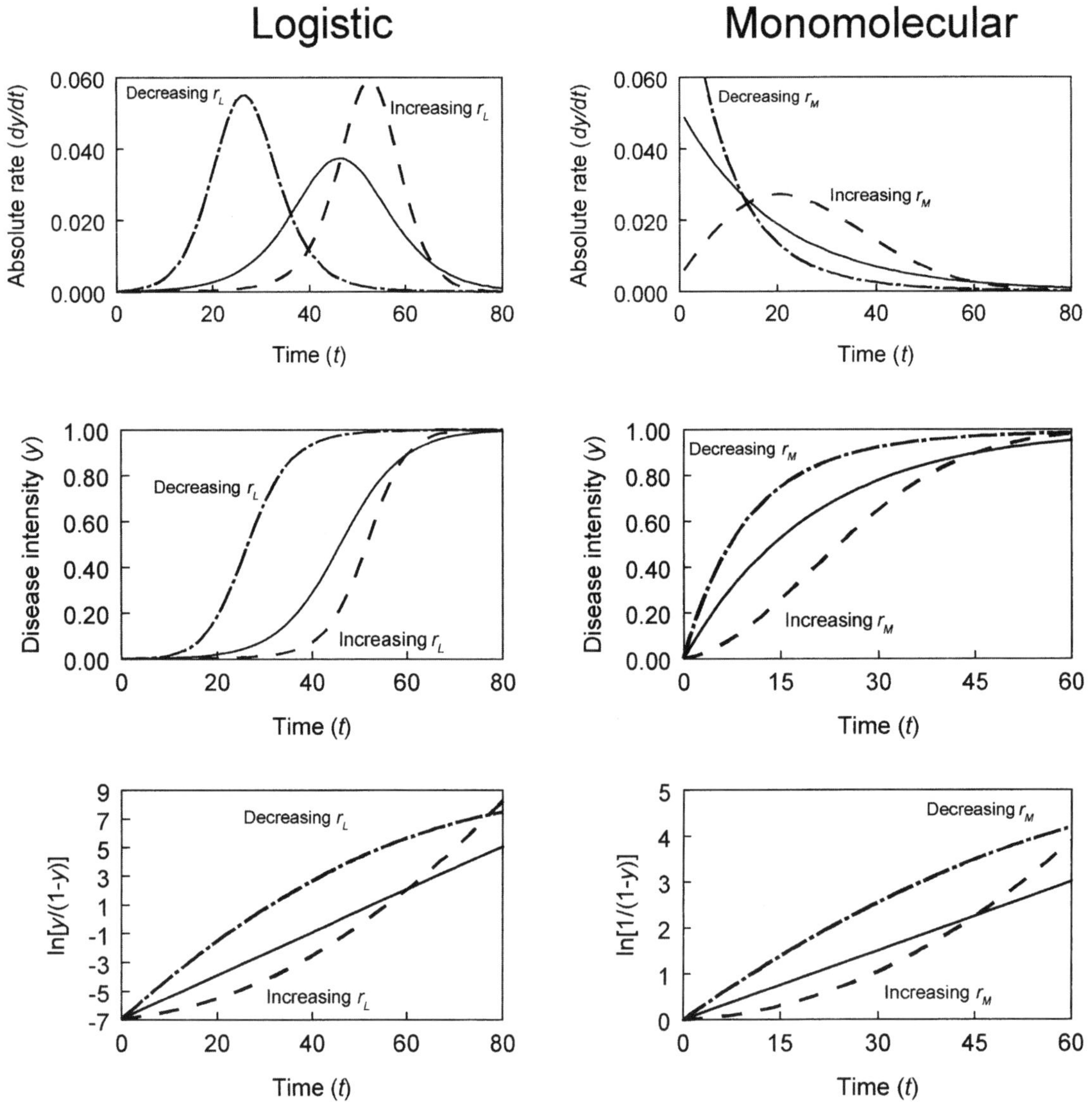

FIG. 4.22. Plots of disease intensity, transformed disease intensity, and dy/dt (absolute rate) versus time based on the logistic (left-hand panels) and monomolecular (right-hand panels) models. Curves are for either the rate term (r_L or r_M) being fixed (solid curves), or increasing or decreasing over time (labeled on the graphs; see equation 4.51).

data exhibits leveling off at large times; section 4.8), or using a time function (e.g., equation 4.51) for $r*$. However, the curvature to y^* versus t does not disappear in the last situation). The mechanisms (population processes) implied by each of these generalizations of dy/dt models are different, of course. It is unlikely that an individual data set will be extensive enough to result in unambiguous evidence as to why a polycyclic epidemic is not described by the simple logistic model (equation 4.15 or 4.16) or monocyclic epidemic is not described by the simple monomolecular model (equations 4.10 and 4.11). This concept in choosing a model was discussed in more detail in section 4.6.2 and elsewhere in the book. Analysis of many disease progress curves for a given pathosystem or use of ancillary evidence can indicate which process (and thus, which model form) is most reasonable for a given disease.

Depending on the function used for $r*(t)$, parameter estimation may be quite straight-forward using linear or nonlinear least squares. For equation 4.52, for instance, it is easy to use *polynomial* regression methods (Neter et al., 1983) to estimate the parameters y_0^*, ζ_0, and $\zeta_1/2$. The use of a straight-line function for $r*(t)$ (giving a quadratic y^* versus t) is probably only reasonable for short-duration epidemics, since it is difficult to hypothesize a continuously increasing or decreasing rate term (with no upper or lower bound, respectively) over long time periods. Much more complicated functions can be considered (Amorim et al., 1993; Hau et al., 1993) that are linear or nonlinear in the parameters. Trigonometric functions could also be used to account for, as an example, cyclical changes in environmental favorability for disease development. Constraints may have to be placed on some parameters (e.g., forcing a parameter to be negative) or nonsensical results could occur (e.g., a predicted decrease in y where only increases (or no changes) are expected or biologically relevant).

We do not elaborate further here because investigators can often use epidemic models with fixed $r*$ or can use more mechanistically based models discussed in the next chapter. There are some circumstances where use of a variable in models is desirable. Interested readers should study Hau et al. (1993), Jeger (1987), or section 9.4 of Campbell and Madden (1990), and references therein.

4.10 Concluding Comments and Prelude to Advanced Topics

In this chapter, we have showed how one can describe and compare epidemics using models such as the logistic and monomolecular. For these and some related models, the absolute rate of disease increase is defined in terms of a single variable (y) and a rate parameter. Integration leads to the addition of one more parameter which can be used to represent disease intensity at $t = 0$. The models could be made more flexible by incorporating a parameter for maximum disease intensity or for the shape of the disease progress curve. Methods were presented for selecting a model, estimating parameters, and comparing epidemics through the model parameters. It was also shown how to use model parameters to calculate synoptic variables, such as AUDPC, and to evaluate control strategies. Fundamental qualitative and quantitative differences were shown between monocyclic and polycyclic diseases.

A fuller biological or mechanistic understanding of plant disease epidemics requires the consideration of additional variables and parameters in the population-dynamic models. This is the subject of the next chapters. To ease the transition between the models and analysis presented so far, and what the reader will find in Chapter 5, an alternative way of describing epidemics with the logistic model is outlined here. We could also do this with the monomolecular (or other) model, but for brevity we focus on the logistic. In all of the models and analysis presented so far, disease intensity was represented as a proportion. This is because most plant pathologists record disease incidence or severity as a fraction of the total number of individuals or total area (Chapter 2). Vanderplank (1963), and others who followed, generally followed this convention and presented models for y.

One can, however, characterize epidemics in terms of the actual *number* of infected individuals or actual *area* of plant (or root) tissue diseased. We use upper case Y to represent this measure of disease intensity and upper case M to represent total number of individuals or total area of plant (root) tissue. Then, $y = Y/M$. Both Y and M can be considered abundances; if they are expressed in terms of a fixed area of land (for instance), then they are densities. In this formulation, $H = M - Y$ is disease-free numbers of individuals or area of plant tissue. For convenience, we call H the 'healthy' area or numbers.

If one multiplies both sides of the logistic differential equation (equation 4.14) by M, one obtains:

$$\frac{d(yM)}{dt} = r_L(yM)(1 - y).$$

Because $Y = yM$ (and $Y/M = y$), one can write the logistic equation as:

$$\frac{dY}{dt} = r_L Y(1 - (Y/M)) = r_L Y\left(\frac{M - Y}{M}\right) = (r_L/M)Y(M - Y).$$

This model represents the change in Y over time as a function of Y and the proportion disease free $[1 - (Y/M)]$, up to a maximum of M. To express both diseased (Y) and disease-free ($H = M - Y$) values in absolute units (rather than as proportions), we can move the M denominator under the rate parameter (see last version of the above equation), and can call this Υ ($=r_L/M$). Then, using H for $M-Y$, the logistic model for Y is:

$$\frac{dY}{dt} = \Upsilon YH \tag{4.53a}$$

Because H decreases directly as Y increases, we can also write the differential equation for H:

$$\frac{dH}{dt} = -\Upsilon YH \tag{4.53b}$$

Equations 4.53a and 4.53b are called *coupled differential equations*. Equations are coupled or linked when one or more variables appear in one or more of the equations. Together, the pair of equations can be used to represent both Y and H over time.

Because in this chapter we considered M to be fixed—in the models, if not (fully) in reality—we have not needed to use a pair of coupled differential equations. That is, if one knows Y, then H is easily determined because $Y + H$ is fixed ($=M$). However, if the host is actively growing or changing in numbers, M is not fixed and one needs to use (at least) two equations to characterize the epidemic. This is considered in subsequent chapters. Even with fixed M, we could have expressed all the models and analyses in terms of H instead of y ($=Y/M$). Integration results in an equation for H as a function of Υ, H_0, and t, and one can plot H (or $h = H/M$) versus t. Disease could then be calculated from $Y = M - H$. As mentioned above, by tradition this is not done. However, with more detailed mechanistic representations of plant disease epidemics (see Chapter 5), all diseased individuals will not be treated the same, and it actually will make a lot of sense to focus on dH/dt as well as other coupled equations for components of disease. Through the use of coupled differential equations, a fuller understanding of the mechanisms involved in epidemics will be achieved, and a theory can be developed for the population dynamics of plant diseases.

4.11 Suggested Readings

Jeger, M. J. 2004. Analysis of disease progress as a basis for evaluating disease management practices. *Ann. Rev. Phytopathol.* 42: 61–82.

Kranz, J. 2003. *Comparative Epidemiology of Plant Diseases*. Springer, Berlin (chapter 5).

Madden, L. V., and Campbell, C. L. 1990. Nonlinear disease progress curves. In: *Epidemics of Plant Diseases: Mathematical Analysis and Modeling* (J. Kranz, editor), 2nd Edn. Springer, pp. 181–229.

Thresh, J. M. 1983. Progress curves of plant virus diseases. In: *Applied Biology* (T. H. Coaker, editor). Vol 8. Academic Press, NY, pp. 1–85.

5

Temporal Analysis II: The Components of Disease

Everything should be made as simple as possible, but not simpler.

Albert Einstein

5.1 Introduction

In the previous chapter, we discussed several types of disease progress curves. Such progress curves are valuable instruments to describe and compare epidemic progress, as we have shown. In this chapter, we will expand on the insight obtained in the previous chapter. Our focus will, however, be different. Although the population biology of plant diseases was discussed at length in the previous chapters, the models presented in Chapter 4 were chosen because of their ability to describe disease progress curves with known dynamics at the level of the population. In this and the next chapter, we will model the mechanisms at the level of the *individual* infection process, including latent period, infectious period, formation of daughter infections, etc., and study what type of dynamics result in the population of plants. Thus, the primary goal of Chapters 5 and 6 is not to choose equations to fit data, *per se*, but to determine the implications of basic biological processes on the population dynamics of disease.

We first develop a series of disease progress models, from very simple caricatures of epidemics, to more complex and realistic ones (section 5.2). Based on the analysis of these models, we discuss when and why epidemics will develop, how fast they will increase and what levels of severity (or incidence) they will reach or approach. An important element of the presentation will be how the more detailed model results here relate to empirical findings shown in the previous chapter. Then in Chapter 6, we consider the complexity of crop growth on epidemics, as well as other advanced topics. In particular, we explore the role of primary infections throughout epidemics, insect transmission of plant pathogens, seasonal disruption of multi-year epidemics (by harvest and subsequent replanting), and other complicating factors on disease dynamics.

The current chapter deals with models for polycyclic diseases with many disease cycles during an epidemic. Such epidemics are generally initiated with a very low level of infection (i.e., low initial disease incidence; y_0). For example, potato late blight epidemics, such as those depicted in Fig. 4.1, start with spores blown into the potato field, where the spores originated from infections that had survived the crop-free period on cull piles, plant volunteers, or debris left from the previous growing season. The amount of this initial disease starting the epidemic is usually negligible compared to the total amount of disease that develops during the epidemic. We will thus consider polycyclic epidemics starting from a very small initial infection. This makes the mathematics a little simpler for presentation purposes. However, all the models presented can be generalized to situations with non-small y_0.

For the models introduced we will study:

1. *Shape of the disease (epidemic) progress curve.* How does disease intensity develop throughout the epidemic? In the previous chapter, most of the epidemic progress curves for polycyclic diseases were *S*-shaped. The models used in the previous chapter were, however, specifically chosen because they are known to produce *S*-shaped curves. The models studied in this chapter do not *a priori* specify that the disease progress curve is *S*-shaped, and we will study which shapes are produced by the (realistic) models.
2. *Threshold for epidemic development.* What characteristics of a disease and of the crop determine whether an epidemic develops or not? In this chapter, we find thresholds expressed in terms of the characteristics of the disease that determine whether an epidemic will develop or not. This is of considerable practical importance in trying to determine the factors that lead to epidemics, and determining how best to prevent epidemics from occurring.
3. *Epidemic development in the initial stages.* Once disease intensity has reached high levels, disease management can hardly be effective in preventing substantial crop losses (Chapter 12; Fry, 1982).

Therefore, effective management of polycyclic diseases has to be targeted at the early stages of an epidemic where disease intensity is low. Information on the epidemic development during these initial stages will be very helpful in developing appropriate disease management strategies. For this reason, throughout this chapter, we will discuss the initial stage of the epidemic for the models to be discussed and show how initial disease increase can be predicted from biologically meaningful parameters.

4. *Final disease level.* What will be the final level or final intensity of disease in an epidemic? By choosing a disease management method to reduce the epidemic development in the initial stage, does this automatically imply that the final levels the epidemic will reach are minimized? We will discuss the final intensity of the disease for each of the models we study in this chapter.

As a final note, we comment on the necessity of complex models. Plant pathologists and other biologists sometimes prefer to incorporate as many aspects or properties of a disease as possible into an epidemic model. The assumption is that a complex model would be more realistic than a simple model, because it has more details, and thus gives more insight into the dynamics of plant disease epidemics. However, even if it was possible to incorporate all/most information available for a disease into a model, the modeler faces the problem of intractability. That is, such complex models are difficult to work with, and mathematical analysis of the model is seldom possible (see Jeger, 1986a, and Chapter 12 in Campbell and Madden, 1990). If all information were incorporated, the "model" would be as complex as reality and thus as difficult to study. As discussed in Chapter 3, we assume that models are simplifications of reality that incorporate the features of reality that are of direct relevance to the investigator. The point of studying models (i.e., studying reality through models) is precisely to isolate the governing aspects of the system from all other aspects, and study the consequences of these important aspects on the dynamics of the system. Once these are understood, additional aspects can be added to the model to see whether they change the dynamics. In this and the following chapter, readers will see that incorporating more aspects of the biology of a disease does not necessarily result in better understanding of the epidemiology of the disease.

A highly relevant question in this regard is: To what extent does a simplification of reality tell us something useful about reality? In the remainder of Chapter 5 we will show, in sequential fashion, that increased realism (i.e., increased model complexity) does not necessarily give us more insight into epidemic development, and that simple models do not necessarily lead to incorrect understanding about the system of interest. The question about how simple models can or should be related directly to the quotation at the start of this chapter. We hope that the reader can be convinced that high model complexity is not always necessary to gain insight in the dynamics of plant disease epidemics.

As discussed in section 4.2 of the previous chapter, deterministic models are used throughout and, with the exception of section 5.2.1 (below), we assume that time is continuous. A distinguishing feature of most of the models in this chapter is that determining disease intensity as a proportion at any given time, $y(t)$, requires numerical integration because there is no analytical solution for the models for all times during an epidemic. This means that the function for $y(t)$ involves an integration symbol. However, as will become apparent in addressing items 2–4 above, analytical solutions (or very good approximations) exist for the early stages (small t) and very late stages (generally, large t) of polycyclic epidemics under the conditions we are concerned about.

5.1.1 Terminology

The models we introduce in this chapter apply to a wide range of diseases. However, specific terminology may vary with the disease system. For example, with a rust disease, we could use the lesion (or pustule) as the unit of disease, and spores as the units of inoculum that are produced, transported, deposited on healthy leaves, and ultimately cause new infections. For viral plant diseases that are systemic, the entire plant could be the unit of disease, and the units of inoculum would be more nebulous (as discussed in the previous chapter), involving viruliferous insects (for example). When describing a model using the terminology of a particular disease, it often becomes difficult to appreciate that the model can also be used for another disease by simply substituting the appropriate terminology for that disease. To help minimize this problem we will mostly use terminology not related to a specific disease. Although this might seem a bit artificial in the beginning, our experience indicates that it helps greatly to understand the generality of the models presented.

In this and the following chapter, we use the following terms:

1. *Individual* for the unit of host crop or disease. As discussed below, an individual can be healthy (disease free in terms of the disease of interest), latently infected, infectious, or removed from the epidemic. The individual could be a plant, a leaf, a fruit, a root, a unit area of a leaf that can contain a lesion, and so on. Even though plant diseases are often measured on a continuous scale (see Chapter 2), it is very awkward to refer to unit areas (e.g., cm^2) of leaf area and individual plants or leaves in the same sentence, for instance. Thus, we generally maintain the term *individual* for the both continuous and discrete units of host and disease for ease of

presentation. The presented models do not depend on discrete units of host or disease.

2. *Infectious unit* is the unit of inoculum that is produced by diseased individuals and is dispersed to other host units where infection may occur. As emphasized in section 4.2, epidemics occur from the contact of inoculum with disease-free (healthy) plant individuals, but the nature of the inoculum varies quite a bit for different diseases. Readers can refer to Chapter 7 for a thorough discussion on infectious units and their dispersal.
3. *Transmission* is used for the transfer of infectious units from individual to individual (e.g., plant to plant). The transmission rate then is the rate at which the disease is transmitted from infectious individuals to healthy individuals. Although the word transmission is often reserved in plant pathology for processes involving insect (or other) biological vectors, we follow Shurtleff and Averre (1997) and use it here in a more general sense. The term is used to include production of inoculum, dispersal, and infection of individuals from the dispersed inoculum.

In many of the models that are discussed in this chapter, diseased individuals are divided into three categories. After the infection takes place, the infected individual first goes through a phase where the disease develops and "grows" in the individual (e.g., fungal colonization of leaves, virus multiplication and movement within plants), but the infected individual does not produce infectious units. This infected individual is in the *latent state*. After a latent period, the infected individual becomes an *infectious* individual, meaning that it now produces infectious units that have the potential to cause daughter infections (i.e., new diseased individuals). The infected individual is in the *infectious state*. For viruses, the infectious state means that vectors can acquire the virus and potentially inoculate a disease-free individual. After a certain length of time, the production of infectious units ceases and the infected individual enters its post-infectious or *removed state*. The individuals are said to be removed from the epidemic (in terms of producing new diseased individuals).

We can define:

1. *Latent period*. This it the length of time between the start of the infection process by a unit of inoculum and the start of production of infectious units (i.e., the beginning of infectivity). Individuals in this period are called latently infected individuals or latent infections. The length of this period is denoted by $\wp$.
2. *Infectious period*. This is the length of time between the start of production of infectious units and the end of the production of infectious units. Individuals in this period are called infectious. The length of this period will be denoted by ι.

For a single individual, it is straightforward to define latent and infectious periods. For a population of individuals, the definitions (or even the concepts) become a little more complicated. For instance, even if the infection process starts at the exact same moment, there will be a distribution of latent periods since the start of the infectious period will not be exactly the same for every individual. Likewise, there will be a distribution of infectious periods in a population. The transition from individual to population is addressed in the relevant models below. Note that ι and $\wp$ are used for *mean* times when representing a population of individuals, unless indicated otherwise.

5.2 Disease Progress Models with Fixed Host Density

5.2.1 A simple discrete-time model

5.2.1.1 Model derivation. In the first model to be discussed, we assume that disease generations are separated in time. By this we mean that infectious individuals produce infectious units that infect healthy individuals during a fixed period of time, after which the infectious individuals stop producing new infections. Then, the new group (generation) of diseased individuals becomes infectious and the process is repeated. We also assume a fixed *total* number of host individuals in the population. The number of infectious individuals in the $n + 1$th generation, I_{n+1}, then is simply the number of new infections produced by the infectious individuals in the previous, i.e., nth, generation, I_n. How many new infections occur is calculated from the number of healthy individuals in generation n, H_n, multiplied by the probability for each healthy individual to be infected by infectious individuals in that generation. Concurrently, the number of healthy individuals in generation $n + 1$, H_{n+1}, equals the number of healthy individuals in the previous generation n, H_n, multiplied by the probability for an individual *not* becoming infected in generation n. In pseudo-equation form, we thus have:

$$\begin{aligned} H_{n+1} &= H_n \begin{pmatrix} \text{probability } \textit{not} \text{ to get infected} \\ \text{in generation } n \text{ by individuals } I_n \end{pmatrix} \\ I_{n+1} &= H_n \begin{pmatrix} \text{probability to get infected} \\ \text{in generation } n \text{ by individuals } I_n \end{pmatrix} \end{aligned} \tag{5.1}$$

Each infectious individual produces, on average, α viable infectious units. The probability that an infectious unit is deposited on a particular host individual (i.e., comes in contact with a particular host) is θ; this means that, on average, a particular host receives $\alpha\theta I_n$ infectious units (that is, on average $\alpha\theta I_n$ units of inoculum come in contact with each host individual). The probability that an infectious unit deposited on a healthy individual is effective, meaning it causes an infection, is denoted by ψ. In summary, this means that when the infectious units are randomly distributed, each individual will, *on average*, receive $\psi\theta\alpha I_n$ effective infectious units. Not every individual will receive exactly this average number of effective infectious units; due to the random nature of the dispersal of infectious units, some individuals will receive less, and some more than this average. Those individuals that receive no effective infectious units at all will be the ones not getting infected in generation n by individuals I_n.

In the situation described here, the probability that an individual receives a certain number, X, of infectious units, when individuals receive on average $\psi\theta\alpha I_n$ infectious units, is a very well known one in statistics. In particular, X has a Poisson distribution with parameter $\psi\theta\alpha I_n$, The reader is referred to Chapter 9 for a detailed description of this probability distribution in other contexts. The only aspect we need to know here is the probability that the individual does not receive any effective infectious units. According to the Poisson distribution, this probability equals $\exp\{-\psi\theta\alpha I_n\}$, which is the term needed in the equation for the number of healthy individuals in equation 5.1. The probability to receive any number of effective infectious units (rather than 0), and thus to become infected, is, of course, equal to one minus the probability not to receive any effective infectious units. Substituting this information in our system of equations 5.1 we have:

$$H_{n+1} = H_n e^{-\beta I_n} \qquad (5.2a)$$

$$I_{n+1} = H_n(1 - e^{-\beta I_n}) \qquad (5.2b)$$

where we simplified the notation by using $\beta = \psi\theta\alpha$. In the general epidemiological literature, β is often called the per capita disease transmission probability, or per capita infection probability (Daley and Gani, 1999; Anderson and May; 1991). The model equations 5.2a and 5.2b have, as far as we know, first been studied by Dr. J. Gielen from Wageningen University (*unpublished*), and the third author of this book has used this model for teaching for many years. In Diekmann and Heesterbeek (2000) a more mathematical discussion can be found.

The model equations 5.2a and 5.2b are very simple to solve numerically. After deciding on the value of the parameter β, we choose a value for the number of healthy and infectious individuals at the beginning of the epidemic (at $n = 0$), H_0 and I_0, respectively. Substituting these in the right-hand side of equations 5.2a and 5.2b we find the number of healthy and infected individuals at $n = 1$. Substituting these numbers again in the right-hand side of the equations we find results for generation $n = 2$, and so on. Plotting the numbers against time (=generation here), we can visualize the epidemic progress curve generated by the model. To enable a comparison with the epidemic curves we studied in Chapter 4 (and later sections of this chapter), we also want to plot the total number of individuals that have been infected (i.e., diseased) since the start of the epidemic, Y_n. Calculating this number is simply a matter of adding up all numbers of infectious individuals in the various generations, from $n = 1$ to n:

$$Y_n = I_1 + I_2 + I_3 + \cdots + I_{n-2} + I_{n-1} + I_n = \sum_{k=1}^{n} I_k \qquad (5.3)$$

In this model, there is no carryover of infectious individuals beyond a single time generation, so total disease is simply the sum of the generation values of I. Note that we do not add the starting number of infectious individuals, I_0, in equation 5.3. In section 5.2.1.5 it will become clear why.

5.2.1.2 Model simulation. Fig. 5.1 shows a series of solutions of the model for different values of the parameter β. All these solutions start with 1000 healthy individuals, $H_0 = 1000$, and one infectious individual, $I_0 = 1$. From these figures we can draw some conclusions about the epidemic dynamics the model generates.

First of all, the disease progress curves, Y_n versus n, are *S*-shaped curves, similar to some of those we studied in Chapter 4. At first sight this might seem logical, but readers are reminded that in Chapter 4 models were chosen that are known to produce an *S*-shape. That is, the *S*-shape was incorporated into the disease progress curve before hand. In the derivation of our present model (equations 5.2a and 5.2b), we did not *a priori* specify that the disease progress curve should be *S*-shaped. The only thing we have put into the model is a biologically reasonable bookkeeping of the transmission of disease from infectious to healthy individuals. That this results in *S*-shaped disease progress curves shows that the biology underlying epidemic development (in this simple

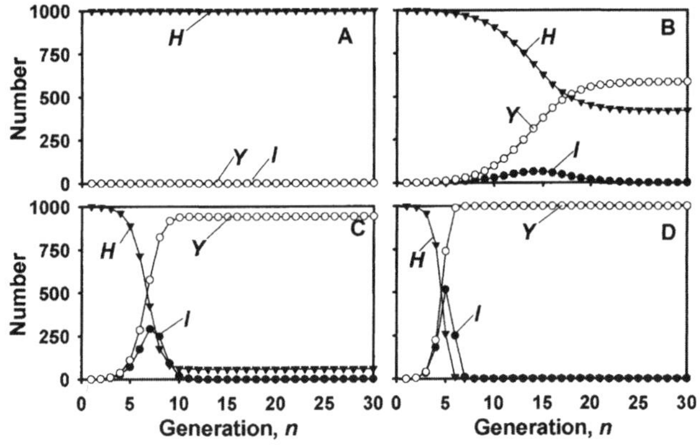

FIG. 5.1. Number of healthy (H) and infectious (I) individuals per unit area in relation to disease generation (n), and total number of individuals who have been infected since the start of the epidemic (Y). The densities are solutions of model equations 5.2a and 5.2b. (**A**) $\beta = 0.0008$; (**B**) $\beta = 0.0015$; (**C**) $\beta = 0.003$; (**D**) $\beta = 0.08$.

scenario) can generate *S*-shaped disease progress curves, and therewith underpins the use of selected models in Chapter 4.

The *initial* increase in the number of infectious individuals for the first few generations, I_n, and the total number of infected individuals, Y_n, seem to resemble an exponential curve as described in Chapter 4 (see Fig. 4.2).

Even though we did not *a priori* incorporate into our model an upper limit for Y, there seems to be one. After the number of infectious individuals per generation has decreased to zero after a long time period, the total number of infected individuals is still smaller than the number of healthy individuals at the start. For example, for $\beta = 0.0015$, only ~60% of the total number of healthy individuals present at the start of the epidemic become infected. The larger the value of the parameter β, the larger the percentage of the population that becomes infected. We will in the following denote the final number of infected or diseased individuals by Y_∞, to denote that this number is reached after the epidemic has petered out (as I_n goes to 0), that is, Y_n for n very large. That not all individuals fall victim to an epidemic is a very well known phenomenon for epidemics in humans, animals and plants (Daley and Gani, 1999). For example, readers are encouraged to think back to the most recent influenza virus strain that caused an epidemic in their country or region. There will definitely be people you know that did not get infected by the end of the epidemic, even though they were likely exposed to the pathogen. This phenomenon is also known from plant disease epidemiology; some of the graphs in Fig. 4.1 demonstrate this leveling off of disease at less than 100%. Here we have shown that even a very simple model only incorporating some aspects of the population biology of infectious diseases shows this phenomenon.

Perusal of Fig. 5.1 shows that for values of β smaller than 0.001, no epidemic develops. This is consistent with the interpretation of this parameter ($\beta = \psi\theta\alpha$). If, for example, the number of infectious units produced by one infectious individual (α) is very small, we expect that the disease is not capable of building up a population of infectious individuals and that the disease will simply die out. Further simulations, not shown in the figure, indicate that the apparent threshold for an epidemic depends on the combination of the initial number of healthy individuals (H_0) and the parameter β. Fig. 5.2 shows a plot of H_0 and β for which an epidemic develops or does not develop. There clearly seems to be a threshold line. For combinations of H_0 and β that fall above this line, an epidemic will develop (based on this simple model), whereas for combinations that fall below the line, no epidemic will develop.

Our model analysis using numerical solutions and graphs thus reveals several interesting phenomena, but it is natural to ask: Are these phenomena the result of the particular parameter values we used, or are they general phenomena of this model irrespective of parameter

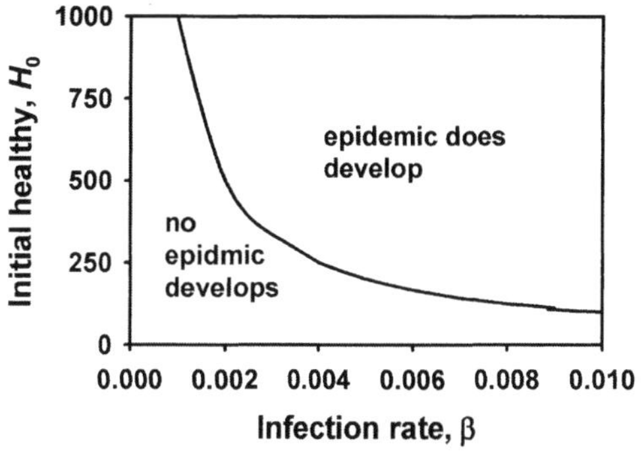

FIG. 5.2. Parameter combinations for which an epidemic develops and does not develop, using equations 5.2a and 5.2b for a non-overlapping generation epidemic model. Horizontal axis: the transmission or infection rate parameter, β. Vertical axis: the number of healthy individuals at generation (n) 0, H_0. Combinations of β and H_0 above the curve result in epidemics, and combinations below the curve do not result in epidemics.

value? For example, would other parameter values result in an epidemic progress curve that is not an exponential type in the initial stage of the epidemic? How exactly does Y_∞ depend on β and H_0? These questions cannot be adequately answered by further model simulation; we can, however, find general answers using analytical methods. These analytical methods will be developed in the following sub-sections where we concentrate on the questions posed in the introduction: (i) when does an epidemic develop; (ii) how fast will the number of infected individuals grow in the initial stages of the epidemic; and (iii) what fraction of the host individuals will be infected at the end of the epidemic?

5.2.1.3 The threshold for epidemic development. Consider the situation where the number of infectious individuals is very small in the initial stages of an epidemic. It is possible to approximate several terms in equations 5.2a and 5.2b by simpler terms which make it easier to analyze the model. First of all, when I_n is very small, the number of healthy individuals, H_n, will remain very close to the initial number of healthy individuals at the start of the epidemic, H_0. So as long as I_n is very small, we can replace the term H_n on the right-hand side of equation 5.2b. by H_0 (i.e., $H_n \approx H_0$) and still have a good description of the dynamics of the epidemic model. Next, as shown in Box 5.1, for small values of I_n the term $1-\exp(-\beta I_n)$ can be approximated by the much simpler term βI_n. This concept was also used in Chapter 3 (section 3.6) to characterize dose-response relations for inoculum density studies. The model equation for the infectious individuals thus becomes:

$$I_{n+1} = \beta H_0 I_n \qquad (5.4)$$

This simplified model—a good approximation to equation 5.2b as long as I_n is very small—has a simple

BOX 5.1

Consider the function $f(I) = \exp(-\beta I)$. Fig. B5.1 shows the graph of this function for $\beta = 0.1$. In the graph we have also drawn the straight line $g(I) = a + bI$ that intersects the graph of $f(I)$ at $I = 0$, and has the same slope as $f(I)$ at that point. This figure shows that the non-linear function $f(I)$ is very well approximated by the linear function $g(I)$ for small values of I. Forcing $g(I)$ to intersect $f(I)$ at $I = 0$ amounts to taking $a = 1$, as then $f(0) = g(0)$. The slope of the function $g(I)$ is b. The slope of the function $f(I)$ at the intersection point equals the derivative of $f(I)$ at $I = 0$. Since $df/dI = -\beta\exp(-\beta I)$, the appropriate value of $b = -\beta$. This shows that for small values of I, the function $f(I)$ can be approximated by $1-\beta I$.

In model equation 5.2b we thus have that $1-\exp(-\beta I_n) \approx \beta I_n$ as long as I_n is small.

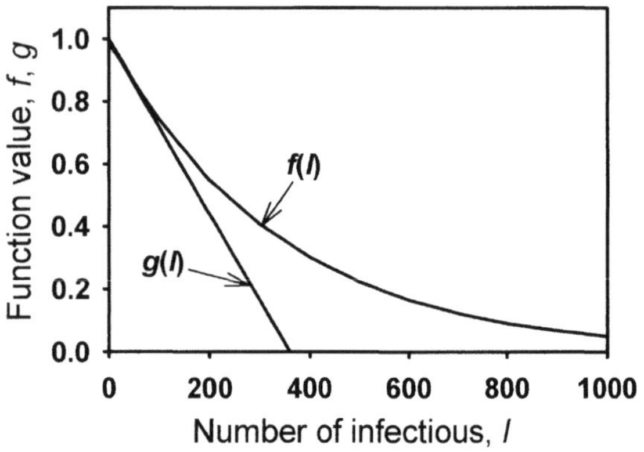

FIG. B5.1. Exponential decay function, $f(I)$, and its approximating linear function, $g(I)$. The functions are $f(I) = \exp(-0.0028I)$ and $g(I) = 1-0.0028I$.

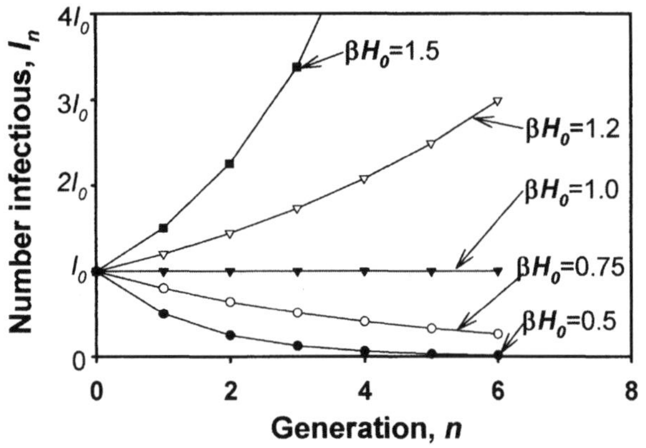

FIG. 5.3. The number of infectious individuals per unit area (I) in relation to disease generation (n) for various values of the basic reproduction number, $R_0 = \beta H_0$. The curves represent solutions of equation 5.5.

solution. Starting with the initial number of infectious individuals I_0, we find after substitution into equation 5.4 that $I_1 = \beta H_0 I_0$. Substituting I_1 again into equation 5.4 we get $I_2 = \beta H_0 I_1$, which can be rewritten as $I_2 = \beta H_0 I_1 = (\beta H_0)^2 I_0$. Again repeating this procedure, we find $I_3 = \beta H_0 I_2 = (\beta H_0)^3 I_0$, and repeating this it become clear that the solution to model equation 5.4 is:

$$I_n = (\beta H_0)^n I_0 \tag{5.5}$$

Equation 5.5 tells us for which values of β and H_0 an epidemic will or will not develop (in terms of whether or not I_n increases). In Fig. 5.3, I_n is plotted as a function of n for a range of values of βH_0. The figure shows that when $\beta H_0 < 1$ the number of infectious individuals decreases in subsequent generations to zero and we conclude that no epidemic develops. When $\beta H_0 > 1$ the number increases and we conclude that an epidemic does develop. The situation where $\beta H_0 = 1$ clearly separates the cases where an epidemic develops from those where no epidemic develops, and the reader can verify that this equation defines the line in Fig. 5.2 that separates epidemic from non-epidemic situations.

The quantity βH_0 has a clear biological interpretation. β is the per capita disease transmission probability, and the quantity βH_0 thus has the interpretation of "the number of new infections (or diseased individuals, in general) in the *next* generation caused by one infectious individual if it is placed into a completely healthy population". Essentially, this is a generation-to-generation multiplication factor. Note that the derivation of equation 5.5 assumed that the number of infectious individuals was very small, and the epidemic is thus in its initial stage. It is intuitively clear that when I_n is small, an epidemic will develop when each infectious individual gives rise to more than one new infectious individual in the next generation, and an epidemic will not develop when each infectious individual gives rise to less than one infectious individual.

The quantity "the number of new infections (or diseased individuals, in general) due to one infectious individual placed in a completely healthy population" is called the basic reproduction number, R_0. When $R_0 > 1$, an epidemic will develop and when $R_0 < 1$ no epidemic will develop. For model equation 5.5 being considered here, we thus found that $R_0 = \beta H_0$. For further reading on the basic reproductive number see Diekmann et al. (1990), Anderson and May (1991), Adler (1992), Jeger and van den Bosch (1994a), and Gubbins et al. (2000).

5.2.1.4 Initial disease increase. Equation 5.5 also shows us what the shape of the epidemic curve in the initial stage of the epidemic is. Equation 5.5 can be written as:

$$I_n = I_0 e^{\ln(\beta H_0)n} \tag{5.6}$$

which is an exponential growth equation similar to equation 4.2 of the previous chapter (which was written for total disease as a proportion, not just infectious disease, and continuous time). The exponential growth rate $r_E = \ln(\beta H_0)$. We thus conclude that independent of the value of β and H_0, if an epidemic is started from a small I_0, there is initially exponential growth of the number of infectious individuals. How long this initial exponential stage will last depends on β and I_0 relative to H_0. The expression for r_E indicates that increasing β (either through increases in α, ψ, or θ), or increasing H_0 will lead to a larger r_E.

5.2.1.5 *Final disease level.* To calculate the final disease intensity (after a long period of time) we will use the full model equations 5.2a and 5.2b since, by definition, final disease level will occur after the initial stages. Starting an epidemic with the initial conditions I_0 and H_0, we find from equation 5.2a for generation $n = 1$:

$$H_1 = H_0 e^{-\beta I_0} \quad (5.7)$$

Substitution this again in equation 5.2a, we get for the next generation $n = 2$:

$$H_2 = H_1 e^{-\beta I_1} = H_0 e^{-\beta I_0} e^{-\beta I_1} = H_0 e^{-\beta(I_0 + I_1)} \quad (5.8)$$

For generation $n = 3$ and 4, respectively, we find:

$$\begin{aligned} H_3 &= H_2 e^{-\beta I_2} = H_0 e^{-\beta(I_0+I_1)} e^{-\beta I_2} = H_0 e^{-\beta(I_0+I_1+I_2)} \\ H_4 &= H_3 e^{-\beta I_3} = H_0 e^{-\beta(I_0+I_1+I_2)} e^{-\beta I_3} = H_0 e^{-\beta(I_0+I_1+I_2+I_3)} \end{aligned} \quad (5.9)$$

From this it is clear that the number of healthy individuals in generation n equals:

$$H_n = H_0 e^{-\beta(I_0+I_1+I_2+I_3+\cdots+I_{n-1})} = H_0 e^{-\beta(I_0+Y_{n-1})} \quad (5.10)$$

where we used the definition of total number of infected individuals from equation 5.3. As we saw from Fig. 5.1 for some specific combinations of β and H_0, the number of healthy individuals, H, and the total number of infected individuals, Y, converges to a final level when n gets indefinitely large (mathematically, $n \to \infty$). The model solution we just derived allows us to let n go to infinity by simply substituting $n = \infty$ in equation 5.10 to arrive at:

$$H_\infty = H_0 e^{-\beta(I_0+Y_\infty)} \quad (5.11)$$

Note that I_0 is kept separate from Y in the exponent. Realizing that the final total number of infected individuals and the number of healthy individuals are related by $Y_\infty = H_0 - H_\infty$ (since the total population size is fixed at H_0 in this model scenario), equation 5.11 can be written as:

$$Y_\infty = H_0\left(1 - e^{-\beta(I_0+Y_\infty)}\right) \quad (5.12)$$

This expression shows us that if we have a value for the parameter β and values for the initial conditions H_0 and I_0, the total number of infected individuals, after the epidemic has ran its course, Y_∞, can be calculated directly. Note that equation 5.12 does not depend on a small value for I_0.

For the further discussion of this final level equation, we restrict ourselves to the situation where the epidemic has started with a very small I_0, as we have done throughout the presentation of the model. When I_0 is very small, $I_0 + Y_\infty \approx Y_\infty$, which is substituted in equation 5.12. Furthermore, we are interested in the fraction of the initial number of healthy individuals that will finally have been infected (diseased) during the epidemic, Y_∞/H_0. This fraction will be denoted by $y_\infty = Y_\infty/H_0$. Equation 5.12 then becomes:

$$y_\infty = 1 - e^{-\beta H_0 y_\infty} \quad (5.13)$$

Note that βH_0 is the basic reproduction number, R_0. Equation 5.12 can thus be written as:

$$y_\infty = 1 - e^{-R_0 y_\infty} \quad (5.14)$$

This final disease level equation has been discussed extensively in the literature and more information can be found in Reddingius (1971), Metz (1978b), Jeger (1986b), Jeger and van den Bosch (1994a), Diekmann and Heesterbeek (2000) and Segarra et al. (2001).

Unfortunately, equation 5.14 cannot be solved explicitly for y_∞. It is however easy to find the solutions numerically. In Box 5.2 one way is shown to calculate solutions. Fig. 5.4 shows the final fraction y_∞ as a function of R_0. The figure shows again the threshold for the development of an epidemic, $R_0 = 1$ (as explained above). Furthermore, it shows that the final fraction of the individuals that get infected increases quickly with increasing R_0. For example, when in the beginning stages of the epidemic, each infectious individual infects three healthy individuals ($R_0 = 3$), more than 90% of all healthy individuals will ultimately get infected during the epidemic.

If the epidemic is not started with a very small initial number of infectious individuals, I_0 has some effect on y_∞. We refer reader to Metz (1978b), Jeger and van den Bosch (1994a), and Segarra et al. (2001) for a discussion of this situation. Furthermore, we point out that there are several related but slightly different definitions of y_∞, some of which include I_0 in the total population size and others that keep I_0 separate from the total population size. The various possibilities are discussed in a plant pathology context in Jeger and van den Bosch (1994a).

5.2.1.6 *Concluding remarks.* We have studied a very simple model describing the dynamics of an epidemic with non-overlapping generations. The formulation introduced some basic features of plant diseases in

Box 5.2

We need to find a solution to the equation $y_\infty = 1 - e^{-R_0 y_\infty}$. First, write this equation in the form $f(y_\infty) = g(y_\infty)$, where $f(y_\infty) = y_\infty$ and $g(y_\infty) = 1 - e^{-R_0 y_\infty}$. The functions f and g are plotted in Fig. B5.2 for $R_0 = 2$. The value for y_∞ where the functions f and g intersect is the value for which $f = g$, and this is the solution for the final level equation 5.14. For different values of R_0, the function g becomes less or more steep and the intersection point is different from the one in Fig. B5.2. Using such a graphical method, y_∞ can easily be found. This graphic method is crude, and only approximate values are found. Using root finders available in many computer packages, the final levels can be calculated very exactly.

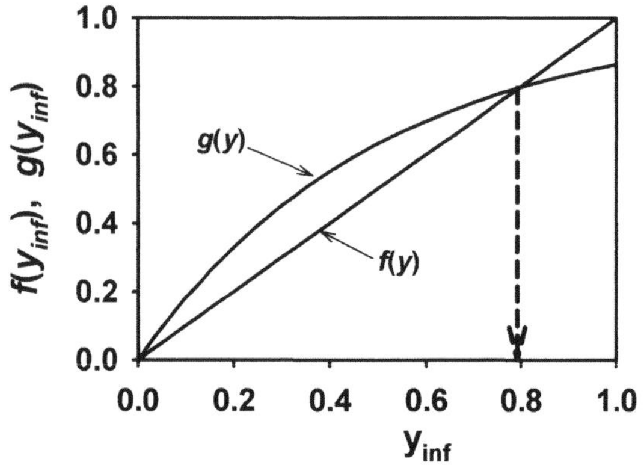

FIG. B5.2. Curves of the functions $f(y) = y$ and $g(y) = 1-e^{-2y}$, where $f(y) = g(y)$ is the final disease level equation 5.14. The intersection point at $y = 0.8$ is the final disease level for $R_0 = 2.0$.

mathematical terms to develop a model for epidemic development. Numerical study of the model, Fig. 5.1, showed some interesting features. These features were further studied using analytical tools. The combination of the two approaches showed that

1. An epidemic will develop when $\beta H_0 > 1$ and no epidemic will develop if this product is smaller than 1. The interpretation of this quantity, the number of new infections resulting from one infectious individual when it is placed in a completely healthy population, is the definition of basic reproduction number R_0.
2. Starting with a small number of initial infectious individuals, the initial stage of the epidemic shows an exponential increase in the number of infectious individuals. The exponential rate parameter, r_E, is a function of R_0.
3. The shape of the full epidemic curve is sigmoid.
4. Not every individual will become infected when an epidemic occurs. The fraction of individuals that will become diseased after the epidemic runs its course depends on R_0, and can be calculated from equation 5.14.

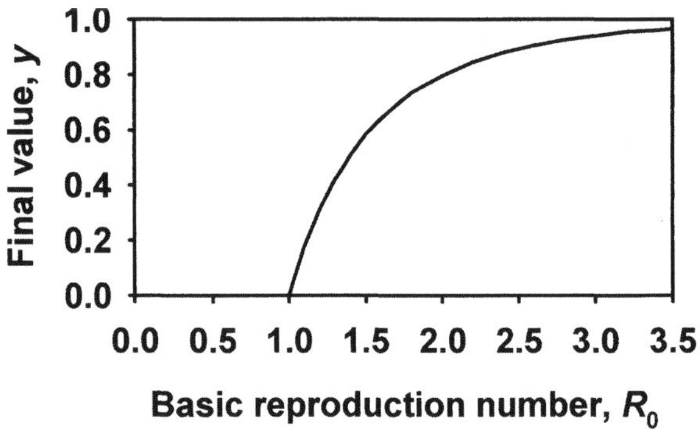

FIG. 5.4. The fraction of plant individuals that will have become infected after the epidemic has run its course, y_∞, in relation to the basic reproduction number, R_0. The curve is the solution of the final level equation 5.14.

The model equations 5.2a and 5.2b are very simple, and the model is a caricature of a real epidemic. In reality, generations are for most diseases not separated. In the model no account is given to latently infected individuals. The infectious period is not mentioned at all. We assumed that infectious individuals in generation n only can cause new infections for one generation. Other discrepancies with a real epidemic could be listed here. The question: "Do these four listed conclusions apply to real epidemics or are they simply artifacts of the simple nature of the model specification?" is a very valid one. In the next sections we will, therefore, consider four other models for plant disease epidemics that incorporate additional aspects of the biology of plant diseases.

5.2.2 The H-I-R epidemic model

The next model we study does not make the unrealistic assumption, as in the previous model, that the disease generations are separated. Now considering time as continuous, the model is formulated as a set of differential equations for the rate of change in the number (or proportion) of individuals in the various states (disease free or healthy, infectious, removed).

In the previous model, we discussed all processes in terms of numbers of individuals. This was done to ease the introduction of the models we consider in this chapter. In practice, one will usually be interested in densities of individuals, such as the number of healthy or diseased plants per square meter. In the next model, as well as the ones that follow, we thus change our vocabulary and discuss densities rather than just numbers.

The reader should be remembered here that in Chapter 2, we discussed three categories of disease intensity: (1) incidence, (2) severity, and (3) count. It is possible to represent incidence, severity and counts per unit area of land, such as 50 diseased leaves per square meter, 0.5 square centimeters of lesion per square meter, and 25 lesions per square meter. In this chapter we use density in this sense.

5.2.2.1 Model derivation. As done in section 5.2.1, we assume that the host population size is fixed. The density of healthy individuals at time t will be denoted by $H(t)$. The density of healthy individuals decreases due to infectious units causing them to become infected (i.e., by the production of infectious units, some of which come in contact with disease-free individuals and cause infections). The density of infectious individuals (i.e., number of individuals in the infectious state per unit area), $I(t)$, increases due to these new infections. Furthermore, the density of infectious individuals decreases due to individuals reaching the end of their infectious periods. These individuals stop being infectious and are called removed. The density of removed individuals, $R(t)$, increases due to infectious individuals reaching the end of their infectious period. Denoting the rate of change in the density of healthy, infectious, and removed individuals by $dH(t)/dt$, $dI(t)/dt$, and $dR(t)/dt$, respectively, the set of pseudo-equations is:

$$\begin{cases} \dfrac{dH(t)}{dt} = -\begin{bmatrix}\text{rate at which healthy}\\ \text{individuals become infected}\end{bmatrix} \\ \dfrac{dI(t)}{dt} = +\begin{bmatrix}\text{rate at which healthy}\\ \text{individuals become infected}\end{bmatrix} \\ \qquad\quad -\begin{bmatrix}\text{rate at which infectious}\\ \text{individuals reach the end}\\ \text{of the infectious period}\end{bmatrix} \\ \dfrac{dR(t)}{dt} = +\begin{bmatrix}\text{rate at which infectious}\\ \text{individuals reach the end}\\ \text{of the infectious period}\end{bmatrix} \end{cases} \tag{5.15}$$

Each infectious individual produces α infectious units per time unit, on average. (Note the difference with the definition of α in the previous model, where α was the number of infectious units produced by one infectious individual during the entire generation.) Consequently, $\alpha I(t)$ is the total density of infectious units produced per time unit per unit area. The probability that an infectious unit comes in contact with a healthy individual in a unit area is θ. That is, the average number of healthy individuals per unit area that are contacted by an infectious unit is given by $\theta H(t)$. Thus, the number of infectious units transmitted from infectious to healthy individuals per time unit per unit area equals $I(t)\theta H(t)$. An infectious unit deposited on a healthy individual has a probability ψ of infecting the individual. We thus have in pseudo-equation format:

$$\begin{bmatrix}\text{rate at which healthy}\\ \text{individuals become infected}\end{bmatrix} = \beta H(t)I(t) \tag{5.16}$$

where we simplified the notation by using $\beta = \alpha\theta\psi$. In the general epidemiological literature, β is often called the per capita disease transmission rate. (Again note the difference with the dimension of the β parameter in the previous model. There it was a value per generation, whereas in the present model it is a rate parameter (value per time unit.)

Each individual has a probability per time unit, μ, to reach the end of its infectious period and thus transfer to the removed category. The total density of infected individuals transferring from infected (or infectious) to removed thus equals $\mu I(t)$, and we have:

$$\begin{bmatrix}\text{rate at which infectious}\\ \text{individuals reach the end}\\ \text{of the infectious period}\end{bmatrix} = \mu I(t) \tag{5.17}$$

Putting the pieces together, the model to be studied in this section is:

$$\begin{aligned} \frac{dH(t)}{dt} &= -\beta H(t)I(t) \\ \frac{dI(t)}{dt} &= \beta H(t)I(t) - \mu I(t) \\ \frac{dR(t)}{dt} &= \mu I(t) \end{aligned} \tag{5.18}$$

The concepts underlying equation 5.18 were first used by Kermack and McKendrick (1927), and this model is still the starting point of many studies in human, animal and plant disease epidemiology (Anderson and May, 1991; Daley and Gani, 1999; Jeger, 1982). In Box 5.3, we show in a little more depth how differential equations result from simple bookkeeping of the densities in the different categories or states of disease. To allow a comparison of the epidemic curves generated by this model with those of the previous model we also want to calculate the total density of individuals that has become infected since the start of the epidemic, $Y(t)$. This density is the sum of the infectious individuals and the removed individuals at a given time, $Y(t) = I(t)+R(t)$. The differential equation describing the dynamics of this quantity is:

$$\frac{dY(t)}{dt} = \beta H(t)I(t) \tag{5.19}$$

which is easily obtained by adding $dI(t)/dt$ and $dR(t)/dt$. $Y(t)$ here and Y_n in the previous model have the same interpretation.

Box 5.3

We consider how the densities of healthy, infectious and removed individuals change in a very small time step of length Δ, between time t and time $t + \Delta$. The density of healthy individuals at time $t + \Delta$, $H(t + \Delta)$, equals its density at time t, $H(t)$, minus the number per unit area that became infected between time t and time $t + \Delta$. To calculate the number per unit area that became infected in this time step, we assume here that the time step, Δ, is very small, implying that the difference between the density of healthy and infectious individuals at the start of the time step, time t, and at the end of the time step, time $t + \Delta$, is very small.

As in the derivation of equation 5.16, each infectious individual produces, on average, α spores per time unit. In the time step Δ, every infectious individual thus produces $\alpha\Delta$ infectious units. Using our assumption that the time step is very small we can approximate the density of infectious individuals during this time step with the density at the beginning of the time step, $I(t)$. Consequently, $\alpha\Delta I(t)$ is approximately the total number of spores produced in the time step Δ per unit area. The average number per unit area of healthy individuals contacted by an infectious unit is given by $\theta H(t)$. (Note that we make use of the approximation of small time step again.) Thus, the number of infectious units transmitted from infectious to healthy individuals equals $\alpha\Delta I(t)\theta H(t)$. An infectious unit deposited on a healthy individual has a probability ψ of infecting the individual. The total density of healthy individuals that becomes infected in the time step between t and $t + \Delta$ is $\psi\alpha\Delta I(t)\theta H(t)$. In equation form:

$$H(t + \Delta) = H(t) - \beta\Delta H(t)I(t) \qquad \text{(B5.3.1)}$$

were, again, $\beta = \psi\alpha\theta$.

The number per unit area of infectious individuals that reach the end of their infectious period between time t and $t + \Delta$ can be calculated in a similar way. Each individual has a probability μ per time unit to reach the end of its infectious period. In the small time step Δ the probability to reach the end of the infectious period thus is approximately $\mu\Delta$, and the density of individuals that reach the end of their infectious period in the time step from t to $t + \Delta$ is approximated by $\mu\Delta I(t)$. The equation for the change in the density of infectious individuals thus becomes:

$$I(t + \Delta) = I(t) + \beta\Delta H(t)I(t) - \mu\Delta I(t) \qquad \text{(B5.3.2)}$$

and for the removed:

$$R(t + \Delta) = R(t) + \mu\Delta I(t) \qquad \text{(B5.3.3)}$$

As discussed previously, differential equations model the rate of change at time t. The rates of change in the densities of healthy, infectious and removed individuals in the time step from t to $t + \Delta$ are given by

$$\frac{H(t+\Delta)-H(t)}{\Delta}, \frac{I(t+\Delta)-I(t)}{\Delta} \text{ and } \frac{R(t+\Delta)-R(t)}{\Delta}$$

respectively. We can rewrite equations B5.3.1–B5.3.3 such that they reflect these rates, by subtracting in each equation the first term on the right-hand side of the equal sign from both sides of the equal sign, and subsequently dividing both sides by Δ. This yields:

$$\begin{aligned} \frac{H(t+\Delta)-H(t)}{\Delta} &= -\beta H(t)I(t) \\ \frac{I(t+\Delta)-I(t)}{\Delta} &= \beta H(t)I(t) - \mu I(t) \qquad \text{(B5.3.4)} \\ \frac{R(t+\Delta)-R(t)}{\Delta} &= \mu I(t) \end{aligned}$$

If we now let the time step Δ become smaller and smaller, the left-hand side of the equations become closer and closer to the instantaneous rate of change at time t (equation 5.18).

5.2.2.2 Model simulations. Equations 5.18 and 5.19 cannot be solved analytically. That is, one cannot derive an equation for $H(t)$ or $I(t)$ (or $Y(t)$) simply as a function of time and parameters that does not involve the integration symbol. There is, however, a range of methods to *numerically* solve the set of differential equations 5.18 plus 5.19. It is beyond the scope of this book to go into detail on this topic, especially as there is a range of computer packages available that can numerically solve such systems of differential equations. To give some idea of such methods, however, we can simply consider the equations B5.3.1–B5.3.3 in Box 5.3. First, take a small value for Δ, choose a value for the parameters β and μ, and choose values for the initial conditions $H(0)$ and $I(0)$. Now substituting these values in the right-hand side of the equations B5.3.1–B5.3.3, we find the densities of healthy, infectious and removed individuals at time $t = \Delta$. Substituting these calculated density values again into the right-hand side of equations B5.3.1, B5.3.2 and B5.3.3, we find the densities of healthy, infectious and removed individuals at time $t = 2\Delta$. Repeating this process over and over again results in the solution of the model equations. Remember that the equations B5.3.1–B5.3.3 are approximations for small values of Δ, the smaller the value of Δ the better the approximation of the solution we find, but at the expense of having to take large numbers of time steps.

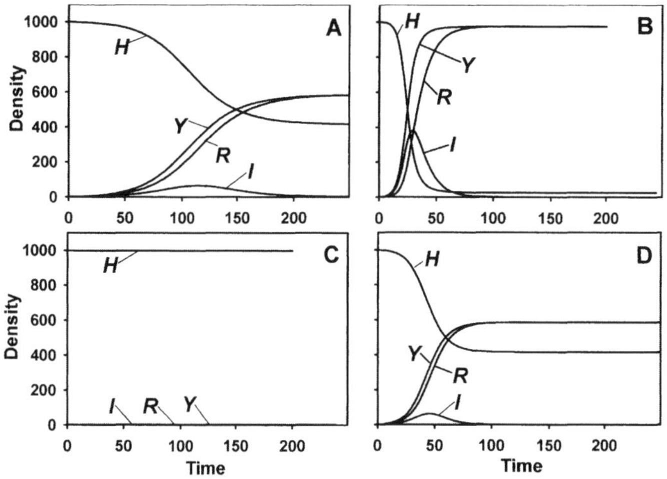

FIG. 5.5. The density of healthy (H), infectious (I), and removed (R) individuals, and the total density of infected individuals ($Y = I + R$), in relation to time since the start of the epidemic (t). Curves represent solutions to model equation 5.18. At the start of the simulation ($t = 0$), $H(0) = H_0 = 1000$, $I(0) = 1$ and $R(0) = 0$. (**A**) $\mu = 0.1$ (implying $\iota = 10$) and $\beta = 0.00015$ per time unit; (**B**) $\mu = 0.1$ ($\iota = 10$) and $\beta = 0.000375$ per time; (**C**) $\mu = 0.25$ ($\iota = 4$); and $\beta = 0.00015$ per time. (**D**) $\mu = 0.25$ ($\iota = 4$) and $\beta = 0.000375$ per time.

There are many more efficient numerical methods to solve differential equations (such as with the programs MATHCAD, MATHEMATICA, and MAPLE). However, the simple method of Box 5.3 can be easily programmed and will give surprisingly accurate solutions for a range of parameter values. Fig. 5.5 shows some numerical solutions of the model equations 5.18 and 5.19. Our initial conditions are again $H_0 = 1000$ and $I_0 = 1$ (together with zero removed individuals). Note that in this model β and μ both have an effect on the shape of the epidemic curves.

As in the previous model, when an epidemic develops, the disease progress curves [i.e., $Y(t)$ versus t] show a sigmoid shape. Also, the initial disease increase shortly after the start of the epidemic closely resembles an exponential curve (as described in Chapter 4). As we have discussed in detail for the previous model, the present model also shows that not every individual becomes infected during the epidemic.

For small values of β and large values of μ, no epidemic seems to develop (Fig. 5.5C). This agrees with our biological intuition about epidemic development. One expects that if an infectious individual transmits the disease at a very low rate (because of low α, θ, and ψ), no epidemic will occur; even when a single infectious individual is surrounded by an entirely healthy population, less than one new infection will result, on average. Similarly, when the probability per time unit that infectious individuals become removed is very large, an individual has a short time to transmit the disease; if this time is very short, one expects that an epidemic may not develop because of the smaller opportunity to produce new infections. Two other trends seen in Fig. 5.5 are biologically intuitive. The initial rate of exponential increase, r_E, is larger when β is larger, and r_E is smaller for larger values of μ.

Comparing Fig. 5.5A and D we see that the epidemic curves strongly differ in the initial rate at which the epidemic increases. Both epidemics do, however, finally approach the same total density of infected individuals, Y_∞. The r_E value thus does not seem to determine the final level of disease intensity, and the simulations do not show what combination of parameters exactly determines the final level of the epidemic.

The simulations suggest that there is a great deal of similarity between the epidemic curves resulting from this model and the model discussed in section 5.2.1 (with continuous time substituted for discrete generations). This is, however, just an impression on basis of visual inspection of graphs. The question on how much the dynamics of these models are the same is answered by analyzing certain aspects of the model. We will in the next sections show, among other things, to what extend the final number of infected individuals, as analyzed in section 5.2.1.5, corresponds to the final level shown by the model equations 5.18 and 5.19.

To aid the interpretation of the analysis, we first consider the parameter describing "the probability per time unit that an individual reaches the end of its infectious period", μ, in some more detail. In the model described here, the probability per time unit to stop producing infectious units is a *constant*, which results in a random variable for time-to-removal (duration of the infectious period for a population of individuals infected at the same time) that has an exponential distribution (see Lawless, 1982). This distribution is discussed in some detail in Chapter 7 in relation to distance of spore flights. For the exponentially distributed infectious period, the *mean* infectious period is the inverse of the rate parameter. We, thus, conclude that for the present model, the mean duration of the infectious period, ι, and the probability per time unit to leave the infectious stage, μ, are related by:

$$\iota = \frac{1}{\mu} \tag{5.20}$$

Note that in this model, the diseased individual starts producing infectious units immediately when it becomes infected because we have ignored the latent period.

5.2.2.3 The threshold for epidemic development. Consider the situation where, in the initial stage of the epidemic, the density of infectious individuals, $I(t)$, is very small. With this condition, the rates of change of the density of healthy individuals, $dH(t)/dt$, is very small. The number of healthy individuals will therefore not change very much during the initial stage of the epidemic and will remain close to its initial value, H_0. We thus approximate the density of healthy individuals in equation 5.18 by its initial condition, $H(t) \approx H_0$, which give us:

$$\frac{dI(t)}{dt} = \beta H_0 I(t) - \mu I(t) = (\beta H_0 - \mu)I(t) \qquad (5.21)$$

This differential equation is the model for exponential growth of $I(t)$ and is equal to equation 4.1 for $Y(t)$ or $y(t)$ with $r_E = \beta H_0 - \mu$. Its solution is:

$$I(t) = I_0 e^{r_E t} = I_0 e^{(\beta H_0 - \mu)t} \qquad (5.22)$$

Referring to the discussion on the exponential growth curve in Chapter 4, we can conclude from equation 5.22, that an epidemic will develop ($r_E > 0$) only when $\beta H_0 - \mu > 0$, and no epidemic will develop when $\beta H_0 - \mu < 0$. These inequalities can also be written as $(\beta H_0/\mu) > 1$ and $(\beta H_0/\mu) < 1$, and $\beta H_0/\mu$ can also be written as $\beta H_0 \iota$. In this format, the inequalities have a simple biological interpretation. Because β is the per capita disease transmission rate, the quantity βH_0 is the number of new infections produced by one infectious individual per time unit when it is surrounded by an entirely healthy population. As discussed above (equation 5.20), the mean infectious period is $\iota = 1/\mu$ time units. If an infectious individual produces new infections at a rate of βH_0 per time unit and it does that for on average $1/\mu$ time units, then the product of these, $\beta H_0/\mu$, has the interpretation of "the mean number of new infections (or diseased individuals) produced by one infected individual during its entire infectious period when placed in an entirely healthy population". One example would be the number of *daughter* lesions resulting from a single *mother* lesion in a field over the entire time period that the mother lesion is producing spores, when the number of lesions is so small that there is no chance that a spore produced by the mother lesion lands on a lesion. Vanderplank (1963) called R_0 the progeny-parent ratio, although he developed the term in the context of a different model (see section 5.2.4). Clearly, when $I(t)$ is small, an epidemic will develop when every infected or diseased individual produces more than one new infection. When every infected individual produces, on average, less than one new infection, no epidemic will develop because the infectious individuals do not replace themselves during the time available for transmission. This result is equivalent to what we found in the previous model. The interpretation of the threshold quantity, $\beta H_0/\mu$, is again equal to the definition of the basic reproduction number, R_0. To be explicit, we write:

$$R_0 = \frac{\beta H_0}{\mu} \qquad (5.23)$$

and state that an epidemic will develop when $R_0 > 1$ and will not develop when $R_0 < 1$. Sometimes the term *invasion* is applied to the situation when an epidemic occurs (Diekmann et al., 1990, 1995; Mollison, 1995). In other words, $R_0 > 1$ is known as the invasion criterion.

5.2.2.4 *Initial disease increase.* As equation 5.22 shows, when an epidemic is started with a small density of infected individuals, in the initial stage of the epidemic the density of infectious individuals will grow exponentially with rate $r_E = \beta H_0 - \mu$. As in the previous model, this initial exponential growth is a general property of the model and does not depend on the parameter value (as long as $R_0 > 1$). The total density of diseased individuals, $Y(t)$, also increases exponentially. This can be seen from model equation 5.19. As discussed above, in the initial stages of the epidemic, the density of healthy individuals can be approximated by its initial value, $H(t) \approx H_0$. Equation 5.19 then becomes:

$$\frac{dY(t)}{dt} = \beta H_0 I(t) \qquad (5.24)$$

and substituting equation 5.22 for $I(t)$ we find

$$\frac{dY(t)}{dt} = \beta H_0 I_0 e^{r_E t} \qquad (5.25)$$

which after integration yields

$$Y(t) = \frac{\beta H_0 I_0}{r_E} e^{r_E t} \qquad (5.26)$$

in which $r_E = \beta H_0 - \mu = \mu R_0 - \mu$. Increases in β (from either increases in ψ, α, or θ) or decreases in μ (increases in ι) lead to larger r_E.

5.2.2.5 *Final disease level.* In the previous model (section 5.2.1), we derived step-by-step the final level equation. For model equation 5.18, the derivation of the final level equation involves solving a linear differential equation. This calculation is developed in Box 5.4. Here we will only discuss the result of the calculation. The fraction of the initial population of healthy individuals that will become infected during the epidemic, $y_\infty = Y_\infty/H_0$ when started with a very small initial infection is calculated from:

$$y_\infty = 1 - e^{-R_0 y_\infty} \qquad (5.27)$$

where, in this case, R_0 is given by equation 5.23. Thus, the final level equation is the same as in our first model.

5.2.2.6 *Concluding remarks.* There is a strong similarity in the epidemic dynamics resulting from the model equations 5.18 and 5.2. Both have a threshold parameter, below which no epidemic will develop and above which an epidemic does develop. This threshold is the basic reproduction number, R_0. Both models show that when

BOX 5.4

This derivation of the final level equation follows that in Murray (1991). We here refer to the three equations in 5.18 as a, b, and c. Dividing equation 5.18b by equation 5.18a we find:

$$\frac{dI(t)}{dH(t)} = -\frac{(\beta H(t) - \mu)I(t)}{\beta H(t)I(t)} = -1 + \frac{\mu}{\beta}\frac{1}{H(t)} \quad \text{(B5.4.1)}$$

Separating variables and integrating both sides we find

$$\int dI(t) = \int\left(-1 + \frac{\mu}{\beta}\frac{1}{H(t)}\right)dH(t) \quad \text{(B5.4.2)}$$

which gives

$$I(t) = -H(t) + \frac{\mu}{\beta}\ln(H(t)) + C \quad \text{(B5.4.3)}$$

where C is the integration constant that is found by considering the state of the system at time $t = 0$

$$I_0 = -H_0 + \frac{\mu}{\beta}\ln(H_0) + C \quad \text{(B5.4.4)}$$

Substituting equation B5.4.4 into equation B5.4.3 and letting time go to infinity we find, realizing that $I_\infty = 0$ (because all diseased individuals will finally become removed in a epidemic with a fixed total number of individuals):

$$0 = -H_\infty + \frac{\mu}{\beta}\ln(H_\infty) + I_0 + H_0 - \frac{\mu}{\beta}\ln(H_0) \quad \text{(B5.4.5)}$$

Rearranging this equation, and substituting the relations $Y_\infty = H_0 - H_\infty$, $y_\infty = Y_\infty/H_0$, and $R_0 = \beta H_0/\mu$ we finally find equation 5.27 (when I_0 is very small).

A similar way to derive this expression is by dividing equation 5.18a by equation 5.18c:

$$\frac{dH(t)}{dR(t)} = -\frac{\beta}{\mu}H(t) \quad \text{(B5.4.6)}$$

Separating variables and integrating both sides we find

$$\int \frac{1}{H(t)}dH(t) = -\frac{\beta}{\mu}\int dR(t) \quad \text{(B5.4.7)}$$

which gives

$$\ln(H(t)) = -\frac{\beta}{\mu}R(t) + C \quad \text{(B5.4.8)}$$

The integration constant C can again be found by considering the state of the system at time $t = 0$

$$\ln(H_0) = -\frac{\beta}{m}R(0) + C \quad \text{(B5.4.9)}$$

At time $t = 0$ there are no removed individuals yet, so $R(0) = 0$. Substituting equation B5.4.9 in equation B5.4.8 and letting time go to infinity we find

$$\ln(H_{¥}) = -\frac{\beta}{\mu}R_{¥} + \ln(H_0)$$

Realizing that $R_\infty = I_0 + Y_\infty$ and rearranging the equation a little we find again equation 5.27 when I_0 is very small (~0).

an epidemic is started with a small density of infectious individuals the initial growth of the density of infectious individuals is exponential in form. The exponential rate parameter, r_E, can be expressed as a combination of the model parameters, βH_0 and μ. Both models have exactly the same final level equation, when expressed in terms of R_0.

5.2.3 The H-L-I-R epidemic model

So far none of the models we have shown included a latent period. This is a serious shortcoming of the models because it is very well known that many diseases have a considerable latent period, some even longer than the infectious period (Vanderplank, 1963). Therefore, we will extend the model from the previous section to include a latent period. Since this model is a little more complicated, we present a schematic form for the model (Fig. 5.6) to help the reader. The boxes refer to the densities of individuals in the different states, and the arrows show the transitions from one state to the next.

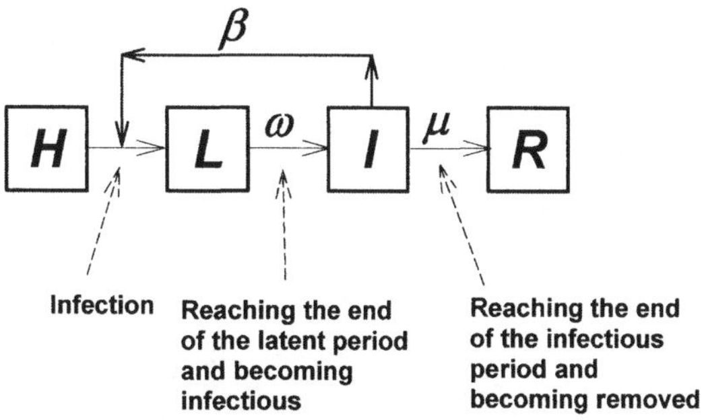

FIG. 5.6. Graphical representation of the H(ealthy)-L(atent)-I(nfectious)-R(emoved) (i.e., H-L-I-R) model of equations 5.30. Boxes represent the state variables H, L, I, and R. The parameter β is the transmission rate, ω is the probability per time unit that a latently infected individual transfers into the infectious category, and μ is the probability per time unit that an infectious individual transfers into the removed category.

5.2.3.1 Model derivation. We use the same notation for the density of healthy, infectious and removed individuals, $H(t)$, $I(t)$ and $R(t)$, respectively, as in the previous model. Adding to this we introduce the density of latently infected individuals as $L(t)$ (i.e., the number per unit area of individuals in the latent state). The main difference with the H-I-R model is that healthy individuals that become infected first enter the latently infected state. The latently infected individuals transfer after their latent period into the infectious category. The pseudo-model equations that describe this are:

$$\begin{aligned}
\frac{dH(t)}{dt} &= -\left[\begin{array}{l}\text{rate at which healthy}\\ \text{individuals become infected}\end{array}\right]\\
\frac{dL(t)}{dt} &= +\left[\begin{array}{l}\text{rate at which healthy}\\ \text{individuals become infected}\end{array}\right] - \left[\begin{array}{l}\text{rate at which latently infected}\\ \text{individuals become infectious}\end{array}\right]\\
\frac{dI(t)}{dt} &= +\left[\begin{array}{l}\text{rate at which latently infected}\\ \text{individuals become infectious}\end{array}\right] - \left[\begin{array}{l}\text{rate at which infectious}\\ \text{individuals reach the end}\\ \text{of the infectious period}\end{array}\right]\\
\frac{dR(t)}{dt} &= +\left[\begin{array}{l}\text{rate at which infectious individuals}\\ \text{reach the end of the infectious period}\end{array}\right]
\end{aligned} \tag{5.28}$$

The only new term in this model is the rate at which latently infected individuals become infectious. We assume that each latently infected individual has a constant probability, ω, per time unit to reach the end of its latent period and become infectious. The total density of latent individuals transferring from latent to infectious per unit time thus equals $\omega L(t)$ and we have:

$$\left[\begin{array}{l}\text{rate at which latently infected}\\ \text{individuals become infectious}\end{array}\right] = \omega L(t) \tag{5.29}$$

Substituting the relevant equations from the previous section and equation 5.29 in equation 5.28 we arrive at the model equations:

$$\begin{aligned}
\frac{dH(t)}{dt} &= -\beta H(t)I(t)\\
\frac{dL(t)}{dt} &= \beta H(t)I(t) - \omega L(t)\\
\frac{dI(t)}{dt} &= \omega L(t) - \mu I(t)\\
\frac{dR(t)}{dt} &= \mu I(t)
\end{aligned} \tag{5.30}$$

To make the epidemic curves generated by this model comparable to the previous model we again want to calculate the total density of individuals that has become infected since the start of the epidemic, $Y(t)$. This density is the sum of the latently infected individuals, infectious individuals and the removed individuals at a given time, $Y(t) = L(t) + I(t) + R(t)$. The differential equation describing the dynamics of this quantity is:

$$\frac{dY(t)}{dt} = \beta H(t)I(t) \tag{5.31}$$

obtained simply from $dL(t)/dt + dI(t)/dt + dR(t)/dt$.

Before discussing the disease progress curves generated by this model let us first consider the relation between the parameter ω, "the probability per time unit that a latently infected individual reaches the end of its latent period and becomes infectious", and the latent period. In the model described here, this probability per time unit is a constant. We are thus in the same situation as with the mean infectious period, equation 5.20 (where latent period was ignored), and we refer to that discussion on the exponential distribution. In short, the time that individuals are in the latent state is a random variable with an exponential distribution. The mean time an individual stays in the latent state for a population of infected individuals with an exponential distribution equals the reciprocal of ω. Thus, we can consider $\wp$ to be the *mean* latent period, given by:

$$\wp = \frac{1}{\omega} \tag{5.32}$$

Now that both latent and infectious periods have been explicitly incorporated into the epidemic model (equations 5.30 and 5.31), it is useful here to compare the model for Y (equation 5.31) with the logistic model for Y given at the end of Chapter 4, $dY(t)/dt = \Upsilon Y(t)H(t)$ (equation 4.51a). For the logistic model, $\Upsilon = r_L/M$, with M being the total host "size" (density here). If initial disease is very low, $H_0 \approx M$, and we can write the rate parameter as $\Upsilon = r_L/H_0$. In one sense, β in the H-L-I-R model is *analogous* to Υ ($=r_L/H_0$) of the logistic model (or βH_0 is analogous to r_L). Increases in r_L or βH_0 result in faster rates of disease increase (dY/dt) overall. However, βH_0 is clearly *not* the same as r_L. This is because in the logistic model, the rate of disease increase is proportional to *total* disease, $Y(t)$, which includes both latent and removed disease as well as infectious disease density. Thus, the logistic model can never fully represent the dynamics of disease development since, for polycyclic diseases, new diseased individuals result only from infectious individuals. As stated by Vanderplank (1963) for a different model (see section 5.4 below), this means that the logistic rate, r_L, depends on the transmission rate (β) as well as on $\wp$ and ι. However, the relationship is quite complicated over the duration of the epidemic. Nevertheless, the logistic model is easy to use for describing and comparing disease progress curves.

When $Y(t)$ or $y(t)$ is low, the exponential model is a good approximation for the logistic, with $r_E \approx r_L$, as shown in Chapter 4. For the two previous models in this chapter, we have already shown that disease increase is also approximately exponential during the initial stages of polycyclic epidemics. We continue to explore the relationship between the exponential model and those based on more biological realism in this and the following sections.

We note that in human epidemiology, disease-free individuals typically are labeled "susceptible" (S), and those in the latent state are labeled "exposed" (E). Thus, the H-L-I-R model is simply referred to as an S-E-I-R model in the literature (Diekmann and Heesterbeek, 2000), even without explanation. Plant pathologists often use terms such as susceptible for other purposes; for instance, susceptible is used as a label for a genetic trait, not the temporary status of an individual or group of individuals within the time scale of an epidemic. Thus, we use the H-L-I-R label in this book.

5.2.3.2 Model simulations. With three parameters, $\beta(=\psi\alpha\theta)$, $\mu(=1/\iota)$, and $\omega(=1/\wp)$, it starts to become more difficult to get a general overview of the effect of parameter values on disease progress curves. Fig. 5.7 shows a number of simulations with various parameter combinations. As in the previous models: (i) the curves are generally S-shaped; (ii) for some parameter combinations, an epidemic develops and for other combinations, no epidemic develops; (iii) the initial phase of epidemic increase seems to be of the exponential type; and (iv) decreased probability per time unit that an infectious individual reaches the end of its infectious period increases the growth rate of the epidemic.

The primary effect of the mean latent period is that the rate of increase appears to be large for small values of $\wp$, which is biologically intuitive—the sooner that infections become infectious, the higher the rate of disease increase. Furthermore, the length of $\wp$ does not seem to determine whether an epidemic does or does not develop, nor does it seem to determine the final disease level. Fig. 5.7 does, however, show only a small range of parameter values and a further mathematical analysis of the model is needed to show the generality of these simulation results.

Analysis to answer our three key questions relating to the epidemic threshold, the initial disease increase rate and the final disease level is still possible for this model. The threshold for epidemic development will be discussed without detailed derivation. The derivation of the final level equation has, as will be discussed, already been shown. To calculate the exponential growth rate we need solutions of a system of linear differential equations. This technique is not too complicated but would need several pages to explain. We will therefore only state and discuss the results of the analysis and sketch its derivation in Box 5.5.

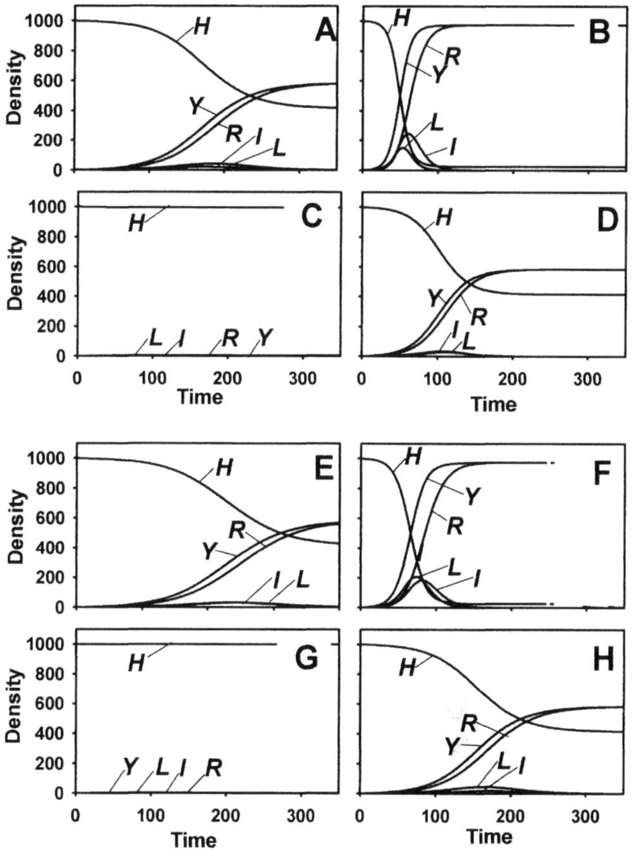

FIG. 5.7. The density of healthy (H), latently infected (L), infectious (I), and removed (R) individuals, and the total density of infected individuals ($Y = L + I + R$), in relation to time since the start of the epidemic (t). Curves represent solutions to model equations 5.30. At the start of the simulation ($t = 0$), $H(0) = H_0 = 1000$, $L(0) = 0$, $I(0) = 1$ and $R(0) = 0$. Upper four frames: short latent period; lower four frames: longer latent period. **(A)** $\mu = 0.1$ (implying $\iota = 10$), $\omega = 0.2$ (implying $\wp = 5$), and $\beta = 0.00015$ per time. **(B)** $\mu = 0.1$ ($\iota = 10$), $\omega = 0.2$ ($\wp = 5$), and $\beta = 0.000375$ per time unit. **(C)** $\mu = 0.25$ ($\iota = 4$), $\omega = 0.2$ ($\wp = 5$), and $\beta = 0.00015$ per time unit. **(D)** $\mu = 0.25$ ($\iota = 4$), $\omega = 0.2$ ($\wp = 5$) and $\beta = 0.000375$ per time unit. **(E)** $\mu = 0.1$ ($\iota = 10$), $\omega = 0.1$ ($\wp = 10$), and $\beta = 0.00015$ per time unit. **(F)** $\mu = 0.1$ ($\iota = 10$), $\omega = 0.1$ ($\wp = 10$), and $\beta = 0.000375$ per time unit. **(G)** $\mu = 0.25$ ($\iota = 4$), $\omega = 0.1$ ($\wp = 10$), and $\beta = 0.00015$. **(H)** $\mu = 0.25$ ($\iota = 4$), $\omega = 0.1$ ($\wp = 10$) and $\beta = 0.000375$ per time unit.

5.2.3.3 The threshold for epidemic development. As before, the threshold for epidemic development is the basic reproduction number, R_0. The formulation for R_0 expressed in the parameters of model equations 5.30 is:

$$R_0 = \frac{\beta H_0}{\mu} \tag{5.33}$$

which is exactly the *same* expression as for the basic reproduction number of the H-I-R model equations 5.18 that did not include a category of latently infected individuals. At first sight this might seem strange. However,

Box 5.5

For the H-L-I-R model, we start with the same approximation as we have used before. In the initial stages of the epidemic we can approximate $H(t)$ by its initial value H_0. Substituting this in equation 5.30 we get for the differential equations describing the density of latently infected and infectious individuals:

$$\frac{dL(t)}{dt} = \beta H_0 I(t) - \omega L(t)$$
$$\frac{dI(t)}{dt} = \omega L(t) - \mu I(t) \qquad \text{(B5.5.1)}$$

This is a system of linear differential equations. These can be solved explicitly. The book by Edelstein-Keshet (1988) describes the method in detail. The solutions are:

$$L(t) = c_1 e^{\lambda_1 t} + c_2 e^{\lambda_2 t} \quad I(t) = c_3 e^{\lambda_1 t} + c_4 e^{\lambda_2 t} \qquad \text{(B5.5.2)}$$

where λ_1 and λ_2 are the eigenvalues of the matrix:

$$J = \begin{pmatrix} -\omega & \beta H_0 \\ \omega & -\mu \end{pmatrix} \qquad \text{(B5.5.3)}$$

which are given by

$$\lambda_1 = \frac{1}{2}\sqrt{(\mu+\omega)^2 + 4\omega(\beta H_0 - \mu)} - \frac{1}{2}(\mu+\omega)$$
$$\lambda_2 = -\frac{1}{2}\sqrt{(\mu+\omega)^2 + 4\omega(\beta H_0 - \mu)} - \frac{1}{2}(\mu+\omega) \qquad \text{(B5.5.4)}$$

The quantities c_i depend on the initial conditions. The eigenvalue λ_2 is negative and with increasing time, the exponential term with this eigenvalue will vanish (go to zero). This implies that the densities of latently infected and infectious individuals is given approximately by:

$$L(t) \cong c_1 e^{\lambda_1 t} \quad I(t) \cong c_3 e^{\lambda_1 t} \qquad \text{(B5.5.5)}$$

From which we conclude that both densities grow exponentially, with growth rate λ_1; thus, $r_E = \lambda_1$ in equation 5.34. Note that λ_1 is only above 0 when $\beta H_0 - \mu > 0$ (or $\beta H_0/\mu > 1$) in equation B5.5.4. Thus, there is only an epidemic when $R_0 > 1$, as demonstrated with other models.

---

in this model, individuals in the latently infected state will eventually all transfer into the infectious category (i.e., we have not included in the model the possibility that a latently infected individual might leave the system due to, for example, death). Therefore, the mean total number of new infections produced by one infectious individual during its entire infectious period when placed in an entirely healthy population does not depend on the latent period. With increasing $\wp$, it will take longer to produce all the new infections, but ultimately the same number is produced.

An other way to see this is to consider the sum of $L(t)$ and $I(t)$ [$=Q(t)$], which is all the non-removed disease. The equation for $dQ(t)/dt$ can be obtained from equations 5.30 by $dQ(t)/dt = dL(t)/dt + dI(t)/dt$. Assuming that $H(t) = H_0$ (as before), because we are concerned here with the early stages of an epidemic when $I(t)$ is very low, the reader is left to confirm that $dQ(t)/dt$ is given by equation 5.21, with Q substituted for I on the left-hand side. $Q(t)$ can only increase if βH_0 is larger than μ, or equivalently $R_0 = \beta H_0/\mu > 1$. In the next chapter, we discuss the basic reproduction number in a situation where latently infected individuals can leave the system due to, for example, death.

5.2.3.4 Initial disease increase. As in the previous two models, it can be shown that when the epidemic is started with a very small initial number of latently infected or infectious individuals, or a very small combination of these densities, infectious disease density will initially increase exponentially as:

$$I(t) = I_0 e^{r_E t} \qquad (5.34)$$

with the exponential growth rate parameter, r_E, given by:

$$r_E = \frac{1}{2}\sqrt{(\mu+\omega)^2 + 4\omega(\beta H_0 - \mu)} - \frac{1}{2}(\mu+\omega) \qquad (5.35)$$

See Box 5.5 for further details. In section 5.2.3.1, we compared the differential equations for the logistic and H-L-I-R models, and pointed out that the logistic rate (r_L)—or for small $y(t)$, the exponential rate (r_E)—incorporates aspects of βH_0, $\wp(1/\omega)$, and $\iota(1/\mu)$. Equation 5.35 here formalizes this relationship mathematically for the early stages of a polycyclic epidemic. In the previous two models, r_E could be expressed as a very simple function of the parameters, and the dependence of r_E on the parameters could simply be understood from its expression. Unfortunately, because of the extra complications of a latent period, equation 5.35 is not very instructive. Drawing graphs of r_E as function of the parameters helps in this respect. To make the interpretation easier and to make the results comparable to the next model we will discuss, we change the parameters from the probabilities per time unit, μ and ω, to the parameters for mean infectious period, ι, and mean latent period, $\wp$, using equations 5.20 and 5.32. Some rearranging of the expression finally yields:

$$r_E = \sqrt{m^2 + \frac{1}{\iota\wp}(\beta H_0 \iota - 1)} - m, \quad \text{where } m = \frac{\wp + \iota}{2\wp\iota} \qquad (5.36)$$

Introducing the expression for R_0 in equation 5.36 we have

$$r_E = \sqrt{m^2 + \frac{1}{\iota \wp}(R_0 - 1)} - m, \quad \text{where } m = \frac{\wp + \iota}{2\wp\iota} \quad (5.37)$$

The reader can confirm that when $\beta H_0 \iota$ ($=R_0$) = 1, both equations 5.36 and 5.37 indicate that $r_E = 0$, as required, and that r_E is only larger than 0 when $R_0 > 1$. That is, there is no epidemic unless $R_0 > 1$. This relationship between r_E and the parameters of the H-L-I-R model serves as an excellent way to contrast the simple growth-curve models of Chapter 4 and the more realistic ones in this chapter. Stating that an epidemic requires that r_E be greater 0 is, in a sense, an empty conclusion or criterion because it does not tell you very much about the population biology of the disease of interest. For instance, with a fungal pathogen, one might think that any sporulation rate per diseased individual (α) above zero, any positive probability of a spore reaching a host plant (θ), or any positive probability of a spore causing an infection when in contact with a healthy plant individual (ψ) could lead to an epidemic. However, as shown in this and the previous sections of this chapter, there is a *nonzero* threshold combination of α, θ, and ψ, expressed by their product, $\beta = \alpha\theta\psi$, that must be exceeded for an epidemic to occur.

We intentionally give these two slightly different models for r_E (equations 5.36 and 5.37) because, as we will see, it changes the *apparent* dependence of r_E on the parameters (when $r_E > 0$). But before going into the difference between equations 5.36 and 5.37, let us discuss some numerical examples. Assume we have done a series of experiments and found that the disease under study has an average latent period of $\wp = 7$ days and an average infectious period of $\iota = 10$ days. Further we have monitored an epidemic that has started in a crop field, and after analyzing the data using the methods discussed in Chapter 4, we conclude that in the beginning stages of the epidemic $r_E = 0.05$/day. What does this tell us about R_0 and the βH_0 transmission rate parameter? To calculate βH_0 we first have to express βH_0 as a function of the other known (or assumed known) parameters from equation 5.36. Some algebraic manipulations show that:

$$\beta H_0 = \frac{\left((r_E + m)^2 - m^2\right)\wp\iota + 1}{\iota} \quad (5.38)$$

Substituting $\wp = 7$ and $\iota = 10$ in $m = (\wp + \iota)/(2\,\wp\iota)$, we find that $m = 0.1214$. Next substituting m, $\wp$, ι, and $r_E = 0.05$ into equation 5.38 we find that the transmission rate is $\beta H_0 = 0.2025$/day. Furthermore, we know that $R_0 = \beta H_0 \iota$, and substituting the calculated or assumed numbers we find that $R_0 = 2.025$. The reader should be reminded here that βH_0 has units of time^{-1}, but that R_0 is unitless.

Suppose we can choose between two disease management methods, one reduces the transmission rate by 50%, the other one increases the mean latent period of the disease by a factor of 2. The first scenario can be achieved using a protective fungicide that reduces the probability that an infectious unit that landed on a healthy individual actually infects the host (ψ, one of three terms making up β). The second scenario can be achieved using a curative fungicide that reduces the development rate of the pathogen. To calculate the effect of these two possibilities we substitute for the first one $\wp = 7$, $\iota = 10$ and $\beta H_0 = 0.2025/2 = 0.10125$ into equation 5.36 and the expression for R_0. We find $R_0 = 1.0125$ and $r_E = 0.0007$/day. In other words, because R_0 is very close to 1, there is very slow increase in disease. For the second possibility we substitute $\wp = 7 \times 2 = 14$, $\iota = 10$ and $\beta H_0 = 0.2025$ into equation 5.36 (with m now given by 0.0857), and find that $r_E = 0.035$/day, lower than before the control, but not changed nearly as much as found when changing β. Clearly, the protective fungicide is the most effective control in this example situation. Note that this is only an example, and we do not advocate that protective fungicides are more effective than curative fungicides, in general—it all depends on the parameter values and on the magnitude of the effect of the fungicide on these parameter values that can be achieved in field applications.

We return to our discussion on the equations 5.36 and 5.37. Before drawing the graphs we have to choose parameter values. We choose values characteristic for two contrasting plant diseases, one highly infectious (large βH_0) with a long infectious period and one moderately infectious with a short infectious period. Stripe rust, caused by *Puccinia striiformis*, is a disease of wheat. The foliar fungal pathogen has a high rate of spore production (α) and long infectious period (ι), characteristic for many rust fungi. Downy mildew, caused by *Peronospora farinosa*, is a disease of spinach. The foliar oomycete plant pathogen has a low spore production rate and short infectious period. In a reanalysis of data published by van den Bosch et al. (1988c), Segarra et al. (2001) estimated for the stripe rust: $R_0 = 55$, $\wp = 10$ days and an infectious period $\iota = 20$ days. In the H-L-I-R model this corresponds to a transmission rate of $\beta H_0 = 2.75$/day. For downy mildew they found: $R_0 = 3.2$, $\wp = 7$ days, and $\iota = 3.4$ days. This corresponds to a transmission rate of $\beta H_0 = 0.94$/day.

Fig. 5.8 shows the exponential growth rate of the epidemic, r_E, as a function of βH_0, $\wp$, and ι, for both diseases. In Fig. 5.8A and B, $\wp$ and ι are fixed at the values mentioned above, and the transmission rate, βH_0, is varied as a parameter on the horizontal axes. Similarly, in Fig. 5.8C and D, βH_0 and ι are fixed, and $\wp$ is varied on the horizontal axis. The asterisk marks the parameter value mentioned above. Fig. 5.8E and F are constructed in a similar way.

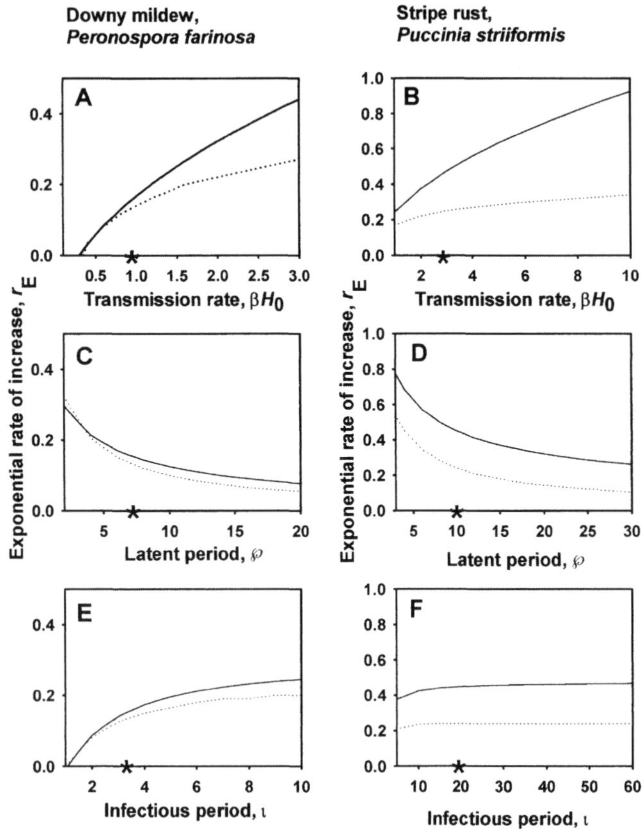

FIG. 5.8. The exponential rate of increase of disease (r_E) as function of (A) and (B) transmission rate; (C) and (D) latent period; and (E) and (F) infectious period, corresponding to values appropriate for *P. farinosa*, left column, and *P. striiformis*, right column. The drawn line shows r_E for the H-L-I-R model, calculated from equation 5.36. The dotted line shows r_E for the Vanderplank model (1963), calculated from equation 5.49a. Parameter values that are not plotted in a given graph are: $\wp$ = 7 days, ι = 3.4 days, and βH_0 = 0.94/day (for *P. farinosa*); and $\wp$ = 10 days, ι = 20 days, and βH_0 = 2.75/day (for *P. striiformis*). Asterisk on the horizontal axis shows the parameter value used as default in the other graphs of this figure.

From the figure we see that the larger the transmission rate, βH_0, the larger the r_E, which can be understood intuitively, since both quantify the increase in disease over time. When the transmission rate of downy mildew is smaller than ~0.3/day, r_E is zero, indicating no epidemic develops. This, of course, corresponds to values of $R_0 < 1$. The relation between βH_0 and r_E is markedly non-linear for small βH_0. For small values of βH_0, the small changes in βH_0 have a considerable effect on the exponential growth rate of the epidemic, whereas for large βH_0 values, changes in βH_0 have little effect on the exponential growth rate. One of the consequences of this relation between transmission rate and exponential growth rate is that disease management targeted at reducing the transmission rate will be more effective for pathogens with a small transmission rate than for pathogens with a high transmission rate.

The value of $\wp$ clearly has a considerable effect on r_E, with long mean latent periods corresponding to small epidemic growth rates, as is intuitively obvious since it takes longer for new infectious units to be first produced. When $\wp$ is short, a change in $\wp$ has a large effect on r_E. When the $\wp$ is long, a change in $\wp$ has much less effect on the exponential growth rate. Again this has considerable implications for disease management. A fungicide or resistant cultivar, for example, that affects the development rate of the pathogen within the host, and therewith increases the mean latent period, will be very effective for pathogens with a short mean latent period but much less effective for pathogens with a long mean latent period.

The exponential growth rate, r_E, increases with the length of the mean infectious period, ι, which again is intuitively obvious. The longer individuals are infectious, the higher the rate of increase in the population since there is more opportunity to cause infections. The dependence of r_E on ι follows the same pattern as the dependence of r_E on βH_0.

Fig. 5.9 is constructed in the same way as Fig. 5.8, but in this case equation 5.37 is used. The dependencies of r_E on R_0 and $\wp$ are very comparable to Fig. 5.8. The main difference between the two figures is in the effect of the mean infectious period, ι, on r_E (Fig. 5.8E and F compared with Fig. 5.9E and F). The trends are the opposite for the two representations of the exponential growth rate in equations 5.37 and 5.36. The increase in r_E with increasing ι, as we discussed in relation to Fig. 5.8, is intuitively obvious, but the decreasing r_E with increasing ι shown in Fig. 5.9 seems entirely counterintuitive. There is, however, a very essential difference between the two representations of the model. In equation 5.36 and corresponding Fig. 5.8, an increase in ι has the effect of increasing R_0 when βH_0 is fixed; in other words, R_0 is not constant in Fig. 5.8. Based on the definition of R_0 ($=\iota\beta H_0$), if we keep βH_0 fixed and increase ι, R_0 must increase. That an increase in the R_0 results in an increase in the exponential growth rate is again obvious.

In equation 5.37 and corresponding Fig. 5.9, we keep R_0 fixed. So in this equation and figure, a change in ι does not change the R_0. This implies that if we decrease ι (without changing R_0), the same total number of new infections per infectious individual (if surrounded by an entirely healthy population) are still produced, but in a shorter time span. This can only happen if βH_0 increases as ι decreases. Likewise, βH_0 decreases as ι increases if R_0 is kept constant, since the new infections are spread out over a longer time period. Thus, it is intuitively clear that r_E is inversely related to ι when R_0 is fixed. We have brought this point up to illustrate the fact that the results and interpretation can depend on the parameterization chosen, and readers should be cautious when choosing a parameterization.

5.2.3.5 Final disease level. The final level equation is exactly the same as equation 5.14 for the simpler H-I-R

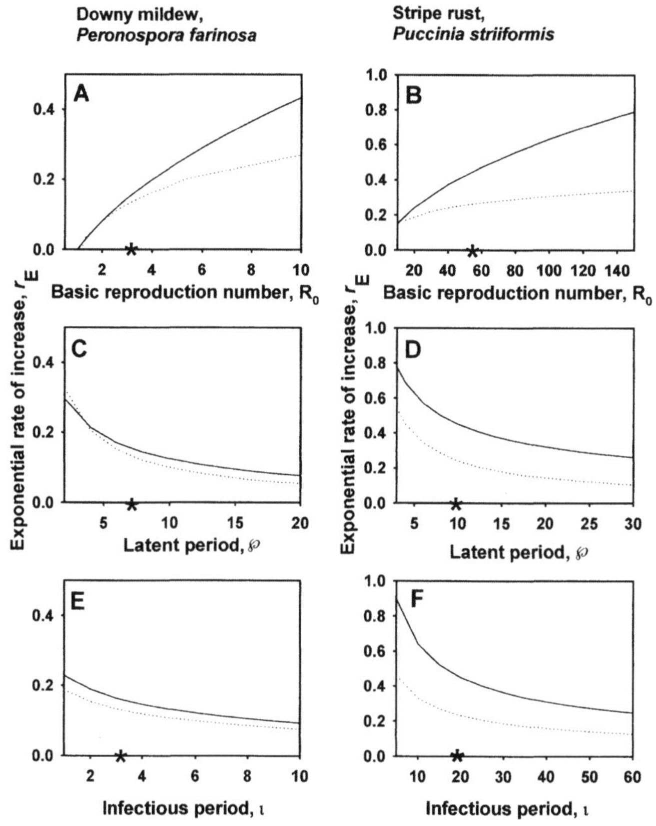

FIG. 5.9. The exponential rate of increase of disease (r_E) as function of (**A**) and (**B**) basic reproduction number; (**C**) and (**D**) latent period; and (**E**) and (**F**) infectious period, corresponding to values appropriate for *P. farinosa*, left column, and *P. striiformis*, right column. The drawn line shows r_E for the H-L-I-R model, calculated from equation 5.37. The dotted line shows r_E for the Vanderplank model, calculated from equation 5.49b. Parameter values that are not plotted in a given graph are: $\wp = 7$ days, $\iota = 3.4$ days, and $R_0 = \iota\beta H_0 = 3.2$ (for *P. farinosa*); and $\wp = 10$ days, $\iota = 20$ days, and $R_0 = \iota\beta H_0 = 55$ (for *P. striiformis*). Asterisk on the horizontal axis shows the parameter value used as default in the other graphs of this figure. Same as Fig. 5.8, except that R_0 is fixed in C–F (instead of βH_0), and that r_E is plotted against R_0 in A and B (instead of against βH_0).

model, with equation 5.33 for R_0 (see Fig. 5.4 for y_∞). Because all diseased individuals ultimately become infectious and then removed (with $\omega > 0$ and $\mu > 0$) with sufficiently long time period (when there is no plant mortality as in this model), all diseased individuals will be in the removed state before 100% of the host population is diseased. This can be seen mathematically from Box 5.4 where the second method of deriving the final level equation (i.e., that based on dH/dR) applies to the model equation 5.30 as well. As can be seen from Fig. 5.4, the effect of R_0 on y_∞ is most pronounced at $R_0 \leq 3$.

Results for the H-L-I-R model can be compared with the logistic (or similar) growth-curve model of Chapter 4. A final level of y less than 1 is easily accommodated in the logistic model by use of a parameter for the upper limit (asymptote, K), as discussed in section 4.8. However, K in equation 4.47 is purely an empirical value that is not defined in terms of the underlying mechanisms of population growth. Now, by modeling the epidemic based on more fundamental parameters, we see that a biological reason for K being less than 1 is that all diseased individuals become removed before all plants are diseased. The basic reproduction number is all that is needed to determine final disease level (when initial disease intensity is low). In a sense, y_∞ is analogous to K. It is important to note that the governing equations for the H-L-I-R model (equation 5.30) do *not* include a parameter for final H (and thus final Y); rather, the final values are the natural *outcomes* of the underlying population dynamics of disease.

5.2.3.6 Some concluding remarks. Again many features of this H-L-I-R model are the same as for the previous two models. Basically, the only real difference between the three is the specific expression for r_E in terms of the parameters of the particular model. The meaning of R_0 and r_E are unchanged, however. The expression for r_E is more complicated because it now accounts for the latent period. Thus, the formula for r_E of the previous simpler model (H-I-R) would only be appropriate to use if $\wp$ was very short.

5.2.3.7 Recapitulation of the model equations—role of latent and infectious periods. When discussing the parameters ι and $\wp$ we referred to them as the *mean* infectious period and the *mean* latent period instead of simply the infectious period and the latent period. Although this might make reading tiring in some places, it was done for a very good reason. In this section we take a closer look at the latent period and the infectious period in the model.

In deriving the model we assumed that there is a constant probability per time unit, ω, that a latently infected individual reaches the end of its latent period and becomes infectious. Assume it would be possible to do an experiment of the following type: We start at time $t = 0$ with a group of L_0 individuals that are newly infected, and thus just entered their latent period. We stop any secondary infections developing and only observe how many of the initial group of latently infected individuals are still in the latent period. These observations are described by the differential equation:

$$\frac{dL(t)}{dt} = -\omega L(t) \tag{5.39}$$

This model describes a simple exponential decay and the solution of this equation is:

$$L(t) = L_0 e^{-\omega t} \tag{5.40}$$

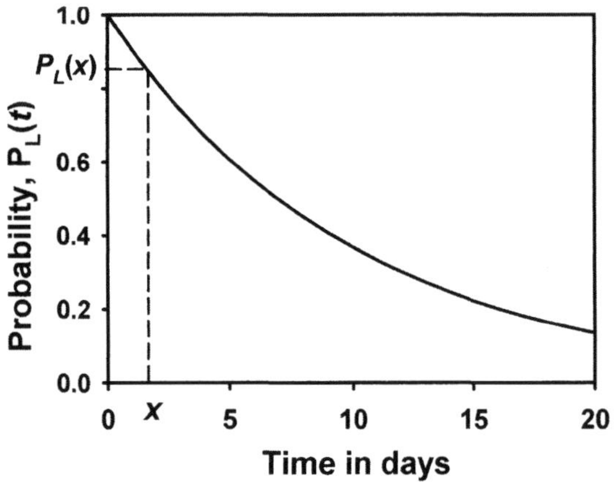

FIG. 5.10. Probability to be in the latent period as a function of time since the individual became infected based on equation 5.41.

This means that the probability that an individual is still latent t time units after it has become infected, $P_L(t)$, is given by the exponential function:

$$P_L(t) = \frac{L(t)}{L_0} = e^{-\omega t} \tag{5.41}$$

Fig. 5.10 shows an example graph of this function. The graph makes clear that some individuals have a very short latent period and others have a long latent period. Since not all individuals will have the same latent period, the description of the latent period incorporated in the model is quite realistic. The model, however, does fall short in the sense that some individuals reach the end of their latent period very shortly after they became infected. For example, in Fig. 5.10, x time units after becoming infected, a small fraction of $1 - P_L(x)$ individuals already moved into the infectious category. Realistically one would expect that for some period of time after becoming infected *all* individuals remain in their latent period (so that $P_L(t) = 1$ for $t > 0$) and then gradually the probability to be in the latent period decreases.

Exactly the same concept holds for the infectious period as incorporated in the model. The probability to be in the infectious period t time units after becoming infectious is given by an exponential distribution function. Again the fact that there is variation in the infectious period between individuals is realistic, but some individuals will have an unrealistically short infectious period in the model.

At this point one might wonder why we expand on this in so much detail, and why there actually is a next section introducing yet another model. All models so far show the same epidemic dynamics: (i) there is a threshold for epidemic development, which is quantified by R_0; (ii) when the epidemic is started with a small initial infection, the number or density of infectious individuals grow exponentially in the initial stage of the epidemic; and (iii) the same final level of disease intensity is reached, defined as a function of R_0. The only difference is in how the basic reproductive number, R_0, and the exponential growth rate, r_E, depend on the details of the model specification. We will, however, see with the next model that this last aspect has some important consequences.

5.2.4 The Vanderplank model

Vanderplank (1963) developed a model for the dynamics of plant disease epidemics that takes account of the latent period and the infectious period. He developed this model in 1963, a year that could be seen as the starting point of theoretical or quantitative plant disease epidemiology (see Chapter 1). In this section we study his general model in the light of the theory we have developed so far.

5.2.4.1 Model derivation. As in the differential equation models of the previous sections, Vanderplank's model describes the rate of change in the density of healthy plants as:

$$\frac{dH(t)}{dt} = -\beta H(t)I(t) \tag{5.42}$$

We now express the density of infectious individuals, $I(t)$, in terms of the density of diseased individuals, $Y(t)$, in the past. To follow the derivation, one has to keep the difference between infected (or diseased) and infectious in mind: infected individuals can be in their latent state, in their infectious state, or be in the removed state. The Vanderplank model assumes that the latent period has a *fixed* duration $\wp$ which is the same for every individual. Further, the infectious period is assumed to have a fixed duration ι which is the same for every individual. This implies that individuals that are infectious at time t became infected between $\wp$ time units ago (the latent period) and $\wp + \iota$ time units ago (the latent plus infectious periods). Individuals infected less than $\wp$ time units ago are still in their latent state and individuals infected more then $\wp + \iota$ time units ago are in the removed category. The individuals that are infectious at time t thus became infected between time $t - \wp - \iota$ and time $t - \wp$. For instance, consider $\wp = 5$ and $\iota = 10$. Infectious individuals at time $t = 40$ became infected between time $t = 25$ and $t = 35$. Any individuals infected *before* time $t = 25$ are in the removed state, and any individuals infected *after* time $t = 35$ are in the latent state. The density of individuals that become infected in the time interval from $t - \wp - \iota$ to $t - \wp$ is simply the difference between the density of infected individuals at $t - \wp - \iota$, $Y(t - \wp - \iota)$, and the density at $t - \wp$, $Y(t - \wp)$. This is illustrated in Fig. 5.11. We thus have the relation $I(t) = Y(t - \wp) - Y(t - \wp - \iota)$ and the Vanderplank model takes the form:

$$\frac{dH(t)}{dt} = -\beta H(t)\left(Y(t - \wp) - Y(t - \wp - \iota)\right) \tag{5.43}$$

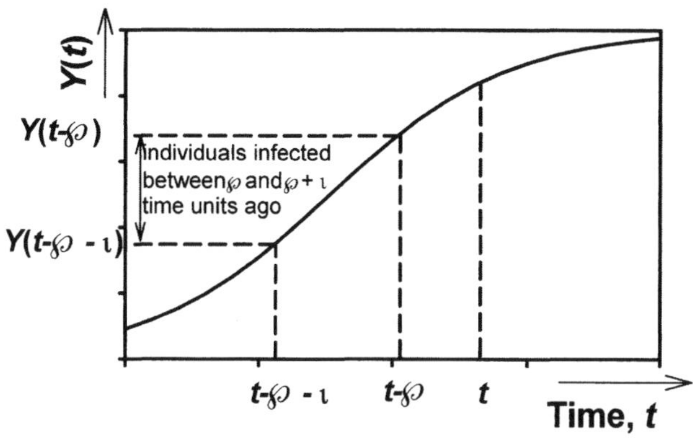

FIG. 5.11. The density of diseased individuals, $Y(t)$, plotted versus time t, for a hypothetical disease progress curve based on the modeling concepts of Vanderplank (1963). When the latent period, $\wp$, and the infectious period, ι, have a fixed duration for all infected individuals, the density of infectious individuals equals the difference between the density of infected individuals at time $t - \wp$, $Y(t - \wp)$, and the density of infected individuals at time $t - \wp - \iota$, $Y(t - \wp - \iota)$.

The model can entirely be expressed in the density of infected (diseased) individuals if we assume that the density of healthy individuals plus the density of diseased individuals adds up to the density of healthy individuals, H_0, at the start of the epidemic, so that $H(t) = H_0 - (Y(t)$. Substituting this in equation 5.43 we get:

$$\frac{dY(t)}{dt} = \beta(H_0 - Y(t))\left(Y(t-\wp) - Y(t-\wp-\iota)\right) \qquad (5.44)$$

The density of individuals in their latent period is given by $L(t) = Y(t) - Y(t - \wp)$, and the density of removed is given by $R(t) = Y(t - \wp - \iota)$.

This completes the derivation of the Vanderplank model. Equation 5.44 is sometimes called a *differential-delay equation* or differential-difference equation. The analysis of differential equations with time delay terms, such as $t-\wp$ and $t-\wp-\iota$, is quite complicated and beyond the scope of this book. We refer the interested reader to Reddingius (1971) and Metz (1978b). However, in the previous sections we have built up some good intuition about the dynamics of epidemics, and we will show that this intuition is sufficient to express the basic reproduction number, R_0, and the exponential growth rate, r_E, as a function of the model parameters. As it turns out, many of the interpretations of equation 5.44 here are the same as for the H-L-I-R model (equation 5.30).

Readers familiar with Vanderplank's (1963) book know that he expressed his differential-delay equation for disease intensity as a proportion, not a density. To show how equation 5.44 can be expressed for proportion diseased $[y(t) = Y(t)/H_0]$, note that $dy(t)/dt = d[Y(t)/H_0]/dt$, $[H_0 - Y(t)]/H_0 = 1 - y$, and $[Y(t - \wp) - Y(t - \wp - \iota)]/H_0 = y(t - \wp) - y(t - \wp - \iota)$. Because one divides the right-hand side of equation 5.44 by H_0 twice to obtain only proportions instead of densities, but the left-hand side is divided by H_0 only once to obtain $dy(t)/dt$, β must be multiplied by H_0 on the right-hand side. This gives $(\beta H_0)[1-y(t)][y(t-\wp)-y(t(t-\wp-\iota)]$ for $dy(t)/dt$. This version of the equation makes it clear that βH_0 (and not just β) is the rate parameter of interest.

5.2.4.2 The threshold for epidemic development. We already know that R_0 is the total number of new infections due to one infected individual during the whole course of its infectious period when it is placed in a completely healthy population. We also know that R_0 determines whether an epidemic will or will not develop. As in the differential equation models discussed in sections 5.2.2 and 5.2.3, in the Vanderplank model, individuals are not removed from the latent state except by becoming infectious and are not removed from the infectious state except by becoming post-infectious (i.e., there is no mortality of diseased individuals prior to being removed). This implies that the latent period will not affect R_0, because all infected individuals ultimately go through all the disease states and become post-infectious.

Consider one infectious individual placed in a host population consisting only of healthy individuals, and a per capita transmission rate of βH_0. Remember that this transmission rate is the number of new infections due to one infectious individual per time unit. Each individual is infectious for ι time units and, thus, the total number of new infections caused before the end of the infectious period is $\iota\beta H_0$. In other words, $R_0 = \iota\beta H_0$, the same value as determined for the H-L-I-R model. Vanderplank called βH_0 the *corrected basic infection rate* and as mentioned before, R_0 the progeny-parent ratio.

5.2.4.3 Initial disease increase: Starting with a very small initial disease level (small $Y(0)$), the density of infected individuals is, in the initial stages of the epidemic, negligible compared to the density of healthy individuals. This implies that $H_0 - Y(t) \approx H_0$. Equation 5.44 then becomes:

$$\frac{dY(t)}{dt} = \beta H_0\left(Y(t-\wp) - Y(t-\wp-\iota)\right) \qquad (5.45)$$

The models in the previous sections showed in the initial stages of the epidemic an exponential increase of the form:

$$Y(t) = Y_0 e^{r_E t} \qquad (5.46)$$

Assuming that for this model the same general result holds, and that equation 5.46 is a solution of model equation 5.45 at very small disease intensity, we substitute the right-hand side of equation 5.46 for $Y(t)$ in equation 5.45 and obtain:

$$\frac{d[Y_0 e^{r_E t}]}{dt} = \beta H_0\left(Y_0 e^{r_E(t-\wp)} - Y_0 e^{r_E(t-\wp-\iota)}\right) \qquad (5.47)$$

which becomes

$$Y_0 r_E e^{r_E t} = \beta H_0 Y_0 e^{r_E t} \left(e^{-r_E \wp} - e^{-r_E (\wp + \iota)} \right) \tag{5.48}$$

Dividing out the term $Y_0 \exp(r_E t)$ on both sides, one obtains

$$r_E = \beta H_0 \left(e^{-r_E \wp} - e^{-r_E (\wp + \iota)} \right) \tag{5.49a}$$

The last equation specifies the exponential growth rate, r_E, in terms of the model parameters of equation 5.45. Equation 5.49a cannot be solved explicitly for r_E. But as we have seen with the equation for the final level of an epidemic, such equations can be solved numerically. Note that equation 5.49a can be written in terms of the basic reproduction number:

$$r_E = \frac{R_0}{\iota} \left(e^{-r_E \wp} - e^{-r_E (\wp + \iota)} \right) \tag{5.49b}$$

Figure 5.8 show the exponential growth rate as function of the transmission rate, βH_0, the latent period, $\wp$, and the infectious period, ι for the Vanderplank model (dotted curves). The solutions of equation 5.49 are plotted in the same graphs as for the solutions with the H-L-I-R model discussed in the previous section. Qualitatively, the relations between the model parameters and r_E are very similar to the ones discussed for the H-L-I-R model and we will not reiterate on this. There is, however, a striking *quantitative* difference in results between the two models, especially for large R_0 and βH_0, as in the example of stripe rust.

For most parameter combinations, the calculated r_E for the H-L-I-R model is larger than for the Vanderplank model, but for some parameter combinations, the Vanderplank model has the largest calculated r_E. The general trend is that for parameter values that give a very small r_E, the Vanderplank model produces a slightly larger r_E, and for parameter combinations that give a high r_E, the H-L-I-R model produces a larger r_E.

5.2.4.4 Final disease level. Various authors (e.g., Reddingius, 1971; Metz, 1978b) have shown that the final level equation is again equation 5.27, and we do not elaborate on the proof. Jeger and van den Bosh (1994a) give a thorough presentation of the derivation of the final-value result for the Vanderplank model for a fuller range of conditions than discussed here. Vanderplank (1975) used equation 5.21 heavily to characterize diseases of perennials over many years.

5.2.4.5 Concluding remarks. From the discussion about Fig. 5.8, we can conclude that many of the qualitative features of the model discussed here are very similar to features of the previous model, the quantitative differences in calculated r_E between models can, however, be large. This is especially relevant for diseases with a large βH_0 or R_0. This shows that if we want to use information on the latent and infectious periods to predict r_E, it is important to utilize an epidemic model that reasonably well represents the actual distribution of latent and infectious periods. In this respect, it may be quite difficult to choose between using the H-L-I-R model or the Vanderplank model to calculate r_E, because both model formulations have the same level of biological detail, and level of realism (or lack of realism). For the H-L-I-R model, too many individuals transfer very early from the latent to the infectious state, and from the infectious to removed state. For the Vanderplank model, all individuals have exactly the same latent period and same infectious period. In the next section we will therefore introduce a model that describes latent and infectious period more accurately, but with the penalty of being more complicated.

5.2.5 The Kermack and McKendrick model

Nearly 80 years ago Kermack and McKendrick (1927) developed a very general model for epidemics that still has a remarkable and profound influence on theoretical epidemiology (Daley and Gani, 1999; Diekmann and Heesterbeek, 2000; Diekmann et al., 1995). One important property of the model is that few assumptions are made regarding the distribution of latent and infectious periods. For the application to plant diseases, van den Bosch et al. (1988b) and Segarra et al. (2001) introduced a quite general model description of latent and infectious period that fits measurements well, and use it to calculate the exponential rate of increase for the early stages of an epidemic.

The full Kermack and McKendrick model and its analysis are beyond the scope of this book. We will, however, show the derivation of the model for the initial stage of the epidemic when the epidemic is started with a very small initial infection level. For further reading on the full model we refer to the extensive literature on this model (Kermack and McKendrick, 1927; Metz, 1978b; Diekmann and Heesterbeek, 2000 and references therein). For an introduction to this model in a plant pathology context we refer to Segarra et al. (2001).

In this section, we introduce the description of "infectiousness of an individual" (or *infectivity* of an individual) and describe the basic reproduction number and the method by which the exponential growth rate is calculated. The model has been used in plant pathology primarily for fungal (or oomycete) foliar plant diseases with spores being the infectious units (units of inoculum). We therefore change our phrasing slightly and often discuss infectious units and the production of infectious units as spores and spore production, respectively. We also assume here that the individual disease unit is the lesion. However, these changes to more specific labels for individuals and infectious units are for ease of expression, because the model can be applied to most if not all plant disease scenarios.

One basic aspect of the Kermack–McKendrick model is in the description of the infectiousness of an individual. In the H-L-I-R and Vanderplank models, all diseased individuals are placed into non-overlapping states (latent, infectious, and removed), and all individuals in a state are treated alike. But in the Kermack–McKendrick model, these categories disappear (in a sense) and a diseased individual is "simply" characterized in terms of time elapsed since the initiation of infection. This time will be denoted by τ and referred to as "time since infection"; infectiousness can vary continuously throughout this time period. One can think of τ as 'age' of the diseased individual.

5.2.5.1 The sporulation curve. Consider an experiment where a group of leafs or plants is inoculated with a spore suspension at one moment in time ($\tau = 0$) and a number of lesions is produced. For a sufficiently long period the development of these lesions is observed. The produced spores are harvested and counted daily. Plotting the number of spores produced per day per lesion, $S(\tau)$, as a function of time since infection, τ, we will find that during the first days no spores are produced because the latent period has not yet ended. Then, spore production starts and the rate of spore production increases to a peak level and then falls off again, and finally decreases to zero. Some examples for a range of fungal foliar pathogens are shown in Fig. 5.12. An example of this was analyzed in section 3.6.2 to demonstrate other modeling concepts (see Fig. 3.9). This sporulation curve, $S(\tau)$, characterizes both the latent period *and* the infectious period in terms of time since infection.

A model that fits most of the data sets on sporulation is the equation

$$S(\tau) = \begin{cases} 0 & \text{if } \tau < \wp \\ \Omega \dfrac{\lambda^w (\tau - \wp)^{w-1} e^{-\lambda(\tau-\wp)}}{\Gamma(w)} & \text{if } \tau \geq \wp \end{cases} \tag{5.50}$$

where $\Gamma(w)$ is the gamma function which is tabulated in several textbooks (e.g. Abramowitz and Stegun, 1965). Equation 5.50 is known as the delayed gamma distribution multiplied by the parameter Ω.

In this description of the sporulation curve, the parameter Ω is the total number of spores produced per lesion during its entire lifetime, and $\wp$ is, as before, the latent period. The parameters λ and w determine the shape of the sporulation curve during the infectious period. In Fig. 5.12 the drawn line shows the model fit to the data which justifies (for these data sets) the use of equation 5.50. Certainly, other equations could be used for $S(\tau)$, including ones that do not require a threshold $\wp$ to be exceeded before sporulation occurs. However,

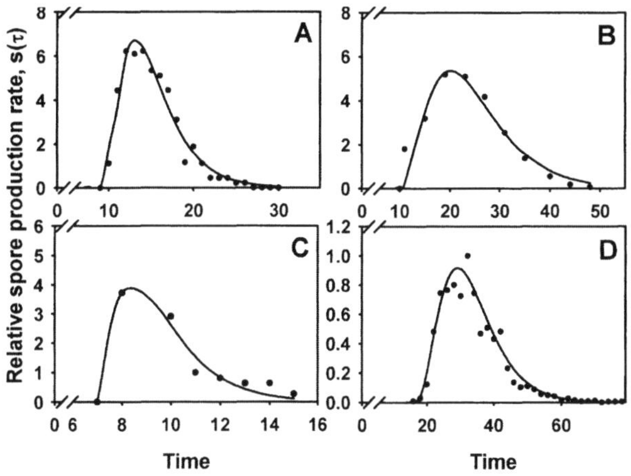

FIG. 5.12. Sporulation (or more generally, infectivity) curves for four foliar diseases. The vertical axis is the relative spore production rate, and the horizontal axis is time since initiation of infection. Points are the measured values; the drawn lines are fit for a delayed-γ distribution function (equation 5.50). (A) *Puccinia recondita*; $\wp = 10$ days, $\lambda = 0.218$, and $w = 2.34$ (data from van den Bosch et al., 1988b, c). (B) *Peronospora striiformis;* $\wp = 10$ days, $\lambda = 0.213$, and $w = 3.15$ (data from van den Bosch et al., 1988b,c). (C) *Peronospora farinosa;* $\wp = 7$ days, $\lambda = 0.91$, and $w = 2.33$ (data from van den Bosch et al., 1988b,c). (D) *Puccinia lagenophora*; $\wp = 16$ days, $\lambda = 0.23$, and $w = 4.00$ (data from Kolnaar and van den Bosch, 2001).

equation 5.50 appears to be a reasonable general model for the infectiousness of individuals.

5.2.5.2 Model derivation. The starting point of the derivation of the Kermack and McKendrick model is very similar to the starting point of the derivation of the Vanderplank model. The number of spores produced per time unit per unit area by all infectious individuals present in the crop is denoted by $\Lambda(t)$. The number of healthy individuals per unit area that will be contacted by an infectious unit is given by $\theta H(t)$, where, as before, θ is the probability that a spore reaches a particular host individual. Thus, the number of infectious units transmitted from infectious to healthy individuals per time unit equals $\Lambda(t)\theta H(t)$. An infectious unit deposited on a healthy individual has a probability ψ of infecting the individual. We thus have for the rate at which individuals become infected:

$$\frac{dY(t)}{dt} = \psi\theta H(t)\Lambda(t) \tag{5.51}$$

In the H-I-R and H-L-I-R models $\Lambda(t)$ was $\alpha I(t)$; that is, all infectious individuals were treated the same (they each had a constant sporulation rate, α). However, here we do not place individuals into states and we have to derive an expression based on the sporulation curve. The differential equation $dY(t)/dt$ is, as before, the number of new infections in the population per time unit per unit area. For ease of explanation we use the notation $b(t)$ for this quantity and use the slightly simpler label 'number of new infections per time unit'.

$$b(t) = \psi\theta H(t)\Lambda(t) \tag{5.52}$$

When the epidemic starts with a very small number of initial infections we can, as before, approximate the density of healthy individuals by its initial density and substitute $H(t) = H_0$.

$$b(t) = \psi\theta H_0\Lambda(t) \tag{5.53}$$

We now need to calculate $\Lambda(t)$, which is the sum (integral) of the rates of spore production of all infected individuals in the population. By definition, all infected individuals present at time t have become infected at some point in the past. For example, the density of individuals that have been infected 10 time units ago is the 'number of new infections per time unit' at time $t - 10$, $b(t - 10)$. In general, the density of individuals at t that have 'time since infection' τ is given by $b(t - \tau)$. As stated above, individuals that are infected τ time units ago produce $S(\tau)$ spores per time unit. Therefore, the total number of spores produced by all individuals that became infected τ time units ago equals $b(t - \tau)S(\tau)$. Calculating $\Lambda(t)$ is a matter of summing up over all possible values of τ. In continuous time, this summing up over all possible values of τ amounts to integrating over τ. We have

$$\Lambda(t) = \int_0^\infty b(t-\tau)S(\tau)d\tau \tag{5.54}$$

Substituting equation 5.54 in equation 5.53 the model becomes:

$$b(t) = \psi\theta H_0 \int_0^\infty b(t-\tau)S(\tau)d\tau \tag{5.55}$$

for the rate of increase of lesions (or diseased individuals, in general). Note that equation 5.55 for the rate of disease increase involves an integration over τ for any given time. To determine $Y(t)$, one then integrates equation 5.55 from 0 to t, which means that one must integrate a function that involves an integral, making this model more complex than the ones discussed in previous sections.

5.2.5.3 The exponential growth rate and derived R_0. The models in the previous sections showed in the initial stages of the epidemic an exponential increase of the form

$$Y(t) = Y_0 e^{r_E t} \tag{5.56}$$

For the current model, we assume that equation 5.56 holds for small Y_0. Equation 5.56 implies that the number of *new* infections per time unit, $b(t)$, also increases exponentially because the derivative of the exponential model (equation 5.56) is still an exponential growth function. Therefore,

$$b(t) = b_0 e^{r_E t} \tag{5.57}$$

with $b_0 = r_E Y_0$.
Substituting this equation into equation 5.55 we find

$$b_0 e^{r_E t} = \psi\theta H_0 \int_0^\infty b_0 e^{r_E(t-\tau)} S(\tau)d\tau \tag{5.58}$$

Dividing out the term $b_0 e^{r_E t}$ on both sides we have

$$1 = \psi\theta H_0 \int_0^\infty e^{-r_E\tau} S(\tau)d\tau \tag{5.59}$$

This equation allows us to calculate the exponential growth rate, r_E, after substituting an appropriate equation for $S(\tau)$ and calculating the integral. In demographics, equation 5.59 is called the Euler equation. A detailed discussion on this equation can be found in Keyfitz (1968) and Roughgarden (1979).

To summarize the model at this point, the total number of spores produced by one infectious individual throughout its infections period equals Ω (equation 5.50), and the number of healthy individuals per unit area that will be contacted by each spore is given by θH_0. Thus, the total number (*not* per unit time) of spores transmitted from one infectious individual to healthy individuals equals $\Omega\theta H_0$. An infectious unit deposited on a healthy individual has a probability ψ of infecting the individual. We can thus conclude that for this model the basic reproduction number is given by $R_0 = \psi\Omega\theta H_0$. This can be compared to the formula derived for other models, $R_0 = \psi\alpha\iota\theta H_0$. In other words, $\alpha\iota$ of the previous models serves the same role as Ω in the current general model. This is as expected, because $\alpha\iota$ is the total number of spores produced per infected individual (=rate per time unit multiplied by time duration of infectiousness) when the sporulation rate is constant over the infectious period.

Let us now, as an example, reconsider the Vanderplank model, in which an infected individual has a fixed latent period, $\wp$ and a fixed infectious period, ι. In terms of the general epidemic model of Kermack and McKendrick, the equation that describes the sporulation curve for the Vanderplank model is:

$$S(\tau) = \begin{cases} 0 & \text{if } \tau \le \wp \\ \alpha & \text{if } \wp < \tau < \wp + \iota \\ 0 & \text{if } \tau \ge \wp + \iota \end{cases} \tag{5.60}$$

Substituting this equation in the Euler equation 5.59, we obtain

$$1 = \psi\theta H_0 \int_{\wp}^{\wp + \iota} e^{-r_E\tau}\alpha\, d\tau \qquad (5.61)$$

Calculating the integral we finally find

$$r_E = \beta H_0 (e^{-r_E\wp} - e^{-r_E(\wp+\iota)}) \qquad (5.62)$$

where we have used, as before, $\beta = \psi\theta\alpha$. Equation 5.62, which is solved numerically to determine r_E, is identical to equation 5.49, which we derived for the Vanderplank equation in section 5.2.4.3.

In Box 5.6 we discuss a very general approximation formula for r_E, which is unfortunately only a valid approximation for very small R_0.

5.2.5.4 The exponential growth rate for sporulation curve 5.50. One advantage of using equation 5.50 for the sporulation curve is that it leads to a fairly simple expression to determine r_E for the initial stages of an epidemic. Substituting the sporulation curve 5.50 (based on the delayed gamma distribution) into the Euler equation 5.59, we find that r_E is calculated from the equation:

$$1 = R_0 e^{-r_E\wp}\left(1 + \frac{r_E}{\lambda}\right)^{-w} \qquad (5.63)$$

where we used $R_0 = \psi\theta\Omega H_0$. As for the Vanderplank model, this equation cannot be solved explicitly to express r_E in terms of the model parameters. An estimate of r_E can be obtained when, after substituting the parameters values for R_0, $\wp$, w and λ, the right-hand side of equation 5.63 is plotted over a wide range of values for r_E and reading of from this graph where the value of the function equals 1. As discussed in Box 5.2, more accurate solutions can be obtained using a numerical root finder in one of the available computer packages.

It is important to point out that other functions lead to different functions for calculating r_E.

5.2.5.5 Final disease level. Various authors (Metz, 1978b; Diekmann and Heesterbeek 2000, and references therein) have shown that the final level equation for the Kermack and McKendrick model is again equation 5.14.

5.2.5.6 Concluding remarks. The Kermack and McKendrik model is a very flexible and general model that can incorporate realistic descriptions of the spore production curve. Contrary to the H-L-I-R and Vanderplank models, investigators do not need to choose which of the very specific types of transitions between infection categories (latent, infectious removed) should be used; rather, on basis of information on the biology of the disease, investigators can derive a function $S(\tau)$ and substitute it in the Euler equation. Because this model has been little used in plant pathology, there are ample opportunities for further research.

Box 5.6

We define

$$s(\tau) = \frac{S(\tau)}{\Omega} \qquad (B5.6.1)$$

as the relative sporulation curve. Referring to equation 5.50 and the discussion following that equation, $s(\tau)$ is a probability density or distribution for sporulation time, implying that integrating $s(\tau)$ over τ from 0 to ∞ equals 1 (by definition). Substituting equation B5.6.1 in equation 5.59 we get

$$1 = R_0 \int_0^{\infty} e^{r_E\tau} s(\tau) d\tau \qquad (B5.6.2)$$

For small values of R_0 the exponential rate of increase, r_E, can be approximated from equation B5.6.2 by:

$$r_E \approx \frac{\ln(R_0)}{\vartheta} \qquad (B5.6.3)$$

where ϑ is the expected value based on the $s(\tau)$ function,

$$\vartheta = \int_0^{\infty} \tau s(\tau) d\tau \qquad (B5.6.4)$$

The interpretation of ϑ is the average time elapsed since start of the infection process at which an infectious unit is produced. For the Vanderplank model, $\vartheta = \wp + (\iota/2)$.

The derivation of this equation can be found in Metz and Diekmann (1980), or Seggara et al. (2001). The equation is only a valid approximation for small values of R_0, where small here means smaller than 5. A slightly more accurate approximation formula is discussed in Segarra et al. (2001). In general these approximation formulae have a limited applicability to plant diseases because plant diseases often have large basic reproduction numbers. When R_0 is low, however, such as when a disease control is *almost* effective enough to prevent an epidemic, one can see the consequence of the achieved R_0 by estimating the exponential rate of increase using equation B5.6.3.

5.3 Discussion

We have introduced five models in this chapter for polycyclic epidemics. The models were developed in sequence, from the very simple caricature one (section 5.2.1) to the very realistic (but much more complex) Kermack and McKenrick one (section 5.2.5). The first, and in our opinion, and most important conclusion to be drawn from the discussion of these models is: if we are interested in *qualitative* features of disease dynamics, like thresholds for epidemic development, initial rate of increase of disease intensity, shape of the disease progress curve, and final intensity of disease (size of the epidemic), a simple model yields the same results as a very complex model. This shows that it is often not necessary, as some biologists and plant pathologists are inclined to think, to incorporate as many aspects (details) as possible of the biology of the pathogen and host under consideration to gain insight into the epidemiology of the disease. However, if we are more concerned about *quantitative* features of epidemics, such as accurate predictions of the exponential growth rate in the initial stage of disease increase, we do need realistic descriptions of the components of the epidemic, especially distribution of latent and infectious periods. In this situation, the Kermack and McKendrick model is a very good tool to arrive at such quantitative estimates.

The models presented here all have their advantages and disadvantages. The simplest model, equations 5.2a and 5.2b, are very easy to solve numerically. Furthermore, equations for the threshold for epidemic development, the exponential growth rate, and the final disease level are easily derived from this model. This is the reason we started with this model. The disadvantage of the model is that it does not include some very important features of plant diseases. The absence of latent and infectious periods, and the restriction to non-overlapping generations, make this model unrealistic. This model is seldom used in animal, human and plant disease epidemiology, but it is an ideal tool to introduce some important qualitative aspects of epidemics of infectious diseases.

The H-L-I-R model, as we showed, is a more realistic description of the population dynamics of plant diseases. This model is still analytically tractable with commonly used mathematical techniques. The disadvantage, as we discussed, is in the restrictive assumptions of the distribution of the latent and infectious periods. Despite this disadvantage, the H-L-I-R model and its variations (not explained here) are extremely popular in theoretical epidemiology, and this model is generally used as a starting point for more model developments to incorporate specific features of the pathosystem of interest. We will discuss examples of such further developments in the next chapter. Some valuable papers in botanical epidemiology based on this general model framework include: Swinton and Gilligan (1996), Chan and Jeger (1994), Gilligan et al. (1997), Jeger (1982).

The H-I-R model is even simpler to work with than the H-L-I-R one. It is thus easy to use the model to explore many qualitative aspects of epidemics, especially the derivation of a basic reproduction number, threshold for an epidemic, and final disease intensity. Its disadvantage is that it produces an unrealistic prediction of r_E, since latent period is not included (or, equivalently, a latent period of 0 time units). As we have seen, however, latent period does not affect the definition of R_0, and corresponding thresholds and final level.

The Vanderplank model incorporates, in some ways, a somewhat more realistic description of the latent and the infectious periods compared with the H-L-I-R model, which is definitely an advantage of this model over the H-L-I-R one. It is, however, much harder to analyze mathematically differential-delay equations than it is to analyze the differential equation models of the H-L-I-R type. It is also very hard to fit such models to data. Nevertheless, because of the strong influence of Vanderplank (1963), botanical epidemiologists have been more interested in using the differential-delay approach than investigators in the rest of theoretical epidemiology (e.g., Jeger, 1984; Jeger and van den Bosch, 1994a). The model also serves as a good foundation for numerical simulation studies (e.g., Zadoks, 1971; Zadoks and Kampmeijer, 1977) since the time delays are handled in a natural way in computer programs with single-day time steps in the model. Because of the many practical advantages of H-L-I-R models, the Vanderplank model is becoming considerably less popular for theoretical studies even in botanical epidemiology. We note that in the ecological literature, differential-delay equations play a useful role in some aspects of the study of population dynamics (Nisbet and Gurney 1982, and references therein).

As we have seen, the Kermack and McKendrick model is the most realistic of the models discussed in this chapter. But it is a difficult model to work with and to fit to data. For these reasons, extensions and applications of this model are not common in plant pathology, with the notable exception of its generalization for studying the spatial spread of plant diseases (see Chapter 8).

We end this discussion on disease progress models with a cautionary note. Considered in more biological detail, the introduced parameters (βH_0, $\wp$, ι, and so on) may be considered to be functions of other properties of diseases, and not just constants. There are two essentially different situations:

(i) *A parameter depends on other parameters*. We have already considered one example of this when we have used β for the product $\psi\theta\alpha$. Each of these component parameters could be functions of other parameters. In these situations, the results, such as equations for R_0, r_E, and final disease level, can simply be applied with no additional difficulty. For example, the probability of a unit of inoculum contacting a host (θ) could be considered to vary with the initial disease-free host density H_0, say, for example, $\theta = \zeta_1 H_0/(\zeta_2 + H_0)$ (where the ζs are positive parameters). Then it is valid to substitute this new definition of θ into the equations for R_0, r_E, and the final level. One especially useful choice of parameters is $\zeta_1 = 1/H_0$ and $\zeta_2 = 0$ (so that $\theta = 1/H_0$). This results in a transmission rate of β (rather than βH_0), and an R_0 of β/μ.

It should be pointed out that most measurements of epidemics involve the proportion of plant area affected by disease (severity) or the proportion of plants (leaves, etc.) that are diseased. That is, actual densities are not known, and more importantly, there have been few studies that quantitatively compared epidemics with different host densities. Thus, whether or not θ (for instance) or other components of β depends explicitly on host density is not necessarily known (as a general rule), and the transmission rate βH_0 should be treated as a single term, not necessarily as two separate (and unrelated) parameters. For instance, βH_0 is analogous to r_L of the logistic model, and not just β, as mentioned above. An implication of this can be addressed in terms of R_0 ($\beta H_0/\mu$). It would appear that host density is a determinate of the basic reproduction number, and directly determines whether or not there is an epidemic. But, as mentioned in the previous paragraph, it is possible that β goes down as H_0 goes up; in this case, R_0 would not vary linearly with host density when other parameters are fixed. What we can say definitively is that R_0 goes up with increases in βH_0. de Jong et al. (1995) nicely describe some ways of deciding from data whether β depends on host density.

(ii) *A parameter depends on variables in the system.* In the second situation it is *not* valid to simply substitute the new expressions in the equations for R_0, r_E, and the final level! For example, if θ is considered to vary with the healthy host density $H(t)$, say, for example, $\theta = \zeta_1 H(t)/(\zeta_2+H(t))$, then it is not valid to substitute this new definition into the equations for R_0, r_E, and the final level. This expression has to be incorporated in the governing model equations (e.g., equation 5.30) and the mathematical analysis has to be redone to see how this new formulation affects R_0, r_E, and the final level. In fact, all qualitative and quantitative features of the model *could* be affected in this scenario. We do not pursue this subject here.

5.4 Suggested Readings

Diekmann, O., and Heesterbeek, J. A. P. 2000. *Mathematical Epidemiology of Infectious Diseases: Model Building, Analysis, and Interpretation.* John Wiley and Son, Ltd, Chichester (A mathematically challenging text, but full of excellent insights on epidemic processes).

Jeger, M. J., and van den Bosch, F. 1994. Threshold criteria for model plant disease epidemics. I. Asymptotic results. *Phytopathology* 84: 24–27.

Segarra, J., Jeger, M. J., and van den Bosch, F. 2001. Epidemic dynamics and patterns of plant diseases. *Phytopathology* 91:1001–1010.

Vanderplank, J. E. 1963. *Plant Diseases: Epidemics and Control.* Academic Press, New York (Chapters 5, 6, and 8).

6

Temporal Analysis III: Advanced Topics

It is the mathematicians' inclination to strive for an abstract and general theory, the hope being, that once such a theory exits, one can make it operational by mere specification and elaboration.

Odo Diekmann and J.A.P. (Hans) Heesterbeek

6.1 Introduction

In the previous chapter, we considered models of increasing complexity for characterizing various aspects of plant disease dynamics. Among other things, we showed that many useful epidemic properties can be elucidated using simple (but well formulated) models. We now consider additional complexities of epidemics that are important in some circumstances. In section 6.2, we introduce crop growth into the models. This section will introduce two main categories of crop growth scenarios. The section serves as an introduction to the many possible types of disease dynamics that can occur when the host population "size" changes.

In section 6.3, we consider in more detail the primary infections that initiate an epidemic. It will be discussed how disease dynamics change when these primary infections occur over the duration of the epidemic rather than just at the start. A special case of the model is the one for a monocyclic disease.

In section 6.4, we use the insight gained from the previous sections to study plant diseases caused by viruses. In particular, we deal with the situation in which the pathogen is transmitted from infected plants to healthy plants by the feeding activity of vectors, usually herbivorous insect species (such as aphids and leafhoppers). In this section, we study the consequences of vectored transmission on the development of epidemics, and use this system to demonstrate how such models can be utilized to evaluate disease control strategies.

In section 6.5, we expand on several of the earlier topics of this and the former chapter and address additional aspects of plant disease epidemics. In particular, we discuss situations where insight into epidemics depends on more than the use of mathematical analysis, and results are obtainable only by the use of computer programs. Finally, in section 6.6, we conclude the chapter by discussing ways of fitting some of the differential-equation models directly (or indirectly) to data.

6.2 Models with Crop Growth

In the previous chapter, we have assumed that the host population density is fixed during the epidemic. Obviously, crops do grow over time, and leaf area index or root mass is clearly dynamic. However, over limited periods of time in which the epidemic occurs, the change in host size may be relatively small, so that the assumption of a fixed size is reasonable. For situations where the entire plant is the host unit (such as for systemic diseases), host size may still be assumed fixed in the sense that total number of plants is not changing (even if leaf area, etc., is changing).

We now discuss models that include the increase and decrease in the density of the crop. As before, we present this in terms of healthy and diseased *individuals*, even when the "individual" refers to unit area of host tissue. The host crop dynamics can be due to planting, plant growth, and harvesting. The literature on models including crop growth is extensive and we can only give a general overview of the subject that will set the scene for reading papers on this subject. For further reading we refer to Gilligan (2002 and references therein) and for the mathematics theory to Edelstein-Keshet (1988 and references therein).

Broadly, there are two categories of models with crop growth. The first category describes crop growth as a continuous process. This is often done by including a growth term in the differential equation for the density of healthy individuals. The dynamics of this type of model can partly be studied using analytical techniques (Jeger and van den Bosch, 1994b), and be further explored numerically. The second category describes planting and harvesting at fixed times during the year, and modeling of the initial disease level at the start of a growing season as a function of disease level at the end of the previous growing season (Madden and van den Bosch, 2002). Such models are often only studied numerically, although, as we will see, thresholds for epidemic development can be derived analytically.

6.2.1 Continuous crop growth

Epidemic models including a dynamic term for the healthy, or sometimes both healthy and diseased, individuals are used by several authors for several purposes. Objectives include:

1. To properly account for change in leaf area index, root mass, or other characteristics of crop growth, in a particular field, or growth in a region where crops are planted around the same time.
2. To approximate continuous cropping systems where crop plants are planted and harvested continuously, as, for example, in several tropical agrosystems. As we will see later in section 6.4, this is, for example, used to describe virus epidemics in cassava crops.
3. To describe perennial crops where part of the plants are harvested every year and replanted with new crop individuals. This, for example, describes tree plantations and nurseries where garden shrubs are grown.
4. As an approximation, to model seasonal cropping over multiple seasons where planting and harvest take place at certain times in each year. Though describing seasonal cropping with a continuous crop growth term is crude, remember that in the previous chapter we obtained a wealth of qualitative insight into plant disease epidemics from simple models, and that these qualitative results carried over to more complex models.

6.2.1.1 Model derivation. We will use as our starting point the H-L-I-R model equation 5.30 of the previous chapter and expand this to incorporate crop growth. Probably the simplest way to incorporate crop growth is to add a growth term to the equation for the density of healthy individuals:

$$\begin{aligned}\frac{dH(t)}{dt} &= \begin{bmatrix}\text{rate of growth of}\\ \text{healthy crop}\end{bmatrix} - \beta H(t)I(t)\\ \frac{dL(t)}{dt} &= \beta H(t)I(t) - \omega L(t)\\ \frac{dI(t)}{dt} &= \omega L(t) - \mu I(t)\end{aligned} \tag{6.1}$$

At this stage we leave out the model equation describing the removed category. In section 6.2.1.3, we will discuss the removed category in some detail. To describe the rate of crop growth, various model terms have been used (Gilligan, 2002 and references therein; see also Jeger and van den Bosch, 1994b). The most commonly used one for theoretical studies may be

$$\begin{bmatrix}\text{rate of growth of}\\ \text{healthy crop}\end{bmatrix} = r_H H(t)\left(1 - \frac{H(t)}{H_{\max}}\right) \tag{6.2}$$

Note that this expression is the same as in the logistic growth curve discussed in Chapter 4. In our present interpretation of crop growth, the parameter $H_{\max}$ represents the maximum crop density in the field and r_H is the growth rate parameter for the crop. The logistic function is chosen here simply to describe continuous increase in total crop density over time. In the absence of disease, the total crop density and healthy crop density is the same, which increases with an *S*-shaped curve towards its maximum value $H_{\max}$. Substituting this description of crop growth into the model we have the model:

$$\begin{aligned}\frac{dH(t)}{dt} &= r_H H(t)\left(1 - \frac{H(t)}{H_{\max}}\right) - \beta H(t)I(t)\\ \frac{dL(t)}{dt} &= \beta H(t)I(t) - \omega L(t)\\ \frac{dI(t)}{dt} &= \omega L(t) - \mu I(t)\end{aligned} \tag{6.3}$$

Accounting for crop growth using a model such as this one will be justified when total host density varies considerably during the epidemic. An example would be an epidemic with low transmission rate parameter β, and disease increase from plant emergence to harvest. Of course, when $r_H = 0$, equation 6.3 reduces to equation 5.30 (fixed host).

Another frequently used crop growth model captures the essentials of the continuous cropping systems as mentioned earlier. Assume that there is continuous planting of new crop, with the density of new crop planted per time unit denoted by σ. The crop growth period equals $1/\eta$ time units. This implies that the probability per time unit that a crop individual is harvested equals η. To understand the relation between harvest rate and growth period see the discussion on the relation between infectious period and probability per unit time to reach the end of the infectious period in section 5.2.2.2. Equation 6.2 then becomes:

$$\begin{bmatrix}\text{rate of growth of}\\ \text{healthy crop}\end{bmatrix} = \sigma - \eta H(t) \tag{6.4}$$

However, note that here not only healthy individuals are harvested (or die) but that latent and infectious individuals are harvested (or die) as well. This feature has to be added to the model, and terms $-\eta L(t)$ and $-\eta I(t)$ have to be added to the equation for the densities of latent and infectious individuals, respectively. The model then takes the form:

$$\begin{aligned}\frac{dH(t)}{dt} &= \sigma - \eta H(t) - \beta H(t)I(t)\\ \frac{dL(t)}{dt} &= \beta H(t)I(t) - \omega L(t) - \eta L(t)\\ \frac{dI(t)}{dt} &= \omega L(t) - \mu I(t) - \eta I(t)\end{aligned} \tag{6.5}$$

This way of representing host growth could also be used for a perennial crop, with continuous emergence of new leafs or shoots and death or senescence of old ones (both diseased and healthy ones).

In the following sections we will analyze model equation 6.3 in some detail. Model equation 6.5 show some similar dynamics and we will only discuss the threshold for epidemic development for this model.

6.2.1.2 Model simulations. In Fig. 6.1 we show the dynamics of healthy, latent and infectious individuals for a range of values of the transmission rate parameter β, starting at time $t = 0$ with a small initial infection. The set of graphs in the Fig. 6.1 shows the range of dynamics that can be found for this model. For other parameters, the densities of healthy, latent and infectious individuals might be different but the qualitative dynamics are the same.

The simulations show that when β is small, no epidemic develops, whereas for larger values of β, an epidemic does develop. There is clearly a threshold for epidemic development. In section 5.2, we also found such threshold behavior, and an interesting question is what the relation is between thresholds in disease progress models with fixed host and thresholds in models including crop growth. We will discuss this in the next section.

In the initial stage of the epidemic, the densities of latent and infectious individuals again seem to increase exponentially. As with the threshold situation, it is interesting to ask what the r_E value is and how it relates to the r_E value for the disease progress models studied in section 5.2.

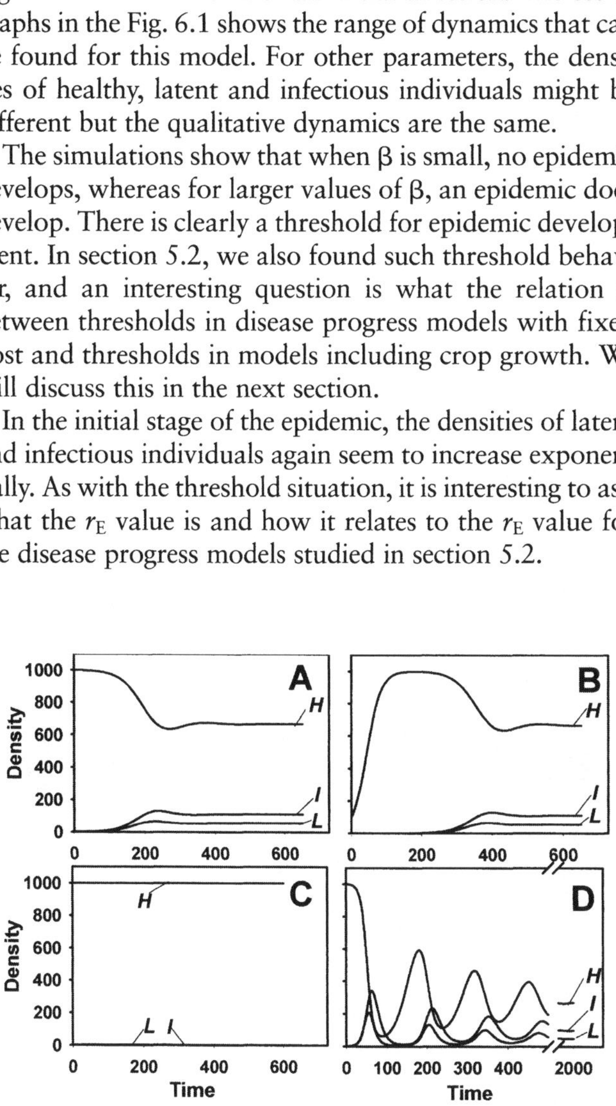

FIG. 6.1. The density of healthy, (H) latently infected (L), and infectious (I) individuals as function of time (t) for the epidemic model with host growth (equation 6.3). $H_{max} = 1000$ for all graphs. (**A**) $H(0) = 1000$, $L(0) = 0$, $I(0) = 1$. $r_H = 0.05$, $\mu = 0.1$ (implying $\iota = 10$), $\omega = 0.2$ (implying $\wp = 5$) and $\beta = 0.00015$ per time unit. (**B**) $H(0) = 99$, $L(0) = 0$, $I(0) = 1$. Same parameters as in (**A**). (**C**) $H(0) = 1000$, $L(0) = 0$, $I(0) = 1$. $r_H = 0.05$, $\mu = 0.1$ ($\iota = 10$), $\omega = 0.2$ ($\wp = 5$) and $\beta = 0.000099$ per time unit. (**D**) $H(0) = 1000$, $L(0) = 0$, $I(0) = 1$. $r_H = 0.05$, $\mu = 0.1$ ($\iota = 10$), $\omega = 0.2$ ($\wp = 5$) and $\beta = 0.000375$ per time unit.

In the simulations shown, the density of healthy individuals first decreases when H_0 is at or near H_{max}, then fluctuates for some time for some parameter values, and finally settles down to a constant level. When H_0 is much less than H_{max}, $H(t)$ initially increases, fluctuates, and then settles down (Fig. 6.1B). In some cases, it may take a very long time to settle down (e.g., Fig. 6.1D). This constant level is called the *steady state* or equilibrium value. Mathematically, the steady state for a variable X is the value at which $dX/dt = 0$. For the latent and infectious individuals, a similar pattern is seen of initial increase, possible fluctuations, and finally settling down at a steady state value. At first sight this might seem similar to the dynamics we found for the disease progress models in section 5.2. There is, however, a very essential difference. In the disease progress models without host growth, the densities of latent and infectious individuals ultimately decreased to zero (because all infected individuals ultimately become removed). In the present model, the densities in the latent and infectious categories do not decrease to zero but converge to a larger-than-zero density. The difference is, of course, due to the crop growth incorporated in the model equation 6.3. The continuous inflow of new healthy individuals sustains the epidemic and causes the density of latent and infectious not to decrease to zero. When there are infectious individuals in the system continuously, new infections will occur continuously (as long as there are disease free individuals).

6.2.1.3 The removed category. The density of the removed state, $R(t)$, and the corresponding total density of infected individuals, $Y(t)$, are less commonly incorporated into the models with continuous crop growth as formulated in equation 6.1. This is because, at first sight, strange phenomena can occur. Consider the situation where we add the standard equation for the density of removed individuals to model equation 6.3 or 6.5. This equation reads:

$$\frac{dR(t)}{dt} = \mu I(t) \qquad (6.6)$$

As we have discussed above, the density of infectious individuals will finally approach a steady state with $I > 0$. This implies that the right hand side of equation 6.6 becomes a constant greater than 0, which in its turn implies that the density of removed individuals will grow continuously without bound. The same holds for total disease $Y(t) = L(t)+I(t)+R(t)$. The reason for this is obvious: as long as new healthy individuals enter the population and infectious individuals are around, healthy individuals will become infected (assuming $\beta > 0$) and finally end up in the removed category.

There are two situations that have to be distinguished in this case. In the first situation, the removed state *is* a relevant state throughout the epidemic, and equation 6.6 is the correct description for this state. For example, the situation applies if we study an epidemic within one growing season in which removed individuals comprise simply dead leaf tissue which remains on the plants till the end of the growing season. At the end of the finite-length growing season, the density of the removed category can be large but is not infinite (because time is limited). The density of the removed category can simply be calculated using equation 6.6. Whether the epidemic has approached its steady state (in terms of H, L, and I) by the end of the growing season is to be decided on basis of the simulation results.

In the second situation, individuals in the removed state can be removed (i.e., eliminated) physically from the system. For example, in our model described by equation 6.5, individuals are harvested including the removed individuals. The differential equation for the density of removed individuals then becomes:

$$\frac{dR(t)}{dt} = \mu I(t) - \eta R(t) \qquad (6.7)$$

In this situation, the density of removed individuals (that have not yet been physically removed from the system) will, in the long term, approach a steady state value itself, which can be calculated from

$$\breve{R} = \frac{\mu}{\eta}\breve{I} \qquad (6.8)$$

(As explained in more detail in the next sub-section, the overstriking of variables indicates steady states.) Another example where removed individuals become eliminated from the system is when leaves drop off the crop plant and the leaf material is degraded by micro-organisms or is consumed by earthworms, insects, or other organisms. In that case, assuming that each removed individual has a constant probability per time unit δ to fall off and be degraded or eaten, the differential equation for the density of removed individuals becomes:

$$\frac{dR(t)}{dt} = \mu I(t) - \delta R(t) \qquad (6.9)$$

and the steady state of this density is

$$\breve{R} = \frac{\mu}{\delta}\breve{I} \qquad (6.10)$$

6.2.1.4 Steady states and thresholds for epidemic development. As mentioned above, a steady state occurs when the densities of healthy, latent and infectious individuals do not change. Lack of change implies that the rate of change in their densities, $dH(t)/dt$, $dL(t)/dt$, and $dI(t)/dt$, respectively, are all zero. The densities are thus not functions of time anymore, and we replace $H(t)$, $L(t)$ and $I(t)$ by $\breve{H}$, $\breve{L}$, and $\breve{I}$, respectively. Substituting these into model equation 6.3 together with 0 for the rates, we find

$$\begin{aligned} r_H\breve{H}\left(1-\frac{\breve{H}}{H_{max}}\right) - \beta\breve{H}\breve{I} &= 0 \\ \beta\breve{H}\breve{I} - \omega\breve{L} &= 0 \\ \omega\breve{L} - \mu\breve{I} &= 0 \end{aligned} \qquad (6.11)$$

This set of equation is easily solved, and shows, after some algebraic manipulations, to have three solutions.

$$\breve{H} = 0, \quad \breve{L} = 0, \quad \breve{I} = 0 \qquad (6.12)$$

$$\breve{H} = H_{max}, \quad \breve{L} = 0, \quad \breve{I} = 0 \qquad (6.13)$$

$$\breve{H} = \frac{\mu}{\beta}, \quad \breve{L} = \frac{r_H\mu}{\beta\omega}\left(1-\frac{\mu}{\beta H_{max}}\right), \quad \breve{I} = \frac{r_H}{\beta}\left(1-\frac{\mu}{\beta H_{max}}\right) \qquad (6.14)$$

The first steady state where all densities are zero is not very interesting for our purpose. It is, however, a real steady state of the system—when there is no healthy host, then there will never develop a population of healthy hosts. (Note that in the selected logistic model for H, there can be no increase in H if H is 0.) If there is no healthy host the disease will not be able to establish and these densities also remain zero.

The second and the third steady state are the interesting ones from our perspective. In the second steady state, the host is at its maximum density and the disease is absent from the system. The third steady state is the one for which an epidemic has developed. Since densities of latent and infectious individuals cannot be negative (by definition), this steady state is only biologically relevant when $\breve{L}$ and $\breve{I}$ are greater than or equal to zero. This implies that the expression in brackets on the right hand side of these steady state equation has to be larger than or equal to zero. This implies,

$$1-\frac{\mu}{\beta H_{max}} \geq 0 \qquad (6.15)$$

Clearly, with this inequality we have found the threshold for epidemic development. There is a steady state with disease present when this quantity is larger than zero and there is no steady state with disease present when this quantity is smaller than zero. The quantity being equal to zero serves as the threshold.

It is natural to ask: How does this compare to the threshold for epidemic development found for the model without crop growth, which is given in equation 5.23?

Rearranging equation 6.15, the relation will become clear. We consider the inequality:

$$1-\frac{\mu}{\beta H_{\max}}>0 \qquad (6.16)$$

which after some rearrangement of the expression becomes:

$$\frac{\beta H_{\max}}{\mu}>1 \qquad (6.17)$$

This last form of the expression is exactly the same as equation 5.23, with H_0 replaced by H_{max}. Without host growth, the initial host density, H_0, *is* the largest possible host density, so there is an equivalence in meaning to H_0 and H_{max}. We will not reiterate and do the step-by-step interpretation of this quantity again, but it is clear that $\beta H_{max}/\mu$ is the basic reproduction number, R_0, of the disease in this situation ($R_0 = \beta H_{max}/\mu$). We thus conclude that the threshold for epidemic development is the same for model equation 6.3 (host growth) and equation 5.30 (fixed host).

Below the threshold, when $R_0 < 1$, no epidemic develops and the system converges to the steady state equation 6.13 (where ultimately all individuals are disease free). The steady states of the system are shown in Fig. 6.2 as function of β. For β values smaller than the threshold value (with the chosen H_{max} and μ approximately 0.00025), no disease is present in the system at the steady state. The density of healthy individuals at the steady state decreases with increasing β (for $R_0 > 1$). For β values larger than but close to the threshold value (and the chosen other parameters of equation 6.14), the density of latent and infectious individuals increases with increasing β. This is due to the depletion of the healthy population by the disease (i.e., replacement of healthy individuals with either latent or infectious individuals). For β values larger than approximately 0.0004 (with the chosen values of the other parameters of equation 6.14), the density of latent and infectious individuals then *decreases* with increasing β. When β is larger than 0.0004 (in this example), the healthy population cannot produce new healthy individuals fast enough to permit increasing L and I with increasing transmission rate at the equilibrium densities. In other words, the density of healthy individuals is small enough that it only supports small densities of latent and infectious individuals.

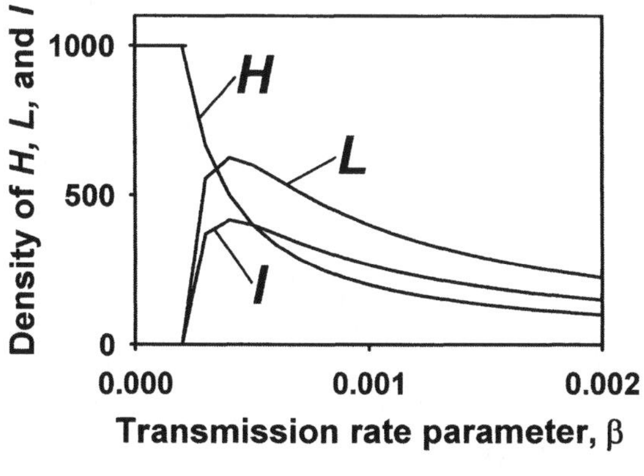

FIG. 6.2. Steady state densities of healthy (H), latently infected (L), and infectious (I) individuals in relation to the transmission rate, β. The steady-state values are calculated from equation 6.14. Parameter values are $H_{max} = 1000$, and $r_H = 0.05$, $\mu = 0.2$ (implying $\iota = 5$) and $\omega = 0.3$ (implying $\wp = 3.33$) per time unit.

The steady state or equilibrium density of healthy individuals can also be expressed as:

$$\breve{H}=\frac{H_{\max}}{R_0} \qquad (6.18)$$

based on the formula for R_0 ($=\beta H_{max}/\mu$). This formula often results from calculating the steady state for the density of healthy individuals. Unfortunately, this expression does not necessarily hold for *all* models of epidemic dynamics with host growth.

It is useful to compare results here with those obtained from the models without host growth (section 5.2 of the previous chapter). Of course, there is no steady state without host growth; instead, $H(t)\ (= H_0 - Y(t))$ approaches an *asymptotic* value as all diseased individuals become removed (final level; H_∞). Equation 5.27 defines the final level of disease as a proportion, but a little algebra can show the value of H_∞ for the H-L-I-R model (which is essentially the same as equation 5.11 for the discrete generation model). In a practical sense, H_∞ for fixed host size is analogous to the steady state $\breve{H}$ for a dynamic host because they are indicators of the long-term number of disease-free hosts when time is not limiting. (However, they are not the same mathematically). For comparisons, consider the following example results:

R_0	H_∞	$\breve{H}$
1.5	420	670
2.0	200	400
3.0	60	333

As is clear with just a few values, with host growth there are more disease-free individuals in the long term than without host growth for a given value of the basic reproduction number. This is logical because with host growth, new healthy individuals are continuously added to the system, but obviously there is no change in the total host density when the host is not growing.

6.2.1.5 *Initial disease increase.* Without showing this, we note here that the initial disease increase, when the epidemic is started with a very small density of infected individuals, is again exponential with rate parameter, r_E. r_E is given by the same equation as used for the H-L-I-R model (equation 5.30) without host growth (see equation 5.36) when H_0 is replaced by the maximum host density, H_{max}.

6.2.1.6 Threshold of epidemic development of model equation 6.5. We now return to the second type of host growth situation, where there are new plantings (or new host growth) coupled with harvesting (or loss of host area through mortality), summarized by equation 6.5. Calculating the steady states of the model using the method discussed in section 6.2.1.3, we find as the steady states where disease is present

$$\breve{H} = \frac{(\omega+\eta)(\mu+\eta)}{\beta\omega}, \quad \breve{L} = \frac{1}{\omega+\eta}\left(\sigma - \frac{\eta(\omega+\eta)(\mu+\eta)}{\beta\omega}\right),$$
$$\breve{I} = \frac{\omega}{(\omega+\eta)(\mu+\eta)}\left(\sigma - \frac{\eta(\omega+\eta)(\mu+\eta)}{\beta\omega}\right) \tag{6.19}$$

The density of latent and infectious individuals is larger than zero only if

$$\sigma - \frac{\eta(\omega+\eta)(\mu+\eta)}{\beta\omega} > 0 \tag{6.20}$$

This expression is most easily interpreted if we rearrange it into the inequality

$$\frac{\omega}{\omega+\eta}\beta\frac{\sigma}{\eta}\frac{1}{\mu+\eta} > 1 \tag{6.21}$$

To understand this, first consider the system in the absence of disease. In this situation the differential equation describing the density of healthy individuals is

$$\frac{dH(t)}{dt} = \sigma - \eta H(t) \tag{6.22}$$

This equation has steady state density of healthy individuals $\breve{H} = \sigma/\eta$ (i.e., at long time, $H(t)$ approaches σ/η). So the quotient σ/η has the same role as the maximum host density, H_{max}, in equation 6.17. One could then actually write the right-hand-side of equation 6.22 as $\eta(H_{max} - H(t))$. Now consider that we place one freshly infected individual in a completely healthy population. Because we included a harvest or death term into the model equation 6.5, not every individual will go through the latent period and become infectious. There is a probability that an infected individual is harvested (or dies) before it reaches the infectious stage. A latently infected individual has a probability per unit time ω to reach the end of its latent period and become infectious, and a probability per unit time η to be harvested (or die). This implies that the probability that an individual reaches the end of its latent period and moves into the infectious category before it is harvested (or dies) equals $\omega/(\omega+\eta)$. The mean time that an infected individual is in the latent stage is $1/(\omega+\eta)$. When the individual does reach the infectious stage, it causes $\beta\sigma/\eta$ new infections per time unit. Since infectious individuals can either reach the end of their infectious period or be harvested or die, the mean time the individual remains infectious equals $1/(\mu+\eta)$. The individual that does reach the infectious stage thus will, on average, produce $\beta(\sigma/\eta)/(\mu+\eta)$ new infections. The quantity on the left hand side of equation 6.21 is thus, again, the average number of new infections cause by one newly infected individual when it is placed in an entirely susceptible population, R_0.

Substituting H_{max} for σ/η we can write R_0 as:

$$R_0 = \frac{\omega}{\omega+\eta}\beta H_{max}\frac{1}{\mu+\eta} \tag{6.23}$$

When $\eta = 0$, this reduces to the expression for R_0 in the model with logistic healthy host growth. The difference is only because here we are specifying host harvest or death in all categories. As with the previous model for host growth, H_{max} has the same role as H_0 when there is no host growth (so that initial healthy host density is the same as maximum healthy host density). A little algebra shows that the steady state value of H is given by $\breve{H} = H_{max}/R_0$ (with R_0 given by equation 6.23), the same as found for logistic host growth (equation 6.18).

It is straightforward to use equation 6.5 for the epidemic coupled with equation 6.7 for the death/harvesting of removed individuals. It is useful to use $\eta(H_{max} - H(t))$ for the equation for $dH(t)/dt$, as discussed above. We leave it as an exercise for the reader to further explore this model. The reader should be able to discover that when the epidemic starts with initial $H + I$ equal to the maximum ($H_0 + I_0 = H_{max} = \sigma/\eta$), that there is *no* net change in the *total* density of the host ($dH(t)/dt + dL(t)/dt + dI(t)/dt + dR(t)/dt = 0$), which means that the total host density is unchanged throughout the epidemic, even though there are new healthy individuals produced (and, of course, mortality/death of individuals in all states). With this situation, one can easily determine steady state disease intensity ($\breve{Y} = H_{max} - \breve{H}$). In particular, based on equation 6.18 for steady-state value of healthy individuals, $\breve{Y} = H_{max}(1-1/R_0)$, and $\breve{y} = \breve{Y}/H_{max} = (1-1/R_0)$. It is easy to show that steady state $\breve{y}$ is less than asymptotic $y(y_\infty)$ when there is no host growth.

6.2.1.7 Concluding remarks. This section on models with continuous crop growth showed some basic features of epidemics with non-constant host size. Some of these features clearly are identical to the ones we found in the previous chapter on epidemic progress models without host growth, but some other features are new. The concept of the basic reproduction number and its role in a threshold for an epidemic is the same for all these models. The densities of healthy, latent and infectious individuals finally reach steady state levels, and we have shown how these steady states can be calculated.

The models discussed here have in common that the density of healthy, latent, and infectious individuals will, after a shorter or longer time, settle down at their steady state densities. This is, however, not the case in general. In some models the densities of these variables do not settle down at a steady state but keep fluctuating forever, either in a regular way with fluctuations of a fixed amplitude and a fixed period, or irregularly (perhaps chaotically). It is beyond the scope of the book to discuss this situation in more detail and we refer to the book of Edelstein-Keshet (1988) for a discussion on such topics.

Steady states of systems of differential equations are often easy to calculate. In the general theoretical literature, when the steady state for L and I are above 0, the disease is said to *persist* (Mollison, 1995). Therefore, the *persistence criterion* is the value of the threshold that must be surpassed for disease to remain in a system. Whether the steady state expression is of relevance for a plant disease epidemic is another matter. As Fig. 6.1 showed, it takes a certain period of time before the system approaches the steady state. Whether the steady state is reached within the time span to which the model applies depends on the parameter values and the relevant time scale of the system of interest. Especially when a model like equation 6.3 is used to describe the growth of an epidemic within a single growing season, it might very well be that at the end of the growing period the epidemic still is far from its steady state values. Under these circumstances, it is more meaningful to determine Y (or H, or I) at particular times—this will generally require numerical solutions to the differential equations.

6.2.2 Seasonal cropping

Many cropping systems are seasonal. The crop is planted and harvested at a certain moment in each year. Planting increases the density of susceptible individuals abruptly. Harvesting abruptly decreases both the density of susceptible individuals and the density of latent, infectious, and removed individuals. In the period between harvest and planting, the disease (pathogen) has to survive on plant material left in the field, in the soil, or in other locations (for example, in infected perennial weeds). This situation leads to a model for the development of an epidemic in a seasonal environment. In this section, we will discuss one of these models in some detail. Very few applicable mathematical methods are available to analyze such systems over multiple seasons, and simulation usually is the main tool to study these models. However, we will show that the threshold for epidemic development can be calculated explicitly for such models. The approach shown is based on a simplified version of the model presented by Madden and van den Bosch (2002). Other approaches have been published by Gubbins and Gilligan (1997), Shaw (1994), and Thrall et al. (1997). As mentioned in Chapter 4, sometimes the term *polyetic* is used to describe epidemics over multiple growing seasons.

6.2.2.1 Model derivation. At the start of the growing season, the crop is planted at density H_0. The crop-growing season has a length of T time units, and during the crop-growing season the epidemic develops. We use the equation 5.18, which we studied in section 5.2.2, as the model for the development of the epidemic during the crop-growing season. Instead of the H-I-R model, we could have used an H-L-I-R model, or even one based on a growing crop, but chose the simpler one for presentation purposes.

At the end of the crop-growing season, at time $t = T$, the crop is harvested. At this time the density of infectious and removed individuals are $I(T)$ and $R(T)$, respectively. At harvest, most of these individuals are removed or destroyed by harvesting, and only a small fraction remains on the field or in the soil as crop residue. During the period when no crop is grown, these remaining individuals experience adverse conditions (for example, low temperatures, humid conditions, etc.) so that from these individuals only a small fraction survives till the start of the next crop-growing season. Moreover, the pathogen may live saprophytically during this time, so that the concept of a diseased individual becomes nebulous (although we can still consider this surviving inoculum as units of diseased individuals for modeling purposes). Often fungal plant pathogens go through a sexual stage on the debris during the time where no crop is grown.

Moreover, some types of infectious units (spores) are only released from infected plants when they decay over time. Verticillium wilt of potato is an example, where the microsclerotia are released into the soil when potato tissue decays (Powelson and Rowe, 1993). Thus, although $R(T)$ may be removed from the current epidemic, these individuals may still contribute to the epidemic the next season.

The infectious and removed diseased individuals survive the crop-free period with probabilities ξ_I and ξ_R, respectively. Thus, the density of infection-producing "individuals" at the start of the *next* epidemic is $\xi_I I(T) + \xi_R R(T)$. (A gradual exponential decline in diseased individuals throughout the crop-free period could be equivalently modeled.) At the start of the next growing season the crop is planted again at density H_0 and the epidemic process is renewed. Each of the $[\xi_I I(T) + \xi_R R(T)]$ units of infection-producing "individuals" produces γ initial infections to start the epidemic in the new growing season. All initially infected individuals in the model are infectious. Thus, initial conditions at the start of the next season are $I(0) = \gamma[\xi_I I(T) + \xi_R R(T)]$, $R(0) = 0$, and $H(0) = H_0$, where the $I(T)$ and $R(T)$ terms are for the *previous* season. The epidemic then proceeds for T time units to reach the end of the next growing season, and the process repeats itself. We add another subscript to I_0 to indicate the season order; $I_{0,0}$, $I_{0,1}$, $I_{0,2}$, and so on.

We maintain the concept of individuals for both between and within-season situations for ease of presentation. As eluded to above, $\xi_I I(T) + \xi_R R(T)$ can be interpreted more broadly (by extending the definitions of the parameters) to represent the total primary inoculum availably at the start of the next season as a function of diseased individuals at the end of the current seasons.

6.2.2.2 *Model simulations.* Fig. 6.3 shows a number of simulations of the model for different values of the within-season transmission rate parameter β. Conceptually, the simulations are not difficult. Within each season equations 5.18 are solved. The only additional aspect is that I_0 each year depends on the densities of I and R at the end of the previous year and the parameters γ, ξ_I, and ξ_R.

For small values of β, no epidemic develops, showing that there is a threshold for epidemic development. When an epidemic develops, the density of infectious and removed individuals increases during the growing season and abruptly change at the end of the growing season at harvest. The larger the value of β, the steeper the epidemic increase and, hence, decline from one growing season to the next.

6.2.2.3 *Threshold for epidemic development.* As we have seen in all models so far, the threshold for epidemic development is the basic reproduction number, R_0. In the seasonal cropping model under consideration, we have to distinguish two different basic reproduction numbers. First of all, as we have seen in section 5.2.2, the model for the within-growing-season epidemic has an R_0 given by equation 5.23. We will use the expression $R_{0\text{within}}$ for this number

$$R_{0\text{within}} = \frac{\beta H_0}{\mu} \tag{6.24}$$

This $R_{0\text{within}}$ value is not the same as the R_0 of our multi-season epidemic model, which will be denoted by $R_{0\text{multi}}$.

The multi-season R_0 can be calculated as follows: Assume we start in year 0 with a very small initial density of infectious individuals, $I(0) = I_{0,0}$. We will explicitly calculate expressions for the initial infection in successive years, $I_{0,0}$, $I_{0,1}$, $I_{0,2}$, $I_{0,3}$, etc. As we have used before, in the situation where the density of infectious individuals is very small in the beginning stages of an epidemic, the density of healthy individuals is very well approximated by its initial density H_0. The within season model equation 5.18 then become

$$\begin{aligned} \frac{dI(t)}{dt} &= (\beta H_0 - \mu) I(t) \\ \frac{dR(t)}{dt} &= \mu I(t) \end{aligned} \tag{6.25}$$

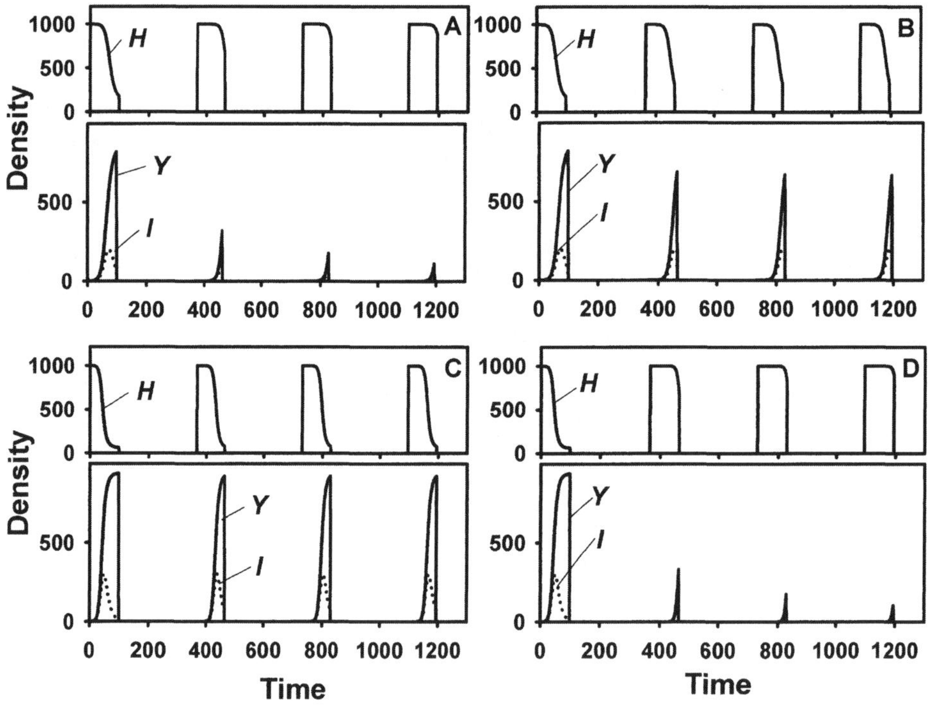

FIG. 6.3. The density of healthy (H), infectious (I), and the total infected (or diseased) (Y) individuals in relation to time (t; days) for an epidemic of an annual crop over multiple years. The model is described in section 6.2.2.1. The growing season is $T = 100$ day (with $365-T$ days between seasons). $H_0 = 1000$ for each growing season. For the first growing season, $I(0) = 1$. (**A**) $\beta = 0.00023$ per day, $\xi_1 = \xi_2 = 0.001$, $\gamma = 0.001$, $R_{0\text{within}} = 2.3$, $R_{0\text{multi}} = 0.78$. (**B**) $\beta = 0.00023$, $\xi_1 = \xi_2 = 0.001$, $\gamma = 0.01$, $R_{0\text{within}} = 2.3$, $R_{0\text{multi}} = 7.8$. (**C**) $\beta = 0.0003$, $\xi_1 = \xi_2 = 0.001$, $\gamma = 0.001$, $R_{0\text{within}} = 3.0$, $R_{0\text{multi}} = 727.7$. (**D**) $\beta = 0.0003$, $\xi_1 = \xi_2 = 0.0001$, $\gamma = 0.00001$, $R_{0\text{within}} = 3.0$, $R_{0\text{multi}} = 0.728$.

Note, we do not write out an equation for $dH(t)/dt$ since we are concerned here with the early stage when $H(t) \approx H_0$ so that $dH(t)/dt \approx 0$. The first equation of equation 6.25 is an exponential growth equation with solution:

$$I(t) = I_{0,0}e^{(\beta H_0 - \mu)t} \qquad (6.26)$$

Substituting this equation into the second equation of equation 6.25 we find

$$\frac{dR(t)}{dt} = \mu I_{0,0}e^{(\beta H_0 - \mu)t} \qquad (6.27)$$

This equation can be solved simply by integrating from 0 to t. Realizing that the initial condition for R is zero, we find for season 0 as solution

$$R(t) = \mu I_{0,0}\frac{1}{\beta H_0 - \mu}\left(e^{(\beta H_0 - \mu)t} - 1\right) \qquad (6.28)$$

Using equations 6.26 and 6.28, the initial condition for the *following* growing season, $I_{0,1} = \gamma[\xi_I I(T) + \xi_R R(T)]$, leads to

$$I_{0,1} = \left(\gamma\xi_I e^{(\beta H_0 - \mu)T} + \gamma\xi_R\frac{\mu}{\beta H_0 - \mu}\left(e^{(\beta H_0 - \mu)T} - 1\right)\right)I_{0,0} \qquad (6.29)$$

Equation 6.29 is obtained simply by multiplying equation 6.26 for $I(T)$ by $\gamma\xi_I$ and equation 6.28 for $R(T)$ by $\gamma\xi_R$. Using the same method we calculate the initial condition for year 2 as:

$$I_{0,2} = \left(\gamma\xi_I e^{(\beta H_0 - \mu)T} + \gamma\xi_R\frac{\mu}{\beta H_0 - \mu}\left(e^{(\beta H_0 - \mu)T} - 1\right)\right)I_{0,1} \qquad (6.30)$$

and, in general, for the initial infection in year $n+1$ we find

$$I_{0,(n+1)} = \left(\gamma\xi_I e^{(\beta H_0 - \mu)T} + \gamma\xi_R\frac{\mu}{\beta H_0 - \mu}\left(e^{(\beta H_0 - \mu)T} - 1\right)\right)I_{0,n} \qquad (6.31)$$

This equation has exactly the same form as equation 5.4, with the density the *next* season being equal to the density *this* season multiplied by a constant. Now, however, a complicated parameter combination is involved in the "constant". Following the same reasoning as done for equation 5.4 in section 5.2.1.3, we conclude that an epidemic will develop when

$$\left(\gamma\xi_I e^{(\beta H_0 - \mu)T} + \gamma\xi_R\frac{\mu}{\beta H_0 - \mu}\left(e^{(\beta H_0 - \mu)T} - 1\right)\right) > 1 \qquad (6.32)$$

and no epidemic will develop when this quantity is smaller than 1. In other words, there is a multi-seasonal epidemic when $I_{0,n+1}/I_{0,n} > 1$. We, thus, have found the expression for the multi-season basic reproduction number, $R_{0\text{multi}}$. To make the relation between the two basic reproduction numbers clearer, we substitute the expression for $R_{0\text{within}}$ (equation 6.24) into the expression for the expression for $R_{0\text{multi}}$.

$$R_{0\text{multi}} = \left(\gamma\xi_I e^{\mu(R_{0\text{within}} - 1)T} + \gamma\xi_R\frac{1}{R_{0\text{within}} - 1}\left(e^{\mu(R_{0\text{within}} - 1)T} - 1\right)\right) \qquad (6.33)$$

Fig. 6.4 shows $R_{0\text{multi}}$ in relation to $R_{0\text{within}}$ for a range of the other parameters.

Fig. 6.4 clearly shows that a $R_{0\text{within}}$ value larger than 1 does not directly imply that $R_{0\text{multi}}$ is larger than 1, although increasing $R_{0\text{within}}$ leads to increasing $R_{0\text{multi}}$. The length of the growing season, T, the mean infectious period, $1/\mu$, the survival of the pathogen between two successive growing seasons, ξ_I and ξ_R, and the efficiency of the initial infection at the start of each subsequent season, γ, all have a crucial effect on the $R_{0\text{multi}}$. When $R_{0\text{within}} < 1$, it takes very large values of $\gamma\xi_1$ and/or $\gamma\xi_2$, meaning that there must be very high between-season survival probability (ξ) and/or very high primary infection efficiency (γ) for there to be increasing I_0 over multiple years. Note that the relationships are highly nonlinear in that a small increase in $R_{0\text{within}}$ can have a large increase in $R_{0\text{multi}}$. Moreover, $R_{0\text{within}}$ could be relatively large (say, >5) and still the disease will die out over multiple years because of low between-season survival or low probability of infection from surviving infected individuals.

It should be emphasized that equation 6.25 and its solution (equations 6.26 and 6.28) are specified when healthy host density is not limiting. But this is precisely the situation that applies when determining thresholds for epidemics. However, Fig. 6.3 was prepared by using the more general equation 5.18 for the within-season dynamics (which adjusts for declining $H(t)$), and not using equation 6.26 and 6.28.

6.2.2.4 *Concluding remarks*. There are several models in the literature with seasonal dynamics incorporating additional aspects of multi-season epidemics (Gubbins and Gilligan, 1997; Gilligan, 2002; Madden and van den Bosch, 2002). Among other things, one can deal with both latent and infectious periods within seasons, and explicit consideration of inoculum in the soil or during the epidemic. Furthermore, the simplistic assumption here that surviving diseased individuals are based on densities at $t = T$ can certainly be relaxed. There may not be, in general, analytical solutions for multi-seasonal final disease values, rate of increase of $I_{0,n}$ over multiple seasons, or other results. However, in some situations

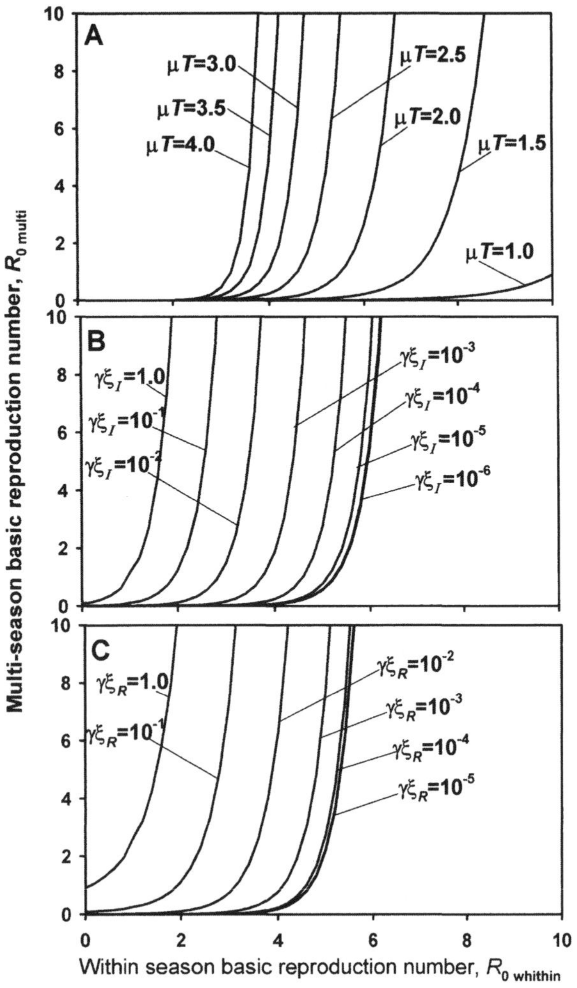

FIG. 6.4. The basic reproduction number of the multi-seasonal epidemic model, $R_{0\text{multi}}$, as a function of the within-season basic reproduction number, $R_{0\text{within}}$. Lines are calculated using equation 6.33. (**A**) $R_{0\text{multi}}$ as function of $R_{0\text{within}}$ for various values of the product of the rate at which infectious individuals become removed, μ, and the duration of the growing season, T. (**B**) $R_{0\text{multi}}$ as function of $R_{0\text{within}}$ for various values of the product of: the probability that an infectious individual at the end of a growing season survives until the next season (ξ_I) and the number of infections at the start of a growing season caused by an infectious individual surviving from the previous season (γ). (**C**) $R_{0\text{multi}}$ as function of $R_{0\text{within}}$ for various values of: the probability that a removed diseased individual at the end of a growing season survives until the next season (ξ_R) and the number of infections at the start of a growing season caused by a removed diseased individual surviving from the previous season (γ).

$R_{0\text{multi}}$ is related to a seasonally recurring final density of Y, $Y(T)$, (Madden and van den Bosch, 2002). That is, when $R_{0\text{multi}} > 1$, an apparently stable end-of-the-season disease intensity, $Y(T)$, can be reached after a few seasons, which can be approximated based on $R_{0\text{multi}}$ (Madden and van den Bosch, 2002). With the short introduction in this section we hope to have provided sufficient overview for those interested to be able to read the key papers. Considerably more research is needed, both theoretical and experimental (observational) to develop a more formal understanding of plant diseases over many individual seasons. Such an understanding should help clarify the development of long term disease management strategies—for instance, would it be more efficient to reduce βH_0 or between-season survival of infectious units for a given disease?

6.3 The Role of Primary Infection

In sections 5.2 and 6.2, each epidemic was started in the model(s) by introducing a number of infected or diseased individuals instantaneously at time $t = 0$. For the H-L-I-R model (equation 5.30), for example, starting the epidemic with I_0 newly infected individuals is specified with the conditions: $H(0) = H_0$, $L(0) = 0$, $I(0) = I_0$, and $R(0) = 0$, and, thus, the density of initial diseased individuals, $Y(0) = Y_0 = L(0) + I(0) + R(0)$, is given simply by I_0. The source of the infectious units that initiate the epidemic may be spores in the soil, migrating viruliferous insect vectors, or infected perennial weeds in the area. This approach to the primary infections that initiate an epidemic is common in most modeling of epidemics for polycyclic diseases in animals, humans, and plants (Anderson and May, 1991; Daley and Gani, 1999). The concept is that, with sufficient time, the secondary infections "overshadow" the initial events, especially if I_0 is very small and β is not very small. However, it may not be realistic in some cases to assume an instantaneous start to the epidemic. It is possible, for instance, that the primary infections occur over an extended period of time, possibly concurrently with the new (secondary) infections occurring due to spread from individual to individual (plant, root, leaf, etc.).

For example, epidemics of light leaf spot of oilseed-rape, caused by the fungus *Pyrenopeziza brassicae,* are initiated from the ascospores produced on the plant debris left in the field from the previous year (the primary infections). The ascospores are released from around September until the following January. The ascospore release rate increases from zero before September to a maximum value and then drops off slowly towards zero around February (Papastamati et al., 2002; Gilles et al., 2000, 2001). Another example is take all of wheat, caused by *Gaeumannomyces graminis* (Bailey and Gilligan, 1999). Primary infections occur as roots grow through the soil and contact inoculum, which occurs for an extended period of time, and secondary infections of roots occur from the contact of healthy roots with diseased roots.

The process of non-instantaneous initiation of the epidemic by primary infections can be incorporated into the models discussed so far in this chapter. As will be seen below, this essentially means combining models for polycyclic and monocyclic epidemics (see section 4.4.2) into a single model formulation. The consequences of such model combinations on predicted disease dynamics is, however, less clear-cut compared to the results obtained for strictly polycyclic or monocyclic epidemics. The approach we introduce here has been studied intensively and extensively. Leading work in this area stems from Gilligan and co-workers (e.g., Truscott et al., 1997; Gilligan and Kleczkowski, 1997; Gilligan, 2002, and references therein). As in some previous sections, we give an introduction to this work, showing some interesting phenomena, and suggest further reading on the subject for those who have interest.

We use the term *primary infectious units* for the infectious units that cause the primary infections. Essentially, primary infectious units comprise the initial (or primary) inoculum. Readers should return to section 4.3 of Chapter 4 for a more complete discussion of these concepts. In the example of light leaf spot, primary infectious units would the ascospores. We can use the term secondary infectious units for the inoculum (broadly defined) that are produced during the current epidemic that can potentially infect other plant individuals.

6.3.1 Model derivation

We base our discussion around the H-L-I-R model we discussed in section 5.2.3. If we assume that all primary infection is brought into the crop *after* $t = 0$ we have as initial conditions $H(0) = H_0$, $L(0) = 0$, $I(0) = I_0 = 0$ and $R(0) = 0$. The primary infections are introduced into the model equation 5.30 in pseudo-equation form as:

$$\begin{aligned}
\frac{dH(t)}{dt} &= -\beta H(t)I(t) - \begin{bmatrix}\text{rate of primary}\\ \text{infection}\end{bmatrix}\\
\frac{dL(t)}{dt} &= \beta H(t)I(t) + \begin{bmatrix}\text{rate of primary}\\ \text{infection}\end{bmatrix} - \omega L(t)\\
\frac{dI(t)}{dt} &= \omega L(t) - \mu I(t)\\
\frac{dR(t)}{dt} &= \mu I(t)
\end{aligned} \tag{6.34}$$

The primary infections are, of course, due to primary infectious units "released" into the system starting at $t = 0$. The term "release" should be considered broadly here to mean the inoculum per unit area or volume per time unit in the soil or on the surface, possibly in neighboring areas, which is available for contact with the healthy host individuals. The inoculum might have all been produced in the previous epidemics (as discussed in section 6.2 for multi-season epidemics), or it could actually be produced (say, from saprophytic growth of the pathogen) between growing seasons or during the current epidemic. The density of available primary infectious units per time unit at t is given by $x(t)$. This function can take many forms. For the example of light leaf spot on oilseed rape as described above, $x(t)$ would be a bell-shaped curve. For many pathogens, one can assume that $x(t)$ is at its highest at the beginning of the epidemic ($t = 0$), and declines towards 0 with increasing t due to mortality of the inoculum. The number per unit area of healthy individuals that are contacted by a primary infectious unit is given by $\theta H(t)$. Thus, the number per unit area of primary infectious units coming in contact with healthy individuals per time unit equals $x(t)\theta H(t)$. A primary infectious unit in contact with a healthy individual has a probability ψ_p of infecting the individual. Note that this infection probability can be different from the infection probability of a secondary infectious unit, so we add a p subscript to ψ. It has, for example, been shown that the primary infectious units for light leaf spot on oilseed rape (ascospores) have a higher probability of infection than the secondary infectious units (conidia) (Gilles et al., 2001). The rate of infection due to the primary infectious units thus is:

$$\begin{bmatrix}\text{rate of primary}\\ \text{infection}\end{bmatrix} = \theta\psi_p H(t)x(t) \tag{6.35}$$

and the model equation 6.34 become

$$\begin{aligned}
\frac{dH(t)}{dt} &= -\beta H(t)I(t) - \theta\psi_p H(t)x(t)\\
\frac{dL(t)}{dt} &= \beta H(t)I(t) + \theta\psi_p H(t)x(t) - \omega L(t)\\
\frac{dI(t)}{dt} &= \omega L(t) - \mu I(t)\\
\frac{dR(t)}{dt} &= \mu I(t)
\end{aligned} \tag{6.36}$$

where, as before, β $(=\theta\psi\alpha)$ is the transmission rate for secondary infections. As we have usually done in this chapter, we keep track of the total density of infections (total disease intensity) since the start of the epidemic, $Y(t) = L(t) + I(t) + R(t)$. The differential equation for this quantity is:

$$\begin{aligned}
\frac{dY(t)}{dt} &= \beta H(t)I(t) + \theta\psi_p H(t)x(t)\\
&= [\beta I(t) + \theta\psi_p x(t)]H(t)\\
&= [\theta\psi\alpha I(t) + \theta\psi_p x(t)]H(t)
\end{aligned} \tag{6.37}$$

The later expression in equation 6.37 makes it clear that $x(t)$ for primary infections plays the same role in the model as $\alpha I(t)$ for secondary infections.

Some special cases can be directly elucidated from equation 6.37 that are worth consideration. If $\beta = 0$, there are no secondary infections, and disease progress is strictly of the monocyclic type (see section 4.4.2). Depending on the value of ψ_p, θ, and $x(t)$, $Y(t)$ can remain low or go to the maximum possible value (H_0) by the end of the epidemic. If the availability of primary infectious units is fixed (so that $x(t) \equiv x_0$), and $\beta = 0$, then the model reduces to: $dY(t)/dt = \theta\psi_p x_0 H(t) = \theta\psi_p x_0[H_0 - Y(t)]$, which is identical to the monomolecular model (equation 4.8) used for monocyclic (simple interest) diseases in Chapter 4 (with $Y = yH_0$). In other words, the r_M parameter of the monomolecular model is equivalent to $\theta\psi_p x_0$. Decreases in the amount of inoculum, the probability of contacting a host, and/or the probability causing infection when in contact with a host all lead to lower r_M. A higher r_M means a steeper slope of $\ln[1/(1-(Y/H_0))]$ versus t. However, if there is enough time, Y can go to H_0 (or y to 1) for any positive value of r_M if $x(t)$ is a constant (x_0).

If $\beta > 0$ and $\theta\psi_p x_0 > 0$, the epidemic consists of a mixture of polycyclic and monocyclic processes. Several years ago Brassett and Gilligan (1988) utilized simplified versions of this mixture model for root disease epidemics. A practical simplification is to use the logistic function (see section 4.4.3) for the polycyclic portion of the epidemic, rather than the more complex H-L-I-R coupled equations here. In other words, using our current notation, one can write: $dY(t)/dt = [\Upsilon Y(t)+r_M]H(t)$, with $\Upsilon = r_L/H_0$ and $r_M = \theta\psi_p x_0$. There is an analytical solution to this model (see also section 9.5 in Campbell and Madden, 1990), allowing for direct (and relatively easy) estimation of the rate parameters from a data set.

Our concern here is with the general model equation 6.37, with the more realistic assumption of a non-fixed availability of primary infectious units, $x(t)$, and the more realistic H-L-I-R structure for the polycyclic component (with $\beta > 0$). Note, for $\theta\psi_p x(t)$ to be constant when $x(t)$ is variable, then either θ or ψ_p must vary over time in such a way that the three-way product is constant. Since this is unlikely (except as an approximation), one can generally expect a more complex pattern to disease progress curves than found for the pure H-L-I-R or the monomolecular models, or the mixture of the logistic and monomolecular (with fixed r_M) models mentioned in the previous paragraph, or the mixture of the H-L-I-R and constant-x monocyclic models.

Of the forms of $x(t)$ that are possible, we only consider the exponential decay equation. In other words, we assume that x declines during the epidemic, from a maximum at $t = 0$. The expression for x is:

$$x(t) = x_0 e^{-\kappa t} \tag{6.38}$$

in which x_0 is the value of x at $t = 0$ and κ is a parameter (with units of 1/time). (Of course, a special case is $\kappa = 0$, which gives $x(t) = x_0$.) The *total* number of primary infectious units that will enter the system per unit area over all possible times, $\Re$, equals the constant

$$\Re = \int_0^\infty x_0 e^{-\kappa t} dt = \frac{x_0}{\kappa} \tag{6.39}$$

The implication of equation 6.39 is that there is a finite amount of primary inoculum available, and once this is released there can be no more individuals infected from the primary inoculum. If $\beta = 0$, as well, the epidemic would cease at this point.

6.3.2 Model simulations

Fig. 6.5 shows a number of simulations of the model equations 6.36 and 6.37, using equation 6.38 for $x(t)$. In all cases, $\kappa = 0.052$/day, a value found by Bailey and Gilligan (1999). This means that $\Re = 19.23x_0$ total units of inoculum are released over a long period of time. For convenience, we use $x_0 = 1$/day. With the exception of Fig. 6.5A, a single nonzero value of $\theta\psi_p x_0$ was used for all model epidemics. In Fig. 6.5A, $\theta\psi_p x_0 = 0$ (either from $\theta = 0$ or $\psi_p = 0$), which means that the epidemic is purely of the polycyclic type. This graph corresponds to Fig. 5.7A already discussed ($H_0 = 1000$, $I_0 = 1$; $\beta H_0 = 0.15$/day, $1/\omega = \wp = 5$, and $1/\mu = \iota = 10$). The R_0 is 1.5. Fig. 6.5B corresponds to the same polycyclic parameters, but with the additional monocyclic component ($\theta\psi_p x_0 = 0.0105$/day). Because of the contributions of the primary infectious units to the epidemic over time, there is higher disease intensity ($Y(t)$)

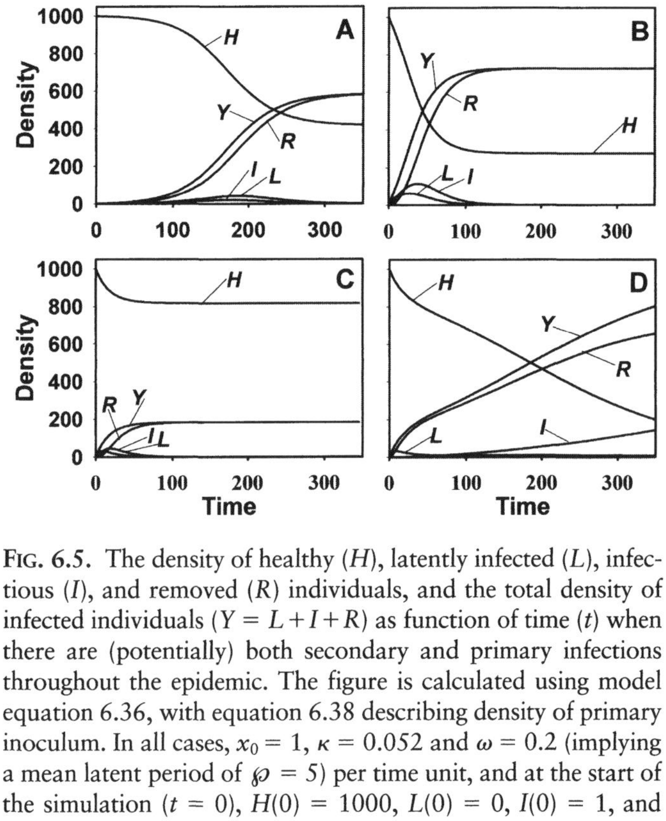

FIG. 6.5. The density of healthy (H), latently infected (L), infectious (I), and removed (R) individuals, and the total density of infected individuals ($Y = L+I+R$) as function of time (t) when there are (potentially) both secondary and primary infections throughout the epidemic. The figure is calculated using model equation 6.36, with equation 6.38 describing density of primary inoculum. In all cases, $x_0 = 1$, $\kappa = 0.052$ and $\omega = 0.2$ (implying a mean latent period of $\wp = 5$) per time unit, and at the start of the simulation ($t = 0$), $H(0) = 1000$, $L(0) = 0$, $I(0) = 1$, and $R(0) = 0$. (**A**) $\beta = 0.00015$, $\theta\psi_p = 0$, $\mu = 0.1$ (implying a mean infectious period of $\iota = 10$) per time unit. (**B**) $\beta = 0.00015$, $\theta\psi_p = 0.0105$, $\mu = 0.1$ ($\iota = 10$) per time unit. (**C**) $\beta = 0$, $\theta\psi_p = 0.0105$, $\mu = 0.1$ ($\iota = 10$) per time unit. (**D**) $\beta = 0.00001$, $\theta\psi_p = 0.0105$, $\mu = 0.001$ ($\iota = 1000$) per time unit.

throughout, and the plot of $Y(t)$ versus t has less of a sigmoid shape than seen for Fig. 6.5A.

Fig. 6.5C depicts the epidemic for the same situation as in Fig. 6.5B, except that $\beta=0$ ($R_0=0$); that is, the graph is for the pure monocyclic epidemic with declining value of $\theta\psi_p x(t)$. As discussed in the previous subsection, final Y is less than H_0 because $x(t)$ declines over time to 0, meaning that $dY(t)/dt$ ultimately declines to zero. For instance, with the chosen value of κ, $x(50)$ is only $0.074x_0$. Note that, technically, there are still latent, infectious, and removed individuals with this epidemic. However, since the transmission rate (secondary infection rate) is 0 here, new infections are not developing from the infectious individuals.

Fig. 6.5D is for the same monocyclic process as depicted in Figs. 6.5B and C, but with much lower βH_0 (0.01/day) and much longer mean infectious-period parameter, $1/\mu$ (=1000), than depicted in Fig. 6.5B for the polycyclic component. The R_0 for the polycyclic component of the epidemic is quite large (=10), meaning that even without the monocyclic component, the $Y(t)$ values would approach H_0 with enough time. Due to the small value of βH_0, the new infections during the first 30–40 days are due mostly to the primary infectious units, and the early stages of the epidemics in Fig. 6.5C and D are very similar. A slight leveling off of the disease progress curve is apparent around day 40 because of the diminishing primary inoculum ($x(t)$). Then there is an apparent increase again in $Y(t)$ as the secondary infections start to have a discernable effect on the disease progress curve. Because of the lack of primary infectious units at times greater than 40 (in this example), disease progress after this time in Fig. 6.5D (as well as in Fig. 6.5B) is almost entirely of a pure H-L-I-R type. Unlike the situation in Fig. 6.5B, the temporary plateau in Y is discernable in Fig. 6.5D because of the slow rate of disease increase due to secondary infections (due to low β).

6.3.3 Discussion

As mentioned in the context of the simulations above, it is straightforward to calculate the basic reproduction number (R_0) for the polycyclic component of a epidemic. Whether or not an epidemic occurs (when there is the possibility of infections from primary and secondary infectious units), however, depends on *both* R_0 and the parameter combination for the monocyclic component of the epidemic. In other words, R_0 could still be less than 1 and there can be an increase in Y over time, as shown in Fig. 6.5C (or in Chapter 4 for monocyclic diseases). Whether the epidemic curve still has an approximate exponential phase, as with the models discussed in section 5.2 for pure polycyclic epidemics, depends on the balance between the primary/monocyclic and secondary/polycyclic components of the epidemic. Certainly, a monocyclic process results in a non-exponential-type increase in Y (see Fig. 4.17C; also see section 4.4.2). But, if the magnitude of this contribution is small relative to the polycyclic component, then there may be an approximately exponential increase. Moreover, if the monocyclic component is short-lived (due to rapid decline in $x(t)$, represented by a high κ in equation 6.38), then $Y(t)$ may still be very low when the polycyclic component starts to dominate; then one would definitely expect an exponential increase until $Y(t)$ becomes larger. The exponential rate of increase for this later situation would be approximately the r_E value calculated for the H-L-I-R model in section 5.2.3.4. Of course, if $Y(t)$ is already relatively high by the extinction of the monocyclic phase of the epidemic (for situations when this phase is short lived [large κ]), then it would be too late for exponential increase to be manifested (even though there may be very high $dY(t)/dt$ if βH_0 is large).

The final level for the epidemic with a monocyclic and a polycyclic component is well defined. Its derivation is beyond the scope of this chapter, but we refer to Metz (1978b) for a derivation in the context of the Kermack and McKenrick model. This derivation is given in a plant pathology context in Segarra et al. (2001). The final level of the epidemic can be calculated from

$$y_\infty = 1 - e^{-R_0 y_\infty - \theta\psi_p \mathfrak{R}} \qquad (6.40)$$

Fig. 6.6 shows the final level of the epidemic as a function of R_0 for various values of $\theta\psi_p\mathfrak{R}$. Note that $\theta\psi_p\mathfrak{R}$ can be considered the total amount of released primary infectious units per unit area per host individual that actually contact and infect host individuals. For the simulations in Fig. 6.5 (except for 6.5A, which had no monocyclic component), $\theta\psi_p\mathfrak{R} = 0.2$.

When $\theta\psi_p\mathfrak{R} = 0$, the epidemic is purely of the H-L-I-R type, and final y given by equation 6.39 is the same as that given by equation 5.27. Final y increases with increasing $\theta\psi_p\mathfrak{R}$ at any given R_0, but the effect depends

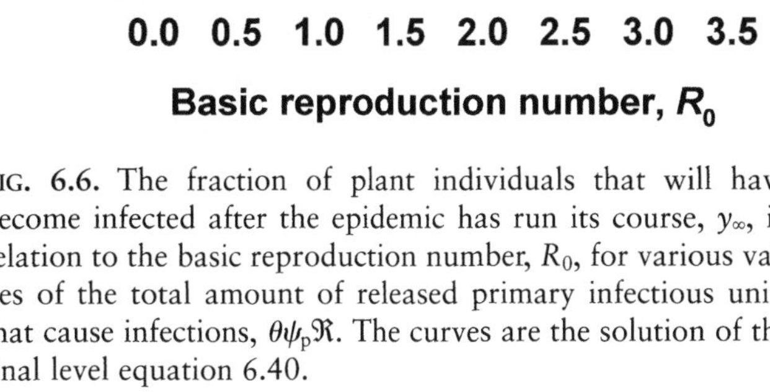

Fig. 6.6. The fraction of plant individuals that will have become infected after the epidemic has run its course, y_∞, in relation to the basic reproduction number, R_0, for various values of the total amount of released primary infectious units that cause infections, $\theta\psi_p\mathfrak{R}$. The curves are the solution of the final level equation 6.40.

on R_0. At $R_0 > 3$, there is just a little increase in y_∞ with increasing primary infections; this is because the polycyclic component dominates at high R_0. The effect of $\theta\psi_p\Re$ is clearly seen at low R_0, when y_∞ would be relatively low without an influx of primary infectious units (Fig. 6.6). Note that y_∞ can be larger than 0 even for values of R_0 from 0 up to 1, due to the influence of the primary infectious units. For example, with $R_0 = 0$ and $\theta\psi_p\Re = 0.2$, about 20% of the individuals become diseased ultimately ($y_\infty \approx 0.2$) in Fig. 6.6, which is in agreement with the simulation results shown in Fig. 6.5C. Likewise, with $R_0 = 1.5$ and $\theta\psi_p\Re = 0.2$, $y_\infty \approx 0.72$, in agreement with the simulation results in Fig. 6.5B.

The final question we address here is: When is it important to consider a mixture of monocyclic and polycyclic processes? There is no formal answer, but guidelines are clear from numerical and some analytical results. For epidemics with a definite polycyclic component, little is gained by the complexity of adding a monocyclic component if βH_0 and/or R_0 are high (higher than in the simulations of Fig. 6.5), unless $\theta\psi_p\Re$ is very high. Moreover, for epidemics with a definite monocyclic component occurring over a long time period (say, the duration of the season), and moderate or higher $\theta\psi_p\Re$, little is gained by the complexity of adding a polycyclic component in the model when R_0 is very low, especially for short-duration epidemics in which there is little time for the compounding effects of polycyclic epidemics to have much of an influence. As one would expect, it is the intermediate situation with moderate R_0 and $\theta\psi_p\Re$ that the mixture model is of great value. Examples include root diseases when the host individual of interest is the root (or root segment), not the entire plant. When entire plants are the units of host being considered, it is much more likely that the epidemic will be of the monocyclic type. Gilligan (2002) thoroughly reviews the topic.

To conclude this section, we point out that the link between the H-L-I-R model (equation 5.30) and the one of this section (equation 6.36) is even stronger than one might think. Using equation 6.38 for $x(t)$, it can be shown mathematically that as κ goes to infinity and $\theta\psi_p$ goes to infinity in such a manner that the ratio $\theta\psi_p/\kappa$ is constant ($=\aleph$), then $\aleph x_0 H_0$ is the initial density of infected individuals ($L(0)$), and $x(t)$ equals 0 immediately after $t = 0$. With a high efficiency of infection but rapid loss of inoculum, the monocyclic phase is very short lived. In other words, consideration of equation 6.36 as a general model with the additional assumptions here, leads to a formal expression for the starting conditions of a pure polycyclic epidemic in terms of initial inoculum density and how this inoculum infects the host individuals.

6.4 Epidemics with Vector Transmission

Many plant pathogens are transmitted from infectious plants to healthy plants by means of vectors. Plant viruses are the largest group of pathogens causing plant diseases that are transmitted by vectors, mostly herbivorous insects, especially aphids, leafhoppers, planthoppers and some beetles (Nault, 1997; Madden et al., 2000b). Phytoplasmas, spiroplasmas, and some other prokaryotes (such as the agent of Pierce's disease of grape) are also transmitted by vectors. Other arthropods (e.g., mites) and other organisms (e.g., nematodes, fungi) may also transmit some viruses, but we restrict attention here to insects as vectors.

When the insect vector feeds on an infectious plant, it can acquire the virus if virus titer is high enough in the plant. For several viral plant diseases, the plant virus has no, or very little, effect on the insect vector (Nault, 1997). When the insect subsequently feeds on a healthy plant the virus can be inoculated into the plant, and a new infection occurs. Viral plant diseases are generally systemic, meaning that once infected, the virus can multiply and move throughout the plant. Because virus diseases are systemic, the natural definition of an individual for these diseases is the whole plant, and in the section we use the term plant rather than individual for ease of presentation.

The latent period in the plant is the time it takes for sufficient virus multiplication and within-plant movement so that the virus can be acquired by a vector individual. The infectious period in the plant is over when there is sufficient reduction in virus titer in the plant so that a vector cannot acquire the virus when feeding. There are four basic transmission types or classes for plant viruses with insect vectors, which are discussed in an epidemiological context by Madden et al. (2000b). The types are based on a combination of: (i) the length of time that a vector remains infective after acquiring the virus from a plant; and (ii) on the physiological mode of transmission (e.g., whether or not the virus multiplies in the vector). In terms of population dynamics, however, the transmission type directly determines the length of the latent period and infectious period in the vector, and the rate at which the virus is acquired from the infectious host plant (or rate at which it is inoculated to a healthy plant). We consider only a small subset of the multiple possibilities.

Since all models are simplifications of reality, it is certainly possible to model and analyze plant virus disease epidemics without explicit consideration of the vector population. Examples of this are in Chapter 4 and elsewhere (e.g., Chan and Jeger, 1994; Madden and Campbell, 1986; Madden et al., 1987b; Nutter, 1997; Zhang et al., 2000a). For instance, if an H-L-I-R model (equation 5.30) was used, the transmission parameter, β, would be a function of, among other things, rates of virus acquisition and inoculation, and the density of the vector population. The functional relationship likely is very complex, depending on the transmission class (Madden and Campbell, 1986). Here we are concerned with expanding the models considered so far to explicitly relate the vector population to the population of diseased plants to develop a better understanding of plant virus epidemics.

As an introduction into the area of virus-vector modeling and analysis, we discuss a version of a general model that has been developed (in various forms) to study management strategies of virus diseases (Chan and Jeger, 1994; Holt and Chancellor, 1996; Holt et al., 1997, 1999; Jeger and Thresh, 1993; Jeger et al., 1998, 2002, 2004; Zhang et al., 2000a,b; Madden et al., 1990b, 2000b). The model is very similar to the first model for virus vector diseases in humans developed by Ross (1911, 1915) to study malaria. Wherever we plot densities of healthy and infected individuals, or the thresholds for the development of an epidemic, we will mostly use parameter values applicable to cassava mosaic virus disease. This disease has attracted a great deal of attention due to the very severe pandemic that has caused dramatic yield losses over the last decade in Africa (e.g., Otim-Nape et al., 2000).

The discussion of the model development here for virus–vector systems also serves the additional purpose of outlining some of the key steps in formulating a model for a plant disease using coupled differential equations, and then exploring the properties of the model, including model predictions for different parameter values.

6.4.1 Model derivation

The host population is divided into healthy and infectious plants. As before, the density of healthy plants is denoted by $H(t)$ and that of infectious plants by $I(t)$. We thus assume that the latent period is 0 and the infectious period is infinite (i.e., the duration of the epidemic or longer). These are obvious simplifications but allow us to consider some long-term dynamics here. This model is an acceptable simplification of reality whenever actual latent period is short and infectious period is long. The vector population is divided into non-viruliferous and viruliferous, where the term viruliferous stands for vectors that have acquired the virus by feeding on an infectious plant and can (potentially) transmit the virus when feeding on healthy plants. The density of non-viruliferous vectors is denoted by $X(t)$ and the density of viruliferous by $Z(t)$. We are thus assuming that there is no (or as a simplification, a very short) latent period in the vector and that the infectious period is very long. The dynamics of the virus-vector system is depicted in Fig. 6.7.

In the model, $H(t)$ increases due to planting of new crop plants and decreases due to harvesting of plants. This part of the model is analogous to the second dynamic-host model of section 6.2 (equation 6.5), applicable for relatively large regions over long time scales (several crops being planted and harvested). The density of healthy plants also decreases due to plants becoming

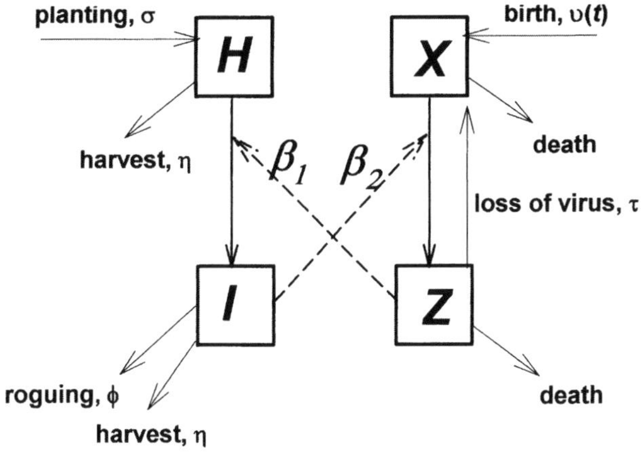

FIG. 6.7. Graphical representation of the virus vector model. Boxes are the state variables, density of healthy plants, H; density of infectious plants, I; density of non-viruliferous vectors, X; density of viruliferous vectors, Z. See text accompanying equations 6.46 and 6.52 for details.

infected with the virus when viruliferous vectors feed on the plant. This is described in the pseudo-equation

$$\frac{dH(t)}{dt} = [\text{planting rate}] - [\text{harvest rate}]_H - \begin{bmatrix}\text{rate at which healthy} \\ \text{plants become infected}\end{bmatrix} \quad (6.41)$$

The H subscript explicitly refers to harvesting of healthy individuals. $I(t)$ increases due to the infections that reduce the density of healthy plants, and decreases due to harvest and due to death of infectious plants. The pseudo-equation describing the dynamics of the density of infectious plants thus is:

$$\frac{dI(t)}{dt} = -[\text{harvest rate}]_I + \begin{bmatrix}\text{rate at which healthy} \\ \text{plants become infected}\end{bmatrix} - \begin{bmatrix}\text{disease-caused} \\ \text{death/roguing rate}\end{bmatrix} \quad (6.42)$$

The I subscript refers to the infectious plant individuals. As in the model in section 6.2.1 we assume that there is a constant planting rate σ and a harvest rate η (a constant), giving the terms

$$\begin{aligned}&[\text{planting rate}] = \sigma, \\ &[\text{harvest rate}]_H = \eta H(t), \text{ and} \\ &[\text{harvest rate}]_I = \eta I(t)\end{aligned} \quad (6.43)$$

Note that η can also be a death rate of plants due to all factors other than disease (so that it is the same for healthy and infected individuals. In fact, η could be the sum of the rates of harvesting and "natural" death/mortality (as long as natural death does not depend on infection status). Infectious plants can die due to the disease, or if sanitation programs are initiated, by removal (roguing) by the grower. We assume that there is a constant probability ϕ per time unit that an infectious plant is removed from the system due to the disease. The death rate term thus is:

$$\begin{bmatrix}\text{disease-caused} \\ \text{death/roguing rate}\end{bmatrix} = \phi I(t) \qquad (6.44)$$

For the rate at which healthy plants become infectious, we assume that there is a constant transmission rate β_1 and that the infection term takes the form

$$\begin{bmatrix}\text{rate at which healthy} \\ \text{plants become infected}\end{bmatrix} = \beta_1 H(t)Z(t) \qquad (6.45)$$

Substituting equations 6.43–6.45 into equations 6.41 and 6.42, the model describing the plant population thus takes the form

$$\begin{aligned}\frac{dH(t)}{dt} &= \sigma - \eta H(t) - \beta_1 H(t)Z(t) \\ &= \eta(H_{\max} - H(t)) - \beta_1 H(t)Z(t) \qquad (6.46) \\ \frac{dI(t)}{dt} &= -\eta I(t) + \beta_1 H(t)Z(t) - \phi I(t)\end{aligned}$$

This completes the specification of the dynamics of the plant population. Note how this equation differs from equation 5.18, where no vector was involved. In the contact rate term for host plant infection, $I(t)$ is replaced by $Z(t)$, the density of viruliferous vectors. We *could* consider the special case of no dynamics to total host density (in the absence of disease). This would mean using $\eta = 0$, so that there is no planting or harvesting. However, for long-term scenarios being of interest here, with new planting and harvesting, this is not a useful simplification, so we only consider $\eta > 0$.

The density of non-viruliferous vectors (i.e., those free of the virus), $X(t)$, increases due to birth (hatching, emergence, and so on) of vectors and decreases due to vector death. Further, the density of non-viruliferous vectors decreases due to acquisition of the virus by individuals feeding on infectious plants and density increases due to *viruliferous* vectors loosing the virus they acquired. This is described by:

$$\frac{dX(t)}{dt} = [\text{birth rate}] - [\text{death rate}]_X - \begin{bmatrix}\text{rate at which} \\ \text{non-viruliferous} \\ \text{vectors acquire} \\ \text{the virus}\end{bmatrix} + \begin{bmatrix}\text{rate at which} \\ \text{viruliferous vectors} \\ \text{lose the virus}\end{bmatrix} \qquad (6.47)$$

Note that we use "viruliferous" here as a synonym for infective or infectious, since we are not considering a latent period in the vector. The number of viruliferous vectors, $Z(t)$, increases due to non-viruliferous vectors acquiring the virus by feeding on infectious plants and decreases due to viruliferous vectors losing the virus and due to vector death. The pseudo-equations for this category is:

$$\frac{dZ(t)}{dt} = -[\text{death rate}]_Z + \begin{bmatrix}\text{rate at which} \\ \text{non-viruliferous} \\ \text{vectors acquire} \\ \text{the virus}\end{bmatrix} - \begin{bmatrix}\text{rate at which} \\ \text{viruliferous vectors} \\ \text{lose the virus}\end{bmatrix} \qquad (6.48)$$

For now we will denote the birth rate of the insect vector by the general function $\upsilon(t)$. Each insect vector has a probability Ξ per time unit to die. This gives us

$$\begin{aligned}&[\text{birth rate}] = \upsilon(t), \\ &[\text{death rate}]_X = \Xi X(t), \text{ and} \qquad (6.49) \\ &[\text{death rate}]_Z = \Xi Z(t)\end{aligned}$$

The viruliferous vector has a probability τ per time unit to lose the virus, which gives

$$\begin{bmatrix}\text{rate at which} \\ \text{viruliferous vectors} \\ \text{lose the virus}\end{bmatrix} = \tau Z(t) \qquad (6.50)$$

Referring to our discussion in sections 5.2.2.2 and 5.2.3.1 on latent and infectious periods in relation to probabilities per time unit to leave a category, we note that this assumption implies that the mean infectivity period of the vector equals $1/\tau$.

For the rate at which vectors acquire the virus we apply a similar term as for the rate at which plants become infected through vector feeding, describing the dependence of acquisition on non-viruliferous vector and infectious plant density as:

$$\begin{bmatrix}\text{rate at which} \\ \text{non-viruliferous} \\ \text{vectors acquire} \\ \text{the virus}\end{bmatrix} = \beta_2 X(t)I(t) \qquad (6.51)$$

Note that the acquisition rate is a function of the product of densities of non-viruliferous vectors and infectious plants. The model equations for the dynamics of the vector population thus are

$$\begin{aligned}\frac{dX(t)}{dt} &= \upsilon(t) - \Xi X(t) - \beta_2 X(t)I(t) + \tau Z(t) \\ \frac{dZ(t)}{dt} &= -\Xi Z(t) + \beta_2 X(t)I(t) - \tau Z(t)\end{aligned} \qquad (6.52)$$

We now need to specify the vector birth (emergence, hatching) rate in more detail. We are considering here only those plant viruses that do not have an effect on the biology of the vector. Madden et al. (2000b), and references therein, cover the issue more broadly. Thus, the total vector density, $X(t) + Z(t)$, does not depend on the virus nor on the density of healthy or infected plants. A simple assumption we can make to derive the birth rate is that every insect vector that dies is replaced by a new vector through birth in the population. This means that we assume the vector population to be of constant size, which we will denote by P. In other words, total vector density is a parameter, not a variable. The total vector population being constant implies that

$$X(t) + Z(t) = P \qquad (6.53)$$

Taking the derivative of equation 6.53 with respect to time yields

$$\frac{dX(t)}{dt} + \frac{dZ(t)}{dt} = 0 \qquad (6.54)$$

when P is a constant. This is because the total rate of change of the vector population must be 0 if P does not change. Of course, the rates of change for non-viruliferous and viruliferous vectors can both be nonzero even when their sum is zero. Substituting the equation 6.52 in equation 6.54 we find:

$$\upsilon(t) - \Xi X(t) - \Xi Z(t) = 0 \qquad (6.55)$$

which can be written as

$$\upsilon(t) = \Xi X(t) + \Xi Z(t) = \Xi(X(t) + Z(t)) = \Xi P \qquad (6.56)$$

We thus find that $\upsilon(t) = \Xi P$, and substituting this in the model equation 6.52, we complete the model specification.

Our assumption about constant vector population density leads to a simplifying consequence: one of the vector equations has become redundant. To see this, rewrite equation 6.53 in the form $X(t) = P - Y(t)$. Substituting this expression for $X(t)$ in the equation for $dZ(t)/dt$ we find

$$\frac{dZ(t)}{dt} = -\Xi Z(t) + \beta_2(P - Z(t))I(t) - \tau Z(t) \qquad (6.57)$$

and based on equation 6.54, $dX(t)/dt$ is obtained as $-dZ(t)/dt$. Thus, the rate of change in the density of viruliferous vectors can be calculated independently of the density of non-viruliferous vectors (i.e., X is not in equation 6.57 for the rate of change in Z). Note that whenever we know the density of viruliferous vectors, $Z(t)$, the density of non-viruliferous vectors is then easily calculated from $X(t) = P - Z(t)$. Putting all the pieces together, the model describing the epidemic dynamics of a virus-vector disease is:

$$\begin{aligned}
\frac{dH(t)}{dt} &= \sigma - \eta H(t) - \beta_1 H(t)Z(t) \\
&= \eta(H_{max} - H(t)) - \beta_1 H(t)Z(t) \\
\frac{dI(t)}{dt} &= -\eta I(t) + \beta_1 H(t)Z(t) - \phi I(t) \\
\frac{dZ(t)}{dt} &= -\Xi Z(t) + \beta_2(P - Z(t))I(t) - \tau Z(t)
\end{aligned} \qquad (6.58)$$

Equation 6.58 is a variation of the epidemic models with host growth (see section 6.2), generalized to account for transmission by a vector. Note that if no roguing (or infection-caused mortality) occurred for infected individuals ($\phi = 0$), then a fixed total plant density of H_{max} ($= H + I$) would be maintained throughout the epidemic if initial host density ($H_0 + I_0$) equaled H_{max}. We do not, however, restrict the model to situations with fixed total host density. It should be further noted that density of individuals that are rogued [$B(t)$] could be explicitly accounted for by using an additional equation in equation 6.58 for $dB(t)/dt$ [$= +\phi I(t)$], but such explicit representation is not needed.

6.4.2 Model simulations

The left-hand column of Fig. 6.8 shows a series of simulation results of model equation 6.58 for different values of the inoculation rate parameter β_1 when there is no roguing (removal) of diseased plants ($\phi = 0$). We assume the acquisition and inoculation rates were the same (which is reasonable for many plant viruses [Madden et al., 2000b]). The epidemics started with $I_0 = 1$ (so, $H_{max} = 1000$), and $P = 50$ and $Z_0 = 0$ (so that all vectors were virus free). The plotted values of plant densities were rescaled to $0-1$ so that they could be shown readily on the same graph with viruliferous vectors. For very small values of β_1, no epidemic develops, whereas for larger values an epidemic does develop. A high percentage of the plants become infected at high β_1, but as found for polycyclic epidemics in general (see sections 5.2 and 6.2), not all individuals become infected. An apparent steady state is seen where loss of infected plants (through harvesting) is balanced by new planting. Furthermore, there is an apparent threshold behavior for epidemic development, which is very similar to the threshold behavior for diseases that are transmitted directly from plant to plant. The simulations also show that after the start of the epidemic, the density of infected plants appears to increase exponentially with time in the initial stages of the epidemic (when β_1 is high enough for an epidemic to occur).

The right-hand column of Fig. 6.8 corresponds to $\beta_1 = \beta_2 = 0.008$/day (same as for Fig. 6.8B), but with $\phi > 0$

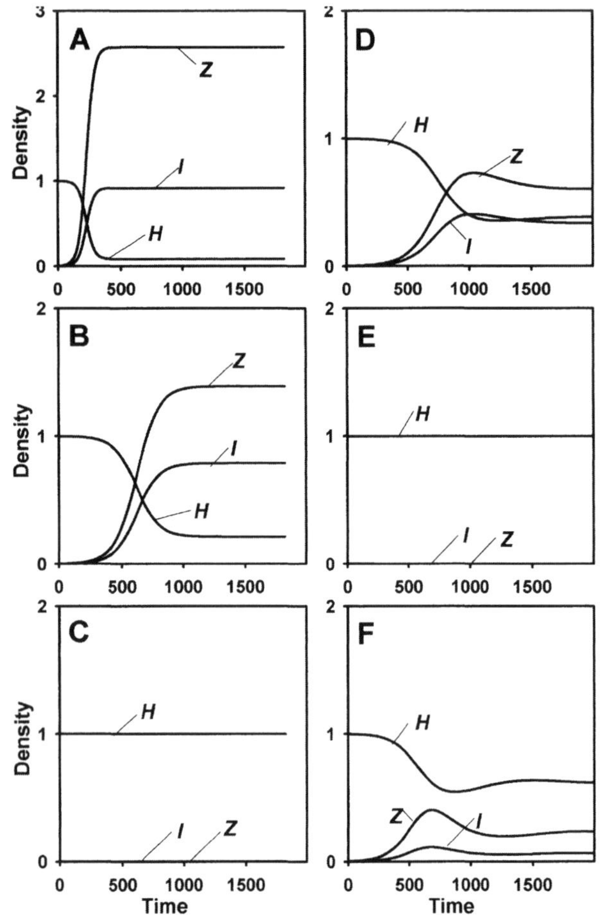

FIG. 6.8. The density of healthy plants, H, infectious plants, I, and viruliferous vectors, Z, as function of time, for a plant disease with an insect vector specified by model equation 6.58. At the start of the epidemics ($t = 0$), conditions were $H(0) = 1000$, $I(0) = 1$ and $Z(0) = 0$. Plant variables (but not the vector variable) were re-scaled between 0 and 1 in the graphs by dividing values by 1000; all parameter values shown are in terms of rescaled disease variables. $H_{max} = 1$ ($\sigma = 0.003$, $\eta = 0.003$), and $\Xi = 0.12$ and $\tau = 0.1$ per time unit for all cases. Left-hand column of graphs: increasing β_1 and β_2 with roguing rate (ϕ) equal to 0. Right-hand column of graphs: increasing ϕ (all above 0) and/or vector density (P) at fixed values of β_1 and β_2. (**A**) $\beta_1 = 0.013$, $\beta_2 = 0.013$, $P = 50$. (**B**) $\beta_1 = 0.008$, $\beta_2 = 0.008$, $P = 50$. (**C**) $\beta_1 = 0.003$, $\beta_2 = 0.003$, $P = 50$. (**D**) $\beta_1 = 0.008$, $\beta_2 = 0.008$, $\phi = 0.0025$, $P = 50$. (**E**) $\beta_1 = 0.008$, $\beta_2 = 0.008$, $\phi = 0.015$, $P = 50$. (**F**) $\beta_1 = 0.008$, $\beta_2 = 0.008$, $\phi = 0.015$, and $P = 100$.

(indicating the removal of infected plants). A small value for ϕ (=0.0025/day) resulted in lower density of diseased plants throughout (Fig. 6.8D) compared to the situation with $\phi = 0$. Also, density of diseased plants increased to a maximum before declining somewhat to an apparent steady state. When the roguing rate was increased to $\phi = 0.015$/day, no epidemic occurred (Fig. 6.8E). However, if vector density was doubled to $P = 100$ at this high roguing rate, an epidemic did occur (Fig. 6.8F), although the density of infected plants was relatively low.

6.4.3 Steady states and thresholds for epidemic development

The steady state densities of healthy plants, infectious plants (same as infected plants in this model, since latent period is 0 and there is no finite infectious period) and of viruliferous vectors can be calculated in the same way as discussed in section 6.2.1.3. Equating all rates of change to zero and replacing $H(t)$, $I(t)$ and $Y(t)$ by $\breve{H}, \breve{I}$ and $\breve{Z}$, respectively, we find that the steady states can be calculated from:

$$\begin{aligned} \sigma - \eta\breve{H} - \beta_1\breve{H}\breve{Z} &= 0 \\ \beta_1\breve{H}\breve{Z} - (\eta+\phi)\breve{I} &= 0 \\ \beta_2(P-\breve{Z})\breve{I} - (\Xi+\tau)\breve{Z} &= 0 \end{aligned} \tag{6.59}$$

Solving this set of equations we find two steady states of the system

$$\breve{H} = \frac{\sigma}{\eta} = K, \quad \breve{I} = 0, \quad \breve{Z} = 0 \tag{6.60}$$

$$\breve{H} = \frac{\sigma}{\eta} - \frac{\eta+\phi}{\eta}\breve{I}, \quad \breve{I} = \frac{\beta_1\beta_2 P\frac{\sigma}{\eta} - (\Xi+\tau)(\eta+\phi)}{\beta_2(\eta+\phi)\left(\beta_1 P\frac{1}{\eta} - 1\right)},$$
$$\breve{Z} = \frac{\beta_2 P\breve{I}}{\beta_2\breve{I} + \Xi + \tau} \tag{6.61}$$

Note that we saved space above by expressing the densities of healthy plants and viruliferous vectors in terms of the density of infectious plants. Equation 6.60 is the steady state when the disease is not present. As discussed in section 6.2.1.5, the density of healthy plants converges to H_{max} when disease is not present for any nonzero starting value of H (this is why $H_{max} = \sigma/\eta$). The second set of steady states correspond to when the disease is present (equation 6.61). Similarly to our discussion in section 6.2.1.3, we note that these steady states are only biologically relevant if all these densities are larger than or equal to zero. Some algebra shows that all three steady state densities are larger than zero when

$$\left(\frac{\beta_1 H_{max}}{\Xi+\tau}\right)\left(\frac{\beta_2 P}{\eta+\phi}\right) > 1 \tag{6.62}$$

with $H_{max} = \sigma/\eta$. We have organized the expression in such a way that it helps the interpretation. Note that there are several algebraic steps involved in obtaining this expression.

The left-hand side of equation 6.62 is the basic reproduction number, R_0, for this virus disease represented by epidemic model equation 6.58. With a dynamic host, the disease will only persist (i.e., $I(t)$ will stay above 0) if $R_0 > 1$. A heuristic explanation follows. Consider the situation where one viruliferous vector flies into the crop, which is disease free but does contain a population of non-viruliferous vectors. This viruliferous vector will infect $\beta_1 H_{max}$ plants per time unit. Referring to the discussion in sections 5.2.2.2 and 6.2.1.6, we know that a viruliferous vector remains viruliferous for, on average, $1/(\Xi + \tau)$ time units (i.e., it either dies or loses the virus). Therefore, the total number of plants one viruliferous vector infects before it reaches the end of its viruliferous period equals the first part on the left hand side of equation 6.62. Next we consider the infected plants (which are also infectious in this model formulation). Each infected plant will be visited by healthy vectors that then can become viruliferous themselves. The vector population is in this stage still entirely healthy (except for this one individual that entered the system) and thus the number of vectors that become viruliferous per time unit per infectious plant equals $\beta_2 P$. Again referring to the discussion in section 5.2.2.2 we know that an infectious plant remains infectious for, on average, $1/(\eta + \phi)$ time units (i.e., either the plant is harvested or is rogued/removed). Therefore, the total number of vectors per infectious plant that acquire the virus before the plant is removed due harvesting and roguing equals the second part on the left hand side of equation 6.62. The product of the two terms thus has the interpretation of the number of viruliferous vectors resulting from one viruliferous vector, with infected plants as intermediate step, when the viruliferous vector is surrounded by an entirely healthy plant population and an entirely non-viruliferous vector population. This explanation can also be reversed, to obtain the number of new infected plants from one infected plant, with viruliferous vectors as the intermediate step, when the infected plant is introduced into an otherwise disease-free plant population with initially no viruliferous vectors. It is clear that if this number of new "cases" per case is larger than unity, an epidemic will develop, whereas if it is smaller than unity, no epidemic will develop. We have thus found the threshold for epidemic development in a virus-vector system.

It is useful to compare the equation here for R_0 with equation 6.23 which was derived for a similar (but not identical) dynamic-host situation in section 6.2. Because there is no latent period, the term with ω in equation 6.23 does not appear here. We also do not consider the infectious period in the current simple model for virus epidemic, which is equivalent to assuming that infected plants remain infectious until harvest. Thus, there is no μ in the current R_0 calculation (equation 6.62). Because we explicitly consider roguing of infected plants, ϕ plays the same role in determination of R_0 here as μ does in equation 6.23. Finally, the βH_{max} term in equation 6.23 is replaced by the four-way product, $\beta_1 \beta_2 H_{max} P$; this is because epidemics are functions of the host and vector densities, and depend on both virus acquisition and inoculation rates.

6.4.4 Some notes on disease management

As mentioned earlier, we have chosen parameter values for the simulations that reflect the situation of cassava mosaic virus transmitted by the whitefly *Bemisia tabaci*. As given in the legend of the figure, the parameters all have days as time unit and m^2 as unit of area. Characteristically, the whitefly density (P) is between 50 and 500 individuals per square meter (see Jeger et al., 2004 and references therein for motivation of these parameter values). When studying virus disease management, three potential strategies are often considered for this and similar diseases (see Jeger et al., 2004, for a general overview of the topic).

- *Roguing*. This was brought up in terms of model development, but we elaborate here. At regular time intervals, fields are inspected and plants showing disease symptoms are removed and destroyed. In our model such sanitation programs are reflected in the parameter ϕ, which is the product of the number of inspection rounds made per time unit and the probability that the inspector (farmer) recognizes and rogues infected plants. The experience of development workers with growers of small cassava plantations in Africa is that they are maximally willing to do one inspection round each month (Fauquet and Fargette, 1990). If they, with certainty, recognize each infected plant, this would mean that the rouging parameter ϕ equals approximately 0.03/day. Since the detection of infected plants is not very easy and many infected plants will be missed during sanitation rounds, we take the value $\phi = 0.02$ as the very maximum possible.
- *Resistant cultivars*. Several organizations are breeding cassava for resistance to cassava mosaic virus. Some of these cultivars are being used at present, but further developments in this area are continuing. The resistance level of the plant to the virus is expressed in β_1 and β_2, the inoculation and acquisition rate by vectors, respectively. Resistant cultivars that have a lower probability to become infected when exposed to a viruliferous vector have a smaller value of β_1. Resistant cultivars that result in a lower probability of a non-viruliferous vector acquiring the virus when exposed to the infectious plant have a smaller value of β_2.
- *Reducing vector density*. In some crops it is possible to reduce vector densities with well-targeted pesticide programs or through changes in the agronomics of the system. Reducing the vector density is reflected in the model parameter P.

Which of these disease management methods will work best in the case of cassava mosaic virus? One way

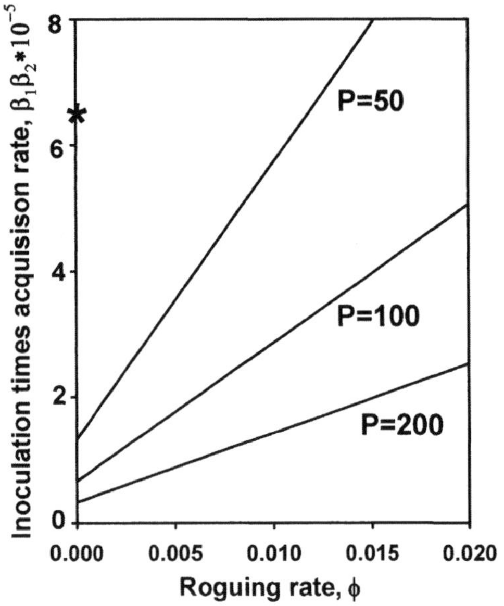

FIG. 6.9. The combination of the product of inoculation and acquisition rate, $\beta_1\beta_2$, and the rouging rate, ϕ, for which the basic reproduction number, R_0, is 1, for various values of the vector density, P. These lines are calculated from equation 6.62. Parameter values are: $H_{max} = 1$ ($\sigma = 0.003$ and $\eta = 0.003$), and $\Xi = 0.12$ and $\tau = 0.1$ per time unit.

to answer this question is to determine if it is possible, in the long run, to eradicate the disease in an area. By 'eradication' here we explicitly mean that an epidemic will not occur (at the temporal/spatial scale of interest). The question can be answered by considering both R_0 (equation 6.62) and the steady state of the densities of healthy and infected plants and the density of the viruliferous vectors.

In Fig. 6.9, we plotted the roguing rate ϕ on the abscissa and the product of the acquisition and inoculation parameters, $\beta_1\beta_2$, on the ordinate. The lines in the figure are where $R_0 = 1$. These lines thus separate parameter combinations for which an epidemic will develop from parameter combinations for which no epidemic will develop. Such lines are drawn in the figure for three different values of the vector density (P). Parameter combinations above the line will lead to the development of an epidemic, but no epidemic will develop for parameter combinations below the line (assuming, of course, that the model is appropriate). For example, consider the situation with $P = 50$, $\beta_1\beta_2 = 6.4 \times 10^{-5}$, and no roguing ($\phi = 0$), which corresponds to the simulation in Fig. 6.8B. This combination of $\beta_1\beta_2$ and ϕ falls above the line; thus, an epidemic will develop when the virus is introduced in the system (as demonstrated numerically in Fig. 6.8). Increasing the roguing rate to above ~0.012/day will cause the parameter combination to fall below the threshold line for $P = 50$, and no epidemic will develop (as shown numerically for $\phi = 0.015$ in Fig. 6.8E). The star on the ordinate depicts the parameter values of the default "standard" system chosen for this example analysis, where susceptible cultivars are used and no roguing is applied.

The figure shows that roguing as *the* disease management measure can only prevent an epidemic when the density of whiteflies, P, is relatively small. For the nominal values of the other parameters (e.g, β_1 and β_2), roguing alone can eradicate the disease when $P = 50$ but cannot control the disease for the other values of P shown in the figure. For instance, with $\beta_1\beta_2 = 6.4 \times 10^{-5}$, an epidemic will occur for all values of the ϕ when P is 100 or higher. A resistant cultivar, if it is to eradicate the disease, should decrease the product of the acquisition and inoculation parameter, $\beta_1\beta_2$, to less than 20% of its chosen nominal value for susceptible cultivars. For instance, with $P = 50$ and $\phi = 0$ (no roguing), $\beta_1\beta_2$ must be reduced from 6.4×10^{-5} to 1.8×10^{-5} or less in order to prevent an epidemic. It seems unlikely that this reduction in the parameters can be achieved through breeding.

We can conclude that individual disease control strategies cannot be expected to eradicate the disease. A combination of the two disease management strategies, however, shows tremendous promise for disease control. Bringing the roguing rate to close to its maximum (0.02/day) and introducing a resistant cultivar that decreases the product of the acquisition and inoculation parameters, $\beta_1\beta_2$, to about 60% of its value for susceptible cultivars (4×10^{-5}) will eradicate the disease even at $P = 100$. With lower P (=50), even a moderate roguing rate ($\phi = 0.01$) will result in disease eradication if $\beta_1\beta_2$ can be reduced to 4×10^{-5}. Experiments have shown that reducing the vector population density using insecticides is very difficult (Palumbo et al., 2001; Dittrich et al., 1990). Thus, although reducing the vector density would be effective in disease control, if used in conjunction of reduction of $\beta_1\beta_2$ and increase in ϕ.

In Fig. 6.10, the density of healthy and infected (=infectious in this case) plants and the density of viruliferous vectors are plotted as function of all four disease management parameters. Graphs such as these are very useful to determine epidemic outcomes when epidemics cannot be prevented ($R_0 > 1$). The nominal values of the parameters not shown on the graphs correspond to those for susceptible cultivars ($\beta_1 = \beta_2 = 0.008$/day), no roguing ($\phi = 0$), and a vector population density of 50 insects per plant. These figures show that the density of healthy plants increases in close to a linear manner with the roguing rate until $\phi \approx 0.01$/day, and then $\breve{H}$ is no longer affected by ϕ (because the steady state equals H_{max}; the disease "dies out") (Fig. 6.10D). Moreover, $\breve{H}$ increases with decreasing vector density, but the relationship is not linear; in fact, there is little change in $\breve{H}$ as P decreases from 100 until about 40 (Fig. 6.10A); at lower P values, there is a rapid change in $\breve{H}$ with change in P. The parameters reflecting host resistance, β_1 and β_2, show a similar relationship to $\breve{H}$ (Fig. 6.10A, B) as that found for P. That is, small reductions in β_1 or β_2 away from the nominal 0.008/day had little impact on $\breve{H}$; the

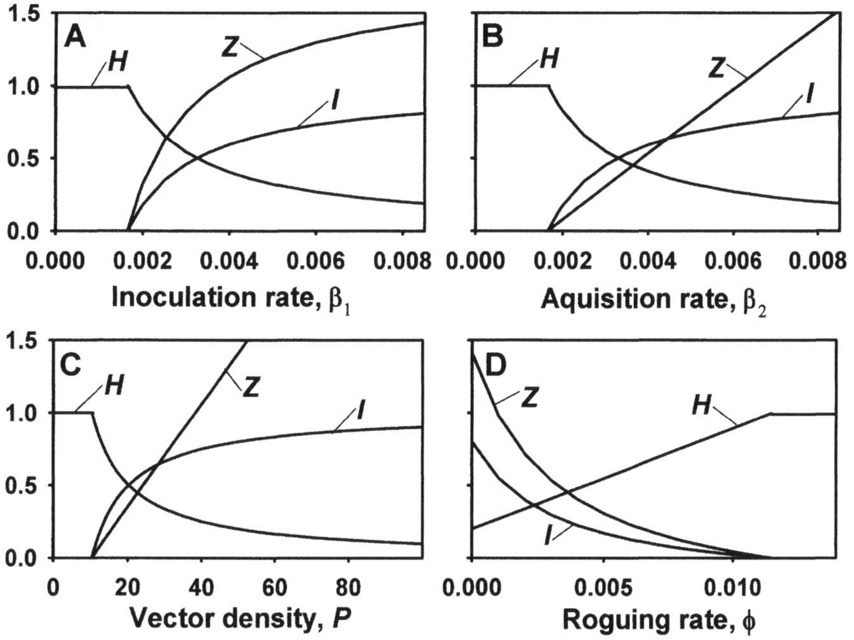

FIG. 6.10. Steady state densities of healthy plants (H) infectious plants (I) and viruliferous vectors (Z) as a function of various parameters for a plant disease with an insect vector specified by model equation 6.58. Parameter values, if not plotted on the horizontal axis, are: $H_{max} = 1$ ($\sigma = 0.003$ and $\eta = 0.003$), $\beta_1 = 0.013$, $\beta_2 = 0.013$, $\phi = 0.0$, $\Xi = 0.12$, and $\tau = 0.1$ per time unit, and $P = 50$.

value of $\breve{H}$ only changed markedly when the parameters decreased by 50% or more (approaching the values that result in eradication). Therefore, in cases where disease is not eradicated, and a *single* control measure has to be used, roguing is more efficient than using resistance. Of course, combinations of controls can be even more effective (not shown here).

6.4.5 Concluding remarks

In this section, we have introduced a simple model for the dynamics in virus–vector systems. Using parameters for the cassava mosaic disease we have demonstrated an application of the model by evaluating the effect of control strategies on epidemics and we were able to draw some general conclusions about disease management. A warning is at place here. It would be unwise to simply implement the results of such a simple model into practice without further work. Other models also describing the cassava mosaic system, but based on different assumptions, should be studied to see whether the conclusions are general and that hold for a larger set of models (as we saw in section 5.2). Several models, some considerably more detailed than the one here, have been studied (e.g., Holt et al., 1997, 1999; Jeger et al., 1998, 2002, 2004; Zhang et al., 2000b, Madden et al., 2000b). The simple model here does serve the purpose of showing how disease dynamic models can be used to evaluate controls, or to suggest the likelihood that a certain control will be effective.

The model we have studied falls short in several aspects. As mentioned at the start of this section, no allowance for latent and infectious periods in the plant host were considered. We also did not consider a latent period in the vector. Both the latent and infectious periods in the vector, together with the parameters β_1 and β_2, are of fundamental importance in classifying plant viruses. Jeger et al. (1998, 2004) and Madden et al. (2000b) show the linkage between these transmission traits and the most effective means of control for the different types of plant viruses.

Lastly, our appraisal of control strategies does not consider the cost of developing the control (e.g., cost of breeding) and the costs of implementing the controls. The issue of control costs and the resulting benefits from the controls are addressed somewhat in Chapter 11. However, the representation of the epidemics is much more superficial in that chapter. Considerably more work is needed to link economic decision theory to population-dynamic modeling of plant disease epidemics.

6.5 Transitional Dynamics and Other Complexities

6.5.1 Models considered so far

In this and the previous chapter we have shown that a considerable understanding of plant disease epidemics can be reached by the use of just a few differential equations and a limited number of well-defined parameters. When host growth and mortality are not major issues during the time period of interest, the models of Chapter 4 for polycyclic diseases are easily expanded to consider latent and infectious periods, allowing for a more realistic model for $dY(t)/dt$ [or $dH(t)/dt$] compared to the logistic and similar models. Essentially, major results can all be expressed in terms of $\wp$, ι, and βH_0, although the latter can be considered the product of multiple parameters ($\alpha\theta\psi$). Additional complexities, such as host growth and mortality, as well as the mixing of polycyclic and monocyclic (primary infection) processes over time and the development of epidemics in annual crops over many years, are readily incorporated into the models, as needed, to satisfy the objectives of the investigator and the type of disease development. Often, a sigmoid disease progress curve is produced by the models, especially for polycyclic diseases when there is no host growth. A fundamental result is that for many scenarios for compound interest diseases, it is possible to define (in terms of a few parameters, or a composite parameter): (1) a threshold for epidemic development; (2) the exponential rate of increase during the early part of the epidemic (*if* there is an epidemic); and (3) a value for the final proportion of diseased individuals (or steady state value of H if the host is dynamic). The basic reproduction number, R_0, is the key composite parameter that is of great value. With proper formulation of the models, results for specific parameter combinations can be found with a calculator, or even by hand. Of course, spreadsheets facilitate the calculations, but are not required.

Researchers sometimes require more than the results for these three numbered issues. In particular, it is often of interest to know the density or proportion of diseased individuals at any given time during an epidemic, and not just at small and very large t. The term *transitional dynamics* is sometimes used when referring to population dynamics between small and very large t (i.e., between small and large $y(t)$). Because the time duration of many crops is not long enough for asymptotic (final level) or steady-state epidemic results to be achieved, calculated values of $Y(t)$ [or other densities, such as $H(t)$] are required for intermediate times (i.e., beyond the initial stages, where exponential-model approximations no longer are reasonable). In general, such calculations require numerical integration techniques since there is typically no analytical solution to the equations in this or the previous chapter (i.e., $Y(t)$ cannot be written as a function of t without the use of the integral symbol). This was once a serious limitation to the use of some of these epidemic models, but the limitation is becoming much less important because of the development of relatively easy-to-use computer programs (such as MATHCAD and MATHEMATICA) for numerically solving coupled differential equations. All the plots of density versus time were obtained based on numerical integration (see Figs. 5.5, 5.7, 6.1, 6.3, 6.5, 6.8).

As pointed out for several of the models in section 5.2, plots of $Y(t)$ versus t are often *S*-shaped (see Fig. 5.7) when the epidemics start with a small density of infected individuals, suggesting that a logistic or similar disease-progress model might be reasonable for describing the epidemics. In fact, a more detailed analysis than presented in these chapters shows that the logistic (or similar) model *can* approximately describe disease dynamics even when the generating models consider the realism of, among other things, a finite infectious period and a nonzero latent period. This should be expected since numerous published disease progress curves are adequately represented by the logistic model (see Chapter 4 for examples), and the models of this chapter would be problematical if they did not duplicate actual observations of plant disease epidemics.

The degree of similarity between the generated disease progress curves based on the H-L-I-R model (as an example) and the logistic model depends on the values of $1/\omega$, $1/\mu$, and β. As shown in Fig. 6.11, plots of logits versus t can be fairly straight over all actual times, indicating a very high similarity in results between simple and more complicated models, or the logit curve can bend to the left or right, even with sigmoid curves of $Y(t)$ versus t.

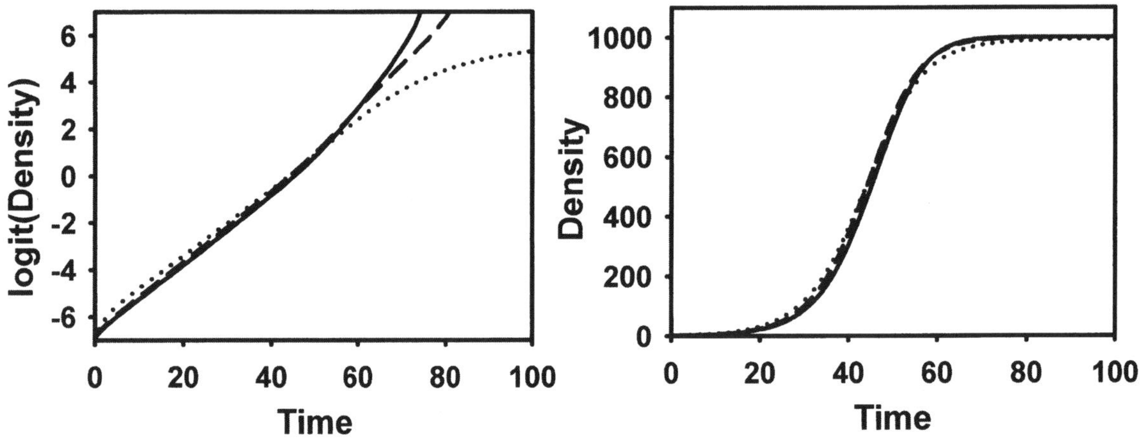

FIG. 6.11. Proportion of diseased *individuals*, $y(t) = Y(t)/H_0$, based on the H-L-I-R model (equation 5.30) and plotted on a logit scale (with an assumed maximum $y(t)$ of $K = 1$). Solid line: $1/\omega = \wp = 5$, $1/\mu = \iota = 1000$, $\beta H_0 = 0.25$ per time unit. Dotted line: $1/\omega = \wp = 14$, $1/\mu = \iota = 6$, $\beta H_0 = 0.90$ per time unit. Dashed line: $1/\omega = \wp = 5$, $1/\mu = \iota = 37$, $\beta H_0 = 0.30$ per time unit.

Note that logits were calculated using $\ln[y(t)/(1-y(t))]$ in the graph, in which $y(t) = Y(t)/H_0$, and parameter choices for the H-L-I-R model were deliberately chosen to emphasize deviation from a logistic relation at high disease density. Most combinations of parameters from H-L-I-R (and related models without host growth) result in logit lines that are fairly straight.

Very long mean infectious period ($\iota = 1/\mu$) typically results in logit curves bending to the left, and very short mean infectious period typically results in logit curves bending somewhat to the right. The curvature typically is an issue only at large $y(t)$. The value of R_0 was above 5 for the three curves in Fig. 6.11, so final disease level as a proportion was near 1 in all cases. If one is interested in approximating a realistic epidemic model when R_0 is small (but still above 1), it is recommended that the more general version of the logistic model [i.e., with a parameter for maximum y (= K)] be utilized (equation 4.47), and that y_∞ from equation 5.27 be used for K (Segarra et al., 2001). Then, the logit is given by $\ln[y(t)/(y_\infty - y(t))]$.

A direct comparison of the equations for $dY(t)/dt$ for the logistic (equation 4.51a) and for the H-L-I-R model (as an example; equation 5.31) shows that the two models can never be equal over the full range of Y (or H) values, because the rate for the logistic model depends on all diseased individuals, $Y(t)$, and for the more realistic H-L-I-R model, the rate depends on just infectious individuals, $I(t)$. This was addressed partly in section 5.2.3.1. $Y(t)$ and $I(t)$ are not equal, unless $\wp = 1/\omega = 0$ and $\iota = 1/\mu = \infty$. Nevertheless, the logistic model can provide a good approximation to more realistic and complex models introduced in this chapter. However, with fairly easy-to-use mathematical programs, obtaining numerical solutions to H-L-I-R and similar or more complex (e.g., host growth) models is now a reasonable (but not trivial) exercise, somewhat removing the need to "force" analysis into a logistic (or similar) framework.

6.5.2 More complicated models

The epidemic models considered in this chapter can all be expanded, as necessary to meet the objectives of the investigator. This is often most directly done by converting parameters into variables, and having these variables be dependent on other parameters. For instance, it is generally assumed, out of convenience, that β (or βH_0) is constant. One or more of the components of β (α, ψ, or θ) may however be functions of H, t, or other variables. This was partly addressed in section 5.3. Just a few of many possible situations are considered here.

In some systems, the susceptibility of the crop decreases over time as plants or plant tissue matures. Among other things, this could mean that the probability that an infectious unit in contact with a healthy plant causes an infection (ψ) decreases over time, so that ψ is replaced by a function $\psi(t)$ (see Kleczkowski et al., 1996; Gilligan, 2002). Also, it is possible that the probability of an infectious unit coming in contact with a host plant (θ) is a function of initial host density (H_0) or even total host density at any time ($M = L+I+R+H$) for situations with a dynamic host crop. Then θ is replaced by functions $\theta(H_0)$ or $\theta(M)$ in the models, if a reasonable relationship between the variables can be formulated. Moreover, rate of production of infectious units (α) could be a function of density of infectious or even total disease intensity, $\alpha(Y)$.

In some cases, it is still possible to analyze these expanded models analytically using the generalizations of the mathematical methods discussed in previous sections of this chapter. However, in general, analysis is based on numerical solutions of the equations, as discussed above for transitional dynamics. It may turn out that the use of constants gives about the same results as more complicated variables in the models. The "parameters" of the simpler models would then be considered types of "averages" of underlying variables (e.g., average of $\psi(t)$ over time). It is important to keep in mind that the "simple" models of this and the previous chapter, with a small number of parameters, *do* provide a realistic description of epidemics. This may be because some variation of the terms "cancel out", on average. For instance, even if ψ, α, and θ all depend on time or densities of diseased individuals, their product (=β) may be reasonably constant (as an approximation) over time, depending on the direction of the underlying dependencies.

There is one obvious "parameter" dependency that does need some discussion in terms of the spatial component of epidemics. Except for some general statements, we have so far ignored the spread of diseased individuals in space as the density increases over time. The probability that an infectious unit contacts a given host individual (θ) obviously depends on the distance (s) between where the infectious unit (e.g., spore) was produced and the location of the given host individual. In general, the greater the distance, the lower the probability of contact. To account more fully for this dependency, *explicit* modeling of the dynamics of Y (or H, etc.) in space *and* time is required. This is addressed in some detail in Chapter 8, after first presenting the concepts and analysis of disease gradients in Chapter 7. This issue is most relevant when the diseased individuals are in a limited number of patches (foci), so that some parts of the field are disease free and other parts have high density of diseased individuals. With this situation, healthy individuals far from the disease will have a very different probability of contact with inoculum than healthy individuals near the focus. With a random pattern of disease (see Chapter 9), however, there is less of an issue in terms of temporal modeling, because there are many distances between diseased and healthy individuals that are "averaged out" (in some circumstances) to permit use of a constant (average) θ in temporal models of epidemics. In other words, disease-free plants would not differ greatly in their distances to diseased individuals.

6.5.3 Computer simulation modeling?

The physical and biotic environment clearly affects the parameters in the models presented in this chapter. For example, the mean length of the latent period ($\wp$), sporulation rate (α), and probability of infection (ψ) are all temperature dependent for fungal diseases, and the latter two are affected by free moisture (e.g., Kolnaar and van den Bosch, 2001; Lalancette et al., 1988a, b). Furthermore, most plant virus epidemics depend on the density and behavior of viruliferous or infective insect vectors. With a few exceptions, the models of this and the previous chapter do not deal with the explicit dependencies of model parameters on these environmental variables. By ignoring any driving variables, the investigator is, in reality, using an "average" value for the parameters (averaged over the different environments). Such an approach is very useful because models with a few parameters often are sufficient to characterize epidemics.

Of course, researchers often are explicitly interested in how the environment affects disease development under specific conditions. Unless one assumes there is systematic variation in these environmental variables over time that can be described by a simple model, or the driving environmental variables do not change (as we assumed in section 6.4 for total vector size in the plant virus model), there are more difficulties in modeling the population dynamics of plant diseases compared with the approaches discussed elsewhere in the chapter. Numerical simulation often is the only possibility. This generally requires utilization of discrete-time analogues of the population-dynamic dynamic models, where new values of the driving environmental variables are input to the model at the time scale of relevance (day, hour, etc.). Results typically are expressed simply as the graphs of disease density versus time (Reynolds and Arneson, 1997).

The topic of environmental variation leads directly into the topic of computer simulation based on the principles of systems analysis. The topic is reviewed in Chapter 12 of Campbell and Madden (1990), and in Teng (1981), Zadoks (1971), and Waggoner (1978). Sometimes the approach is called *simulation modeling* or computer simulation modeling to emphasize that the approach essentially requires a computer program. The models may simply be called simulators. Readers should be cautious when seeing the term simulation, however. After all, the term simulation could be applied to the method of generating the disease progress curves in this chapter since analytical solutions do not exist for all times. Here we are specifically referring to models with numerous linked equations (perhaps in the hundreds), with numerous parameters (perhaps dozens or hundreds or more) that attempt to describe disease development over time, often in relation to the environment or other external factors (e.g., fungicide applications).

This type of systems-based simulation modeling is very important historically in plant disease epidemiology because the early model by Waggoner and Horsfall (1969), and those that followed in the 1970s and later, helped demonstrate the quantitative nature of the discipline and showed that one could explore and understand disease dynamics with the use of models. A reductionist approach has generally been taken in constructing the models, which means that epidemics are broken down into multiple sub-parts (components), models are written for each component, and then the simulation model is constructed by linking the individual models within a computer program. The terminology is heavily influenced by systems analysis (de Wit and Rabbinge, 1979).

Such models are most useful when researchers are interested in *specific* answers to *specific* questions. That is, a computer simulation model can be useful for determining disease intensity with certain hourly or daily profiles of temperature, relative humidity, surface moisture, and rainfall over a growing season when a crop of a particular susceptibility is grown and fungicides are applied on an irregular basis in response to identified risk factors. Some excellent simulation models include those published by Bruhn and Fry (1981), Andrade-Piedra et al. (2005), Gilligan (1994), Gilligan et al. (1994), and Irwin et al. (2000). These types of models are not useful for developing a general understanding of epidemics, and ascertaining such population-dynamic features such as thresholds for epidemics and final disease levels in relation to underlying processes. In particular, none of the principles of epidemics discussed in this and the previous chapter were elucidated using these complex simulation models.

The authors of this book are great believers in the principle of model *parsimony*—keeping the model(s) as simple as possible to meet the objectives of the investigator. We agree with the philosophy advanced by Mollison (1995) that "we aim to find a small set of model components that determine the dynamics, and to describe these as far as possible in terms of simple parameters with clear ecological interpretations." In Chapter 4, we showed that a single equation is sufficient for many purposes when quantifying rates of disease increase and comparing epidemics for different treatments. In this chapter, we showed that the use of two-to-four components (linked differential equations), with just a few parameters (e.g., β, $\wp$, and ι), leads to a very rich understanding of epidemic processes that is applicable to many pathosystems, and allows for predictions of disease density at certain stages of epidemics. The general approach taken is to start with the simplest model possible, and then to expand it just enough to answer the questions of interest to the researcher. A thorough discussion of simulation models in comparison to the ones of this chapter (often called analytical models) is given by Jeger (1986a).

6.5.4 Stochasticity

As explained in the first part of Chapter 4, we focus on the use of deterministic models in representing plant disease epidemics because of the large population sizes in typical fields or plots. When population sizes are low, or the interest is in the individual rather than in the sum or average of many individuals, then stochastic models can be of benefit. The models of this chapter can be formulated as stochastic ones, but at a cost of additional model complexity (see Daley and Gani, 1999; Renshaw, 1991). Moreover, the threshold and final level results can be constructed within a stochastic framework. For those interested in this subject, Gibson et al. (1999) and Filipe and Gibson (1998) show how stochastic models can be used for understanding temporal or spatio-temporal dynamics of epidemics.

The extra complexity of stochastic models was not needed to make the general points of this chapter, so we do not find it necessary to deal with the topic. As shown by Diekmann and Heesterbeek (2000), deterministic models can be used to address many of the issues of interest in studying disease dynamics, and deterministic models can result in a rich diversity of disease progress curves. However, stochastic models are of value for some applications, such as determining if there is a critical size or density of the *pathogen* population (in contrast to the host population, H_0) for an epidemic to occur, and the mean time for the disease to go extinct in a dynamic system (Daley and Gani, 1999; Shaw, 1994).

Stochasticity and variability arise in a different context later in this book under the topic of spatial pattern analysis (Chapter 9).

6.6 Parameter Estimation

Although the emphasis in this chapter is on determining the epidemic outcomes from known parameters (i.e., from assumptions about the pathogen-host interaction), for a given set of starting conditions, one can take the approach emphasized in Chapter 4 and explicitly estimate parameters from observed disease progress curves. To inform the reader on this subject, we conclude the current chapter with some discussion on parameter estimation. For brevity, we restrict attention to the H-L-I-R type model (i.e., no host growth considered, and primary infections only at $t = 0$).

6.6.1 Estimating parameters without direct curve fitting

A practical advantage of the realistic models of this chapter is that, in principle, parameters can be estimated (or approximated) separate from an observed disease progress curve. There are three relevant parameters in the differential equations of the H-L-I-R model, $\wp$ (or $1/\omega$), ι (or $1/\mu$), and β, although the later is a product of three other parameters ($\beta = \alpha\theta\psi$). Integration of the coupled differential equations (equation 5.30) results in a constant of integration (as discussed for the population growth models of Chapter 4), which can be considered an additional parameter to be estimated. However, we will consider y_0 to be known here. Both $\wp$ and ι can be determined directly from controlled studies (e.g., see Kolnaar and van den Bosch, 2001; Chapter 8 in Vanderplank (1963), and references therein), or from observations of infected individuals during epidemics of interest. That is, one can continually monitor inoculated plants (leaves, etc.) to determine when spores are produced in new lesions. Further monitoring can determine when older lesions no longer produce new spores. The procedure could be somewhat more tedious for other pathogens. For plant viruses with insect vectors, the methods would be more complicated, since one would expose inoculated plants at various times after inoculation with non-viruliferous vectors (and then expose the vectors to healthy plants for possible transmission) to determine when vectors first acquire the virus and when vectors no longer can acquire the virus from diseased plants. There is more information on latent than on infectious periods, because researchers seem less interested in determining ι directly. This could be because of the long time required before individuals are no longer infectious. It is possible that ι is the life span of the crop plant.

Determining β (or βH_0) generally involves consideration of the three component parameters ($\alpha\theta\psi$). Sporulation rate per unit area of infected host provides an estimate of α. An estimate of ψ is obtained by inoculating plants with different densities of inoculum, and determining the number of lesions divided by the number of spores (as long as inoculum density is not too high). In some respects, estimating θ may be the most challenging. As will become clear after reading Chapters 7 and 8, spore dispersal studies can be used to determine the probability of spore contact with a host in relation to distance from an infected individual.

As mentioned in section 6.5 and elsewhere, all of the "constants" in the models are likely "averages" of underlying variables that may or may not be measurable, and all depend on the environment. The transmission parameter, β, may be the most dependent on specific environmental conditions, meaning that it may not be possible to use prior information on sporulation rate, infection efficiency, and so on, to determine a value for representing a particular disease progress curve. Even if some components of β were known with some certainty (say, ψ), other components may be unknown (say, θ); thus, β would only be known up to a proportionality constant. Fitting the H-L-I-R model to a disease progress curve could then provide a direct estimate of the realized β under a given set of circumstances.

A transmission rate is also of relevance for pathogens with vectors (and other complicated scenarios not discussed here). For convenience, we only showed in this chapter a virus-vector model when there is also host growth (section 6.4). An expanded H-L-I-R type model with fixed host density could also be used (see Madden et al., 2000b). The parameter product $\beta_1\beta_2$ (see equation 6.58) plays an analogous role to β in the simpler standard H-L-I-R model. The parameters for acquisition and inoculation probability can also be determined (or determined up to a proportionality constant) by determining in controlled studies the proportion of vectors that acquire the virus from infectious plants with a fixed feeding time (β_2) or the proportion of plants that become infected when exposed to viruliferous vectors for a fixed feeding time (β_1). Jeger et al. (1998) and Madden et al. (2000b) provide more details.

6.6.2 Fitting models to data

It was once very difficult to directly fit models such as equation 5.30 to disease progress curves. Model fitting requires a computer program for estimating parameters of nonlinear models, which is an iterative process (see Chapter 3), which can *also* calculate numerical solutions to linked differential equations at each step of the iterative process. In other words, parameter estimation is a doubly iterative process. With programs such as PROC MODEL of SAS (part of the SAT/ETS package) and fast computers, this is less of an obstacle than it was in the past. However, it is our experience that the general problems in performing nonlinear least squares or maximum likelihood (as discussed in Chapters 3 and 4) are magnified when the nonlinear models are only definable as numerical solutions to differential equations (see Madden et al., 1987b, for an early use of such methodology for a different set of differential equations).

We demonstrate direct fitting of equation 5.30 to data below, but first we emphasize a practical alternative. As mentioned in section 6.5.1, disease progress curves for polycyclic diseases often are reasonably described by the logistic model with parameter r_L. It was shown in Chapter 4 that when $y(t)$ is low, $r_L \approx r_E$, and in several places in section 5.2 we showed how r_E can be determined from the parameters of the more realistic models (e.g., equations 5.35–5.37). Thus, one can: (i) fit the logistic model to the data, using standard linear or nonlinear techniques (see sections 4.62–4.64); (ii) use the estimate of r_L (for the entire data set) as the estimate of r_E (for low disease intensity); and (iii) use the relevant equation for r_E to estimate βH_0, given assumed values of $\wp$ and ι, and use the parameter combination to obtain an estimate of R_0; and (iv), if desired, use simulation to generate a disease progress curve for the H-L-I-R model (equation 5.30) based on the estimated βH_0 determined from r_E.

The procedures are demonstrated for the tobacco etch virus disease epidemic considered in Chapter 4 (see Fig. 4.13). The pathogen, tobacco etch virus (TEV), is transmitted by aphids in a non-persistent manner. This means that the latent and infectious periods are both very short in the vector, with aphids continually acquiring and losing the virus from their stylets. As shown by Madden et al. (2000b), this simplifies the modeling because temporal periods need not be considered in the vector population. When not directly considering the vector population (as here), β of equation 5.30 reflects the joint probability of the virus being acquired from a diseased plant and then inoculated to a healthy plant, and the density of the vector population. As done in Madden and Campbell (1990), we assume $\wp = 7$ days (or $\omega = 1/7$). There is no direct evidence about the length of the infectious period in the plants, but it is possible that once infectious, plants remain infectious for the duration of the epidemic. We, thus, first assume that $\iota = 50$ days (or $\mu = 1/50$).

The logistic model with $K = 1$ provided an excellent fit to the disease progress curve. We use the estimates obtained from nonlinear weighted least squares: $\hat{y}_0 = 0.0006$ (s.e. = 0.00035), $\hat{r}_L = 0.314/\text{day}$ (s.e. = 0.0247). Based on the good fit of the logistic model, we can assume that $\hat{r}_E = 0.314/\text{day}$ when $y(t)$ is low. Using equation 5.38 (with $m = (7 + 50)/(2 \times 7 \times 50) = 0.081$), we obtain an estimate of βH_0 of 1.03/day. This gives an estimate of R_0 (equation 5.33) of 51.5. One can see in Fig. 6.12 that the numerical solution to equation 5.33 with these parameters provides a very good fit to these data. One can ask: what if the estimate of μ is very far from the true value. It turns out in this case that the estimate of βH_0 is robust to errors in μ. For instance, using $\mu = 1/25$ in equation 5.38, one obtains 1.05/day, and with 1/100 for μ, one obtains 1.02/day for the estimate of βH_0.

We then used the MODEL procedure of SAS to directly fit equation 5.33 to the tobacco etch data (with, once again, $\omega = 1/7$ and $\mu = 1/50$). We used the estimate of y_0 from the logistic model as the starting intensity of disease. The estimate of βH_0 was 0.85/day (s.e. = 0.017), just a little below the estimate obtained without direct fitting of the model (above). With this estimate of the transmission

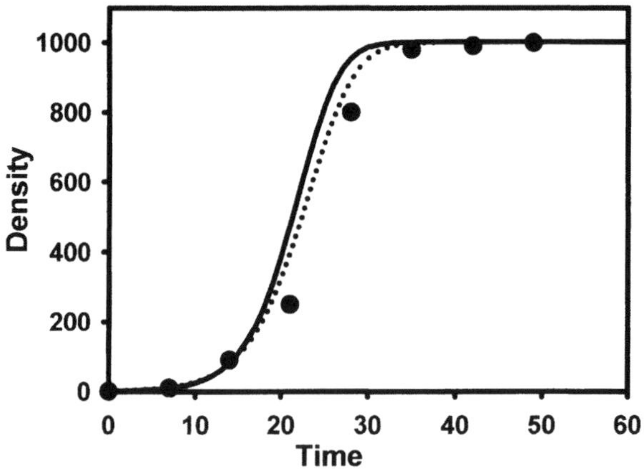

FIG. 6.12. Proportion of pepper plants infected with tobacco etch virus (as shown in Fig. 4.13) and predicted proportion based on numerical solution equation 5.30 (with $1/\mu = 50$ days and $1/\omega = 7$ days). Solid line was determined by calculating βH_0 from equation 5.38, and the dotted line was determined by fitting the model to the disease progress data (using of PROC MODEL of SAS).

parameter, there also was a good fit to the data (see dotted line in Fig. 6.12. The coefficient of determination (R^2) was above 0.999. The estimate of R_0 here was 42.5. Although less than the 51.5 determined above, both values are far above the number that gives $y_\infty \approx 1$.

Attempting to simultaneously estimate βH_0, μ, and ω using the MODEL procedure is problematical with the number of data points available. Our experience is that one should have more than eight assessments of disease in order to estimate all three parameters of equation 5.33 through nonlinear optimization methods. Further adding to the challenge is that with a final disease intensity of ~1, a wide range of βH_0 and μ combinations can result in the same high final level. This is because $y_\infty \approx 1$ is obtained with any R_0 above 5 or so with the H-L-I-R model. Thus, there was little information in the data set (with a third of the observations corresponding to $y \geq 0.98$) to, for example, distinguish between an R_0 of 10, 20, 50, or higher. In fact, assuming $\mu = 0$ resulted in a transmission rate of 0.80/day, little different from when $\mu = 1/50$.

It is now becoming more common to use nonlinear least squares or maximum likelihood to directly estimate parameters of H-L-I-R or other coupled-differential-equation models for population processes. This is opening up new opportunities for linking theoretical epidemiological studies to observations of disease progress in annual and perennial crops. Gilligan (2002, and references therein) nicely demonstrates some of these linkages. However, the general problems of using nonlinear model fitting methods compared with linear fitting methods (see discussion in Chapters 3 and 4) are magnified when the nonlinear model consists of a collection of differential equations. Clearly, applied statistical research is needed to further assess the distributional properties of parameters estimated using these sophisticated and powerful methods. One important outcome of new statistical research (involving direct or indirect estimation of parameters) is an appraisal of how uncertainty in model parameters (e.g., α), determined from other epidemics or the current epidemic, affects the accuracy and precision of predicted R_0.

6.7 Suggested Readings

Gilligan, C. A. 2002. An epidemiological framework for disease management. *Adv. Bot. Res.* 38: 1–64.

Gubbins, S., and Gilligan, C. A. 1997. Persistence of host–parasite interactions in disturbed environments. *J. Theor. Biol.* 188: 241–258.

Jeger, M. J., Holt, J., van den Bosch, F., and Madden, L. V. 2004. Epidemiology of insect-transmitted plant viruses: modeling disease dynamics and control interventions. *Physiol. Entomol.* 29: 1–14.

7

Spatial Aspects of Epidemics—I: Pathogen Dispersal and Disease Gradients

Everything is related to everything else, but near things are more related than distant things.

W. R. Tobler (*Tobler's Law of Geography*)

7.1 Introduction

Epidemics are population processes that occur in time and space (see Chapter 1). For the previous three chapters, however, we have mostly ignored the spatial component of epidemics so that a general framework could be developed to characterize and analyze disease development over time. Plant diseases generally do spread in space as they increase in time, and an understanding of plant disease epidemics would be incomplete without consideration of this aspect of disease development.

Disease spread is the result of inoculum dispersal. *Dispersal* is the movement of infectious units (e.g., spores) of a pathogen from one place to another or the movement of infectious units from the place they were formed to other locations (Campbell and Madden, 1990). We consider the terms propagules, propagative units, and units of inoculum as being equivalent to infectious units in this and other chapters. Because propagative units mostly move short distances, new infections primarily occur near other infections and, therefore, there usually is higher disease intensity in some locations than others. If the inoculum initially is concentrated in one location, one can study dispersal by measuring the spore deposition on surfaces (e.g., leaves, flowers) at a range of distances from the inoculum source. Likewise, if diseased individuals are initially concentrated in a focus in one area, disease spread can be determined by measuring disease intensity at a range of distances from the initial disease source. These types of situations are considered in this and the next chapter. However, if there are many inoculum sources, or diseased individuals are located at multiple locations at the start of an epidemic, then disease spread may not be easily measured (although the same biological process is occurring). This is because it will not be known where a deposited spore originated or which inoculum source produced the spore that resulted in a new diseased individual at a given location. In this situation, different spatial methods typically are used for characterizing epidemics (see Chapter 9).

By re-addressing the models and analysis of the previous chapters, one can easily see the importance of the spatial component of epidemics. For instance, for monocyclic diseases, the probability per unit time that a unit of inoculum will come in contact with a host individual and cause an infection (see section 4.4.2) is a function of the distance between the inoculum and the host. Also, the probability that a spore produced on a lesion during the epidemic of a polycyclic disease will reach another host individual is a function of distance between the diseased individual and the "other" host individual. By treating these terms as constants, one is assuming that the epidemic is homogeneous in space, in that there are many inoculum sources and/or infected individuals evenly positioned throughout the host population. Such an assumption is not realistic when the epidemic starts with a single (or a few) spatially-restricted inoculum sources.

We now explicitly consider the spatial component of epidemics. In particular, this chapter deals directly with characterizing disease spread from a single (original) inoculum source. Relatively simple models are presented and their properties discussed. It is shown how to estimate model parameters, and how to compare the spread of two or more epidemics based on the fitted models. It is assumed here that the reader has already digested the information in Chapter 4 on the temporal characterization of epidemics, especially the concepts related to statistical model fitting and comparisons. Thus, where the concepts are the same, the detailed statistical methods presented in Chapter 4 will not be repeated here.

In the next chapter, a much more mechanistic approach is taken to understand the spread of disease through the movement of inoculum. There, we expand on the rate equations (dy/dt) for epidemics in Chapter 5, by explicitly incorporating distance functions into terms for the probability of a spore starting at one location coming in contact with a host at another location. Then, in Chapter 9, we take a different approach and show how to characterize spatial patterns of epidemics without explicit reference to the source(s) of inoculum.

7.2 Dispersal Gradients, Disease Gradients, and Disease Spread

7.2.1 Concepts

The study of spore dispersal is very important historically in plant disease epidemiology. In fact, before the concepts of Vanderplank (1963) started to take hold, "epidemiology" often was used to describe the study of pathogen spores in the atmosphere, including the study of spore movement to new infection sites, and how the environment affected the production, movement, and germination of spores. Gregory (1968, 1973) was a pioneer in the study of spore dispersal, and many of the concepts and principles he developed are still valid today and are reflected in this chapter. Gregory's work was even instrumental in the ultimate formation of an entire scientific discipline—*aerobiology*—that focuses on all aspects of spores, pollen, and other organisms in the atmosphere. Although plant disease epidemiology is now considered to involve a lot more than dispersal, this process is clearly a key component of disease development.

Dispersal of propagative or infectious units (e.g., spores) involves liberation (takeoff), transport (flight), and deposition (landing). The terminology is heavily influenced by the study of spore movement in the atmosphere, but dispersal certainly occurs in the soil as well, although the distances involved are generally much shorter. Inoculum also needs to be considered more broadly than just spores in order to account for the movement of bacterial cells and pathogens in arthropods or other vectors. As discussed in Chapter 4, we deliberately use a loose definition of "inoculum" in this book that can accommodate the movement of vectors that have acquired plant viruses, spiroplasmas, phytoplasmas, and other pathogens.

Liberation of inoculum involves removal of spores or cells from diseased individuals and their release into the transport medium (i.e., air, water, or soil). Transport involves movement of inoculum through a turbulent medium such as air or water. Deposition involves the removal of the inoculum from the transport medium and placement on a surface (e.g., leaf). Overall, dispersal is both a physical and biological process, and research on the topic typically is interdisciplinary in nature. We will not cover the physics of inoculum dispersal in this book, but refer the interested reader to several thorough references (Aylor 1999, and references within; Ferrandino, 1993; Madden, 1992; McCartney and Fitt, 1985).

Instead of focusing on the biophysical details of inoculum movement, we summarize the dispersal processes by the *disease gradient*—the change (generally, decline) in disease intensity or number of infections over distance from an inoculum source. A note of caution should be mentioned here. Gradients of disease can be due to multiple causes. For instance, there can be a gradient of root disease incidence from one side of a field to the other due to an *environmental gradient* such as decreasing soil moisture. We do not deal with such environmental gradients here, but limit our discussion (for the most part) on gradients of disease that are caused by dispersal. Often, the term *dispersal gradient* is used to describe a disease gradient that is caused by variation in deposition of inoculum in relation to distance from an inoculum source. By moving a little earlier in the chain of events, one can deal directly with a *deposition gradient*, which is a measure of spores per unit area or length in relation to distance from an inoculum source. Multiplying this spore (or infectious unit) density by an infection-efficiency factor produces number of infections in relation to distance (i.e., the disease gradient).

Typically, observed gradients may be caused by a mixture of inoculum dispersal and environmental gradients, and researchers should be careful in over-interpreting disease gradient data if they do not have some evidence that dispersal is primarily responsible for the observed systematic change in disease over space. Because this is a book about diseases in plant populations, we focus mainly on disease gradients, and assume in this chapter that they are the result of dispersal. That is, we assume that the disease gradient is a dispersal gradient. One can categorize the dispersal gradient as either primary or secondary. In a *primary gradient*, all infections are due to spores (or other infectious units) originating at the inoculum source. Essentially, primary gradients are due to primary spread of disease (from initial inoculum, which, in this situation, is located at a well-defined inoculum source). In a *secondary gradient*, infections are due to inoculum produced beyond the original source. Essentially, secondary gradients occur when there is secondary spread of disease during an epidemic, which occurs for polycylic diseases (see Chapter 4).

Typically, diseased plants do not move. Thus, plant disease spread is not exactly the same as, for example, the spread of insects from one field to another. The inoculum does move, and as mentioned in the Introduction (section 7.1), spread can be considered the result of inoculum dispersal. In other words, the area occupied by diseased individuals, and the intensity of disease at a given distance from the source, increase due (at least in part) to the dispersal process. In this sense, disease spread is analogous to the population expansion of animals where the parents do not move but the offspring move out to new areas.

Example gradients are shown in Fig. 7.1. These were chosen to show a wide range of pathogen and host systems, and to show that distances of disease spread can be short or long. In most studies, disease gradients are measured over a few meters; thus, most disease gradients are quantified for distances less than 10 m. However, in some cases, gradients are determined over greater distances (>1 km), but this is most easily done when there is a known single inoculum source. In Fig. 7.1, disease intensity is measured or expressed in various ways, such

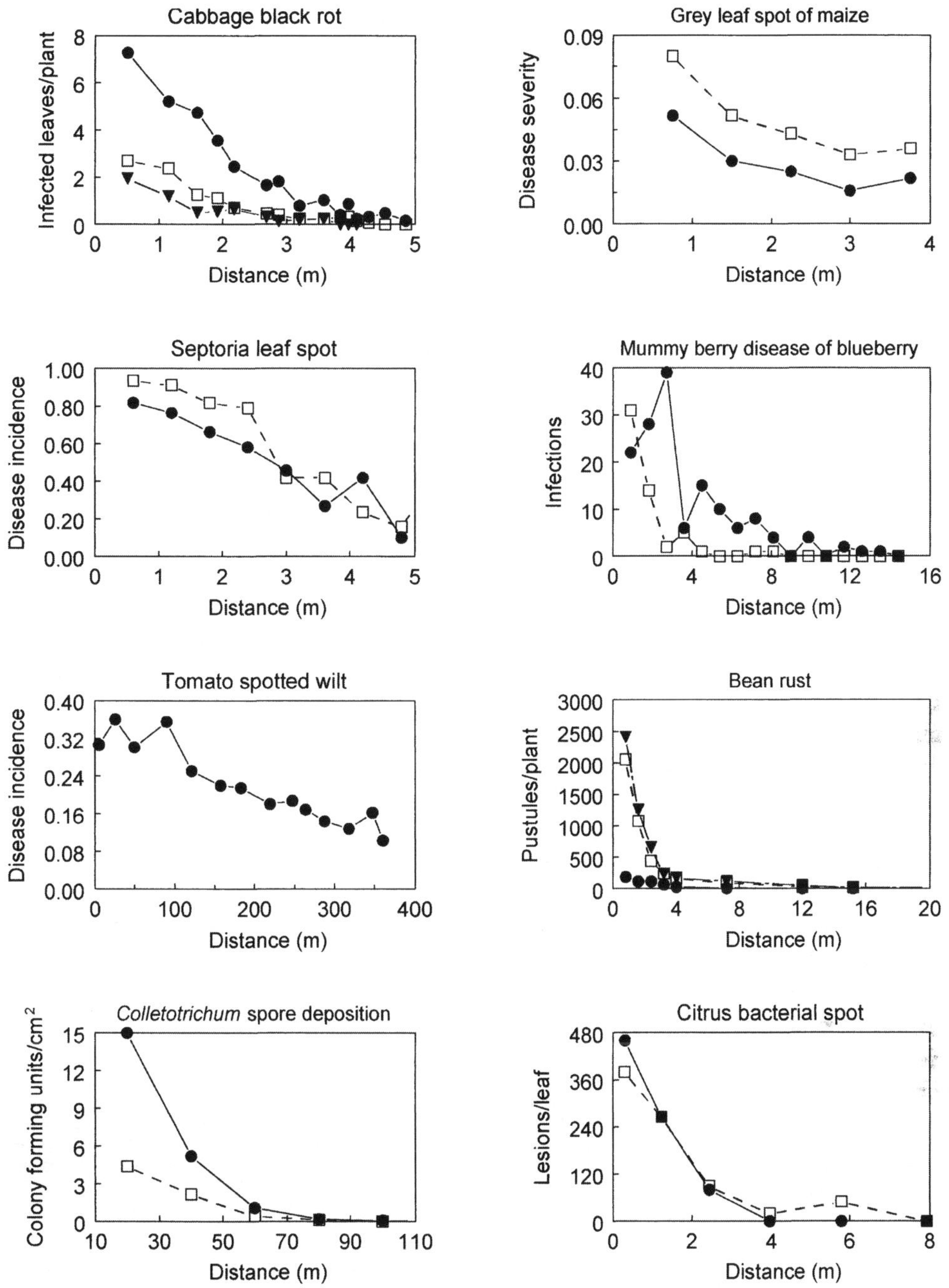

FIG. 7.1. Exampled dispersal and disease gradients. *Left-hand column*: Number of cabbage leaves infected by *Xanthomonas campestris* pv. *campestris*, cause of black rot, from a point source (Kocks and Ruissen, 1996). Incidence of Septoria leaf spot of tomato, caused by *Septoria lycopersici*, in relation to distance from a point source (Parker et al., 1997). Lines are for two different years. Incidence of tomato spotted wilt of peppers, caused by tomato spotted wilt virus, from an area source in Georgia (Gitaitis et al., 1998). Number of spores of *Colletotrichum acutatum* deposited over 1 min at 5 and 15 min into a generated rain over a plastic mulch surface (Yang et al., 1991). *Right-hand column*: Severity of grey leaf spot of maize, caused by *Cercospora zeae-maydis*, in relationship to an area source (Asea et al., 2002). Lines are for two different seasons. Number of shoots of blueberry infected by *Monilinia vaccinii-corymbosi*, cause of mummy berry disease, from a point source (Cox and Scherm, 2001). Lines are for up- and down-wind from the source. Number of bean rust pustules, caused by *Uromyces appendiculatus*, in relation to distance from a line source (Aylor and Ferrandino, 1989). Lines are for different times. Number of lesions/leaf of citrus bacterial spot, caused by *Xanthomonas axonopodis* pv. *citrumelo*. Lines are for two treatments at one assessment time (Gottwald et al., 1997).

as number of infected or diseased leaves, rust pustules per plant, proportion of individuals infected (incidence), and proportion of area with lesions (severity). An example of a spore deposition gradient is also shown for comparison purposes.

The disease gradients have, for the most part, a characteristic shape that is determined largely by the physical process of dispersal. Disease intensity is high at and near the inoculum source (especially if the inoculum source is at or near the soil surface), and generally declines with increasing distance from the source. The rate of decline is usually large at small distances from the source, especially early in the epidemic, and small at large distances from the source. When disease intensity is moderately high near the source (e.g., considerably more than 10% of the maximum possible), intensity only declines slowly at small distances, and then a steeper decline occurs before the decline becomes small again at large distances (see *Septoria* example). In some cases, there is an increase in disease intensity before the general pattern of decline in disease with distance is observed (see mummy berry disease example). This is more typical when the inoculum source is elevated (i.e., in the upper parts of the host plants).

The disease/dispersal gradients shown in Fig. 7.1 may be quantified and compared using the models presented in this chapter.

7.2.2 Inoculum sources

We have used the term "inoculum source" so far without really defining it. For a polycyclic disease, any infected (and infectious; see Chapter 5) individual is an inoculum source for other healthy individuals. For a monocyclic disease, an inoculum source can be infected individuals from another location or a previous year, or the source can be overwintering spores in the soil. For the purposes of characterizing disease spread, however, we tend to think of an *inoculum source* as a spatially-restricted concentration of infectious units, such as spores, bacterial cells, or even diseased individuals that can be producing the infectious units. That is, the source is separate from, or on the "edge" of, the area over which disease spread is measured, at least at the start of the epidemic. The inoculum source can also be called a disease *focus*, but we mostly use this term in the following chapters. Generally, the source covers an area far less than the area where disease is measured, although this is not a requirement.

It is important to note that this spatially-restricted concept of an inoculum source is, in part, a matter of convenience. That is, in many (most) epidemics, the inoculum source is not separate from the host individuals, but is part of the system where disease is developing. With polycyclic diseases, every lesion (for example) can be an inoculum source. Likewise, with monocyclic diseases, overwintering spores may be found throughout the soil where the host crop is grown. To *study* disease gradients and spread, in practice, one often must use systems where the (original) inoculum source is well-defined in space. Exciting new statistical approaches have been developed in recent years, where such as restriction is not needed, as will be discussed in section 9.10 (Gibson, 1997a, b).

There are three general types of inoculum sources based on geometry and relative size: point, line, and area sources. The distinction is important for the characterization of physical movement of propagules across a plane. Even if the physical process is not of direct concern, parameters in gradient models will depend on source type; thus, one must be cognizant of the source type when comparing results among systems. Although a point is mathematically dimensionless, in the context of disease spread, the term *point source* is used for a very small source relative to the area where spread is being studied. A single infected plant with sporulating lesions in the center or at one corner of a field is an example of a point source. For practical purposes, the diameter (or, more generally, width) of a point source is about 1% of the distance over which spread is being measured. One can think of a *line source* as a point source stretched out across the entire width of a field or plot, perpendicular to the direction of measured disease spread. An example could be a row of infected plants along one edge of a field, or across the center of a field. Fig. 7.2 shows idealized examples of a point and line source, but these are placed at difference locations for clarity of presentation.

An *area source* is bigger than a point or line source, but the meaning is relative to the size of the area over which dispersal or spread is measured. A planting of cedars infected by the cedar apple rust pathogen, *Gymnosporangium juniperi-virginianae*, is an area source for a neighboring apple orchard. One can think of an area source as a line source with depth, so that there is, for example, more than one row of infected (and infectious) plants. One can also think of an area source as an expanded point source, such that its

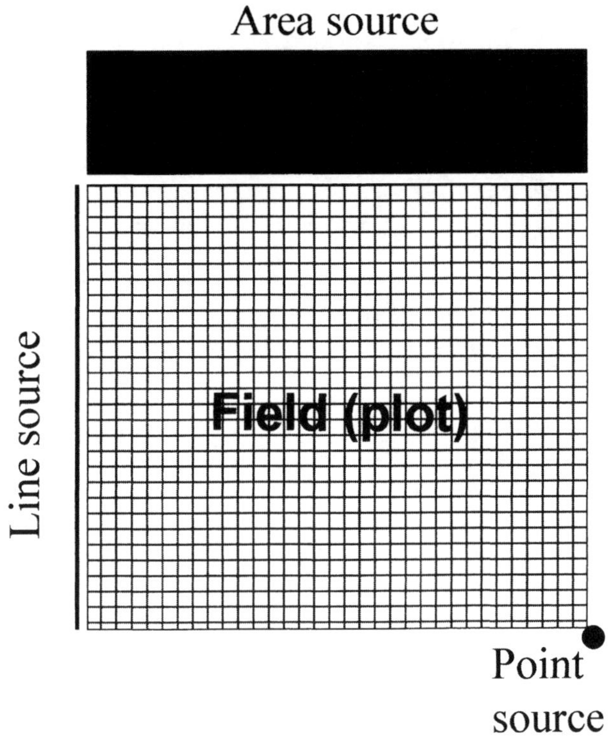

FIG. 7.2. Schematic of a field of host plants with a point, line, and area inoculum source.

diameter is considerably more than 1% of the distance over which spread is quantified. A potato field with late blight may be considered an area source for an adjacent field but a point source for a region, county, or province. Fig. 7.2 shows diagrammatically one version of an area source.

7.3 Models

There are many ways of modeling dispersal and disease spread. Although not presented here, spore dispersal can be characterized by various types of models based on the physics of particle movement in different types of media (Aylor, 1999; Ferrandino, 1993; McCartney and Fitt, 1985). There are also many relatively simple nonlinear models for representing spore deposition, infections, or disease intensity in relation to distance from a source (Jeger, 1983; Maffia and Berger, 1998; Minogue, 1986). These typically are correlative or descriptive (see Chapter 3), but they can be consistent with known mechanisms of dispersal. There are also quite elaborate theoretical mathematical representations of dispersal (and movement of individuals, in general) that account for many population attributes of the biological processes involved without explicitly dealing with physics of particle transport (Othmer et al., 1988; Pielaat and van den Bosch, 1998; van den Bosch et al., 1990a).

For the most part, we focus our attention on some fairly simple nonlinear models of disease gradients that have been proposed. Being consistent with the models used in previous chapters, the ones used here throughout most of the chapter are deterministic and treat distance (s, as in "space") as a continuous variable. Although descriptive in nature, they are also generally consistent with more mechanistic models of dispersal and can accurately characterize observed gradients. These models also have stochastic versions that are useful for understanding certain aspects of the dynamics of disease development over time and space.

As before, we use Y to represent disease in absolute units, such as number of lesions, number of infected leaves, or area of lesions, and y for disease intensity as a proportion ($y = Y/M$, where M is number of individuals or area of host that can be infected). For convenience, we can refer to Y as number of infections, because if intensity is not too high, one can count lesions and consider these to represent individual infections. Assuming that a fixed proportion (ψ) of deposited spores (or other infectious units) cause infections, then Y/ψ represents the number of deposited spores when Y is not high. For completeness, we can write $Y(s)$ for Y at distance s from the inoculum source, but typically we do not need the extra notation in this chapter. Moreover, because space is explicitly dealt with here, Y represents infections per unit area (such as infections per square centimeter of the surface where infections are counted or measured), or infections per unit length (such as lesions per centimeter distance along a radius of the circle centered on a point source).

For the physical purists, our presentation below is strictly valid for a line source of inoculum, with spread perpendicular to that source. In other words, we present model just for a one-dimensional spatial system. In Chapter 8, we discuss some of the issues regarding spread in one versus two dimensions.

7.3.1 Exponential

The first model we consider is the exponential, sometimes called the negative exponential, which has been used to represent gradients since at least 1940's (Frampton et al., 1942; Gregory and Read, 1949). Here, Y is proportional to an exponential function of distance from the source:

$$Y = a_E \exp(-b_E s) \qquad (7.1)$$

in which a_E and b_E are parameters. The parameter a_E represents the (theoretical) level of Y at the inoculum source ($s = 0$), because at the inoculum source one writes equation 7.1 as:

$$Y = a_E \cdot e^0 = a_E 1 = a_E$$

This parameter is thus an indication of *source strength*—the amount of inoculum or inoculum-producing disease intensity at $s = 0$. The parameter b_E, sometimes known as the *spread parameter*, measures the steepness of the gradient and has units of 1/distance. The inverse of b_E (i.e., $1/b_E$) has units of distance and is an indicator of the spatial scale over which spread (or dispersal) is occurring. More will be said about this later. The lower limit for each of the two parameters is 0 (when Y is non-increasing with distance) and there is no upper limit (at least in the model). Obviously, if $a_E = 0$, there are no diseased individuals (infections), and no dispersal is being assessed ($Y = 0$ at all s). When $b_E = 0$, there is no decline in Y with increasing distance. This would be an indication that there was no true inoculum source at $s = 0$; it also could be because disease intensity was measured after a long time and that all plant individuals were diseased. Example realizations of equation 7.1 are shown in Fig. 7.3. As expected, increasing b_E results in a steeper decline in Y over s, and increasing a_E results in an increase in the height of the curves. Equation 7.1 can also be used for disease intensity as a proportion (y). If M is the number of sites that can be infected at a given distance (e.g., maximum possible number of lesions), then $y = Y/M$. Dividing both sides of equation 7.1 by M gives an equation for y instead of one for Y.

The exponential model can be linearized by taking natural logs of both sides of the equation, producing:

$$\ln(Y) = \ln(a_E) - b_E s \qquad (7.2)$$

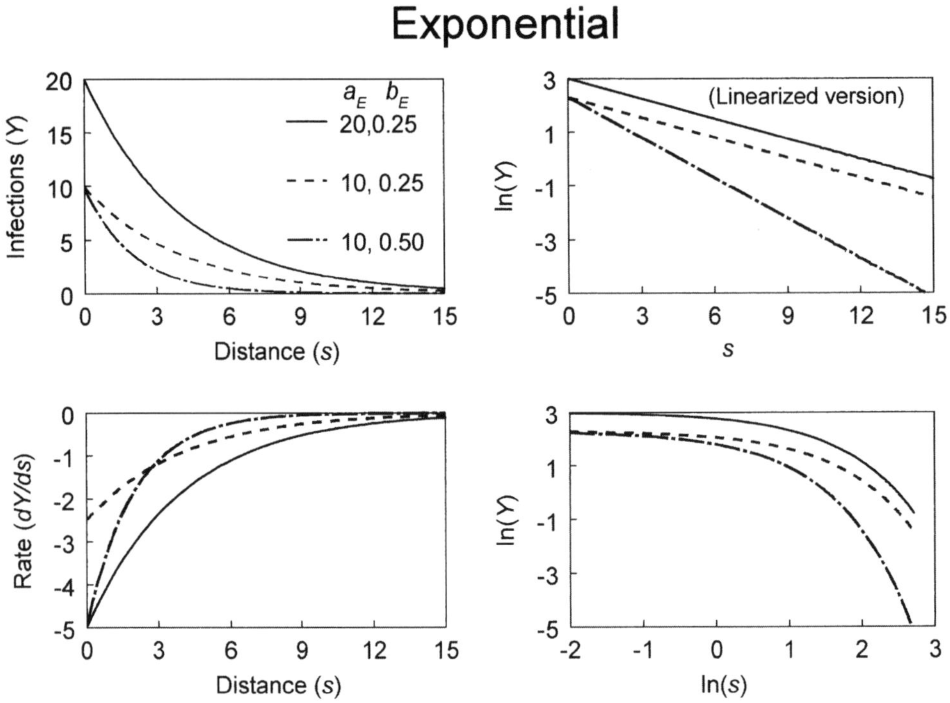

FIG. 7.3. Dispersal gradients (Y versus s) characterized by the exponential model (equation 7.1).

Equation 7.2 is a straight-line relationship between $\ln(Y)$ and s, with slope of $-b_E$ and intercept $\ln(a_E)$ (linearized version frame in Fig. 7.3). In this transformed scale, it can be easier to see the effects of a_E and b_E on the gradients. For instance, two gradients with the same b_E but different a_E values are seen as parallel lines when $\ln(Y)$ is plotted versus distance. With different b_E values, two gradient lines would not be parallel, and the lines could cross depending on the values of a_E. On an untrasformed scale (i.e., Y versus s), it can be very hard to see the effects of different spread parameters on the gradients. If one only had observations of Y at two distances, $Y(s_1)$ and $Y(s_2)$, one could determine b_E using: $b_E = [\ln\{Y(s_2)\} - \ln\{Y(s_1)\}]/(s_2 - s_1)$.

Equation 7.2 is useful for visualizing disease gradients and for parameter estimation (see below). However, the actual exponential gradient of disease is a curve and not a straight line (equation 7.1), meaning that the steepness varies continuously with distance from the source. This can be seen by writing the exponential gradient model in differential-equation form:

$$\frac{dY}{ds} = -b_E Y \qquad (7.3)$$

in which dY/ds is the rate of change in Y over distance from the source. We could have introduced the exponential model with equation 7.3 instead of equation 7.1 to emphasize the dynamics of disease spread in space. However, because these gradient models often are justified based on their ability to describe observed gradients and not based on dynamics arguments, we have elected to present the models first in their integrated form. Readers should refer to Chapter 4 for detailed explanations of differential equations and rates of change.

The rate dY/ds is a direct measure of the gradient steepness. As shown in the bottom-left frame of Fig. 7.3, the rate is a large negative value close to the source (near $s = 0$), and approaches zero at large s (depending on the value of the spread parameter). It can be shown that $dY/dt = -a_E b_E$ at $s = 0$ (because Y at $s = 0$ is given by a_E). The steepness of the gradient is a function of only one variable, Y, and one parameter, b_E. The spread parameter can, therefore, be thought of as a summary of all the conditions and factors (e.g., environment) that influence disease gradients other than the actual level of disease intensity. Two exponential gradients that have the same b_E can be said to have equal gradients, no matter what average disease intensity is found for a data set, because they will have the same dY/ds at any common Y.

7.3.2 Power model

The second model considered in this chapter for representing disease gradients is known as the power, power law, or inverse-power model. It has also been called the *Gregory model*, because he was an early proponent of its use to describe gradients and he utilized it extensively to summarize dispersal results (Gregory, 1968, 1973). For the power model, Y is proportional to a power of distance:

$$Y = a_P s^{-b_P} \qquad (7.4)$$

in which a_P and b_P are parameters. These parameters have some similarity to the parameters of the exponential model, but there are also some fundamental differences. Note that when $s = 1$, equation 7.4 reduces to $Y = a_P 1^{-b_P} = a_P 1 = a_P$. Thus, a_P represents theoretical Y at a distance of 1 unit from the source [i.e., $a_P = Y(1)$], not at the source. In fact, Y becomes indefinitely large as s approaches 0 (if b_P is positive), which is biologically meaningless. However, just because the model gives a meaningless result at one particular distance does not mean that it is not useful at all (or most) other distances. Moreover, if the distance between plants (or between the source and the nearest plant) is, say, x cm, it does not matter if the model generates unrealistic values at distances less than x cm. What matters is that the model generates realistic values at distances of x, $2x$, and any actual distance between the source and a plant.

Because the meaning of a_P is linked to a non-zero distance, its physical meaning depends on the units for recording s. For instance, if equation 7.4 was appropriate for representing the gradients in Fig. 7.1, and distance was recorded in meters, then a_P would be the (theoretical) value of Y at 1 m from the source. However, if the distances were recorded in centimeters, then a_P would be the value of Y at 1 cm from the source. Thus, one must always keep the distance units in mind when interpreting the specific value of a_P. If one wants to use a_P as an approximate measure of Y at (more precisely, "very near") the source, then the distance units should be chosen so that $s = 1$ is a lot closer to the source than to the first observed Y value. For example, if the closest distance to the source with observed Y was at 0.5 m, then recording distance in centimeters (and not meters or kilometers) would be appropriate.

In equation 7.4, b_P is also known as a spread parameter. However, it is a unitless term, unlike b_E, which has units of 1/distance. Thus, there is a fundamental difference in the physical interpretation of the exponent parameter of the two models. A rescaling of distance (say, from kilometers to centimeters) will change the value of b_E but not of b_P. (In contrast, remember that a rescaling of distance will change the value of a_P but not of a_E).

Because the value of Y at $s = 0$ is not defined with the power gradient model, the model often is modified by adding a small positive constant (λ) to s:

$$Y = a_P (s + \lambda)^{-b_P} \qquad (7.5)$$

We call this the *modified power model* or *modified power-law model*. Here, λ has the same units as s (i.e., units of distance). When $s = 0$, equation 7.5 reduces to $Y = a_P \lambda^{-b_P} = Y(0) = Y_0$, which is a positive constant for positive values of a_P and b_P. In other words, $a_P \lambda^{-b_P}$ is the source-strength term for the power model. The interpretation of a_P can be determined by inserting $1-\lambda$ for s in equation 7.5. This gives $Y = a_P 1^{-b_P} = a_P$. In other words, a_P is the theoretical Y at a distance of $1-\lambda$ from the center of the inoculum source. Mundt (1989) made the interesting argument that λ can be considered a measure of the physical "size" (diameter, width) of the inoculum source.

Of course, when $\lambda = 0$, equation 7.5 reduces to equation 7.4. For generality, we refer to equation 7.5 as the power model for gradients, with the understanding that often it is used with an assumed value of 0 for λ. A plot of Y versus s (Fig. 7.4) for this model has some similarity to the plot for the exponential model (Fig. 7.3). However, for the power model, the curve is steeper close to the inoculum source and shallower a large distances.

Equation 7.5 can be linearized (at least in terms of the a_P and b_P parameters) by taking logs of both sides:

$$\ln(Y) = \ln(a_P) - b_P \ln(s + \lambda) \qquad (7.6)$$

This equation describes a straight-line relation between $\ln(Y)$ and $\ln(s + \lambda)$ (see insert of lower right frame of Fig. 7.4). In fact, for this model, a plot of $\ln(Y)$ versus $\ln(s)$ is very close to a straight line except close to the source (i.e., where s is approaching λ). This kind of situation could arise if the true relation was equation 7.5 or 7.6 with $\lambda > 0$, but one fitted equation 7.4 (i.e., $\lambda = 0$) to the data. On the other hand, a plot of $\ln(Y)$ versus s (no transformation), which is the standard scaling for the exponential gradient model, exhibits a large amount of curvature (Fig. 7.4).

The (modified) power model can be expressed in differential-equation form:

$$\frac{dY}{ds} = \frac{-b_P Y}{s + \lambda} \qquad (7.7)$$

As with the exponential model, dY/ds is proportional to Y, meaning that the gradient is most steep (most negative) at the highest values of disease (which are typically near the source). However, dY/ds is also indirectly proportional to s (or $s + \lambda$, if λ is non-zero). Thus, at a given level of Y, the gradient is steeper close to the source than farther away. The lower left frame of Fig. 7.4 shows some typical dY/ds curves for the power model. These curves are similar to those for the exponential model, except the steepness is greater near the source for the power model. In fact, if $\lambda = 0$, dY/ds becomes indefinitely large and negative as s approaches 0 (or, for positive λ, dY/ds becomes indefinitely large and negative as $s + \lambda$ approaches 0). For positive λ, dY/ds has a finite value at all s values, and equals $-a_P b_P \lambda^{-(b_P+1)}$ at $s = 0$. This can be seen in the inset of the lower left frame of Fig. 7.4.

In some ways it is more difficult to compare power-model gradients than exponential gradients. This is because the steepness of the gradient in the former case depends on both Y *and* s (see equation 7.7) (see Minogue (1986) for a valuable discussion). Thus, two gradients with the same b_P values do not necessarily have the same gradient steepness (dY/ds) at a given (chosen) Y. Nevertheless, it is still common to compare gradients

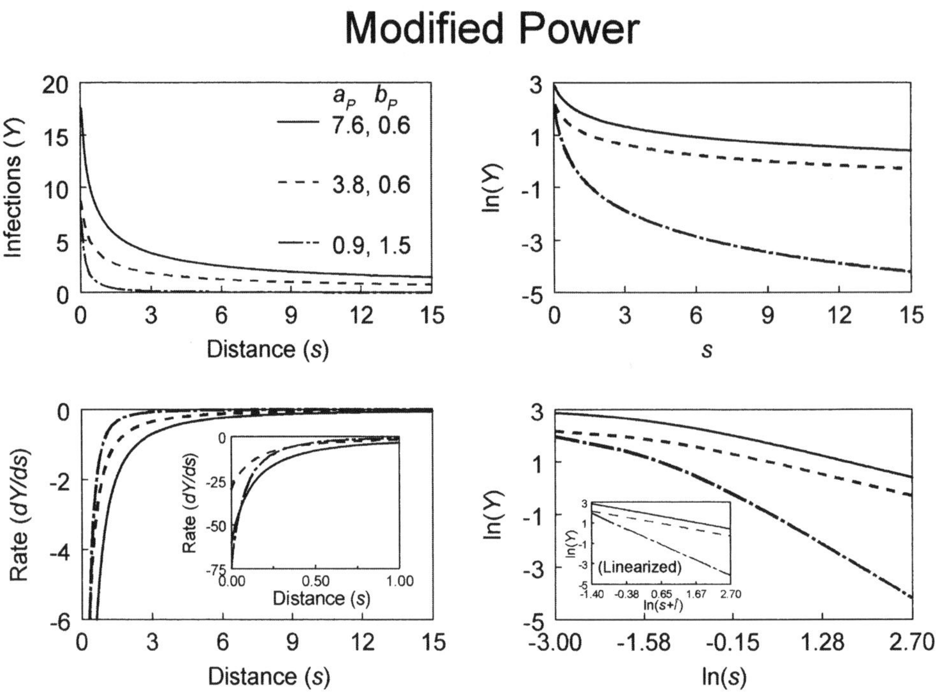

FIG. 7.4. Dispersal gradients (Y versus s) characterized by the (modified) power model (equation 7.5), with $\lambda = 0.2$. Insert in dY/ds graph shows the rate at very close to the source ($s = 0$). Insert in $\ln(Y)$:$\ln(s)$ graph shows $\ln(Y)$ versus $\ln(s + \lambda)$.

through their (estimated) b_P values. In comparative studies, investigators must be aware that two gradients with identical b_P values will have equal changes in $\ln(Y)$ with unit changes in $\ln(s)$ [i.e., $d\ln(Y)/d\ln(s) = b_P$ (on average); see equation 7.6], but not equal changes in actual measures of disease intensity with unit changes in distance at a given Y.

7.3.3 Power versus exponential model

The exponential and (modified) power models of gradients are, by far, the most common for characterizing disease (dispersal) gradients. In fact, these two models, or similar ones, do provide a very good representation of many observed gradients (e.g., Fitt et al., 1987). Moreover, these two models serve as the basis for further model expansions that are used to represent spatio-temporal dynamics of plant disease development (Gibson and Austin, 1996; Xu and Ridout, 1998). Thus, it is beneficial to discuss these two models, and the gradients they characterize, in a little more detail. The differences in the concept (or meaning) of gradient equality for the two models was already discussed at the end of the previous sub-section. This discussion here expands on some points already made, and covers some of the similarities and differences in the model behavior and interpretation.

Both models represent a situation with a nonlinear decline in Y with s (Fig. 7.5, top frame). For the power model, the gradient is steeper near the source and shallower far from the source, compared with the exponential model. The model difference is even more striking when plotting $\ln(Y)$ versus s (Fig. 7.5): the exponential model results in a straight line but the power model can show considerable curvature. This graph is especially useful to keep in mind in conceptualizing the meaning of the two models. For the power model, at large distances from the source, $\ln(Y)$ changes very little with unit change in distance (relative to the exponential model), and there remains a small (but non-trivial) level of disease even at very large distances. In terms of dispersal, this means there is a small (but non-trivial) chance of spore deposition even at very large distances.

As discussed in several articles (Aylor, 1999; McCartney and Fitt, 1998), the two models are representative of different physical aspects of spore dispersal in the air. Specifically, propagule dispersal is characterized by:

- the rapid decrease in the aerial spore concentration, and
- the rate of deposition of spores on plants with distance away from the source.

The rapid decrease in spore concentration is due to *filtration* of spores by plants (or other structures) and to *dilution* by mixing of the air, which leads to *escape* of spores from the canopy (Aylor, 1999). When air turbulence is low, the tendency for spores to be deposited is much greater than the tendency to be diluted and escape, and spore density decreases exponentially, on average, with distance (corresponding to equation 7.1). This situation also would occur with splash-dispersed pathogens where air turbulence is less of a factor (Pielaat et al., 1998; Yang et al., 1991). Conversely, when air turbulence

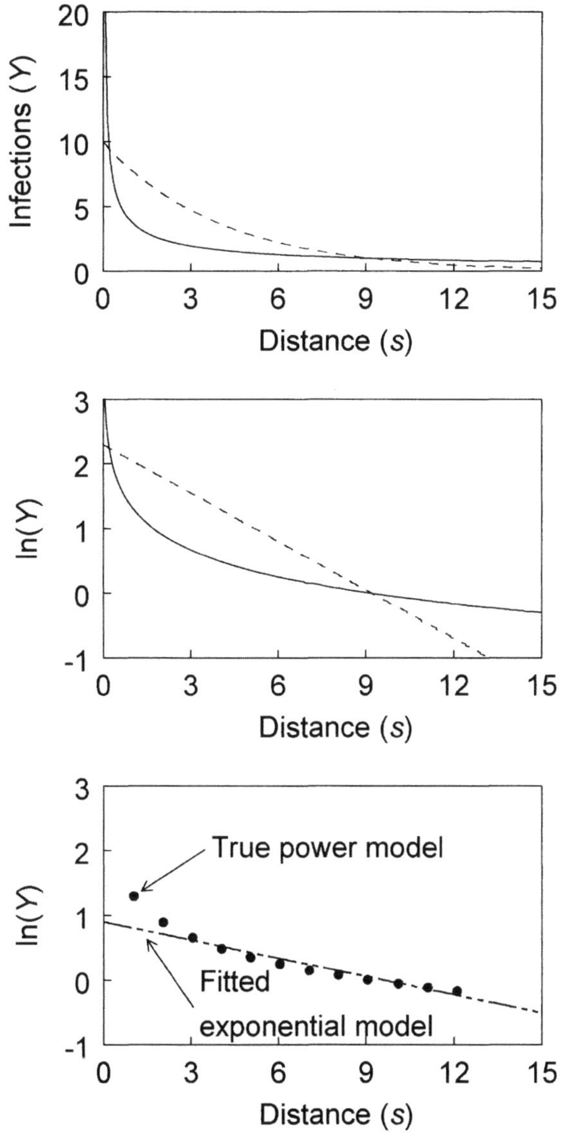

FIG. 7.5. Direct comparison of exponential (equation 7.1; dashed line) and (modified) power (equation 7.5; solid line) dispersal functions, on linear and log scales. In bottom frame, exponential model is fitted to true power-model values.

is high, the tendency for spores to be diluted and escape the canopy is much greater than the tendency to be deposited. In this case, spore density decreases, on average, according to a power model (corresponding to equation 7.4).

Of course, it is quite possible for neither dilution/escape nor deposition to dominate in an actual dispersal event; then, more complex models (see section 7.3.5 and Aylor, 1999) may be needed to adequately characterize the process. In fact, Aylor (1999) points out that deposition and dilution/escape may dominate over different distances. So, in principle, if one could observe distances at a very fine spatial resolution (say, every centimeter), starting at the source and extending for several kilometers, one could observe power-type and exponential-type gradient curves over different distance segments. Fortunately, such refinement in the modeling is not needed for many purposes, and models of the simple exponential or (modified) power types are sufficient for representing a high percentage of observed gradients.

7.3.4 Contact distributions

So far, we have taken a mostly descriptive approach to characterize disease or dispersal gradients (Fig. 7.1), and have indicated how the power and exponential models are consistent with known physical processes of spore dissemination. It is sometimes helpful to characterize dispersal with probabilistic (i.e., statistical) models rather than deterministic ones. Assume that all spores originate at a single (line) source ($s = 0$). Now suppose that the probability of spore deposition in a very small distance increment (Δs) is constant no matter how far the spore has moved, and that the probability is also the same for all spores in a population originating at the source. Call this probability $\zeta\Delta s$ (with ζ having units of 1/distance). Using standard methods for working with continuous distributions, it can be shown that the probability density function (*pdf*) for the random variable s (distance of spore movement until deposition) is given by:

$$D(s) = \zeta \exp(-\zeta s) \tag{7.8}$$

Note that the contact distribution is utilized again in Chapter 9 for certain types of spatial pattern analysis, and is a key component of all the modeling in Chapter 8. The integration of $D(s)$ from s to $s + \Delta s$ is the probability of a spore being deposited over this distance (when starting at distance 0); equivalently, $D(s)$ is the probability per unit distance of a spore being deposited at ("immediately around") s. Borrowing terminology from general theoretical epidemiology (Diekmann and Heesterbeek, 2000), $D(s)$ can be called the *contact distribution*, because, in a sense, it describes the probability of "contact" between a diseased individual at one location [i.e., the source in this case ($s = 0$)] and an individual at another location (the location identified by s). More precisely here, in terms of plant diseases, the contact distribution characterizes the probability of an infectious unit (e.g., spore) originating $s = 0$ coming in contact with a host individual at distance s (determined from the integration of $D(s)$ from s to $s + \Delta s$). Because $D(s)$ is a *pdf*, its integration from $s = 0$ to $s = \infty$ is 1, by definition. Biologically, this means for spore dispersal that a spore must end up somewhere. Additional refinements to the concept of a contact distribution will be found in the next chapter.

It should be pointed out that a spore being deposited at s is not quite the same thing as a spore contacting a host individual at s—there may not be a host individual exactly at distance s (for instance). For practical purposes, both disease gradients and spore deposition gradients are utilized to determine the appropriate (or reasonable) contact distribution. When appropriate (see Chapter 8), one can multiply a contact distribution based on spore transport by a constant to reflect the fraction of deposited spores at s that actually contact a host.

Equation 7.8 is directly analogous to the exponential disease gradient model (equation 7.1), with $\zeta = b_E$. The difference is that equation 7.8 is scaled so that the integration over all distances is 1. Thus, the "source strength" parameter here is simply ζ. The expected value (mean) and the variance of a random variable with equation 7.8 as its *pdf* equal $1/\zeta$ and $1/\zeta^2$, respectively. The median s ($s_{50\%}$) equals $\ln(2)/\zeta$.

Now, suppose the probability of deposition per unit distance (ζ) is not constant, but is a random variable (with a possibly different value for each spore). This means that each spore has a different probability of being deposited, possibly because of its size or flight taken. The gamma distribution (Evans et al., 2000) is one useful distribution for ζ because it can take on any positive value. Using standard methods for mixing distributions, the following expression for flight distance is obtained:

$$D(s) = \Psi\lambda^{\Psi}(s+\lambda)^{-(\Psi+1)} \qquad (7.9)$$

Equation 7.9 is the *pdf* of a Pareto distribution of the second kind (Johnson et al., 1994), often simply called a Pareto distribution (Minogue, 1989). It is directly analogous to the modified power model for dispersal gradients (equation 7.5), with $b_P = \Psi + 1$. The only difference is that scaling is used here so that the integration of equation 7.9 over all distances equals 1. In contrast, there is no scaling of spore numbers (or infections) when typically using equation 7.4, and a_P gives the height of the Y curve. It is interesting to note that the mean and variance for the Pareto distribution are only defined for certain (large) values of Ψ. Minogue (1989) gives a very useful discussion of this property. If $\Psi > 1$ (i.e., $b_P > 2$), then the mean s is $\lambda/(\Psi - 1)$. The median s, which is defined for all Ψ, is $s_{50\%} = \lambda(2^{1/\Psi} - 1)$.

The differences between equations 7.8 and 7.9 are most apparent at very small and very large s values (near and far from the source), where the Pareto function gives larger values than the exponential. This is, essentially, the same concept as shown in the middle frame of Fig. 7.5 for the deterministic dispersal models. A plot of $\ln[D(s)]$ versus s has the same shape as a plot of $\ln(Y)$ versus s. When $\ln[D(s)]$ declines linearly with s, it is said that the distribution is *exponentially bounded*. Thus, the exponential distribution has this property. More details on the concept of exponential bounds for distributions are given in Chapter 8 and in Minogue (1989). Chapter 8 also describes generalizations of the distributions considered here.

With reference to the predicted values at large s, the Pareto distribution is not exponentially bounded, and therefore has a so-called "fatter tail" than the exponential. The implications of this difference can be seen in a simple simulation of spore flights. Using a random number generator, 2000 spore flight distances (from random positions along a line segment) were generated for an exponential and Pareto distribution. To make them equivalent in one sense, a median s ($s_{50\%}$) of 0.6 was chosen for both distributions. As shown in Fig. 7.6, points are clustered close to the source, and the density of points drops off fairly rapidly, for both models. With the exponential distribution, there is a very smooth drop off in density, with the largest distance around 8. However, with the Pareto model (i.e., power-model type dispersal), there is more irregularity in the drop off of point density, and there are points found at a very wide range of distances. In fact, 26 distances (out of 2000) were actually greater than the upper limit of $s = 1000$ used in the graph. It should be re-emphasized that half of the points in both graphs are for distances below 0.6.

One other feature of Fig. 7.6 is worth noting. For the Pareto distribution, not only are spores deposited far beyond the source, but some clumping or clustering of points at various distances are somewhat apparent. Thus, for pathogens with a Pareto-type distribution for spore dispersal, establishment of new foci of infection beyond are original source, and clustering of disease around the new infections, are standard features. These topics are dealt with in later chapters.

The exponential and Pareto contact distributions are emphasized in this sub-section because these correspond directly to the popular and useful exponential and power disease gradient models. Other distribution functions have been found to be useful, however. For instance, the normal, (half-)Cauchy, and other distributions have been used for $D(s)$ (Shaw, 1995; van den Bosch et al., 1988b, c; Pielaat and van den Bosch, 1998; Xu and Ridout, 1998). These all vary in their general shape, especially in terms of the "heaviness" or "fatness" of their tails at large distances (i.e., how fast $D(s)$ declines with s at large distances). The implication of tail fatness will become even more apparent later in this chapter and the next chapter when spatio-temporal disease development is discussed.

7.3.5 Some other dispersal models

As with temporal models of disease progress, there are other models than can be used to represent disease gradients away from an inoculum source. As mentioned above, spore dispersal is determined by a mixture of deposition and dilution (and escape) processes (Aylor, 1999), and one or the other may be more important under different conditions. It is possible that neither process will dominate, and that an observed gradient shows characteristics of both deposition and dilution. Lambert et al. (1980) made an initial step at expanding

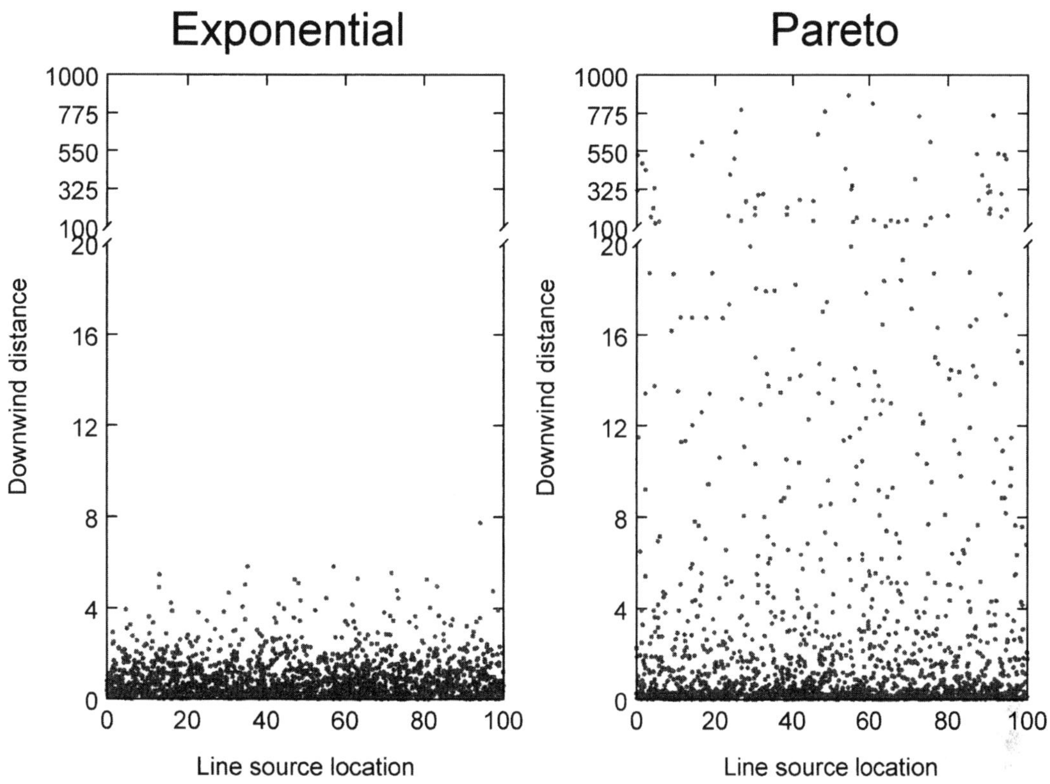

FIG. 7.6. Simulated dispersal of 2,000 spores (or other units of inoculum) from a line source (distance, s, of 0) according to the exponential (equation 7.8) contact distribution, with $\zeta = 1.15$, and Pareto distribution (equation 7.9), with $\lambda = 0.2$ and $\Psi = 0.5$. Both distributions have the same median, $s_{50\%} = 0.6$, meaning half of all points on the graphs are less than 0.6.

on the above models by developing a more general gradient model. Their function can be written as:

$$Y = a_1 \exp(-b_1 s^{\eta}) \tag{7.10}$$

where a_1, b_1, and η are parameters. When the unitless shape parameter, η, equals 1, equation 7.10 reduces to the exponential model (equation 7.1). When η approaches 0, equation 7.10 reduces in the limit to the power model (equation 7.4). Other values of η allow the model to take on other shapes.

In some ways, the generalization of equation 7.10 may not be needed. This is because the modified power model is, in fact, a generalized dispersal gradient model if λ can take on any value. In this situation, λ and b_P work together to indicate the shape of the gradient. This can be seen by the equation 7.7 for dY/ds. At large λ (for example, much larger than the largest distance considered), s in the denominator has very little influence on the rate. In other words, if λ is large, increasing s does not change the denominator to any extent, and the model can be written as $dY/ds \approx (b_P/\lambda)Y$, which is the formulation for the exponential model (with $b_E = b_P/\lambda$). In more mathematical terms, when λ becomes indefinitely large, with b_P/λ held to a finite value (meaning that b_P also becomes indefinitely large), the modified power model reduces to the exponential model.

A different generalization is possible by combining both exponential and power functions in the same model (Maffia and Berger, 1998). One formulation is:

$$Y = a_1 s^{-b_2} \exp(-b_1 s) \tag{7.11a}$$

in which a_1, b_1, and b_2 are parameters. This model is consistent with some physical models for that have been developed for spore dissemination (see equation 15 in Aylor, 1999). When $b_2 = 0$ (in principle, when dilution/escape of spores is of no or little importance relative to deposition processes), equation 7.11a reduces to equation 7.1. Likewise, when $b_1 = 0$ (in principle, when deposition is of little importance relative to dilution/escape processes), equation 7.11a reduces to equation 7.4. It is possible, however, for both b parameters to be non-zero. The equation can be linearized to:

$$\ln(Y) = \ln(a_1) - b_2 \ln(s)(-b_1 s) \tag{7.11b}$$

which facilitates in parameter estimation (see below). The model can be further generalized by adding a constant (λ) to s, as done with the power model (equation 7.5).

Another possible model is based on a deterministic version of the half-Cauchy statistical distribution for $D(s)$. One way that this model can be written is:

$$Y = \frac{a_1}{(1 + (b_1 s)^2)} \tag{7.12}$$

Interestingly, Aylor and Ferrandino (1990) developed this same model for spore deposition near an inoculum source based on assumed mechanisms of spore trajectories. Like the power model (equation 7.4 or 7.5), this model involves raising s (actually, a function of s) to a power; thus, it may be appropriate as an alternative to the power model for gradients that are very steep near the source ($s \approx 0$) and flat at large distances. Unlike the power model (without modification; equation 7.4), however, equation 7.12 has a finite value at $s = 0$, namely $Y(0) = a_1$. As a contact distribution, the half-Cauchy distribution has been used heavily in simulation and theoretical studies (Shaw, 1995; Xu and Ridout, 1998), but has not gained much popularity (yet) for characterizing observed gradients. The interested reader can verify that with equation 7.12, $1/Y$ is a linear function of s^2.

Some other contact distributions are discussed in Chapter 8.

7.3.6 Some calculations

Readers who have learned how to perform calculations with the empirical disease progress models of Chapter 4 (see equation 4.4.6) will have no trouble performing similar operations with gradient models. It is sometimes convenient to consider that the two most common models (exponential and power) can be written generically as:

$$\ln(Y) = a_* - b_* s^* \tag{7.13}$$

where a_* represents either $\ln(a_E)$ or $\ln(a_P)$, b_* is either b_E (units of 1/distance) or b_P (unitless), and s^* is either just s (for exponential) or $\ln(s + \lambda)$ for (modified) power (with $\lambda = 0$ in many cases).

Just a couple of calculations are demonstrated here. Suppose that distance is measured in centimeters from a source, Y represents (mean) rust pustules per plant, and that $a_E = 500$ and $b_E = 0.02$/cm. *What is Y at 4 m (400 cm)?* Using equation 7.13, $\ln(Y) = \ln(500) - 0.02 \times (400) = 6.215 - 8 = -1.875$. Back-transformation gives a Y of 0.168. Such a simple calculation could be done directly with equation 7.1. Now, with the same parameters, *at what distance is $Y = 1$?* Rearranging equation 7.13, leads to:

$$s^* = \frac{-[\ln(1) - \ln(500)]}{0.02} = \frac{-[0 - 6.215]}{0.02}$$
$$= 310.8 \text{ cm } (\sim 3.1 \text{ m}).$$

The interested reader should be able to determine *how much a_E or b_E would have to be changed so that Y declines to 1 pustule/plant not at 3.1 m but at 1 m (100 cm)*. Answer: either new $a_E = 7.39$ (a reduction in source strength of ~496) or new $b_E = 0.062$/cm (an increase of gradient steepness of 0.042/cm).

When performing similar calculations with the power model, it is important to remember that one is dealing with a logarithmic transformation of distance. For instance, with $\lambda = 1$ cm, suppose that $b_P = 1.2$ and that $a_P = 500$ (i.e., that Y at the source is $500 \times 1^{-1.2} = 500$ pustules/plant). The distance at which $Y = 1$ can be determined from: $s^* = -[\ln(1) - \ln(500)]/1.2 = 5.18$. Given that $s^* = \ln(s + 1)$ here, s is calculated as: $\exp(5.18) - 1 = 176.7$ cm (~1.8 m).

7.4 Model Fitting

7.4.1 Choosing a model

As with disease progress curves (Chapter 4), a combination of interrelated graphic and statistical methods can (or should) be used to select a proper model for describing a disease gradient. As with all methods, the model selection process is constrained by the available data. That is, any conclusion regarding the 'best' model may be limited due to measurement error (see Chapter 2), sampling error (Chapter 10), and the range of distances over which Y is assessed. Nevertheless, choosing a reasonable model is essential for quantifying dispersal and disease spread (see later), and in comparing results for different treatments, locations, years, pathogens, etc.

The general methods presented in Chapter 4 can typically be applied to observed dispersal gradients to select a reasonable model for the data. We demonstrate the process with data on number of pustules per plant for rust of beans (Aylor and Ferrandino, 1989). A line source was used, and results for the first assessment time in 1986 in the south direction were read from the graph. As expected, there was a decline in Y with increasing s (Figs. 7.7 and 7.8), which is consistent with either exponential or power-type models for dispersal. The sharp decline in the first 3 m or so, and the very flat curve at large distances are suggestive of a power model, but this could be misleading if data are not recorded for very small and large s.

A plot of dY/ds versus s is of importance for theoretical distinctions between the two model processes (dY/ds goes rapidly to very large negative values for power, and less rapidly to finite negative values for exponential) (see Figs. 7.3 and 7.4). However, the plot of $\Delta Y/\Delta s$ typically is not of practical value in model selection. This is because it is generally unrealistic to have a sufficient number of sampling distances very close to the source to see if dY/ds is declining fast enough to be representative of the power model. Since estimated dY/ds is a ratio, there can be considerable imprecision in its estimate,

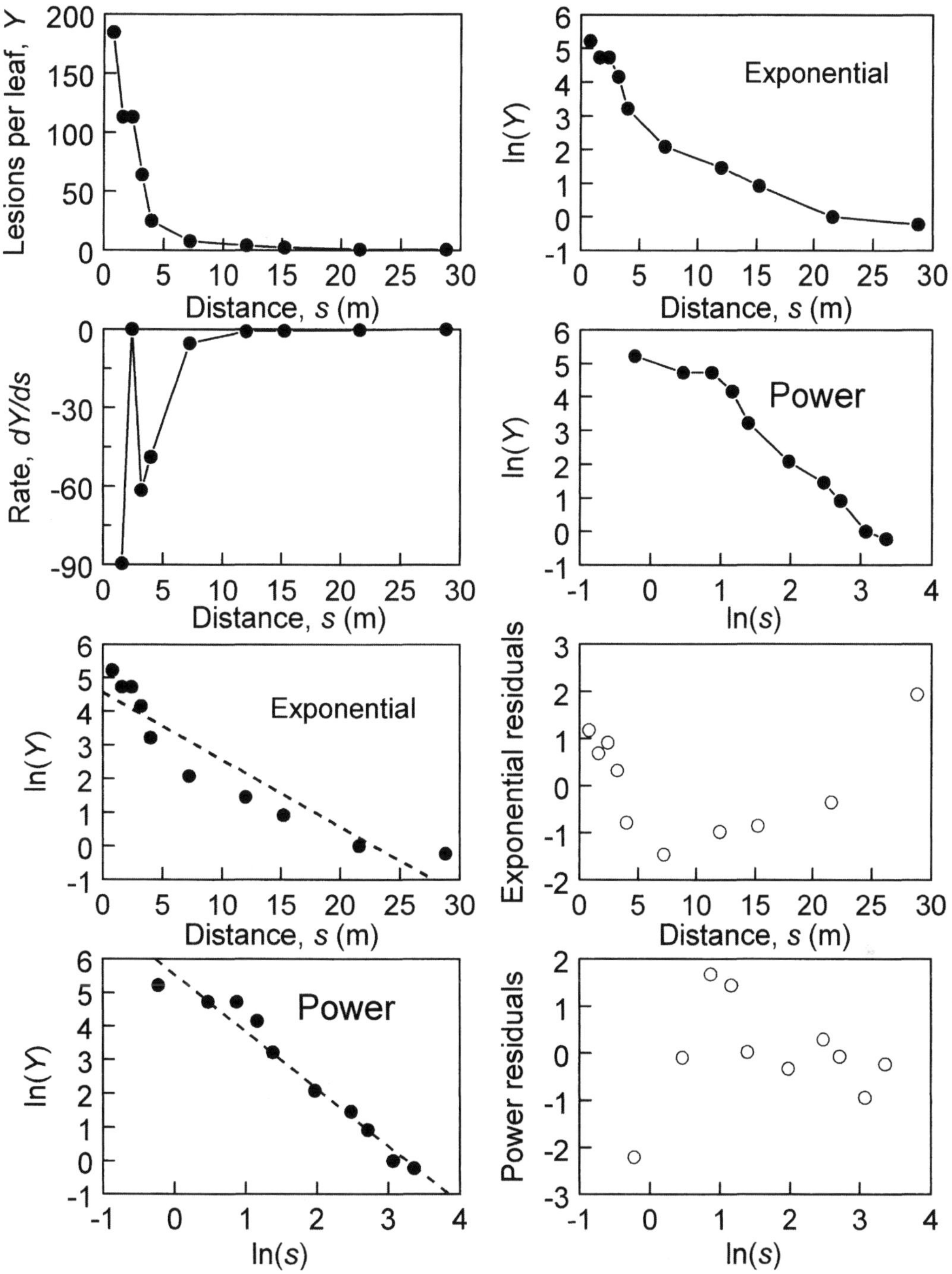

FIG. 7.7. Number of lesions (pustules) per leaf in relation to distance from a line source, for bean rust (from Aylor and Ferrandino, 1989). Upper four frames: data are shown on linear and transformed scales. Lower four frames: fit of exponential (equation 7.2) and power (equation 7.6) models to the transformed data, together with the residuals.

resulting in large scatter of points. This can be seen for the bean rust example: although the trend clearly is towards large negative numbers as s goes towards 0, there is also a wide range of rate estimates over the first three-to-four distances (Fig. 7.7).

Plots of $\ln(Y)$ versus s and $\ln(s)$ are especially useful for selecting a model. These graphs correspond to the linearized versions of the two dispersal models (equations 7.3 and 7.6)—a straight line results if a particular model is appropriate. For the bean rust example (Fig. 7.7), the plot of $\ln(Y)$ versus s shows considerable curvature, indicating that the exponential model would not be appropriate. The type of curvature is typical when power-model values are plotted on this scale (see Fig. 7.3). On the other hand, a plot of $\ln(Y)$ versus $\ln(s)$ (Fig. 7.7) was a fairly straight line, except very close to the source, where some curvature of the line was revealed. As discussed above, this curvature at small $\ln(s)$ occurs when a λ is greater than 0 in equation 7.6 (see Fig. 7.4). Using a value of λ equal to half the distance between the source and the first sampling distance (0.4), reduced this curvature somewhat, but as we show later, the impact of this addition was minor on model fitting. Overall, the power model seems more reasonable than the exponential model for describing the example data set.

As with all evaluations of dispersal/disease gradients, the range of sampling distances can greatly influence the results (Ferrandino, 1996). Over a limited distance range, the model predictions can be very similar, even though there are fundamental differences in the two model types at very large and small distances from the source. This can be seen with the bottom frame of Fig. 7.5. The solid points represent disease values corresponding exactly to the power model (equation 7.4), but the data are plotted in the form of $\ln(Y)$ versus s—the standard scale for the exponential model (equation 7.1). The straight line represents one fit of the exponential model to these values. The fit is quite good, except close to the source, where the exponential model underpredicts the true Y values. If the first observation (at $s = 1$) was not present (e.g., when the first row of plants was 2 m from the source), a researcher would have little evidence that the exponential model was inappropriate. If data were available far beyond $s = 12$, the differences between the two models could become noticeable—the decrease in true $\ln(Y)$ with distance would continue to decline for this power-model relation (i.e., the curve would become flatter), whereas the exponential model would indicate a continuous linear decline in $\ln(Y)$.

The main point from this discussion is that the investigator should never over-interpret the results from a single data set. Given that different physical processes may dominate over different distance ranges (Aylor, 1999), and that observed gradients generally represent only a portion of an entire dispersal gradient, even a perfect fit of a simple model to a small data set does not indicate that this model would necessarily hold at all distances or for all dispersal events for the same pathogen/host combination. One *does* need to have reasonable model choices to quantify properly the spread of disease, and to compare treatment results. However, conclusions on mechanisms must always be based on cumulative evidence, and not on a limited dataset. Ferrandino (1996) further elaborates on the challenges in distinguishing between exponential-like and power-law-like gradients when disease is assessed over limited spatial ranges.

7.4.2 Estimating parameters—linear methods

A wide range of statistical methods can be used to fit disease-gradient models to data. It is assumed here that the reader has digested the presentation on model fitting for disease progress curves (section 4.6), so that considerably less detail is needed here. Least squares and maximum likelihood (as well as some other statistical methods) can be used for model fitting and parameter estimation. We focus in this chapter primarily on least squares methods.

As shown with equation 7.13, both the exponential and modified power models (with λ assumed known) can be expressed as linear models. Readers can see Chapter 3 for discussion of the meaning of models that are linear in the parameters. Statistical versions of these mathematical dispersal models can be written as:

$$\ln(Y)_j = \ln(a_E) - b_E s_j + \varepsilon_j \tag{7.14a}$$

$$\ln(Y)_j = \ln(a_P) - b_P \ln(s + \lambda)_j + \varepsilon_j \tag{7.14b}$$

in which the j subscript indicates the j-th data point or distance ($j = 1, \ldots, n$), and ε is the error (residual) term, a random variable (difference between observed transformed Y and that determined by the model). The intercept for the models is $\ln(a_E)$ or $\ln(a_P)$, and the slope is b_E or b_P. Equation 7.14b is not, of course, linear in terms of the parameter λ, because λ appears only as a logarithmic function of $s + \lambda$. One needs to pre-specify a value for λ to use linear statistical methods. The most common approach is to assume that $\lambda = 0$, which results in the simpler regular power dispersal-gradient model. However, one could specify a value roughly equal to the "size" (diameter, depth) of the source (Mundt, 1989).

To use linear least squares, the standard assumptions about the distribution of the error term (ε), and hence the response variable [$\ln(Y)$ in this situation], are made (see equation 4.6.2, and Neter et al., 1983). Normality usually is not overly important, except under extreme circumstances, but independence of the observations and constant variance are very important assumptions. With disease progress curves, the response (transformed disease intensity) is cumulative, in that Y at $t = 20$ equals Y at $t = 10$, plus any new disease that develops between times 10 and 20. Thus, it is very important to check for serial correlations with disease progress modeling. However, Y [or $\ln(Y)$] for disease gradients, as typically expressed, is not a cumulative value over space; for example, Y at $s = 20$ is *not* the accumulation of lesions (or infections) from distance values of 0 to 20. In this sense, one might not expect to find correlations of Y values for disease gradients. In fact, because Y *is* declining with distance (Fig. 7.1), there is (spatial) autocorrelation (lack of independence) because of this spatial trend; that is, Y is serially correlated over distance. In most cases when there is a strong gradient of disease from a (single) source, the correlation of Y values can be "removed" by fitting a gradient model to the data. For instance, $\varepsilon = Y - a_E \exp(-b_E s)$ may not be correlated (if the exponential model is reasonable) even if the Y values are correlated. Using time series descriptions, the trend is removed by fitting a particular model to gradient data. It is easy to check for independence by assessment of the residuals, as discussed in Chapter 4, just to be sure.

If Y equals number of deposited spores or number of lesions, it is unlikely that $var(Y)$ is constant. In general, $var(Y)$ can be expected to increase with mean Y (Schabenberger and Pierce, 2002). It turns out that for discrete data in the form of counts, the logarithmic transformation does a reasonable job of stabilizing variances. That is, the variable $\ln(Y)$ may have close to a constant variance for practical purposes, and weighting based on assumed (or calculated) variances is not necessary. Thus, we assume that $\ln(Y)$ does have a constant variance for model-fitting purposes here. Later in this chapter we deal with situations where the response variable (whether transformed or not) cannot be assumed to have a constant variance.

Fig. 7.8 shows the data points plotted in Fig. 7.7 in the form of a SAS program for least squares analysis. Fitting the exponential model (equation 7.14a) to the bean rust data resulted in the following equation:

$$\ln(\hat{Y}) = 4.58(0.352) - 0.20(0.027)s$$

in which the numbers in parentheses are estimated standard errors. The coefficient of determination (R^2) was 0.878, and the residual variance was estimated as 0.579. The estimate of a_E can be obtained by back-transformation of 4.58 as exp(4.58) = 97.5, which is considerably below the observed Y at the first distance ($s = 0.8$). It should be noted that the estimate of $\hat{b}_E$ has units of 1/distance; that is, $b_E = 0.20$/m. Evaluation of the predicted line and the residuals in the lower frames of Fig. 7.7 reveal clearly that this model did not provide an appropriate fit over the range of observed data. In particular, the (standardized) residuals showed a very strong pattern. Of course, if the observations within 2 m of the source and beyond 25 m were not made, a good fit would have been obtained.

Fitting the modified power model (equation 7.14b, with $\lambda = 0$) to the data resulted in the following equation:

$$\ln(\hat{Y}) = 5.56(0.246) - 1.70(0.119)\ln(s)$$

```
/* Dispersal gradient analysis  */
data a;
            input s Y @@; /*pairs of (s,Y) strung across records */
            lnY=log(Y); lns=log(s); lns4=log(s+.4);
            datalines;
0.8  184.9   1.6 113.3   2.4 113.3    3.2  64.1
4.0   25.0   7.2 8.0     12.0  4.3    15.2  2.5
21.6 1.0     28.8  0.8
;
proc reg data=a;
            title 'Linear regression analysis';
            /* residual plot specified for each model */
            model lnY = s;  plot r.*s;  *exponential model;
            model lnY = lns;  plot r.*lns; *power model;
            model lnY = lns4; plot r.*lns4; *modified power model (c=0.4);
proc nlin data=a;
            title 'Power model (fixed c = 0)';
            parameters a=1000 b=2; /* Initial guesses of parameters */
            bounds a >= 0; bounds b >=0;
            c=0.0;
            model lnY = log( a*(s+c)**(-b) );
proc nlin data=a;
            title 'Modified power (with c unknown)';
            parameters a=1000 b=2 c=.4;
            bounds a >= 0; bounds b >= 0; bounds c >= 0;
            model lnY = log( a*(s+c)**(-b) );
proc nlin data = a;
            title 'Modified power (c = 0.4)';
            parameters a = 1000 b = 2;
            bounds a >= 0; bounds b >= 0;
            c=0.4;
            model lnY = log( a*(s+c)**(-b) );
run;
```

FIG. 7.8. Input program file of SAS to read in dispersal data and fit dispersal models, in linear and nonlinear forms. Note: in version 9.1 of SAS, the NLIN procedure no longer displays the total sum of squares, SST (see section 3.5.2), which is needed to calculate the pseudo-R^2. SST can be obtained by determining the variance of the response variable and multiplying by number of observations minus 1.

which had an R^2 of 0.962 and an estimated residual variance of 0.179. The slope here is unitless. The back-transformation of the intercept is given as exp(5.56) = 259.8, which is an estimate of Y at s = 1 m, *not* at the source. One could re-scale distance from meters to centimeters, as an example, by multiplying distance by 100, and then taking the log of s. Fitting the power model (with λ = 0) produces the same results, except for the intercept term which is now estimated as 13.38 (standard error = 0.766). Back-transformation gives exp(13.38) = 646,934, obviously a very large number, for the predicted Y at 1 cm. Re-scaling to millimeters would give an even larger number. This is because Y becomes indefinitely large as s goes towards 0 with this model (and a_P is the value of Y at s = 1). Using the results for the centimeter distance scale, predicted Y at 100 cm (=1 m) is given by:

$$13.38 - 1.70 \ln(100) = 13.38 - 7.83 = 5.55$$

which is the same as the original intercept with distance in meters (except for rounding error).

Graphic evaluation of the predicted line and residuals showed that this model was clearly more appropriate than the exponential for the bean rust data set. The greatest errors were near the source, where the slight curvature on the log-log scale was previously identified. Improvement in fit could be accomplished by consideration of a value of λ greater than 0. Readers can evaluate a range of λ values, but we show results for just λ = 0.4 m. This is half the distance between the first sampling distance and the center of the source, which may be a reasonable choice if λ is interpreted physically as a measure of size of the source. Least squares resulted in the following equation:

$$\ln(\hat{Y}) = 6.10(0.228) - 1.88(0.108) \ln(s + 0.4)$$

which had an R^2 of 0.974, slightly larger than when λ = 0. The slope parameter was slightly larger (1.88 versus 1.70), and the intercept was also larger. Back-transformation of the intercept, exp(6.1) = 445.9, is an estimate of a_P, equaling the predicted Y when s + 0.4 = 1 m (or s = 0.6 m). One can determine the predicted Y at s = 1 m, with $Y(1) = a_P(1 + \lambda)^{-b_P} = 445.9(1.4)^{-1.88} = 236.9$. This estimate is close to the 259.8 value obtained for a_P with the simpler power model above (λ = 0), which is, by definition, estimated Y at s = 1 when λ = 0.

Other than possibly improving the fit to observed data, the advantage of the modified power model (with $\lambda > 0$) is that one can directly estimate Y at the source, $Y(0)$, as a measure of source strength. Using the formula given after equation 7.5, one obtains:

$$Y(0) = 445.9(0.4)^{-1.88} = 2497.$$

This value is considerably larger than the source-strength estimate based on the exponential model ($\hat{a}_E$ = 97.5). This $Y(0)$ value, however, is considerably below the non-sensical estimate of Y at 1 cm [i.e., $Y(0.01)$], given above when the regular power model (λ = 0) was fitted to the data. The reader is reminded that the exponential model provided a rather poor fit to these data close to the inoculum source, and that the regular power model breaks down as s approaches 0.

In summary, for the example data set, the (modified) power model provided a better fit than the exponential model. This will not necessarily be true for other data sets. Using a value of λ greater than 0 resulted in a better fit than the use of λ = 0. However, it is not necessarily clear that one needs to use the more general model for these data. The improvements in R^2 and residual plot (not shown) were marginal when using λ = 0.4 m. The reader can determine that the fit can be even better with higher values for λ. If the goal of the research was to obtain a precise estimate of the "source strength" and the gradient of disease close to the source, then the use of a more complex model is certainly warranted. But if the goal was to provide a reasonably simple overall description of the gradient (starting around 0.7 m), to facilitate comparisons among groups (e.g., treatments, cultivars, years), then the more general model would not be needed. In fact, the use of the more general model where it is not fully needed may be disadvantageous because model fitting would be more complicated.

It must be pointed out that a direct comparison of the fit of the exponential and modified power models is, in a sense, not completely fair if λ is allowed to take on any positive value. In this situation, the modified power model contains three unknown parameters, whereas the exponential model has only two unknowns. Generally, a 3-parameter model will provide a better fit than a 2-parameter one, especially if the 2-parameter version is a special case of the 3-parameter model. This was seen by fitting a Richards model (with unknown η parameter) to disease progress data (Chapter 4). When λ is determined directly from the observed data, the modified power model becomes a flexible (general) gradient model that can take on multiple shapes (Minogue, 1989). One cannot directly estimate λ using linear least squares because the model (equation 7.24b) is not linear in terms of λ. Regressing $\ln(Y)$ on $\ln(s + \lambda)$ for a large number of possible λ values (say, λ = 0, 0.1, 0.2, ..., 1), and selecting the λ with the highest R^2 is an indirect way of estimating this parameter. We show in the next sub-section how to estimate λ with nonlinear methods.

Specifying a value of λ greater than 0 in the power model *is* important if one is trying to quantify the gradient very near the source. The obvious question then arises from this discussion: how can one compare the fit of the exponential model and the modified power model (with non-zero λ) without giving unfair advantage to the latter? One answer to this is to choose a λ independent of the gradient. This could be done before collecting the data or before looking at a graph. If λ can be considered a measure of the size of the source, then this value can certainly be approximated. The rule followed above was to use a λ equal to half of the distance between the center of the source and the first data point. Other possibilities exist. However, investigators should be aware that estimates of a_P and b_P are always conditional on the choice of λ.

7.4.3 Estimating parameters—nonlinear methods

The standard models for dispersal gradients are all nonlinear in the parameters. In common practice, as discussed above, the models are linearized so that linear least squares can be utilized to estimate parameters. Nonlinear estimation methods offer great flexibility, in that investigators are not forced to use models that can be linearized. There is a penalty associated with nonlinear methods, however, because the methods are more complicated, may fail or may produce biased estimates of parameters, and require a considerable number of data points compared with linear methods. Experienced users, however, can benefit greatly from nonlinear estimation methods. Interested readers should refer back to sections 3.5 and 3.6 of Chapter 3. We present some analyses here for the bean rust data set. Relevant code for fitting is given in Fig. 7.8.

A dispersal model can be expressed generally as: $Y_j = f(s_j; \Theta)$, where Θ represents a vector of parameters. Two ways of expressing the model in statistical form are:

$$Y_j = f(s_j;\Theta) + \varepsilon_j \qquad (7.15a)$$

and

$$Y_j = f(s_j;\Theta)\varepsilon_j \qquad (7.15b)$$

in which the j corresponds to the j-th observation (distance from the source). The comments made in Chapter 4 (section 4.6.3) regarding appropriateness of the different statistical forms of disease progress models are applicable here. In short, if equation 7.15a is correct, only nonlinear methods can be used, because this statistical model cannot be linearized (but see section 3.5), even though a transformation of $f(\bullet)$ is linear. On the other hand, if $f(\bullet)$ represents the exponential or power models (and some others), equation 7.15b can be linearized—by taking logs of both sides. It is not clear based on our understanding of the population biology of diseases whether the error term in a nonlinear statistical model should be additive or multiplicative.

To avoid excessive repetition of ideas, we only discuss one assumption about the ε, and hence Y, values. Because spore density or lesions per plant varies greatly with distance from a source (see Fig. 7.1), it is very unlikely that the variance of ε (or Y) will be constant. Rather, $var(Y)$ likely increases with increasing mean Y (or some monotonically increasing function of Y). One approach for incorporating this property of Y into the analysis is to use weighted least squares (or maximum likelihood). Reasonable choices for the weights are $1/Y$ or $1/Y^2$.

It can be shown that if $var(Y)$ is proportional to Y^2, then $\ln(Y)$ has an approximate constant variance (Neter et al., 1983). Given that the linearized forms of the exponential and (modified) power models involve the log of Y, unweighted linear least squares for model fitting can be justified. A similar approach is to use nonlinear methods, but with the model in log-transformed form. For instance, with equation 7.15b (but not 7.15a), one can write:

$$\ln(Y)_j = \ln\left[f(s_j;\Theta)\right] + \ln(\varepsilon)_j \qquad (7.15c)$$

in which $\ln(\varepsilon) = \varepsilon^*$ is a "new" error term, equivalent to the ε in equation 7.14. For the exponential and power models, algebraic manipulation of $\ln[f(\bullet)]$ results in a linear equation. However, equation 7.15c can be used directly with nonlinear parameter estimation methods with any model in which Y has a variance proportional to the square of Y.

In Fig. 7.8 we show how to fit a few versions of the modified power model to the rust data, with the models in the form of equation 7.15c. Many other statistical programs could be used. $\ln(Y)$ is created in the data step ($\ln Y$), and the right-hand side of equation 7.15c is given in the model statement. The error term is implied, and not written. The parameter statement identifies unknown parameters and gives initial guesses for their values. Bounds could have been placed on the parameters, and an initial search of parameter values for starting values in the optimization method could have been specified. Estimates of the parameters and coefficients of determination are given here, with standard errors in parentheses:

Power:	$\hat{a}_P = 260(64)$,	$\hat{b}_P = 1.70(0.12)$,
	$\lambda = 0$,	$R^2 = 0.96$
Modified Power:	$\hat{a}_P = 447(102)$,	$\hat{b}_P = 1.88(0.11)$,
	$\lambda = 0.4$,	$R^2 = 0.97$
Modified Power:	$\hat{a}_P = 1993(2462)$,	$\hat{b}_P = 2.36(0.38)$,
	$\hat{\lambda} = 1.7(1.1)$,	$R^2 = 0.983$.

Note that we use a hat over λ when it is estimated directly from the data (third case) and do not use a hat when a specific value of λ is simply inserted in the model (first two cases here).

The model with λ fixed at either 0 or 0.4 (first two cases) gave the same results as found with linear least squares (see above) because the log transformation results in a linear model. For nonlinear models in general, one definitely cannot expect to find this correspondence between linear and nonlinear models. Note that $\ln(a_P)$ is estimated with linear models, and a_P is directly estimated here. By allowing λ to take on any value (third case), a considerable improvement in the coefficient of determination occurred, with nearly a perfect fit to the data. The estimate of λ ($\hat{\lambda} = 1.7$) was about equal to the distance from the source to the *second* sampling location. Thus, the concept of λ as a measure of source size probably does not hold in this case. Rather, the role of λ in influencing the shape of the dispersal gradient likely is dominant here. As expected, the estimates of a_P and b_P depended on the value of λ (either fixed or estimated).

The estimate of a_P for the general case (=1993) is so different from the other estimates, that one might question the validity of the estimate. One must recall, however, that a_P is the predicted Y at $s = 1 - \lambda$. With $\lambda = 1.7$, this means that a_P represents Y at a distance of -0.7 m, $Y(-0.7)$, which is *not* a physical location. It is more interesting to determine the predicted Y at $s = 0$ m. This is indeterminate when $\lambda = 0$. In the modified power model, $Y(0)$ is predicted as $\hat{a}_P\hat{\lambda}^{-\hat{b}_P}$. With λ fixed at 0.4, the source strength is determined as $447(0.4)^{-1.88} = 2503$; with λ estimated, the source strength is determined as $1993(1.7)^{-2.36} = 570$, a value smaller than found with the small (fixed) λ value. As discussed previously, the power model becomes more exponential like as λ gets very large, so it is not surprising that the $Y(0)$ predictions becomes smaller here as λ increases. In one sense, $s = 0$ is a hypothetical (theoretical) location for this data set in that the first data point occurs nearly a meter from the inoculum source. For the actual observed data points, all versions of the model fit well.

One interesting result when fitting the modified power model with unknown λ was the large estimated standard errors for estimated a_P and λ. This suggests a fair degree of imprecision in the estimates of these two parameters here. These two estimates also were highly and positively correlated, because of the nature of the model formulation. For instance, as shown in the formula for estimated $Y(0)$ in the previous paragraph, as $\hat{\lambda}$ increases, $\hat{\lambda}^{-\hat{b}_P}$ must decrease (for positive b_P). So, to achieve the same, $\hat{Y}(0)$, $\hat{a}_P$ must increase with increasing $\hat{\lambda}$. Ratkowsky (1990) indicates that in models of the modified power type, a_P and $\hat{\lambda}$ parameters have very nonlinear behavior (see Chapter 3 for some discussion). One way to improve the precision of the estimates, and reduce their bias, is to re-parameterize the model using the so-called "expected value" parameter in place of a_P. The issue is very important for those working on nonlinear modeling of dispersal gradients. The modeling presentation in Chapter 5 of Schabenberger and Pierce (2002) is very helpful in this regard.

Based on use of the nonlinear estimation procedure, other dispersal models described in section 7.3.5 can be fitted. For example, the estimate of the shape parameter in the Lambert et al. model (equation 7.10) was $\hat{\eta} = 0.24$ (s.e. = 0.12), close to, but slightly above the limiting value for a pure power model ($\eta = 0$). This suggests that the power model is more appropriate than the exponential, which agrees with the results already presented. Fitting the joint power-exponential model (equation 7.11a) to the data resulted in: $\hat{b}_1 = 0.05$ (s.e.= 0.034) for the exponential component, and $\hat{b}_2 = 1.31$ (s.e.= 0.271) for the power component. Theoretically, if the gradient was purely of the power type, b_1 would be 0. The estimated b_1 was not significantly different from 0, which indicates the appropriateness of the power model. This can also be tested based on the reduction in residual sum of squares between the fit of the power model and equation 7.11a. However, such a simple interpretation is hindered by the fact that s and $\ln(s)$ (see equation 7.11b) are correlated, and the correlation can be above 0.9 in many cases. Another way of looking at this is that equation 7.11a involves two functions of the *same* variable (s). Because of this, tests involving the parameters can be ambiguous. Ferrandino (1996) developed an ingenious approach to replace s and $\ln(s)$ with two uncorrelated functions, and test whether or not the power and exponential models equally fit the data. Readers should check the original article for details.

It is worth pointing out that equation 7.12 provided the best fit of all the models to the example data in terms of smallest residual sum of squares and largest coefficient of determination. This is noteworthy, because this model has only two parameters, and the next best models had three parameters (modified power with estimated λ). This suggests that more work is needed to assess the appropriateness of the Cauchy-like model for additional dispersal gradients.

7.5 Disease Gradients—Correcting for Maximum Intensity

As discussed at the beginning of the chapter, dispersal of propagules typically is assessed by quantifying deposition of spores (or other infectious units) at various distances from a source, or, more commonly, by measuring disease intensity (such as number of lesions or pustules) at each distance. The latter is relevant when number of spores is directly and linearly proportional to number of lesions, which is reasonable when density of lesions is not too high. But it may be difficult to specify with great precision what is meant by "too high". With increasing spore deposition, there is a decrease in the fraction of spores that cause countable lesions (Vanderplank, 1975). Spores may compete with each other, or spores may land on previously infected sites on plants. As shown with disease progress curves, one must account for the fact that disease intensity has an upper limit in an epidemic.

In this section, we address in detail how to make adjustments in models to account for an upper limit to disease intensity at a given location. The need for such an adjustment has been known for decades (Gregory, 1973), yet surprisingly, papers are still published in plant pathology where no adjustments are made, leading to some questionable interpretations of results. We start with an *ad hoc* approach to altering the models. Then, we consider a more formal approach, similar to the protocol taken for disease progress, where the exponential model (equation 4.1) is expanded by the incorporation of a disease-free term. Because disease gradients are often determined for more than one time during an epidemic, we show how to quantify and compare gradients obtained at multiple times. This approach is general, and can be used to compare any gradients, not just those corresponding to different times. The developed models form the basis for additional theoretical characterization of disease in time and space.

7.5.1 Simple adjustment

Several decades ago, it was pointed out that a modification must be made to the common dispersal models if disease intensity is used as the dependent variable and intensity is not low (Gregory, 1973). So far in this chapter, we have operated on the assumption that Y in the exponential (equation 7.1), (modified) power (equation 7.4), and similar models is a random variable with 0 as a lower bound and *no* upper bound. However, Y really has M (number of plant individuals or host area that could be diseased) as a physical upper bound, which corresponds to an upper bound for y of 1 ($y = Y/M$). In many cases, the actual upper bound for y, K, may be even less than 1 ($K < 1$ or $KM < M$; see section 4.4.8), but we restrict our attention to the situation with $K = 1$. The simple no-upper-bound assumption, although reasonable if Y or y is not too high, is unrealistic when Y or y is moderate or high. The recommendation of Gregory is to define Z as the multiple infection transformation of disease intensity. If y represents disease on a 0–1 scale, then one replaces Y with $\ln[1/(1-y)]$; that is, one uses $Z = \ln[1/(1-y)] = \ln[M/(M-Y)]$ for Y in the above models. The rationale here is that there is close to unit increase in y or Y with unit increase spore deposition at low disease intensity, but the increase in y or Y with increase in spore deposition becomes less and less at higher disease levels (because spores land on lesions, or spores compete for infection sites). The linear relation between Z and spore deposition would be maintained at all disease densities, in principle. This transformation can be obtained directly from a consideration of the discrete statistical distributions that are discussed in Chapter 9.

This approach results in:

$$\ln\left(\ln\left(\frac{1}{1-y}\right)\right) = \ln(a_E) - b_E s \qquad (7.16a)$$

for the linearized version of the exponential model (compare with equation 7.2), and:

$$\ln\left(\ln\left(\frac{1}{1-y}\right)\right) = \ln(a_P) - b_P \ln(s+\lambda) \qquad (7.16b)$$

for the linearized version of the (modified) power model (compare with equation 7.6). Note that the left-hand side of the equations involves a double-log transformation, which is known as the complementary log-log transformation of y, CLL(y), in statistics. This transformation can also be written as: $\text{CLL}(y) = \ln[-\ln(1-y)]$.

The multiple infection transformation commonly is used to represent disease development over time for monocyclic diseases (Chapter 4). Thus, it *might* appear that equations 7.16a and 7.16b are primarily of use for monocyclic diseases. This multiple-infection-transformation correction is, in fact, used for both monocyclic- and polycyclic-disease gradients. Under some circumstances, the CLL transformation *is* obtained from more mechanistic models of monocyclic disease development in space and time (see Minogue, 1986, and section 8.3.3). However, the reader is reminded that the pure monocyclic growth model results in $\ln[1/(1-y)]$ on the left hand side of the temporal equation (equation 4.11), not CLL(y).

At a minimum, investigators should use CLL(y) rather than $\ln(y)$ or $\ln(Y)$ in linear models whenever they are characterizing disease gradients and y is greater than ~0.2. We show in the next sub-section that possibly more appropriate transformations can be derived for describing gradients based on a more formal analysis of dy/ds.

7.5.2 Generalizations of the exponential and power models

For disease progress curves, one can model disease on an absolute scale, Y, or on a proportion scale, y ($=Y/M$). For dispersal studies, it is more common to work with Y than with y, but this is an arbitrary choice in many cases. In Chapter 4, it was shown that a direct approach to expanding the differential equation model for unlimited population growth was to multiply the level of disease intensity (y) by the level of disease-free intensity ($1-y$). This same idea can be used here to develop models for disease gradients. In the following sub-sections, we apply appropriately expanded models to an actual data set. Because we now focus directly on disease intensity, and not just dispersal processes, we explicitly call the following equations disease gradient models.

One can divide both sides of the dY/ds equations for the exponential and (modified) power models (equations 7.1 and 7.7) by M to obtain:

$$\frac{dy}{ds} = -b_E y \qquad (7.17a)$$

and

$$\frac{dy}{ds} = \frac{-b_P y}{s} \qquad (7.17b)$$

for the rate of change in y with change in s. Just to simplify presentation, we assume that $\lambda = 0$ for the modified power model for the rest of the section. Based on the arguments made in Chapter 4, equations 7.17a and 7.17b can produce totally unrealistic values for dy/ds at large y, leading to possibly impossible predictions of y (e.g., $y > 1$) at small s. Multiplying the right-hand side of these rate equations by $1 - y$ can result in more realistic and interpretable representation of spread over the entire range of disease intensities. As in section 7.5.1, we assume that $K = 1$. The exponential model for spread becomes:

$$\frac{dy}{ds} = -b_L y(1-y) \qquad (7.18)$$

after multiplication by the proportion of host that is disease free. This is, actually, a logistic model for a disease gradient. We use b_L for the gradient parameter (units of 1/distance) to be a reminder that this parameter is not quite the same as b_E (equation 7.17a); at low y, however, equations 7.17a and 7.18 are indistinguishable, in general (with $b_L \approx b_E$). Integration leads to:

$$y = \frac{1}{1 + A_L \exp(b_L s)} \qquad (7.19)$$

in which A_L is the constant of integration, equaling $(1 - y_0)/y_0$, where y_0 is (theoretical) disease intensity at the inoculum source (i.e., at $s = 0$). Readers should note that the exponent in the denominator is positive, which is different from the temporal version of the logistic model (equation 4.15b). This is because of the negative sign in the differential equation (equation 7.18). An equivalent way to write equation 7.19 is:

$$y = \frac{1}{1 + \exp(-(\ln[y_0/(1-y_0)] - b_L s))}.$$

This can be compared with equation 4.15a for temporal disease progress, but now there is a negative sign before the spread parameter.

Algebraic manipulation of equation 7.19 results in the following linear equation:

$$\ln\left(\frac{y}{1-y}\right) = -\ln(A_L) - b_L s \qquad (7.20)$$

Because $-\ln(A_L) = -\ln[(1-y_0)/y_0] = +\ln[(y_0/(1-y_0)]$, equation 7.20 can be written as:

$$\ln\left(\frac{y}{1-y}\right) = \ln\left(\frac{y_0}{1-y_0}\right) - b_L s.$$

which is an equation for a straight line with slope $-b_L$ and intercept given by the logit of disease intensity at the source (see equation 4.16 for comparison).

Example disease gradients corresponding to equation 7.19 are shown in Fig. 7.9 for selected values of y_0 (or A_L) and a single b_L value. These curves represent five different times in an epidemic; the choice of A_L values appears arbitrary here, but is explained in a later subsection. The curves are mirror images of logistic disease progress curves (because of the negative sign before b_L), with the highest y values at the smallest s values. Unlike the situation for no upper limit to the response variable, the curves level off at high y (or low s). For cases where y is not too high (e.g., <20%), the curves look the same as for the exponential dispersal model (equation 7.2).

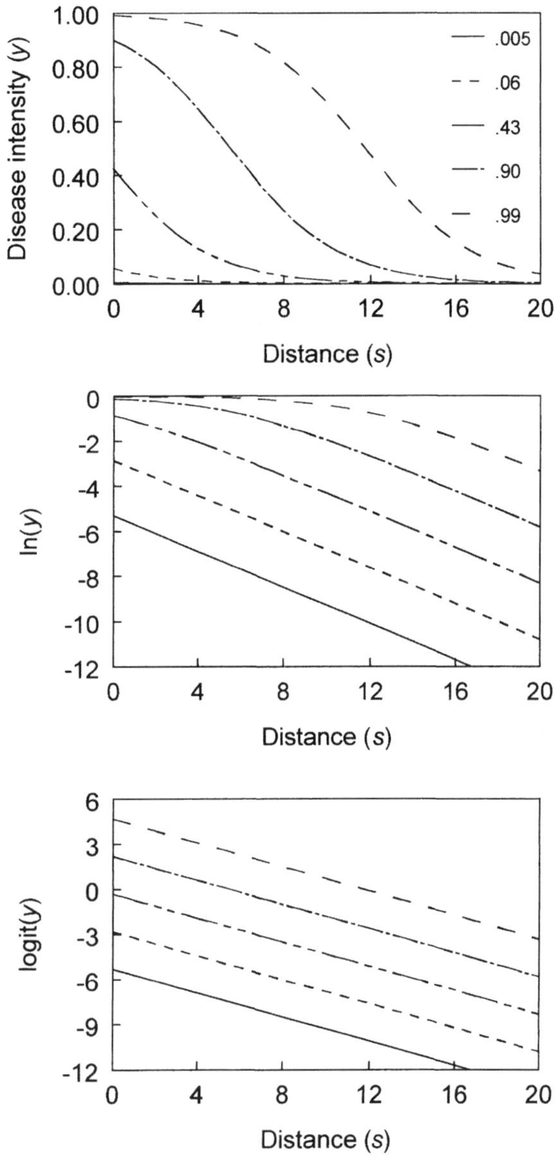

FIG. 7.9. Disease gradients according to the logistic model (equation 7.19), for five different levels of disease intensity at $s = 0$ (y_0; labeled in upper frame), all with the same spread parameter, b_L. Gradients shown on a linear and two transformed scales.

This is because the exponential is a good approximation of the logistic model at low y.

If one plotted the gradients on a log-scale (middle frame of Fig. 7.9), it appears that the gradients flatten over time, especially at distances relatively near the source ($s < 12$, here), even though a single spread parameter value was used. The idea of dispersal gradients flattening over time with polycyclic diseases has a long history in plant pathology. Gregory (1968) indicated that secondary spread flattens a primary gradient, which makes sense when untransformed disease intensity is used as the measure of infections (see top frame of Fig. 7.9). With an upper limit to disease intensity, the rate of change of y with change in s must slow down (with all other things being equal) as y increases, because there are fewer opportunities for infections when most of the plant host is already infected. However, Zadoks and Kampmeijer (1977), van den Bosch et al. (1988a–c), and Minogue and Fry (1983b), among others, posited that infection gradients do not necessarily flatten over time during an epidemic. We only point out a few issues here related to gradient flattening.

To determine whether or not there is flattening of the infection gradient, one must use the appropriate transformation of y (or a model that makes the proper adjustment for maximum y). If equation 7.18 is appropriate, then one must plot logits versus distance, as shown in the bottom frame of Fig. 7.9. Here, the correction for maximum disease is accounted for graphically, and one can see that the gradients are stable in the sense that there is no change in the slope for the different times. Because logit(y) is very similar to ln(y) at low y, one can also see the stable gradients at low y in the ln(y) : s graph in Fig. 7.9 if one just looks at the right-hand side of the graph. Of course, the steepness of the gradients does not change here precisely because equation 7.19 was used with a *fixed* (single) value of b_L at all values of y_0. If the gradients were becoming less steep over time, then b_L of equation 7.19 (or equation 7.20) would be getting smaller (closer to 0) with increasing time (if equation 7.19 was appropriate at *all* times). Tests of b_L, therefore, become a basis for evaluations of gradients over time.

The power model can also be easily generalized to accommodate a limitation to disease. The model can be written in differential-equation, integrated, and linearized forms as:

$$\frac{dy}{ds} = \frac{-b_{PL}y(1-y)}{s} \quad (7.21)$$

$$y = \frac{1}{1 + A_{PL}s^{b_{PL}}} \quad (7.22)$$

$$\ln\left(\frac{y}{1-y}\right) = -\ln(A_{PL}) - b_{PL}\ln(s) \quad (7.23)$$

in which A_{PL} equals $(1 - y_1)/y_1$, where y_1 is (theoretical) disease intensity at a distance of 1 [$=y(1)$]. The intercept term in equation 7.23 is equal to $+\ln[y_1/(1 - y_1)]$, the logit of disease at $s = 1$. Note that we label the gradient parameter as b_{PL} (for "power-logistic"; unitless) because the numerator in equation 7.21 is of the logistic form, but there is also a denominator of s, reflecting a power-type gradient. The model can also be referred to as the *log-logistic*. The link between this model and the logistic disease gradient model may be seen more clearly by an alternative expression for y of equation 7.22:

$$y = \frac{1}{1 + \exp(-(\ln[y_1/(1-y_1)] - b_{PL}\ln(s)))}$$

The log-logistic model, with a different parameterization, is used in section 3.5 of Chapter 3.

Disease gradients described by equation 7.22 are shown in Fig. 7.10, corresponding to five different values of y_1 (or A_{PL}), for a fixed (single) value of b_{PL}. Because of the nature of power-model-type gradients, in which dy/ds is a large negative number at small s, the decline in y with increasing s is relatively large near the source even when overall disease intensity is high (e.g., top curve in Fig. 7.10). Nevertheless, the curves show some appearance of becoming flatter near the source as y increases. This is expected since there is an upper limit to disease (=1) in this model. The flattening of the gradient is more pronounced when one plots ln(y) versus ln(s) (middle frame of Fig. 7.10), which is the proper scaling for the simpler power model (equation 7.6), but not the proper scaling when a correction for maximum disease is made. When one looks just at the right hand side of this graph, however, corresponding to relatively low y, one sees parallel lines (i.e., no flattening). When logit(y) is plotted against ln(s) (bottom frame of Fig. 7.10), which gives the proper scaling for the model introduced here (equation 7.23), lines with the same slope are found for each time, an indication of stable gradients. As with the logistic model above, the gradients *must* be stable in this theoretical example because they were generated with a single spread parameter in the equation. One can ask the question for observed gradients of disease: does steepness change (as measured by b_{PL}) over time? In other words, are gradients stable? As discussed in the next paragraph, the answer is more complicated than for an exponential-type gradient.

The discussion of the power model (equation 7.7) in terms of the concept of gradient steepness also applies to any model in which distance is in the denominator of the dy/ds equation. In short, there is always some ambiguity of the meaning of the equality (or inequality) of two gradients in which dy/ds (or dY/ds) involves both y and s. For the simpler situation of the exponential or logistic models of spread, dy/ds only involves y and a parameter. Then, if two gradients have the same b_L, steepness (as defined here

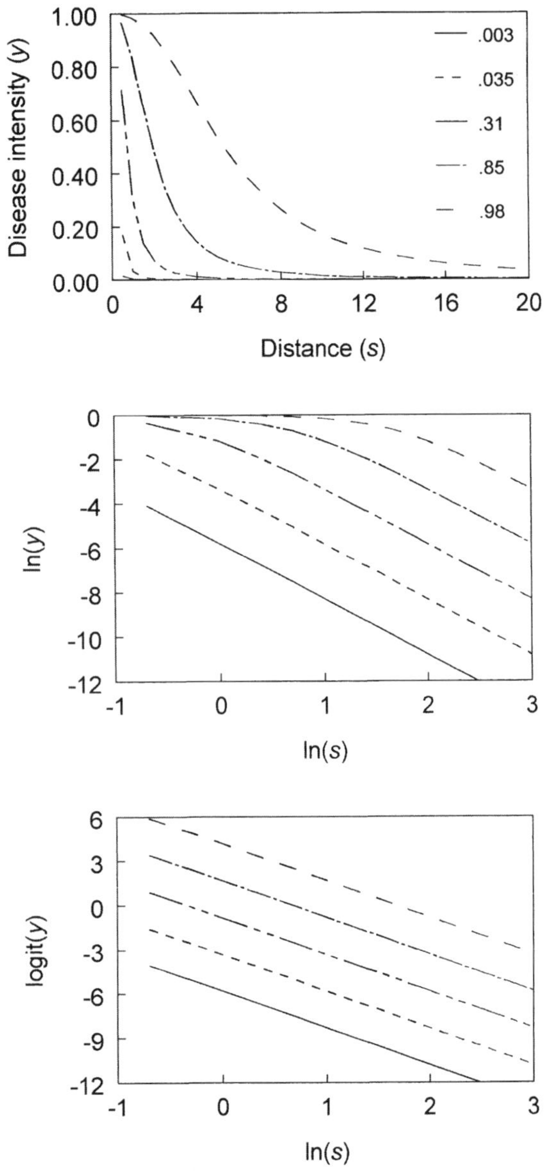

FIG. 7.10. Disease gradients according to the power-logistic model (equation 7.22), for five different levels of disease intensity at $s = 1$ (y_1; labeled in upper frame), all with the same spread parameter, b_{PL}. Gradients shown on a linear and two transformed scales.

by *dy/ds*) is the same at a given value of *y*. However, for equation 7.21, if two gradients have the same b_{PL}, steepness is not necessarily the same at a given *y*, because steepness depends also on distance from the source. Comparing estimates of b_{PL} is, nevertheless, common and useful, but one should be cautious in not over-interpreting the meaning of the parameter differences.

7.5.3 Other models

In addition to the models in the above sub-section, various other generalizations and expansions of the dispersal and disease gradient models can be utilized. Just as the logistic model can be modified to account for asymmetry in *dy/dt* over time (see section 4.4.4), one can use functions of $1 - y$ [e.g., $\ln(1) - \ln(y)$ for the Gompertz model (see equation 4.19)] rather than $1 - y$ directly in the equations for *dy/ds*. This approach is occasionally used in practice (e.g., Maffia and Berger, 1998). One could also expand the other gradient models in section 7.3.5 to adjust for maximum disease. We do not pursue these approaches here, although they can be valuable when one has the goal of a very accurate description of observed gradients.

One can also consider simplifications of the models presented for disease gradients, at least for comparison purposes. For instance, suppose that *dy/ds* is not affected by *y*, but there is still a maximum possible disease intensity equal to 1. If distance does not affect the gradients directly, then one obtains:

$$\frac{dy}{ds} = -b_M(1-y) \tag{7.24}$$

in which b_M is now the spread parameter (units of 1/distance). This is analogous to the monomolecular model (equation 4.8) for temporal disease progress, and we use the M subscript for this reason. If distance does affect the gradient, then one obtains:

$$\frac{dy}{ds} = \frac{-b_{PM}(1-y)}{s} \tag{7.25}$$

in which the spread parameter (b_{PM}) is unitless. These and other models were developed and considered by Jeger (1983). They have occasionally been evaluated (e.g., Jeger et al., 1983; Madden et al., 1990a). Solutions are easy to obtain, as either nonlinear of linearized equations.

7.5.4 Model fitting

7.5.4.1 General comments. As with the simpler exponential and power models, one can use various model fitting methods utilizing either the nonlinear or linear versions of the disease gradient models discussed here.

One issue that needs some attention, however, concerns the variance of the response (dependent) variable in the models. For a discrete count variable with no upper limit, *var*(*Y*) can be expected to increase with increasing *Y*. We pointed out that the transformation used to linearize the nonlinear models for *Y*, ln(*Y*), also does a good job of stabilizing variances. With disease intensity on a proportion (0–1) scale, *var*(*y*) is a more complex function of *y*. As discussed in detail later in Chapter 9, disease incidence most likely has a binomial or more general discrete distribution. If so, *var*(*y*) is proportional to $y(1 - y)$. In fitting the nonlinear models (with *y* as the dependent variable) to data, one could then use weights equal to $1/[y(1 - y)]$.

When fitting the linearized versions of these models to data, the variance of the transformed *y*, y^*, is the relevant quantity. In section 4.6.2, we showed how to estimate the variance of y^*, $var(y^*)$, when $var(y)$ was known (or assumed). For the logit transformation, $y^* = \ln[y/(1-y)]$, so $var(y^*) = [y(1-y)]^{-1}$. Thus, to fit either equation 7.20 or 7.23 to data (with an assumed additive error term, ε), one could use $y(1-y)$ as weights.

When dealing with disease severity data, it is not clear which statistical distribution is most appropriate. However, one can generally assume that $var(y)$ is still proportional to $y(1-y)$, and use the same weights as discussed for disease incidence.

For either disease incidence or severity, there may or may not be statistical evidence from the model fit that weights are required if there is one observation per distance. A residual plot can certainly provide strong evidence that variances are unequal, and hence the need for weights. However, with a limited number of data points, the residual plot may be ambiguous. Then the decision regarding weights is more subjective. One could argue that weights are always necessary for a variable that ranges between 0 and 1, especially if there are observed values close to either limit. On the other hand, if the empirical evidence from model-fit diagnostics (e.g., residual plot) does not give clear evidence of unequal variances, then there is not direct evidence to use any weights. In the latter situation, it could be argued that weighting is counter-productive, because one is obtaining a poorer fit (in terms of residual sum of squares) than without weights. In this chapter we use weights (or appropriate variance-stabilizing transformation in previous sections), but we do not take an inflexible view on this issue, and accept the merit of both arguments in cases where there is no clear evidence of unequal variances for the data being analyzed.

7.5.4.2 Example—graphical evaluation. We consider another bean rust example here, based on one of the data sets described in Maffia and Berger (1998). A line source of infected plants was used, and data originated in the spring-planting. Disease intensity was measured in various ways, but we only discuss results for leaf disease incidence, that is, the proportion of leaves with pustules. Results were obtained for six times, $t = 1, 8, 15, 22, 29$, and 36 days, in which $t = 1$ corresponded to the first day with observed pustules.

This example is typical of many in which gradients are observed at several times in the epidemic. For the first and second assessment times, there was a fairly steep decline in y with increasing s near the source (Fig. 7.11). At distances of 5 or more meters, incidence was very low. By later times, there was a considerable flattening of the gradients on the original scale (y), and y was close to 1 near the original inoculum source at the later times. Moreover, at these later times, y was still very high at the largest distance considered (8.5 m). Ideally, one would want to use large enough distances so that the "full" gradient could be observed at each time, but this is totally impractical in most field studies. In this case, one would need to use at least a four-fold increase in field size to observe low values of y at large distances from the source.

When one ignores the fact that there is an upper limit to y, and plots $\ln(y)$ versus s or $\ln(s)$, one sees a flattening of the disease gradients with time, especially at $s < 5$ m. Also note that incidence is so low for the first time that it is difficult to compare these results for large distances with results for the other times. We, thus, (mostly) focus on $t \geq 8$.

Considering that the upper limit of y (=1) is approached here, it is appropriate to utilize the logistic and power-logistic gradient models (equations 7.20 and 7.23) to represent these data. Plots of logit(y) versus s and $\ln(s)$ are shown on the right-hand side of Fig. 7.11. In general, a clear choice between the two models is not possible. There is slight curvature when logit(y) is plotted versus s, with decreasing slope at increasing distance. This is indicative of a power-type model. The plot of logit(y) versus $\ln(s)$ does produce a straight line, except for large distances (high s), where a slight downward bending of lines could be seen. Residual plots from linear regression analysis show some pattern with both model choices (not shown). Because of the straight lines obtained for logit(y) versus $\ln(s)$ at $s < 6$ m [$\ln(s) < 1.6$], we select the power-logistic as the primary model to quantify these gradients. This is consistent with other studies that found a power-law type relation for dispersal gradients of rust spores (Aylor and Ferrandino, 1989; Mundt, 1989). However, data are insufficient in this example to interpret the gradients in terms of dispersal mechanisms. This would require other data sets and larger fields (larger maximum s) in the studies. In particular, for the latter issue, it would be informative to know if that slight curvature of the logit line at $\ln(s) > 1.6$ (Fig. 7.11) would continue at much larger distances.

7.5.4.3 Example—linear regression. For the interested reader, the actual data points are given in Fig. 7.12 in the form of a SAS program for regression analyses based on equations 7.20 and 7.23. Before showing results for the logistic power model, we present some results that are obtained if one (incorrectly) ignored the fact that y is limited from above by 1. That is, we fitted the regular power model (equation 7.6, with y, rather than Y, used for disease intensity) to the data and compared slopes (estimates of b_L) among times. This analysis is not specified in the SAS program. Estimated slopes were 0.69, 0.41, 0.20, 0.12, 0.10, and 0.07 for times 1 through 36, based on weighted linear regression of $\ln(y)$ on $\ln(s)$, with weight equal to $y/(1-y)$ [see section 4.6.2 for variance of $\ln(y)$]. Results indicate a strong tendency for decreasing slope over time. In fact, the slope for day 8 was about six times larger than the slope for day 36 (0.41/0.07).

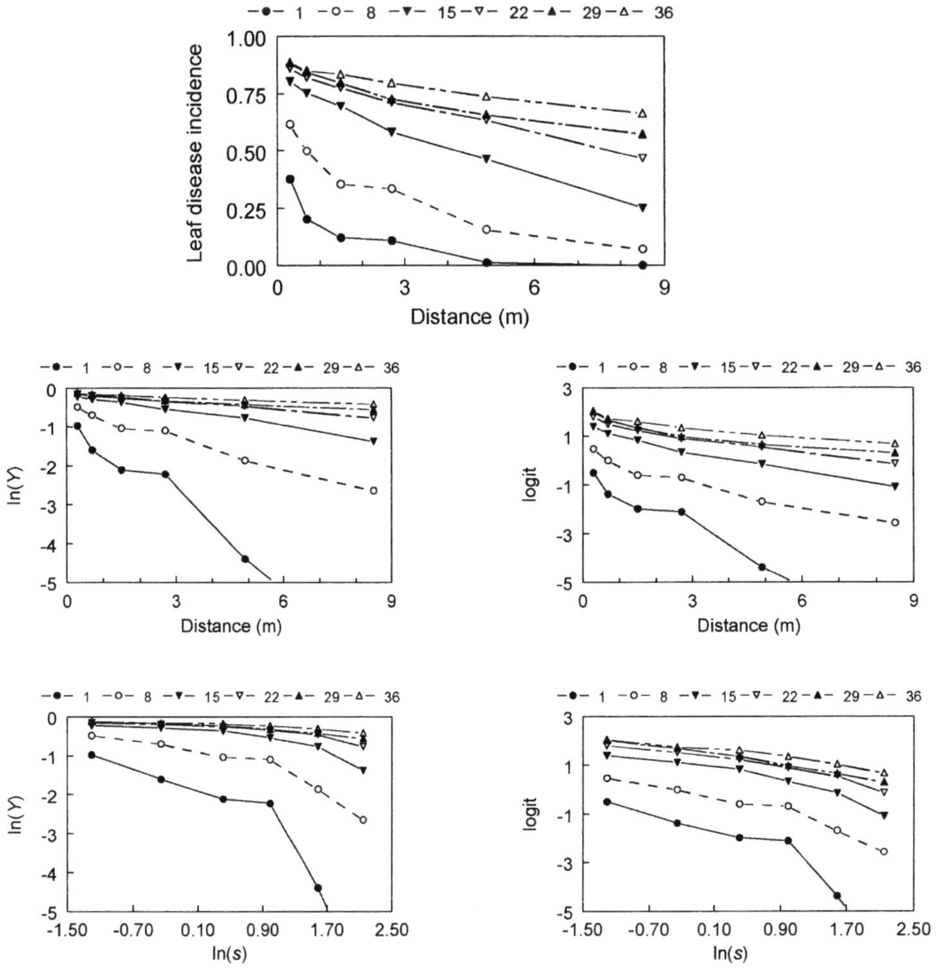

FIG. 7.11. Disease gradients at six assessment times for bean rust (Maffia and Berger, 1998). Observations of leaf disease incidence (proportion of leaves with pustules) in relation to distance from a line source were taken for the Fall planting of the study. Data shown on a linear and various transformed scales.

For later purposes, it is helpful to indicate explicitly that the statistical form of power-logistic model (equation 7.23) can be written as:

$$\ln\left(\frac{y}{1-y}\right)_j = -\ln(A_{\mathrm{PL}}) - b_{\mathrm{PL}}\ln(s)_j + \varepsilon_j \qquad (7.26a)$$

or

$$y_j^* = A_{\mathrm{PL}}^* - b_{\mathrm{PL}}\ln(s)_j + \varepsilon_j \qquad (7.26b)$$

in which the j subscript refers to the j-th distance (e.g., the first distance is $s_1 = 0.3$ m), y^* is the symbol for the transformation that produces a linear model (logit here), and $A_{\mathrm{PL}}^* = -\ln(A_{\mathrm{PL}}) = \ln[y_1/(1-y_1)]$ for the power-logistic model. Fitting equation 7.26b to the data for each time, using weighted least squares [with $y(1-y)$ as weight], produces the results in Table 7.1.

The error degrees of freedom (df) equals 4 for each time. As shown in the top frame of Fig. 7.13, the fitted lines are very reasonable summaries of the gradients on a logit:ln(s) scale. The $\hat{A}_{\mathrm{PL}}^*$ values increased with time, as expected, since y increased over time. Back-transformation of estimated A_{PL}^* can be used to estimate y at 1 m. For instance, at $t = 29$ days, $\hat{y}_1 = 1/\left(1+\exp(-A_{\mathrm{PL}}^*)\right) = 0.81$. The estimates of b_{PL} do become smaller over time, but the trend is much weaker than when the simpler power model (with no correction for maximum disease) was used. Although the R^2 values were high, there was some pattern to the residual plot, indicating that the power-logistic model was not ideal. For comparison, some results from fitting the statistical version of the logistic gradient model (equation 7.20) are also shown in Table 7.1. The estimates of b_{L} declined faster with time than the estimates of b_{PL}. This and other models resulted in undesirable residual plots, although the R^2 was very high for some assessment times with the logistic model. As discussed above, we concluded that the power-logistic model was a reasonable (but not ideal) choice for analysis.

7.5.4.4 Example—comparing parameter estimates. When one wants to compare two gradients, one can test the equality of the spread parameters or the intercepts (transformation of the constant of integration) using

```
data rust;
            input t s y @@; * several grous of t-s-y per record;
            logit=log(y/(1-y)); lns=log(s);
            wt = y*(1-y);  * weight for linear regression of logits;
            datalines;
 1 0.3 0.377   8 0.3 0.616  15 0.3 0.802  22 0.3 0.860  29 0.3 0.883  36 0.3 0.886
 1 0.7 0.202   8 0.7 0.500  15 0.7 0.751  22 0.7 0.820  29 0.7 0.845  36 0.7 0.851
 1 1.5 0.122   8 1.5 0.355  15 1.5 0.695  22 1.5 0.775  29 1.5 0.798  36 1.5 0.835
 1 2.7 0.109   8 2.7 0.334  15 2.7 0.582  22 2.7 0.711  29 2.7 0.726  36 2.7 0.796
 1 4.9 0.012   8 4.9 0.156  15 4.9 0.463  22 4.9 0.635  29 4.9 0.658  36 4.9 0.738
 1 8.5 0.0009  8 8.5 0.071  15 8.5 0.251  22 8.5 0.467  29 8.5 0.574  36 8.5 0.664

 ;
 proc sort data=rust out=rust; by t;
 proc reg data=rust;
            title 'Bean rust, leaf disease incidence; logit; WEIGHT; separate times';
            by t;  /*separate weighted least squares regression analysis for each time*/
            weight wt;
            model logit = s; plot r.*s;  *plot of residual (r.) versus distance (s);
            model logit = lns; plot r.*lns;
proc mixed data=rust;
            title 'Bean rust, leaf disease incidence; logit; WEIGHT; COVARIANCE';
            weight wt;
            class t; /*covariance analysis, with separate intercepts & slopes*/
            model logit = lns t lns*t / solution ddfm=satterth;
proc reg data=rust;
            title 'Bean rust, leaf disease incidence; logit; WEIGHT; SPATIO-TEMPORAL';
            weight wt;
            model logit = lns t; plot r.*p.; *residuals vs. predicted;
run;
```

FIG. 7.12. Input program of SAS to read in data shown in Fig. 7.11 and fit power-logistic (equation 7.23) models to the data using linear (weighted) least squares. Program also performs a covariance analysis to determine the effect of time on gradients, and fits the power-logistic spatio-temporal model (equation 7.36) to the data.

TABLE 7.1. Fit of the logistic (equation 7.20) and power-logistic (equation 7.23) models to the bean rust data shown in Fig. 7.12, together with the estimated distance ($\hat{s}'$) where disease intensity equals 0.6, for the different disease assessment times.

	Power logistic				*Logistic*			
t	$\hat{A}^*_{PL}$	$\hat{b}_{PL}$	R^2	$\hat{s}'$	$\hat{A}^*_L$	$\hat{b}_L$	R^2	$\hat{s}'$
1	−1.6 (0.15)	0.88 (0.17)	0.88	–	−0.5 (0.23)	0.72 (0.15)	0.85	–
8	−0.3 (0.12)	0.77 (0.12)	0.91	0.4	0.3 (0.15)	0.37 (0.05)	0.92	–
15	0.8 (0.15)	0.71 (0.12)	0.90	1.7	1.3 (0.09)	0.29 (0.02)	0.98	3.1
22	1.3 (0.11)	0.58 (0.07)	0.93	4.7	1.6 (0.09)	0.21 (0.02)	0.97	5.7
29	1.5 (0.04)	0.53 (0.03)	0.99	7.9	1.7 (0.16)	0.18 (0.03)	0.89	7.2
36	1.7 (0.06)	0.41 (0.12)	0.96	23.5	1.9 (0.09)	0.15 (0.02)	0.94	10.0

various methods. The gradients may be from disease assessments made at different times in a single epidemic (as here), or correspond to contemporaneous assessments of disease for different treatments (e.g., cultivars, fungicide applications). The approaches discussed in Chapter 4 are applicable here, and we directly address only a limited number of approaches.

The simplest approach is to use a t-test to compare the parameter estimates. For instance, if one had two gradients, each characterized by the power-logistic model (equation 7.26a), one can test the null hypothesis:

$$H_0: b_{PL1} = b_{PL2}, \text{ or } H_0: b_{PL1} - b_{PL2} = 0$$

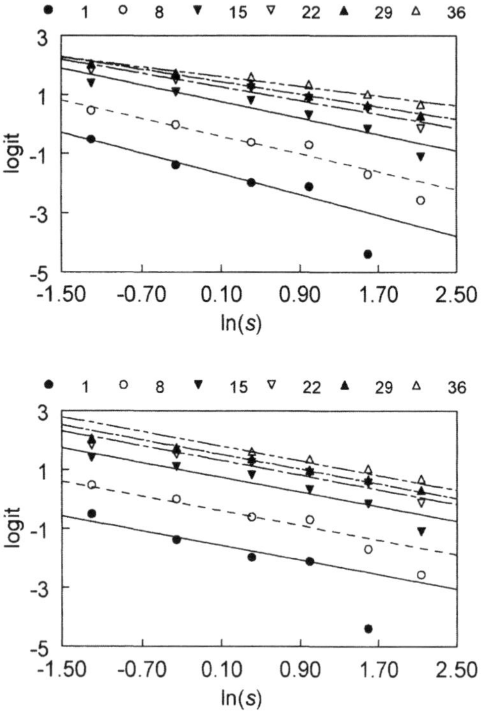

FIG. 7.13. Fit of the power-logistic model (equation 7.23) to the bean rust gradient data (see Fig. 7.11) for each of six assessment times. Upper frame: a separate intercept and slope for each time. Lower frame: fixed slope for all times.

versus the alternative hypothesis that the parameters are not equal. The t-statistic can be written as:

$$t = \frac{\hat{b}_{\mathrm{PL}1} - \hat{b}_{\mathrm{PL}2} - 0}{s(\hat{b}_{\mathrm{PL}1} - \hat{b}_{\mathrm{PL}2})}$$

in which the denominator is the standard error of the difference of the two estimated slopes. This can be estimated (for uncorrelated slopes) as:

$$s(\hat{b}_{\mathrm{PL}1} - \hat{b}_{\mathrm{PL}2}) = \left(s^2(\hat{b}_{\mathrm{PL}1}) + (s^2(\hat{b}_{\mathrm{PL}2})\right)^{1/2}$$

in which the two terms on the right hand side are the squares of the standard errors of the estimated slopes. For the comparisons of estimated slopes for days 8 and 15 of the bean rust example, the t-statistic is given by:

$$t = \frac{0.77 - 0.71}{(0.12^2 + 0.12^2)^{1/2}} = \frac{0.06}{0.17} = 0.35.$$

The degrees of freedom for the t-test is equal to the sum of the error degrees of freedom for the two disease progress curves ($df = 4 + 4 = 8$). This achieved t-statistic is much below the critical value for 8 df (2.31) to be significant at $\alpha = 0.05$. Thus, the slopes are not significantly different (at $\alpha = 0.05$) for these two gradients.

As an aside, readers should note that s is used for both the distance variable and for an estimated standard error or standard deviation, and that t is used for the time variable as well as the t-statistic. Other than using unusual symbols for time and space variables (or for standard errors and t-statistics), this double interpretation cannot be avoided. The meaning of the symbols should be apparent based on context of the material.

The two intercepts can be compared in the same way as the comparisons of the slopes. Then, other pairs of gradients can be compared. The reader should be able to determine that the intercepts clearly are different between many times (as expected), and that some slopes are also different from each other. This approach is very useful when there are only a small number of gradients, or the investigator only has information on parameter estimates and standard errors (say, from a published paper). With multiple gradients, it may be difficult to gain a general understanding of disease spread based on pairwise comparisons. Thus, it is generally more efficient to use a covariance analysis to test whether the heights of the lines (intercepts) or their slopes vary among groups (Milliken and Johnson, 2002). This approach also allows the investigator to incorporate features of the experimental design, such as blocking, into the analysis.

Assuming that the power-logistic model is appropriate, we can expand equation 7.26b to:

$$y_{ij}^* = A_{\mathrm{PL}i}^* - b_{\mathrm{PL}i}\ln(s)_{ij} + \varepsilon_{ij} \qquad (7.27a)$$

in which the i subscript represents the i-th group (time in the example). There is an i subscript on $\ln(s)$ (in addition to the j subscript) because it is possible that the distances are not the same for each group (although this is not the case here). Thus, the potential for each group to have a different intercept (A^*) and slope (b_{PL}) is incorporated into this statistical model. An alternative way of writing this equation is:

$$y_{ij}^* = \gamma + A_{\mathrm{PL}i}^* - \iota\ln(s)_{ij} - b_{\mathrm{PL}i}\ln(s)_{ij} + \varepsilon_{ij} \qquad (7.27b)$$

in which γ and ι represent overall intercepts and slopes, respectively. In this formulation, $\gamma + A_i^*$ is the intercept for the i-th group, and $\iota + b_{\mathrm{PL}i}$ is the slope for the i-th group. Readers should compare this with equation 4.38b, and reread the discussion about use of this type of equation. The formulation in equation 7.27b is most useful in a covariance analysis for testing whether or not there is a significant effect of group on the intercept and slope. This model is fitted to the bean rust data, with time as the group, using the MIXED procedure of SAS in Fig. 7.12. Note that time (t) is given as a class (category or factor) variable in the program. An overall intercept (γ) is implied by the model statement and is not identified; t specifies the effect of time on the intercept [a "main" effect term ($A_{\mathrm{PL}i}^*$)]. The overall slope and the group effect on the slope are specified with lns (ι) and lns*t ($b_{\mathrm{PL}i}$), respectively. That is, in the latter case, the meaning of a group-dependent slope is that the slopes are not the same (interaction of group and the continuous variable).

Fitting this model to the data showed the following: a highly significant effect of distance ($F = 246.6$; $df = 1, 24$; $P < 0.001$), meaning that ι was different from 0; a highly significant effect of time ($F = 103.0$; $df = 5, 24$; $P < 0.001$), indicating that (A^*_{PLi}) varied with time; and a marginal effect of the interaction of time and distance ($F = 2.6$; $df = 5, 24$; $P = 0.05$), meaning some evidence that b_{PLi} was not the same for all times. In this model formulation, the actual estimates of the intercepts and slopes for each time can be obtained by the estimates of γ, ι, and the group-specific parameter estimates (using the solution option). For instance, the slope for group three (third time) would be $\hat{\iota} + \hat{b}_{PL3}$ and the intercept would be $\hat{\gamma} + \hat{A}^*_{PL3}$. A more direct way of obtaining these estimates would be to use the following model statement:

```
model logit = t t *ln s/noint solution;
```

where noint option eliminates γ from the model ("no intercept"). Because there is also no individual lns term specified (thus, no ι), one is essentially fitting equation 7.27a to the data. The F tests displayed for group and the interaction of group and the continuous variable, however, are no longer applicable for testing group and interaction effects when these changes to the computer code are made (even though the estimates of the slopes and intercepts, and their standard errors are appropriate). Using this approach, as an example, one obtains the following parameter estimates for 15 days:

$$t = 15: \quad \hat{A}^*_{PL3} = 0.85(0.11), \quad \hat{b}_{PL3} = 0.71(0.08)$$

which agrees with the individual regression results shown above. The agreement does not have to be this good, since the covariance analysis, as done here, is assuming a single error variance for all the data, not a separate one for each gradient (each assessment time, in this case).

Since there is only marginal effects of group on the slope (marginal interaction), it is natural to consider fitting a covariance model with a single (group independent) slope. Equation 7.27a can be simplified to:

$$y^*_{ij} = \gamma + A^*_{PLi} - \iota \ln(s)_{ij} + \varepsilon_{ij} \qquad (7.27c)$$

Note this formulation indicates that $b_{PLi} = 0$ for all i (all groups). In other words, ι now represents a single slope (i.e., $\iota \equiv b_{PL}$, with no i subscript). This model can be fitted using the following code:

```
model logit = t ln s/solution;
```

With this analysis, the (single) estimated slope was: $\hat{b}_{PL} = 0.62$ (s.e.= 0.043), intermediate among the individual slopes for the different times. Estimates of the different intercepts can be obtained from $\hat{\gamma} + \hat{A}^*_{PLi}$. By giving the noint option as well, one directly obtains $\hat{A}^*_{PLi}$ (because $\gamma = 0$). The estimated intercepts for the six times were: −1.5, −0.3, 0.8, 1.4, 1.6, and 1.8, similar to those obtained when analyzing data for each time separately. The residual variance for this model was 0.0125, which is not a lot larger than the residual variance found for the fit of equation 7.27b (0.0098), indicating that a reasonable fit of this multi-time data set is obtainable with a single spread parameter. As shown in the bottom frame of Fig. 7.13, a single slope is reasonable for describing these data, although the fit is slightly better with time-dependent slopes (top frame).

To conclude this section, analysis of the bean rust gradients shows that logistic and power logistic models could be used, but latter was chosen based on common use of a power-type model for rust diseases. On the original y scale, there was clear flattening of the gradients with increasing time, especially near the inoculum source. There also was some evidence that the gradients flattened over time on a logit(y) : ln(s) scale, but a constant gradient steepness (b_{PL}) over time was a reasonable approximation for the data. Through the use of the logit transformation (in this case), the limitation to disease intensity (i.e., $y < 1$) was accounted for, so that the relationship between (transformed) disease intensity and (transformed) distance could be characterized.

7.6 Spatio-Temporal Dynamics of Disease Spread

7.6.1 General comments

By characterizing disease gradients at several times in an epidemic in the previous section, we have been casually considering the spatio-temporal dynamics of disease development in host populations. That is, we considered how y changed with s, in the form of disease gradients, and how y changed with t, through the change in gradient height and (potentially) gradient steepness with increasing time. Here we start to formalize the relationship between y and both t and s.

As mentioned in section 7.1, in previous chapters we mostly considered a plant disease epidemic as a temporal process, characterized by dy/dt, even though we recognized that space was important. With monocyclic diseases, for instance, we indicated that $dy/dt = r_M(1 - y)$ (equation 4.8), where r_M equals the product of inoculum density or abundance (x'), and the probability per unit time that a unit of inoculum causes an infection (ϕ = the probability per unit time that a unit of inoculum contacts a host individual multiplied by the probability that the unit of inoculum in contact with a disease-free host actually infects the host). If x' is clustered at a single location (line, point, etc.), then ϕ is unlikely to be constant, but rather would be function of distance from x', $\phi(s)$. This function, $\phi(s)$, is actually analogous to a contact distribution (see section 7.3.4). For polycyclic diseases, the situation is (considerably) more complicated because any infected individual can then serve as an inoculum source, at least during the diseased individual's infectious period (see Chapter 5). Instead of a constant probability of any produced spore coming into contact with a host individual (θ; see sections 5.2.2 and 5.2.3), a spore produced by an infected individual at location ξ has a much greater chance of contacting a host individual at location s if the locations are close together ($s - \xi$ is small) than if the locations are far apart ($s - \xi$ is large). In other words, θ is actually a function of $s - \xi$, $\theta(s - \xi)$. Thus, the logistic model (equation 4.14) and the more complex models of Chapter 5 for y (e.g., equation 5.30) are inadequate for representing epidemics in space and time when the epidemics start from a well-defined inoculum source (or sources). Although the temporal models of the previous chapters were quite useful for describing and understanding epidemics, the next level of quantification requires a formal consideration of space and time in disease development. The approach presented in this section is the first step towards developing a better understanding of the spatio-temporal processes within an epidemic. Through the models developed, the concept of disease spread hopefully becomes clearer, and empirical approaches for characterizing spread are obtained that are natural consequences of the developed models. However, the specific models shown here should be considered a stepping stone towards the fuller characterization of spread that is presented in Chapter 8.

7.6.2 Two spatio-temporal models

Previously, we sometimes found it useful to use the notation $y(t)$ instead of just y to represent disease intensity at time t. With the simultaneous consideration of space and time, we can use the notation $y(t, s)$ to refer to disease intensity at location s and time t. Representation in two spatial dimensions can be done by replacing s with s_1 and s_2 (for instance) to represent row and column locations, $y(t, s_1, s_2)$. The latter approach is not considered here. A general formulation to represent $y(t, s)$ can be given as:

$$y(t, s) = f(t, s; \Theta). \qquad (7.28a)$$

where Θ is a vector of parameters. It is quite possible that $f(\bullet)$ is not a simple (or even complicated) single equation, but can only be written as an integral

$$f(t, s; \Theta) = \iint \Xi \, d\hat{s} d\hat{t} \qquad (7.28b)$$

in which Ξ is an arbitrary general mathematical expression, $\hat{s}$ and $\hat{t}$ are specific locations (between 0 and s) and times (between 0 and t), respectively, and the integrations are over space and time. In this situation, numerical solutions would be required, although approximate analytical solutions may exist for certain circumstances (see Chapter 8). Here we consider situations where a solution to the integration exists, so that $f(\bullet)$ is a single equation.

Disease intensity is now a function of two continuous independent variables, s and t. Mathematically, partial derivatives ($\partial\bullet$) are used instead of ordinary derivatives to represent rates of change in models with more than one independent variable. One can take the partial derivatives of equation 7.28a to determine rates of change of y with respect to time and space. In particular, we write $\partial y(t, s)/\partial t$ for the absolute rate of change in y with time at location s, and $\partial y(t, s)/\partial s$ for the absolute rate of change in y with space (distance) at time t. For some presentation circumstances, these can be simplified to $\partial y/\partial t$ and $\partial y/\partial s$, where the dependence on s and t, respectively, is assumed. For this approach to work, there is a mathematical requirement that $f(\bullet)$ is a continuous function.

For polycyclic diseases, a logistic model is often an appropriate description of disease progress over time (Chapters 4 and 5). Thus, it is natural to consider a form of equation 7.28 in which the partial derivative with respect to time is given by:

$$\frac{\partial y}{\partial t} = r_L y(1-y) \qquad (7.29)$$

Other functions could be considered here, such as the Gompertz, for temporal disease progress. Likewise, a H-L-I-R-type model (Chapter 5), with the first y term on the right hand side of equation 7.29 replaced by infectious disease, could be used for temporal progress. For monocyclic diseases, it is natural to consider a form of equation 7.28 in which the partial derivative for disease with time is given by:

$$\frac{\partial y}{\partial t} = r_M(1-y) \qquad (7.30)$$

We have shown in previous sections that the rate of change in disease with distance can be reasonably represented by the logistic or power-logistic models. Thus, we consider forms of equation 7.28 in which partial derivatives with respect to s can be written as:

$$\frac{\partial y}{\partial s} = b_L y(1-y) \qquad (7.31)$$

and

$$\frac{\partial y}{\partial s} = \frac{b_{PL} y(1-y)}{s}. \qquad (7.32)$$

Of course, other models could also be considered. This is a good place to point out that a logistic (or power-logistic) gradient model does not necessarily mean that the disease is polycyclic (compound interest) in nature.

The next step is to consider *pairs* of partial differential equations, one for space and one for time. For instance, consider logistic equations for temporal (equation 7.29) and spatial (equation 7.31) dynamics of disease. Solution leads to the following equation:

$$y = \frac{1}{1 + A_{LL}\exp(b_L s - r_L t)} \qquad (7.33)$$

in which A_{LL} is the constant of integration, equal to $(1-y_0)/y_0$, where y_0 is the theoretical y at $t = 0$ and $s = 0$. That is, $f(\bullet)$ of equation 7.28a is given by the right hand side of equation 7.33. Another way of looking at this is that the partial derivatives of equation 7.33 equal equations 7.29 and 7.31. A little algebra leads to the following linear equation:

$$\ln\left(\frac{y}{1-y}\right) = -\ln(A_{LL}) - b_L s + r_L t \qquad (7.34)$$

The intercept term can also be written as $+\ln[y_0/(1 - y_0)]$. Thus, equations 7.33 or 7.34 can be used to describe disease in space and time.

Fig. 7.14 shows one example epidemic that follows equation 7.33, with $r_L = 0.25$/day and $b_L = 0.4$/m. The 3D graph in the top frame visually shows disease development in the two dimensions. Note, for visual clarity, the time axis runs into the page (at a slight angle), with the smallest times in the foreground. At any selected distance (e.g., $s = 4$ m), there is a logistic increase in disease over time, with low y at $t = 0$, and y near 1.0 at large time (when s is small). The temporal curves are sigmoid in shape, although high disease intensity is not reached at large distances (within the time frame considered here). At $s = 0$, one can see the logistic increase in y at the inoculum source.

Reading the 3D graph from left to right, one can see the disease gradients over distance. These are reverse-sigmoid in shape as well, because equation 7.31 was specified for the gradient. The middle frame of Fig. 7.14 shows disease gradients for selected (labeled) times, although the curve for $t = 0$ days barely rises above the 0 axis. At early times, the gradients are exponential in shape, as expected, but they flatten close to the source because of the limitations to disease. The bottom frame of the figure shows that logit-transformed disease intensity declines linearly with distance (see equation 7.34). Moreover, because the times selected were equally spaced (every 10 days), the vertical difference of the logits is constant over all distances ($10 \cdot r_L$). It should be noted that the theoretical values plotted in Fig. 7.14 are the same as those in Fig. 7.9. The values of y_0 chosen in the earlier figure were simply theoretical disease intensities at $s = 0$ for 10-day increments with the model parameters given here.

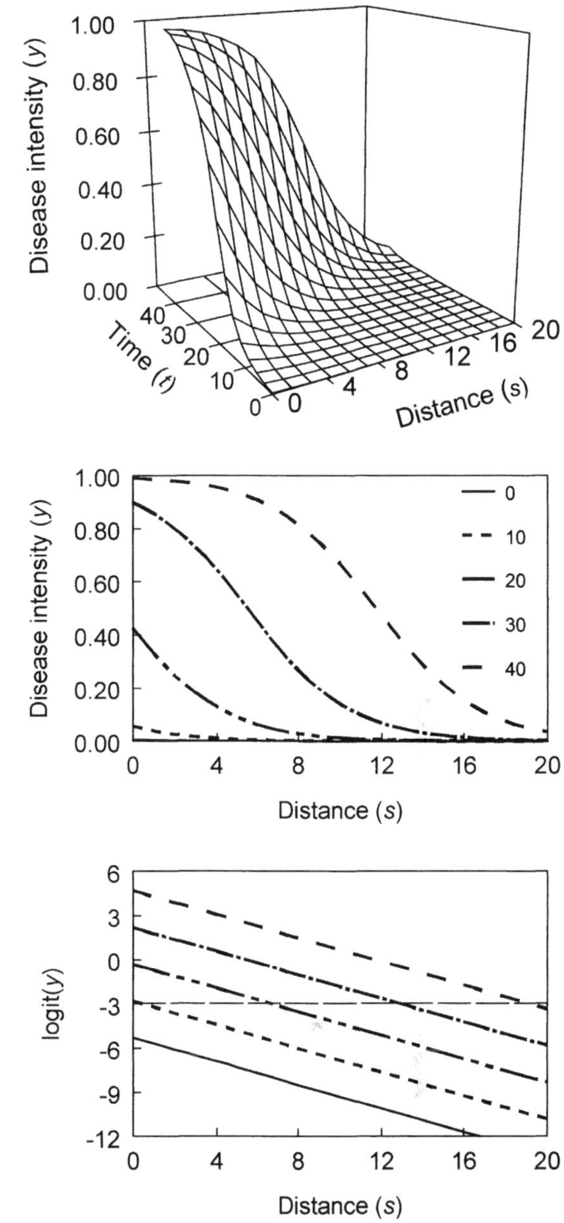

FIG. 7.14. Spatio-temporal development of disease intensity based on logistic equation 7.33. Disease gradients for six labeled times are shown on linear and transformed scales in the middle and lower frames. Horizontal line is isopath of $y' = 0.05$.

Another natural pairing of partial differential equations is the logistic one for temporal dynamics (equation 7.29) and the power-logistic one (equation 7.32) for spatial dynamics. This corresponds to the following equation for y:

$$y = \frac{1}{1 + A_{LPL} s^{b_{PL}} e^{-r_L t}} \qquad (7.35)$$

where A_{LPL} is the constant of integration, $(1 - y_1)/y_1$, where y_1 is y at $t = 0$ and $s = 1$. In other words, the right hand side of equation 7.35 equals $f(\bullet)$ of equation 7.28a. Equation 7.35 can also be written as:

$$y = \frac{1}{1 + A_{LPL}\exp(b_{PL}\ln(s) - r_L t)}$$

Algebra leads to the linear equation:

$$\ln\left(\frac{y}{1-y}\right) = -\ln(A_{LPL}) - b_L \ln(s) + r_L t \qquad (7.36)$$

which indicates straight-line relations between logits and both t and $\ln(s)$. For equations 7.34 and 7.36, we can write A^*_{LPL} for $-\ln(A_{LPL})$.

An example corresponding to equation 7.35, with $r_L = 0.25$/day and $b_{PL} = 2.5$, is shown in Fig. 7.15. There is a sigmoid-type increase in y at all distances beyond 0, because of the logistic nature of $\partial y/\partial t$, with y approaching 1 near the source (over the time span considered here). Compared with the example in Fig. 7.14, there was a sharper decline in y as s increases, because of the s term in the denominator of $\partial y/\partial s$ (see middle frame of Fig. 7.15). Moreover, logits decreased in a linear manner with increasing $\ln(s)$, with the same slope for all times, as required by equation 7.36 (bottom frame of Fig. 7.15) in this example. In particular, the vertical difference between the lines is constant at all distances because the difference in times for the lines was 10 days ($10 \cdot r_L$). As pointed out for the previous figure, the values in Fig. 7.15 are the same as those in the previous sections of the chapter and shown in Fig. 7.10. What looked like arbitrary choices for y_1 (or A) in Fig. 7.10 were actually theoretical values for y at $s = 1$ at the selected time values.

Many other models for $y(t, s)$ can be specified by considering pairs of partial differential equations for space and time dynamics. Before discussing the usefulness of some of these possibilities, we must first present an additional concept of great importance when characterizing spatio-temporal dynamics.

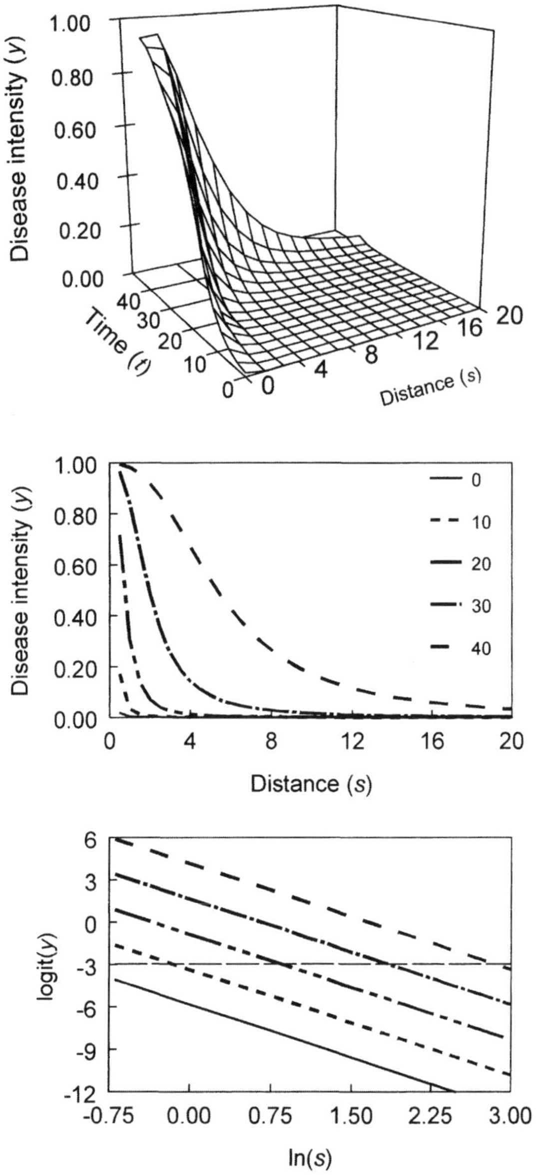

FIG. 7.15. Spatio-temporal development of disease intensity based on power-logistic equation 7.35. Disease gradients for six labeled times are shown on linear and transformed scales in the middle and lower frames. Horizontal line is isopath of $y' = 0.05$.

7.6.3 $\partial s/\partial t$

7.6.3.1 Isopaths. Suppose a pathogen is introduced into a previously disease-free (perennial) host population at a single location (point, line) and the disease becomes established (see Chapter 5). With invasive diseases, it is natural to ask: How fast is disease spreading beyond the point of introduction? For this thought exercise, suppose further that one has absolute knowledge of all diseased individuals at the disease assessment times (no missed individuals due to sampling errors and no misdiagnoses of individuals for disease status). Given that the time of introduction is defined as $t = 0$, suppose that the farthest distance (s) from the initial inoculum source with a diseased individual at various times (weeks) was found to be:

$$t = 2 \text{ wk}, \quad s = 14 \text{ m};$$
$$t = 4 \text{ wk}, \quad s = 28 \text{ m};$$
$$t = 6 \text{ wk}, \quad s = 42 \text{ m}.$$

The distances can be considered to represent the front of the epidemic at the listed times. As in Chapter 4, we use Δ here to indicate a calculated difference between successive values. The calculated rate of spatial spread of disease with time ($\Delta s/\Delta t$) can be determined by dividing the difference of distances by differences in times.

Between weeks 2 and 4, the disease spread 14 m; that is, disease was 14 m farther away from the source at week 4 compared with week 2. Thus, $\Delta s/\Delta t$ = 14/2 = 7 m/wk, or 1 m/day. With the hypothetical example, this rate is the same for all pairs of times. If this same rate continued indefinitely, one could calculate when disease would be, for example, 1 km from the original source. The answer is 1000/7 = 142.9 wk, or ~2.7 years. Likewise, one could ask: how far will the disease spread in the next 5 years? Assuming a continuous crop in all directions and no seasonal effects, and given that there are 260 wk in 5 years, the answer is 7×260 = 1,820 m or ~1.8 km.

This rate of spatial spread of disease over time tells us nothing about the intensity of disease at any distance from the original source, unless other assumptions are made or additional information is available (see below). Moreover, it is often impractical (because of costs and time) to assess disease at great enough distances from a source to find the farthest diseased individual. For instance, early in the bean rust example (Fig. 7.11), assessing leaves for pustules over 9 m from the source was satisfactory, but by 3 wk into the epidemic, about 50% of the leaves were infected at 9 m. Thus, it is quite possible that the front of the epidemic is unknown.

The practical approach to determining $\Delta s/\Delta t$ is to choose a (generally low) fixed value of y, y', and determine the distance where y' is found at selected times. It may be more practical to work with transformed disease intensity, and select a specific y^*, $y^{*\prime}$, for calculations. Consider, again, Fig. 7.14 for logistic-type disease gradients over time. In the bottom frame, a thin horizontal line is shown, corresponding to $y^{*\prime} = -2.944$, or $y' = 0.05$. A line (or curve) of fixed levels of disease intensity is called an *isopath*. The horizontal line is not shown in the middle frame because it is hard to see (being so close to 0). The points where the horizontal line intersects the gradient lines (or curves) are the distances where 5% disease intensity are found for the different times on the graph. At $t = 0$, there is no distance where 5% intensity is found. For the other shown times, the following distances are found (assuming that time is measured in days and distance in meters):

$$t = 10\,\text{days}, \quad s' = 0.38\,\text{m};$$
$$t = 20\,\text{days}, \quad s' = 6.63\,\text{m};$$
$$t = 30\,\text{days}, \quad s' = 12.88\,\text{m};$$
$$t = 40\,\text{days}, \quad s' = 19.13\,\text{m}.$$

Based on these specific results, one calculates $\Delta s/\Delta t$ as 0.625 m/day. For instance, between days 30 and 40, one obtains: (19.13−12.88)/10 = 0.625. The distances here are shown to two decimal places, although with observed gradients (e.g., Fig. 7.13), one will not be able have such precision (later, we show how we obtained such precise values for distance here). Plus, given that observations of disease are made at selected (fixed) values of distance and time, it is very possible that exactly 5% intensity is not found at any distance. This requires interpolation or use of model descriptions.

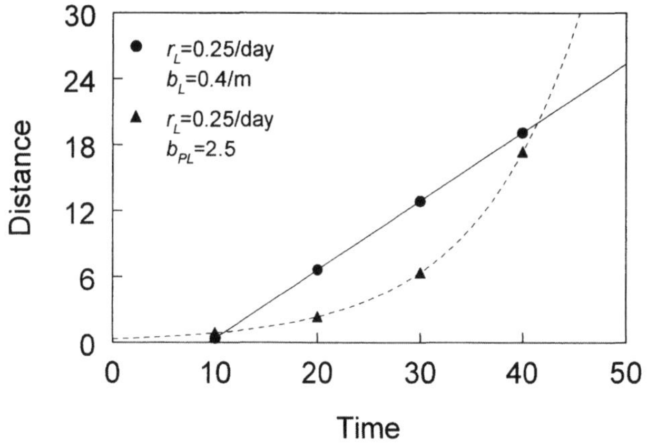

FIG. 7.16. Distance where $y' = 0.05$ is found in Figs. 7.14 (logistic spatio-temporal model) and 7.15 (power-logistic spatio-temporal model). Lines are solutions to equations 7.38 (solid) and 7.39 (dashed).

Before discussing model results, we reconsider Fig. 7.15 for the power-logistic type of gradients. The horizontal line is, as before, $y^{*\prime} = -2.944$ (for $y' = 0.05$). The values of log of distance, $\ln(s')$, can be read from the intersection points, and s' determined from $\exp(\ln(s'))$:

$$t = 10 \text{ days}, \quad \ln(s') = -0.145, \quad s' = 0.86 \text{ m};$$
$$t = 20 \text{ days}, \quad \ln(s') = 0.855, \quad s' = 2.35 \text{ m};$$
$$t = 30 \text{ days}, \quad \ln(s') = 1.850, \quad s' = 6.39 \text{ m};$$
$$t = 40 \text{ days}, \quad \ln(s') = 2.855, \quad s' = 17.38 \text{ m}.$$

Although the difference in $\ln(s')$ between times is constant here, the difference in s' increased with each successive time period. That is, the farther the distance from the source, or the longer the time since the start of the epidemic, the greater the $\Delta s/\Delta t$. For instance, between 30 and 40 days, $\Delta s/\Delta t$ equaled 1.1 m/day, but between 10 and 20 days, $\Delta s/\Delta t$ was only 0.15 m/day. The distance values for the two cases (Figs. 7.14 and 7.15) are shown in Fig. 7.16, together with lines that will become apparent shortly.

Since we are now dealing with changes in distance where a *specific* non-zero y (y') is found, it would be more exact to represent the rate of change as $\Delta s'/\Delta t$. This is because the result could depend on the specific disease intensity chosen. However, with a few exceptions, such additional notation is not needed.

7.6.3.2 *Two models.* Using the models for temporal ($\partial y/\partial t$) and spatial ($\partial y/\partial s$) disease development in section 7.6.2, it is straightforward to characterize the rate of spatial spread of disease over time. As discussed in section 7.6.2 above, partial derivatives are used with the deterministic models for rates of change of a variable with two "independent" variables (t and s). It can be shown (Jeger, 1983) that the rate of isopath movement, $\partial s/\partial t$, is given by:

$$\frac{\partial s}{\partial t} = -\frac{\partial y / \partial t}{\partial y / \partial s} \quad (7.37)$$

for a general spatio-temporal process such as that represented by equation 7.28a. It is useful to think of $\partial s/\partial t$ as a *velocity* since it has units of distance per time. Specifically, $\partial s/\partial t$ is the velocity (or speed) of isopath movement, how fast a given level of disease moves from one location to another, or the rate at which the disease focus expands. The $\Delta s/\Delta t$ value calculated above for specific times and distances is replaced by the partial derivative, $\partial s/\partial t$. Because we are assuming that $y(t, s)$ is represented by a continuous function, one can think of a $\Delta s/\Delta t$ calculation as an estimate of $\partial s/\partial t$. With the adoption of a particular model form for equation 7.28a—which is equivalent here to selection a pair of functions, $\partial y/\partial t$ and $\partial y/\partial s$—one can directly define spatial movement or spread using equation 7.37.

Using the logistic equation 7.29 for $\partial y/\partial t$ and logistic equation 7.31 for $\partial y/\partial s$ (with solution given by equation 7.33), one obtains the very simple expression for the velocity based on equation 7.37:

$$\frac{\partial s}{\partial t} = \frac{r_L}{b_L} \quad (7.38)$$

Since both the numerator and denominator are constants, equation 7.37 indicates that $\partial s/\partial t$ is a constant ($C = r_L/b_L$), with units of distance per time. This means that any given fixed level of disease (y') or its transformation ($y^{*\prime}$) moves a fixed distance over a fixed time. For the example in Fig. 7.14, the horizontal difference between successive lines or curves (which are for 10-day increments) is the same, and equal to C multiplied by the time difference: $(0.25/0.4) \times 10 = 6.25$ m. Although the example was worked through for $y' = 0.05$, other isopaths could have been chosen. The actual distances will be different (e.g., $y' = 0.5$ will be closer to the source than $y' = 0.01$), but the change in distance over time will be the same (for this model).

Integration of equation 7.38 leads to the straight-line expression $s_0 + (r_L / b_L)t$, where s_0 is a constant of integration, corresponding to the distance where y' is found at $t = 0$. For the example, y at $t = 0$ is less than 0.05 at all distances, even at the source. Thus, s_0 does not necessarily have a simple physical interpretation. The straight line (with a slope of r_L/b_L) is shown on Fig. 7.16 with the previously obtained specific distances. For a selected y' less than 0.05, the points and the line will be higher (because the low disease intensity is found farther from the source); for a selected y' greater than 0.05, the points and the line will be lower (because the high disease intensity will be found closer to the source).

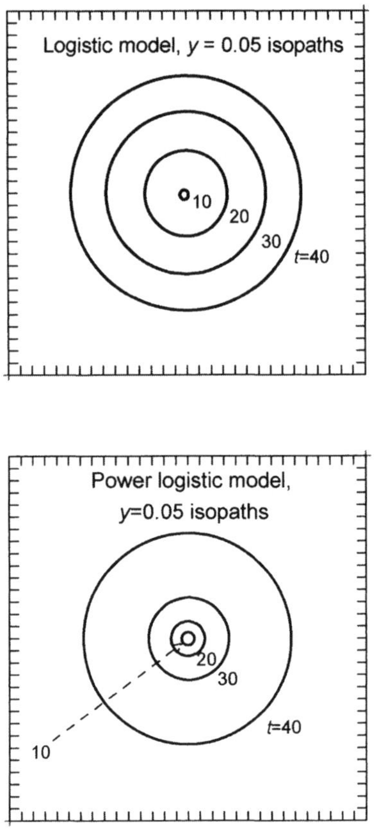

FIG. 7.17. Theoretical radial expansion of the $y' = 0.05$ isopath for a logistic (equation 7.33) and power-logistic (equation 7.35) spatio-temporal model at labeled time points.

When $\partial s/\partial t$ is constant, it is said that disease spreads in a *wave-like* manner (Diekmann, 1978; Thieme, 1977), and that epidemic develops as a traveling wave with velocity C. This can be seen by the top frame of Fig. 7.17, which corresponds to the same values in Fig. 7.16, but specifically for a point source (at the center of a field). The isopaths of $y' = 0.05$ increase from the center in a uniform manner at rate $C = 0.625$ m/day.

Now consider the situation where the power-logistic model describes the disease gradients (equation 7.32) and logistic model for temporal development (equation 7.29). In other words, we consider equation 7.35 as the solution for $y(t, s)$ in equation 7.28a. Utilization of equation 7.37 produces:

$$\frac{\partial s}{\partial t} = \frac{r_L s}{b_{PL}} = (r_L / b_{PL})s = \upsilon s \quad (7.39)$$

for the rate of isopath movement. Note that because b_{PL} is unitless, υ ($= r_L/b_{PL}$) has units of 1/time; thus, $\upsilon \cdot s$ is the velocity, and not just υ. Equation 7.39 is similar to the expression found in equation 7.38, except that the velocity is directly proportional to distance. That is, all other things being equal, $\partial s/\partial t$ (velocity) increases as s increases. This is exactly the situation found for the situation in Fig. 7.15. Solution to this equation is given by:

$$s_0\exp[(r_L/b_{PL})t],$$

where the constant of integration (s_0) is the theoretical distance at $t = 0$ where y' occurs. Taking the derivative of this equation with respect to time gives another way of expressing the disease expansion velocity:

$$C(t) = (r_L/b_{PL})s_0\exp[(r_L/b_{PL})t]$$

Note that we explicitly labeled the velocity as $C(t)$ to emphasize that the implication of equation 7.39 is that velocity increases with time since the start of the epidemic. Because of the dependence of the velocity on distance, the isopaths increase exponentially with *time*, as can be seen by the broken curve in Fig. 7.16. For an idealized situation of a single point source in Fig. 7.17, one can see that the differences of the circle radii increase with time.

Epidemics characterized by equation 7.39 are said to spread in a non-wave-like manner because of the acceleration of isopath movement (Minogue, 1989). Ferrandino (1993) refers to this as *dispersive wave* spread.

It is important to note for both equations 7.38 and 7.39, $\partial s/\partial t$ is directly proportional to r_L. This means that the velocity of spatial spread increases with the temporal increase in disease (where the disease is present). For instance, if infection efficiency is higher, latent period shorter, or sporulation rate per lesion higher for disease 1 compared with disease 2, then $\partial s/\partial t$ will also be higher for disease 1 than for 2. On the other hand, $\partial s/\partial t$ is inversely proportional to b_* (b_L or b_{PL}) for both models. Since b_L is a measure of the gradient steepness, it is logical that isopath velocity increases as b_L decreases. For instance, if spores travel farther (on average) for pathogen 1 than for 2, then $1/b_L$ will be larger for 1 than for 2, producing a larger $\partial s/\partial t$ for 1 than for 2. For the unitless b_{PL} parameter, interpretation is a little more difficult, as discussed for the b_P in the dispersal gradient models and contact distribution models. However, since b_P characterizes the steepness of the gradient on a log distance scale, one can (in this situation) give roughly the same interpretation to b_P as to b_L, in terms of velocity of spread (in logs).

7.6.4 Other models

The spatio-temporal models of equations 7.33 and 7.35 were chosen for expository purposes in the previous section, because they built on common models for temporal and spatial components of epidemics and they capture some of the essential features of velocity of disease movement. Of course, other pairings of partial differential equations can (or should) be considered as the components of equation 7.28a. For instance, consider the situation where disease progress over time is of the monomolecular form (equation 7.30) and that the disease gradients are represented by $\partial y/\partial s = b_{PM}(1-y)/s$ (a power-monomolecular type). Utilizing equation 7.37, one obtains:

$$\frac{\partial s}{\partial t} = \frac{r_M s}{b_{PM}} = \left(\frac{r_M}{b_{PM}}\right)s$$

which is identical to equation 7.39, except that different symbols are used for the parameters. Likewise, pairing the monomolecular temporal model with a pure monomolecular model for the gradients $[b_M(1-y)]$ results in a constant velocity (r_M/b_M).

However, what if temporal disease progress was of the monomolecular form (equation 7.30) and that the gradients were of the logistic type (equation 7.31)? This results in:

$$\frac{\partial s}{\partial t} = \frac{r_M}{b_L y} = \frac{(r_M/b_L)}{y} \qquad (7.40)$$

This velocity equation has some essential characteristics in common with the others we considered: $\partial s/\partial t$ increases with increasing temporal disease increase (r_M) and with decreasing steepness of the gradient ($1/b_L$). However, the velocity is also related to the reciprocal of disease intensity. That is, $\partial s/\partial t$ is a function of y, with greater $\partial s/\partial t$ at low intensity than at high intensity. For this situation, one would obtain a different disease expansion rate for every specific disease value (y'). For this circumstance, one could be more explicit and refer to the velocity as $\partial s'/\partial t$.

To understand disease-dependent velocities, consider a graph similar to the middle frame of Fig. 7.14 corresponding to the solution of equation 7.40. If one drew horizontal lines at different y' values, and calculated $\Delta s'/\Delta t$ for each, one would find greater $\Delta s'/\Delta t$ at low y' than at high y'.

Jeger (1983) showed some other interesting pairings of equations for the temporal and spatial components of epidemics. In general, however, the pairings of partial differential equations may not even lead to an analytical solution for $y(t, s)$ (equations 7.28a, b). The situation can be even more complicated. For instance, in this chapter we assumed that the population terms for temporal and spatial rates of change (r_* and b_*) are constants. But this is not necessarily so. For instance, even if y increased logistically everywhere in the population, r_L might not be fixed but depend on distance from (original) inoculum source; in particular, r_L could either increase or decrease with s. One would write this as $r_L(s)$ to show that r_L is a function of s. This could be directly assessed by analyzing the disease progress curves at *each* sampling location (e.g., the vertical slices of y in Fig. 7.11) in the field to see if estimates of r_L changed systematically with distance. To add to the complexity, r_L could also vary with time, such that r_L values for the early and late parts of the epidemic were different. This could be written as $r_L(t, s)$. It should be noted that if r_L did vary systematically with time, the disease progress curve likely would appear to be non-logistic in shape.

Additionally, disease gradients might be logistic (or power-logistic) in nature at all times, but with a b_L (or b_{PL}) term that is a function of time, $b_L(t)$ [or $b_{PL}(t)$]. We assessed this directly for the power-logistic example in Fig. 7.11 and saw some evidence that the slope did decrease with time. Moreover, b_L (or b_{PL}) could vary with distance, so that the decline in y near the source has a different rate than far from the source. This could be written as $b_L(s)$ [or $b_{PL}(t, s)$]. This latter situation would probably appear as a non-logistic (or non-power-logistic) type of disease gradient at individual times.

The main points here are that the spatio-temporal dynamics of disease can be very complicated, and it may be very challenging to find a relatively simple model (equation 7.33 or 7.35) for representing the dynamics of disease in space and time that can be justified based on population biology. Even when the purely temporal dynamics and disease gradients are well described by standard models (e.g., logistic), characterizing the resulting spatio-temporal dynamics is a difficult endeavor. Ultimately, diffusion and/or integral modeling is necessary, as presented in the next chapter. Nevertheless, the models presented here do provide a reasonable starting point for understanding epidemics and for describing and comparing their development in space and time.

7.6.5 Analysis

As demonstrated above, one can (empirically) determine the velocity of isopath movement for a given epidemic, $\Delta s/\Delta t$, by choosing a specific value of y, y' (or $y^{*\prime}$), and determine the distance where this value occurs at the different assessment times. The change in distance divided by the change in time is a measure of the velocity. Generally, one should use a small value for y', because one is usually most interested in the movement of the front of the epidemic. A reasonable choice is $y' = 0.05$ or $y' = 0.10$. Much smaller values can lead to some misleading results because of the low precision of the estimated y (see Chapter 10). To determine whether or not $\Delta s'/t$ is affected by disease intensity, one could perform the calculations for other y' values (e.g., 0.25, 0.5).

In attempting to apply this protocol to the example data set in Fig. 7.11, one can see some immediate difficulties. Using $y' = 0.05$ (or $\text{logit}(y') = -2.94$), one can only determine $\Delta s/\Delta t$ between times 1 and 8 (with interpolation), since all y values are above 0.05 at later times. If one chose $y' = 0.6$, then the isopath movement rate could be determined between several times. However, if the rate depends on y, then it may be less interesting to know the movement of the 0.6 isopath than the 0.05 isopath (i.e., nearer the front).

One can use the model fitted to the gradient data to determine the distance where y' is obtained. For instance, assuming that the power-logistic gradient model is appropriate (equation 7.23), and using the parameter estimates shown above for the data in Fig. 7.13, one can determine that the distance corresponding to $y^{*\prime} = -2.94$ at $t = 8$ days as:

$$\hat{s}' = \exp\left(\frac{y^{*\prime} - \hat{A}^*_{PL}}{-\hat{b}_{PL}}\right) = \exp\left(\frac{-2.94 - (-0.3)}{-0.77}\right)$$
$$= \exp(3.43) = 31\text{m}$$

Note that the exponential function is used here because simple rearrangement of equation 7.23 produces an expression for $\ln(s')$. This value can also be obtained by rearranging nonlinear equation 7.22 [where one uses $\hat{A}_{PL} = \exp(-\hat{A}^*_{PL}) = 1.35$]. This actually is the type of calculation used to obtain the s' values to two decimal places in section 7.6.3.1 for the idealized disease gradients of Figs. 7.14 and 7.15. If this calculation is done for the logistic spatio-temporal model (equation 7.34), one would not perform the exponential operation.

The calculated distance here is considerably beyond the largest distance used in the study (Fig. 7.13), and given the uncertainty about the appropriateness of the model at large distances, may not be reliable. The reason the estimated s' is so large is because of the very shallow gradient (small dy/ds) at large distances for power-type model processes. For the data in Fig. 7.13, the researcher who wishes to estimate disease expansion over time will be much better off using disease intensity values above 0.25 for y'. This is because the calculated distances will be (mostly) within the range of distances used in the study.

Using calculated s' values for each time, based on the chosen disease-gradient model parameter estimates, one can then calculate $\Delta s/\Delta t$ values. In other words, the calculated s' values from the equation rearrangements are treated as observed distances. If we use $y' = 0.6$ (so that $y^{*\prime} = 0.405$), and the parameter estimates obtained when there was a different slope for each time, one obtains the s' estimates shown in Table 7.1. The differences in s' between times were: 1.3, 3.0, 3.2, and 15.6 m over the last five assessment times. One divides these by 7 days (the difference in time between the disease assessments on which the gradients are based) to obtain the following velocity values: 0.19, 0.41, 0.46, and 2.22 m/day. Clearly, there is an increasing velocity for $y' = 0.6$. This is partly because of the power-type relation (s in the numerator for $\partial y/\partial s$) and also because b_{PL} was decreasing somewhat over time.

Using the models developed in section 7.6.2, one does not need to base isopath calculations on observed gradients or estimated gradients for individual times. In particular, if one can assume a specific spatio-temporal model, or find empirical evidence that such a model is reasonable, one can directly use model parameters for determining (or estimating) $\partial s/\partial t$. For the bean rust example (Fig. 7.15), we fitted the power-logistic model (equation 7.36) to the data using least squares. The statistical model can be written as:

$$y_{ij}^* = A_{LPL}^* + r_L t_i - b_{PL} \ln(s)_j + \varepsilon_{ij} \tag{7.41}$$

where y_{ij}^* is the logit of disease intensity for the i-th time and j-th distance, and ε_{ij} is the error. The SAS code for fitting this model is shown at the bottom part of Fig. 7.12. This model should be compared with equations 7.27a,b where a separate intercept and slope (on a transformed scale) were specified for *each* group, in which "group" corresponded to time (a class/factor variable). Here, there is one intercept term, both time and space are continuous variables, and there is no interaction of time and ln(s).

Fitting the model resulted in:

$$\hat{y}^* = -0.97(0.150) + 0.088(0.007)\ t - 0.564(0.071)\ \ln(s)$$

where the numbers in parentheses are standard errors of the parameter estimates. The R^2 value was 0.84. The estimated residual variance was 0.034. Using the parameter estimates, estimated velocity of isopath movement ($\partial s/\partial t$) was $(0.088/0.564) \cdot s = 0.156s$ m/day. For instance, the estimated velocity at 1, 3, 9, and 18 m equals 0.156, 0.47, 1.4, and 2.8 m/day. Distance of a fixed y at time t is estimated by:

$$\hat{s}' = \hat{s}_0 \exp\left(\frac{\hat{r}_L}{\hat{b}_{PL}} t\right)$$

where estimated s_0 (distance at $t = 0$ where y' is found) can be estimated by:

$$\exp\left(\frac{\hat{A}_{LPL}^* - y^{*\prime}}{\hat{b}_{PL}}\right)$$

in which $y^{*\prime}$ is the logit of transformed y'. This expression for predicted s_0 can be found by algebraic rearrangement of equation 7.36 (with $t = 0$). Using the parameter estimates and $y' = 0.6$ ($y^{*\prime} = 0.405$), one obtains:

$$\hat{s}' = 0.087 \exp(0.156t)$$

For days 8, 15, and 22, 29, and 36, one obtains estimated distances of: 0.3, 0.9, 2.7, 8.0, and 24.0 m. Based on an assessment of the data in Figs. 7.11 and 7.13, it would appear that these values are little too small for the early times. The problem lies with the fit of equation 7.36 to these data. In particular, at each distance, disease progress is not logistic, at least not logistic with $K = 1$. This can be seen by the plot of logit(y) versus ln(s). Since disease was assessed every 7 days, a logistic increase would correspond to *equal* vertical separation of the transformed gradients. For example, at $s = 0.7$ m, a purely logistic increase would mean equal changes in logits between times 1 and 8, 8 and 15, and so on. If r_L and b_{PL} were fixed for the entire epidemic, the vertical differences ($=r_L \cdot \Delta t$) would be the same at each distance on the logit scale. The interested reader can construct disease progress curves for each distance by taking the vertical slices of logits and plotting these versus time. These data exhibit a slowing down of the epidemic as time increased, which corresponds to decreasing vertical differences of the logits at each time. A residual plot of the multiple regression confirmed the logistic model for temporal progress was not ideal here.

One could also fit the logistic-logistic model (equation 7.34) to the data to describe the epidemics. Then, one would use:

$$\hat{s}' = \hat{s}_0 + (\hat{r}_L / \hat{b}_L)t$$

to represent isopath values. Here, s_0 is estimated as $[\hat{A}_{LL}^* - y^{*\prime}]/\hat{b}_L$. However, this model also would not do any better than the power-lotgistic because of the declining temporal increase. As an alternative, one could use a different temporal model (e.g., monomolecular, Gompertz), but it is difficult to combine either of these with a power-logistic (or logistic) disease gradient model. This is, in part, because $\partial s/\partial t$ is then a function of y. As an alternative to using a different temporal model, one could specify a temporal model where $\partial y/\partial t$ is inversely proportional to time, which results in logit(y) being a linear function of ln(t). This is the same as specifying a logistic model but with r_L a function of time, $r_L(t)$, rather than a constant. Jeger (1983), Campbell and Madden (1990), and Madden et al. (1990b) show how to use this type of temporal model.

The above discussion underlines the challenges faced with spatio-temporal modeling and analysis of epidemics. Analytical solutions to partial differential equations for temporal and spatial development (equation 7.28a) may only be obtainable by making strong assumptions about the process being modeled (e.g., pairs of partial differential equations), which may or may not be reasonable. When the assumptions are reasonably met, the solutions make it quite convenient to describe, compare, and understand epidemics. That is, models such as equations 7.33 and 7.35, and their linearized forms, offer an elegant way of characterizing disease progress in space and time. As will become apparent in the next chapter, these models, especially for polycyclic diseases, can only be considered an intermediate step in developing and understanding of epidemics in time and space.

7.7 Disease Management

Knowledge of dispersal gradients can be of great value in developing and testing disease controls (McCartney and Fitt, 1998). For instance, one can decide how far a host crop must be separated from an inoculum source (e.g., an alternative host) in order to maintain disease intensity below a critical value. Or, one can determine the effects of resistant cultivars, cultivar mixtures, and other control treatments on the level of disease at, say, s' meters from an inoculum source. By having knowledge of disease development in time and space, one can determine how long it will take for disease to reach a certain critical value (say, y') under a given set of crop and environmental conditions. Then, through experiments, one can determine which treatments will delay the occurrence of the critical disease value.

The evaluation of controls on dispersal and disease spread is done through the appraisal of (estimated) model parameters. Calculations done using example parameters throughout this chapter are, thus, of value in evaluation of controls, and we do not repeat them here. With primary gradients, the effect of many controls will be to reduce a_* but not b_* (Minogue, 1986). That is, if a control (e.g., a resistant cultivar) reduces disease intensity by a constant fraction throughout a field (because, for instance, the probability of infection by a deposited spore is reduced by 50% compared with another cultivar), then the only difference in gradients is in the height of the curves (a_*). Thus, the spread parameter has no relation to control in this case. However, b_* may or may not be affected by some agricultural practices that can have an effect on disease intensity at any time, for instance, planting density (see, e.g., Madden and Boudreau, 1997). Certainly, the increased number of plants or leaves filters out spores due to barrier effects; however, spores would only have to travel short distances to reach, and possibly infect, other plants. Moreover, the crop environment may change with change in plant population density, leading to a change in the probability of infection by deposited spores.

For secondary spread, it is more difficult to make generalizations about the height of the disease gradient or the steepness of the gradient in reference to the original inoculum source. This is because the number of infections at a given location could be due to inoculum produced at many locations in a field, not just at the original source. If the control affects temporal disease development (r_*), then the control does affect spread of disease ($\partial s/\partial t$; see equations 7.38 and 7.39), whether or not b_* is affected.

Despite the documented evidence that dispersal is a key factor in epidemic development, Jeger (1999) has observed that knowledge of dispersal has been inadequately used in developing and implementing actual integrated pest management programs. Thus, there are considerable opportunities to improve disease control strategies and tactics by the utilization of pathogen dispersal and disease spread information obtained from epidemiological research.

7.8 Concluding Comments and a Prelude to the Following Chapters

In this chapter, we have considered the dispersal of spores and other infectious units (including inoculative vectors) from one location to another. More than for any other topic in botanical epidemiology, the study of dispersal involves a combination of principles and methods from physics and biology (Aylor, 1999; Madden, 1992). As is often done in epidemiology, in this chapter we condensed the complex physical models for spore transport to (relatively) simple ones for spore deposition (per unit area) or infection (e.g., pustule number per leaf) as a function of distance from an inoculum source—the dispersal gradient. These models are also scaled versions of the contact distribution, which specifies the probability of a spore (or any infectious unit) originating at location 0 arriving (being deposited) at location s.

Dispersal gradients are often of an exponential or (modified) power-law type (equations 7.1 and 7.4). Both model types can produce very similar predictions over a narrow-to-moderate distance range, and it can be difficult to choose between the models with many data sets (Ferrandino, 1996; Fitt et al., 1987). At distances very close to the inoculum source and far away, gradients corresponding to these two model types can deviate substantially. In particular, a power model is characterized by steeper declines in Y near the source and shallower declines in Y far from the source, relative to an exponential model (Fig. 7.5). Both linear and nonlinear statistical methods can be used to fit these models to data. Through the estimated parameters, one can compare gradients and evaluate the effects of imposed treatments or uncontrolled factors (e.g., differences between years).

When one is characterizing dispersal gradients through measurements of disease intensity (number of infections), one must be careful to adjust for the maximum possible disease whenever disease intensity is moderate or high. That is, if infections are represented on a proportion scale (y: 0 − 1), one should not use the simple gradient models without modification. The modification typically involves multiplying the expression for dy/ds by $1 - y$ (or a function of $1 - y$). The resulting disease gradient models can be used to determine, among other things, how disease gradients change over time in an epidemic, and how far disease spreads over time. It is especially useful to determine the rate of movement (velocity: C or $C(t)$) of a fixed amount of disease from a source, which depends on the form of the dispersal gradient or contact distribution. One can characterize the simultaneous spatio-temporal development of disease by linking partial differential equations for disease progress in time and disease gradient over distance. For some situations, analytical solutions that simplify epidemic descriptions are possible.

Ultimately, more elaborate modeling is required to represent the spatio-temporal epidemic process, especially for polycyclic diseases. For instance, in the derived models for disease intensity (equations 7.33 and 7.35), y at a given time is related only to distance from the *original* source. Once the primary spread of disease has occurred, however, every infected individual in the field can serve as an inoculum source—once the latent period has been passed and before the infectious period has ended. Thus, there is no unique (individual) distance of relevance, but many distances to consider. Infected individuals near location s are more likely to cause infections at s than infected individuals very far from s. This complicates the matter dramatically, as is shown in the next chapter.

Adding to the complexity of the quantification problem is spread of disease from multiple foci at the start of the epidemic rather from one inoculum source. Other than through simulation (as discussed partly in Chapter 8), the more common approach is to take a different perspective to the analysis, and to quantify the pattern of disease at one or more assessment times during an epidemic. This is dealt with in detail in Chapter 9.

7.9 Selected Readings

Aylor, D. E. 1999. Biophysical scaling and the passive dispersal of fungus spores: relationship to integrated pest management strategies. *Agric. For. Meteorol.* 97: 275–292.

Jeger, M. J. 1999. Improved understanding of dispersal in crop pest and disease management: current status and future directions. *Agric. For. Meteorol.* 97: 331–349.

McCartney, H. A., and Fitt, B. D. L. 1998. Dispersal of foliar fungal plant pathogens: mechanisms, gradients, and spatial patterns. In: *Plant Disease Epidemiology* (D. G. Jones, editor). Kluwer, London, pp. 138–160.

Pielaat, A., Madden, L. V., and Gort, G. 1998. Spores splashing under different environmental conditions: a modeling approach. *Phytopathology* 88: 1131–1140.

8

Spatial Aspects of Epidemics—II: A Theory of Spatio-Temporal Disease Dynamics

Although this may seem a paradox, all exact science is dominated by the idea of approximation.

Bertrand Russell

8.1 Introduction

The previous chapter dealt with the basic concepts of dispersal and disease gradients and how gradients change with time. Some simple ways of characterizing the simultaneous expansion of disease intensity in space and increase in time were discussed based on descriptive models. In this chapter, we study the spatio-temporal dynamics of plant disease epidemics using a more mechanistic modeling approach.

The chapter starts with two examples of the expansion of plant diseases, and based on these examples, some key questions will be posed. These questions will be answered using a model for the spatial expansion of plant diseases. Just as we expanded the simple logistic model in Chapter 4 to the H-L-I-R or similar models in Chapter 5, we here expand on the models in Chapter 5 for temporal increase and Chapter 7 for disease gradients to obtain a more satisfactory understanding of spatial expansion of epidemics. Through the developed models, we expand on some of the finer points of the previous chapters to develop a better overall understanding of epidemics. Finally, we then show an example of how to apply the derived models to evaluate a disease control strategy.

8.1.1 Large scale spread: the case of potato late blight

Plant diseases newly emerging in an area can have devastating consequences for the crop and the human population (Schumann, 1991). Potato late blight, caused by *Phytophthora infestans*, is one of the best documented examples. Information on the epidemiology of this disease has been used in several parts of this book. Here we provide a synopsis on the emergence and long-range spread of this disease based on published material (Bourke, 1964; Stevens, 1933; Schumann, 1991).

The disease, previously unknown, was first reported in 1843 in the United States around the large harbors on the east coast. Diseased specimens were possibly brought in by shipping from South America (the Andes region) where potato is an endemic plant species. Between 1843 and 1845 the disease spread in a *wave-like* fashion, from the east coast inwards, at an average rate of 295 km/year (Stevens, 1933). Fig. 8.1, which is redrawn from Bourke's map, shows the spatial expansion of the disease in these years. In 1845, the disease reached the geographical limits of potato production. Yield losses of up to 40% were reported for many places in the country. The social consequences were, however, limited due to the fact that several other crops were grown in sufficient quantities. Up to the present day, the disease continues to be a constraint to potato production.

In Europe the situation was much different. The disease was first found at the end of June 1845, near the west coast in Belgium (Bourke, 1964), presumably brought in by a ship from America. Within one season, the disease spread in a wave-like fashion all over Europe. Its rate of expansion was estimated by Bourke (1964) to be around 13 km *per day*. Yield losses of up to 60% were reported, and losses continued to be severe during the following years (Bourke, 1964). Fig. 8.2, which is redrawn from Bourke (1964), shows the expansion of the epidemic.

Because of the reliance on a single crop for local food consumption, coupled with a very unstable political situation, the losses were devastating to the country, both economically, politically, and socially. Many farmers in Ireland who could not pay their rents to the English landlords were evicted from their land and homes, often resulting in death due to starvation. From the eight million people in Ireland, approximately one million died due to starvation and malnutrition related diseases. Another 1.5 million people managed to emigrate, most to the United States and Canada (Schumann, 1991).

This continental expansion of an epidemic is just one well documented example of disease spread. Many more such examples are described in the literature and we refer the interested reader to Heesterbeek and Zadoks (1987, and references therein), Zadoks (2001), and Hengveld (1989, and references therein).

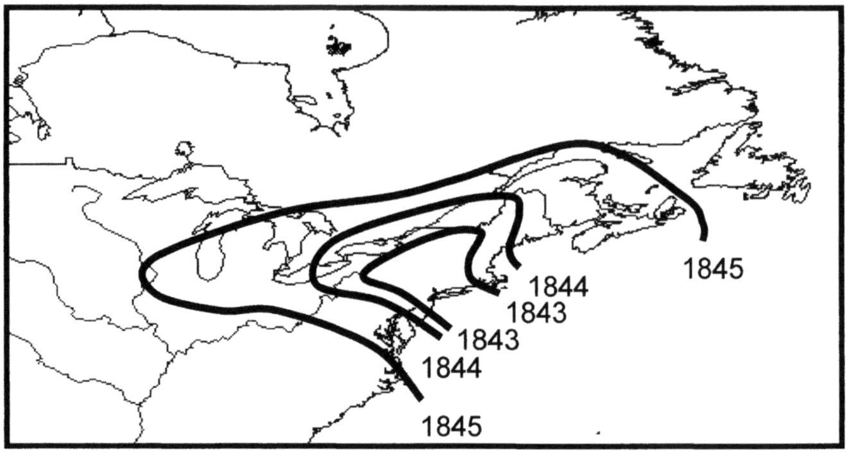

FIG. 8.1. Spatial expansion of potato late blight in the United Sates and Canada. Lines represent boundaries of disease observations. Figure is drawn from information in Bourke (1964).

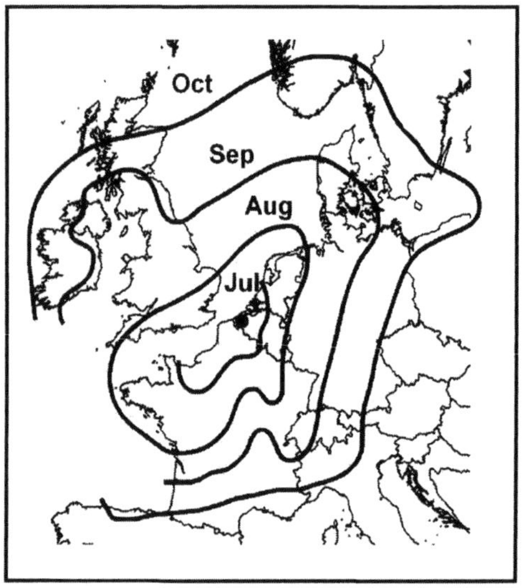

FIG. 8.2. Spatial expansion of potato late blight in Europe. Dot represents first observed infected field. Lines represent boundaries of disease observation. (Redrawn from Heesterbeek and Zadoks, 1987, and based on information in Bourke, 1964.)

8.1.2 Small scale, focus expansion

On a different spatial scale are the disease foci developing and expanding in single fields during epidemics. Such foci originate at the point of infection, where disease intensity subsequently expands radially (but not necessarily symmetrically) due to production of infectious units (e.g., spores), dispersal of the spores (or any infectious units), and new infections caused by some of the dispersed spores. Disease foci of this type were first described by Duhamel in 1728 (Zadoks, 1981)! Without knowing it at that time, he was dealing with an infectious disease of saffron crocus caused by the soil-borne fungus, *Rhizoctonia violacea* (telemorph: *Helicobasidium* sp.). His work led him to recommend the containment of such foci by means of a circular trench because he realized that the disease was expanding from the center outward. Duhamel was far ahead of his time with his research and it took centuries before the research on focal disease expansion was taken further.

One of the best studied examples of focal disease expansion in crop fields is stripe rust disease of wheat, caused by the fungal pathogen *Puccinia striiformis*. Research on foci of this plant pathogen system was done by Zadoks and co-workers (Zadoks and Schein, 1979; van den Bosch et al., 1988c, 1990b), as well as by Emge and Shrum (1976). Early work showed that the rate of focus expansion was as low as 8 cm/day (Zadoks and Schein, 1979). More recent research, based on large-size field experiments, shows that the rate of expansion for stripe rust actually increases in time under many circumstances (Cowger et al., 2005; Sackett and Mundt, 2005a, b) (Fig. 8.3).

Many other diseases are known to spread through crop fields as expanding foci. Rates of expansion vary from just a few centimeters per day, such as for downy mildew on spinach, caused by *Peronospora farinosa* (van den Bosch et al., 1988c), to close to a meter per day for rust on oats caused by *Puccinia coronata* (Berger and Luke, 1979). In Chapter 7 we covered the basic ways of quantifying the focus expansion through the use of observed disease (dispersal) gradients at multiple times, as well as with the use of simple descriptive spatio-temporal models for epidemics. The reader should review the material in section 7.6 before proceeding here. Those with more interest in the subject can read Zadoks and Schein (1979), Zadoks and van den Bosch (1994), Heesterbeek and Zadoks (1987), Jeger (1989), and references therein.

8.1.3 Common features of spatial disease expansion

Although the examples given above discuss spatial expansion of disease on widely different spatial scales, these epidemics have some basic features in common. First of all, the epidemics, whether expanding on a continental scale

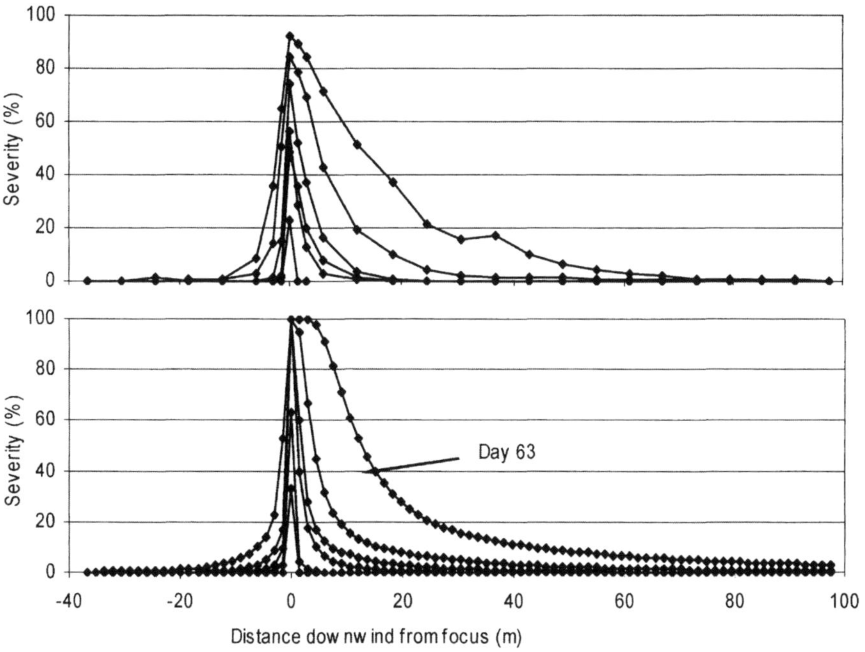

FIG. 8.3. Gradients of the severity of stripe rust of wheat at multiple times for (**A**) observed epidemics and (**B**) computer simulations. Differences in curves for different assessment times at a fixed value of severity increase with time, typical for increasing velocity of expansion. (Reprinted from Sackett and Mundt, 2005b.)

or on a field scale, showed this common wave-like expansion characteristic of many biological invasions. Often the rate of expansion of such epidemic waves is constant and can thus be expressed as single rate value (change in distance divided by change in time). In other cases (Ferrandino, 1993; Aylor, 1999; Frantzen and van den Bosch, 2000), it seems as if the wave increases its rate of expansion continuously, till it runs into the boundaries, either physical or climatological, of the area where disease can occur. Fig. 7.16 demonstrates both of these scenarios.

Underlying epidemic expansion both on a continental scale and on a field scale, and all scales in between, are two basic epidemiological processes we have studied in previous chapters. In Chapters 4, 5, and 6 we studied how epidemics develop in time. In particular, we have studied how epidemic increase depends on mean latent period ($\wp$) and the basic reproduction number (R_0) of the disease, the latter being determined by the per-capita transmission rate (βH_0) and the mean infectious period ($\iota = 1/\mu$) (see equation 5.33). As will be shown, *reproduction* [in the general sense, including transmission (from production of infectious units and probability of infectious units causing infection)] is one underlying processes of epidemic expansion. The disease can spread in space (i.e., occupy new area in the crop, region, and so on) because of the production of daughter infections over time. In Chapter 7 we discussed the dispersal of infectious units, such as spores, around sources of infected individuals, and the description of the resulting disease gradients with various nonlinear models. If all infections are originally at one location, then the disease gradient represents the new infections *beyond* the mother infection(s), an indicator of spread. Thus, in addition to reproduction, dispersal is the other underlying process of epidemic expansion. The greater the reproduction and the greater the distance traveled by infectious units, the greater the epidemic expansion. The basic question is how the combination of reproduction (multiplication) and dispersal of the pathogen around infectious individuals adds up to the development of spatially expanding disease epidemics over time. There are several other related questions.

- How specifically does the rate of epidemic expansion (in, e.g., meters per day) depend on R_0?
- How does the rate of expansion depend of the particular form of the contact distribution, as discussed in section 7.3.4?
- Under what circumstances does the epidemic expand with a constant rate and when does the rate of expansion increase (or change) continuously?

All such questions are difficult, if possible at all, to investigate with just observations of disease at various times and locations. We will, therefore, in the next section develop a model to describe the spatial expansion of disease epidemics based on the underlying mechanisms of reproduction and dispersal. Our understanding of the underlying mechanisms is based on observations and analysis of disease increase in time and dispersal of the pathogen (as presented in Chapters 4, 5, and 7). Analyzing several variants of the model we will be able to theoretically answer the questions posed above. This understanding of spread can then guide us in how we monitor and quantify actual epidemics in time and space.

Our concern in this chapter is mostly on polycyclic diseases, where there is spread from plant to plant, or individual to individual. However, in one section we discuss some of the key issues involved in the spread of monocyclic diseases.

In the next section we utilize the terminology developed in Chapter 5, section 5.1. We suggest that readers review that material before proceeding further.

8.2 Models for Spatial Population Expansion

8.2.1 Introduction

In this section we will study the rate of epidemic expansion using a model. As in Chapter 5, we describe a model that captures the bare essentials behind spatial and temporal dynamics of the epidemic. We assume here that the epidemic does not run into so-called *density dependence* in terms of the healthy host. This amounts to saying that the density of healthy crop at any place and any time can be approximated by the density the crop had at the start of the epidemic (i.e., density of disease-free individuals, H_0). The model thus describes a *density independent* situation. Although this assumption may appear too unrealistic, it turns out, as will be discussed later, that this assumption is fully warranted for assessing expansion rates. This is because the density of infected individuals is generally low (or H_0 is high) at the forefront or leading edge of the epidemic wave. This can be seen for several observed disease gradients in Fig. 7.1, where Y (or $y = Y/H_0$) is small at large distances from the inoculum source.

The assumption about density independence implies that each infected individual (such as a plant, leaf, or unit area of leaf surface) causes, on average, a total of R_0 new infections over its entire infectious period (i.e., life-time). These infections are distributed through space in some fashion. In Chapter 7, we discussed disease gradients and contact distributions and we will use the concept of contact distribution to describe the pattern of the R_0 new infections around an infectious individual. Further simplifying assumptions we use here are: (i) that disease generations are separated in time, as we also assumed for the discrete-time (non-overlapping generation) model at the start of Chapter 5 (section 5.2.1); and (ii) that space is one-dimensional. For practical purposes, the second assumption is consistent with a line source for the initially infected individuals (such as a row of diseased plants). A schematic of a line source is given in Fig. 7.2. In Chapter 7 we assumed that dispersal or spread was from the "edge" of the area of interest, so that distance (s) was exclusively a non-negative number ($s \geq 0$). Here we assume that spread can be to the "left" and "right" of the line source; in other words, we assume that the original and subsequent infections can be in any location in the area. With this assumption, s can be a positive or negative value.

In section 8.3 we will discuss all these assumptions, describe how (or if) they affect the results we will obtain for the simple model, and refer to the relevant literature dealing with these issues. These assumptions can all be relaxed as needed.

8.2.2 Model derivation

To derive the model we show how the density of infectious individuals at position s in disease generation $n + 1$, $Y_{n+1}(s)$, is related to the density of infectious individuals in generation n at *all* possible positions. We use the symbol ξ for any possible (arbitrary) location and s for the specific location of interest. With this notation, $Y_n(\xi)$ is the density of infectious individuals at location ξ at generation n. With a one-dimensional process, $|s - \xi|$ then is the distance between location s and location ξ Note that here we assumed all diseased (infected) individuals are infectious. Thus, $Y = L + I + R = 0 + I + 0 = I$. For ease of comparison with later sections, we express the models in terms of Y (although this is equivalent to I here).

First we calculate the contribution of infectious individuals at location ξ in one generation to the density of infectious individuals in the next generation at distance s. We will use for this the notation $Y_{n+1}(s)\backslash Y_n(\xi)$. Because we are considering the situation in which density dependence does not play a role (e.g., that each spore has no chance to land on an infected individual), each infected individual causes R_0 new infections in the next disease generation; these new infections can be anywhere in the system. The total density of new infections in generation $n + 1$ caused by all infectious individuals present at position ξ in generation n thus is given by $R_0 Y_n(\xi)$. We now have to calculate how many of these infectious individuals will develop at position s. The *contact distribution*, as defined in section 7.3.4, is a scaled version of the dispersal-gradient model that describes how the new infectious individuals in generation $n + 1$ are distributed around the place where the infectious individual was in generation n. Equivalently, the contact distribution, $D(s - \xi)$, multiplied by a scaling constant, is interpreted as the probability of an infectious unit produced at location ξ coming in contact with a host individual at s. (Strictly speaking, $D(s - \xi)$ is the *fraction* per unit area of the *new* infections (i.e., in generation $n + 1$) at position s due to an infectious individual at position ξ). The continuous distribution can also be called a probability density function (*pdf*). Fig. 8.4 represents the process for a hypothetical case of infected individuals at two locations, ξ_1 and ξ_2. The simplest situation would be for the infectious individuals being at one location (e.g., $\xi = 0$) at generation n.

We can thus conclude that the density of infectious individuals in generation $n + 1$ at position s due to all

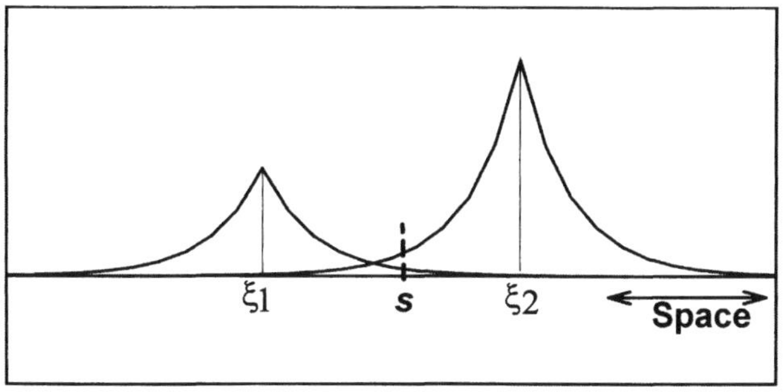

FIG. 8.4. Schematic of spatial spread of new infections around inoculum sources at positions ξ_1 and ξ_2. Number of new infections at position s is the sum of the contributions of new infections from the two sources. See text for further explanation.

infectious individuals in generation n at position ξ is given by:

$$Y_{n+1}(s) \backslash Y_n(\xi) = R_0 Y_n(\xi) D(s-\xi) \qquad (8.1)$$

To calculate the total density of infectious individuals at position s in disease generation $n + 1$, we have to add up all contributions to infections at s from *all* individuals in generation n at *all* possible locations ξ. Adding up over a continuous variable, in this case continuous space, amounts to integrating over ξ. The model equation, therefore, takes the final form:

$$Y_{n+1}(s) = R_0 \int_{-\infty}^{\infty} Y_n(\xi) D(s-\xi)\, d\xi \qquad (8.2)$$

This is a spatial extension of the purely temporal model for disease increase over discrete generations discussed in Chapter 5 (equation 5.4) for the situation when healthy host density is not limiting. Note that infected individuals at any location, including at location s, can contribute to infections at s. The magnitude of the contribution from each location depends on the distance separating s and ξ, as quantified by $D(s - \xi)$.

As with the discrete-time model for the temporal dynamics of an epidemic derived in Chapter 5 (equations 5.2a and 5.2b), model equation 8.2 is, in principle, easily solved, although the computations are tedious. Substituting the contact distribution and the value of R_0, it is possible to calculate the density of infectious individuals in generation $n = 1$, $Y_1(s)$ if we know the initial condition $Y_0(s)$. For instance, $Y_0(s)$ could be zero at all ξ values except one (say, at $\xi = 0$). The integral might be difficult to work with, but numerical procedures are available in several computer packages. To calculate the second generation, $Y_2(s)$, we substitute $Y_1(\xi)$ again in equation 8.2. This is continued for multiple generations.

The reader is reminded here that this is a density independent variant of a model first studied by Aronson and Weinberger (1978) and Weinberger (1978, 1982). Recently this model was studied by Kot et al. (1996) and versions including more aspects of demography by Neubert and Caswell (2000).

8.2.3 Rates of expansion in relation to contact distributions

In this section we will consider examples of rates of disease expansion based on four different contact distributions. The four distributions are given in Table 8.1. We will discuss each of these expansion-rate results in sequence. Each epidemic is started with Y_0 individuals at position $s = 0$; that is, $Y_0(s) = 0$, except for $Y_0(0) > 0$.

Before discussing spatial dynamics based on the model, we first discuss some properties of the selected contact distributions that are relevant for the interpretation in the next sections. It is convenient to discuss these distributions in terms of spore flight distances, although the contact distribution applies to more than just spore movement (Chapter 7). In the previous chapter, we exclusively considered situations with $s \geq 0$, and mostly dealt with two contact distributions, the exponential (equation 7.8) and Pareto (equation 7.9). Now, we allow s to be positive or negative ($-\infty \leq s \leq +\infty$), to allow for, for example, spore arrival from both "sides" of the location of interest (as shown schematically in Fig. 8.4).

Fig. 8.5 plots realizations of the four contact distributions, with parameter values chosen such that the predicted values had a fair degree of overlap (although total overlap is impossible). Other distributions could also be used, but these four encompass a sufficiently wide range of possibilities. The *double exponential distribution*, sometimes known as the Laplace distribution, is a generalization of the exponential when s can take on any value from positive to negative infinity. Likewise, the *modified power-law distribution* is a generalization of the Pareto distribution, which was used in Chapter 7, for the case of positive and negative s. The *Gaussian distribution* is the classical normal distribution for the random variable s. Finally, the *root distribution* is a nonstandard model for the distribution of s that serves a useful purpose in the comparison of contact distributions, as will become clearer below.

Statistical distributions are often characterized by their first two moments, that is, the mean (expected value) and the variance. However, all distributions do not have finite or definable means and variances, as discussed by Minogue (1989) in one particular context. In Chapter 7 it was pointed out that the Pareto distribution (i.e., for the case with the distance variable being nonnegative) only has a finite mean and variance with suitably large Ψ. The distributions in Table 8.1 are specifically defined here to be symmetrical around distance 0, so they all have means of 0. The variance of the Gaussian, double exponential, and root distributions are σ^2, σ^2, and $\Gamma(6)/\Psi^6$, respectively (where $\Gamma(\bullet)$ is the gamma function). The modified power law distribution

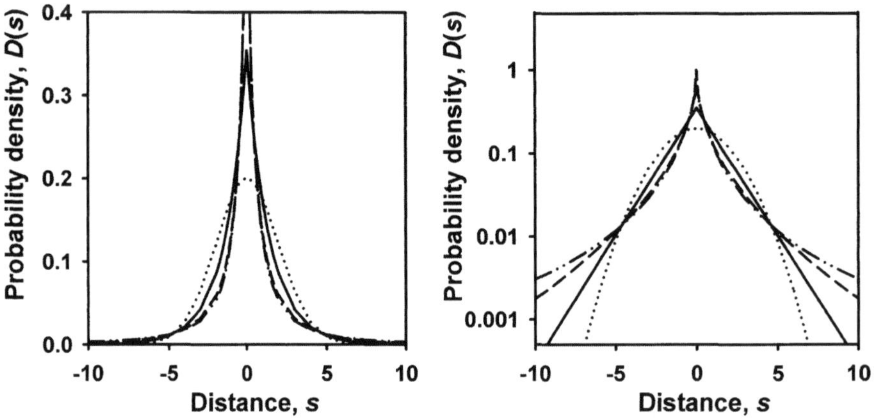

FIG. 8.5. Realizations of four contact distributions, $D(s)$ (or $D(s - \xi)$, with $\xi = 0$). Horizontal axis is distance from source, s, and vertical axis is either the probability density function for the contact distribution listed in Table 8.1 (left-hand graph), or the probability density on a log scale (right-hand graph). Dotted line is for the Gaussian, solid line is for the double exponential, dashed line is for the root, and dash-dotted line is for the modified power-law distribution.

TABLE 8.1. Four contact distributions $[D(s)]^a$ of relevance in the spread of plant diseases.

Common name	*D(s) =*[b]	*ln[D(s)] =*	*Disease expansion rate [C_n or C(t)]*[c]
Gaussian	$\frac{1}{\sigma\sqrt{2\pi}}\exp\left(-\frac{s^2}{2\sigma^2}\right)$	$\ln\left(\frac{1}{\sigma\sqrt{2\pi}}\right) - \frac{s^2}{2\sigma^2}$	Approaches constant (C)
Double exponential	$\frac{1}{\sigma\sqrt{2}}\exp\left(-\frac{\lvert s\rvert\sqrt{2}}{\sigma}\right)$	$\ln\left(\frac{1}{\sigma\sqrt{2}}\right) - \frac{\sqrt{2}\lvert s\rvert}{\sigma}$	Approaches constant (C)
Root	$\frac{\Psi^2}{4}\exp\left(-\Psi\sqrt{\lvert s\rvert}\right)$	$2\ln(\Psi) - \ln(4) - \Psi\sqrt{\lvert s\rvert}$	Increases linearly
Modified power law	$\frac{\Psi}{2}\lambda^{\Psi}(\lambda + \lvert s\rvert)^{-(\Psi+1)}$	$\ln(\Psi/2) + \Psi\ln(\lambda) - (\Psi + 1)\ln(\lambda + \lvert s\rvert)$	Increases exponentially

[a]Distributions written with $\xi = 0$. For a more general version of the models, one can replace s by $s - \xi$ throughout. The range for s is negative to positive infinity.
[b]Parameter units: distance for σ in double exponential and Gaussian distributions; distance$^{-1/2}$ for Ψ in the root distribution; unitless for Ψ, and distance for λ in the modified power law distribution. All parameters are positive.
[c]Change in distance per generation or per time period where a low value of Y (Y') is found. See equation 8.6 for general formula.

only has a finite variance at $\Psi > 2$, and its form is so cumbersome as to be intuitively uninformative. For the four distributions here, however, σ, $1/\Psi$ (root), and λ/Ψ (modified power law) have the same general (but not exact) interpretation in terms of contact and dispersal: larger values of σ, $1/\Psi$ (root), and λ/Ψ (modified power law) mean greater distances of spore movement or larger spatial ranges of contact between host individuals.

The σ parameter of the Gaussian and double exponential distributions, the Ψ parameter of the root distribution, and the Ψ and λ parameters of the modified power-law distribution were chosen for Fig. 8.5 so that $D(s)$ values all intersect at the same s value. The vertical axis is on a linear scale (left-hand side) or a logarithmic scale (right-hand side). On a linear scale, the four distributions look quite different near the center (at distances near 0), but appear quite different in the tails (i.e., values relatively far from $s = 0$).

The differences between the tails of the four contact distributions become very clear when using a log scale for the ordinate (Fig. 8.5). The double exponential distribution consists of straight lines. The tail of the Gaussian distribution decreases faster with distance from the origin than the exponential distribution. Both the double exponential and the Gaussian distribution are said to be *exponentially bounded* (see section 7.3.4), meaning that the tails fall off (declines) faster than, or as fast as, an exponential function. In contrast, the root and the modified power-law distributions clearly decrease less than exponentially with distance. In particular, the rate of decline in $\ln[D(s)]$ decreases with increasing distance from the origin (Fig. 8.5). This is *not* dependent on the chosen values for Ψ and/or λ. Distributions with this property are said to be non-exponentially bounded, "fat-tailed", or "heavy tailed".

As discussed in Chapter 7, it was emphasized that there are many examples of dispersal gradients described by the exponential gradient model and many other examples of gradients described by the power-law gradient model. This means there are numerous examples of contact distributions that are either exponentially bounded or not exponentially bounded. In the next sections we show that there is an essential difference between the spatial expansion of the epidemic for an exponentially bounded and for a non-exponentially bounded contact distribution. These mechanistic results add to the more descriptive results given in the previous chapter.

8.2.3.1 Gaussian contact distribution. Model equation 8.2 with the Gaussian contact distribution (Table 8.1) can be solved explicitly (N. L. Johnson et al., 1994, 1995) for disease density at location s in generation n. The solution is:

$$Y_n(s) = Y_0 R_0^n \frac{1}{\sigma\sqrt{2\pi n}} \exp\left(-\frac{s^2}{2\sigma^2 n}\right) \qquad (8.3)$$

Fig. 8.6 shows an example realization of the solution of this model. How will the front of the epidemic wave move through space over time (i.e., between generation)? To calculate this we determine from equation 8.3 the position of a pre-specified density of infected individuals, Y', at time (generation) n, relative to the starting point of the epidemic, $s_{Y',n}$. To simplify the presentation of the equations and other calculations, we simplify the notation somewhat and use s'_n for $s_{Y',n}$. In Fig. 8.6A this concept is illustrated by the identification of Y' on the ordinate and the position of this value of disease for each generation on the abscissa. The line of fixed disease level is known as an *isopath* (see section 7.6.3), so s'_n is an isopath. Calculating the position of the small density of infectious individuals requires solving the equation $Y' = Y_n(s'_n)$. Using Y' and s'_n results in:

$$Y' = Y_n(s'_n) = Y_0 R_0^n \frac{1}{\sigma\sqrt{2\pi n}} \exp\left(-\frac{(s'_n)^2}{2\sigma^2 n}\right) \qquad (8.4)$$

and after some algebraic manipulation we find:

$$s'_n = \sqrt{2\sigma^2 n \left\{\ln(Y_0) - \ln(Y') - \ln\left(\sigma\sqrt{2\pi n}\right) + n\ln(R_0)\right\}} \qquad (8.5)$$

Fig. 8.7 shows s'_n as function of generation, n. The figure shows that after a very short initial phase, the distance of Y' to the starting point of the epidemic increases

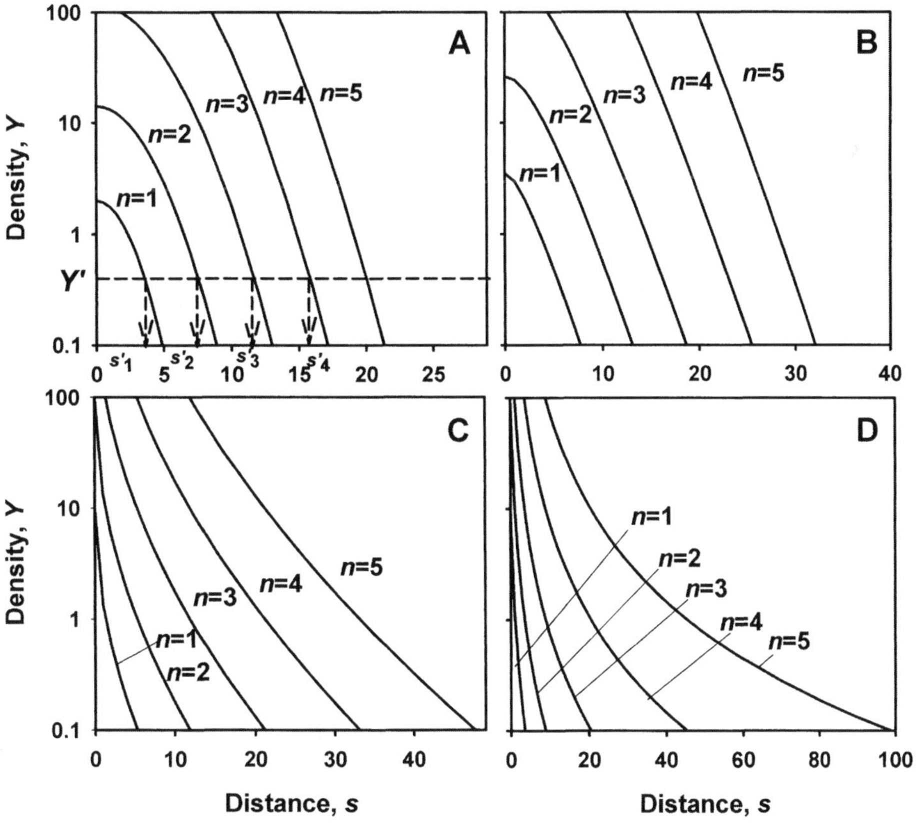

FIG. 8.6. Spatial expansion of disease density, $Y_n(s)$, as described by equation 8.2, corresponding to the four contact distributions depicted in Fig. 8.5 and listed in Table 8.1 It is assumed that the original inoculum source is at distance 0. Vertical axis is density of diseased individuals, and horizontal axis is distance from the starting point of the epidemic. Only positive distances are shown, although spread can be in both directions. (**A**) Gaussian contact distribution; (**B**) double exponential contact distribution; (**C**) root contact distribution; and (**D**) modified power law contact distribution.

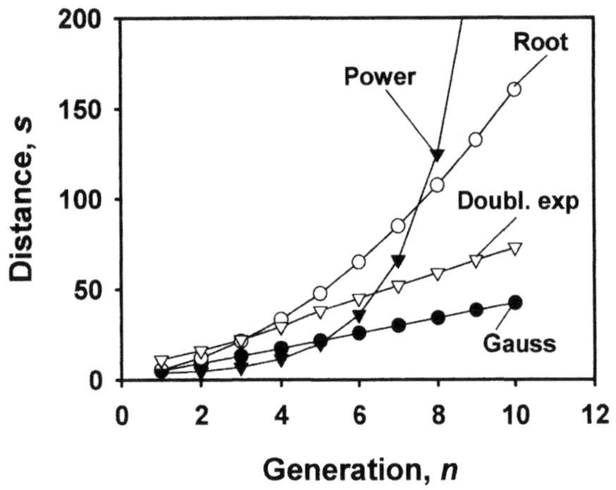

FIG. 8.7. Distance between the starting point of the epidemic (original inoculum source) and the position of a specified low density of disease, Y' (see Fig. 8.6A), for a sequence of non-overlapping generations. Results based on solving equation 8.2 for distance, s'_n ($=s_{Y',n}$), where Y' is found, for each of the contact distributions in Table 8.1.

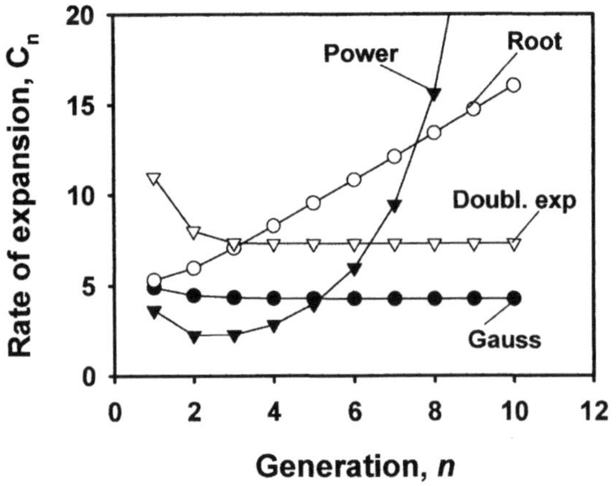

FIG. 8.8. Mean rate (or velocity) of epidemic expansion, C_n, between the beginning of the epidemic (generation 0) and the current generation (n), based on distances shown in Fig. 8.7. Results obtained using equation 8.6 for each of four contact distributions.

linearly with time. This implies that the rate of expansion is a constant. To calculate the rate we will consider the average rate of expansion between the start of the epidemic, $n = 0$, and the present generation n. This average rate of expansion, denoted by C_n, is calculated from:

$$C_n = \frac{s'_n}{n} \tag{8.6}$$

Substituting equation 8.5 in this expression we find as the rate of epidemic expansion:

$$C_n = \frac{s'_n}{n} = \sqrt{\frac{2\sigma^2\left\{\ln(Y_0) - \ln(Y') - \ln\left(\sigma\sqrt{2\pi n}\right)\right\}}{n} + 2\sigma^2\ln(R_0)} \tag{8.7}$$

In Fig. 8.8 the average rate of epidemic expansion is plotted as a function of time, for the same parameters and initial condition as those used in Fig. 8.6. C_n is the rate (or velocity or speed) of isopath movement (as in section 7.6.3.2). We see from this figure that the rate of expansion reaches a constant rate, C, very soon after the start of the epidemic. This result also holds for other parameter values than the ones used to calculate Fig. 8.8. This constant rate of expansion can be calculated mathematically from equation 8.7 by letting n go to infinity (i.e., increase indefinitely). We immediately see from this equation that the whole first term under the square-root sign vanishes. The constant rate of expansion thus is:

$$C = \sigma\sqrt{2\ln(R_0)} \tag{8.8}$$

Fig. 8.9A shows the dependency of C on the parameter values R_0 and σ. C increases linearly with the standard deviation of the Gaussian distribution; that is, increasing σ, which means a shallower gradient, results in a higher C. Moreover, C increases with R_0 meaning that increasing the number of new infections resulting from each infection also results in a faster velocity of isopath movement. For large values of R_0, say $R_0 > 10$, C approximately increases linearly with the logarithm of R_0. The per unit increase in C with R_0 is higher when R_0 is low than when it is high.

Before assessing the model (equation 8.2) with the other contact distributions, we discuss some numerical examples. Suppose we have done a series of experiments and found that the contact distribution of the disease under consideration is Gaussian and has a standard deviation of 10 cm. This value could be obtained, for instance, with a fungal disease dispersed by spores, by introducing infectious plants in one row in a field (for generation 0), and counting new infections at several distances from the origin the equivalent of one generation later (at $n = 1$). As long as $Y_1(s)$ is not high, one can regress $\ln[Y_1(s)]$ on s^2, and the slope is an estimate of $1/(2\sigma^2)$ (see Table 8.1). Furthermore, suppose that an experiment on focus expansion of this disease, as discussed in sections 8.1.2 and 7.6.5, shows an expansion rate at around 12 cm/day over a few generations. This can be done by identifying a particular density of disease (say, $Y' = 1$), determining where this density is found for each generation (s'_n), and then using equation 8.6. What does this tell us about the value of R_0? Solving R_0 from equation 8.8 we find:

$$R_0 = \exp\left(\frac{1}{2}\left(\frac{C}{\sigma}\right)^2\right) \tag{8.9}$$

Substituting $\sigma = 10$ and $C = 12$ we find that $R_0 = 2.05$. Suppose we can choose between two focus expansion

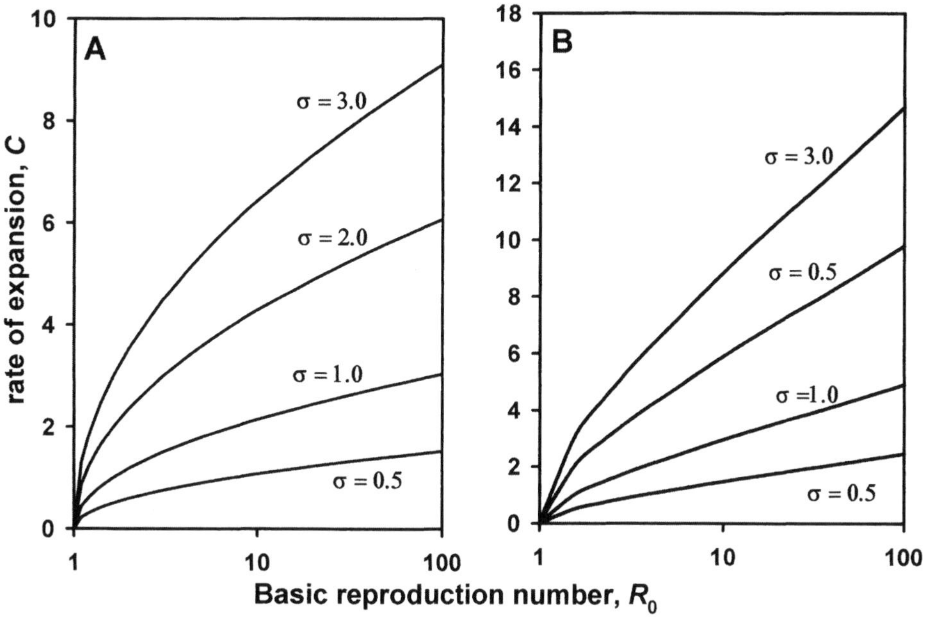

FIG. 8.9. The final (or asymptotic) rate of disease expansion, C, for epidemics with a (**A**) Gaussian contact distribution, and (**B**) double exponential contact distribution, in relation to the basic reproduction number (R_0; log scale) and standard deviation of the contact distribution (σ).

management methods. First, that it is possible to reduce σ by 50%. This can be done, for example, by planting a different crop variety whose plant architecture is such that the crop is more dense, implying less air movement in the crop and consequently infectious units being dispersed over shorter distances. Substituting this into equation 8.8 we find that the rate of disease expansion is reduced to 6 cm/day. We are assuming here that R_0 is not affected by this management tactic. Our second disease management method involves the use of fungicides that reduces R_0 by 50%. Substituting this new R_0 into equation 8.8, we find that the rate of focus expansion is reduced to 2.2 cm/day. For small values of R_0 it is more efficient (at least for this Gaussian contact-distribution example) to restrict disease expansion by reducing the number of new infections from each mother infection than it is to change the length scale of the contact distribution (σ).

Instead of using equation 8.6, one could obtain estimates of expansion velocity for each *successive* pair of generations. For a given Y', we can define the generational rate (or velocity) as:

$$C'_n = s'_{n+1} - s'_n \tag{8.10}$$

where s'_n and s'_{n+1} are the distances where Y' is found in generations n and $n+1$, respectively. One could plot these, and if no trend is apparent (as expected here), then an average of the C'_n values could be an estimate of C. Equation 8.10 produces "instantaneous" values of velocity (as also utilized in Chapter 7), whereas equation 8.6 produces an overall average from the first to nth generation. We emphasize the use of equation 8.6 here, since the biological interpretation is the same as that obtained with the mean of equation 8.10.

8.2.3.2 Double exponential contact distribution. Model equation 8.2 with the double exponential contact distribution (Table 8.1) can be solved explicitly (N. L. Johnson et al., 1994, 1995), although the solution is very messy:

$$Y_n(s) = Y_0 R_0^n \frac{1}{\phi} \frac{e^{-|s|/\phi}}{2^{2n-1}(n-1)!} \sum_{j=0}^{n-1} \frac{2^j(2n-j-2)!}{j!(n-j-1)!} \left|\frac{s}{\phi}\right|^j \tag{8.11}$$

where $\phi = \sigma/\sqrt{2}$. Fig. 8.6B shows a solution of the model for the same parameter values as used for the solution of the model with the Gaussian contact distribution. The figure shows that the solutions for density of the infected individuals are very similar to those using the Gausian density as contact distribution, except that the rate of expansion of the model with the exponential contact distribution seems to be higher.

Unfortunately, equation 8.11 is so complicated that it turns out not to be possible to explicitly derive an equation for s'_n nor for the average rate of expansion C_n, as was done for the model with the Gaussian contact distribution. From equation 8.11, distance s'_n and the rate C_n *can* be calculated numerically (i.e., iteratively), and Figs. 8.7 and 8.8 show the results of this calculation. As for the model with the Gaussian contract distribution, we see from the figure that C_n approaches a constant value (C) just a few disease generations after the start of the epidemic. Without further proof we mention that this is not due to the particular set of parameter values chosen. For each possible set of parameter values the rate of expansion approaches a constant.

Fig. 8.9B shows C as a function of the logarithm of R_0 for various values of σ of the double exponential distribution. This figure shows that the rate of expansion for the model with the double exponential distribution is larger than for the model with the Gaussian distribution. This is due to the fact that the exponential distribution has thicker (heavier, fatter) tails than the Gaussian distribution (but still exponentially bounded), resulting in more new infections further away from an infected individual. For values of R_0 larger than approximately 2, the rate of expansion is close to linear in the logarithm of R_0.

8.2.3.3 Root contact distribution. When using the root contact distribution in model equation 8.2, it is not possible to solve the model explicitly. Kot et al. (1996) have, however, introduced a very useful method to calculate approximations to the solution. Applying their method we find as approximate solution:

$$Y_n(s) \cong Y_0 R_0^n \frac{\Psi^2}{4} e^{-\Psi\sqrt{|s|}} \tag{8.12}$$

Fig. 8.6C shows a solution of the model. The figure suggests that the rate of epidemic expansion does not apparently converge to a constant. In fact, the rate of expansion keeps increasing over time (i.e., generation) as indicated by increasing distances between low Y values from generation to generation. This is a different qualitative behavior than found with the Gaussian and the double exponential contact distributions.

The average rate of disease expansion, C_n, can be calculated using the same procedure as for the model with the Gaussian contact distribution. This procedure yields:

$$\begin{aligned} C_n &= \frac{s'_n}{n} \\ &= \frac{1}{\Psi^2}\left[\frac{\ln(Y_0) - \ln(Y') + \ln(\Psi^2/4)}{\sqrt{n}} + \sqrt{n}\ln(R_0)\right]^2 \end{aligned} \tag{8.13}$$

For large values of n, this becomes:

$$C_n = n\frac{1}{\Psi^2}(\ln(R_0))^2 \tag{8.14}$$

This last expression shows that the rate of epidemic expansion indeed increases linearly with time (generation), with a slope of $[\ln(R_0)]^2/\Psi^2$. Specifically, the rate of expansion increases per generation with the square of the logarithm of the basic reproduction number and with the square of the distances traveled by spores (related to $1/\Psi$). The plot of s'_n versus n (Fig. 8.7) shows the expected upward curvature, indicating increasing rate of disease expansion. The plot of C_n versus n (Fig. 8.8) confirms the *linear* increase in the expansion rate after a few generations. The implication of equation 8.14 is that the farther the distance from the inoculum source, the greater the expansion rate of the epidemic.

It was pointed out before that, all other things being equal, C_n (or C) is higher with a double exponential than with a Gaussian distribution for contact in the epidemic model. However, there is no simple overall comparison of C_n obtained with a root and the previous contact distributions. This is because the expansion rate increases with time (generation) for the root distribution. Depending on the value of Ψ, C_n may be lower or higher than the corresponding rates for the previous two contact distributions when n is not large. With enough time, however, the expansion rate for a root distribution will (must!) surpass that obtained with the previous two distributions, since C_n is directly proportional to n. With sufficient time, C_n for a root distribution can be considerably larger than that obtainable with the Gaussian or double exponential distribution.

8.2.3.4 Modified power law contact distribution. Using the modified power law contact distribution in equation 8.2, it is not possible to solve the model explicitly. Applying the method of Kot et al. (1996) again we find as approximate solution:

$$Y_n(s) \cong Y_0 R_0^n \frac{\Psi}{2} \lambda^{\Psi} (\lambda + |s|)^{-(\Psi+1)} \tag{8.15}$$

Fig. 8.6D shows a solution of the model for one set of parameters. The figure suggests that, as for the model with the root contact distribution, that the rate of epidemic expansion does not converge to a constant, but that the rate of expansion keeps increasing over time (as demonstrated by the increasing distance between a selected low disease density for successive pairs of generations). In fact, with this example, it appears that the rate of expansion is increasing faster between generations than with the root contact distribution. How does the rate of epidemic expansion, C_n precisely increase with time? To answer this question, C_n was calculated using the same procedure as for the model with the Gaussian contact distribution. This procedure yields:

$$C_n = \frac{s'_n}{n} = \lambda\left(\frac{Y_0\Psi}{2\lambda Y'}\right)^{(1/(\Psi+1))}\left(\frac{R_0^{(n/(\Psi+1))}}{n}\right) - \frac{\lambda}{n} \tag{8.16}$$

For large times (generations), equation 8.16 approaches:

$$C_n = A\frac{R_0^{(n/(\Psi+1))}}{n} = \frac{A}{n}\exp\left(\frac{n}{\Psi+1}\ln(R_0)\right) \tag{8.17}$$

where the constant A is given by:

$$A = \lambda\left(\frac{Y_0\Psi}{2\lambda Y'}\right)^{1/(\Psi+1)} \tag{8.18}$$

From equation 8.17 we see that the rate of epidemic expansion increases as an exponential function of time (n). However, the relationship is a little more complicated, since n also appears in the denominator outside the exponential or power function in equations 8.16 and 8.17. This means that at small n (i.e., when equation 8.16 applies), the expansion rate *could* initially decrease with increasing time (depending on the other parameters), but the exponential function of n will dominate after a few disease generations, giving an increasing C_n.

Figs. 8.7 and 8.8 demonstrate how s'_n and C_n, respectively, are related to time (generation) with a particular combination of R_0 and contact-distribution parameters. The distance where Y' is found (s'_n) is clearly seen to increase over all times as the focus expands (Fig. 8.7), although there is a slight decrease in the rate of disease expansion for the first two generations, followed by an increasing rate (with no upper bound) (Fig. 8.8). Unlike the case for a root contact distribution, the increase in disease expansion rate is more than linear. That is, the change in C_n ultimately becomes larger with every successive pair of generations.

As discussed for the root distribution, there is no simple comparison of C_n between models with the modified power-law contact distribution and with the Gaussian and double exponential distributions. During the early stages of an epidemic, C_n for a modified power-law distribution may be smaller or larger than the C_n for a Gaussian or double exponential distribution (depending on the other parameters). However, with sufficient time, with the modified power-law distribution will be larger, possibly much larger, than the others.

8.2.4 Comparisons

Most plant diseases do not increase in discrete and non-overlapping generations (see Chapters 4 and 5). However, we showed in section 5.2.1 that use of discrete-generation models can readily lead to a general understanding of temporal dynamics of diseases. In fact, key results obtained with such a simple model—thresholds for epidemics, final disease levels, exponential increase early in the epidemic—all carry over to more realistic (but more complicated models) (sections 5.2.2–5.2.5). Here we utilize the simplicity of a discrete generation model to explore the consequences of different forms of the contact distribution (a scaled version of the disease gradient when healthy host density is not limiting) on spatio-temporal dynamics of disease. As required, the increase in disease density from generation to generation is dependent on R_0. Moreover, we found that the expansion rate of a disease focus (C_n or C) is also dependent on R_0 (the descriptor of reproduction, in the general sense) for all the contact distributions considered. The functional relationship *does* depend on the contact distribution, but typically C_n is proportional to $\ln(R_0)$. Since $\ln(R_0)$ is only positive when $R_0 > 1$, it is clear that for C_n to be positive in the long term (e.g., Gaussian distribution), or for the increase in C_n to be positive in the long term (e.g., root and modified power-law distributions), then R_0 must exceed 1.

The results of the previous sections show that there are two very distinct qualitative classes of disease expansion in epidemics. The first one is the class of *traveling epidemic waves*, where it can be said that spread is in a wave-like manner (see section 7.6.3). A traveling wave occurs when the contact distribution is Gaussian, double exponential, or any distribution that has an exponentially bounded tail. Mollison (1972) expressed this as a more formal rule based on the moment generating function of the contact distribution. The traveling waves approach a constant rate of expansion. As far as our understanding goes, this constant rate is approached very quickly, but this does depend on the parameter values. Traveling waves for populations have been described by mathematicians as early as 1937 (Fisher, 1937). The concept has been applied to: ecology for population expansion (Skellam, 1951; Hengveld, 1989; van den Bosch et al., 1992; Kot et al., 1996; Shigesada and Kawasaki, 1997; Neubert and Caswell, 2000); animal epidemiology for disease expansion (Diekmann, 1978, 1979; Thieme, 1977, 1979; Mollison, 1972); and to genetics for the rate of expansion of a gene (Fisher, 1937). Traveling epidemic waves also have been modelled for plant disease epidemic models (Kampmeijer and Zadoks, 1977; Zadoks and Kampeijer, 1977; Minogue and Fry, 1983a, b; van den Bosch et al., 1988a–c, 1990a, b, 1999).

The second class of disease expansion is of *dispersive epidemic waves*. A dispersive wave occurs with the root and modified power-law contact distributions, or with any distribution that does not have an exponentially bounded tail (see also section 7.6.3). Waves of this type continuously keep increasing in their rate of expansion (C_n never approaches a constant). Mechanistic, statistical, and meteorological arguments for this type of disease expansion were provided by Ferrandino (1993), Minogue (1989), and Aylor (1999). Recently, the idea of dispersive waves has also received attention both in plant pathology (Aylor, 1999; Cowger et al., 2005; Frantzen and van den Bosch, 2000) and in ecology (Kot et al., 1996). As our results for the root and modified power-law distributions demonstrate, the rate of expansion of dispersive waves can increase with time in various ways. We do not know of any thorough investigation about the formal relation between the shape of contact distributions and the way the rate of epidemic expansion increases for dispersive waves, and much still has to be discovered about this.

Traveling waves and dispersive ways were introduced in the previous chapter (section 7.6.3) in terms of observed epidemics and using a more descriptive model. We discuss some other features of the spatio-temporal model of Chapter 7 in more detail later, but here we point out an aspect of the spatial component of that model. Equations 7.33 and 7.35 were not based on *explicit* consideration of the contact distribution, but dealt directly with the disease gradient (Y or y in relation to s), with correction for a maximum disease level of 1. Moreover, although the equations of the previous chapter could describe gradients of disease in relation to the original inoculum source at multiple times, they could not adequately represent spread of disease between any two arbitrary locations away from the original inoculum source. Basically, of the possible models that could have been used for the spatial component, in order to provide general agreement with observed gradients at several times in an epidemic, equations were chosen specifically to obtain either a constant expansion rate or one that increased with time. Using a model for spread in this chapter, where contact between host individuals is directly incorporated at the level of each infected and healthy individual, we showed how traveling and dispersive wave can (must) naturally develop.

The more realistic spatial component here also shows that more complicated spread behavior can occur during the very early stages of polycyclic epidemics, before the spread "settles down" into its long term behavior of either a constant expansion rate or accelerating rate. Interestingly, in the simpler modeling of Chapter 7, we showed that the expansion rate was a function of the temporal rate parameter (e.g., r_L for the logistic model). Moreover, in Chapter 5 we showed that r_E (same as r_L when y is low for a logistic epidemic) is related to R_0. Thus, the results of the previous chapter are broadly consistent with the ones here, where C is seen to be a function of $\ln(R_0)$.

8.3 Some Extensions

8.3.1 One-dimensional versus two-dimensional epidemic expansion

Model equation 8.2 describes an epidemic in a one-dimensional space. Epidemics generally develop in two dimensions, and we need to know how the rate of epidemic expansion in one dimension relates to the rate of epidemic expansion in two dimensions. If the epidemic develops from a line source which is long compared to the distance infectious units disperse (see Chapter 7), the density of infectious units (and resulting infected host individuals) at a given distance perpendicular to this line source will be the same (on average) for every position along the line source. For instance, in Fig. 8.10, the density of infected individuals at point A will be the same as the density of infected individuals at point B, and the same will hold for any point that is a perpendicular distance d from the line source. This implies that we are dealing with a one-dimensional spatial epidemic, at least in terms of the model, where the space direction is perpendicular to the line source. Only at the edges of the line source we will see differences. If the line source is long compared to the standard deviation of the contact distribution (where there is a definable standard deviation, as with the Gaussian distribution), then the complications of an edge effect will not be of concern.

Now consider an epidemic developing from a point source, and assume that the two-dimensional space is homogeneous, meaning that all environmental circumstances are the same at every location, and the density of the crop is the same at every point. Under this assumption, the epidemic will be *rotationally symmetric* at every moment in time (at least in the model). Fig. 8.10 shows the epidemic developing from a point source (similar to Fig. 7.17). Contour lines for a fixed density of infected individuals are drawn for various points in time. These contour lines are concentric circles. When these concentric circles are large enough, locally they look like a straight line. That is, in the vicinity of any point on a circle, the circle appears a straight line in the same way as any uninformed observer would consider the world to be a flat disk based on local observations. This implies that such an epidemic can *eventually* be approximated (after enough time so that the circle circumference is not too small) by an epidemic expanding in one dimension, in which the direction is parallel to the circle. The one-dimensional model of this chapter thus applies to epidemics that develop from a point source as well, if we take care to use the proper form of the contact distribution, as will be discussed below.

There is, however, still the question left of what the general relation is between the two-dimensional contact distribution around an infectious individual and the one-dimensional contact distribution perpendicular to a line of infectious individuals. Consider a two-dimensional contact distribution $D(s_1, s_2)$. This distribution gives the scaled probability per unit area of an infectious unit originating at $(0, 0)$ being at (s_1, s_2). Equivalently, $D(s_1, s_2)$ is the fraction per unit area of the new infections at position (s_1, s_2) due to an infectious individual at $(0, 0)$. We could have just as easily taken any other position as the origin (ξ_1, ξ_2); then the contact distribution is written as $D(s_1 - \xi_1, s_2 - \xi_2)$. Let us assume a line source is parallel to the ordinate of our coordinate system. Fig. 8.10 shows an example in the lower right corner. Such a line source can be thought of as composed of a linear series of point sources. Drawn in Fig. 8.10 is a series of contour lines depicting $D(s_1, s_2)$ distributions around arbitrary points on the line. The contact distribution for the line source in *just* the s_1 direction, $D(s_1)$, is the scaled probability of infectious units on a line perpendicular to the line source. The figure

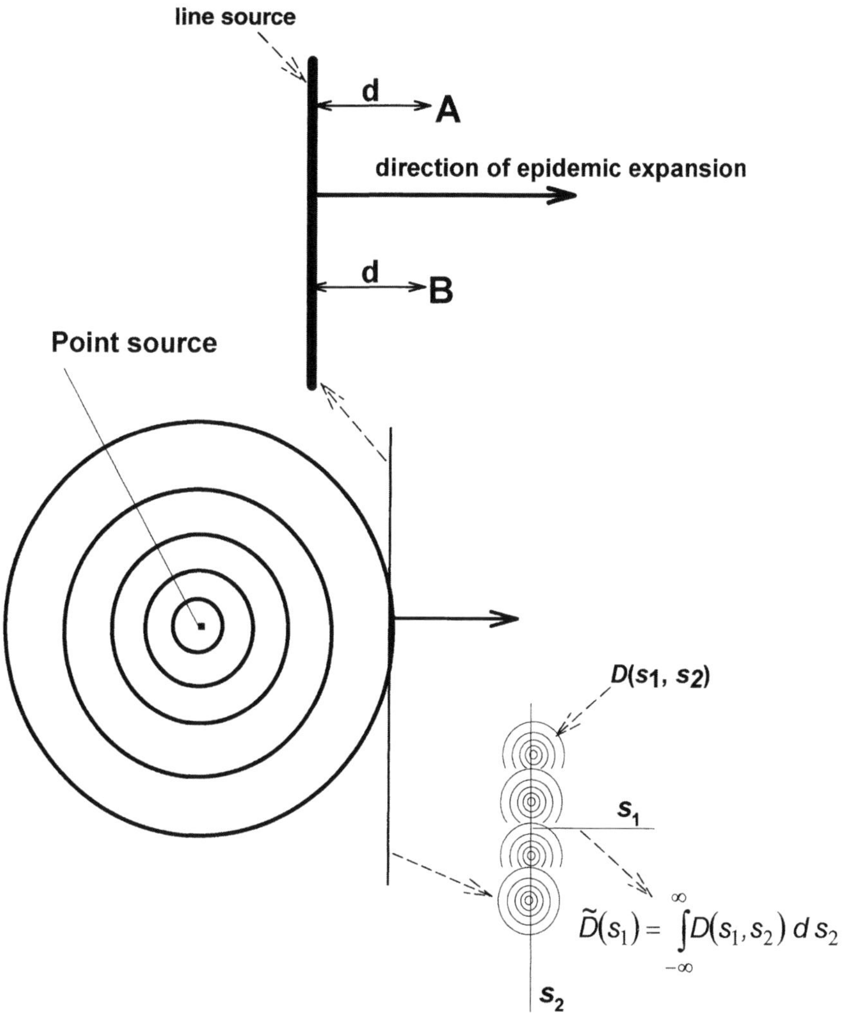

FIG. 8.10. Schematic for the relation between one-dimensional and two-dimensional disease spread in epidemics. See text for explanation.

attempts to make clear that the one-dimensional contact distribution can be calculated by adding up all contributions of all point sources, which is equivalent to the summation over all possible values of contact probability across the second spatial dimension (s_2). Assuming continuous space, we use integration instead of summation and write:

$$\tilde{D}(s_1) = \int_{-\infty}^{\infty} D(s_1, s_2)\, ds_2 \qquad (8.19)$$

$\tilde{D}(s_1)$ is known in statistics as the *marginal distribution* of the bivariate distribution $D(s_1, s_2)$.

Each contact distribution of Table 8.1 can be considered the marginal distribution of some bivariate distribution. Although one can, in principle, determine the marginal distribution for a given bivariate distribution, the opposite is not necessarily true—one cannot necessarily determine the bivariate distribution for a given marginal distribution. For the Gaussian distribution, the relationship is straightforward—the marginal and bivariate forms are both Gaussian (see Johnson and Kotz, 1970).

Some papers that deal with two-dimensional models for contact distributions include: Aronson and Weinberger (1978) and Weinberger (1978, 1982). We do not further work with explicit two-dimensional contact distributions in this chapter.

8.3.2 Continuous time and more realistic temporal dynamics

8.3.2.1 Model and simulations. Model equation 8.2 can be extended to represent epidemics with: overlapping disease generations (i.e., with continuous time); latent and infectious periods; dependence of the rates on healthy host density (i.e., for the density-dependent situation, when $H(t)$ is not considered fixed at H_0); and healthy host-density growth in the absence of disease. These extensions are similar to the way our simplest epidemic progress model (equations 5.2a and 5.2b) was extended in Chapter 5. The H-I-R, H-L-I-R, Vanderplank differential-delay, and the Kermack-McKendrick models can all be extended to allow for an explicit spatial component (Diekmann, 1978, 1979; Thieme, 1977, 1979).

For brevity, we consider only a spatial extension of the H-L-I-R model discussed in section 5.2.3. The densities of healthy, latent, infectious and removed individuals are now a function of both time, t, and position (location) in space, s, $H(t, s)$, $L(t, s)$, $I(t, s)$, and $R(t, s)$, respectively. We now assume that t and s are both continuous variables. The model is given in pseudo-equation form as:

$$\begin{aligned}
\frac{\partial H(t,s)}{\partial t} &= -\begin{bmatrix}\text{rate at which healthy}\\ \text{individuals become infected}\end{bmatrix}\\
\frac{\partial L(t,s)}{\partial t} &= \begin{bmatrix}\text{rate at which healthy}\\ \text{individuals become infected}\end{bmatrix} - \omega L(t,s)\\
\frac{\partial I(t,s)}{\partial t} &= \omega L(t,s) - \mu I(t,s)\\
\frac{\partial R(t,s)}{\partial t} &= \mu I(t,s)
\end{aligned} \tag{8.20}$$

where the $\partial/\partial t$ notation means the *partial* derivative with respect to time, with the same interpretation as in the differential equation models of Chapter 5. The rate of infection term at position s now depends, as in the discrete-generation equation 8.2, on infectious units (e.g., spores) that are produced by infectious individuals at all possible positions ξ. The density of infectious individuals at time t at position ξ is represented by $I(t, \xi)$. The probability that an infectious unit (e.g., spore) produced by an infectious individual at ξ will contact a host individual at position s is now given by the contact distribution, $D(s - \xi)$, multiplied by θ'. Spores may be deposited between plants, for instance. Thus, θ' reflects the fact that some infectious units never contact a host anywhere. If each infectious individual produces α spores per time unit, the total influx of spores at position s per time unit due to all infectious individuals at position ξ is given by $\alpha I(t, \xi)D(s - \xi)\theta'$. To calculate the total influx rate of spores at position s due to all infectious individual in the crop we need to add up all contributions of all infectious individuals at all possible positions ξ. This amounts to integrating over ξ and we have:

$$\begin{bmatrix}\text{rate of spore influx}\\ \text{at postition } s\end{bmatrix} = \alpha\theta' \int_{-\infty}^{\infty} I(t,\xi)D(s-\xi)\,d\xi \tag{8.21}$$

Further, noting that each infectious unit in contact with a host has a probability ψ to cause an infection, and following the same reasoning as for the derivation of the infection term in the H-L-I-R model in section 5.2.3.1, we find that:

$$\begin{bmatrix}\text{rate at which healthy}\\ \text{individuals become infected}\end{bmatrix} = \alpha\psi\theta' H(t,s) \int_{-\infty}^{\infty} I(t,\xi)D(s-\xi)\,d\xi \tag{8.22}$$

The model equations become:

$$\begin{aligned}
\frac{\partial H(t,s)}{\partial t} &= -\alpha\psi\theta' H(t,s)\int_{-\infty}^{\infty} I(t,\xi)D(s-\xi)\,d\xi\\
\frac{\partial L(t,s)}{\partial t} &= \alpha\psi\theta' H(t,s)\int_{-\infty}^{\infty} I(t,\xi)D(s-\xi)\,d\xi - \omega L(t,s)\\
\frac{\partial I(t,s)}{\partial t} &= \omega L(t,s) - \mu I(t,s)\\
\frac{\partial R(t,s)}{\partial t} &= \mu I(t,s)
\end{aligned} \tag{8.23}$$

We can simplify the expression by defining the transmission rate as $\beta = \alpha\theta'\psi$, as done in Chapter 5, and use β in equation 8.23. It is important to note that θ' may not always be explicitly used in the model formulation. That is, any scaling of the contact distribution simply becomes an implicit part of the transmission parameter (together with sporulation rate and probability of infection).

Note that this model *does* include density dependence in terms of disease-free host density. Minogue and Fry (1983a, b) analyzed a model including similar extensions of a purely temporal model to a spatio-temporal model, but they used the Vanderplank specification of latent and infectious periods (see equation 5.44). Models for the spatial expansion of epidemics also have been developed to include the more general sporulation ("infectivity") curve approach of Kermack and McKendrick (1927) for the latent and infectious periods as described in section 5.2.5. These models were first developed and analyzed by Diekmann (1978, 1979) and Thieme (1977, 1979) and were subsequently operationalized by van den Bosch et al. (1988a–c, 1990a, b, 1992, 1999) to analyze biological invasions and the expansion of plant disease foci.

By summing the partial differential equations for latent, infectious, and removed disease states, one can determine the equation for the change in total disease at a given time and location:

$$\frac{\partial Y(t,s)}{\partial t} = \beta H(t,s)\int_{-\infty}^{\infty} I(t,\xi)D(s-\xi)\,d\xi \tag{8.24}$$

When density of diseased individuals is low, $H(t, s)$ is near the initial total host density, H_0 (and we assume here that H_0 does not depend on location). We can thus assume that $H(t, s) \cong H_0$. Then, equation 8.24 can be approximated by:

$$\frac{\partial Y(t,s)}{\partial t} = \beta H_0\int_{-\infty}^{\infty} I(t,\xi)D(s-\xi)\,d\xi \tag{8.25}$$

It is helpful to compare equation 8.24 with the one for a purely temporal epidemic given in the previous chapter (section 5.2.3) for the H-L-I-R model. In equation 5.31, the density of new infections at a given time is based on the density of disease-free individuals and infectious individuals for the entire population, $H(t)$ and $I(t)$, respectively, with $\beta H(t)I(t)$ as the transmission term. In equation 8.24, the density of new infections at a given time at a given location depends on healthy host density at that location, $H(t, s)$, and on infectious units over the entire area (all possible ξ values) in which the epidemic occurs. By the nature of contact distributions (see Fig. 8.5), infectious individuals at or near location s contribute more to new infections than infectious units farther away.

The basic reproduction number for this model is, as with the non-spatial situation, defined as $R_0 = \beta H_0/\mu$. This is because one infected individual in an otherwise disease-free population of H_0 individuals produces a total of α/μ $(=\alpha \cdot \iota)$ spores (for the example of a spore-producing fungus) during its infectious period, with $\alpha\theta' H_0/\mu$ spores contacting a host (somewhere), and with $\alpha\theta'\psi H_0/\mu$ resulting infections. In other words, the contact distribution does not affect the calculation of R_0.

There is no analytical solution to equation 8.23; rather, numerical solutions must be used, based on use of a computer program such as MATHCAD or MATHEMATICA. Because the epidemic model now involves partial derivatives, the required numerical integration methodology is more involved than that used for numerically solving the ordinary differential equations of Chapter 5. An example solution is shown for selected times as the heavy (thick) curves in Fig. 8.11, based on the double-exponential contact distribution (see Table 8.1). The pattern of disease gradients is very similar to those found for the more descriptive logistic-logistic spatio-temporal model of Chapter 7 (see Fig. 7.14) and, for values of Y not close to H_0, for the non-overlapping generation (and density-independent) epidemic model (Fig. 8.6B). Density of diseased individuals approaches the final disease level, which depends on R_0. When the ordinate is plotted on a log scale, a straight-line relation with distance is seen, except for densities above 100 or so, where the density dependence is most apparent. The simulation also shows an apparent equal distance between the gradient lines at low values of $Y(t, s)$, implying that the rate of disease expansion is constant (with this contact distribution). Because we are now dealing with continuous time, we refer to the velocity of disease expansion as $C(t)$ instead of as C_n, which is equivalent to $\partial s/\partial t$ discussed in the previous chapter (change in distance where a given level of disease is found with change in time).

When the modified power law distribution is used, a different pattern is found for the gradients (Fig. 8.12). In particular, even when $Y(t, s)$ is plotted on a log scale, there is strong curvature, especially at large distances from the source. The pattern is very similar to that found for the non-overlapping generation epidemic model with the same contact distribution (Fig. 8.6D), except for large Y. As seen in Fig. 8.6D, there is a very apparent increasing distance between the gradient curves at low values of $Y(t, s)$.

8.3.2.2 Disease expansion rates—traveling waves. In the temporal models discussed in Chapter 5, we took great care to describe the density of healthy individuals and how this density affects the number of new infections

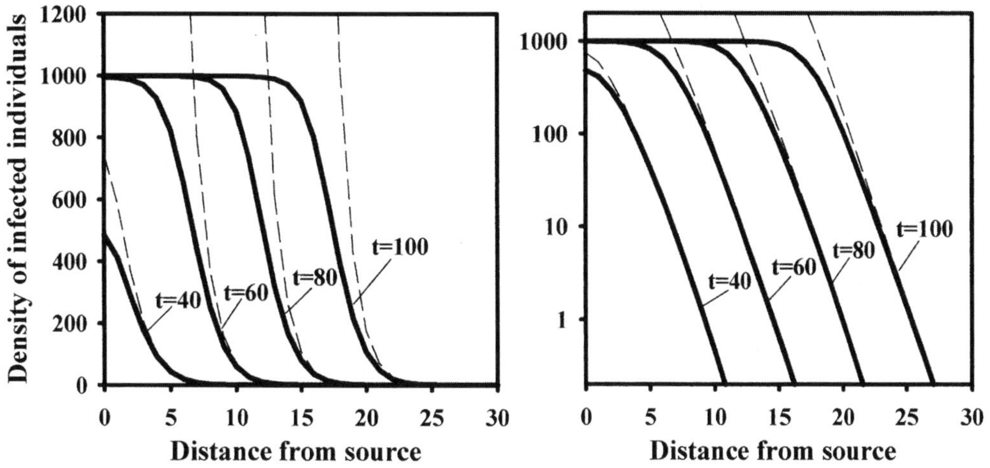

FIG. 8.11. Numerical solution of the spatial H-L-I-R epidemic model, based on use of a double-exponential contact distribution (with $\sigma = 1$) (see Table 8.1). Vertical axis is disease density, $Y(t, s)$, and horizontal axis is distance from the original inoculum source. Thick solid lines are gradients of disease for the density-dependent situation (equation 8.23), at selected times, t. Thin dashed lines are disease gradients for the density-independent situation (equation 8.23, with $H_0 = 1000$ used instead of $H(t, s)$). Note that in the front of the epidemic (large distances with low $Y(t, s)$), the two sets of gradients merge. Disease spread was in both directions from the original source (i.e., for positive and negative s), but only positive distances are shown. Other parameter values were: $\beta H_0 = 0.6$, $\omega = 1/7$, and $\mu = 1/10$.

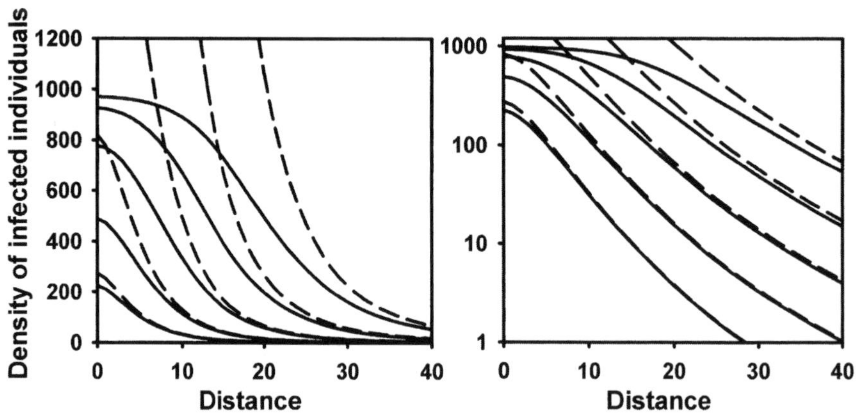

FIG. 8.12. Numerical solution of the spatial H-L-I-R epidemic model, based on use of a modified power law contact distribution (with $\Psi = \lambda = 3$) (see Table 8.1). Vertical axis is disease density, $Y(t, s)$, and horizontal axis is distance from the original inoculum source. Thick solid lines are gradients of disease for the density-dependent situation (equation 8.23), at selected times, t. Thin dashed lines are disease gradients for the density-independent situation (equation 8.23, with $H_0 = 1000$ used instead of $H(t, s)$). Note that in the front of the epidemic (large distances with low $Y(t, s)$), the two sets of gradients merge. Disease spread was in both directions from the original source (i.e., for positive and negative s), but only positive distances are shown. Other parameter values were: $\beta H_0 = 0.6$, $\omega = 1/7$, and $\mu = 1/10$.

produced per unit area per time unit. This is, of course, for the obvious reason that if the density of healthy individuals decreases during the epidemic, the number of new infections (for a given density of infectious individuals) will automatically also decrease. In other words, there is less opportunity for new infections as $H(t)$ declines (e.g., more and more spores land on diseased individuals). The effect of such density dependence on the rate of epidemic expansion has been studied by mathematicians in quite some detail. The work on this goes back to Kolmogorov et al. (1937) and Bramson (1983). Density dependent extensions of model equation 8.2 have been studied in detail by Weinberger (1978, 1982), and continuous-time versions of the model have been studied by Diekmann (1978, 1979), Thieme (1977, 1979) and Mollison (1972).

These studies considered the situation where the tail of the contact distribution falls off faster than exponentially with distance from the source. No particular contact distribution is used, so that general results are valid for *any* contact distribution with exponentially bounded tails. As we have discussed in section 8.2 for non-overlapping generations, this type of contact distribution implies that a traveling wave develops when there is no density dependence. A remarkable achievement of these cited theoretical studies is that *the rate of epidemic expansion in the density-dependent models is the same as the rate of expansion calculated from the density-independent models.* In other words, the results show that $C(t)$ approaches a constant velocity (C) and this velocity is the same in density dependent and density independent models. All the studies use a similar type of density dependence in the model formulation, which specifies that: the number of new infections produced per infectious individual at places and times where the density dependence matters is always less than when density dependence does not matter. In other words, $\partial Y(t, s)/\partial t$ decreases as $H(t, s)$ decreases for a given density of infectious individuals (since there are fewer opportunities for infection at location s). Moreover, the mathematical work assumed that the form of the contact distribution is not influenced by the density of healthy, infected and removed individuals. (Actually these authors use an even more general assumption, which we do not deal with here.)

The highly significant result can easily be understood intuitively by returning to Fig. 8.11. Superimposed on the disease density curves generated from density-dependent equation 8.23 are the disease gradients for the density independent case (i.e., with H_0 for $H(t, s)$ in equation 8.23), drawn as thin broken lines. Put another way, the thick (heavy) lines are the solutions of more complex equation 8.24, and the thin lines are the solutions of the simpler equation 8.25 for $\partial Y(t, s)/\partial t$. All parameter values are the same for both epidemic models. As expected, the use of the density-independent model ultimately results in meaningless results close to the original inoculum source when, in fact, there is an upper limit to disease density (H_0). In particular, prediction of $Y(t, s)$ can exceed H_0, given sufficient time. However, the fronts of the epidemics (i.e., Y values relatively far from the origin) are virtually identical for the two situations.

It may be helpful to the reader to refer back to the theory of polycyclic epidemics described in Chapter 5. Even for models of temporal disease progress for polycyclic diseases, an exponential growth model provided a good approximation when disease density was low. In a general sense, this approximation at low Y carries over to spatio-temporal models, for both the time and space dimensions. That is, because $H_0 \approx H(t, s)$ at small $Y(t, s)$, simplifications are possible. So, no matter how complicated the model is for dynamics at moderate or high disease density, much simpler models are applicable far from the original inoculum source. The expanding epidemic is drawn forward through space by the infectious individuals at the forefront of the epidemic. It is exactly in the front of the epidemic where the density of infected individuals is so small that the dynamics of the epidemic is not affected by any (healthy-host) density-dependent process.

Several important mathematical results are obtainable based on the use of a density-independent model (Diekmann, 1978, 1979; Thieme, 1977, 1979) for the epidemic. Without mathematical proof, we emphasize the key ones related to the velocity or rate of disease expansion, $C(t)$. A major result is that after an initial period of time, $C(t)$ approaches an asymptotic constant velocity (C) when the contact distribution is exponentially bounded. As with the model with discrete generations, and as demonstrated in Fig. 8.11 for one set of parameters, an exponentially bounded contact distributions leads to a traveling epidemic wave. In particular, C can be written as an approximate function of R_0:

$$C \cong A + B\ln(R_0) \tag{8.26}$$

in which A and B are constants. These constants depend on other model parameters, such as σ and ω ($1/\wp$), but the exact functional dependencies will be influenced by the contact distribution used. We have not derived the relations for various contact distributions here but merely used the close-to-linear relation between C and $\ln(R_0)$ seen in Fig. 8.9. Based on equation 8.26, one can see that the farther spores (or other infectious units) travel, as indicated by σ, and the greater the new infections per each infectious unit, as indicated by R_0, the greater the velocity of focus expansion. From a control perspective, one can limit the invasion of a disease into new areas by reducing σ (perhaps with barriers) or by reducing R_0.

An important point is that it does take some time for the constant velocity, C, to be approached. The time period is not easy to define exactly (van den Bosch et al., 1990b), but depends on all model parameters. No estimates of convergence time exist and further work in this area is needed. Observed epidemics confirm that there is a time period before a constant expansion rate is found (Minogue and Fry, 1983b; van den Bosch et al., 1988c).

8.3.2.3 Disease expansion rates—dispersive traveling waves. It appears that the density-independent version of equation 8.23 (H_0 used for $H(t, s)$) accurately describes the front of disease expansion also when the contact distribution is not exponentially bounded. This is demonstrated in Fig. 8.12, where the thin broken lines correspond to numerical solutions of the non-density dependent version of equation 8.25. However, the mathematical theory has not been adequately developed for these types of contact distributions to give more detailed results. By referring back to the non-overlapping generation model, one should expect that the increase in $C(t)$ is proportional to the scale of contact distance ($1/\Psi$ in root, and λ/Ψ in modified power law distributions). Based on continuous-time results for exponentially-bounded distributions, one would also expect that $C(t)$ is inversely related to mean latent period. More research is clearly needed here, because many diseases are often characterized by power-law type gradients (e.g., Sackett and Mundt, 2005a), as discussed and demonstrated in the previous chapter. Consistent with a power law type of gradient, calculated velocities often do increase with time (Cowger et al., 2005; Madden et al., 1990a; Sackett and Mundt, 2005b).

8.3.2.4 Multi-seasonal epidemic expansion. van den Bosch et al. (1999) developed a model to describe both the within growing season and the multi-seasonal expansion of plant disease epidemics on a continental scale. We do not give the theory here, but the model describes the dynamics of a polycyclic epidemic during the crop growth season and combines this with a simple description of pathogen survival through the season where no crop host is present.

8.3.3 Disease expansion with monocyclic diseases

The basic concepts used here for continuous time and for discrete non-overlapping generations can easily be applied to monocyclic diseases, in which the only infections occur from the primary inoculum (the primary infectious units, such as overwintering spores in the soil or spores produced on weeds on the edge of a field). We consider only the continuous time scenario, and build on a portion of the model introduced in section 5.4 for the situation where primary infectious occur throughout the epidemic. Readers should re-study that section before reading the following material.

We can define $x(t, s)$ as the density of primary infectious units available for infection ("released spores") at time t and location s (with units of 1/time). We can consider $x(t, s)$ to be fixed, or to change over time. $\theta' D(s - \xi)$ is the probability that a primary infectious unit originating at ξ comes in contact with a host individual at s, and ψ_p is the probability that an infectious unit in contact with a host causes an infection. If there are no secondary infections ($\beta = 0$), one can write:

$$\frac{\partial Y(t, s)}{\partial t} = \theta' \psi_p H(t, s) \int_{-\infty}^{\infty} x(t, \xi) D(s - \xi) d\xi \tag{8.27}$$

A comparison of this equation with equation 8.24 shows that $x(t, s)$ is used here instead of spores produced by infectious diseased individuals [$\alpha I(t, \xi)$]. For the simpler situation where there is a single inoculum source, at $s = 0$, the equations simplifies to:

$$\frac{\partial Y(t, s)}{\partial t} = \theta' \psi_p H(t, s) x(t, 0) D(s) \tag{8.28}$$

No integration is necessary over space, since inoculum arriving at s can only be coming from *one* location (s units away). An even simpler model is possible if one can assume that inoculum density does not change, so that $x(t, 0)$ is a constant ($\equiv x_0$). Obviously, a fixed density of available inoculum is not to be expected exactly, but it may be quite a reasonable assumption over the time period of concern. Then one can write:

$$\frac{\partial Y(t, s)}{\partial t} = \theta' \psi_p x_0 D(s) H(t, s) \tag{8.29}$$

Equation 8.29 is a spatial expansion of the monocyclic model of Chapter 4 (equation 4.8), in terms of Y (rather than y), where $\theta' \psi_p x_0$ is analogous to the simple-interest rate parameter r_M. Minogue (1986) showed how to use and interpret this model in terms of disease expansion. Interestingly, there is an analytical solution for this model with a single inoculum source.

$$Y(t, s) = H_0(1 - \exp(-\theta' \psi_p x_0 D(s) t)) \tag{8.30}$$

Equation 8.30 assumes that $Y(0, s)$ ($= Y_0$) is 0 everywhere and the epidemic starts with all inoculum at $s = 0$. A numerical example is shown in Fig. 8.13 for the double-exponential contact distribution (see Table 8.1) and one set of parameters. An interesting result is that the rate of disease expansion ($C(t)$) appears to decline over time; that is, the distance between the curves at a single (low) value of Y declines over time.

Because there is an analytical solution for this model, one can readily determine the rate of disease expansion. With the double-exponential contact distribution, a little algebraic manipulation of equation 8.30 results in:

$$\ln(-\ln(1 - y(t, s))) = \ln\left(\frac{\theta' \psi_p x_0 t}{\sigma\sqrt{2}}\right) - \frac{\sqrt{2}}{\sigma}|s| \tag{8.31}$$

in which $y(t, s) = Y(t, s)/H_0$. The left-hand side of equation 8.31 is the so-called complementary log-log transformation (CLL), which is used more extensively in Chapter 9 for other purposes. The CLL declines with distance, with a slope of $\sqrt{2}/\sigma$, and increases with the log of time. To determine $C(t)$, one can first define a fixed value of $y(t, s)$, say y', and find the value of s that gives y' ($s_{y'(t)}$, or s' for short). This results in an equation for the isopath in relation to time:

$$s' = \frac{\sigma}{\sqrt{2}}\left(\ln\left(\frac{\theta' \psi_p x_0 t}{\sigma\sqrt{2}}\right) - \ln(-\ln(1 - y'))\right) \tag{8.32}$$

The rate of disease expansion ($C(t)$) can be obtained by determining the partial derivative of s' with respect to t ($\partial s'/\partial t$). For equation 8.32, the partial derivative is given by:

$$\frac{\partial s'}{\partial t} = C(t) = \frac{\sigma}{t\sqrt{2}} \tag{8.33}$$

Based on equation 8.33, one can see that $C(t)$ is directly proportional to σ, as found for polycyclic diseases (see section 8.2)—the longer the distance traveled by spores, for example, the greater the rate of disease expansion.

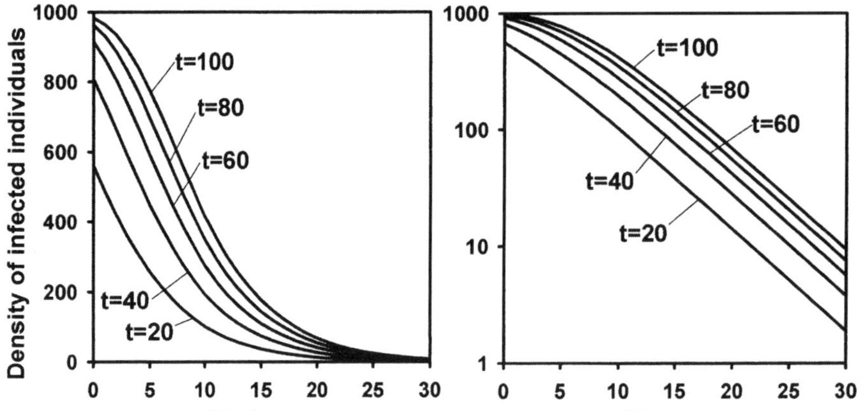

FIG. 8.13. Numerical solution of the spatial monocyclic epidemic model for a single inoculum source at distance 0 (equation 8.30), based on use of a double-exponential contact distribution (with $\sigma = 7$) (see Table 8.1). Vertical axis is disease density, $Y(t, s)$, and horizontal axis is distance from the inoculum source. Solid lines are gradients of disease at selected times, t. Other parameter values: $H_0 = 1000$, and $\theta' \psi_p x_0 = 0.4$.

Moreover, $C(t)$ is inversely proportional to time since the start of the epidemic. This declining expansion rate makes intuitive sense, because infections do not lead to further infections with monocyclic diseases, and diseased individuals at the front of the epidemic do not move the disease forward in space. In other words, when all infections beyond the origin originate exclusively from inoculum at the origin, there is a decline in expansion over time (even with a fixed inoculum density at the source).

Equation 8.30 is applicable for any of the contact distributions used in this chapter. However, equations 8.31–8.33 are specific for the double-exponential distribution. Other contact distributions will lead to different formulations for s' and $C(t)$, sometimes with more complicated equations for $C(t)$.

The results here should be contrasted with those obtained in Chapter 7. It was found in section 7.6.4 to be challenging to link the monocyclic temporal model with a logical disease-gradient model and obtain an easy-to-use, realistic, and interpretable combined model. By being more realistic in the model formulation, equation 8.30, and using contact distributions consistent with observed dispersal gradients, an interpretable spatio-temporal model is readily obtained.

8.3.4 Multiple foci and temporal dynamics

The previous sections of this chapter showed that a considerable understanding of the spread of disease can be obtained using a range of models that capture the essential biological features of epidemics in time and space. Depending on the model formulation, exact or approximate equations can be developed to predict the rate of disease expansion from a single focus. Of course, many epidemics are initiated from several foci instead of one, but, fortunately, the principles of spread expressed in this chapter hold for any number of foci. With multiple original foci, the concept of a simple single expansion rate for the epidemic no longer applies. There is, in fact, expansion of each disease focus, possibly with mergers of originally distinct foci, as well as a resulting "global" expansion rate at the edge of the full spatial extent of the epidemic.

Models such as the spatial H-L-I-R one in equation 8.23 can be numerically solved using mathematics computer programs for any set of initial conditions $[Y(0, s)]$. For example, infectious individuals can be located at any number of locations over the spatial domain. With multiple foci, graphs of observed disease gradients or predicted disease gradients from equation 8.23 (such as in Figs. 8.11 and 8.12) are of less direct interpretation. This is because there is no longer an unambiguous $s = 0$ point in terms of the starting conditions with multiple original foci. For this situation, maps of (observed or predicted) disease density at selected times are often produced instead of simple curves of density versus distance. Maps of disease intensity will be considered in detail starting in the next chapter.

One can also use spatio-temporal models (such as equation 5.23) to determine disease progress curves for the *entire* field or region in which the epidemic is occurring, not just for specific locations, based on any starting conditions. As done in Chapters 4 and 5, we can define $Y(t)$ as the total number of diseased individuals over all locations at time t. With explicit consideration of space, $Y(t)$ is given by:

$$Y(t) = \int_{-\infty}^{\infty} Y(t, s)\, ds .$$

One can then explore the effects of temporal parameters [e.g., β $(=\alpha\theta'\psi)$, ω $(=1/\wp)$, and μ $(=1/\iota)$], spatial parameters (σ for certain contact distributions), and initial conditions on the disease progress curves. This is demonstrated as line 3 (l_3) in Fig. 8.14 for an example

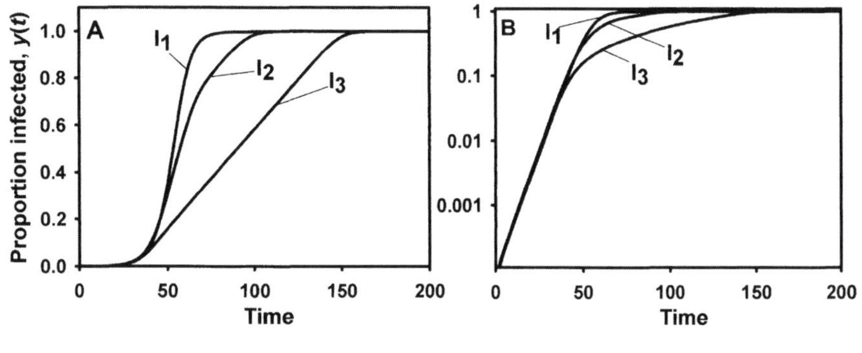

FIG. 8.14. Total proportion of diseased individuals across all locations, $y(t)$ $[=Y(t)/H_0]$, versus time, t, for a spatial H-L-I-R model (equation 8.23) with a double-exponential contact distribution ($\sigma = 1$; see Table 8.1), or aspatial model (equation 5.30). (**A**) arithmetic scale and (**B**) log scale. Line l_1: realization for equation 5.30, which corresponds to homogeneous mixing. Line l_2: realization for equation 8.23, with three evenly inoculum line sources. Line l_3: realization of equation 8.23, with a single inoculum line source in the center of the spatial range. Total density of diseased individuals at $t = 0$ was the same for all three epidemics. Other parameter values were: $\beta H_0 = 0.6$, $\omega = 1/7$, and $\mu = 1/10$.

epidemic where s ranges from 0 to 60, with one initial focus in the field center. Results are expressed as proportions, $y(t) = Y(t)/H_0$. The double exponential contact distribution was used, and the parameters are the same as used in generating the gradients in Fig. 8.11. Line 1 (l_1) is corresponds to the $y(t)$ values for predicted based on the aspatial H-L-I-R model of Chapter 5 (equation 5.30) with the same initial density of infected individuals and the same model parameters. With the aspatial model, homogeneous mixing of the infectious units throughout the host population is implied (i.e., contact does not depend on position).

As seen in the graph, aggregation of disease into a single initial focus results in a slower increase in total disease density (not necessarily at any given location) than when there is homogeneous mixing of inoculum, at least for the chosen value of σ. This is because, for the typical situation of a contact distribution that declines rapidly with increasing distance, infectious units produced by an infectious individual are likely to come in contact with a previously infected individual when the diseased individuals are mostly near each other to start with. Careful evaluation of Fig. 8.14 shows that the difference between the two cases is virtually nonexistent at low $y(t)$, since there are many disease-free host individuals at all locations to come in contact with inoculum. Then, the differences in the two cases become apparent as the effect of the declining disease-free hosts is manifested. Ultimately, the same final disease density is reached for both cases. It should be noted that a larger value of σ would diminish the effect shown in the figure. This is because infectious units travel greater distances, on average, resulting is a lower likelihood that an infectious unit will come in contact with a diseased host.

The line 2 (l_2) in Fig. 8.14 corresponds to the same situation as the l_3 line, but with three evenly separated foci at time 0, instead of one focus. Initial total density of diseased individuals at time 0 was the same. The rate of disease increase was, again, lower than for the aspatial (homogeneous mixing) case (l_1), but the difference was not as great as between the aspatial case and the single-focus case. With well separated initially diseased individuals, there is less likelihood that spores will contact diseased host individuals than when all diseased individuals are initially together.

Various efforts have been made to account for some aspects of the spatial component of epidemics using strictly temporal models of disease progress (Gubbins and Gilligan, 1997; Hughes et al., 1997; Liu et al., 1987; Madden and Campbell, 1990; Waggoner and Rich, 1981; Yang and TeBeest, 1992; Zhang et al., 2000b). In particular, the proposed models account for, in principle, the non-homogeneous mixing of the infectious units without explicit consideration of the spatial location of individuals or the contact distribution. Sometimes the equations are called *spatially implicit models* of disease progress. The basic approach is to replace $I(t)$ and $H(t)$ in the transmission term ($\beta I(t)H(t)$) in temporal models (e.g., in equation 5.30) with nonlinear functions of infectious (or more simply, infected [$Y(t)$]) and/or healthy host density, $f_I(I(t))$ and $f_H(H(t))$, respectively. The equation component for $dH(t)/dt$ in the H-L-I-R model would then be written as:

$$\frac{dH(t)}{dt} = -\beta^* f_I(I(t)) \cdot f_H(H(t)) \tag{8.34}$$

This type of modification can also be used in models with a dynamic host crop in the absence of disease (see section 6.2 of Chapter 6), as well as in models that include vector dynamics (see section 6.4). The "transmission parameter" is written as β^* because it is no longer simply interpreted just as $\alpha\theta'\psi$. Power functions are common choices for $f_I(I(t))$ and $f_H(H(t))$, such as I^a and H^b (Liu et al., 1987). Then, a special case is when $a = b = 1$, where the standard (homogeneous mixing) temporal model is obtained. Values of a or b less than 1 are indicative of non-homogeneous mixing. For instance, with $b = 0.8$, the reduction in healthy plants per unit time (or the increase in infected plants) will be less than when $b = 1$ (homogeneous mixing), since $H^{0.8} < H$. This would be, in principle, accounting for the fact that there are fewer contacts between infectious units and healthy plants with non-homogeneous mixing than with homogeneous mixing.

More complicated functions can also be used in the transmission expression (see equations 6 and 7 in Hughes et al., 1997). Although equation 8.34 and similar models can provide good fits to observed or simulated disease progress curves, there is no formal quantitative linkage between a and b (or more generally, the explicit functions of I and H) and the underlying contact-distribution parameters and the initial spatial conditions (i.e., locations of initially diseased individuals). Thus, this approach should only be considered an approximation to represent the increase in density of diseased individuals for explicit spatio-temporal epidemic processes.

As will be shown in Chapter 9, there are useful alternatives to the methods discussed in this and the previous chapter for characterizing and quantifying the spatial component of epidemics. These approaches are based on *statistical* models of patterns of disease intensity rather than mathematical models of contact distributions and disease gradients. These statistical models also lead to modifications of temporal epidemic models that are in the same format as general equation 8.34 (Hughes et al., 1997; McRoberts et al., 1996; Waggoner and Rich, 1981) although we do not give more details in this book. The general conclusion from this work is that increasing aggregation of disease leads to slower rates of increase in $Y(t)$ over time, as found here based on explicit consideration of the spatial component of the epidemic (Fig. 8.14).

Modification of an H-L-I-R (or similar) model with power functions is analogous to modification of the

$1 - y$ term in the logistic equation (equation 4.14) to obtain the Richards growth model (see equation 4.22). By adding one or two parameters to the model, one would expect better fits to the data than obtained without the new parameters. This is, indeed, generally found with experimental data (e.g., Gubbins and Gilligan, 1997; Zhang et al., 2000b). The next research step will be to develop spatially implicit models (which could include modifications of the ones currently in use) that can be more closely related to the parameters of the contact distribution and the initial spatial conditions.

8.4 An Application

The effect of mixing two or more cultivars to control plant diseases is well documented (Browning and Frey, 1969; Mundt and Browning, 1985; Wolfe, 1985). In this section, we consider the rate of epidemic expansion in cultivar mixtures to demonstrate an application of the above model. We consider a mixture of a susceptible and a completely resistant cultivar and study how the fraction of the crop consisting of the susceptible cultivar, φ, influences the rate of expansion (C). We use results for two well-studied pathosystems, stripe rust (caused by *Puccinia striiformis*) of wheat, and rust (caused by *Uromyces appendiculatus*) on bean (van den Bosch et al., 1988c, 1990a; Assefa et al., 1995). For these pathosystems we assume here that the spatial expansion of the disease may be described by a traveling wave (which is consistent with some of the published studies).

In this section, we will derive an equation that relates the fraction of susceptible crop, φ, in a mixture of susceptible and completely resistant cultivars, to the rate of focus expansion. It turns out that the other model parameters, such as R_0, σ, and $\wp$ can be condensed into two constants, and that their numerical values are not needed to either derive our hypothesis about the relation between φ and C, nor to test the hypothesis we derive.

As indicated above, C for both the Gaussian and the double exponential contact distributions can be predicted by equation 8.26. Suppose that our mixture is a random mixture of susceptible and resistant plants. The infectious units produced by an infectious individual are then distributed over the two cultivars in proportion to their fractions in the mixture. This implies that a fraction φ of the infectious units end up on the susceptible cultivar and a fraction $1 - \varphi$ on the resistant cultivar. Only the infectious units on the susceptible cultivar cause infections and, therefore, the basic reproduction number realized in the mixture is:

$$R_0 = \varphi \tilde{R}_0 \tag{8.35}$$

where we use $\tilde{R}_0$ to represent the basic reproduction number of the disease when the crop consists only of susceptible cultivar. Substituting equation 8.35 into 8.26 gives:

$$C = A + B\ln(\varphi \tilde{R}_0) \tag{8.36}$$

which can also be written as:

$$C = \tilde{C} + B\ln(\varphi) \tag{8.37}$$

where $\tilde{C} = A + B\ln(\tilde{R}_0)$, the (asymptotic) constant rate of expansion when the crop consists of susceptible cultivar only.

Equation 8.37 suggests that the rate of disease expansion is linearly related to the logarithm of the fraction of susceptible cultivar in the crop. This is a hypothesis that can be tested experimentally. Plots have to be established with a range of seed mixtures to cover the range of proportions, from $\varphi = 1$ down to $\varphi \approx 0$; a plot with φ equal to 0 provides no information since then there is no epidemic. The center plant of each plot (the center line could also be used) is inoculated by the pathogen to serve as the initial source of infectious units, and the developing epidemic is observed and disease gradients are determined over the time period the disease is spreading through the plot. Using the empirical methods in section 7.6.3 of the previous chapter, one can determine the distance where a small value of disease is found at each time, $Y'(t, s')$. In brief, this can be done by drawing a horizontal line through a gradient graph (as done in Fig. 8.6) at Y', and finding the distance where this value occurs at each time. To avoid interpolation, one could also fit an empirical gradient model to the data at each time [making sure to use a gradient model that adjusts for a maximum y of 1 (see section 7.5.2)], and rearrange the prediction equation for distance (s'). From these maps the rate of epidemic expansion is calculated. If the hypothesis is correct we will find a straight line relation between the rate of expansion and the logarithm of the fraction of susceptible crop (assuming that the contact distribution is exponentially bounded, and that the contact distribution does not depend on φ.

Fig. 8.15 shows the results of the experiments of van den Bosch et al. (1999) and Assefa et al. (1995). From the figure we see that the experimental results agree with the hypothesized relation, although there is considerable variation in C values for a given fraction of susceptibles in the mixture. For applications, it is helpful to have rules of thumb telling how certain epidemiological quantities, such as the rate of epidemic expansion, varies with the fraction of susceptible plants in the crop. Generalizing from the theoretical considerations in this section and the experiments done on two different plant pathogen systems we can state the rule of thumb: In a mixture of susceptible and resistant plants, the rate of epidemic expansion increases linearly with the logarithm of the fraction of susceptible plants in the crop.

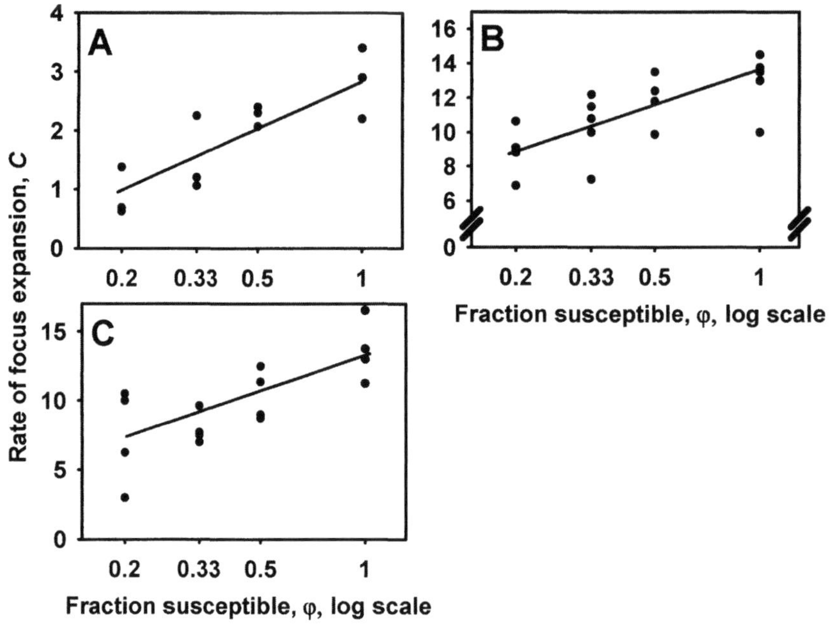

FIG. 8.15. Observations on the rate of disease expansion in relation to the fraction of susceptible crop in a mixture. Drawn line is the fit of equation 8.36 based on least squares regression. (**A**) stripe rust of wheat, caused by *Puccinia striiformis* (data from van den Bosch et al., 1990b). (**B**) and (**C**) bean rust, caused by *Uromyces appendiculatus* in (**B**) 1990 and (**C**) 1991 (data from Assefa et al., 1995).

Complicating the issue is the fact that there is evidence from other studies (done in other locations) that power law type contact distributions may best describe stripe rust and bean rust disease gradients (Maffia and Berger, 1998; Sackett and Mundt, 2005a, b; see also Figs. 7.11 and 7.13). In particular, Cowger et al. (2005) recently showed that the rate of focus expansion increased exponentially with time for stripe rust, which is consistent with a power-law contact distribution (equation 8.17). Thus, the experimental data supports the epidemiological theory for their pathosystem.

8.5 Concluding Remarks

In this chapter we derived a simple model formulation for the spatial expansion of polycyclic (compound interest) plant disease. The model is an integration of: (i) reproduction (in the general sense)—i.e., infections from each infected individual, including the production of infectious units and their probability of causing an infection when in contact with a disease-free host individual—and (ii) the contact distribution, the measure of dispersal of infectious units (equation 8.2 or 8.23). Analyzing the model we found that there are two essentially different types of expansion rates for the front of the epidemic, or for a specified isopath of low disease intensity (Y'). The categories are determined entirely by the form of the contact distribution. A traveling wave, in which the front moves at constant rate, is found for contact distributions with exponentially bounded tails. A dispersive wave, in which the front moves out with increasing velocity, is found for contact distributions with non-exponentially bounded tails. For dispersive waves, the rate of expansion of dispersive waves (acceleration) can be linear or nonlinear (e.g., exponential) functions of time. In all situations, the expansion rate is a positive function of the basic reproduction number, R_0.

The basic results for polycyclic diseases were fairly easy to show using a simple non-overlapping-generation model with no (host) density dependence (equations 8.2 and 8.3). The mathematical complexity increases greatly with consideration of latent and infectious periods, overlapping generations, and density dependence, but various authors have shown that the basic results hold for expansion of the epidemic (Diekmann, 1978, 1979; Thieme, 1977, 1979). In contrast to polycyclic diseases, the mathematical formulation is much simpler for monocyclic (simple interest) diseases, even when there is density dependence. Depending on the contact distribution, it can be shown that the expansion rate may decline with time.

Direct fitting of spatio-temporal models such as equation 8.23 to data is more difficult than fitting of the equivalent purely temporal models. This is because of the partial derivatives instead of ordinary derivatives that must be numerically integrated. For practical purposes, it is probably most useful to estimate model parameters from ancillary studies and then use these parameters to predict $Y(t, s)$ for a specific situation using numerical methods. For instance, gradients of disease intensity, especially early in the epidemic, provide the required information for calculating contact-distribution parameters (e.g., σ). Furthermore, C can be estimated by measurements of disease intensity at the front of the epidemic at several times (see Chapter 7). Depending on the contact distribution, R_0 can then be estimated from σ and C.

There are several other approaches to model spatial dynamics of populations. Ecologists, theoretical biologists and sometimes plant pathologists use diffusion-type models (e.g., Gilligan, 1995; White and Gilligan, 1998). The basic assumption underlying diffusion equations is that the individuals, or the entities being dispersed (e.g., spores), move randomly through space (Minogue, 1989). This assumption may be a reasonable approximation, at least over certain spatial and temporal scales of interest, especially for animal species. This approach could, for example, be useful to model virus diseases with arthropod vectors, where the vector can reasonably be assumed to move randomly through the crop. A good introduction to the use of diffusion models is found in Okubo (1988). As a general rule, however, a diffusion-type model is unrealistically simplistic for many plant disease epidemics (see discussion in Minogue, 1986; and also Mollison, 1972), so we did not discuss development and applications of this model in the chapter.

Spatial structure can also be incorporated into epidemic models using spatially explicit simulation models (Kampmeijer and Zadoks, 1977; Sackett and Mundt, 2005b). As discussed in regards to temporal models (see section 5.6.3), simulation is especially useful for determining specific outcomes under specific conditions, such as in studying particular profiles of environmental conditions on spatio-temporal dynamics. Simulation is also valuable to explore disease outcomes at all locations and times when there are multiple initial inoculum sources, and also when researchers wish to study stochastic aspects of epidemics. The spatio-temporal stochastic model of Xu and Ridout (1998) is especially useful in this regard. For a more mathematical treatment of spatio-temporal dynamics for a lattice of plants, using stochastic modeling, readers should see Filipe and Gibson (1998) and Filipe and Maule (2003, 2004). Among several other results, Filipe and Maule (2004) developed a spatially implicit epidemic model (see equation 8.34) that is different from the power-function expression explained in section 8.3.4. Keeling (1999) presents an alternate way of addressing the problem of spatial structure of disease in temporal dynamics of epidemics without the need for simulation. Additional valuable references include Durrett and Levin (1993), Park et al. (2001), Stacey et al. (2004), and Gilligan (2002).

Although the rate of focus expansion is determined by the contact distribution, it appears that different contact distributions may be appropriate for the same disease under different circumstances (Assefa et al., 1995; Maffia and Berger, 1998; Sackett and Mundt, 2005a, b; van den Bosch et al, 1988c, 1990a). Results on realized disease gradients, and corresponding contact distributions, may reflect the scale of the experiments used to quantify disease spread. As discussed in Chapter 7, disease intensity must be measured at relatively large distances from the inoculum source in order to clearly distinguish among contact-distribution types. Moreover, based on Aylor (1999), based on physical principles of spore transport, it could be argued that different contact-distribution types may be of most relevance at different distance ranges from the source for a given pathogen. Clearly more research is needed in this area since results have a major consequence on predictions of disease spread into new areas.

Finally, a subject that is widely acknowledged for its importance in the spatial dynamics of populations, but is very little studied, is heterogeneity of the environment. Shigesada and Kawasaki (1997) and Shigesada et al. (1986) are excellent references on this topic.

8.6 Selected Readings

Aylor, D. E. 1986. A framework for examining inter-regional aerial transport of fungal spores. *Agric. For. Meteorol.* 38: 263–288.

Kot, M., Lewis, M. A., and van den Driessche, P. 1996. Dispersal data and the spread of invading organisms. *Ecology* 77: 2027–2042.

Shigesada, N., and Kawasaki, K. 1997. *Biological Invasions: Theory and Practice.* Oxford Series in Ecology and Evolution. Oxford University Press, New York.

van den Bosch, F., Metz, J. A. J., and Zadoks, J. C. 1999. Pandemics of focal plant disease, a model. *Phytopathology* 89: 495–505.

Zadoks, J. C., and van den Bosch, F. 1994. On the spread of plant disease: a theory on foci. *Annu. Rev. Phytopathol.* 32: 503–521.

9

Spatial Aspects of Epidemics—III: Patterns of Plant Disease

Whenever they appear, I turn to look and my looking excites the patterns.

Pat Cadigan ("Patterns", 1989)

9.1 Why We Look at Spatial Patterns

When plant pathologists look at patterns of pathogens, disease, vectors or other components of pathosystems, it is the outcomes of dispersal processes and the effects of environmental heterogeneity that are being observed. Although it is not usually a good idea to draw conclusions about processes on the basis of outcomes, it is nevertheless true that one of the reasons why so much attention has been devoted to the study of patterns of plant disease is that these patterns are regarded as realizations of underlying dispersal processes of pathogens (see Chapters 7 and 8). Thus, in one of the first epidemiological studies of field-scale spatial patterns of disease, Bald (1937) monitored the incidence of tomato plants showing symptoms of infection with *Tomato spotted wilt virus* (TWSV). A program of intensive mapping of diseased plants in field plots was undertaken, with the objective of conducting: "*... an ecological survey of incidence, to find out something about the natural dissemination of spotted wilt.*"

The first step in understanding ecological processes is to identify patterns (Fortin et al., 2002). However, when the objective of an analysis of spatial pattern is an explanation of the underlying processes, statistical methods alone may not suffice. Instead, such methods may be used to formulate biological hypotheses that can be investigated by non-statistical means (Diggle, 1983). For example, both Bald (1937) and Cochran (1936), discussing the same TSWV data set, offered biological interpretations of their statistical analyses of pattern in terms of the behavioral ecology of the vector of the virus (see section 9.9). Here, we distinguish between such studies, carried out primarily with the objective of finding a biologically based explanation for observed spatial patterns of disease, and studies in which the primary objective is to make statements about disease intensity, where the pattern of disease is of interest in that it affects the sampling distribution of the intensity estimator (Diggle, 2003). Methods for both types of study are discussed in this chapter.

Spatial analysis in ecology is largely concerned with the identification of patterns at multiple scales, in pursuit of the processes underlying those patterns. Plant pathologists have borrowed freely from the statistical methodology that drives progress in spatial analysis in ecology, and will no doubt continue so to do. However, this chapter does not attempt to review the methodology of spatial analysis in ecology, nor even, in any comprehensive way, provide a catalog of phytopathological applications. For readers interested more generally in spatial analysis in ecology, Ludwig and Reynolds (1988, part II) is a good introduction to the methodology. Fortin et al. (2002) provide an overview of methods for identifying and characterizing the way in which observations from nearby locations are more likely to have similar magnitude than by chance alone, and Dale (1999) covers similar ground in much more detail. Volume 25 of the journal *Ecography* contains a series of useful articles devoted to spatial analysis in ecology (Liebhold and Gurevitch 2002, Dale et al., 2002, Perry et al., 2002, Legendre et al., 2002, Keitt et al., 2002, Dungan et al., 2002) that, as well as being of interest in their own right, provide an extensive insight into the literature.

The scales of phenomena can only be studied using a range of scales of observation and analysis (Dungan et al., 2002). Inferences about patterns may change dependent on the scales of observation and analysis (e.g., Horne and Schneider, 1995), and most plant pathologists are well aware of this (e.g., Ferrandino, 2005). For example, Hudelson et al. (1997) developed cyclic sampling as a means of obtaining information about pattern at several scales. However, we are not always concerned with the identification of patterns at multiple scales and the underlying processes that give rise to them. For studies in which the primary objective is to make statements about disease intensity in the context of, for example, disease management, there will usually be a particular scale at which disease intensity is of interest that will be the deciding factor in the choice of a scale of observation and analysis. In such cases, the scale-dependence of patterns of disease and of characterizations of those patterns from data need not be a major concern. The pattern of disease is of interest insofar as it affects the sampling distribution of disease intensity at the chosen scale of observation and analysis. Analyses based on the use of statistical probability distributions to provide empirical descriptions of patterns are helpful here.

For plant pathologists, investigation of patterns at different scales often forms part of an attempt to infer the dispersal processes that result in observed patterns. Some of the statistical methods used in an exploratory manner to formulate biological hypotheses are briefly outlined. Of greater interest in this context, perhaps, are recent attempts to use stochastic simulation and modeling to make inferences about dispersal processes from patterns of disease. To some extent, these methods circumvent the need for protracted discussions of scale dependence of spatial statistics by focusing on the qualitatively different kinds of patterns that arise from different dispersal processes.

9.2 Terminology

Following Campbell and Madden (1990), we adopt Pielou's (1977) suggestion that "*To avoid ambiguity ... it is most desirable to use the word distribution in its statistical sense only. Then a variate has a distribution whereas a collection of organisms has a pattern.*" The term distribution is then reserved for use in the context of statistical probability distributions, while the term *pattern* is used to describe the dispersion in space of pathogens, diseased or infected plants or other components of pathosystems.

We also need some terminology to be able to refer to the different kinds of pattern that we may encounter. The nature of pathogen dispersal processes means that we frequently observe patterns that are non-random. Specifically, it is often the case that where, say, a plant is diseased, the chance that neighboring plants will also be diseased is increased. We refer to the kind of pattern that results in such cases as *aggregated*. Thus the term aggregation (Ord, 1982) is used to refer to clustering in space, and its use here is synonymous with the terms heterogeneity, overdispersion, clustering, clumping and contagion, when these terms are used in a spatial context in plant disease epidemiology. The kind of pattern that results if the chance that a plant is diseased is independent of the disease status of its neighbors is a *random* pattern. *Regular* patterns—rarely encountered in spatial studies of plant disease—may occur if the chance that a plant is diseased is reduced by proximity to other diseased plants.

Most statistical analyses of spatial pattern begin with a test of spatial randomness (Diggle, 2003). The hypothesis of spatial randomness acts as a useful dividing hypothesis to distinguish between patterns that are more aggregated or more regular than random, so a test of randomness can guide the formulation of more interesting hypotheses about pattern. Where the null hypothesis of spatial randomness is not rejected, further formal statistical analysis of pattern is seldom warranted (Diggle, 2003).

Recall that there are different ways in which we can record disease intensity (Chapter 2). The distinctions between different types of data obtained are important when it comes to the analysis of spatial patterns. When whole plants, or parts of plants, are assessed as either "diseased" or "healthy", the data reflect disease incidence. The term disease incidence is usually used to refer to the number of plants or plant parts diseased expressed as a proportion of the total number assessed. Disease incidence is therefore a discrete variable. When expressed as a proportion, the values of both the numerator and denominator are known. The term disease severity is used to refer to the proportion of host plant tissue that exhibits visual symptoms of disease. Thus, like disease incidence, disease severity is a proportion. However, unlike incidence, severity is a continuous variable, and the value of the denominator of the proportion is not usually known. Disease assessments that are made on the basis of counts of propagules, or of countable symptoms (such as lesions) are, in effect, assessments of population density. Often, assessments of counts made in this way are not distinguished from assessments of disease severity in the phytopathological literature. However, if we assess, say, the number of lesions per leaf, we have a discrete variable taking values 0, 1, 2, ... without (at least in theory) a specified upper limit. Such a variable has some different statistical properties to those of disease severity as defined above. For many diseases, we can think of severity as the number of lesions per unit area of leaf multiplied by the average area of an individual lesion.

9.3 Spatial Plant Disease Data

In order to collect spatial plant disease data, we must have in mind an appropriate *sampling unit*. The simplest situation is when the sampling unit is an *individual* (plant or plant part), and the presence or absence of disease or an assessment of the amount of disease is recorded. In the latter case, the amount of disease might comprise a visual assessment of the proportion of plant tissue diseased or a count of visible symptoms (such as lesions). To conduct a *census*, all possible sampling units in the *population* of interest must be assessed. Only rarely will this be a practical proposition. Usually, therefore, results will be based on a *sample* of data recorded from an appropriate number of sampling units drawn from the population comprising all possible sampling units. That said, it is undoubtedly the case that published data sets that are, in effect, censuses of plant disease (albeit over a relatively small area) have been instrumental in the development of methods for the analysis and interpretation of spatial plant disease data. Since we are concerned here, for the most part, with relatively small-scale patterns, the population will usually be defined by the scale of a field experiment or perhaps of a commercial crop of particular interest.

A sampling unit may comprise a number of individuals (sometimes generically referred to as elements). For example, a quadrat is a sampling unit containing a number of adjacent plants that may be assessed for disease; a plant may be thought of as a sampling unit containing a number of adjacent leaves (or stems, or roots, or other plant parts) that may be assessed. The choice of an appropriate sampling unit will reflect a number of things, including the objectives of the observer, issues relating to practicability, and the resources available for data collection.

Experiments that comprise one or more plots, in which the disease status of the adopted sampling unit is recorded at one or more times, probably provide the largest body of field-scale spatial data on plant disease epidemics. Adopting the terminology of Schabenberger and Pierce (2002), these may be observational studies involving only the recording of disease in space and time, without imposition of experimental treatments on the plots; or designed experiments, involving the assignment of treatments to plots (using some sort of randomization procedure). There is a special case among designed experiments in which the spatial pattern of diseased plants is itself a treatment. Such experiments typically involve the artificial inoculation of some plants in pre-specified spatial patterns, to enable study of the effects of pattern of disease on dependent variables. In fact, a study involving disease gradients in relation to an introduced inoculum source is a good example of a designed experiment with pattern as the treatment (see Chapter 7). Other examples include experiments on the effects of pattern of disease on crop yield (or yield loss) (Chapter 12; Loomis and Bennett, 1966; Johnson, 1991) and temporal disease progress (for example, Rodriguez-de Gonzalez et al., 1995). For such designed experiments, in which spatial pattern of diseased plants is one of the treatments, it is the effect of the imposed patterns that is of most interest, rather than the patterns themselves.

Although spatial plant disease data may comprise records of incidence, severity, or counts, probably the majority of examples in the phytopathological literature are for disease incidence data. It is also noteworthy that the relative ease with which disease incidence can be assessed has led to its use in making indirect assessments of other measures of disease intensity. Mean count per sampling unit (e.g., lesions per plant) or mean severity may be assessed by collecting disease incidence data and then using a previously established relationship either between mean disease incidence and mean count per sampling unit or between mean disease incidence and mean disease severity, as appropriate, to obtain the required assessment of disease intensity.

9.3.1 Data collection

Like Diggle (2003), we distinguish between spatial data collected by sparse sampling and those collected by intensive mapping. For the former, only the disease status of a restricted number of sampling units is recorded. For the latter, both the disease status and location of sampling units are recorded. In the context of epidemiological studies of plant disease, intensive mapping sometimes involves recording the disease status and location of every sampling unit in the population of interest (census data). Methods for the analysis of both types of data were developed in connection with the earliest quantitative studies of spatial pattern of plant disease (Bald, 1937; Cochran, 1936).

The most important distinction between sparse sampling and intensive mapping methods is that in the case of the former, the spatial arrangement of the sampling units is not recorded and so cannot be taken into account in any subsequent data analysis. Information on the location of sampling units cannot be obtained retrospectively from sparsely sampled data. Data collected by intensive mapping, in contrast, include details of the spatial arrangement of the sampling units. This means that data collected by intensive mapping may be analyzed either by methods specifically appropriate for intensively mapped data or, if desired, by methods for sparsely sampled data. However, data collected by sparse sampling may only be analyzed by methods specifically appropriate for such data. Note that in some cases, particularly when plant parts (such as leaves, branches or shoots) serve as the sampling units, it may not be practicable (or even possible) to collect intensively mapped data (see, for example, Madden et al., 1995a).

It is relatively easy to see the spatial component of data collected by intensive mapping. The data collected contain explicit information on spatial locations of sampling units that enables us, if required, to reproduce a representation of the study area in the form of a map. Analyses often entail characterization of the degree of association of disease intensity among neighboring sampling units. However, data collected by sparse sampling contain no information on the spatial locations of sampling units, so we might question the extent to which they can be regarded as spatial data. The analysis of data collected by sparse sampling often entails characterization of the extent of variability in the mean level of disease intensity per sampling unit. Generally, this is of interest where the variability exceeds that predicted on the basis of a baseline model generated on the assumption of spatial randomness. Many studies of aggregation have adopted a sparse sampling approach (Perry, 1994). Although the data are not explicitly spatial, characterizing such excess variability has an important place in the study of plant disease epidemics and in disease management.

There are two main, interrelated approaches to the analysis of spatial pattern with sparsely sampled data: testing the goodness-of-fit of statistical probability distributions to data and calculating indices of aggregation (Gilligan, 1988). The details of data analysis differ for disease incidence (section 9.4) and disease assessments based on counts (section 9.5).

9.4 Analysis of Sparsely Sampled Incidence Data

9.4.1 Summary statistics

To a large extent, patterns of disease result from the way that the disease status of a plant or plant part depends on the disease status of neighboring plants or plant parts. Spatial data, therefore, arise naturally when disease incidence data are collected from groups of neighboring plants or plant parts, even if the location of the groups is not known or recorded. In the statistical literature, these procedures come under the heading of *cluster sampling* (Chapter 10). In this terminology, the sampling units are the clusters. Thus, the term "cluster", as used in this context, has no mechanistic connotations. Sometimes it *is* the case that a natural cluster (say, leaves on a branch) makes an appropriate sampling unit, but there is no requirement that this should necessarily be so. The clusters that make up the sampling units in cluster sampling are artifacts of the sample design. Analyses of sparsely sampled cluster sampling data relate to spatial pattern at the within-sampling-unit scale. As Nicot et al. (1984) and Ferrandino (1998) have pointed out, convincingly if unsurprisingly, such analyses do not tell us about pattern at a scale larger than that of the sampling unit. To characterize such larger-scale patterns, we require additional information in the form of intensively mapped data for the locations of the sampling units.

The analysis of cluster sampling data is more straightforward if the number of individual elements (denoted n) is the same in each sampling unit, and if all n elements in each sampling unit are assessed. For the moment we continue under the assumption that this is the case. Some procedures that are useful when this is not so are dealt with later (section 9.4.9 and Chapter 10). The elements (i.e., plants or plant parts) of each sampling unit are assessed as either "healthy" or "diseased". The proportion of diseased plants or plant parts in each sampling unit is given by $y_i = Y_i/n; i = 1, 2, \ldots, N$. N is the total number of sampling units inspected (*not* the total number of elements inspected) and Y_i (the number diseased in the ith sampling unit) may take integer values $0, 1, \ldots, n$. Then disease incidence, the *mean proportion* of diseased plants or plant parts, is estimated by:

$$\bar{y} = \frac{\sum_i y_i}{N} \tag{9.1}$$

and an unbiased estimate of the variance of the y_i values is:

$$s_y^2 = \frac{\sum_i y_i^2 - \frac{\left(\sum_i y_i\right)^2}{N}}{N-1} \tag{9.2}$$

The estimated variance of the *mean* proportion, $\bar{y}$, is $s_{\bar{y}}^2 = s_y^2/N$, as long as the number of sampling units, N, is small relative to the total number of sampling units in the population of interest. When this is not the case, the finite population correction should be applied (see section 10.2.5). The denominator of the variance formula in equation 9.2 ($N-1$) is the degrees of freedom (*df*). If the number of observations (N) is used as the denominator instead, the maximum likelihood estimate of the variance of the y_i values is obtained. The estimated standard error of the mean proportion is $\sqrt{s_{\bar{y}}^2}$.

Note that the mean *number* of diseased plants or plant parts per sampling unit is estimated by $\bar{Y} = n\bar{y}$, the estimated variance of the number per sampling unit is $s_Y^2 = n^2 s_y^2$, and the estimated variance of the *mean* number is $s_{\bar{Y}}^2 = s_Y^2/N$. As noted above, the finite population correction should be applied if N is not small relative to the total number of sampling units in the population of interest. When the mean is expressed as a number rather than as a proportion, the estimated standard error is $\sqrt{s_{\bar{Y}}^2}$.

It can be shown (Madden and Hughes, 1995) that the maximum variance of the y_i values occurs when all diseased individuals are in the smallest possible number of sampling units, with no diseased individuals in the other sampling units. If, for example, the number of sampling units is $N = 100$, with a total of $n = 25$ plants in each, there can be no more than 25 diseased plants in any sampling unit. When the total number of diseased plants is 125, the maximum variance of the y_i values occurs when five sampling units each have 25 diseased plants, and the remaining 95 sampling units have no diseased plants. The maximum variance is given by $s_{y(\max)}^2 = [N/(N-1)]\bar{y}(1-\bar{y})$. When N is large, this is approximated by $s_{y(\max)}^2 \approx \bar{y}(1-\bar{y})$. The minimum variance of the y_i values occurs when there is exactly the same number of diseased plants in each sampling unit. In this case, the minimum variance is zero. However, if (as in this example) the number of sampling units is not an exact (integer) multiple of the total number of diseased plants, it is not possible to have the same number of diseased plants in each sampling unit. In this case, the minimum variance the y_i values is (slightly) greater than zero.

9.4.2 The binomial distribution

Consider a cluster sampling scenario, as follows. For each of N separate sampling units, the number of diseased plants is recorded. At this stage, we are still confining attention to the case where all the sampling units contain the same total number of individual elements, denoted n. If the probability that an individual plant is diseased, p, is constant over the area in question, the statistical probability distribution that should provide a basis for describing the frequency of diseased plants in a sampling unit of n plants is the *binomial distribution*:

$$\Pr(Y) = \binom{n}{Y} p^{Y}(1-p)^{n-Y} \tag{9.3}$$

The notation:

$$\binom{n}{Y} = \frac{n!}{Y!(n-Y)!}$$

is shorthand for the number of different combinations of n items taken Y at a time, and $n!$ is read as "factorial n" and denotes $n \times (n-1) \times (n-2) \times \cdots \times 3 \times 2 \times 1$, n being a positive integer. $\Pr(Y)$ denotes the probability that a sampling unit contains Y diseased plants out of a total of n, ($Y = 0, 1, \ldots, n$). Disease incidence on a proportion scale, $\bar{y}$, is usually taken as an estimate of the unobserved probability p. The binomial distribution is appropriate when the disease status of one plant in a sampling unit is independent of the disease status of the other plants in the same sampling unit. This is what we take as a working definition of a random pattern of diseased plants at the within-sampling-unit scale. Thus we adopt the binomial distribution as the reference distribution appropriate for the description of a random pattern of disease incidence within sampling units.

The following example data are taken from Snedecor and Cochran (1989). In $N = 40$ sampling units (quadrats), each containing $n = 9$ plants, the numbers of diseased plants were Y_i = 2, 5, 1, 1, 1, 7, 0, 0, 3, 2, 3, 0, 0, 0, 7, 0, 4, 1, 2, 6, 0, 0, 1, 5, 4, 0, 1, 4, 2, 6, 0, 2, 4, 1, 7, 3, 5, 0, 3 and 6. Disease incidence on a proportion scale is $\bar{y} = 0.275$ (equation 9.1). For a binomial distribution, an unbiased estimate of the variance of the y_i values is $s^2_{\text{bin}} = \bar{y}(1-\bar{y})/(n-1)$. Notwithstanding this, the slightly biased estimate:

$$s^2_{\text{bin}} = \frac{\bar{y}(1-\bar{y})}{n} \tag{9.4}$$

is the one usually used in practice. Multiplying this variance by n^2 produces the estimated binomial variance in terms of numbers rather than proportions. For the data above, $s^2_{\text{bin}} = (0.275 \times 0.725)/9 = 0.022$. Note, however, the variance of the y_i values as calculated by equation 9.2 is $s^2_y = 0.067$, considerably larger than obtained by use of the binomial formula. Such *extra-binomial variation* is typical of aggregated plant disease incidence data. The analysis of this variation has applications in the description of pattern for sparsely sampled disease incidence data and in the development of methods of sampling for disease incidence. Indeed, the ratio s^2_y/s^2_{bin} is an example of the "design effect" (or *deff*), an empirical heterogeneity factor used in the sampling literature to characterize the excess variation that arises from non-independence among elements of a sample (Kish, 1995).

Use of the binomial formula to provide an estimate the variance of the y_i values should be preceded by a test of goodness-of-fit to show that the binomial distribution is indeed an appropriate description of the observed frequency distribution of disease incidence. Most widely used spreadsheet programs have a built-in function for calculation of binomial probabilities based on equation 9.3. These probabilities can be converted to expected binomial frequencies by multiplication of $\Pr(Y)$ by N (the number of sampling units). Fig. 9.1 shows the observed (O) and expected (E) binomial frequencies for four TSWV disease epidemics reported by Bald (1937). Visually, there appears to be reasonable agreement between the observed and expected frequencies. This can be tested more formally with a χ^2 test of goodness-of-fit, calculated as follows:

$$\chi^2 = \sum \frac{(O-E)^2}{E} \tag{9.5}$$

The expected values should not be too small in any frequency class, or the test will not be valid (Snedecor and Cochran, 1989). This situation is usually avoided by pooling frequency classes in the tails of the distribution. In general, the guidelines for pooling offered by Snedecor and Cochran (1989) and Mead et al. (1993) have been adopted, with details of specific implementations given in context. In this case, agreement between the observed and expected values is confirmed by the goodness-of-fit tests (Table 9.1). This is evidence to support the view that the pattern of disease incidence is, in this case, indistinguishable from random.

9.4.3 The index of dispersion

For disease incidence data collected by cluster sampling, a simple test of deviations from randomness is provided by calculating the *index of dispersion*, D (Fisher, 1925):

$$D = \frac{s^2_y}{\bar{y}(1-\bar{y})/n} \quad \left(= \frac{s^2_Y}{n\bar{y}(1-\bar{y})} \right) \tag{9.6}$$

Note that this index is a ratio of two variances, the observed variance (estimated from data for proportions or numbers per sampling unit) and the variance estimated on the assumption that data have a binomial distribution. Thus, a value $D = 1$ indicates the two variance estimates are equal, that is to say, the observed variance of the y_i values is equal to the theoretical variance

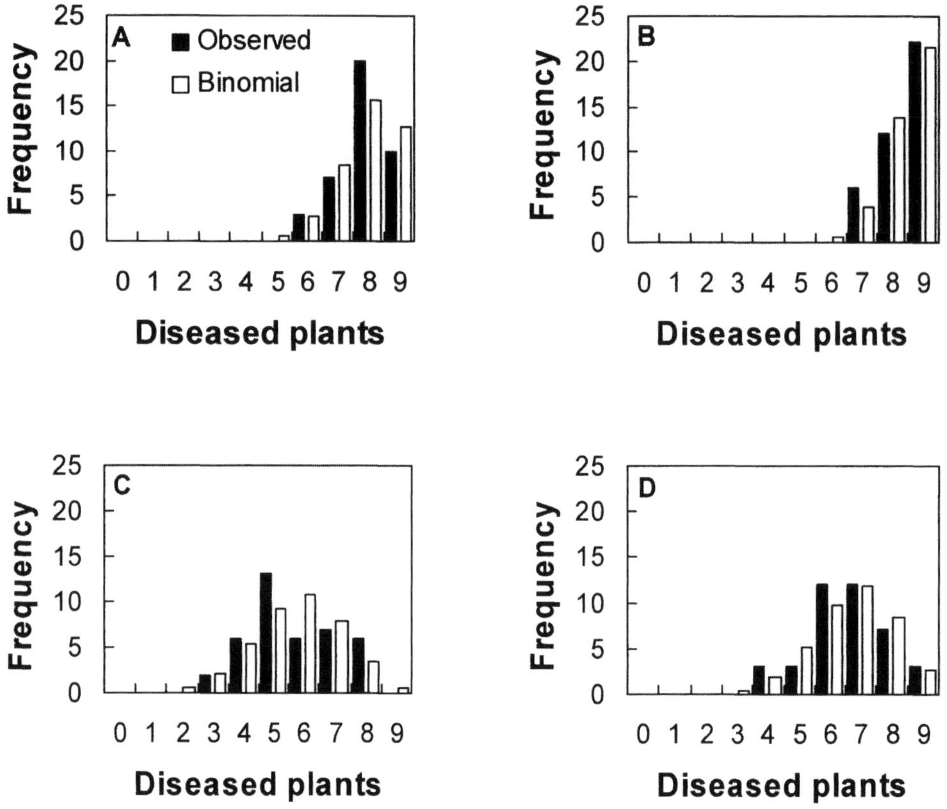

FIG. 9.1. Frequency distributions of tomato plants with symptoms of *Tomato spotted wilt virus* disease. The assessments of disease incidence were made on 12 December 1928 and reported by Bald (1937). Four rectangular plots, each containing 360 plants, were divided into 40 quadrats containing nine plants (three rows by three columns). The solid bars represent observed frequencies and the open bars represent expected binomial frequencies. (**A**) Cultivar Burwood Prize watered by overhead sprays. (**B**) Cultivar Burwood Prize watered by trenches. (**C**) Cultivar Early Dwarf Red watered by overhead sprays. (**D**) Cultivar Early Dwarf Red watered by trenches. Goodness-of-fit statistics are given in Table 9.1.

TABLE 9.1. Analysis of the *Tomato spotted wilt virus* disease incidence data shown in Fig. 9.1. Each plot comprised $N = 40$ quadrats containing $n = 9$ plants.

	Binomial distribution: tests of goodness-of-fit				*Index of dispersion*		
Plot[a]	$\bar{y}$[b]	χ^2[c]	*df*	*P*	*D*	χ^2 *(39 df)*	*P*
BP-OHS	0.881	2.15	2	0.34	0.78	30.4	0.84
BP-T	0.933	1.32	1	0.25	0.99	38.6	0.49
EDR-OHS	0.633	4.64	4	0.33	1.01	39.4	0.45
EDR-T	0.739	2.47	4	0.65	0.96	37.5	0.54

[a]BP, cultivar Burwood Prize; EDR, cultivar Early Dwarf Red; OHS, watered by overhead sprays; T, watered by trenches.
[b]$\bar{y}$ is estimated from the data, so that the number of degrees of freedom for χ^2 is 2 fewer than the number of frequency classes after pooling to ensure that expected frequencies are at least 1.
[c]χ^2 goodness-of-fit statistic with listed degrees of freedom. The significance level is given by P (conventionally, $P > 0.05$ indicates an acceptable fit).

of the reference distribution for a random pattern of diseased individuals within sampling units. When s_y^2 is at its maximum (see section 9.4.1), $D = n$. Conversely, when $s_y^2 = 0$, $D = 0$ (its smallest possible value).

The index of dispersion (equation 9.6) provides a statistical significance test that is of use in the analysis of sparsely sampled data collected by cluster sampling. The value $(N - 1)D$ has a χ^2 distribution when the null hypothesis $D = 1$ is true. To test deviations from randomness, $(N - 1)D$ is compared with the tabulated χ^2 distribution at $N - 1$ *df*. Table 9.1 shows results for the index of dispersion test for the four TSWV data sets from Fig. 9.1. All test results indicated that the pattern of disease incidence at the within-quadrat scale was indistinguishable from random. Sun and Madden (1997) give an alternative method of testing the null hypothesis $D = 1$, using a standard normal distribution. For most practical purposes, however, the χ^2 test shown here is satisfactory, and sometimes superior.

The two χ^2 tests outlined above (index of dispersion and goodness-of-fit) are alternatives. It would not normally be necessary to carry out both tests on the same set of data: this is done here for the sake of illustration. The index of dispersion test is appropriate whether or not the data are given in the form of a frequency distribution, and is especially useful for small numbers of sampling units. The goodness-of-fit test is appropriate when the data are given in the form of a frequency distribution and so can only be used when n does not vary between sampling units.

After compiling a frequency distribution, it is sometimes necessary to pool classes (i.e., when some expected frequencies are small) before carrying out the test. Pooling can be a rather arbitrary procedure, so the fact that this is avoided by the index of dispersion test is an advantage of that procedure. The goodness-of-fit test is most useful when N is large and n is small, so that there are many observations for each class and pooling is not needed.

Readers should note that the two tests are not exactly the same conceptually, although the distinction is rather subtle. The goodness-of-fit test directly evaluates how close the observed frequencies are to theoretical values for a binomial distribution. The test of D compares the estimated (empirical) variance of the y_i values with the corresponding estimated (theoretical) variance of a binomial distribution. Presumably, there could be other statistical probability distributions (of no particular relevance in the present context) with the same variance as that of the binomial distribution. Thus, the goodness-of-fit test is a stricter test of randomness (as defined in terms of a binomial distribution). However, this test can only be performed when all sampling units contain the same total number of individual elements (a requirement for the calculation of expected frequencies).

9.4.4 Intra-cluster correlation

We may think of aggregation, as characterized by the analysis of cluster sampling data, as the tendency for elements in a sampling unit to have the same disease status more frequently than we would expect on the basis of spatial randomness. More formally, we can express this in terms of the *intra-cluster correlation coefficient*, ρ (Mak, 1988; Ridout et al., 1999). Mak (1988) showed that the probability (p_s) that any two members of the same sampling unit have the same status is given by $p_s = 1 - 2p(1-p)(1-\rho)$. Thus, for any given value of the probability (p) that an individual is diseased, p_s increases with ρ.

Conceptually, ρ may be thought of as Pearson's correlation coefficient calculated over all possible pairs of observations that can be constructed within sampling units, though this is not usually the best way to obtain an estimate. Ridout et al. (1999) review methods for estimation of the intra-cluster correlation coefficient from observed data.

If data are collected on the disease status of plants in quadrats (the quadrats being the sampling units, or clusters) an estimated value of the intra-cluster correlation coefficient $\hat{\rho}$ of zero indicates that the disease status of any plant in a quadrat is unaffected by the disease status of other plants in the same quadrat. The upper limit of $\hat{\rho}$ is one. Positive values $0 < \hat{\rho} \leq 1$ (subject to the usual conventions of statistical significance testing) are indicative of a tendency for plants in the same quadrat to have the same disease status.

9.4.5 The β-binomial distribution

When plants or plant parts in the same sampling unit tend to have the same disease status, the *β-binomial distribution* is more likely to provide a description of the frequency distribution of diseased individuals per sampling unit than the binomial distribution. The β-binomial distribution arises as follows (Skellam, 1948). The assumption of constant probability that an individual is diseased is relaxed, and instead this probability is allowed to be a variable, π, with a beta probability density given by $f(\pi) = (\pi^{\alpha-1}(1-\pi)^{\beta-1})/(\mathrm{Be}(\alpha, \beta))$ in which $0 < \pi < 1$, the parameters α and β are positive constants, and $\mathrm{Be}(\alpha, \beta)$ is the beta function. Then, using rules for mixing of distributions (N. L. Johnson et al., 1996), one can write:

$$\Pr(Y) = \int \binom{n}{Y} \pi^{Y}(1-\pi)^{n-Y} f(\pi)\mathrm{d}\pi$$

or:

$$\Pr(Y) = \binom{n}{Y} \frac{\mathrm{Be}(\alpha + x, \beta + n - x)}{\mathrm{Be}(\alpha, \beta)}$$

As previously, $\Pr(Y)$ denotes the probability that a sampling unit contains Y diseased plants out of a total of n, ($Y = 0, 1, ..., n$). A widely used alternative parameterization of the β-binomial distribution is $p = \alpha/(\alpha + \beta)$ and $\theta = 1/(\alpha + \beta)$, in which p is interpreted as the average probability of a plant being diseased (instead of the constant probability of a plant being diseased in the binomial distribution, as above) and θ is an index of aggregation (aggregation increases with increasing θ). Then:

$$\Pr(Y) = \binom{n}{Y} \frac{\prod_{j=0}^{Y-1}(p + j\theta) \prod_{j=0}^{n-Y-1}(1 - p + j\theta)}{\prod_{j=0}^{n-1}(1 + j\theta)} \tag{9.7}$$

in which j is an index variable and Y takes the values $0, 1, 2, ..., n$.

Estimates of β-binomial parameters p and θ based on summary statistics ("moment estimates") are, respectively, $\bar{y}$ and:

$$\hat{\theta} = \frac{s_y^2 - \bar{y}(1-\bar{y})/n}{\bar{y}(1-\bar{y}) - s_y^2} \quad (9.8)$$

in terms of proportions, or $\bar{Y}$ and:

$$\hat{\theta} = \frac{s_Y^2 - n\bar{y}(1-\bar{y})}{n^2\bar{y}(1-\bar{y}) - s_Y^2} \quad (9.9)$$

in terms of numbers of diseased individuals per sampling unit.

It is informative to consider the form of the moment estimate of the parameter θ. The numerator is the difference between the observed variance of the y_i values (equation 9.8) or Y_i values (equation 9.9) and the corresponding theoretical variance for a binomial distribution. The denominator is the difference between the maximum possible variance of the y_i values (equation 9.8) or Y_i values (equation 9.9) and the corresponding observed variance. The value of $\hat{\theta}$ becomes indefinitely large as the observed variance of the y_i values (equation 9.8) or Y_i values (equation 9.9) approaches the maximum possible variance (i.e., as the denominator of the $\hat{\theta}$ formula approaches zero). When the observed variance of the y_i values (equation 9.8) or Y_i values (equation 9.9) is zero, the minimum value of $\hat{\theta}$ occurs, which is $\hat{\theta} = -1/n$.

Computer software is available for maximum likelihood estimation of β-binomial parameters and calculation of expected β-binomial frequencies (Smith, 1983; Madden and Hughes, 1994). One advantage of maximum likelihood estimation is that estimated standard errors of the parameter estimates can be obtained. These are useful for comparing two or more data sets. Fig. 9.2 shows the observed and expected binomial and β-binomial frequencies for data from the Dogwood Anthracnose Impact Assessment Program in 1990 and 1991, reported by Zarnoch et al. (1995). The calculations were carried out using the BBD computer program described by Madden and Hughes (1994). In this case, there appears to be poor agreement between the observed and binomial expected values, an impression confirmed by χ^2 tests of goodness-of-fit (1990: χ^2 = 495.7 (4 *df*), $P < 0.001$; 1991: χ^2 = 1651 (6 *df*), $P < 0.001$). There is much better agreement between the observed and β-binomial expected values, both visually (Fig. 9.2) and on the basis of χ^2 tests of goodness-of-fit (Table 9.2). These results are consistent with the view that in this case, the pattern of disease incidence within sampling units is aggregated.

To calculate expected β-binomial probabilities manually, first calculate the zero term Pr(Y=0), which is the probability that no plants are diseased (out of n in a quadrat):

$$\Pr(Y = 0) = \prod_{j=0}^{n-1} \frac{1 - p + j\theta}{1 + j\theta}$$

(substituting the appropriate estimated values for parameters p and θ). Then set (1) $\alpha = p/\theta$; (2) $\beta = (1-p)/\theta$, and, for Y = 1, 2, ..., n, calculate recursively:

$$\Pr(Y) = \left(\frac{n+1-Y}{Y}\right)\left(\frac{\alpha - 1 + Y}{n + \beta - Y}\right)\Pr(Y-1).$$

The resulting probabilities can be multiplied by N, the number of sampling units, to give the expected β-binomial frequencies. Compared with expected frequencies based on the binomial distribution with the same mean, β-binomial expected frequencies are higher in the tails of the distribution and lower near the mean, and this tendency increases with increasing θ (Fig 9.3).

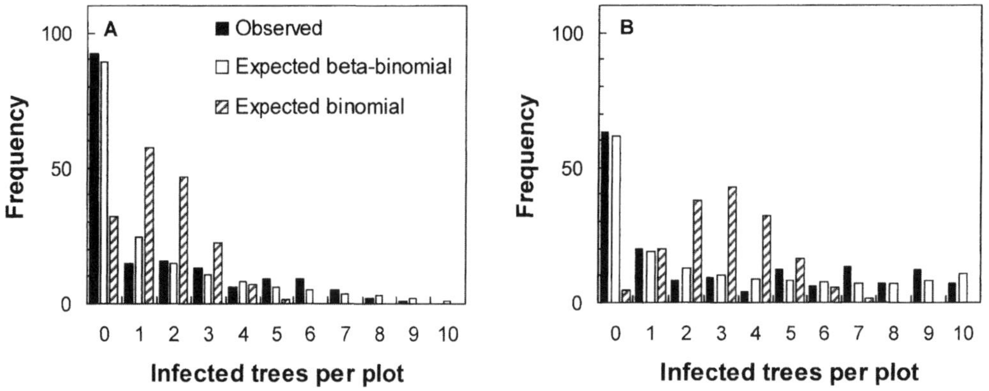

FIG. 9.2. Frequency distributions of dogwood trees with symptoms of dogwood anthracnose. The assessments of disease incidence were based on plots containing ten dogwood trees and reported by Zarnoch et al. (1995). The solid bars represent observed frequencies, the open bars represent expected β-binomial frequencies and the hatched bars represent expected binomial frequencies. (A) 1990 and (B) 1991. Goodness-of-fit statistics are given in Table 9.2.

TABLE 9.2. Analysis of the dogwood anthracnose disease incidence data shown in Fig. 9.2. Estimates of β-binomial parameters p and θ and their standard errors were obtained by maximum likelihood using the program BBD. Analyses are based on $N = 168$ (1990) and $N = 161$ (1991) plots, each containing $n = 10$ trees.

	Parameter estimate (standard error)		*Goodness-of-fit*			*Index of dispersion*		
Year	$\hat{p}$	$\hat{\theta}$	$\chi^{2\,a}$	*df*	P^b	D	χ^2	P
1990	0.153 (0.016)	0.522 (0.087)	12.1	8	0.15	3.85	642.7 167 *df*	<0.001
1991	0.299 (0.025)	0.998 (0.136)	14.8	8	0.06	5.60	895.4 160 *df*	<0.001

[a] $\hat{p}$ and $\hat{\theta}$ are estimated from the data, so the number of degrees of freedom for χ^2 is three fewer than the number of frequency classes (no pooling was required).
[b] The significance level of the χ^2 goodness-of-fit statistic with listed degrees of freedom is given by P. Conventionally, $P > 0.05$ indicates an acceptable fit.

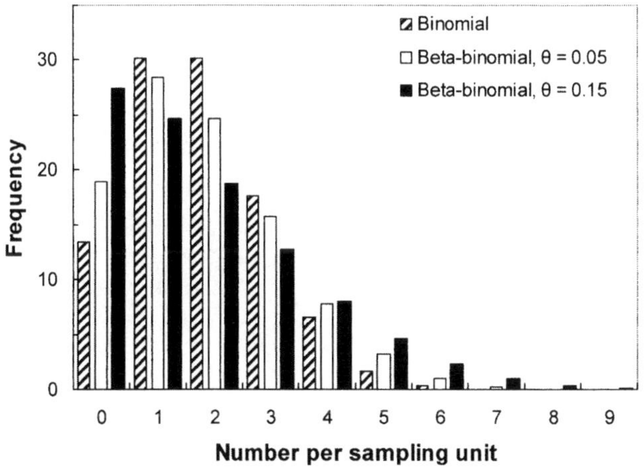

FIG. 9.3. Binomial and β-binomial frequency distributions. For all three distributions, $p = 0.2$, $n = 9$ and $N = 100$. The hatched bars represent binomial frequencies, the open bars represent β-binomial frequencies for $\theta = 0.05$ and the solid bars represent β-binomial frequencies for $\theta = 0.15$.

As an alternative to the index of dispersion test and the χ^2 goodness-of-fit test, Tarone (1979) provides a test of the binomial distribution against the specific alternative of the β-binomial, referred to as the $C(\alpha)$ test. The $C(\alpha)$ test is one-sided (in that the alternative is aggregation, not just "non-randomness") and is based on calculation of a test statistic, z, that has an asymptotic standard normal distribution under the null hypothesis. When all the sampling units contain the same total number of plants, denoted n, the test statistic is calculated from:

$$z = \frac{s - Nn}{[2Nn(n-1)]^{1/2}}$$

in which $S = n(N-1)D$ and D is the index of dispersion (equation 9.6). The BBD computer program described by Madden and Hughes (1994) can be used to calculate $C(\alpha)$ test statistics. For the data in Table 9.1, the outcomes of $C(\alpha)$ tests (separately for each plot) support acceptance of the hypothesis of the binomial distribution. For the data in Table 9.2, the outcomes of $C(\alpha)$ tests (separately for each year) support the rejection the hypothesis of the binomial distribution, in favor of the β-binomial alternative.

For a β-binomial distribution, the variance of the y_i values is estimated by:

$$s^2_{\beta\,\mathrm{bin}} = \left(\frac{\bar{y}(1-\bar{y})}{n}\right)\left(\frac{1+n\hat{\theta}}{1+\hat{\theta}}\right) \quad (9.10)$$

This variance is the product of two terms: the binomial variance of the y_i values (equation 9.4) and a "heterogeneity factor" that depends on both $\hat{\theta}$ and the number of individuals per sampling unit (n). When $\hat{\theta}$ is zero, the β-binomial variance of the y_i values ($s^2_{\beta\mathrm{bin}}$) is equal to the corresponding binomial variance (s^2_{bin}). When $\hat{\theta}$ is greater than zero, $s^2_{\beta\mathrm{bin}}$ is larger than s^2_{bin}. At a given value of mean disease incidence, the β-binomial variance of the y_i values increases with increasing $\hat{\theta}$. Thus, aggregation is indicated when the variance of the y_i values is larger than that of the binomial distribution with the same mean. The β-binomial distribution can be bimodal, with peaks in the 0th and nth frequency classes. The binomial distribution can never be bimodal, so can never be a good description of observed frequencies where there is a pronounced tendency for members of the same sampling unit have the same disease status.

Another parameterization of the β-binomial in frequent use is $p = \alpha/(\alpha + \beta)$ (as before) and $\rho = 1/(\alpha + \beta + 1)$ (equivalent to $\rho = \theta/(\theta + 1)$). Where data are described by the β-binomial distribution, ρ can be interpreted directly as the intra-cluster correlation coefficient. As θ increases indefinitely, ρ approaches a value of one. The smallest possible ρ (obtained when $\theta = -1/n$) is $\rho = -1/(n-1)$. So, although ρ can be thought of as a correlation coefficient, its smallest value is not -1 (as with continuous data), but a value that depends on n. In fact, the smallest possible ρ for a given set of conditions (i.e., given n and N) may be even larger (i.e., closer to zero) than $-1/(n-1)$ because, as discussed in section 9.4.1, it may not be possible to obtain an exactly equal number of diseased individuals in each sampling unit (Ridout and Xu, 2000). An estimate of the β-binomial variance of the y_i values can now be written in terms of $\hat{\rho}$:

$$s^2_{\beta\text{bin}} = \left(\frac{\bar{y}(1-\bar{y})}{n}\right)(1+\hat{\rho}(n-1)) \qquad (9.11)$$

Again, the β-binomial variance formula can be seen as the product of two terms: the binomial variance of the y_i values (equation 9.4) and a "heterogeneity factor" that depends, in this case, on both $\hat{\rho}$ and the number of individuals per sampling unit (n).

The first phytopathological applications of the β-binomial distribution seem to have been in the description of the frequency distribution of rice plants infected with yellow dwarf disease (caused by a mycoplasma-like organism) (Shiyomi and Takai, 1979; Takai and Shiyomi, 1980; Shiyomi, 1981) and in a methodological example given by Qu et al. (1990). Unfortunately, these studies appear to have had little impact on the plant disease epidemiology literature. Hughes and Madden (1993) showed that the use of the β-binomial distribution to characterize the frequency distribution of diseased plants per sampling unit was consistent with an empirical power law relationship between variances (section 9.4.7). Subsequently, a number of studies in which plant disease incidence was recorded (e.g., Gottwald et al., 1998; Madden and Hughes, 1994; Madden et al., 1995a, b; Roumagnac et al., 2004; Shah and Bergstrom, 2002; Scott et al., 2003; Shah et al., 2002; Tanne et al., 1996; Turechek and Madden, 1999a, b; Xu et al., 2001, Zarnoch et al., 1995) show that in cases where the binomial distribution provides an inadequate description of the frequency distribution of diseased plants or plant parts per sampling unit because the data are aggregated, the β-binomial usually provides a significant improvement (see also *Example 9.3*, section 9.9).

The binomial and β-binomial distributions have an important place in the analysis of disease incidence data, but they are not the only statistical probability distributions that are useful in this context. For example, the logistic-normal-binomial distribution is a possible alternative to the β-binomial distribution for the description of aggregated disease incidence data (Hughes et al., 1998; Hughes and Samita, 1998). The hypergeometric distribution has important applications in sampling, particularly when sampling from small populations (Chapters 10 and 11).

9.4.6 The index of dispersion revisited

For cluster sampling data, the variance of the y_i values (s^2_y) can always be written as the corresponding binomial variance (s^2_{bin}) multiplied by a constant. That is to say, by the incorporation of one additional parameter (representing an appropriate heterogeneity factor), the variance s^2_y can be determined from the binomial variance. Substituting for s^2_y (equation 9.2) with the expression for the β-binomial variance of the y_i values from equation 9.11, the following formula for D is obtained:

$$D = \frac{s^2_y}{s^2_{\text{bin}}} = \frac{\left(\frac{\bar{y}(1-\bar{y})}{n}\right)(1+\hat{\rho}(n-1))}{\left(\frac{\bar{y}(1-\bar{y})}{n}\right)} = 1+\hat{\rho}(n-1) \qquad (9.12)$$

In fact, this ratio of variances can be derived without reference to the β-binomial distribution (see section 6.2 in Collett, 2003). Equation 9.12, $D = 1+\hat{\rho}(n-1)$, allows several useful results to be summarized. When $D = 1$ (i.e., when the observed and binomial variances of the y_i values are equal), $\hat{\rho} = 0$, as required. When $D = n$ (i.e., when s^2_y is at its maximum possible value, $\bar{y}(1-\bar{y})$), $\hat{\rho} = 1$, reflecting perfect within-sampling-unit agreement in the disease status of plants. When $D = 0$ (i.e., when $s^2_y = 0$, its minimum possible value), $\hat{\rho} = -1/(n-1)$. Finally, note that for cluster sampling data described by the β-binomial distribution, the index of dispersion D (equation 9.12) is a version of the deff (section 9.4.2), as used in the sampling literature (Kish, 1995).

9.4.7 A power law relationship between variances

Suppose that, instead of a single disease assessment, resulting in a single frequency distribution of disease incidence, we had a number of such assessments over a range of disease incidences. A typical scenario would be the collection of cluster sampling data at multiple times during an epidemic. One possibility would be to analyze the data from each assessment separately, for example by means of the methods outlined previously in section 9.4. Alternatively, we could start by summarizing the characteristics of the data set as a whole, and relate this subsequently to the properties of statistical probability distributions.

Hughes and Madden (1992) made an empirical investigation of the relationship between the observed variance (s^2_y) and mean ($\bar{y}$) of the y_i values for 16 tobacco virus disease epidemics described by Madden et al. (1987a) and 13 maize dwarf mosaic virus disease epidemics described by Madden et al. (1987b). Good descriptions of the data were provided by the relationships:

$$s^2_y = a[\bar{y}(1-\bar{y})]^b \qquad (9.13)$$

and

$$s_y^2 = a\bar{y}^{b_1}(1-\bar{y})^{b_2} \quad (9.14)$$

in which a and either b (equation 9.13) or b_1 and b_2 (equation 9.14) are parameters to be estimated from data. The more general equation 9.14 allows for asymmetry in the curve describing the relationship between s_y^2 and $\bar{y}(1-\bar{y})$. Logarithmic transformation of equations 9.13 and 9.14 produces the linear models:

$$\log_{10}(s_y^2) = \log_{10}a + b\log_{10}(\bar{y}[1-\bar{y}]) \quad (9.15)$$

and

$$\log_{10}(s_y^2) = \log_{10}a + b_1\log_{10}(\bar{y}) + b_2\log_{10}(1-\bar{y}) \quad (9.16)$$

respectively. Since the logarithm of an estimated variance may have a reasonably symmetric sampling distribution with a constant variance, linear least-squares regression may be used to estimate the parameters of equations 9.15 or 9.16. Alternatively, non-linear estimation methods may be used, employing a procedure that allows specification of a non-normal distribution for the response variable. Schabenberger and Pierce (2002) argue that the gamma distribution provides a good description for estimated observed variances.

For the most part, equation 9.13 (usually shown as the linear version, equation 9.15) has provided a satisfactory description of observed disease incidence data. As a result, little effort has gone into evaluating the more general equations 9.14 and 9.16. Equation 9.13 can be rewritten explicitly as a relationship between the observed variance of the y_i values and the corresponding variance on the assumption of a binomial distribution:

$$s_y^2 = A(s_{\text{bin}}^2)^b \quad (9.17)$$

in which $s_{\text{bin}}^2 = \bar{y}(1-\bar{y})/n$, $A = an^b$ (disease incidence $\bar{y}$ being used as an estimate of the binomial parameter p, and n being the total number of plants, or plant parts, per sampling unit, and a and b as in equation 9.13). Again, the parameters A and b can most easily be estimated after logarithmic transformation to $\log_{10}(s_y^2) = \log_{10}A + b\log_{10}(s_{\text{bin}}^2)$. Equations 9.13–9.17 represent alternative formulations of a "power law", analogous to Taylor's (1961, 1984) power law for count data (section 9.5.5), taking account of the statistical characteristics of disease incidence data. Because incidence at the scale of the individual (e.g., plant) is binary (a plant is either diseased or disease-free; see Chapter 2), the model in equation 9.17 is often referred to as the *binary power law* (Madden and Hughes, 1995).

Many published cases show that the power law relationship represented by equation 9.17 seems to provide an appropriate empirical description of incidence data (e.g., Madden and Hughes, 1995; Madden et al., 1995a, b; Hughes et al., 1996, 2002a; Nault and Kennedy, 1996; Hughes and Gottwald, 1998, 1999; Turechek and Madden, 1999a, b; Shah et al., 2001, 2002; Turechek et al., 2001; Shah and Bergstrom, 2002; Bassanezi et al., 2003; Roumagnac et al., 2004) (Fig. 9.4). Note that when the data used to formulate such power law relationships are based on variable size sampling units, the variance formulae given in section 9.4.9 are required. Detailed stochastic simulations confirm the applicability of the power law relationship between variances for a wide range of epidemiological (e.g., dispersal distance, sporulation rate) and sampling (e.g., quadrat size and shape) conditions (Xu and Ridout, 2000).

9.4.8 How the power law is related to statistical probability distributions

For incidence data described by a β-binomial distribution, an estimate of the aggregation parameter θ can be written in terms of the power law parameters A and b (equation 9.17) as follows:

$$\hat{\theta} = \frac{A - n^{b-1}f(\bar{y})}{n^b f(\bar{y}) - A}$$

in which $f(\bar{y}) = [\bar{y}(1-\bar{y})]^{1-b}$. Alternatively, an estimate of the aggregation parameter ρ can be written in terms of the power law parameters A and b (equation 9.17) as follows:

$$\hat{\rho} = \left(\frac{n}{n-1}\right)\left(\frac{A}{n^b f(\bar{y})} - \frac{1}{n}\right).$$

When $b = 1$ and $A = 1$ (equation 9.17), $f(\bar{y}) = 1$, and then $\hat{\theta} = 0$ ($\hat{\rho} = 0$) for all $\bar{y}$. These parameter values are thus consistent with the binomial distribution as a description of variability within sampling units at all incidence values. These values of b and A are also consistent, on average, with a value $D = 1$ of the index of dispersion (equation 9.6).

When $b = 1$ and $a > n^{-1}$ (equation 9.13) (i.e., $A > 1$, equation 9.17), $\hat{\theta}$ (or $\hat{\rho}$) is constant (>0) for all $\bar{y}$. For instance, $\hat{\rho} = (A-1)/(n-1)$ for all $\bar{y}$ (on average). When data are characterized by $b = 1$ and $a > n^{-1}$ ($A > 1$), the observed variance of the y_i values exceeds the corresponding binomial variance for all data sets, and, on average, the index of dispersion $D = A$.

When $b > 1$ and $A > 1$ (equation 9.17), $\hat{\theta}$ (or $\hat{\rho}$) is not constant, but increases to a maximum at $\bar{y} = 0.5$, then decreases. Most of the published examples of equation 9.17 have estimated values of $A > 1$ and $1 < b < 2$, and thus illustrate a situation in which the pattern of disease incidence is aggregated to greatest extent around $\bar{y} = 0.5$, while the pattern when $\bar{y}$ is close either to 0 or

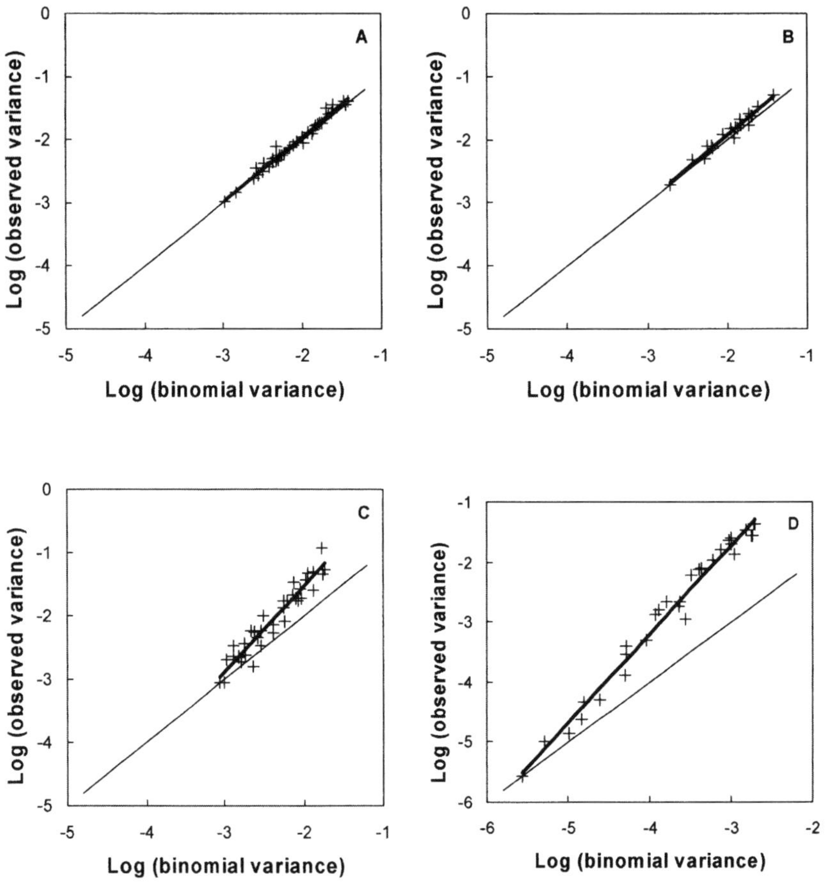

FIG. 9.4. Examples of the linear relationship, on logarithmic axes, between observed and binomial variances, for disease incidence data (based on equation 9.17). (**A**) *Citrus tristeza virus* disease (Hughes and Gottwald, 1998), $\log_{10}(s_y^2) = 0.10 + 1.03 \log_{10}(s_{\mathrm{bin}}^2)$. (**B**) *Plum pox virus* disease (Hughes et al., 2002a), $\log_{10}(s_y^2) = 0.20 + 1.06 \log_{10}(s_{\mathrm{bin}}^2)$. (**C**) Grape downy mildew (Hughes et al., 1997), $\log_{10}(s_y^2) = 1.15 + 1.35 \log_{10}(s_{\mathrm{bin}}^2)$. (**D**) Bacterial blight of onion (Roumagnac et al. 2004), $\log_{10}(s_y^2) = 2.71 + 1.48 \log_{10}(s_{\mathrm{bin}}^2)$.

to 1 is indistinguishable from random. If $b > 1$, then the value of D is not a constant, but depends on the estimated disease incidence ($\bar{y}$). This can be seen on dividing both sides of the equation $s_y^2 = A(s_{\mathrm{bin}}^2)^b$ by s_{bin}^2, the binomial variance of the y_i values, which results in $D = A(\bar{y}(1-\bar{y})/n)^{b-1}$.

Example 9.1. Here, we demonstrate relationships between the power law model represented by equation 9.17 and the intra-cluster correlation coefficient. The examples presented have been selected for the purpose of illustration, so no parameter estimation from data is involved.

The power law relationship represented by equation 9.17 is illustrated in Fig. 9.5A, with $n = 9$, $b = 1.4$ and $A = 13$. The binomial line ($b = 1$, $A = 1$) and the maximum-variance line ($b = 1$, $A = n$) are also shown. It can be seen that the power law and binomial lines cross at a low value of the binomial variance. Two values of disease incidence $\bar{y}$ (the mean proportion of diseased plants) give this variance. These values can be found by solving the quadratic equation $\bar{y}^2 - \bar{y} + nA^{1/(1-b)} = 0 (b > 1)$. In this example, the two solutions are (correct to 3 decimal places) $\bar{y} = 0.015$ and $\bar{y} = 0.985$. At both these values of $\bar{y}$, the binomial variance of the y_i values and the corresponding variance given by the power law model (equation 9.17) are equal (with a value $\approx 1.64 \times 10^{-3}$). In this example, the power law model (equation 9.17) predicts a higher variance than that of the corresponding binomial distribution for disease incidences $0.015 < \bar{y} < 0.985$. At disease incidences $\bar{y} < 0.015$ and $\bar{y} > 0.985$ the power law predicts a lower variance than that of the corresponding binomial distribution. Fig. 9.5B shows relationships between the intra-cluster correlation coefficient ρ and disease incidence $\bar{y}$, corresponding to the variance relationships in Fig. 9.5A. When $b > 1$, the line for ρ corresponding to the power law line in Fig. 9.5A is a curve with a maximum value at $\bar{y} = 0.5$ and which intersects the line $\rho = 0$ at the values of $\bar{y}$ that are the solutions of $\bar{y}^2 - \bar{y} + nA^{1/(1-b)} = 0$.

For any point along the binomial line in Fig. 9.5A, a binomial $(n, \bar{y})$ frequency distribution can be calculated. For any point along the power law line in Fig. 9.5A, a β-binomial $(n, \bar{y}, \rho)$ frequency distribution can be calculated. β-Binomial distributions with $-1/(n-1) \le \rho < 0$ (i.e., with a variance of the y_i values smaller than that of the binomial distribution with the same mean) can be constructed (e.g., Prentice, 1986)

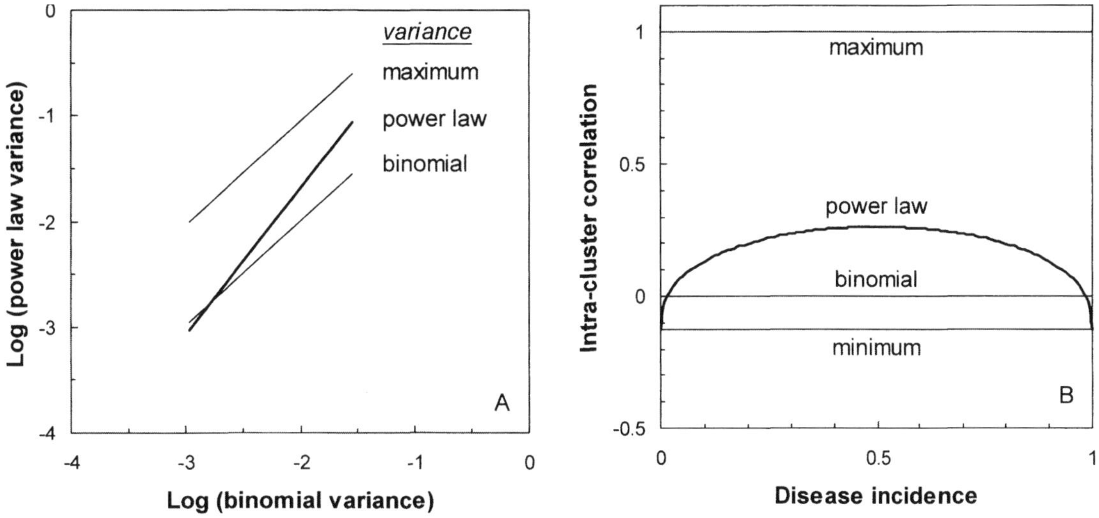

FIG. 9.5. (**A**) The power law relationship represented by equation 9.17 is illustrated, with n = 9, b = 1.4 and A = 13 (plotted on logarithmic axes). The binomial line (b = 1, A = 1) and the maximum-variance line (b = 1, A = n) are also shown. (**B**) Relationships between the intra-cluster correlation coefficient ρ and mean disease incidence $\bar{y}$, corresponding to the variance lines shown in A. The relationship corresponding to the maximum variance line is ρ = 1. The relationship corresponding to the power law variance line is $\rho = (n/(n-1))\left[A/(n^b(\bar{y}(1-\bar{y}))^{1-b}) - (1/n)\right]$. Alternatively, calculate the index of dispersion D (in this case D is a variable) by dividing the power law variance by the binomial variance, then ρ = (D − 1)/(n − 1). The relationship corresponding to the binomial variance line is ρ = 0. The relationship corresponding to the minimum variance line is ρ = −1/(n − 1). The minimum-variance line could not be shown on the logarithmic axes used in A. See *Example 9.1*.

but are of little practical significance in plant disease epidemiology.

9.4.9 Unequal size sampling units

In the cluster sampling scenarios considered so far, it has been assumed that all sampling units have the same number of plants or plant units (denoted n). It is convenient when this actually is the case, but important to be able to proceed with an analysis when it is not. Suppose now that data have been collected on the number of diseased plants in N quadrats, and that in the ith quadrat (i = 1, 2, ..., N) there were n_i plants, of which Y_i were diseased. The proportion of diseased plants in the ith quadrat is then $y_i = Y_i / n_i$. Disease incidence, as the mean proportion of diseased plants over all N quadrats, is given by $\bar{y} = \Sigma_i Y_i / \Sigma_i n_i$. The mean number of plants per quadrat is $\bar{n} = \Sigma_i n_i / N$. In this case, there is not an (move) exact formula for the variance of the y_i values. The following approximate formulae (Cochran, 1977) are useful. The observed variance of the of the y_i values is estimated by:

$$s_y^2 \approx \frac{\sum n_i^2 (y_i - \bar{y})^2}{\bar{n}^2 (N-1)} \qquad (9.\ 18)$$

For a binomial distribution, the variance of the y_i values is estimated by:

$$s_{\text{bin}}^2 = \frac{\bar{y}(1-\bar{y})}{\bar{n}} \qquad (9.19)$$

Equations 9.18 and 9.19 are used to formulate power law relationships (section 9.4.7) when the size of the sampling unit varies. The observed variance of the number of diseased plants or plant parts per sampling unit is estimated by:

$$s_Y^2 \approx \frac{\sum (Y_i - n_i \bar{y})^2}{N-1}.$$

The index of dispersion D (section 9.4.2) can now be calculated from:

$$D = \frac{s_Y^2}{\bar{n}\bar{y}(1-\bar{y})} \left(\text{or} \quad D = \frac{s_y^2}{\bar{y}(1-\bar{y})/\bar{n}} \right).$$

The $C(\alpha)$ goodness-of-fit test of the binomial distribution against the specific alternative of the β-binomial (section 9.4.5) can be calculated from:

$$z = \frac{S - N\bar{n}}{\left[2\sum n_i(n_i - 1)\right]^{1/2}}$$

in which $S = \bar{n}(N-1)D$.

Although these summary statistics can be calculated, it is no longer possible to plot the observed data in the form of a frequency distribution, and generate expected frequencies on the basis of the binomial or β-binomial distribution. Distributional analysis *is* still of interest, however, if we think in terms of parameter estimation rather than just comparing observed with expected frequencies (Madden and Hughes, 1994; Zarnoch et al., 1995). We return to this topic in Chapter 10.

9.4.10 Two-stage sampling

Two-stage cluster sampling is the name given to cluster sampling when there is subsampling from the clusters, so that not all the individual elements in each sampling unit are assessed. Cochran (1977) gives a phytopathological example, which is probably based on data that appear in Cochran (1936), and later in this chapter in *Example 9.3*. A similar example follows.

Suppose that a field containing 1440 individual plants is divided into $M = 160$ quadrats each containing $m = 9$ plants. A total of $N = 40$ quadrats is inspected, and in each quadrat, $n = 3$ plants are assessed for disease. This procedure leads to a frequency distribution of disease incidence.

Number of diseased plants per quadrat (/3 assessed)	*Frequency*
0	21
1	16
2	2
3	1

The estimated mean proportion of plants diseased per quadrat is $\bar{y} = 0.192$ (equation 9.1). The estimated variance of the y_i values has two components: the variance among quadrat means and the variance among individuals within quadrats. The latter arises because only a sample of individuals in each quadrat is assessed. The variance among quadrat means is estimated by $s_1^2 = 0.0563$ (equation 9.2) (summations in equations 9.1 and 9.2 are for $i = 1, 2, ..., N$). The variance among individuals within quadrats is estimated by

$$s_2^2 = [n/(N(n-1))]\sum_i y_i(1-y_i)$$

(Cochran, 1977). Here, $s_2^2 = 0.150$. Let f_1 and f_2 be the sampling fractions N/M and n/m, respectively. Here, $f_1 = 40/160 = 1/4$, and $f_2 = 3/9 = 1/3$. The variance of the mean proportion of plants diseased per quadrat can then be estimated by

$$s_{\bar{y}}^2 = (1-f_1)\frac{s_1^2}{N} + f_1(1-f_2)\frac{s_2^2}{nN}$$

(Cochran, 1977). Here:

$$\begin{aligned} s_{\bar{y}}^2 &= \left(1-\frac{1}{4}\right)\frac{0.0563}{40} + \left(\frac{1}{4}\right)\left(1-\frac{1}{3}\right)\frac{0.150}{3\times 40} \\ &= 0.001056 + 0.000208 \\ &= 0.001265. \end{aligned}$$

The mean proportion diseased is 0.192 with an estimated standard error equal to 0.036 $(=\sqrt{0.001265})$.

Cochran (1977) gives details of calculations for further levels of subsampling (multistage sampling) and of subsampling when sampling units are of unequal size (which makes things rather more complicated). Note that when only a small fraction of the sampling units in the population is sampled (i.e., when N/M is close to 0) or when a large fraction of the plants in each sampling unit is sampled (i.e., when n/m is close to 1), the within-quadrat variance (s_2^2) has little effect on the variance of the mean proportion diseased (because $f_1(1-f_2) \approx 0$).

9.5 Analysis of Sparsely Sampled Count Data

In a plant pathology context, disease data in the form of counts usually relate to the number of infections per sampling unit. Count data are more frequently collected in entomology and weed science (where population densities are probably the most common type of data collected for estimating populations of harmful organisms) than in studies of plant disease. Accordingly, plant pathologists have often applied methods developed in these disciplines (e.g., Bliss, 1958; Marshall, 1988) for characterizing patterns in disease data relating to population counts. Disease data in the form of counts arise in two main ways. Typically, such data are collected by looking at a sequence of individual plants, or plant units, and counting the number of lesions on each. The sampling unit is, in this case, the individual plant or plant unit (often a leaf in the case of foliar pathogens). For example, Waggoner and Rich (1981) reported data relating to number of lesions per plant, per leaf, and per leaflet for a variety of diseases caused by fungal pathogens. Sometimes it is the pathogen population that is assessed, rather than disease symptoms. For example, Charest et al. (2002) measured airborne ascospore concentration in a study of apple scab (caused by *Ventura inaequalis*).

9.5.1 Summary statistics

Suppose that the leaf is the sampling unit of interest and that in a sample comprising a total of N leaves, Y_i lesions are counted on the ith leaf ($i = 1, 2, ..., N$). The mean number of lesions per leaf is estimated by:

$$\bar{Y} = \frac{\sum_i Y_i}{N} \qquad (9.20)$$

and an unbiased estimate of the variance of the number of lesions per leaf is given by:

$$s_Y^2 = \frac{\sum_i Y_i^2 - \frac{\left(\sum_i Y_i\right)^2}{N}}{N-1} \qquad (9.21)$$

The estimated variance of $\bar{Y}$ is $s_{\bar{Y}}^2 = s_Y^2 / N$, if the number of sampling units, N, is small relative to the total number of sampling units in the population of interest (when this is not the case, the finite population correction should be applied). Then, the estimated standard error of the mean, $\bar{Y}$, is $\sqrt{s_{\bar{Y}}^2}$.

9.5.2 The Poisson distribution

As in section 9.5.1, suppose that the leaf is the sampling unit of interest and that counts of lesions per leaf are made. If the presence of a lesion on a leaf makes it no more or less likely that other lesions will occur (i.e., lesion occurrences are independent events), the *Poisson distribution* may be used to provide a description of the frequency distribution of lesions per leaf:

$$\Pr(Y) = \frac{\mu^Y e^{-\mu}}{Y!} \tag{9.22}$$

in which e is the base of natural logarithms (e = 2.718, correct to three decimal places). Pr(Y) denotes the probability of a sampling unit containing Y infections, Y = 0, 1, 2, ... without a specified upper limit. The observed mean number of infections per sampling unit, $\bar{Y}$, is usually taken as an estimate of the parameter μ. The estimated variance of the counts for a Poisson distribution is also equal to the mean, that is:

$$s_P^2 = \bar{Y} \tag{9.23}$$

In general, we adopt the Poisson distribution as the reference distribution appropriate for the description of a random pattern of disease data in the form of counts of infections within sampling units.

To calculate expected Poisson probabilities corresponding to observed counts, first calculate the zero term Pr(Y = 0), which is the probability that a sampling unit contains no infections:

$$\Pr(Y = 0) = e^{-\mu}$$

(substituting the appropriate estimated value of μ). Then, for Y = 1, 2, 3, ... calculate recursively:

$$\Pr(Y) = \frac{\mu}{Y}\Pr(Y-1).$$

In practice, these probabilities can usually be calculated using a spreadsheet with a built-in function for calculation of Poisson probabilities. Probabilities can be converted to expected Poisson frequencies by multiplication by N, the number of sampling units. Agreement between observed and expected Poisson frequencies is tested by a χ^2 test of goodness-of-fit (equation 9.5). When μ has been estimated from the data (that is, the calculation of the Poisson probabilities was based on $\bar{y}$), the number of degrees of freedom is two fewer than the number of frequency classes after any pooling to ensure that expected frequencies are not too small.

9.5.3 The negative binomial distribution

Again suppose that the leaf is the sampling unit of interest and that counts of lesions per leaf are made. If the presence of a lesion on a leaf makes it more likely that other lesions will occur (i.e., lesion occurrences are not independent events), there will be an aggregated pattern of lesions per leaf. A number of statistical probability distributions have been applied by ecologists to provide descriptions of aggregated population data in the form of counts. Southwood (1978) and Krebs (1998) provide useful overviews. The *negative binomial distribution* has been widely used in plant disease epidemiology to describe aggregated patterns of disease data in the form of counts of infections within sampling units. An early example is provided by Strandberg (1973), who fitted several distributions, including the Poisson and the negative binomial, to data for the number of lesions per plant of cabbage black rot, caused by *Xanthomonas campestris* pv. *campestris* infection of cabbage plants.

There are a number of different formulations of the negative binomial distribution. For example:

$$\Pr(Y) = \left(\frac{\Gamma(k+Y)}{\Gamma(k)\Gamma(Y+1)}\right)\left(\frac{\mu}{k}\right)^Y\left(1+\frac{\mu}{k}\right)^{-(k+Y)} \tag{9.24}$$

in which, as in equation 9.22, Pr(Y) denotes the probability of a sampling unit containing Y infections, Y = 0, 1, 2, ... without a specified upper limit, and the mean number of infections per sampling unit, $\bar{Y}$, is usually taken as an estimate of the parameter μ. The parameter k is an aggregation parameter ($k > 0$, aggregation at the within-sampling-unit scale increases as *k decreases*). The moment estimate of k is $\hat{k} = \bar{Y}^2 / (s_Y^2 - \bar{Y})$. $\Gamma(\bullet)$ represents the gamma function. If k is an integer, equation 9.24 may be written:

$$\Pr(Y) = \binom{Y+k-1}{Y}\left(\frac{\mu}{k}\right)^Y\left(1+\frac{\mu}{k}\right)^{-(k+Y)} \tag{9.25}$$

Maximum likelihood estimation of negative binomial distribution parameters can be obtained using statistical software such as SAS.

When k is non-integer, mathematical software such as MATHCAD facilitates the calculation of expected negative binomial probabilities based on equation 9.24. To calculate expected negative binomial probabilities using equation 9.25, let $q = 1 + (\mu / k)$ (substituting the appropriate estimated values of the parameters μ and k), then calculate $\Pr(Y = 0)$, the probability that a sampling unit contains no infections:

$$\Pr(Y = 0) = q^{-k}.$$

Then, for $Y = 1, 2, 3, \ldots$, calculate recursively:

$$\Pr(Y) = \frac{(Y + k - 1)(q - 1)}{Yq}\Pr(Y - 1).$$

Using a spreadsheet with a built-in function for calculation of negative binomial probabilities usually requires k to be rounded to an integer value. Expected probabilities obtained from either equation 9.24 or 9.25 are multiplied by N to obtain expected negative binomial frequencies.

As the value of k increases, expected negative binomial frequencies increasingly resemble expected Poisson frequencies. Agreement between observed and expected negative binomial frequencies may be tested by a χ^2 test of goodness-of-fit (equation 9.5). When μ and k have been estimated from the data (that is, the calculation of the negative binomial probabilities was based on $\bar{Y}$ and $\hat{k}$), the number of degrees of freedom is three fewer than the number of frequency classes after any pooling to ensure that expected frequencies are not too small. Two further tests of goodness-of-fit are available. These compare the observed and expected negative binomial variances and the observed and expected third moments of the negative binomial distribution (Anscombe, 1950; Southwood, 1978; Krebs, 1998). The third moment is a measure of skewness of a distribution.

For a negative binomial distribution, the variance of the counts is estimated by:

$$s_{\text{nb}}^2 = \bar{Y} + \frac{\bar{Y}^2}{\hat{k}} \qquad (9.26)$$

From equation 9.26, it can be seen that the negative binomial variance decreases as the estimate of the aggregation parameter increases. Because of this, k^{-1} is sometimes referred to as the aggregation parameter (so that aggregation increases with increasing k^{-1}). Thus, when referring to the negative binomial aggregation parameter, it is helpful to be explicit as to whether k or k^{-1} is meant. We use k throughout.

Campbell and Noe (1985) and Gilligan (1988) provide useful discussions of the application of statistical probability distributions in the analysis of spatial pattern of disease assessments in the form of counts within sampling units, with particular reference to soil-borne pathogens. In addition to the Poisson and the negative binomial, reference is made to the Polya-Aeppli and Neyman type A distributions, two other discrete distributions that may be used as alternatives to the negative binomial for describing aggregated disease data in the form of counts within sampling units. Xu and Madden (2002) provide an example of the Neyman type A distribution used to characterize patterns of apple powdery mildew lesions (caused by *Podosphaera leucotricha*) over multiple years.

9.5.4 The index of dispersion for counts

The index of dispersion test is, as before, based on a comparison of the observed variance of counts with the corresponding theoretical variance of the reference distribution appropriate for the description of a random pattern of disease data in the form of counts of infections within sampling units. The former is given by s_Y^2 (equation 9.21), the latter, for disease data in the form of counts, is the Poisson variance, which is estimated by $\bar{Y}$ (equation 9.23). Thus, in this case:

$$D = \frac{s_Y^2}{\bar{Y}} \qquad (9.27)$$

(Fisher, 1925). This is sometimes referred to as the variance-to-mean ratio. Numerous indices of aggregation for count data, based on the relationship between the variance and the mean, have been developed and discussed extensively in the ecological literature. Indeed, the relationship between the variance and the mean has been a cornerstone of the analysis of aggregated count data by population ecologists. Patil and Stiteler (1974), Myers (1978), Taylor (1984), Campbell and Madden (1990), Garson and Moser (1995), Hughes and Madden (1995) and Shiyomi and Yoshimura (2000) are among those who have reviewed work in this area. However, for our main purpose here—the identification of deviations from randomness in connection with the analysis of epidemiological surveys and experiments—the index of dispersion will serve (see Diggle, 2003, p. 32).

Example 9.2. Fig. 9.6 shows a frequency distribution representing the observed number of lesions (Y) on each of N=100 leaves. For the data in Fig. 9.6, $\bar{Y} = 3.35$ and $s_Y^2 = 13.99$. The statistical software SAS was used to calculate maximum likelihood parameter estimates ($\log(\bar{Y}) = 1.209(\text{SE} = 0.055)$, $\hat{k}^{-1} = 1.026(\text{SE} = 0.207)$). Expected Poisson and negative binomial frequencies (Fig. 9.6) were calculated using the mathematical software MATHCAD (the negative binomial aggregation parameter was rounded to $k = 1$ for this purpose). There is an excess of values in the tails ($Y \leq 1$ or $Y \geq 8$) of the observed distribution and a deficit near the mean, compared with the Poisson expected frequencies. Such *extra-Poisson variation* is typical of aggregated disease data in the form of counts. Visually, the negative binomial distribution appears to provide a better description of the observed distribution. Support for these impressions is provided by χ^2 tests of goodness-of-fit (equation 9.5). For the Poisson distribution, $\chi^2 = 205.74$ with 7 df ($P < 0.001$). For the negative binomial distribution,

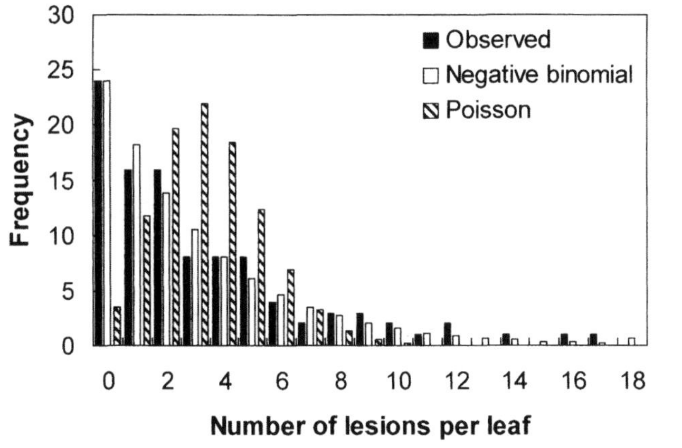

FIG. 9.6. Frequency distributions representing the number of lesions (Y) on each of $N = 100$ leaves. The solid bars represent observed frequencies, the open bars represent expected negative-binomial frequencies and the hatched bars represent expected Poisson frequencies. For the expected frequencies, frequency class 18 represents frequencies ≥ 18. Goodness-of-fit statistics are given in *Example* 9.2.

$\chi^2 = 3.41$ with 8 *df* ($P = 0.91$). It was appropriate in both cases to pool frequency classes until no expected frequencies were less than two. For the Poisson distribution, the number of degrees of freedom is two fewer than the number of frequency classes after pooling. For the negative binomial distribution, the number of degrees of freedom is three fewer than the number of frequency classes after pooling. A negative binomial distribution with a mean of 3.35 and aggregation parameter $k = 1$ (and $N = 100$) provides a good description of the observed data.

For the observed data in Fig. 9.6, the value of the index of dispersion $D = 13.99/3.35 = 4.176$ (equation 9.27). To test the statistical significance of deviations from randomness, $(N - 1)D = 99 \times 4.176 = 413.4$ is compared with the tabulated χ^2 distribution at $N - 1$ *df*. This gives a significance probability $P < 0.001$, indicating that the assumption of randomness is not supported in this case. The available evidence from analysis of data from Fig. 9.6 supports the view that there is aggregation of lesions within leaves (the adopted sampling units).

9.5.5 Taylor's power law

Suppose that (as in section 9.4.7, but now thinking of disease data in the form of counts rather than disease incidence), instead of a single disease assessment, resulting in a single frequency distribution of counts per sampling unit, we had a number of such assessments over a range of values of mean count. One possibility would be to analyze the data from each assessment separately, for example by means of the methods outlined in sections 9.5.3 and 9.5.4. Taylor (1961, 1984) provides an alternative approach.

Taylor (1961) described an empirical relationship (which has come to be known as "Taylor's power law") between the observed variance and mean for count data, in which the variance is proportional to a power of the mean:

$$s_Y^2 = a\bar{Y}^b \tag{9.28}$$

At least part of the motivation for Taylor's original work was to provide a transformation that would render count data (arising mainly from entomological studies of pest population density) suitable for use of analysis of variance. Subsequently there has been discussion of the biological significance (or otherwise) of equation 9.14 (see, e.g., Anderson et al., 1982; Taylor et al., 1983; Taylor 1984; Keeling, 2000). However, in the present context, this discussion is of little concern. Taylor's power law provides a useful empirical description of aggregated count data that has been extensively applied by entomologists, although less so by plant pathologists. After logarithmic transformation, equation 9.28 becomes

$$\log_{10}(s_Y^2) = \log_{10}(a) + b\log_{10}(\bar{Y}) \tag{9.29}$$

from which estimates of a and b may be found by linear regression analysis. Alternative estimation procedures are discussed by Perry (1981) and Schabenberger and Pierce (2002, pp. 393–396). Boivin et al. (1990, Fig. 1) provide a phytopathological example of Taylor's power law from a study of leaf blight of carrots caused by *Cercospora carotae*.

For fairly obvious reasons, Taylor's power law is often referred to as a variance–mean relationship. However, it is also possible to interpret Taylor's power law as a relationship between two variances. Because $\bar{Y}$ is the estimate of the Poisson variance, equation 9.28 represents a relationship between the observed variance of the counts and the corresponding variance of the reference distribution appropriate for the description of random count data: $s_Y^2 = a(s_P^2)^b$. Recall that in section 9.4.7, the equation $s_y^2 = A(s_{bin}^2)^b$ was used to represent a relationship between the observed variance and the corresponding variance of the reference distribution appropriate for the description of random disease incidence data. It appears that a power law relationship between the observed variance and the variance of the reference distribution appropriate for the description of random disease data—usually plotted on logarithmic axes to provide a straight line graph—provides a good empirical description of aggregated disease data, both for incidence and for counts.

Because the variance is equal to the mean for a Poisson distribution, a Taylor's power law analysis of count data characterizing a pattern that was indistinguishable from random should lead to estimates of both a and b equal to one (within the conventional limits of statistical significance testing). The index of dispersion (equation 9.27) is given by $D = a\bar{Y}^{b-1}$ when Taylor's power law holds. Thus, when $b = 1$, $D = a$; i.e., the index of dispersion is constant, on average, over the entire range of mean count per sampling unit. If $b > 1$, D must vary with mean count per sampling unit. For a large majority of published analyses, the estimate of b is greater than one, and (when the data are plotted on logarithmic axes) the regression line corresponding to equation 9.14 crosses the Poisson line ($s_Y^2 = \bar{Y}$) from below at some very low value of mean count per sampling unit at which most sampling units are empty. Aggregated count data lie above the Poisson line. A value of $b > 1$ is often considered to be indicative of aggregation across all the data sets analyzed.

While there is a simple distributional interpretation of the power law description of aggregated disease incidence data (section 9.4.8), the distributional interpretation of aggregated data that follow Taylor's power law is less straightforward (see, e.g., Kemp, 1987; Perry and Taylor, 1988). Different statistical probability distributions may describe the frequency distribution of population density at different points along a Taylor's power law relationship (Taylor, 1984, Fig. 4.I). The Adès distribution (Perry and Taylor, 1985; Holgate, 1989) was introduced specifically in an attempt to provide a statistical probability distribution with Taylor's power law properties, but as yet has received relatively little attention from plant pathologists. For ad hoc handling of count data, the negative binomial distribution is a reasonable option (Taylor, 1984). If equation 9.28 is substituted into the expression for the negative binomial aggregation parameter, $\hat{k} = \bar{Y}^2/(s_Y^2 - \bar{Y})$, it can be seen that Taylor's power law is consistent with the negative binomial distribution if the aggregation parameter changes with the mean according to $\hat{k} = \bar{Y}^2/(a\bar{Y}^b - \bar{Y})$ (subject to the restriction that $\hat{k} > 0$) (Taylor et al., 1979).

9.6 Relationships between Distributions

The characteristics of four key statistical probability distributions for discrete data have been outlined in this chapter. The binomial and β-binomial distributions are applicable when disease incidence data collected by cluster sampling are being described. The Poisson and negative binomial distributions are applicable when disease data in the form of counts within sampling units are being described. Binns et al. (2000, Chapter 4) give a useful account of the same four distributions as a prelude to discussion of their application in crop protection. A full account of the statistical properties of these distributions can be found in N. L. Johnson et al. (1996).

The four distributions are linked (Leemis, 1986; Qu et al., 1990) (Fig. 9.7). It has already been noted that the β-binomial distribution (appropriate for aggregated incidence data) approaches the binomial (appropriate for random incidence data) as the aggregation parameter $\theta \to 0$ ($\alpha + \beta \to \infty$), and that the negative binomial distribution (appropriate for aggregated data in the

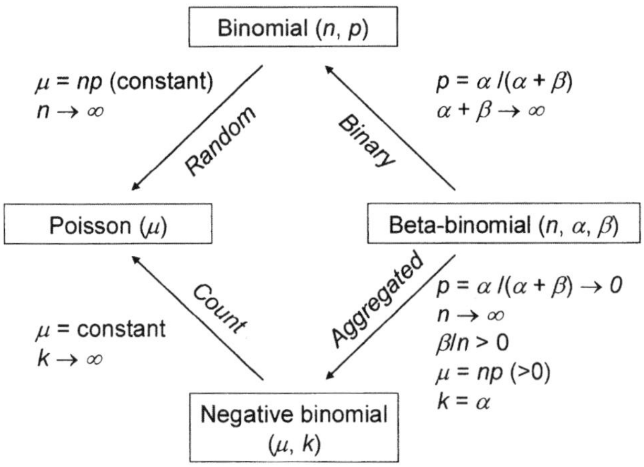

FIG. 9.7. Relationships among four discrete statistical probability distributions.

form of counts) approaches the Poisson (appropriate for random count data) as the aggregation parameter $k \to \infty$. In addition, the binomial distribution approaches the Poisson (with the same mean, i.e., $np = \mu$) as the size of the sampling unit $n \to \infty$. The β-binomial distribution approaches the negative binomial (with the same mean, i.e., $np = \mu$, and negative binomial aggregation parameter $k = p/\theta$), as the size of the sampling unit $n \to \infty$.

Aggregation at the within-sampling-unit scale is characterized by an observed variance larger than that of the reference distribution for random data for either count data or incidence data, as appropriate. The variance of the negative binomial distribution ($k > 0$) exceeds that of the Poisson distribution with the same mean and the variance of the β-binomial distribution ($\theta > 0$) exceeds that of the binomial distribution with the same mean.

9.7 Spatial Hierarchies

There is a further way in which the statistical probability distributions for discrete data that have been introduced in this chapter may be used in the analysis of spatial pattern, one that has particular application in sampling for disease. Whether the data collected are disease incidence (from cluster sampling) or in the form of counts, they provide information on the magnitude of disease intensity within sampling units. Typically, this is summarized by estimation of the mean, using either equation 9.1 or 9.20 as appropriate. The same data also provide information on whether or not disease is present in sampling units. That is to say, sampling units may be classified as "healthy" if no disease is present or "diseased" if disease is present. This, of course, is an assessment of disease incidence at the scale of the adopted sampling unit. Viewed in this way, the data as a whole comprise a spatial hierarchy (Hughes et al., 1997).

9.7.1 Disease incidence in a spatial hierarchy

The probability that an individual element (e.g., plant, leaf) in a sampling unit is diseased is denoted p_{low}. The subscript "low" indicates reference to the within-sampling-unit scale. For simplicity, we consider here the case where the s ampling units each comprise the same number of elements, denoted n. If the binomial distribution (equation 9.3) is regarded as an appropriate characterization of the frequency distribution of disease incidence at the within-sampling-unit scale, the probability that a sampling unit has no diseased elements is given by the zero term of the distribution, $\Pr(Y = 0) = (1 - p_{\text{low}})^n$. If the sampling unit is a quadrat containing n plants, this is the probability that no plants are diseased out of n in a quadrat. The probability that a sampling unit contains at least one diseased element is then $1-\Pr(Y = 0)$:

$$p_{\text{high}} = 1-(1-p_{\text{low}})^n \tag{9.30}$$

in which the subscript "high" indicates reference to the sampling unit scale.

If the β-binomial distribution (equation 9.7) is regarded as an appropriate characterization of the frequency distribution of disease incidence at the within-sampling-unit scale, the probability that a sampling unit has no diseased elements is

$$\Pr(Y = 0) = \prod_{j=0}^{n-1}(1 - p_{\text{low}} + j\theta)/(1 + j\theta)$$

(i.e., the zero term of the distribution), in which θ is the β-binomial aggregation parameter. The probability that a sampling unit has at least one diseased element is then:

$$p_{\text{high}} = 1-\prod_{j=0}^{n-1}\frac{1-p_{\text{low}}+j\theta}{1+j\theta} \tag{9.31}$$

Fig. 9.8 shows relationships between p_{high} and p_{low} based on equations 9.30 and 9.31 for $n = 9$ and (for the latter equation) a range of values of θ. When p_{high} is plotted against p_{low}, the β-binomial curves (equation 9.31) fall below the binomial curve (equation 9.30). This shows that when there is aggregation at the within-sampling-unit scale, p_{high} is smaller than would be expected for a random pattern, at any specified level of p_{low}. Suppose, for example, we make disease assessments based on recording the individual plants as "healthy" or "diseased" in quadrats each containing n plants. Aggregation of disease incidence at the plant scale results in more quadrats containing no diseased plants than for a random pattern with the same mean proportion of diseased plants per quadrat. Examples of the application of equations 9.30 and 9.31, including use of the power law $s_y^2 = A(s_{\text{bin}}^2)^b$ (equation 9.17) to describe variation of the β-binomial aggregation parameter θ, as described in section 9.4.8, are given in Hughes et al. (1997). Graphical plots of relationships between p_{high} and p_{low} do not provide a useful means of estimating β-binomial or power law aggregation parameters from data.

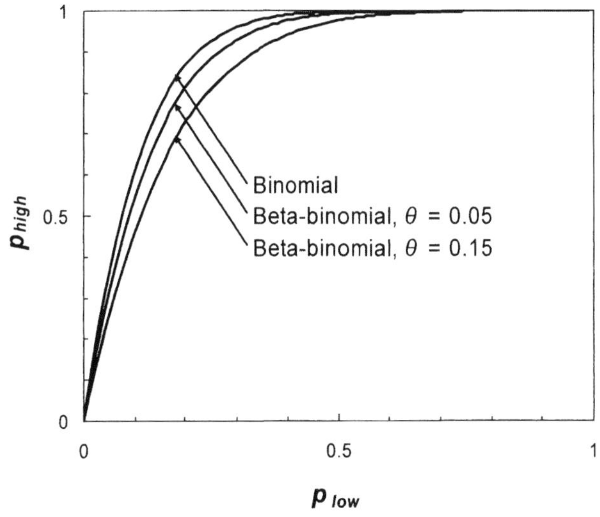

FIG. 9.8. Disease incidence in a spatial hierarchy. Relationships between p_{high} and p_{low} based on equations 9.30 (from the binomial distribution) and 9.31 (from the β-binomial distribution) for $n = 9$ and (for the latter equation) two different values of the aggregation parameter θ.

As an alternative to the use of equation 9.31 to describe aggregated disease incidence in a spatial hierarchy, Hughes and Gottwald (1999) used:

$$p_{\text{high}} = 1-(1-p_{\text{low}})^{\nu} \tag{9.32}$$

This can be written in the form $\text{CLL}(p_{\text{high}}) = \ln(\nu) + \text{CLL}(p_{\text{low}})$, in which CLL(•) denotes the complementary log-log transformation, $\text{CLL}(\bullet) = \ln(-\ln(1-\bullet))$, and ν is a parameter. From this, an estimate of ν may be made from data, provided that it is reasonable to regard the coefficient of $\text{CLL}(p_{\text{low}})$ as being equal to one. Values of ν (equation 9.32) less than the corresponding number of individuals per sampling unit n (equation 9.30) characterize curves that fall below the binomial curve and so are representative of aggregation of disease at the within-sampling-unit scale. Hughes and Gottwald (1999) gave an example with $n = 4$, $\nu = 3.3$, in a study of *Citrus tristeza virus* (CTV) incidence when the brown citrus aphid (*Toxoptera citricida*) was the main vector species. Madden and Hughes (1999b) discussed in some detail the use of equation 9.32 and the interpretation of ν as an effective sample size, and provided a graphical method of approximating ν. Hughes et al. (2002a) applied this graphical method for ν in the development of a surveillance program for *Plum pox virus* (PPV) incidence, and reported $\bar{n} = 3.7$, $\nu = 3.2$.

9.7.2 Counts in a spatial hierarchy

Recall that disease data in the form of counts are usually measured as number of infections per sampling unit.

If the Poisson distribution (equation 9.22) is regarded as an appropriate characterization of the frequency distribution of infections per sampling unit, the probability that a sampling unit contains no infections is given by the zero term of the distribution, $\Pr(Y = 0) = e^{-\mu}$ in which μ is the mean number of infections per sampling unit. The probability that a sampling unit contains at least one of whatever is being counted as an infection is then $1 - \Pr(Y = 0)$:

$$p_{\text{high}} = 1 - e^{-\mu} \tag{9.33}$$

This device—essentially, working out the probability that something has occurred by calculating "one minus the probability that it has not occurred"—was introduced to the study of relationships between disease incidence and disease data in the form of counts by Gregory (1948), citing Thompson (1924) as a precedent in the entomological literature. Equation 9.33 is also Wilson and Room's (1983) "Model 3". From rearranging equation 9.33:

$$\ln(1 - p_{\text{high}}) = -\mu \tag{9.34}$$

Thus, the relationship between $\ln(1-p_{\text{high}})$ and μ is a straight line with a slope of -1 for this Poisson-based relationship (see Fig 9.9). Equation 9.34 may be empirically modified as follows:

$$\ln(1 - p_{\text{high}}) = -a\mu$$

in which a is a parameter. This is equivalent to writing:

$$p_{\text{high}} = 1 - e^{-a\mu}.$$

This is Wilson and Room's (1983) "Model 4". Consider the effect of a on the relationship $p_{\text{high}} = 1 - e^{-a\mu}$. Appealing to the Poisson distribution, we know that $a = 1$ corresponds to a random pattern. If $a < 1$, then fewer sampling units are infected (i.e., p_{high} is smaller) for any given value of μ than when $a = 1$. That is to say (for example), that when $a < 1$, the same number of lesions occupy a smaller number of leaves than if the pattern of lesions per leaf were random, and more leaves have no lesions. Small values of a are thus indicative of deviations from randomness in the direction of aggregation.

Observed data for $(\hat{p}_{\text{high}}, \bar{Y})$ may used to estimate a. A graphical plot of $1 - \hat{p}_{\text{high}}$ against $\bar{Y}$ is indicative of a random pattern of number of infections per sampling unit when the slope of this relationship is -1 (with the $1 - \hat{p}_{\text{high}}$ axis on a ln scale) or -0.4343 ($= -\log_{10}e$) (with the $1 - \hat{p}_{\text{high}}$ axis on a $\log_{10}$ scale). Analysis of data for late blight of potato (caused by *Phytophthora infestans*) gave $\hat{a} = 1.35$ (Gregory, 1948). Thus the slope of the linear relationship ($=-1.35$) is close to -1, indicating conformity with the Poisson distribution. Note that if the $1 - \hat{p}_{\text{high}}$ axis had a $\log_{10}$ scale, the slope of the relationship would be given by $-\log_{10}(e^{1.35}) = -0.586$, close to -0.4343. In contrast, analysis of data for cedar-apple rust of apple (caused by *Gymnosporangium juniperi-virginianae*) gave $\hat{a} = 0.076$ (Gregory, 1948). Thus the slope of the linear relationship ($=-0.076$) is less steep than -1, indicating that the data do not conform to the Poisson distribution. Note that if the $1 - \hat{p}_{\text{high}}$ axis has a $\log_{10}$ scale, the slope of the relationship is given by $-\log_{10}(e^{0.076}) = -0.033$, which is less steep than -0.4343. This is indicative of deviation from randomness in the direction of aggregation.

Consider the further empirical modification of Equation 9.34 by inclusion of a second parameter b, as follows:

$$\ln(1 - p_{\text{high}}) = -a\mu^{b}.$$

This can be written:

$$\text{CLL}(p_{\text{high}}) = \ln(a) + b\ln(\mu)$$

where, as before, $\text{CLL}(p_{\text{high}})$ denotes the complementary log-log transformation of p_{high}. This is a linear relationship on a plot of $\text{CLL}(p_{\text{high}})$ against $\ln(\mu)$. It corresponds to the relationship:

$$p_{\text{high}} = 1 - e^{-a\mu^{b}}$$

on a plot of p_{high} against μ (Nachman, 1981). Thus, this represents a two-parameter generalization of the Poisson model (equation 9.33). de Jong (1995) used this model, in a simulation study of sampling for detection of leek rust (caused by *Puccinia allii*), to characterize the relationship between proportion of diseased leek plants and the mean number of uredosori of *P. allii* per plant. The two-parameter model reduces to Gregory's (1948) one-parameter model when $b = 1$, and to the Poisson model itself when $a = 1$ and $b = 1$.

If the negative binomial distribution (equation 9.11) is regarded as an appropriate characterization of the frequency distribution of infections per sampling unit, the probability that a sampling unit contains no infections is given by the zero term of the distribution, $\Pr(Y = 0) = (1 + (\mu / k))^{-k}$, in which μ is the mean number of infections per sampling unit and k is the negative binomial aggregation parameter. The probability that a sampling unit contains at least one of whatever is being counted as an infection is then:

$$p_{\text{high}} = 1 - \left(1 + \frac{\mu}{k}\right)^{-k} \tag{9.35}$$

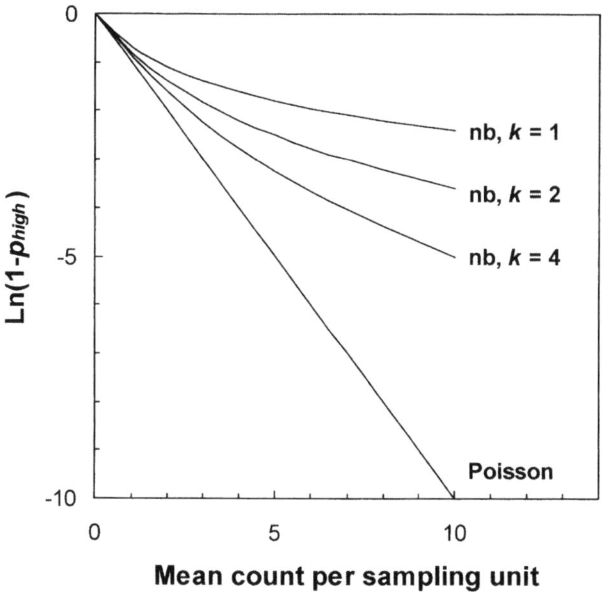

FIG. 9.9. Counts in a spatial hierarchy. Relationships between $\ln(1 - p_{\text{high}})$ and mean count per sampling unit, μ, based on equations 9.34 (from the Poisson distribution) and 9.36 (from the negative binomial distribution) and (for the latter equation) three different values of the aggregation parameter k.

Equation 9.35 is (in a re-parameterized form) Wilson and Room's (1983) "Model 1". From rearranging equation 9.35:

$$\ln\left(1 - p_{\text{high}}\right) = -k \ln\left(1 + \frac{\mu}{k}\right) \qquad (9.36)$$

This relationship between $\ln(1-p_{\text{high}})$ and μ is a curve rather than a straight line (as it is for the Poisson-based relationship) (see Fig. 9.9). There is no simple graphical method of finding an estimate of k from equation 9.36.

When $\ln(1-p_{\text{high}})$ is plotted against μ, the negative binomial curves are above the Poisson line (Fig. 9.9). This shows that when there is aggregation at the within-sampling-unit scale, p_{high} is smaller (because $1-p_{\text{high}}$ is larger) than would be expected for a random pattern, at any specified level of μ. Suppose, for example, that the leaf is the sampling unit of interest and that we make disease assessments based on recording the number of lesions on each of N individual leaves. For an aggregated pattern of lesions, the same number of lesions occupies a smaller number of leaves than if the pattern of lesions per leaf were random, and more leaves have no lesions. A precedent from the entomological literature for this analysis is given by Ingram and Green (1972). Waggoner and Rich (1981) and Seem (1984) provide a phytopathological viewpoint.

The negative binomial model can be generalized by using Taylor's power law to describe variation in the negative binomial aggregation parameter k. This is Wilson and Room's (1983) "Model 2". First, the parameters in equation 9.35 are replaced by their estimates:

$$\hat{p}_{\text{high}} = 1 - \left(1 + (\bar{Y}/\hat{k})\right)^{-\hat{k}}.$$

Then, substitute for $\hat{k}$ using $\hat{k} = \bar{Y}^2/(s_Y^2 - \bar{Y})$, and substitute for s_Y^2 in the result, using $s_Y^2 = a\bar{Y}^b$ (equation 9.28). Then (after some rearrangement):

$$\hat{p}_{\text{high}} = 1 - e^{-\bar{Y}(a\bar{Y}^{b-1}-1)^{-1}\ln(a\bar{Y}^{b-1})}$$

is obtained, which can be written:

$$\ln(1 - \hat{p}_{\text{high}}) = -\bar{Y}(a\bar{Y}^{b-1} - 1)^{-1}\ln(a\bar{Y}^{b-1}).$$

Boivin et al. (1990) used this model to describe the relationship between the proportion of diseased carrot leaves and the mean number of cercospora leaf blight (caused by *Cercospora carotae*) lesions per leaf.

It is apparent that most of the relationships described here in the context of plant pathology have a precedent in the entomological literature. In economic entomology, relationships between incidence and pest population density form the basis of "binomial sampling", in which an estimate of pest population density is based on collection of data relating only to a binary assessment (either presence/absence or relative to a tally threshold) of the pest in question from sampling units (Chapter 10).

9.8 Sparsely Sampled Disease Severity Data

Just as statistical probability distributions may be fitted to sparsely sampled data for disease incidence and for disease assessments in the form of counts, the same kind of analysis could, at least in principle, be applied to sparsely sampled disease severity data. In fact, little attention has been devoted to the description and analysis of disease severity data in this way. There is probably not a single reason for this, but the fact that there are many different ways of recording severity data (Chapter 2) must be a contributory factor. Much more attention has been paid to the formulation of empirical relationships between disease severity and disease incidence.

9.8.1 The severity–incidence relationship—regression models

McRoberts et al. (2003) reviewed regression models for severity–incidence relationships. The most common approach to the analysis of relationships between disease severity and disease incidence is to collect contemporaneous severity and incidence data and prepare a graphical plot of these, before embarking on a statistical

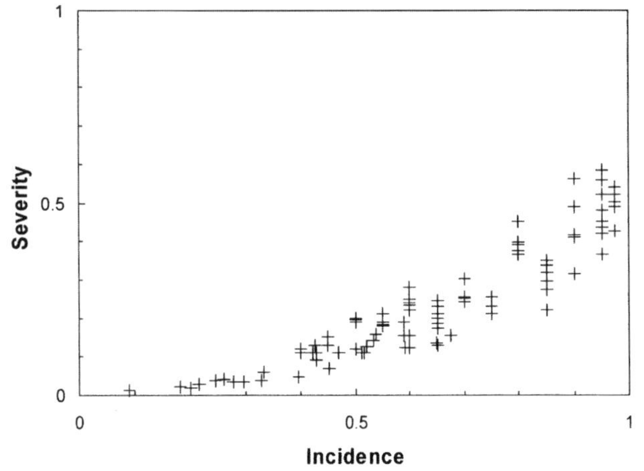

FIG. 9.10. Severity and incidence data for Fusarium head blight on wheat, from a study by Paul et al. (2005).

analysis. Since neither severity nor incidence can be incontrovertibly said to be the independent (explanatory) variable (Snedecor and Cochran, 1989) for the purpose of regression analysis, there is no general rule about which of the variables appears on which of the axes of graph. Examples in both orientations appear in the literature. Over a wide enough range of severity and incidence values, a graphical plot of the raw data, either as severity against incidence (Fig. 9.10) or incidence against severity, usually shows a curved relationship.

When the severity-incidence data follow a curve, common practice is to seek a transformation of one or both variables that results in a linear graphical plot of the data, as a precursor to least-squares linear regression analysis (see Chapter 3). It is worth bearing in mind at this stage that transformations that provide satisfactory descriptions of observed data may not provide an appropriate description of a severity–incidence relationship if extrapolated beyond the range of the data (see, e.g., Zahareiva et al., 1984). Notwithstanding this, many different formulations of severity–incidence relationships appear in the literature.

In the notation adopted for this discussion, estimated mean severity is denoted $\bar{y}_{sev}$ and the corresponding estimated mean incidence is $\bar{y}_{inc}$. To obtain a value of $\bar{y}_{sev}$ (using an example of severity at the leaf scale), visual assessments of severity (as a proportion) could be recorded for a number of leaves, and the mean value calculated. The corresponding value of $\bar{y}_{inc}$ would then be the proportion of leaves diseased (i.e., showing any level of symptoms that can be detected visually). Estimation of the relationship between severity and incidence of course requires an appropriate number of pairs of $\bar{y}_{sev}$ and $\bar{y}_{inc}$ data. In the following discussion, arbitrary constants are denoted *a*, *b*, *c* (as required) for relationships showing severity as a function of incidence, and *a*′, *b*′, *c*′ (as required) for relationships showing incidence as a function of severity.

If logarithmic transformation of both the $\bar{y}_{sev}$ and the $\bar{y}_{inc}$ data results in a satisfactory straight line plot, a power curve of the form:

$$\bar{y}_{inc} = a'(\bar{y}_{sev})^{b'}$$

is an appropriate description. An example is given by Chuang and Jeger (1987) in a study of banana leaf spot (caused by *Mycosphaerella fijiensis* var. *difformis*). The authors point out that such a model arises directly if severity and incidence are both increasing exponentially, but at different rates. Of course, disease does not increase indefinitely in an exponential manner (see Chapter 4), but exponential increase may be a reasonable assumption during the early stages of an epidemic. Price and Williams (1990) (working with both incidence and severity as percentages, rather than proportions) used a power curve of the form:

$$(\bar{y}_{sev} + 1) = a(\bar{y}_{inc} + 1)^{b}$$

in a study of paspalum leaf blight (caused by *Ascochyta paspali*). Presumably, $\bar{y}_{sev} + 1$ and $\bar{y}_{inc} + 1$ were adopted by the authors in order to avoid problems with the logarithmic transformation (employed for the purpose of parameter estimation by linear least squares regression analysis) when the observed values of severity and/or incidence (on a percentage scale) were zero. For the data in Fig. 9.10, the power curve $\bar{y}_{sev} = 0.47(\bar{y}_{inc})^{1.8}$ has an R^2 value of 0.91 (R^2 values for this and subsequent fits are given for reference, but should not be taken as an indication of the superiority or otherwise of any particular model, based as they are on a single data set).

From the general shape of the relationship between severity and incidence in Fig. 9.10, it appears that a polynomial equation in which $\bar{y}_{sev}$ is related to, say, $(\bar{y}_{inc})^2$ might provide a satisfactory description of the data. Such an example is given by Silva-Acuña et al. (1999) in a study of severity—incidence relationships for coffee rust (caused by *Hemileia vastatrix*):

$$\bar{y}_{sev} = a + b(\bar{y}_{inc}) + c(\bar{y}_{inc})^2.$$

For the data in Fig. 9.10, the polynomial $\bar{y}_{sev} = 0.0234 - 0.0559(\bar{y}_{inc}) + 0.5442(\bar{y}_{inc})^2$ has an R^2 value of 0.88. Pataky and Headrick (1988) showed examples of relationships between $\bar{y}_{sev}$ and $\bar{y}_{inc}$ for common maize rust of sweet corn (caused by *Puccinia sorghi*) described by higher-order polynomials. Filajdic and Sutton (1992) showed relationships between $\bar{y}_{inc}$ and $(\bar{y}_{sev})^{0.5}$ for Alternaria blotch of apple (caused by *Alternaria mali*). In this case, severity was measured on a six-point scale, at integer intervals from 0 to 5.

James and Shih (1973) discussed use of the restricted (or negative) exponential equation:

$$\bar{y}_{inc} = 1 - e^{-a'(\bar{y}_{sev})}$$

to describe the relationship between incidence and severity of powdery mildew (caused by *Blumeria graminis* f.sp. *tritici*) and leaf rust (caused by *Puccinia recondita*) of wheat. In this case, disease incidence increases from a lower value (usually taken to be zero) towards an upper value (usually taken to be 100 on a percentage scale or one on a proportion scale) with increasing severity. After some rearrangement of their relationship between $\bar{y}_{inc}$ and $\bar{y}_{sev}$, James and Shih (1973) obtained the relationship:

$$\bar{y}_{sev} = a\ln(1-\bar{y}_{inc}).$$

For the data in Fig. 9.10, the relationship $\bar{y}_{sev} = -0.165 \ln(1-\bar{y}_{inc})$ has an R^2 value of 0.81.

Now consider a two-parameter generalization of James and Shih's (1973) restricted exponential equation:

$$\bar{y}_{inc} = 1 - e^{-a'(\bar{y}_{sev})^{b'}} \tag{9.37}$$

After some re-arrangement, this can be written:

$$\mathrm{CLL}(\bar{y}_{inc}) = \ln(a') + b'\ln(\bar{y}_{sev})$$

Writing this instead as a relationship between severity and incidence (on the transformed scales), we obtain:

$$\ln(\bar{y}_{sev}) = \ln(a) + b\,\mathrm{CLL}(\bar{y}_{inc}) \tag{9.38}$$

For the data in Fig. 9.10, the relationship $\ln(\bar{y}_{sev}) = -1.742 + 1.022\,\mathrm{CLL}(\bar{y}_{inc})$ has an R^2 value of 0.88.

9.8.2. The severity–incidence relationship—a mathematical model

The starting point for this mathematical model is equation 9.30, based on the binomial distribution. The notation used for this discussion is based on that used in section 9.7.1. In this application, the leaf is the higher of the two spatial scales of interest. The probability that a leaf is diseased is denoted p_{leaf}. Consider a leaf as comprising a number of sites (n), each of which may either be fully occupied by a cluster of pustules or unoccupied. If each leaf has the same total area and all sites are the same size, n is constant. The probability that a site is occupied is denoted p_{site}. Assuming that the distribution of occupied sites per leaf is binomial, the probability that a leaf has no occupied sites is then:

$$p_{leaf} = 1-(1-p_{site})^n \tag{9.39}$$

which may be re-arranged as:

$$p_{site} = 1-(1-p_{leaf})^{1/n} \tag{9.40}$$

These are relationships between incidence in a spatial hierarchy, of the type discussed by Hughes et al. (1997) and discussed in section 9.7.1.

The assumptions about constant leaf and site size imply that the proportion of leaf surface visibly diseased is the same as the proportion of sites visibly diseased. Consider a particular value of p_{site}, denoted X ($0 \le X \le 1$). The corresponding severity, denoted here $\bar{y}_X$, may be thought of as the area under a standard uniform [0, 1] distribution between 0 and X. The corresponding value of p_{leaf} may be calculated from equation 9.39. This gives a severity ($\bar{y}_X$)–incidence (p_{leaf}) relationship discussed by Daamen (1986). Daamen (1986, Fig. 1) calculated graphs of $\bar{y}_X$ against p_{leaf} for $0 \le X \le 1$ and $n = 2^i$, $i = 0, 1, ..., 6$.

Daamen's (1986) severity-incidence model has been discussed further by Hughes et al. (2004), as follows. The beta probability distribution is a continuous distribution defined for a random variable x, the possible values of which are between 0 and 1 (see Olkin et al., 1980, for details). The two parameters of the beta probability distribution, denoted α and β, are positive numbers. The cumulative distribution function, referred to as the incomplete beta function, is denoted $I_X(\alpha, \beta)$ and written:

$$I_X(\alpha,\beta) = \frac{\int_0^X x^{\alpha-1}(1-x)^{\beta-1}\,dx}{\int_0^1 x^{\alpha-1}(1-x)^{\beta-1}\,dx} \tag{9.41}$$

($0 \le X \le 1$). When α and β are both positive integers, the denominator of the expression on the right-hand side of equation 9.41 simplifies to $[(\alpha-1)!(\beta-1)!]/(\alpha+\beta-1)!$, and then simplifies further to β^{-1} if $\alpha = 1$. As integrals can be interpreted as areas, the format of equation 9.41 indicates that $I_X(a, b)$ represents an area expressed as a proportion of the whole. Thus the incomplete beta function may be thought of as a statistical analogue of disease severity.

As above, assume that the proportion of leaf surface visibly diseased is the same as the proportion of sites visibly diseased. X again denotes a particular value of p_{site}, with a corresponding value of severity $\bar{y}_X$ and a corresponding value of p_{leaf} calculated from equation 9.39. Now, note that the standard uniform distribution is a special case of the beta probability distribution, with parameters $\alpha = 1$ and $\beta = 1$ (Olkin et al., 1980), so $\bar{y}_X = I_X(1, 1)$. The incomplete beta function $I_X(1, 1)$ may therefore be used to calculate a graph of $\bar{y}_X$ against p_{leaf} for $0 \le X \le 1$ and any given value of n, replicating Daamen's (1986) analysis. A practical difficulty with this model is that the value of n, the number of sites per leaf, is unlikely to be countable in any straightforward manner. In practice, we are more likely to have estimates of p_{leaf} and p_{site} and to treat n as a parameter to be estimated from data. The relationship between $\bar{y}_X$ and p_{leaf}

is calculated via the assumption that sites may only be occupied fully or unoccupied, and the use of either equation 9.39 or 9.40 (both involving p_{site}).

As an alternative, we can set $n = 1$, so that a leaf is always thought of as comprising a *single* site that may be partially occupied (i.e., visibly diseased to some extent). The starting point is now a beta probability distribution with parameters $\alpha = 1$ (for simplicity) and $\beta = q$ ($0 < q \leq 1$). Now severity, the proportion occupied, is modeled by $\bar{y}_X = I_X(1, q)$, in which q is a parameter to be estimated from data and X is an index variable ($0 \leq X \leq 1$). The relationship between severity and incidence is then calculated directly from a graph of $\bar{y}_X$ against p_{leaf}, which has the form:

$$\bar{y}_X = 1 - (1 - p_{leaf})^q \qquad (9.42)$$

Then:

$$\text{CLL}(\bar{y}_X) = \ln(q) + \text{CLL}(p_{leaf}) \qquad (9.43)$$

Fitting equation 9.43 to data (Hughes et al., 2004) provides a method of estimating the parameter q of the incomplete beta function $I_X(1, q)$. The use of the incomplete beta function to represent disease severity has the merit that the statistical description matches the biological variable of interest: both are continuous variables describing an area as a proportion of the whole.

9.8.3 Another regression model

An obvious question, but one still of interest, is what happens when equation 9.43 is used as the basis for a regression model for a severity–incidence relationship. Reverting to the notation of section 9.8.1 (and allowing for the coefficient of $\text{CLL}(\bar{y}_{inc})$ to be estimated), we can write this model as:

$$\text{CLL}(\bar{y}_{sev}) = \ln(a) + b\text{CLL}(\bar{y}_{inc}) \qquad (9.44)$$

and use observed data for severity and incidence at the leaf scale to estimate a and b. Using this approach with severity and incidence data for Fusarium head blight on wheat (caused by *Fusarium graminearum*), Paul et al. (2005) reported that equation 9.44 was a consistently good description of the observed data. For the data in Fig. 9.10, the relationship $\text{CLL}(\bar{y}_{sev}) = -1.6055 + 1.144\text{CLL}(\bar{y}_{inc})$ has an R^2 value of 0.89. The estimated value of the coefficient b is slightly greater than one, which appears to be a typical outcome.

9.8.4 Overview of the severity-incidence relationship

It would probably be unwise to make too much of one type of severity-incidence relationship or another being preferable on theoretical grounds. In essence, the relationships described can serve as devices for summarizing certain epidemiological data in a way that has found a number of useful applications (Seem, 1984; McRoberts et al., 2003). The main issues in making a choice of one model or another are often likely to be how to obtain estimates of model parameters and how good is the fit of the model to the data under consideration.

Extracting information about patterns of disease from a severity–incidence relationship is less straightforward than for the spatial hierarchies described in section 9.7. Suppose, for example, that an assessment of leaves indicated a mean severity of 25%. Notionally, this value could arise if all leaves were 25% diseased (incidence is 100%) or if 25% of the leaves were 100% diseased (incidence is 25%). These extremes imply very different patterns of disease severity. Of course, the value of incidence corresponding to a particular value of severity will likely lie somewhere between the possible extremes. Linking pattern to the severity–incidence relationship for observed data is an interesting problem (McRoberts et al., 2003). The analysis of severity–incidence relationships specifically in the context of spatial patterns of plant disease deserves further investigation.

9.9 Analysis of Intensively Mapped Disease Data

Data collected by intensive mapping include details of the spatial arrangement of the sampling units. There are two main types of intensely mapped plant disease data. For the first, each sampling unit consists of an individual element, such as a plant, that is classified as either disease-free (denoted, for example, "0" or "−" or "H") or diseased (denoted, for example, "1" or "+" or "D"). Thus, in this case, the variable to be analyzed is binary in form (see Chapter 2). For the second type of intensely mapped disease data, each sampling unit consists of either a count, such as the number of diseased plants out of a total of n (with $n > 1$) or the number of lesions, or a measurement of some continuous-scale variable, such as visual estimate of disease severity. Methods for both types of intensely mapped disease data are discussed, beginning with the analysis of binary data.

Bald (1937) reported the results of monitoring epidemics of tomato spotted wilt virus (TSWV) disease in experimental crops of tomato near Adelaide in Australia. Fig. 9.11 shows some of the intensively mapped data from this study, as originally presented by Cochran (1936) (i.e., without reference to experimental treatments). Data from two disease assessments are shown, with plants diseased at the first count marked "×" and those diseased at the second count, two weeks later, marked "+". In relation to the data from the first count, Cochran (1936) concluded that: "... in addition to a gradual increase in the degree of infectivity from the top to the bottom of the field, there is a tendency for

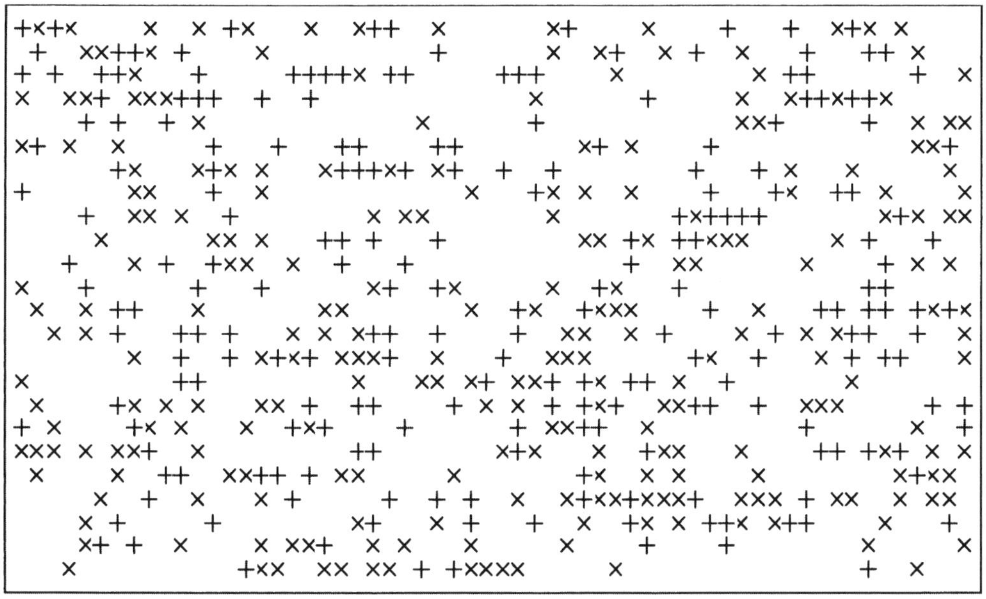

FIG. 9.11. A field map of *Tomato spotted wilt virus* disease incidence, reported by Cochran (1936). Sowing date was 26 November 1929. × indicates diseased at first assessment (18 December 1929), and + indicates diseased at second assessment (31 December 1929).

diseased plants to congregate in small patches." Considering further data from the same experiments, the absence of large, aggregated groups of TSWV-infected plants along rows led Bald (1937) to rule out dispersal of the virus by mechanical transmission during crop husbandry operations. TSWV was carried into the plots by adult thrips that had acquired the virus as nymphs, while feeding on infected plant material outside the plots. The existence of small clusters of infected plants indicated, perhaps, that after landing in a plot, infective adult thrips took flight again, but only for a short distance. Secondary spread of the virus within plots was rare, as indicated by the lack of large clusters of diseased plants. There was thought to be little chance of disease spreading from one tomato plant to another in a plot, as thrips larvae do not move from plant to plant, and adults were not thought to become infective by feeding on diseased tomato plants (Bald, 1937). The statistical analyses underlying these biological interpretations represent the first major contribution to the quantitative study of spatial patterns of plant disease. Methods that applied to the data in both sparsely sampled and intensively mapped formats were used.

Example 9.3. Methods for sparsely sampled data described in this chapter so far can be applied to intensively mapped data. In doing so, information on the spatial arrangement of the sampling units is ignored. (Later in this chapter we show how to obtain a finer resolution in our quantification of spatial pattern for this data set.) The frequency distribution of observed number of TSWV-infected tomato plants out of $n = 9$ in a sampling unit (Fig. 9.12) was obtained by dividing the first-count data from the mapped area shown in Fig. 9.11 into 160 quadrats of three columns by three rows and counting the infected plants (marked "×" in Fig. 9.11) in each quadrat.

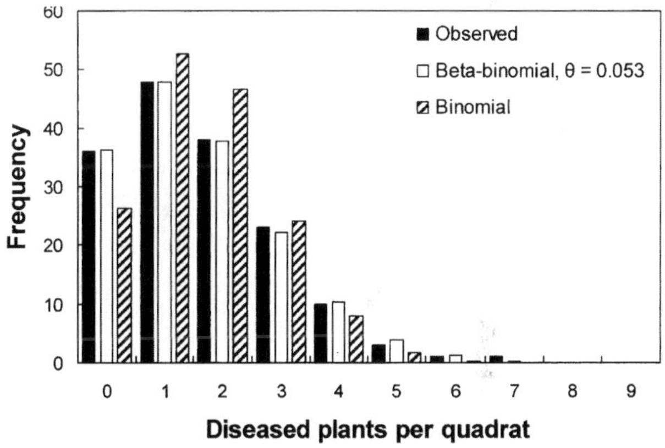

FIG. 9.12. Frequency distribution of tomato plants with symptoms of *Tomato spotted wilt virus* disease incidence. Disease incidence was assessed by visual symptoms on individual plants, made on 18 December 1929 and reported by Cochran (1936) (see Fig. 9.11). The solid bars represent observed frequencies, the open bars represent expected β-binomial frequencies and the hatched bars represent expected binomial frequencies. Goodness-of-fit statistics are given in *Example 9.3*.

The BBD computer program (Madden and Hughes, 1994) was used to calculate maximum likelihood parameter estimates ($\bar{y} = 0.181$(SE = 0.012), $\hat{\theta} = 0.053$ (SE = 0.020)) and expected binomial and β-binomial frequencies (Fig. 9.12). There is an excess of values in the tails ($Y = 0$ or $Y \geq 4$) of the observed distribution and a deficit near the mean, compared with the binomial expected frequencies. Such results are typical of aggregated disease incidence data. Visually, the β-binomial

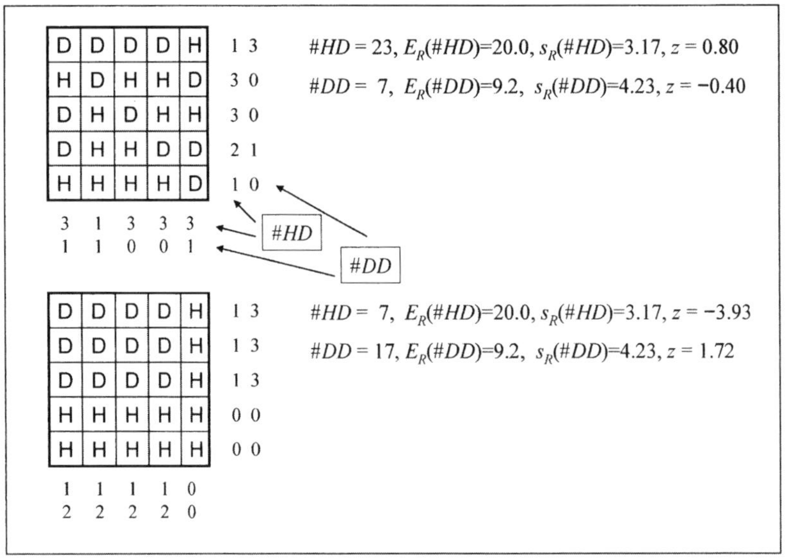

FIG. 9.13. Two maps of diseased (D) and disease-free (H) plants. (A) The upper 5 × 5 array is taken from the top left-hand corner of Fig. 9.11. The lower 5 × 5 array has the same numbers of D and H plants re-arranged in an obviously aggregated pattern. The numbers of H–D joins (#HD) and D–D joins (#DD) for the rows and columns of each map are shown, along with the join-count statistics for each of the maps.

distribution appears to provide a better description of the observed distribution. Support for these impressions is provided by χ^2 tests of goodness-of-fit (equation 9.5). For the binomial distribution, $\chi^2 = 10.22$ with 4 *df* ($P = 0.037$). For the β-binomial distribution, $\chi^2 = 0.44$ with 4 *df* ($P = 0.98$). In both cases, frequency classes were pooled until no expected frequencies were less than one. For the binomial distribution, the number of degrees of freedom is two fewer than the number of frequency classes after pooling. For the β-binomial distribution, the number of degrees of freedom is three fewer than the number of frequency classes after pooling. A β-binomial distribution with a mean of 0.181 and aggregation parameter $\theta = 0.053$, with $n = 9$ (and $N = 160$) provides a very close description of the observed data.

For the observed data in Fig. 9.12, the value of the index of dispersion is $D = 1.42$ (equation 9.8). To test the statistical significance of deviations from randomness, $(N - 1)D$ is compared with the tabulated χ^2 distribution at $N-1$ *df*. Here χ^2 (159 *df*) = 225.55 ($P < 0.001$). Thus the result here is indicative of aggregation. The value of the $C(\alpha)$ test statistic is 3.89, which is larger than the (one-sided) 5% point of the standard normal distribution. Conventionally, then, we would reject the hypothesis of the binomial distribution for these data in favor of the specific alternative of the β-binomial, on the basis of the $C(\alpha)$ test.

The available evidence from analysis of the first-count data from Fig. 9.11, using methods for sparsely sampled data, supports the view that there is aggregation of disease within the adopted sampling units. Now, because we have information on the spatial locations of the diseased plants, we have access to additional methods of analysis that cannot be applied to sparsely sampled data.

9.9.1 Join-count statistics

Join-count statistics (Cliff and Ord, 1981; Upton and Fingleton, 1985) may be used to analyze spatial associations for disease incidence data (de Jong and de Bree, 1995; Gumpertz, 1997). As in the previous section, this approach is appropriate when each individual plant is a sampling unit. If healthy plants are coded "H" and diseased plants coded "D", two adjacent (proximal, contiguous) plants may be classified by the type of "join" linking them: H–H, D–D or H–D. In an array of plants, both the type of join to be counted and the orientation(s) of interest are specified. One can consider joins in different orientations: along rows, across rows, diagonally, or a combination of these. Then one simply counts the number of joins of the specified type in the orientation(s) of interest.

This is demonstrated with an example in Fig. 9.13. The 5 × 5 array in the upper part of Fig. 9.13 is taken from the upper left-hand corner of Fig. 9.11 (first- and second-count data combined). Looking at the first row, there is only one H–D join, between the fourth and fifth plants. There are, however, three D–D joins (between plants one and two, two and three, and three and four). Looking at the first column, there are three H–D joins and just one D–D join. The total number of joins of each type is counted (separately for H–D and D–D joins) for the orientation(s) of interest. Here we are interested in pattern along and across rows, so the number of joins of each type is summed over both the rows and columns of the diagram. Using an analogy with the game of chess, this is usually called the "rook" proximity (or contiguity) case

(Cliff and Ord, 1973). We define #HD and #DD, respectively, as the observed number of H–D and D–D join counts. As shown in the upper part of Fig. 9.13, there were 23 H–D joins (#HD = 23) and 7 D–D joins (#DD = 7).

The question then is whether the observed join-count is large (or small) relative to that expected for a random pattern of Hs and Ds. For large sample sizes, the expected number of join-counts can be defined under randomness [denoted E_R(#HD), E_R(#DD)] and the corresponding standard errors of these expected values can be estimated [s_R(#HD), s_R(#DD)] (see Upton and Fingleton, 1985; Gumpertz, 1997). The formulae for expected values depend on some implicit assumptions, and we show here only the equations for the "free sampling" case, which simply means that the number of diseased plants in an array of a given size is assumed to be random, with a fixed probability that any individual is diseased (Gumpertz, 1997). Assuming that the data set is a rectangular array, comprising *#r* and *#c* columns, with no missing values, one first defines:

$$S_0 = 2(2\#r\#c - \#r - \#c)$$
$$S_1 = 2S_0$$
$$S_2 = 8(8\#r\#c - 7\#r - 7\#c + 4).$$

Note that these constants depend only on the physical dimensions of the field and not on the pattern of disease, and that the definitions given here apply *only* for a rectangular array with no missing values, with the "rook" definition of proximity. Formulae are available for other kinds of array and for other definitions of proximity (Upton and Fingleton, 1985; Gumpertz, 1997). For the diagram in the upper part of Fig. 9.13 (with $\#r = \#c = 5$), $S_0 = 80$, $S_1 = 160$, and $S_2 = 1072$. Expected values and their estimated standard errors are given by:

$$E_R(\#HD) = S_0\bar{y}(1-\bar{y})$$
$$s_R(\#HD) = \frac{\sqrt{S_2\bar{y}(1-\bar{y}) + 4(S_1 - S_2)\bar{y}^2(1-\bar{y})^2}}{2}$$
$$E_R(\#DD) = \frac{S_0\bar{y}^2}{2}$$
$$s_R(\#DD) = \frac{\sqrt{S_1\bar{y}^2 + (S_2 - 2S_1)\bar{y}^3 + (S_1 - S_2)\bar{y}^4}}{2}$$

where $\bar{y}$ is, as before, the observed proportion of diseased plants in the array (our estimate of the probability that an individual plant is diseased).

If there is an aggregated pattern of diseased plants, #HD will be less than E_R(#HD) because there will be fewer H–D (and D–H) joins than expected on the basis of randomness. With a large enough sample size, this can be tested with a standard normal distribution test statistic:

$$z = \frac{\#HD - E_R(\#HD) - 0.5}{s_R(\#HD)}.$$

The subtraction of 0.5 from the numerator is a correction for continuity, applied because join-counts are discrete. A large negative value of the test statistic indicates aggregation of diseased plants. As shown in the upper part of Fig. 9.13, #HD was actually larger than E_R(#HD), so the standard normal test clearly indicates randomness in this example. For comparison, consider the array in lower part of Fig. 9.13, which has the same proportion of diseased plants as the array in the upper part, but with all the diseased plants contained in a single patch. In this case, #HD was considerably less than E_R(#HD), and the standard normal test statistic was large and negative, indicating an aggregated pattern of diseased plants ($P < 0.05$). The test for aggregation is one-sided, so values of $z < -1.64$ (i.e., more negative) are conventionally taken as a basis for rejection of the hypothesis of randomness at $P = 0.05$.

For an aggregated pattern of diseased plants, #DD will be greater than E_R(#DD), because there will be many D–D joins. The following standard normal statistic is used to test for aggregation:

$$z = \frac{\#DD - E_R(\#DD) + 0.5}{s_R(\#DD)}.$$

Here, the correction for continuity involves addition of 0.5 to the numerator (Gumpertz, 1997). For the example in the upper part of Fig. 9.13, there were actually fewer D–D joins than expected, indicating randomness. For the example in the lower part of Fig. 9.13, with all the diseased plants in a single patch, #DD was considerably larger than expected, producing a test statistic $z > +1.64$, indicative of aggregation of diseased plants ($P < 0.05$).

The standard normal test is intended for use with a large number of observations and when disease incidence is not very close to 0 or 1. With small numbers of observations, such as with the 5×5 arrays given as examples in Fig. 9.13, randomization methodology (Upton and Fingleton, 1985) is a better procedure. Keeping the same proportions of Ds and Hs as for the observed data, a large number of random arrangements of Ds and Hs are generated to determine how likely the observed arrangement is, compared with typical random patterns. Gumpertz (1997) provides a SAS macro by means of which this randomization methodology may be implemented.

In its simplest form, the join-count statistic only quantifies spatial association between nearest neighbors (adjacent plants, with a specified orientation), thus allowing no assessment of association between more distant neighbors. However, this is not an inherent limitation of the method, and it is possible to define joins to exist between more distant neighbors (Cliff and Ord, 1981). Pethybridge and Madden (2003) and Pethybridge et al. (2004) provide examples using join-counts to analyze spatial scales in their study of spatio-temporal dynamics of virus spread in Australian hop gardens. Ferrandino (1998) also discusses the analysis of spatial associations for disease incidence data with pairs of plants that are not immediately adjacent, based on a different methodology.

The manual calculation of join counts is relatively easy for small data sets such as in those given in Fig. 9.13, but with moderate-sized data sets (such as shown in Fig. 9.11) the analysis would be impracticable without the use of computer software. Fortunately, Gumpertz (1997) wrote *SAS* programs for many of the required calculations. An alternative, spreadsheet-based approach is described by Sawada (1999). Using a modification of the program written by Gumpertz (1997), we analyzed the full data set mapped in Fig. 9.11 (first assessment), based on the "rook" proximity criterion. There were 798 observed H–D joins (#HD = 798), which was smaller than the expected number under randomness (E_R(#HD) = 829.8), suggesting aggregation of disease. With the estimated standard error of s_R(#HD) = 39.56, the standard normal statistic was calculated to be $z = -0.79$. This gives an achieved significance level of $P = 0.21$. Thus, there was insufficient evidence to reject the null hypothesis of randomness with this analysis.

There is a relationship between join-count analysis and ordinary runs analysis (see, e.g., Madden et al., 1982; Madden and Campbell, 1986). An "ordinary run" is a sequence either of diseased or of healthy plants. For example, the numbers or ordinary runs (#O) in the five rows in the upper part of Fig. 9.13 are 2, 4, 4, 3, and 2. Each of these values exceeds by one the corresponding #HD value. Thus, it turns out that ordinary runs analysis is a special case of H–D join-count analysis for a one-dimensional array (with #O = #HD + 1). The derivation is different, but the outcome is the same. For ordinary runs analysis of a two-dimensional array, the data must first be transformed into a one-dimensional array, by combining the rows end-to-end. However, there is little reason to use this method, because join-count statistics provide a basis for a two-dimensional analysis that can be calculated directly.

9.9.2 The cross-product statistic

Join-count statistics are just one of a number of spatial analyses that can be derived from a general cross-product statistic, denoted here Θ. Mantel (1967) provides the starting point for epidemiological methodology based on the cross-product statistic. Legendre and Legendre (1998) provide a more recent overview of ecological applications. The general cross-product statistic takes the form:

$$\Theta = w\sum_i\sum_j W_{ij}U_{ij} \tag{9.45}$$

which we explain by reference to the join-count statistic, and in particular to the 5 × 5 array shown in the upper part of Fig. 9.13. The variable W_{ij} is used to indicate whether two locations in the array are neighbors, according to whatever definition of proximity has been adopted. In its simplest form, $W_{ij} = 1$ when two locations are defined as neighbors, otherwise $W_{ij} = 0$ (Table 9.3A). The variable U_{ij} is used to indicate whether or not the disease status Y of two locations in the array is the same. Let $Y = 1$ for a location denoted D in the upper part of Fig. 9.13, and $Y = 0$ for a location denoted H. If U_{ij} is defined as $U_{ij} = (Y_i - Y_j)^2$, then $U_{ij} = 0$ when both locations have the same disease status, and $U_{ij} = 1$ when they differ (Table 9.3B). The factor w is a scaling factor, used to control the range of Θ. In this case, $w = 0.5$, because the arrays in Tables 9.3A and 9.3B are both symmetrical about the main diagonal. All the required information about proximity and agreement or otherwise of disease status is contained in the upper off-diagonal elements (the lower off-diagonal elements are actually redundant). As we now have the appropriate W_{ij} values for the adopted definition of proximity, the appropriate U_{ij} values for counting locations with differing disease status, and an appropriate value of w, equation 9.45 can be used to calculate Θ. Evaluation of the term $\sum_i\sum_j W_{ij}U_{ij}$ requires that the corresponding W_{ij} and U_{ij} values, respectively in Tables 9.3A and 9.3B, are multiplied together *element-by-element* (matrix multiplication is *not* required), and the sum of the resulting values is then calculated (the result is 46). Then with $w = 0.5$, $\Theta = 23$. Because the definition of U_{ij} was appropriate for counting pairs of locations with differing disease status, we have found Θ = #HD as obtained previously in the analysis of join counts.

Other definitions of U_{ij} are possible. For example, still with $Y = 1$ for a location denoted D in the upper part of Fig. 9.13 and $Y = 0$ for a location denoted H, let $U_{ij} = (Y_i \times Y_j)$. Now, $U_{ij} = 0$ if both locations are H, or if one location is H and one is D, and $U_{ij} = 1$ if *both* locations are D (Table 9.3C). Using equation 9.45 to calculate Θ, with the W_{ij} values from Table 9.3A and $w = 0.5$ (both as previously), but now with the U_{ij} values from Table 9.3C, gives $\Theta = 7$. Because the definition of U_{ij} in this case was appropriate for counting pairs in which both locations are denoted D, we have found Θ = #DD as obtained previously in the analysis of join-counts.

When $U_{ij} = (Y_i - Y_j)^2$, Θ increases if there are more H–D joins (within the adopted definition of proximity) and decreases if there are more H–H and D–D joins. When $U_{ij} = (Y_i \times Y_j)$, Θ increases if there are more D–D joins (within the adopted definition of proximity) and decreases if there are more H–H and H–D joins. If a different proximity definition were adopted, the calculations would be carried out using a different array of W_{ij} values.

The formula for Θ is simple in form, but direct usage is tedious, mostly because of the way that W_{ij} and U_{ij} variables are used. For instance, for a map of disease consisting of 1440 plants (Fig. 9.11), there are $1{,}440^2$ = 2,073,600 values of both W_{ij} and U_{ij}. Every plant is

TABLE 9.3A. Values of W_{ij} for the "rook" definition of proximity in a 5 × 5 array. Neighboring locations are indicated by $W_{ij} = 1$, otherwise $W_{ij} = 0$.

	Row	*1*	*1*	*1*	*1*	*1*	*2*	*2*	*2*	*2*	*2*	*3*	*3*	*3*	*3*	*3*	*4*	*4*	*4*	*4*	*4*	*5*	*5*	*5*	*5*	*5*
Row	*Column*	*1*	*2*	*3*	*4*	*5*	*1*	*2*	*3*	*4*	*5*	*1*	*2*	*3*	*4*	*5*	*1*	*2*	*3*	*4*	*5*	*1*	*2*	*3*	*4*	*5*
1	*1*	0	1	0	0	0	1	0	0	0	0	0	0	0	0	0	0	0	0	0	0	0	0	0	0	0
1	*2*	1	0	1	0	0	0	1	0	0	0	0	0	0	0	0	0	0	0	0	0	0	0	0	0	0
1	*3*	0	1	0	1	0	0	0	1	0	0	0	0	0	0	0	0	0	0	0	0	0	0	0	0	0
1	*4*	0	0	1	0	1	0	0	0	1	0	0	0	0	0	0	0	0	0	0	0	0	0	0	0	0
1	*5*	0	0	0	1	0	0	0	0	0	1	0	0	0	0	0	0	0	0	0	0	0	0	0	0	0
2	*1*	1	0	0	0	0	0	1	0	0	0	1	0	0	0	0	0	0	0	0	0	0	0	0	0	0
2	*2*	0	1	0	0	0	1	0	1	0	0	0	1	0	0	0	0	0	0	0	0	0	0	0	0	0
2	*3*	0	0	1	0	0	0	1	0	1	0	0	0	1	0	0	0	0	0	0	0	0	0	0	0	0
2	*4*	0	0	0	1	0	0	0	1	0	1	0	0	0	1	0	0	0	0	0	0	0	0	0	0	0
2	*5*	0	0	0	0	1	0	0	0	1	0	0	0	0	0	1	0	0	0	0	0	0	0	0	0	0
3	*1*	0	0	0	0	0	1	0	0	0	0	0	1	0	0	0	1	0	0	0	0	0	0	0	0	0
3	*2*	0	0	0	0	0	0	1	0	0	0	1	0	1	0	0	0	1	0	0	0	0	0	0	0	0
3	*3*	0	0	0	0	0	0	0	1	0	0	0	1	0	1	0	0	0	1	0	0	0	0	0	0	0
3	*4*	0	0	0	0	0	0	0	0	1	0	0	0	1	0	1	0	0	0	1	0	0	0	0	0	0
3	*5*	0	0	0	0	0	0	0	0	0	1	0	0	0	1	0	0	0	0	0	1	0	0	0	0	0
4	*1*	0	0	0	0	0	0	0	0	0	0	1	0	0	0	0	0	1	0	0	0	1	0	0	0	0
4	*2*	0	0	0	0	0	0	0	0	0	0	0	1	0	0	0	1	0	1	0	0	0	1	0	0	0
4	*3*	0	0	0	0	0	0	0	0	0	0	0	0	1	0	0	0	1	0	1	0	0	0	1	0	0
4	*4*	0	0	0	0	0	0	0	0	0	0	0	0	0	1	0	0	0	1	0	1	0	0	0	1	0
4	*5*	0	0	0	0	0	0	0	0	0	0	0	0	0	0	1	0	0	0	1	0	0	0	0	0	1
5	*1*	0	0	0	0	0	0	0	0	0	0	0	0	0	0	0	1	0	0	0	0	0	1	0	0	0
5	*2*	0	0	0	0	0	0	0	0	0	0	0	0	0	0	0	0	1	0	0	0	1	0	1	0	0
5	*3*	0	0	0	0	0	0	0	0	0	0	0	0	0	0	0	0	0	1	0	0	0	1	0	1	0
5	*4*	0	0	0	0	0	0	0	0	0	0	0	0	0	0	0	0	0	0	1	0	0	0	1	0	1
5	*5*	0	0	0	0	0	0	0	0	0	0	0	0	0	0	0	0	0	0	0	1	0	0	0	1	0

compared with every other plant, both for proximity and disease status. This means that other approaches to the required calculations are often adopted, while retaining the cross-product statistic as a conceptual basis for the calculation of either a measure of the agreement of disease status in neighboring locations, or a measure of the disagreement. For example, spatial autocorrelation coefficients (section 9.9.3) and semivariances (section 9.9.4) are versions of cross-product statistics.

9.9.3. Spatial autocorrelation

If a plant being diseased makes it more (or less) likely that neighboring plants will be diseased, diseased plants are exhibiting *spatial autocorrelation*. For intensively mapped disease incidence data where each individual plant is a sampling unit (such as those shown in Fig. 9.11), join-count statistics provide a basic measure of spatial autocorrelation.

More generally, the principles of spatial autocorrelation analysis have been presented by Cliff and Ord (1981) and Upton and Fingleton (1985). The use of this type of analysis is not restricted to disease incidence data, where each individual plant is a sampling unit. Provided that the spatial locations of the sampling units are known, any type of disease intensity data can be analyzed using spatial autocorrelation techniques. For instance, the sampling units may be individual plants, and the disease variable may be the number of diseased leaves or number of lesions. Alternatively, the sampling units may be quadrats containing a number of plants, and the disease variable may be the number of diseased plants. Note also that the methods described in this section are applicable to the analysis of continuous data such as disease severity. The minimum requirement for spatial autocorrelation analysis is that the spatial locations and disease status of the sampling units be known. Although data typically consist of a two-dimensional array of sampling units and their disease status, the methodology can easily be illustrated by application to a one-dimensional array, representing data from a transect across a field.

Example 9.4. The following data represent a transect of sampling units extracted from Table 11.3 of Campbell and Madden (1990), and show the number Y_i ($i = 1, \ldots, N$) of *Macrophomia phaseolina* propagules per 10 g air-dry soil recorded in $N = 16$ contiguous quadrats across a field in Edgecombe County, North Carolina.

i	Y_i
1	41
2	60
3	81
4	22
5	8
6	20
7	28
8	2
9	0
10	2
11	2
12	8
13	0
14	43
15	61
16	50

From these data the following quantities may be estimated. The mean:

$$\bar{Y} = \frac{\sum_{i=1}^{N} Y_i}{N} = 26.75 \qquad (9.46)$$

and the variance

$$\hat{C}(0) = \frac{\sum_{i=1}^{N} (Y_i - \bar{Y})^2}{N} = 641.94 \qquad (9.47)$$

We introduce some new notation here for the variance in order to be consistent with that often adopted in spatial studies that use autocorrelation methods. Note that the denominator used in equation 9.47 is N, the number of observations, rather than $N-1$, the degrees of freedom. Thus, equation 9.47 gives the maximum likelihood estimate of the variance.

The autocovariance between two values of Y_i a distance (or *lag*) j apart is estimated by:

$$\hat{C}(j) = \sum_{i=1}^{N-j} \frac{(Y_i - \bar{Y})(Y_{i+j} - \bar{Y})}{N_j} \qquad (9.48)$$

where N_j is the number of *pairs* of sampling units j units apart (i.e., the number of pairs of Y values on which the calculation of $\hat{C}(j)$ is based). For instance, when the data comprise a one-dimensional array, $N_j = N - j$, and for first-order spatial lags (i.e., immediately adjacent sampling units), $N_j = N - 1$. The $C(\bullet)$ symbol is used to represent covariance. Since the variance is just the covariance of a variable with itself (i.e., the covariance

TABLE 9.3B. Values of U_{ij} defined as $U_{ij} = (Y_i - Y_j)^2$, with $Y = 1$ for a location denoted D in the upper part of Fig. 9.13, and $Y = 0$ for a location denoted H. Then $U_{ij} = 0$ if both locations have the same disease status, and $U_{ij} = 1$ if they differ.

	Row	*1*	*1*	*1*	*1*	*1*	*2*	*2*	*2*	*2*	*2*	*3*	*3*	*3*	*3*	*3*	*4*	*4*	*4*	*4*	*4*	*5*	*5*	*5*	*5*	*5*
Row	*Column*	*1*	*2*	*3*	*4*	*5*	*1*	*2*	*3*	*4*	*5*	*1*	*2*	*3*	*4*	*5*	*1*	*2*	*3*	*4*	*5*	*1*	*2*	*3*	*4*	*5*
1	*1*	0	0	0	0	1	1	0	1	1	0	0	1	0	1	1	0	1	1	0	0	1	1	1	1	0
1	*2*	0	0	0	0	1	1	0	1	1	0	0	1	0	1	1	0	1	1	0	0	1	1	1	1	0
1	*3*	0	0	0	0	1	1	0	1	1	0	0	1	0	1	1	0	1	1	0	0	1	1	1	1	0
1	*4*	0	0	0	0	1	1	0	1	1	0	0	1	0	1	1	0	1	1	0	0	1	1	1	1	0
1	*5*	1	1	1	1	0	0	1	0	0	1	1	0	1	0	0	1	0	0	1	1	0	0	0	0	1
2	*1*	1	1	1	1	0	0	1	0	0	1	1	0	1	0	0	1	0	0	1	1	0	0	0	0	1
2	*2*	0	0	0	0	1	1	0	1	1	0	0	1	0	1	1	0	1	1	0	0	1	1	1	1	0
2	*3*	1	1	1	1	0	0	1	0	0	1	1	0	1	0	0	1	0	0	1	1	0	0	0	0	1
2	*4*	1	1	1	1	0	0	1	0	0	1	1	0	1	0	0	1	0	0	1	1	0	0	0	0	1
2	*5*	0	0	0	0	1	1	0	1	1	0	0	1	0	1	1	0	1	1	0	0	1	1	1	1	0
3	*1*	0	0	0	0	1	1	0	1	1	0	0	1	0	1	1	0	1	1	0	0	1	1	1	1	0
3	*2*	1	1	1	1	0	0	1	0	0	1	1	0	1	0	0	1	0	0	1	1	0	0	0	0	1
3	*3*	0	0	0	0	1	1	0	1	1	0	0	1	0	1	1	0	1	1	0	0	1	1	1	1	0
3	*4*	1	1	1	1	0	0	1	0	0	1	1	0	1	0	0	1	0	0	1	1	0	0	0	0	1
3	*5*	1	1	1	1	0	0	1	0	0	1	1	0	1	0	0	1	0	0	1	1	0	0	0	0	1
4	*1*	0	0	0	0	1	1	0	1	1	0	0	1	0	1	1	0	1	1	0	0	1	1	1	1	0
4	*2*	1	1	1	1	0	0	1	0	0	1	1	0	1	0	0	1	0	0	1	1	0	0	0	0	1
4	*3*	1	1	1	1	0	0	1	0	0	1	1	0	1	0	0	1	0	0	1	1	0	0	0	0	1
4	*4*	0	0	0	0	1	1	0	1	1	0	0	1	0	1	1	0	1	1	0	0	1	1	1	1	0
4	*5*	0	0	0	0	1	1	0	1	1	0	0	1	0	1	1	0	1	1	0	0	1	1	1	1	0
5	*1*	1	1	1	1	0	0	1	0	0	1	1	0	1	0	0	1	0	0	1	1	0	0	0	0	1
5	*2*	1	1	1	1	0	0	1	0	0	1	1	0	1	0	0	1	0	0	1	1	0	0	0	0	1
5	*3*	1	1	1	1	0	0	1	0	0	1	1	0	1	0	0	1	0	0	1	1	0	0	0	0	1
5	*4*	1	1	1	1	0	0	1	0	0	1	1	0	1	0	0	1	0	0	1	1	0	0	0	0	1
5	*5*	0	0	0	0	1	1	0	1	1	0	0	1	0	1	1	0	1	1	0	0	1	1	1	1	0

TABLE 9.3C. Values of U_{ij} defined as $U_{ij} = (Y_i \times Y_j)$, with $Y = 1$ for a location denoted D in the upper part of Fig. 9.13, and $Y = 0$ for a location denoted H. Then $U_{ij} = 0$ if both locations are H or if one is D and one is H, and $U_{ij} = 1$ if both locations are D.

	Row	*1*	*1*	*1*	*1*	*1*	*2*	*2*	*2*	*2*	*2*	*3*	*3*	*3*	*3*	*3*	*4*	*4*	*4*	*4*	*4*	*5*	*5*	*5*	*5*	*5*
Row	*Column*	*1*	*2*	*3*	*4*	*5*	*1*	*2*	*3*	*4*	*5*	*1*	*2*	*3*	*4*	*5*	*1*	*2*	*3*	*4*	*5*	*1*	*2*	*3*	*4*	*5*
1	*1*	1	1	1	1	0	0	1	0	0	1	1	0	1	0	0	1	0	0	1	1	0	0	0	0	1
1	2	1	1	1	1	0	0	1	0	0	1	1	0	1	0	0	1	0	0	1	1	0	0	0	0	1
1	3	1	1	1	1	0	0	1	0	0	1	1	0	1	0	0	1	0	0	1	1	0	0	0	0	1
1	4	1	1	1	1	0	0	1	0	0	1	1	0	1	0	0	1	0	0	1	1	0	0	0	0	1
1	5	0	0	0	0	0	0	0	0	0	0	0	0	0	0	0	0	0	0	0	0	0	0	0	0	0
2	*1*	0	0	0	0	0	0	0	0	0	0	0	0	0	0	0	0	0	0	0	0	0	0	0	0	0
2	2	1	1	1	1	0	0	1	0	0	1	1	0	1	0	0	1	0	0	1	1	0	0	0	0	1
2	3	0	0	0	0	0	0	0	0	0	0	0	0	0	0	0	0	0	0	0	0	0	0	0	0	0
2	4	0	0	0	0	0	0	0	0	0	0	0	0	0	0	0	0	0	0	0	0	0	0	0	0	0
2	5	1	1	1	1	0	0	1	0	0	1	1	0	1	0	0	1	0	0	1	1	0	0	0	0	1
3	*1*	1	1	1	1	0	0	1	0	0	1	1	0	1	0	0	1	0	0	1	1	0	0	0	0	1
3	2	0	0	0	0	0	0	0	0	0	0	0	0	0	0	0	0	0	0	0	0	0	0	0	0	0
3	3	1	1	1	1	0	0	1	0	0	1	1	0	1	0	0	1	0	0	1	1	0	0	0	0	1
3	4	0	0	0	0	0	0	0	0	0	0	0	0	0	0	0	0	0	0	0	0	0	0	0	0	0
3	5	0	0	0	0	0	0	0	0	0	0	0	0	0	0	0	0	0	0	0	0	0	0	0	0	0
4	*1*	1	1	1	1	0	0	1	0	0	1	1	0	1	0	0	1	0	0	1	1	0	0	0	0	1
4	2	0	0	0	0	0	0	0	0	0	0	0	0	0	0	0	0	0	0	0	0	0	0	0	0	0
4	3	0	0	0	0	0	0	0	0	0	0	0	0	0	0	0	0	0	0	0	0	0	0	0	0	0
4	4	1	1	1	1	0	0	1	0	0	1	1	0	1	0	0	1	0	0	1	1	0	0	0	0	1
4	5	1	1	1	1	0	0	1	0	0	1	1	0	1	0	0	1	0	0	1	1	0	0	0	0	1
5	*1*	0	0	0	0	0	0	0	0	0	0	0	0	0	0	0	0	0	0	0	0	0	0	0	0	0
5	2	0	0	0	0	0	0	0	0	0	0	0	0	0	0	0	0	0	0	0	0	0	0	0	0	0
5	3	0	0	0	0	0	0	0	0	0	0	0	0	0	0	0	0	0	0	0	0	0	0	0	0	0
5	4	0	0	0	0	0	0	0	0	0	0	0	0	0	0	0	0	0	0	0	0	0	0	0	0	0
5	5	1	1	1	1	0	0	1	0	0	1	1	0	1	0	0	1	0	0	1	1	0	0	0	0	1

for a spatial lag of zero), C(0) is used for the variance. If the sampling units are not set out in a regular array, the number of pairs of Y values separated by a lag of exactly j may be small. In such circumstances the lag j is usually taken to be a mean distance calculated over some appropriate interval. From equation 9.47, the corresponding autocorrelation coefficient is then estimated by:

$$\hat{r}(j) = \hat{C}(j)/\hat{C}(0) \tag{9.49}$$

Dividing the covariance by the variance has the effect of scaling the autocovariances, so that for continuous data, one obtains a correlation coefficient that lies between +1 and −1. Disease intensity in each sampling unit is often compared with that in the immediately adjacent unit(s), in which case $j = 1$ and the coefficient is $\hat{r}(1)$ obtained. In general, spatial autocorrelation analysis provides a quantitative assessment of whether a large value of disease intensity in a sampling unit makes it more (positive autocorrelation) or less (negative autocorrelation) likely that neighboring sampling units tend to have a large value of disease intensity.

Example 9.4 continued. For the transect data, the calculated values of $\hat{r}(j)$ for $j = 1$, 2, and 3 respectively are 0.625, 0.144, and −0.041. As might be expected, the smaller the distance between sampling units, the stronger is the correlation between their Y_i values. The method used in this example comes from Moran (1950) and is usually referred to as "Moran's *I*". Cliff and Ord (1981) give full details of statistical significance testing. As with join-count statistics, there is a standard normal test for use with a larger numbers of observations, but with small numbers of observations, randomization methodology is preferable. Sawada's (1999) ROOKCASE software provides both tests for $\hat{r}(1)$. In this case, the null hypothesis of no spatial autocorrelation is rejected ($P < 0.05$).

Gilligan (1988) and, in more detail, Gumpertz and Ristaino (1997) discuss the two-dimensional version of Moran's *I* in a phytopathological context. This coefficient is another version of the general cross-product statistic discussed in section 9.9.2. As with the join-count statistic, a two-dimensional analysis requires specification of the orientation(s) in which spatial associations are to be investigated. This allows investigation of *isotropy*. Spatial patterns that do not depend on direction are said to be isotropic. Spatial patterns that depend on direction are anisotropic. Plant pathologists often perform separate directional spatial autocorrelation analyses (e.g., along and across agronomic rows) in order to investigate whether or not patterns are isotropic (e.g., Gottwald and Graham, 1990).

Gottwald et al. (1992) and Reynolds and Madden (1988) have described computer software for spatial autocorrelation calculations. Both programs have been widely used in plant pathology. The program by Gottwald et al. (1992) is based on methodology described by Modjeska and Rawlings (1983), while the program by Reynolds and Madden (1988) is based on the theoretical work of Bennett (1979). Normally, results are presented as a graphical plot of autocorrelation coefficients against spatial lag, which is referred to as an *autocorrelogram*. Sometimes this diagram is simplified to show only the statistical significance probability (usually as $P \leq 0.05$ or $P > 0.05$) of the autocorrelation coefficient at each lag (e.g., Campbell and van der Gaag, 1993). Techniques for modeling the autocorrelogram generally borrow directly from methods developed for time-series analysis (see Madden et al., 1988; Reynolds and Madden, 1988; Reynolds et al., 1988a; Hudelson et al., 1993).

An intensively mapped spatial data set often comprises a single realization of the process underlying the observed pattern of disease. Interpretation of an observed pattern then requires some assumption about the *stationarity* of the underlying process. In carrying out a spatial autocorrelation analysis and interpreting the results, second-order stationarity is usually assumed. Essentially, this requires that the mean and variance of the spatial process under investigation do not vary over the study area (i.e., estimated mean and variance do not depend on the location of the sampling units), and that the autocovariance depends only on the distance (and perhaps direction, unless isotropy has been established) between the sampling units. It can be seen from equations 9.47 and 9.48, respectively, that a constant mean has been assumed in calculating the variance and the autocovariance, and that a constant variance has been assumed in calculating the autocorrelation. Autocorrelation coefficients can be calculated for any data set with spatially referenced sampling units, but interpretation of the results, especially in terms of a stochastic modeling process (see Madden et al., 1988; Reynolds et al., 1988a) is based on the assumption of second-order stationarity.

9.9.4 Semivariance

Statistical methods for the study of spatial variation first developed in the earth sciences, and hence collectively known as geostatistics, have since found an increasing range of application in the life sciences. The main area of application for geostatistics, at least in their original context, was for continuously distributed spatial data such as soil characteristics. In this way, geostatistics differs from spatial autocorrelation, where distances are usually measured in discrete spatial lags. Geostatistical methods were introduced to the study of spatial patterns of plant disease by Chellemi et al. (1988). Gumpertz et al. (1997) provide a more recent methodological account. A key quantity is the *semivariance*.

Example 9.4 continued. Continuing the illustrative example in one-dimensional space, the semivariance at distance j is estimated by:

$$\hat{\gamma}(j) = \frac{\sum_{i=1}^{N-j} \left(Y_{i+j} - Y_i\right)^2}{2N_j} \qquad (9.50)$$

where, once again, N_j is the number of pairs of sampling units distance j apart (and for one-dimensional data, $N_j = N - j$). Thus, the semivariance is one half of the variance of the differences $(Y_{i+j} - Y_i)$, and is another version of a cross-product statistic. For the transect data, calculated values of $\hat{\gamma}(j)$ for j = 1, 2, and 3 respectively are 258.8, 533.0, and 576.6. For these data, we can see that as the semivariance increases, the autocorrelation (section 9.8.2) decreases. This is generally to be expected, because the geostatistical approach is to look at variability, which is the opposite of correlation. When the distance between two sampling units is small, the sampling units are close together and, usually, variability is low. As the distance increases, so (usually) does the variability.

Normally, results are presented as a graphical plot of semivariance against distance, which is referred to as a *semivariogram* (or sometimes just a variogram). As in the use of spatial autocorrelation analysis, plant pathologists often decline to assume isotropy and instead present results in the form of directional semivariograms (e.g., Xiao et al., 1997). Under conditions of second-order stationarity, the semivariance is related to the autocovariance by $\hat{\gamma}(j) = \hat{C}(0) - \hat{C}(j)$, and the semivariogram and autocorrelogram are mirror images. Thus, one can write $\hat{\gamma}(j) = \hat{C}(0)(1 - \hat{r}(j))$ and $\hat{r}(j) = (\hat{C}(0) - \hat{\gamma}(j))/\hat{C}(0)$. However, calculation and interpretation of the semivariance does not require the same assumption about stationarity as for spatial autocorrelation analysis. Instead, the assumption is that the variance of the difference between disease intensities in two sampling units depends on the distance between them. If the assumption of stationarity is likely to be problematic, an analysis based on semivariances is a useful alternative to spatial autocorrelation.

The semivariogram has a terminology all of its own, used to classify generic types of plots that are often obtained from semivariance analysis, and to relate the different parts of the plot to the characteristics of the spatial data from which it was derived. Burrough (1987) provides a comprehensive guide. The main characteristics of the semivariogram of interest are the *nugget*, the *range* and the *sill* (Fig. 9.14). Calculated directly from equation 9.50, the value of $\hat{\gamma}(j)$ when $j = 0$ would be equal to zero, because the numerator of the expression on the right-hand side of equation 9.50 is zero when $j = 0$. However, if the analysis is approached from the point of view of modeling the semivariogram, it is possible that the curve representing the relationship between semivariance $\hat{\gamma}(j)$ and distance j may indicate a small positive value

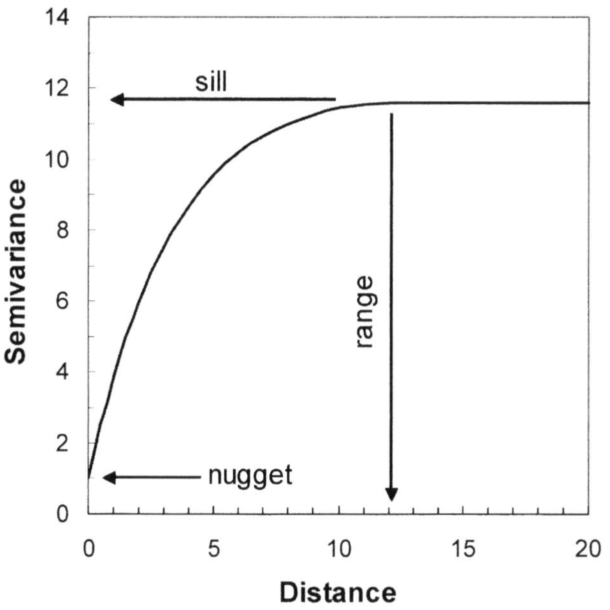

FIG. 9.14. The main characteristics of interest for a model semivariogram are the nugget, the range, and the sill.

when $j = 0$. This value is termed the nugget. It may be thought of as indicating the existence of spatial variability at scales smaller than the distance between sampling units, or variability arising from measurement errors.

The distance j at which $\hat{\gamma}(j)$ no longer continues to increase with increasing j is termed the range. Sampling units separated by distances greater than the range may be treated as spatially independent. For the value of j that characterizes the range, the corresponding value of $\hat{\gamma}(j)$ is termed the sill. Various statistical models have been used to describe generic types of semivariogram. The approach is entirely different to that usually used for modeling autocorrelograms. Estimates of the nugget, range and sill are usually based on an appropriate (nonlinear) model fitted to the data points representing the semivariogram (see Chapter 3 for more details on nonlinear modeling). However, not all generic types of semivariogram exhibit all three of these characteristics.

Once a suitable model has been fitted to the data, it can be used to derive interpolated values of attributes for unsampled locations, a process known as *kriging*. Munkvold et al. (1993), Larkin et al. (1995), Xiao et al. (1997) and Wu et al. (2001) are among those reporting examples of semivariance analysis applied in plant pathology. Gottwald et al. (2002) and Roumagnac et al. (2004) have used kriging to identify disease foci. Detailed accounts of geostatistical methods are given by Cressie (1993, Part I) and Schabenberger and Pierce (2002, Chapter 9).

9.9.5 Spatial analysis by distance indices

Spatial analysis by distance indices, usually referred to by the acronym SADIE, provides an alternative approach to the autocorrelation- and semivariance-based methods of analysis for intensively mapped spatial data outlined in sections 9.9.3 and 9.9.4. The SADIE literature is increasing rapidly, but the properties of the analysis have not yet been as thoroughly characterized as those of autocorrelation- and semivariance-based methods. A useful introduction to the main aspects of the methodology is provided by Perry (1998) and Perry et al. (1999). The former article provides a brief comparison of SADIE methodology with the geostatistical approach to spatial analysis. More details on some of the mathematical properties can be found in Xu and Madden (2005). Phytopathological applications of SADIE include Turechek and Madden (1999a,b), Shah et al. (2001), Thackray et al. (2002), and Dallot et al. (2003). Here, the use of SADIE methodology for the analysis of counts at spatially referenced sampling units (e.g., quadrats) is outlined. Spatial associations between two sets of count data can also be tested using SADIE (Perry and Dixon, 2002). Phytopathological examples include Turechek and Madden (2000), Thackray et al. (2002) and Pethybridge and Turechek (2003). Downloadable software is available, and has been used for the SADIE examples calculated here.

The starting point for the analysis is usually a two-dimensional array of sampling units, for each of which the co-ordinates locating it in the array and the associated count of individuals have been recorded. Intuitively, it can be seen that if the counts for individual sampling units are scattered widely around the mean value, the pattern will be more irregular than if most sampling units have counts close to the mean count per sampling unit. The basis for the SADIE method is to quantify pattern by calculating the minimum total distance, denoted D_a, that individuals must be moved from the starting point defined by the observed counts to the end point at which there is the same number of individuals in each sampling unit, and the pattern of counts per sampling unit is therefore as regular as possible.

Since the application here is for observed counts of diseased plants, there is no actual movement of individuals; rather, this is a conceptual device used to quantify the extent of non-randomness in terms of the minimum "distance to regularity". If the original data are highly aggregated, the distance to regularity will be large, because many moves will be required in order to reach regularity. If the data are close to regular to start with, the distance to regularity will be smaller. The total distance moved depends not only on the pattern of observed counts per sampling unit, but also on number of individuals (i.e., diseased plants) and the size of the intensively mapped array of sampling units. Because of this, D_a is standardized, using a randomization procedure. For each of a large number of random rearrangements of the observed counts per sampling unit, the total distance to regularity is determined. The mean of these values, denoted E_a, is then calculated. Because it is the counts per sampling unit that are randomized (*not* the individual diseased plants), the within-sampling-unit scale characteristics (mean, variance, aggregation indices) are the same for each randomization as for the observed data set. Thus, statistics calculated using the *SADIE* methodology have the desirable property of being conditioned on the small-scale pattern.

Although it is the distance to regularity that is calculated, the null hypothesis of interest is that the observed pattern is random. Regularity is used as the basis for the calculation because this can be unambiguously defined in terms of numbers of individuals per sampling unit. An index of aggregation denoted I_a, is defined as D_a/E_a. When $I_a = 1$, randomness is indicated. A value $I_a > 1$ indicates aggregation and a value $I_a < 1$ indicates regularity. Hypothesis testing is based on the randomization procedure. For example, when $I_a > 1$, the proportion of randomizations in which the distance to regularity is greater than the value of D_a calculated for the original data can be determined. This proportion is an estimate of the significance probability for a null hypothesis of randomness, with an alternative hypothesis of aggregation.

Perry et al. (1999) extended the SADIE methodology to be able to quantify the contribution of each sampling-unit count to the observed pattern. Of interest is whether a particular large count occurs in isolation or is located close to other large counts, and whether a particular small count occurs in isolation or is located close to other small counts. In the terminology of SADIE, a "cluster" may be either a region of relatively large counts close to one another (termed a "patch") or a region of relatively small counts close to one another (termed a "gap"). SADIE methodology permits the calculation of a dimensionless clustering index for each sampling unit.

For a given sampling unit with more than the mean number of diseased plants for the data set as a whole, the number of (conceptual) moves to regularity involves a net outflow of individuals. The clustering index is determined by first calculating a weighted average distance of outflow from the sampling unit, the weights being the magnitudes of the individual outflows. This distance value is then standardized, again using the randomization procedure, to produce a dimensionless clustering index (υ_{Pi}, by convention a positive number). The index υ_{Pi} measures the degree to which a particular sampling unit contributes clustering as a member of a patch. For a sampling unit containing less than the mean number of diseased plants, a clustering index is calculated in corresponding fashion. In this case, the number of (conceptual) moves to regularity involves a net inflow of individuals. The resulting clustering index ($\upsilon_{\hat{P}i}$, by convention a negative number) measures the degree to which a particular sampling unit contributes to clustering as a member of a gap. Values of the clustering indices greater than 1.5 in absolute value are considered large (Perry et al., 1999). Mean values of these indices can be calculated for a data set. For a random pattern, $\bar{\upsilon}_{Pi} = -\bar{\upsilon}_{Gi} = 1$. Tests of spatial randomness are obtained by determining the proportion of randomizations with $\bar{\upsilon}_{Pi}$ greater than the $\bar{\upsilon}_{Pi}$ for the observed data, or the proportion of randomizations with $\bar{\upsilon}_{Gi}$ less than the $\bar{\upsilon}_{Gi}$ for the observed data. Of potentially greater value than such tests is a plot of the clustering indices for individual sampling units in a contour map format. Patches and gaps may be more easily recognized and quantified than from a map of the individual sampling units themselves.

Example 9.4 continued. Although the starting point for a SADIE analysis is usually a two-dimensional array of sampling units, the analysis can also be conducted for a one-dimensional transect of observations. For illustration of the method, a SADIE analysis of the transect data previously used to illustrate spatial autocorrelation and semivariance analyses gives $I_a = 1.49$. A value of I_a of 1.5 or higher is regarded as fairly large, but in this case the significance probability (based on 5967 randomizations of the data) was $P = 0.11$, conventionally a value not small enough to reject the hypothesis of randomness. With only 16 sampling units in the example, the power of the test for non-randomness is expected to be low. Mean cluster indices were $\bar{\upsilon}_{Pi} = 1.55$ and $\bar{\upsilon}_{Gi} = -1.33$ ($0.10 < P < 0.15$). The transect counts are repeated here, together with the cluster indices for individual sampling units (positive values denote υ_{Pi}, negative values denote υ_{Gi}).

i	Y_i	υ_i
1	41	2.0
2	60	2.8
3	81	1.4
4	22	−1.1
5	8	−0.7
6	20	−1.0
7	28	1.0
8	2	−2.0
9	0	−1.8
10	2	−2.2
11	2	−1.0
12	8	−1.3
13	0	−0.7
14	43	1.2
15	61	1.0
16	50	1.5

The gap around locations 8–13 is easily identified with negative individual clustering index values, and the patches near the edges are identified with positive individual clustering index values. It should be emphasized that the individual clustering index values indicate more than just magnitude of the counts (otherwise, the counts themselves would be just as satisfactory for representing the pattern). Consider the two zero counts, for sampling units 9 and 13. Sampling unit 9 has a υ_{Gi} much larger (in absolute value) than the sampling unit at position 13. This is because sampling unit 9 is located within a gap (i.e., surrounded by small counts), whereas sampling unit 13 is adjacent to a patch (with relatively large counts). Likewise, the sampling units in positions 1 and 14 had similar counts (41 and 43, respectively), but had rather different cluster indices. The smaller index occurred at position 14, where a relatively high count occurred next to a gap. The use of clustering indices in this way is more advantageous with analyses of spatially mapped data in a two-dimensional array of counts.

Example 9.5. The top left part of Fig. 9.15 shows an array representing a single realization of a model epidemic produced by a stochastic simulator (Xu and Ridout, 2000). The array shows the number of diseased plants (out of a total of n = 100) in each of N = 144 spatially mapped quadrats in a field. The model epidemic started with four randomly placed diseased plants. Because the data are counts with a natural denominator, summary statistics as described in section 9.4 can be calculated, as follows: $\bar{Y} = 27.85$, $\bar{y} = 0.279$, $s_Y^2 = 954$, D = 47.4, and $\hat{\rho} = 0.47$. These are indicative of aggregation at the within-sampling-unit scale.

In this example, the quadrat data from the simulated epidemic, shown in the array in the top left part of Fig. 9.15, are used to provide a comparative demonstration of the spatial analyses discussed in sections 9.9.3 – 9.9.5. Two cautionary notes precede these analyses. First, slightly different results may be obtained, dependent on the choice of computer software, if different programs employ different estimation procedures. Second, although it is true that it can be useful to employ more than one method when analyzing spatial data (Madden, 1989; Perry et al., 2002) that is not the point here. This example is an illustration of three methods that extract similar information from the data.

For the quadrat data in the array in the top left part of Fig. 9.15, a semivariogram was calculated using the VARIOGRAM procedure in SAS (shown in the top right part of Fig. 9.15). The default condition, under which variability in all directions is considered, was retained. The $\hat{\gamma}(j)$ values increased with distance j, up to about j = 8. The increases in $\hat{\gamma}(j)$ were greatest over distances 1 – 4. At the larger distances ($j \geq 8$), there was a scatter of values. Although there is an apparent tendency for decreasing semivariance at distances $j \geq 9$, we avoid giving detailed interpretations of slight increases and decreases in $\hat{\gamma}(j)$ when $\hat{\gamma}(j)$ is large. This is because the variance of the estimated semivariance is (as an approximation) proportional to the square of the semivariance

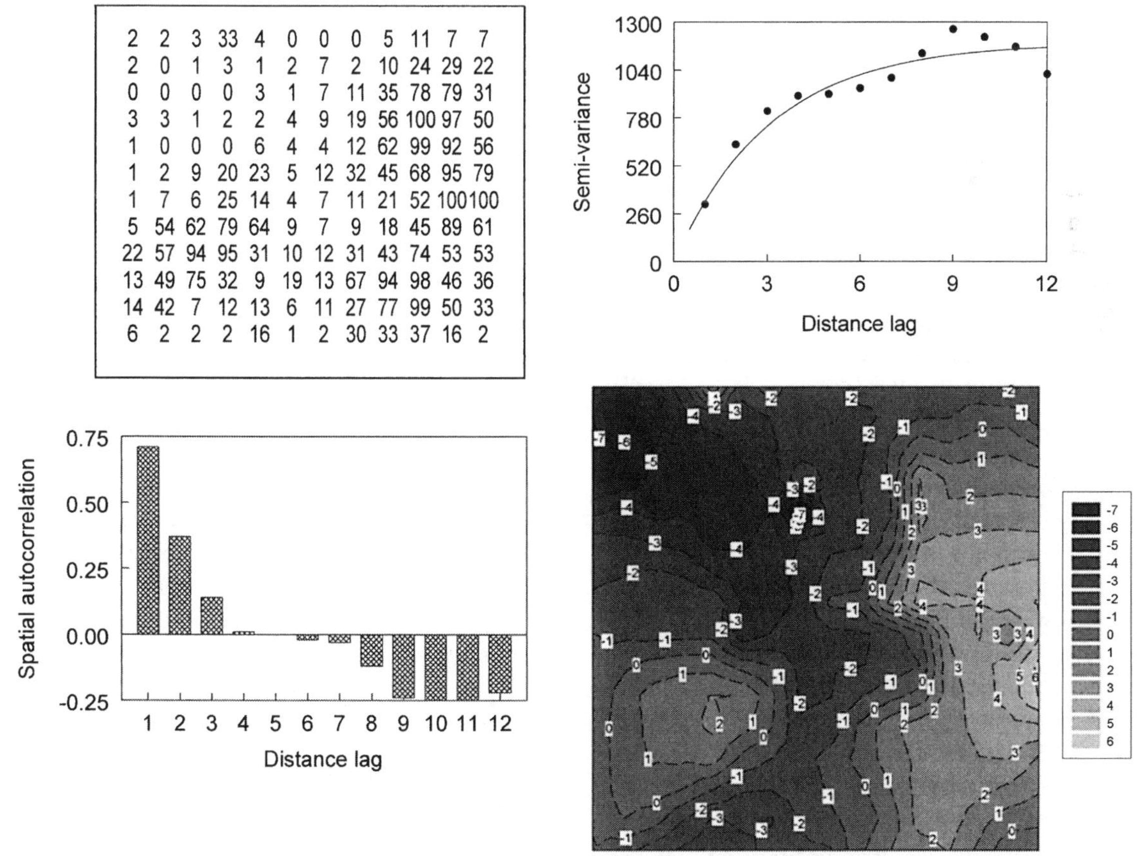

2	2	3	33	4	0	0	0	5	11	7	7
2	0	1	3	1	2	7	2	10	24	29	22
0	0	0	0	3	1	7	11	35	78	79	31
3	3	1	2	2	4	9	19	56	100	97	50
1	0	0	0	6	4	4	12	62	99	92	56
1	2	9	20	23	5	12	32	45	68	95	79
1	7	6	25	14	4	7	11	21	52	100	100
5	54	62	79	64	9	7	9	18	45	89	61
22	57	94	95	31	10	12	31	43	74	53	53
13	49	75	32	9	19	13	67	94	98	46	36
14	42	7	12	13	6	11	27	77	99	50	33
6	2	2	2	16	1	2	30	33	37	16	2

FIG. 9.15. Top left: the data are number of diseased plants per quadrat (out of a total of n = 100 plants in each) in N = 144 quadrats, based on an epidemic generated using the stochastic simulator of Xu and Madden (2004). Top right: a plot of semivariance against distance lag for the simulated epidemic data, with a fitted exponential model. Bottom left: spatial autocorrelations for 12 distance lags for the simulated epidemic data. Bottom right: map of cluster indices for the simulated epidemic data, calculated using SADIE software. See *Example 9.5*.

and inversely proportion to N_j (which declines with increasing j), and the precision of the estimated semivariance is therefore lower at larger distances.

The approach most frequently adopted for characterizing spatial pattern based on the semivariogram is to fit a statistical model. Here, an exponential model of the form:

$$\gamma(j) = \alpha_n + (\alpha_s - \alpha_n)(1 - \exp(-3j/\alpha_r)$$

was fitted to the data ($R^2 = 0.92$), in which α_n, α_r, and α_s are the nugget, range, and sill, respectively. Distance, j, is not restricted to integer values. Because $\hat{\gamma}(j)$ is a random variable (an estimate of the true semivariance), there is uncertainty about its value. The formula for the variance of $\hat{\gamma}(j)$ is complicated, but as an approximation this variance can be estimated by $2\hat{\gamma}^2(j)/N_j$. The reciprocal of this variance is the appropriate weight for use when fitting the model by weighted least squares. The fitted model is shown as a curve in top right part of Fig. 9.15. For the exponential model, the sill is approached asymptotically; thus, there is no exactly determined range. For this model, α_r is considered to be the *practical* range, the distance at which the predicted semivariance is 95% of the sill. In this example, the estimated practical range was 9.4 (arbitrary distance units) (SE = 1.34). Thus (assuming that the model is appropriate), the data are considered to be spatially independent at distances of 9.4 or larger. The estimate of the nugget was 0. The estimate of the sill was 1189 (SE = 70.2).

The spatial autocorrelation analysis of the quadrat data in the array in the top left part of Fig. 9.15 was also based on the VARIOGRAM procedure in SAS. This means that spatial lags were interpreted as distances (i.e., continuous), as in the semivariance analysis, rather than as being discrete, as would be the case if the spatial autocorrelation analysis were calculated using, for example, LCOR2 (Gottwald et al., 1992). The default condition of considering variability in all directions was retained. The nearest analogous proximity rule for a spatial autocorrelation analysis based on discrete spatial lags is the "queen" case, again using a chess-based analogy (Cliff and Ord, 1981). However, the difference between spatial lags calculated as a continuous distances and spatial lags calculated as discrete numbers of sampling units means that the correspondence is not exact. Use of the SAS VARIOGRAM procedure provides estimates of autocovariance over all distances. Dividing these by the variance gives the autocorrelogram in the lower left part of Fig. 9.15, showing $\hat{r}(j)$ values for 12 distance lags (j = 1, ..., 12). For the first distance lag, $\hat{r}(1) = 0.71$, $s(\hat{r}(1)) \approx 0.045$. The standard error estimate given here is based on an approximation, $s(\hat{r}(j)) \approx N_j^{-1/2}$, that is valid only for large numbers of observations. In this example, there were N_1 = 506 pairs of sampling units 1 distance unit apart.

The semivariogram and the autocorrelogram based on the quadrat data in the array in the top left part of Fig. 9.15 yield the same conclusion. The rapid increase in the former over distance lags 1 − 4 is mirrored by a similarly rapid decrease in the latter. Both analyses indicated that adjacent sampling units tended to have similar numbers of diseased plants, either high (as on the right-hand side and in the lower left-hand corner) or low (as in the upper left-hand corner and the center) in magnitude. This similarity between sampling units decreased with increasing distance between them.

A SADIE analysis of the quadrat data in the array in the top left part of Fig. 9.15 was indicative of aggregation, with I_a = 2.83 ($P < 0.001$), and mean clustering indices of $\bar{\upsilon}_{Pi} = 2.48$ and $\bar{\upsilon}_{Gi} = -2.58$ ($P < 0.001$) (based on 5967 randomizations of the data). A map of the individual clustering indices (in the lower right part of Fig. 9.15) shows the patch of high disease intensity along the right-hand side and a minor patch in the lower left-hand corner (much smaller than anticipated from the raw counts). The gaps in the upper left-hand corner and the center of the map are also apparent. The map of clustering indices is useful for the identification of patches and gaps. However, particularly for large data sets, summary statistics on which comparisons can be based are also required.

The overall results from the three spatial analyses are in good agreement, as expected. Turechek and Madden (1999a, 2000) have also shown agreement between results of spatial autocorrelation and SADIE analyses of pattern for strawberry leaf blight (caused by *Phomopsis obscurans*) and strawberry leaf spot (caused by *Mycosphaerella fragariae*). Simulation studies by Xu and Madden (2003, 2004) confirm that SADIE statistics and spatial autocorrelations measure some common properties of patterns, but also reveal differences. Recall that the starting point for a SADIE analysis is usually a two-dimensional array of sampling units, for each of which the co-ordinates locating it in the array and the associated count of individuals has been recorded. Xu and Madden (2003, 2004) show that SADIE statistics are affected not just by the relative positions of the sampling-unit counts (i.e., using spatial autocorrelation terminology, the spatial lags), but also by the actual positions of counts in the array (i.e., nearer the center or nearer an edge). Hence care should be taken when comparing I_a (or related) values across data sets.

9.10 Spatial Patterns and Dispersal Functions

9.10.1 Simulation models

As mentioned at the outset of this chapter, making inferences about dispersal has been a recurring theme from the beginnings of spatial pattern analysis in plant disease epidemiology (Cochran, 1936; Bald, 1937) to the present (e.g., Del Ponte et al., 2003; Ristaino and Gumpertz, 2000; Thackray et al., 2002). Dispersal gradients are typically characterized by a graphical plot of disease intensity, or of some assessment of the pathogen population density, against distance from a localized inoculum source (Chapter 7). Two empirical descriptions have been widely used to describe such graphical plots; one in which disease intensity or pathogen population density decreases exponentially as distance from the source increases, and one in which disease intensity or pathogen population density is inversely proportional to some power of distance from the source (Minogue, 1986; Fitt et al., 1987). These two descriptions explicitly correspond to the (negative) exponential and inverse power law dispersal functions of Chapter 7 (see equations 7.1 and 7.4, respectively), and the corresponding exponential and Pareto contact distributions (equations 7.8 and 7.9, respectively). At large distances from the source, an inverse power dispersal function decreases more slowly than a negative exponential dispersal function with the same mean, resulting in a higher probability of longer-distance dispersal (Minogue, 1989; Shaw, 1994,1995) (see sections 7.3.3, 7.3.4, and 8.2.2).

With experimental data, it is sometimes difficult to choose between these empirical descriptions of dispersal functions solely on the basis of how well a regression line fits the observations. Notwithstanding this, the choice between a negative exponential dispersal function and an inverse power dispersal function has a large impact on the outcome of stochastic simulations that generate spatial patterns of epidemics (Minogue, 1989; Shaw, 1994,1995). This is illustrated by Fig. 9.16, which shows the outcome of two stochastic simulations produced using the DISP_PROJECT software written by M.W. Shaw. Fig. 9.16A shows the outcome of a simulation based on a negative exponential dispersal function. Dispersal has essentially resulted in a single expanding focus of disease. Longer-range dispersal, far beyond this focus, is not in evidence. Fig. 9.16B shows the outcome of a simulation based on an inverse-square dispersal function. In this case, a single expanding focus is not apparent. Instead, disease is widely spread in an apparently heterogeneous pattern. Patterns resulting from stochastic simulations of disease based on dispersal functions using an inverse power formulation "look 'natural'" (Shaw, 1995) and were described as "impressively realistic" by Minogue (1989).

In Chapters 7 and 8, we also showed that the form of the dispersal function or contact distribution has a qualitative effect on the rate of disease focus expansion. For an exponential-type of dispersal function, focus expansion is constant ("wave-like"), but for a power-type of function, the rate increases with time. In the latter case, the further disease spreads from an original inoculum source, the greater the rate of spread into new (previously disease free) areas. The stochastic simulation

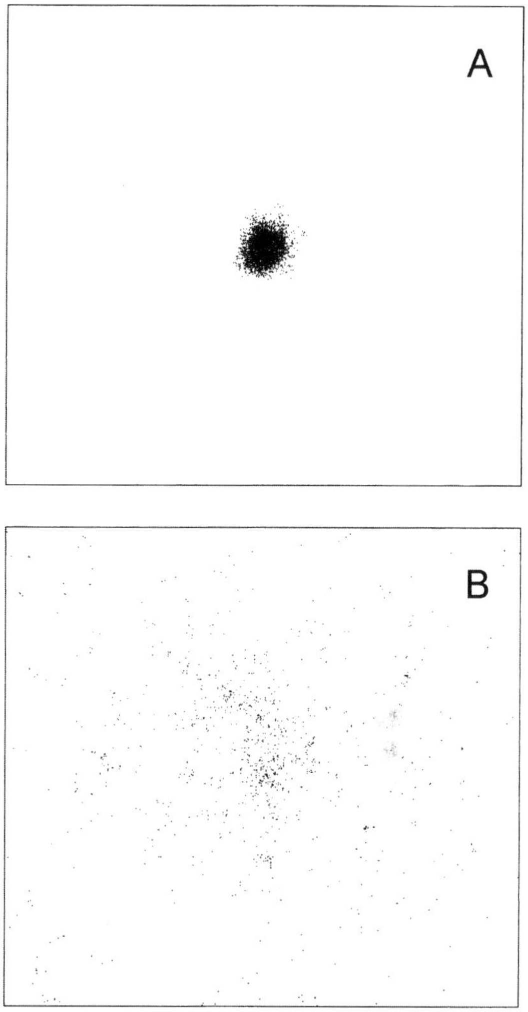

FIG. 9.16. Two stochastic simulations produced using the DISP_PROJECT software written by M. W. Shaw. Both simulations ran for 120 generations and produced a final population size of about 3500. The three model parameters (latent period, infectious period and median dispersal distance) were the same for both simulations and both realizations are drawn to the same scale. (**A**) Simulation based on a negative exponential dispersal function and (**B**) Simulation based on an inverse-square dispersal function.

results in Fig. 9.16 demonstrate a possible mechanism for the increasing expansion rate: formation of "daughter" foci at relatively large distances from the source.

9.10.2 Inference of dispersal from pattern using stochastic models

Instead of starting with a dispersal function, generating a spatial simulation, and then facing the problem of characterizing the extent to which the outcome is realistic/natural, think instead of starting with observed data, such as those in Fig. 9.11. Disease incidence increased between the first and second counts. What, if anything, is possible to infer about dispersal from such data? Gibson and Austin (1996) developed a likelihood-based method for fitting spatio-temporal stochastic models for the spread of plant disease to field-scale experimental observations of patterns. Here, the starting point is a temporal sequence showing intensively mapped disease incidence on at least two occasions. The time interval between "maps" needs to be long enough to allow secondary dispersal, so that newly infected individuals could have become infected as a result of dispersal of inoculum from those already infected. Gibson and Austin (1996) used an example taken from Marcus et al. (1984) showing the pattern of trees infected with CTV in 1981 and 1982. The objective of Gibson and Austin (1996) was to compare models based on different contact distributions (see section 7.3.4). In "model 1", the "infective pressure" (of plants already infected on those previously disease-free) decreased exponentially with distance. Infective pressure is essentially just a scaled version of the contact distribution, in which the probability of inoculum contact with a host and the probability of infection are combined. In "model 2" the infective pressure decreased in inverse proportion to a power of distance. On the basis of simulated patterns, Gibson and Austin (1996) were able to conclude that model 2 was more likely than model 1 "to produce an epidemic which exhibits aggregation similar to the CTV epidemic, given the assumption that an epidemic starts from a small number of randomly placed infections." In this case, the comparisons of simulated patterns with patterns based on observed data were based both on visual impressions of maps and on analysis of maps: join-count statistics (section 9.9.1), intra-cluster correlation coefficients (section 9.4.3), and β-binomial aggregation parameters (section 9.4.4) were calculated. The importance of Gibson and Austin's (1996) work is that it provides a direct method of inferring dispersal from the spatial pattern of disease, using information extracted from temporal sequences of intensively mapped disease incidence data.

Gibson (1997a, b) developed the methodology further. The following account is a brief descriptive outline of the analysis and its interpretation. Consider a population of plants set out in a rectangular array (as in Fig. 9.11). Here, equal spacing between the rows and columns of the array is assumed, and distances are normalized so that this distance is equal to one. Plants in the array are either healthy or infected, and those that are healthy are susceptible to infection. The model is stochastic, so the acquisition of disease by susceptible plants is probabilistic. Stochastic models are especially useful when one is concerned with the individuals within a population and their characteristics, such as, in the present context, the disease status of individual plants, and the spatial locations of the diseased and healthy individuals in a field. For a plant at location i that is susceptible at time t, the probability of becoming infected in the following (short) time interval depends on the length of the interval and the locations of previously infected plants in the population. In the model, the probability that the plant at location i becomes infected in the following time interval is proportional to $a_1 + \sum |j-i|^{-2a_2}$, in which $|j - i|$ represents Euclidean distance between locations j and i (and so between the plants at those locations) and a_1 and a_2 are parameters (both ≥ 0).

Parameter a_1 characterizes the probability with which an uninfected plant acquires disease from "background" sources (i.e., outside the host population represented by the plants in the array). For the purpose of parameter estimation (and presentation of results), the transformation $a_1' = [\ln(1+a_1)]^{1/2}$ is applied. As a_1 (or a_1') increases, background sources increasingly become the dominant factor in the acquisition of disease by susceptibles. Thus, when a_1 (or a_1') is large, the position of newly infected plants is effectively independent of the position of previously infected plants in the population, and the pattern of diseased plants appears random. Parameter a_2 characterizes the probability with which an uninfected plant at location i acquires disease from "local" sources (i.e., previously infected plants in the host population). In most published examples of this type of analysis, the infective pressure exerted by these local sources is assumed to decrease in inverse proportion to a power of distance between the susceptible and the source (as shown above), but exponential decrease of infective pressure with distance is available as an alternative assumption. The overall rate at which an uninfected plant at location i acquires disease from local sources is obtained by summing over all previously infected plants in the host population. Low values of a_2 characterize shallow dispersal gradients, increasing the chance that disease can be spread over long distances. If $a_2 = 0$, any diseased plant in the array affects all susceptibles equally, regardless of distance, and the pattern of diseased plants appears random. As a_2 increases, the dispersal gradient becomes steeper, in which case disease spread is increasingly localized to susceptible near-neighbors. At large values of a_2 ($a_2 \geq 3$), disease spread is restricted almost entirely to immediately adjacent susceptible plants.

Following parameter estimation, results are typically shown as a contour plot of (normalized) parameter densities. The region of highest density on the contour plot indicates the most probable values of the parameters, given the data. An example is given in Fig. 9.17. Fig. 9.17A shows a rectangular array representing CTV-infection of citrus plantings in eastern Spain over three sequential assessments. The data were first described by Gottwald et al. (1996), then analyzed further by Gibson (1997a). The corresponding plot of parameter densities (Fig. 9.17B) is strongly indicative of dispersal involving both long-range background transmission and short-range local transmission of infection. The parameter values that maximize the density in Fig. 9.17B are $a_1' = 1.0$, $a_2 = 3.5$. The region of parameter space rejected at a significance probability P corresponds to the region outside the contour for the Pth percentile on the normalized density plot (as discussed by Gibson, 1997a).

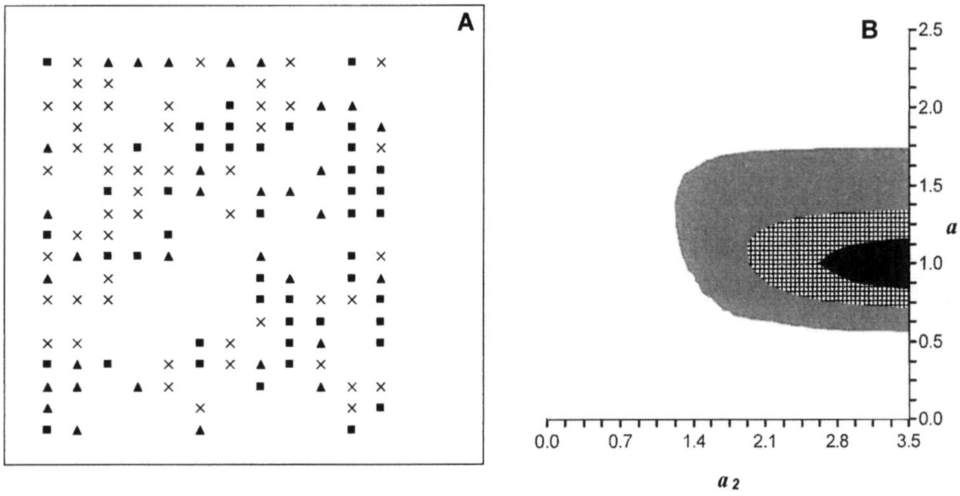

FIG. 9.17. Inference about dispersal from observed pattern. (A) Map of *Citrus tristeza virus* disease incidence at three assessments (1st, ■; 2nd, ▲; 3rd, ×) (see Gibson, 1997a,b; Gottwald et al., 1996). (B) Contour plots of parameter densities over a 21 × 21 grid of parameter values. Contours estimated by interpolation, with densities delineated as follows: white, $P \leq 0.05$; grey, $0.05 < P \leq 0.25$; cross-hatched, $0.25 < P \leq 0.5$; black, $P > 0.5$.

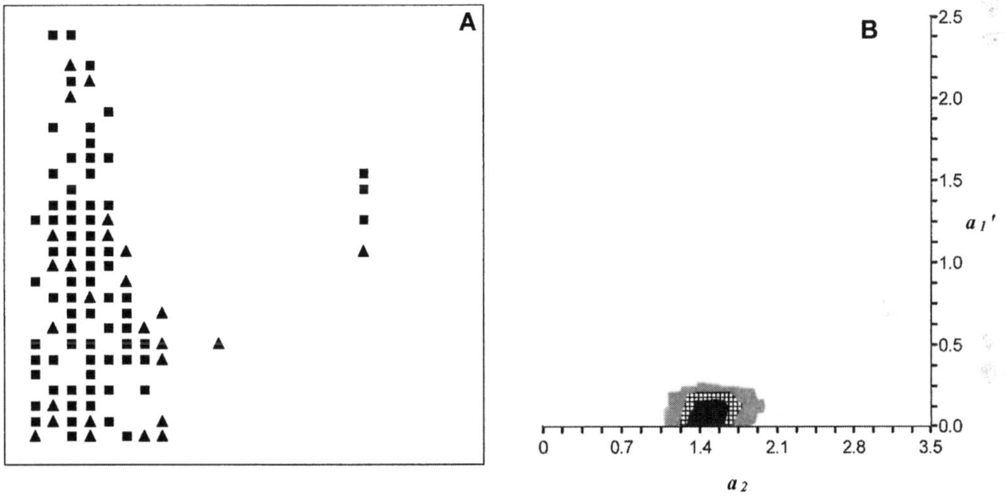

FIG. 9.18. Inference about dispersal from observed pattern. (A) Map of bacterial blight of onion incidence at two assessments (1st, ■ and 2nd, ▲) (see Roumagnac, et al., 2004). (B) Contour plots of parameter densities over a 21 × 21 grid of parameter values. Contours estimated by interpolation, with densities delineated as follows: white, $P \leq 0.05$; grey, $0.05 < P \leq 0.25$; cross-hatched, $0.25 < P \leq 0.5$; black, $P > 0.5$.

A second example is given in Fig. 9.18. Fig. 9.18A shows a rectangular array representing incidence of bacterial blight of onion (caused by *Xanthomonas axonopodis* pv. *allii*) in part of an experimental plot in La Réunion. The data originate from experimental work reported by Roumagnac et al. (2004). The corresponding plot of parameter densities (Fig. 9.18B) is indicative of dispersal involving only local transmission of infection between plants. The parameter values that maximize the density in Fig. 9.18B are $a_1' = 0.0$, $a_2 = 1.4$. In this case, the contour plot of parameter densities provides strong evidence against long-range transmission from background sources of infection.

Further examples of this kind of analysis can be found in Gottwald et al. (1999) (for CTV epidemics in Costa Rica and the Dominican Republic), and Pethybridge and Madden (2003) and Pethybridge et al. (2004) (for viruses in Australian hop gardens). Gottwald et al. (1999) were able to demonstrate convincing differences in model parameters estimated from two data sets originating from CTV-infected citrus plots between which there were major differences in the assemblage of aphid vector species. Pethybridge and Madden (2003) found evidence of local transmission of virus infection to susceptible near-neighbors combined with varying degrees of background transmission.

Fig. 9.19 shows a contour plot of parameter densities obtained by applying the analysis described by Gibson (1997a) to Cochran's (1936) intensively mapped TSWV data (Fig. 9.11). For computational reasons, randomly

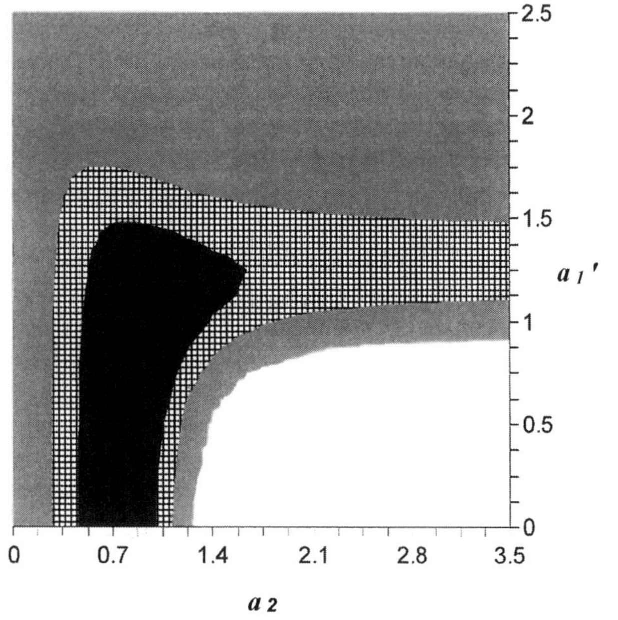

FIG. 9.19. Inference about dispersal from observed pattern. Contour plots of parameter densities over a 21 × 21 grid of parameter values, estimated from the map of *Tomato spotted wilt virus* disease incidence in Fig. 9.11. Contours estimated by interpolation, with densities delineated as follows: white, $P \leq 0.05$; grey, $0.05 < P \leq 0.25$; cross-hatched, $0.25 < P \leq 0.5$; black, $P > 0.5$.

sampled areas from the map shown in Fig. 9.11, rather than the whole map, were analyzed. Fig. 9.19 is a representative example of the results obtained. The example shows that a straightforward interpretation of results is not always possible. The parameter values that maximize the density in Fig. 9.19 are $a_1' = 1.150$, $a_2 = 0.875$, suggesting a role for both local and background transmission of infection. As the interval between the two assessments recorded in Fig. 9.11 was short, the mechanism of local spread is most likely to have been short-range movement of viruliferous vectors within the plot. This would be in agreement with the way that Bald (1937) characterized the dispersal process (see section 9.9). Overall, however, the contour plot of parameter densities by itself does not, in this case, provide a basis for unequivocally characterizing the mechanism(s) of disease spread. Either local transmission with a shallow dispersal gradient coupled with a low rate of background infection $(a_2 < 1,\ a_1' \approx 0)$, or local transmission with a steeper dispersal gradient coupled with a higher rate of background infection $(a_2 > 1,\ 1 < a_1' < 1.5)$ are viable explanations for the observed pattern. Nor is it possible to rule out (at $P = 5\%$) purely random patterns (a_2 low or a_1' high) of TSWV-infection. However, the low parameter density in the bottom right-hand corner of the contour plot (a_2 high, a_1' low) is strong evidence against a combination of local transmission with a steep dispersal gradient and a low rate of background infection.

The methodology described by Gibson (1997a, b) requires at least two sets of intensively mapped disease incidence data in order to permit estimation of parameters relating to spatial spread of disease. Keeling et al. (2004) outline an alternative approach to the estimation of spatial parameters that requires only a single intensively mapped disease incidence data set, provided that one is prepared to make some assumptions regarding the extent to which a single "snapshot" of an epidemic provides a comprehensive description of the spatial dynamics. Keeling et al. (2004) illustrate their analysis using spatial data taken from Marcus et al. (1984), previously analyzed by Gibson and Austin (1996).

9.11 Distance-Based Methods

9.11.1 Events and intervals

So far, the description of spatial patterns of disease has focused on "events". That is to say, descriptions of pattern have been based on observations of diseased host plants or of pathogen population density. Here, instead, descriptions of pattern based on intervals in space (i.e., distances) between events are considered. For simplicity of illustration, we begin with a one-dimensional space. To make it seem more realistic, the one-dimensional space in this example could be thought of as an agronomic row in a field, although for the purpose of the theoretical development, it has to be an indefinitely long such row. If the events of interest occur at random, it is expected that the number of events per unit length of row would follow a Poisson distribution (section 9.5.2). What, then, is the corresponding distribution of intervals between events?

In this section, the symbol d is used to refer to distance, in order to avoid the notation clash that would have occurred had s (used to denote distance in Chapter 7, but used extensively to denote estimated standard errors in the current chapter) been adopted. Let the true mean number of events in a length of row d (with appropriate distance units) be $\mu = \lambda d$, in which λ is the true mean number of events per unit length of row. In the terminology of distance-based spatial analysis, λ is the *intensity* of the pattern (not to be confused with disease intensity). From the Poisson distribution, the probability that no event occurs within a distance d of a chosen event is $\exp(-\mu) = \exp(-\lambda d)$. This is the same as the probability that the distance from a chosen event i to the first neighboring event (the distance denoted d_i) is greater than d. Thus, the probability that the distance from a chosen event to the first neighboring event is less than or equal to d is given by $\Pr(d_i \leq d) = 1 - \exp(-\lambda d)$. This is a cumulative distribution function, from which the probability density $f(d)$ is obtained as follows:

$$f(d) = \frac{\mathrm{d}}{\mathrm{d}d}\left(1 - \exp(-\lambda d)\right) = \lambda \exp(-\lambda d) \quad (d \geq 0).$$

This is the *exponential distribution*, which describes the probability distribution of intervals between successive independent events. Because the intervals can be of any length, this is a continuous distribution (unlike the Poisson, which is discrete). This exposition here gives a link between discrete distributions for organisms and contact distributions for (continuous) distance of spread of diseases that was considered in detail in previous chapters. The expected value of the mean interval between two successive events is obtained from:

$$\int_0^\infty d\lambda \exp(-\lambda d)\, \mathrm{d}d = \frac{1}{\lambda}$$

which is consistent with the definition of λ given above. The lower limit of integration is zero in this case, because we are only considering intervals in one direction in the one-dimensional space. As pointed out by Minogue (1989), the exponential distribution of intervals underlies the empirical use of decreasing exponential functions as descriptions of dispersal gradients (see Chapter 7). In the present context, however, there is no assumption required that a particular individual is the source of infection from which, following dispersal, others are infected.

9.11.2 Neighbors

In practice, the starting-point for use of distance-based methods of pattern analysis is usually an intensively mapped area showing the locations of all individuals of interest (not just the locations of sampling units (e.g., quadrats each containing a number of individuals)). This is often referred to by statisticians as a *spatial point pattern*, sometimes by epidemiologists as a "dot map" (Barreto, 1993). Ecological studies of point patterns based on distances between individuals began with Clark and Evans' (1954) analysis of distance to nearest neighbor. Good introductions to the methodology are provided by Krebs (1989, Chapter 4) and in Upton and Fingleton (1985). Essentially, these are accounts of methods used to describe spatial point patterns, usually in terms of deviations from spatial randomness. This is also the focus of the outline that follows. There is, in addition, an extensive literature on the statistical methodology for investigating and modeling the stochastic mechanisms (or spatial point *processes*) that underlie observed patterns. This topic is not pursued here. Cressie (1993, Part III) and Diggle (2003) provide excellent statistically orientated accounts.

Using the same notation as above (section 9.11.1), but now with d as the radius of a circle rather than distance along a line and intensity λ having units of number of events per unit area, the probability that the distance d_i from a chosen event i to the first neighboring event is less than or equal to d is given by $\Pr(d_i \le d) = 1 - \exp(\lambda \pi d^2)$, from which the probability density $f(d)$ is obtained as follows:

$$\begin{aligned} f(d) &= \frac{\mathrm{d}}{\mathrm{d}d}\left(1 - \exp(-\lambda\pi d^2)\right) \\ &= 2\lambda\pi d \exp(-\lambda\pi d^2)\ (d \ge 0). \end{aligned}$$

The expected value of the mean distance to nearest neighbor is now obtained from:

$$\int_0^\infty 2\lambda\pi d^2 \exp(-\lambda\pi d^2)\, \mathrm{d}d = \frac{1}{2\sqrt{\lambda}}.$$

The variance of the distance to nearest neighbor is $(4-\pi)/4\pi\lambda$. These formulae, derived by Clark and Evans (1954), provide a description of nearest-neighbor distances for a random pattern of events. In this context, "random" means that the intensity of the pattern does not vary over the study area and that there are no interactions between neighbors.

The intensity λ may be estimated from the number of events in the study area divided by the size of the study area (in appropriate units). Observed nearest-neighbor distances may be obtained directly from the mapped data and averaged. Edge effects can influence this calculation, because the nearest neighbors of some individuals in the study area may not themselves be located in the study area. This can be overcome by including a boundary strip around the study area for the purpose of correctly locating the nearest neighbors of those individuals that are located in the study area, but near its extremities (Krebs, 1989). The ratio of the expected value of distance to nearest neighbor for a random pattern to the observed mean distance will be equal to one for a random pattern of individuals, less than one (with a lower limit of zero) for an aggregated pattern and greater than one (with an upper limit of about 2.15) for a regular pattern. Clark and Evans (1954) provide a test of randomness based on calculation of a standard normal deviate. In addition to distance to nearest neighbor, distances to second, third, and further neighbors may also provide useful information about pattern. Thompson (1956) provides the required generalization of Clark and Evans' (1954) nearest-neighbor approach, and Krebs (1989) again gives a useful summary. From a practical point of view, the use of distances to neighbors beyond the nearest is likely to increase the influence of edge effects, unless an appropriately wide border strip is included around the study area.

9.11.3 The *K* (distance) function

Average distances to nearest and further neighbors are not, by themselves, an adequate summary of pattern. It is not difficult to imagine two quite different patterns for which the average nearest-neighbor distance is the same. For example, a pattern comprising patches, within which there is a high density of individuals and among which there are a few widely scattered individuals, may have a combination of short and long nearest-neighbor distances leading to the same *average* distance as for a random pattern. In addition to expected mean distances between neighbors, the extent of spatial dependence between neighbors also requires investigation. One way to investigate spatial dependence between neighbors is to consider the entire distribution of distances (Bartlett, 1964) rather than just the mean. We could, for example, compile the cumulative frequency distributions of observed distance to nearest neighbor, second-nearest neighbor, third-nearest neighbor and so on, and examine their properties. An alternative (or additional) procedure is to calculate the function $K(d)$ (Ripley 1976, 1977, 1979), now a widely used method for the analysis of spatial point patterns. Dixon (2002) provides a useful introductory account.

The function $K(d)$ is a sort of cumulative distribution function. Consider an arbitrary (i.e., randomly chosen) event in a point pattern. The expected mean number of events within a distance d of this event is equal to $\lambda K(d)$. A random pattern of events is the realization of a homogeneous Poisson (i.e., stationary, isotropic) process. In this particular case, the expected mean number of events within a distance d of an arbitrary event is equal to $\lambda \pi d^2$, since the Poisson parameter λ is the mean number of events per unit area and πd^2 is the area of the circle, radius d, centered on the arbitrary event. Then:

$$K(d) = \pi d^2 \, (d \geq 0).$$

A useful linear transformation is:

$$L(d) = \sqrt{\frac{K(d)}{\pi}} \qquad (d \geq 0)$$

(see discussion following Ripley, 1977). For a random pattern of events, a graphical plot of $L(d)$ against d is a straight line with a slope of one, passing through the origin. This provides a useful baseline for a test of randomness. An alternative to this is the graphical plot of $L(d) - d$ against d (Fig. 9.20), which is a horizontal line equal to zero for a random pattern of events.

For a point pattern where N is the observed number of events and $\hat{\lambda}$ is the estimated mean intensity, an estimate $\hat{K}(d)$ can be calculated from:

$$\hat{K}(d) = \frac{\sum_{i}^{N} \sum_{j \neq i}^{N} I_d(d_{i,j})}{\hat{\lambda} N}$$

in which $d_{i,j}$ is the distance between the ith and jth events and I_d is an indicator function, $I_d = 1$ for $d_{i,j} \leq d$; $I_d = 0$ otherwise. With this estimator, edge effects arise when events within a distance d of an event in the study area are not themselves in the study area. This could be overcome by mapping a boundary strip around the study area

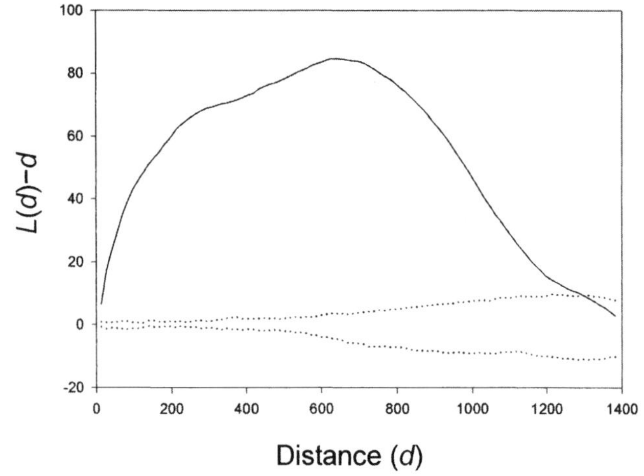

FIG. 9.20. Spatial pattern of oak mortality, Marin County, California, 2000. On a graphical plot of $L(d) - d$ against d (in this case the distance scale is calibrated in meters), a random pattern of mortality is represented by a horizontal line equal to zero. Upper and lower bounds for this line were generated using Monte Carlo methods to produce a simulation envelope representing a 95% confidence interval (shown as dotted lines). The estimated values of $\hat{L}(d) - d$ (shown as a solid line) were greater than zero (and beyond the upper bound of the simulation envelope) at all distances up to about 1200 m, indicating aggregation of oak mortality at these distances. Figure courtesy of N. M. Kelly and Q. Guo.

(see discussion following Ripley, 1977), but more usually the calculation of $\hat{K}(d)$ is adjusted to take account of edge-effects (Ripley, 1976; Haase, 1995; Ward and Ferrandino, 1999). Once a satisfactory adjustment has been applied to the calculation of $\hat{K}(d)$, $\hat{L}(d) = \sqrt{\hat{K}(d)/\pi}$ can be calculated.

It is of interest to compare the estimate $\hat{L}(d)$ with the expected value for a random pattern, $L(d) = d$. For an aggregated pattern, for example, an excess of short distance observations would be expected, leading to $\hat{L}(d) > L(d)$ when d is small. This can be seen as $\hat{L}(d) - d > 0$ on the graphical plot of $L(d) - d$ against d (Fig. 9.20). Significance testing of such deviations from randomness is usually based on randomization procedures. Estimation and significance testing are facilitated by the availability of specialist computer software such as S + SPATIALSTATS. As mentioned above (section 9.11.2), a simple test of deviations from randomness could be followed by the formulation and fitting of an appropriate stochastic model, consistent with observed deviations from randomness. In this way, it is possible to go from description of an observed pattern to inference of a possible underlying spatial process.

The more or less regular layout of most field crops (as in Fig. 9.11) lends itself to the use of quadrat-based methods of analysis. As a result, epidemiologists whose interests are either clinical (Elliot et al., 2000; Lawson, 2001) or veterinary (Pfeiffer and Morris, 1994; Durr and Pfeiffer, 2002 and the following articles in Volume 56 of the journal *Preventive Veterinary Medicine*) have used distance-based spatial methods more than plant pathologists have. The notable exception is in the study of landscape-scale pathology (see Holdenrieder et al., 2004; Lundquist, 2005), where the impact of forest diseases is often a major concern (Reich and Lundquist, 2005). In cases where patterns over a wide range of spatial scales may be of interest, the availability of global positioning system (GPS) receivers that can process satellite signals has greatly facilitated the collection of the necessary data. Geographical information systems (GIS) have similarly enhanced our ability to extract and present the required information subsequent to data collection (Kelly and Tuxen, 2003). Studies by Brooks (2002) of brown root rot disease (caused by *Phellinus noxius*) in the tropical rain forests of American Samoa, by Gottwald et al. (2002) of citrus canker (caused by *Xanthomonas axonopodis* pv. *citri*) in the citrus tree population of metropolitan Miami, and by Kelly and Meentemeyer (2002) of Sudden Oak Death (caused by *Phytophthora ramorum*) in the oak population of China Camp State Park in California illustrate potential applications of the $K(d)$ function in plant disease epidemiology.

9.12 Conclusions

Patterns of disease are complex and multifaceted. The simulation studies of X.-M. Xu and colleagues (Xu and Ridout, 1998, 2000, 2001; Xu and Madden, 2004) make clear the nature of the challenge facing plant pathologists who seek to describe and explain observed patterns of disease. Spatial statistics may be influenced by many factors related to sampling, environment, and initial epidemic conditions. To meet the challenge, a range of statistical techniques, mostly borrowed from related disciplines such as ecology and entomology, is available to plant pathologists. The choice of technique for any particular analysis must take into account the nature of the available spatial data and the objectives of the analysis. When there are many questions, it is most unlikely that a single spatial summary statistic will provide all the answers.

Many phytopathological studies have been carried out with the aim of making inferences about the dispersal mechanism (i.e., the spatial process) that underlies an observed spatial pattern. Such studies involve the collection of intensively mapped spatial data. Statistical methods are often used in an exploratory manner to arrive at a *description* of pattern; then, a biologically based *explanation* of how such a pattern could have arisen is developed. The work of Gibson (1997a, b) represents a real advance for the investigation of dispersal mechanisms on the basis of observed patterns, although the lack of widely available software is a restriction on the uptake of this methodology. Ripley's $K(d)$ function has been used to describe deviations from randomness in patterns of diseased trees over a range of spatial scales. However, in a case where the pattern of all trees (diseased and healthy) in the population of interest could not be assumed to be homogeneous, such analysis would reveal rather little. Diggle and Chetwynd (1991) discuss an alternative, specifically motivated by the analysis of spatial data in an epidemiological context, in which the test is essentially whether the diseased individuals can be regarded as a random sample from the population of all individuals. There is then the possibility of being able to characterize the pattern of diseased individuals as more aggregated (or perhaps more regular) than that of the whole population.

Sparsely sampled spatial data are most frequently associated with applications in disease management. A statistical description of the frequency distribution of quadrat counts underlies the development of sampling strategies (Krebs, 1989), as discussed further in Chapter 10. We re-iterate the importance of matching the type of spatial data collected to the properties of the statistical probability distribution used as a basis for description.

Analysis of intensively mapped data on the basis of either an autocorrelogram or a semivariogram provides information on patterns at scales above that of the sampling unit. Analysis on the basis of fitting statistical probability distributions to quadrat count data provides information on pattern at the within-sampling-unit scale. The two kinds of analysis characterize patterns in terms of spatial correlation or spatial variability (depending on one's point of view) of disease *between* sampling units or *within* sampling units. The outcome is a descriptive summary of the extent to which the occurrence of disease at a location influences the occurrence of disease at other specified locations. When the mapped units are individual plants, additional analyses are available to provide an even finer resolution of the pattern of disease status over a continuum of spatial scales (Ferrandino, 1998; Pethybridge and Madden, 2003). Madden (1989) discusses the use of a combination of methods to characterize patterns of disease intensity between sampling units and between individuals within sampling units. Since no single method can be expected to identify all of the spatial characteristics in data, the use of several statistical techniques—as recommended by Perry et al. (2002)—is becoming more common. Madden et al. (1995b), Gottwald et al. (1998), Turechek and Madden (1999a, b; 2000), Xu et al. (2001) and Scott et al. (2003) provide examples of the application of this approach in particular plant pathosystems. Care is required: it is not enough merely to subject a data set to the catalog of spatial analyses for which computer software has been made available, in the hope that the results will somehow be more meaningful because more effort has been expended. We need to be clear about the purpose of each analysis in the overall scheme of an investigation.

The analysis of pattern is an active area of statistical research. New methods should be given a cautious welcome by those interested in phytopathological applications. The primary concern relating to the application of spatial statistics should be the extent to which the scope for answering interesting questions about plant disease epidemics is extended.

9.13 Suggested Readings

Binns, M., Nyrop, J. P., and van der Werf, W. 2000. Distributions. *Sampling and Monitoring for Crop Protection Decision Making*. Chapter 4. CAB International, London.

Cochran, W. G., 1936. The statistical analysis of field counts of diseased plants. Supplement to *J. R. Stat. Soc.* 3: 49–67.

Gottwald, T. R., Gibson, G. J., Garnsey, S. M., and Irey, M. 1999. Examination of the effect of aphid vector population composition on the spatial dynamics of citrus tristeza virus spread via stochastic modeling. *Phytopathology* 89: 603–608.

Madden, L. V., and Hughes, G., 1995. Plant disease incidence: distributions, heterogeneity, and temporal analysis. *Annu. Rev. Phytopathol.* 33: 529–564.

Ristaino J. B., and Gumpertz M. L. 2000. New frontiers in the study of dispersal and spatial analysis of epidemics caused by species in the genus *Phytophthora*. *Annu. Rev. Phytopathol.* 38: 541–576.

10

Estimating Plant Disease by Sampling

Statistics are facts; the rest is opinion.

Craig Brown (in an interview; *The Guardian*, 16 Nov. 1999)

10.1 Why We Sample for Epidemiological Data

Sampling provides the data that link epidemiological theory to the solution of problems involving plant disease in fields, forests, and elsewhere. Without sampling, there are no data, and without data, our theories can only provide us with qualitative pictures of the dynamics of plant disease in time and space. Such qualitative pictures are useful in providing a strategic overview, but the extent to which this can contribute to practical problem-solving is limited if it is not eventually accompanied by quantitative data.

At the outset we need to specify whether we are sampling to estimate the value of a population parameter, such as mean disease incidence, or to classify a population relative to a threshold level of some epidemiological significance. The current chapter is concerned with various aspects of sampling for estimation. Sampling for classification in the context of crop protection decision making has been advanced by recent research in economic entomology (Pedigo and Buntin, 1994; Binns et al., 2000). We return to this topic in Chapter 11.

What follows is only a summary of some aspects of sampling likely to be of interest to those involved in collecting plant disease data. Texts by Cochran (1977) and Kish (1995) are essential reference sources for those who require further details of statistical aspects of sampling theory. Other useful resources include Southwood (1978, Chapter 2) and Krebs (1999, Chapters 7–9), both of which provide discussions on sampling in an ecological context. Perry (1994) provides an interesting overview of the development of sampling applications in pest and disease management.

10.2 Sampling Preliminaries

10.2.1 Terminology

The idea of the *sampling unit* was introduced in section 9.3. To reiterate briefly, a sampling unit contains the individual elements—often whole plants or parts of plants—on which (in the present context) disease assessments are made. Sampling units must not overlap, so individual elements may only belong to one sampling unit. The *sample* comprises those sampling units that are actually inspected, drawn from the *population* comprising all possible sampling units of interest. The sample data are used as a basis for calculating *estimates* of population characteristics. Since the whole population is not assessed, there is uncertainty attached to these estimates. In some circumstances, the amount of uncertainty can be controlled by calculating a sample size that will provide estimates with a specified degree of *reliability*. The term "reliability" is used here to refer to the extent of the deviations that would occur if a quantity were to be repeatedly estimated using the same sampling procedure. Cochran (1977) refers to this as *precision*. *Accuracy* refers to deviations between the true value of a quantity and estimates of that value based on sampling. Sample estimates that consistently deviate from the true value are *biased*. These definitions of the terms reliability and accuracy are consistent with their usage in the context of survey sampling. For details on the use of these terms in measurement science, see Chapter 2.

10.2.2 Sample size

Sample size may be calculated before sampling, based on some pre-specified requirement for reliability. Alternatively, it is possible to sample until a pre-specified "stop" condition has been met. The latter approach is discussed in sections 10.11 and 10.12.

The main issues to be addressed when calculating a sample size ahead of sampling are summarized in the following brief example. For a sample mean $\bar{y}$, assumed normally distributed (at least to a reasonable approximation), the 95% confidence interval is $\bar{y} \pm 2\sqrt{\sigma^2/N}$. The quantity $\sqrt{\sigma^2/N}$ is the standard error (SE) of $\bar{y}$, in which σ^2 is the variance of the population of the individual y values, N is the number of observations on which the calculation of $\bar{y}$ is based and ± 2 is a reasonable approximation of the values cutting off 2.5% in both the upper and the lower tails of the standard normal distribution.

To be precise, a two-sided $100(1 - \alpha)\%$ confidence interval is constructed using the $100(1 - \alpha / 2)$th percentile of the standard normal distribution, which is 1.96 when $\alpha = 0.05$. For $N < 30$, one would normally use the $100(1 - \alpha / 2)$th percentile of the *t*-distribution with $N - 1$ degrees of freedom (*df*), in place of the value from the standard normal distribution. Specifying the size of the confidence interval is one way of defining reliability. Let *h* be the half-width of this interval, in which case $h = 2\sqrt{\sigma^2 / N}$, and so $N = 4\sigma^2 / h^2$. This tells us that in order to calculate N, the required sample size in this case, we need to specify *h* and have an initial estimate of σ.

In general, in order to calculate sample sizes, we need to adopt a suitable definition of reliability and to obtain a description of variability in the quantity of interest. Often the latter will take the form of an appropriate statistical probability distribution and estimates of the parameter(s) for the situation in which sampling is to be performed (section 10.2.4). The need to supply anticipated values of the unknown parameter(s) of interest is often a stumbling block, since it may appear that would-be samplers are being asked to supply information that they wish to obtain from the survey, as a prerequisite for properly conducting the survey. Krebs (1999) provides some guidance on this. Essentially, the required information may be obtained:

- by sampling in two steps and using the information from the first step to estimate the required parameters (and then the size of the second step),
- by conducting a pilot survey,
- by reference to previous sampling of a similar population, and
- by informed guesswork.

"Informed guesswork" is the use of information that a subject-matter specialist might be able to provide, based on their expertise and experience in relation to the plant pathosystem of interest. The problem is how to formulate this implicit expert knowledge in such a way that it will be of use in calculating sample sizes. The process of constructing a statistical description of implicit expert knowledge is known as *elicitation*. Elicitation may involve a subject-matter specialist, a psychologist and a statistician. The first has the required information, but not in a form appropriate for the objectives at hand (here, sample size determination); the second is concerned with how best to obtain that information; and the third is concerned with processing the information obtained so that it is rendered available in an appropriate form. Winkler (1967) discusses many of the issues involved in these processes. In practice, elicitation procedures are not always elaborate. However, even simple procedures need to observe "good practice". A useful list of guidelines is provided by Kadane and Wolfson (1998). Hughes and Madden (2002) demonstrate some elicitation methods in the context of sampling for plant disease.

The statement, at the beginning of the example, that the sample mean was "assumed approximately normally distributed" was an appeal to the central limit theorem. In particular, as sample size increases, the distribution of sample means increasingly resembles a normal distribution. Thus, if sample sizes are large enough, the estimated means will be approximately normally distributed. Leaving aside for now the question of how large is large enough, use of an appropriate sample design (section 10.2.3) based on probability sampling provides a formula based on sampling theory for estimating the standard error of the sample mean. This statistic is instrumental in sample size calculations.

10.2.3 Sample design

Sample design is the procedure for working out how sampling units will be selected for inspection. The purpose of sample design is to ensure that the sample is properly representative of the population from which it is drawn. The number of available designs is large, and a comprehensive discussion is beyond the scope of this chapter. We concentrate on a few designs that have proved useful in plant disease epidemiology. Reference sources (Cochran, 1977; Kish, 1995) should be consulted for details of other designs and the circumstances in which their use may be appropriate.

Wherever possible, some form of *random sampling* is recommended. Use of random selection of sampling units is the means by which the large body of theory concerned with probability sampling may be accessed as a basis for analysis of sample data. The essential idea underlying probability sampling is that each sampling unit is assigned a probability of selection. The sample is then drawn from the population, using a procedure for random selection of sampling units that is consistent with the pre-assigned probabilities of selection. Calculation of estimates from the sample data then takes into account the pre-assigned probabilities of selection. In order to be able to accomplish all this, we must, at least in principle, be able first to make a list of all possible sampling units in the population of interest. In the sampling literature, this list is known as the frame. Those who solicit public opinion via survey sampling may make use of lists such as electoral registers or telephone directories as their frame. While it is not difficult to see that such lists may be inaccurate both in terms of their original purpose and as representations of a particular population of interest, it is probably the case that relatively few plant pathologists have earned the right to criticize by constructing a good frame prior to their own field sampling. Random sampling is attractive in that it removes the possibility of bias, conscious or otherwise, on the part of whoever carries out the sampling. Without a frame, however, even the intuitively straightforward prospect of selecting plants at random in an agricultural field may prove to be a difficult practical proposition.

If all sampling units are assigned an equal chance of selection and the sampling unit is the individual (plant, for example), the sample design is referred to as *simple random sampling*. If all sampling units are assigned an equal chance of selection but the sampling unit is a group of neighboring individuals, the sample design is referred to as *cluster sampling*. Different procedures are required for analysis because individuals in the same cluster (plants in the same quadrat, for example) cannot be regarded as independent. If cluster sampling involves assessment of all the individuals in each sampling unit selected for inspection, the sample design may be referred to as single-stage cluster sampling. If cluster sampling is carried out, but there is subsampling from clusters so that not all individuals in selected sampling units are inspected, the sample design is referred to as two-stage cluster sampling (section 9.4.10).

It is not a requirement that all sampling units have an equal chance of selection. The population of all sampling units may be divided into non-overlapping sub-populations. These sub-populations are referred to as strata, and the sample designs associated with this approach are known as *stratified sampling*. Each stratum is sampled separately, using an appropriate form of probability sampling within strata. Different probabilities of selection may be assigned to the sampling units in the different strata. As long as these different probabilities are known, they can be compensated for by the assignment of appropriate weights when calculating estimates relating to the whole population from the sample data from the different strata. Snedecor and Cochran (1989) and Krebs (1999) give formulae and illustrative examples.

The requirement for some form of random selection of sample units is often at odds with the practicalities of sampling for disease. In agricultural fields, for example, it is inevitable that sampling units are selected along a path traversed through the field by the sampler. There have been comparative investigations of various sampling paths. Field data may be collected along different paths and the results compared (e.g., Basu et al., 1977). Alternatively, different paths may be simulated on field data (Hau et al., 1982; Punja et al., 1985; Mihail and Alcorn, 1987) or on simulated data (Lin et al., 1979). A simulation approach provides the chance to compare a result based on sampling with the population value. As might be expected, the results indicate that there is not a single type of sampling path that is the best for every situation. Among the few generalities that seem to come out of these studies is that sampling paths that cause the sampler to draw sampling units from sites more widely distributed over a field appear to perform relatively well, compared both with paths where sites are less widely distributed and with sites selected at random, particularly when the spatial pattern of disease is patchy.

If a starting point for a sampling path is selected at random, but sampling units thereafter are selected at regular intervals along the path (the interval chosen so that a sample of a particular size is drawn), this is known as *systematic sampling*. There is no doubt that this procedure violates the assumptions of probability sampling, the most serious consequence of which is that there is no sampling theory leading to a reliable method of estimating the standard error of the sample mean (Snedecor and Cochran, 1989). The practical implications of this violation are less clear. Krebs (1999, Chapter 8), Binns and Nyrop (1992) and Binns et al. (2000, Chapter 6) have all discussed this issue and concluded, broadly speaking, that there is a place for systematic sampling of biological populations. Krebs (1999) provides the following helpful advice: "...if you have a choice of taking a random sample or a systematic one, always choose random sampling. But if the cost and inconvenience of randomization are too great, you may lose little by sampling in a systematic way." Sampford (1962, Chapter 5) has a detailed discussion of the advantages and disadvantages of systematic sampling.

10.2.4 Variability

In Chapter 9, disease incidence data (sections 9.4.3 and 9.4.5) and disease data in the form of counts (sections 9.5.2 and 9.5.3) were described by reference to statistical probability distributions. These distributions, and related power law relationships between variances (sections 9.4.7 and 9.5.4), provide descriptions of variability needed for developing sample size formulae. Initial values for the parameters of these statistical probability distributions or of power law relationships are required—these are not obtained from the sample that is to be drawn, they are a prerequisite to drawing a sample of the appropriate size. These initial values are themselves estimates. We follow current practice in sampling plant populations damaged by pests, pathogens or weeds, by not correcting sample size calculations for uncertainty in the initial estimate of variability. For a brief discussion of this topic, see Julius (2004).

10.2.5 Population size

Recall that in section 10.2.2 the sample size formula was based on specification of a 95% confidence interval by $h = 2\sqrt{\sigma^2 / N}$. In practice σ^2 will be replaced by a previously obtained sample estimate s^2. The formula is rearranged so that N is a function of reliability (defined by h) and variability (defined by s^2). We now regard the sample size calculated from this as an initial value N_0, i.e., $N_0 = 4s^2 / h^2$. If the initial value N_0 is greater than 10% of the population size, the estimated standard error $\sqrt{s^2 / N}$ should be replaced by $\sqrt{(s^2 / N)}\sqrt{1-f}$, in which f is the sampling fraction (i.e., the sample size expressed as a proportion of the population size). The factor $\sqrt{1-f}$ is the *finite population correction* (Cochran, 1977). Now, writing M for population size,

$h = 2\sqrt{s^2/N}\sqrt{1-(N/M)}$, and solving this for N gives (after some rearrangement):

$$N = \frac{N_0}{1+\left(\frac{N_0}{M}\right)}$$

It can be seen that when M is large relative to sample size, N_0 will be a good approximation of N. When M is comparatively small, the effect of ignoring the finite population correction is to overestimate the standard error and so inflate the required sample size. For the sake of simplicity of presentation, the finite population correction is not routinely included in sample size formulae, but it should still be applied wherever appropriate. Conventionally, this is done when the required sample size is more than about 10% of the population size.

10.2.6 Reliability of the estimated sample mean

Karandinos (1976) gave definitions of reliability considered meaningful in practice and showed how they could be used in developing sample size formulae. Reliability was defined either in terms of coefficient of variation of the sample mean, CV:

$$CV = \frac{\sqrt{s^2/N}}{\bar{y}}$$

or in terms of the half-length of the required confidence interval for the sample mean, $z_{\alpha/2}\sqrt{s^2/N}$ in which $z_{\alpha/2}$ is the $100(1 - \alpha/2)$th percentile of the standard normal distribution, used to construct a two-sided $100(1 - \alpha)\%$ confidence interval. If reliability is defined by setting the half length of the required confidence interval of the sample mean equal to a fixed *proportion* of the mean, H:

$$H\bar{y} = z_{\alpha/2}\sqrt{\frac{s^2}{N}}$$

If reliability is defined by setting the half length of the required confidence interval of the sample mean equal to a fixed positive number h:

$$h = z_{\alpha/2}\sqrt{\frac{s^2}{N}}$$

(using a previously obtained sample estimate of the variance, s^2, instead of the population value σ^2 as specified in section 10.2.2). Thus, all three definitions of reliability are based on the estimated standard error of the sample mean, $\sqrt{s^2/N}$. Note, again, that the variance estimate s^2, used to characterize variability, must be supplied before calculating sample size. In practice, it is usually preferable to control the size of $\sqrt{s^2/N}$ relative to $\bar{y}$ rather than in absolute terms. On this basis, the coefficient of variation of the estimated mean, CV, is often used in sample size calculations. For instance, to achieve $CV = 0.20$ (i.e., 20%), $\sqrt{s^2/N}$ would need to be 0.1 when $\bar{y} = 0.5$ and 0.01 when $\bar{y} = 0.05$. Values of CV of 0.2 or 0.1 are typically considered as representative of appropriate levels of reliability for field studies in plant disease epidemiology.

The development of sample size formulae as described by Karandinos (1976) involves rearranging the formulae given above so that N is on the left-hand side and the quantities characterizing reliability and variability are on the right-hand side. On carrying out this rearrangement, the variance appears in the numerator of the right-hand side, and the mean (when a definition of *relative* reliability is adopted) appears in the denominator. This tells us that in general, the sample size required for a given level of relative reliability increases with increasing variability and decreases as the mean increases.

10.3 Simple Random Sampling for Disease Incidence Data

In simple random sampling, the sampling units are individual plants or plant parts. A specified number of such units is selected from the population in such a way that every unit has an equal chance of being chosen. The selection of a unit does not affect the chances of selection of other units.

The binomial distribution is the appropriate model when sampling *with replacement*. The scenario is as follows. A sample comprising N individual elements (plants, for example) is randomly drawn from a population size M in which X plants were "diseased" and $M - X$ were "healthy". Sampling with replacement means that M and X do not change as the sample is drawn, plant by plant. The binomial probability of Y, the number of plants diseased (out of a total of N), is given by:

$$\Pr(Y) = \binom{N}{Y}\left(\frac{X}{M}\right)^{Y}\left(1-\frac{X}{M}\right)^{N-Y} \qquad (Y = 0, 1, \ldots, N)$$

Since, in practice, sampling is usually *without replacement* (i.e., no sampling unit is allowed to be observed more than once), use of the binomial distribution requires a brief justification. The hypergeometric distribution is the appropriate model when sampling without replacement. Sampling without replacement means that the population size and the number of diseased plants both change as the sample is randomly drawn, plant by plant. So if a plant is randomly drawn without replacement from a population size M (in which X plants are diseased and $M - X$ are healthy), the population size for the second plant to be drawn is $M - 1$. If that first plant drawn were diseased, the population size $M - 1$ would comprise $X - 1$ diseased plants and $M - X$ healthy plants.

The hypergeometric probability of Y, the number of plants diseased (out of a total of N), is given by:

$$\Pr(Y) = \frac{\binom{X}{Y}\binom{M-X}{N-Y}}{\binom{M}{N}} \quad (Y = 0, 1, \ldots, N)$$

The two distributions have the same mean (=X/M). This is estimated by the proportion of sampling units assessed as diseased (i.e., mean disease incidence):

$$\bar{y} = \frac{\sum_i Y_i}{N}$$

in which N is the total number of sampling units, i is an index variable for the sampling units (i = 1, 2, …, N) and Y_i = 0 for a sampling unit assessed as healthy or Y_i = 1 for a sampling unit assessed as diseased. The variance of $\bar{y}$, the sample mean, is estimated as follows:

$$s_{\bar{y}}^2 = \frac{\bar{y}(1-\bar{y})}{N} \quad \text{for the binomial distribution, and}$$

$$s_{\bar{y}}^2 = \frac{\bar{y}(1-\bar{y})}{N}\left(\frac{M-N}{M-1}\right) \quad \text{for the hypergeometric distribution.}$$

Thus, the estimated standard error of $\bar{y}$ based on the binomial distribution differs from the estimated standard error of $\bar{y}$ based on the hypergeometric distribution by a factor of $\sqrt{(M-N)/(M-1)} \approx \sqrt{(M-N)/M} = \sqrt{1-(N/M)}$, which we recognize as the finite population correction (section 10.2.5). The finite population correction is close to one if N is small relative to M. In such circumstances we may use the binomial distribution even if sampling without replacement. If N is not small relative to M, we may still use the binomial distribution, but should apply the finite population correction. Preference for use of the binomial distribution over the hypergeometric is based largely on the fact that the former is easier—often much easier—to handle than the latter in some calculations.

10.3.1 Sample size calculations

We are now in a position to develop formulae for sample size calculation based on the binomial distribution, as described by Karandinos (1976). For a simple random sample comprising a total of N sampling units, the estimated standard error of the sample mean is $s_{\bar{y}} = \sqrt{\bar{y}(1-\bar{y})/N}$. If reliability is defined by the coefficient of variation of the sample mean, CV:

$$CV = \frac{\sqrt{(\bar{y}(1-\bar{y})/N)}}{\bar{y}} \qquad (10.1)$$

If reliability is defined by setting the half length of the required confidence interval of the sample mean equal to a fixed proportion of the mean, H:

$$H\bar{y} = z_{\alpha/2}\sqrt{(\bar{y}(1-\bar{y})/N)} \qquad (10.2)$$

If reliability is defined by setting the half length of the required confidence interval of the sample mean equal to a fixed positive number h:

$$h = z_{\alpha/2}\sqrt{(\bar{y}(1-\bar{y})/N)} \qquad (10.3)$$

Table 10.1 shows each of equations 10.1–10.3 rearranged so as to provide an expression for N, the sample size. As noted in section 10.2.2, all sample size calculations require a suitable definition of reliability and an appropriate description of variability. In this case, the variance of the distribution is completely specified by the mean, so an initial estimate of disease incidence, $\bar{y}$, is all that is needed to fulfill the requirement for a description of variability. Fig. 10.1 shows how N varies with $\bar{y}$, using CV = 0.2 as a definition of the required level of reliability. N decreases with increasing $\bar{y}$.

Inverse sampling (section 10.11) or sequential sampling (section 10.12), in which the sample size is not fixed ahead of sampling, may be preferable if the initial estimate of disease incidence, $\bar{y}$, is small. Alternatively, Rahme and Joseph (1998) describe an approach to sample size calculation for the binomial, avoiding use of the normal approximation.

TABLE 10.1. Formulae for sample size calculations for disease incidence data collected by simple random sampling, with three different definitions of reliability[a].

	Reliability defined by:		
		Half-width of confidence interval equal to:	
Description of variability	*Coefficient of variation, CV*	*Proportion of mean, H*	*Fixed positive number, h*
Random (binomial distribution)	$N = (1-\bar{y})/(\bar{y}CV)$	$N = [(1-\bar{y})/\bar{y}](z_{\alpha/2}/H)^2$	$N = \bar{y}(1-\bar{y})(z_{\alpha/2}/h)^2$

[a] N is the number of sampling units required for a pre-specified level of reliability, $\bar{y}$ is an initial estimate of disease incidence and $z_{\alpha/2}$ is the 100(1 − $\alpha/2$)th percentile of the standard normal distribution. The formulae omit the finite population correction.

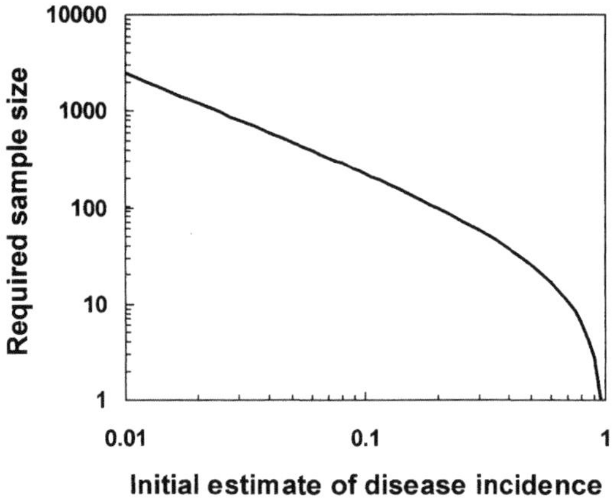

FIG. 10.1. Simple random sampling for disease incidence data. Required sample size (N) varies with the initial estimate of disease incidence ($\bar{y}$), using $CV = 0.2$ as a definition of the required level of reliability (see Table 10.1).

Example 10.1. It is required to estimate disease incidence in a batch of $M = 480$ seedlings, using $CV = 0.2$ as definition of the required level of reliability. An initial estimate indicates that about 15% of the seedlings are diseased. From Table 10.1, $N = 0.85/(0.15 \times 0.2^2) = 142$. Since 142 represents about 30% of the population in this case, the finite population correction is applied. Then $N_0 = 142$, and the revised N is given by

$$N = \frac{N_0}{1 + \left(\frac{N_0}{M}\right)} = 110 \text{(see section 10.2.5).}$$

10.3.2 Inspection errors

Thus far we have made the unstated assumption that the disease status of the inspected individuals is determined without error. However, there exists the possibility that assessments of disease incidence may result in some diseased individuals being recorded as healthy, and some healthy individuals being recorded as diseased. For laboratory-based tests, procedures for defining a threshold on the scale of the test output variable in such a way as to control the rates at which these errors are made are discussed by Sutula et al. (1986). For visual assessments of disease incidence, especially where a number of assessors are involved, acceptable rates for these errors could be defined and assessors trained so as to be able to achieve these rates as a minimum performance standard. Either way, suppose we now can provide values (between zero and one) for ζ (the proportion of diseased individuals erroneously recorded as healthy) and δ (the proportion of healthy individuals erroneously recorded as diseased). Thus, the quantities $(1 - \zeta)$ and $(1 - \delta)$ represent, respectively, the proportion of diseased individuals correctly determined, and the proportion of healthy individuals correctly determined. In the present context, we note that what we record as $\bar{y}$ is the frequency of positive tests (i.e., the result of the test is "diseased"):

$$\bar{y} = p(1-\zeta) + (1-p)\delta \qquad (10.4)$$

in which p is the true disease incidence. This value of $\bar{y}$ can be a badly biased estimate of p (Rogan and Gladen, 1978). Instead, the estimator

$$p^* = \frac{\bar{y} - \delta}{1 - \zeta - \delta} \qquad (10.5)$$

is preferable to the use of $\bar{y}$ as an estimator of p. Recalling that p lies in the interval [0, 1], it can be seen from equation 10.4 that equation 10.5 provides an estimator of p that is applicable for $\delta < \bar{y} < 1 - \zeta$. Otherwise, if $\bar{y} \le \delta, p^* = 0$ and if $\bar{y} \ge 1 - \zeta, p^* = 1$. Rahme and Joseph (1998) consider the implications of using equation 10.5 for sample size calculations. If reliability is defined by setting the half the length of the required confidence interval of the sample mean equal to a fixed positive number h, the sample size N is given by:

$$N = \frac{\bar{y}(1-\bar{y})}{(1-\zeta-\delta)^2}\left(\frac{z_{\alpha/2}}{h}\right)^2 \qquad (10.6)$$

where $\bar{y}$ is given by equation 10.4, based on ζ, δ and p. Since p is unknown, in practice one must calculate values of N for a range of likely values of p in order to select an appropriate final sample size. From equation 10.6, it can be seen that if disease assessments are error-free (i.e., ζ and δ are both equal to zero) the binomial sample size formula is retrieved (Table 10.1). For $0 < (\zeta + \delta) < 1$, the required sample size is larger than that calculated from the binomial formula. If $\zeta + \delta = 1$, the required sample size is indefinitely large and so the required confidence interval cannot be constructed, however many data are collected. If we do not know (or assume) values of ζ and δ, and $\hat{\zeta}$ and $\hat{\delta}$ have instead been estimated from data, dealing with inspection errors is less straightforward. Rogan and Gladen (1978) give details of the required procedure in this eventuality.

10.3.3 Exact binomial confidence intervals

In section 10.3.1, the simplest approximate confidence interval, $\bar{y} \pm z_{\alpha/2}\sqrt{(\bar{y}(1-\bar{y})/N)}$ was used in sample size calculations. The basis for this was mentioned in section 10.2.2: if samples are large enough, the means estimated from them are approximately normally distributed, whatever the distribution of the observations. So, what is large enough? Unfortunately, there is no simple answer.

The easiest way to see the deficiencies of the approximate binomial confidence interval based on the normal distribution is by comparison with an alternative. Here, we compare this approximate confidence interval with the so-called "exact" confidence interval based on calculations of tail areas for the binomial probability distribution

(Clopper and Pearson, 1934; Santner and Duffy, 1989). The term "exact" is qualified because for discrete statistical probability distributions it is not always possible to calculate exact values that cut off, say, 2.5% of the area under the distribution function in the upper and lower tails (as it is in the case of the normal distribution). Because of this, the construction of confidence intervals for discrete distributions such as the binomial is not a straightforward matter (Blaker, 2000). The Clopper–Pearson intervals for binomial proportions, although widely used, are actually rather conservative. A number of different guidelines exist for deciding what is an appropriate confidence interval for a binomial proportion (Samuels and Lu, 1992; Newcombe, 1998). Here, we restrict our attention to Clopper–Pearson intervals, values of which can conveniently be obtained from tables (Diem and Lentner, 1970; Fisher and Yates, 1963) or by use of statistical software such as MINITAB.

Suppose that one diseased plant is observed in a sample of 10. Then disease incidence $\bar{y} = 0.1$, and the resulting approximate normal two-sided 95% confidence interval given by MINITAB is −0.086 to 0.286. This interval is calculated using $z_{\alpha/2} = 1.96$; if the corresponding value from the *t*-distribution with 9 *df* (= 2.26) were used instead, the resulting interval would be −0.115 to 0.315. Either way, this confidence interval is obviously unsatisfactory, since we know that the true disease incidence cannot be negative. Setting the lower limit to zero would avoid that problem, but if we have to arbitrarily adjust the lower limit, might we not also question the authenticity of the upper limit? In this case, the Clopper–Pearson 95% interval (the default calculation in MINITAB) is 0.0025–0.445. The lower limit is positive, as required, and the upper limit is considerable larger than that given by the approximate normal formula (even if the *t*-distribution is used to calculate the interval). The interval is asymmetric about $\bar{y} = 0.1$. The interpretation is that there is a 95% probability that the calculated interval contains the population parameter p.

Now consider a sample of N plants, in which no diseased plants are observed. Obviously, disease incidence is zero, but in this case the resulting approximate normal confidence interval (±0) is not helpful. The lower limit to the interval, equal to zero, is what we would expect. However, the fact that disease incidence in the sample is zero does not necessarily mean that disease incidence in the population is zero, and we would like the upper limit to the interval to reflect this. In this situation, a one-sided Clopper–Pearson $100(1-\alpha)\%$ interval can be calculated, for which the lower limit is zero and the upper limit is denoted p_U, such that $\Pr(Y=0) = \alpha = (1-p_U)^N$. Then:

$$p_U = 1 - \alpha^{1/N}$$

(Selvin, 1996). For example, if no diseased plants are observed in a sample of $N = 10$, $p_U = 1 - 0.05^{1/10} = 0.259$.

For the situation in which no diseased plants are observed in a sample, size N, and a one-sided 95% Clopper–Pearson interval is required, a useful approximation for the upper limit is provided by the "Rule of Three":

$$p_U \approx 3/N$$

(Selvin, 1996; Jovanovic and Levy, 1997). The derivation is based on a Taylor series expansion of $\alpha^{1/N}$, leading to $p_U \approx -\ln(\alpha)/N$, and the further approximation $-\ln(0.05) \approx 3$. Similar approximate rules for other one-sided Clopper–Pearson intervals are $p_U \approx 2.3/N$ (one-sided 90% interval), $p_U \approx 4.6/N$ (99%), $p_U \approx 5.3/N$ (99.5%) and $p_U \approx 7/N$ (99.9%) (Selvin, 1996; Jovanovic and Levy, 1997). The accuracy of these approximations increases with increasing N. Interested readers will find more information on a Taylor series expansion in Chapter 12 (section 12.7.3).

Jovanovic and Zalenski (1997) give:

$$p_U \approx 3/(N+1)$$

for the Rule of Three. The reasoning behind this version of the rule relies on a Bayesian approach to the problem (as discussed in Jovanovic and Levy, 1997), but since numerical results show that the outcome is a uniformly better approximation than the original version of the rule, non-Bayesians can choose to adopt it on these grounds, without feeling guilty.

Bayesian analysis of one-sided confidence intervals for empty samples, based on binomial probabilities, is discussed by Louis (1981), Nicholson and Barry (1995), Jovanovic and Levy (1997) and Winkler et al. (2002). We do not pursue this here, except to state one useful result for combining results from sequence of samples. In its simplest form, this depends on the adoption of a uniform distribution (a special case of the beta probability distribution) as the prior distribution for p. If the first sample, size N_1, is an empty sample, $p_U = 1 - \alpha^{1/(N_1+1)}$. Following a second empty sample, size N_2, $p_U = 1 - \alpha^{1/(N_1+N_2+1)}$, and after K empty samples:

$$p_U = 1 - \alpha^{1/\left(\sum_{i=1}^{K} N_i + 1\right)}$$

The corresponding Rule of Three approximation for the upper limit of a one-sided 95% confidence interval after K empty samples is:

$$p_U \approx \frac{3}{\sum_{i=1}^{K} N_i + 1}.$$

If the summation term in the denominator is large (e.g., following a long sequence of empty samples):

$$p_U \approx \frac{3}{\sum_{i=1}^{K} N_i}$$

is usually adequate.

10.4 Simple Random Sampling for Count Data

Formulae for calculation of sample size for use in simple random sampling of count data are given by Karandinos (1976), Ruesink (1980), Wilson and Room (1982), Ives and Moon (1987), and Campbell and Madden (1990). The Poisson distribution, the negative binomial distribution, and Taylor's power law, all introduced in Chapter 9, provide descriptions of variability for disease assessments based on counts. Initial estimates of parameters are required as prerequisites for sample size calculations. Reliability is defined in section 10.2.6, based on the estimated standard error of the mean.

10.4.1 The Poisson distribution

The single parameter of the Poisson distribution (section 9.5.2) is μ, estimated by $\bar{Y}$, the mean count per sampling unit (for example, mean number of lesions per leaf). For a Poisson distribution, the variance of the counts is also estimated by $\bar{Y}$. Thus, for a sample comprising a total of N sampling units, the estimated standard error of the sample mean is $\sqrt{\bar{Y}/N}$. If reliability is defined by the coefficient of variation of the sample mean, CV:

$$CV = \frac{\sqrt{\bar{Y}/N}}{\bar{Y}} \tag{10.7}$$

If reliability is defined by setting the half the length of the required confidence interval of the sample mean equal to a fixed proportion of the mean, H:

$$H\bar{Y} = z_{\alpha/2}\sqrt{\frac{\bar{Y}}{N}} \tag{10.8}$$

If reliability is defined by setting the half the length of the required confidence interval of the sample mean equal to a fixed positive number h:

$$h = z_{\alpha/2}\sqrt{\frac{\bar{Y}}{N}} \tag{10.9}$$

Table 10.2 shows each of equations 10.7–10.9 rearranged so as to provide an expression for N, the sample size. Sample size calculation requires choice of a definition of reliability and an initial estimate of the mean, $\bar{Y}$ (since this specifies the variance).

10.4.2 The negative binomial distribution

The two parameters of the negative binomial distribution (section 9.5.3) are denoted μ and k. The parameter μ can be estimated by $\bar{Y}$, the mean number of counts per sampling unit (for example, mean number of lesions per leaf). For a negative binomial distribution, the variance of the counts can be estimated by $\bar{Y} + (\bar{Y}^2/\hat{k})$. Thus, for a sample comprising a total of N sampling units, the estimated standard error of the sample mean is $\sqrt{(\bar{Y} + (\bar{Y}^2/\hat{k}))/N}$. If reliability is defined by the coefficient of variation of the sample mean, CV:

$$CV = \frac{\sqrt{(\bar{Y} + (\bar{Y}^2/\hat{k}))/N}}{\bar{Y}} \tag{10.10}$$

If reliability is defined by setting the half the length of the required confidence interval of the sample mean equal to a fixed proportion of the mean, H:

$$H\bar{Y} = z_{\alpha/2}\sqrt{\frac{\bar{Y} + (\bar{Y}^2/\hat{k})}{N}} \tag{10.11}$$

If reliability is defined by setting the half the length of the required confidence interval of the sample mean equal to a fixed positive number h:

$$h = z_{\alpha/2}\sqrt{\frac{\bar{Y} + \bar{Y}^2/\hat{k}}{N}} \tag{10.12}$$

TABLE 10.2. Formulae for sample size calculations for count data collected by simple random sampling, with three different definitions of reliability[a].

	Reliability defined by:		
		Half-width of confidence interval equal to:	
Description of variability	*Coefficient of variation, CV*	*Proportion of mean, H*	*Fixed positive number, h*
Random (Poisson distribution)	$N = 1/(\bar{Y}CV^2)$	$N = (1/\bar{Y})(z_{\alpha/2}/H)^2$	$N = \bar{Y}(z_{\alpha/2}/h)^2$
Aggregated (negative binomial distribution)	$N = ((1/\bar{Y}) + (1/\hat{k}))(1/CV^2)$	$N = ((1/\bar{Y}) + (1/\hat{k}))(z_{\alpha/2}/H)^2$	$N = (\bar{Y} + (\bar{Y}^2/\hat{k}))(z_{\alpha/2}/h)^2$
Aggregated (Taylor's power law)	$N = \hat{a}\bar{Y}^{\hat{b}-2}/CV^2$	$N = \hat{a}\bar{Y}^{\hat{b}-2}(z_{\alpha/2}/H)^2$	$N = \hat{a}\bar{Y}^{\hat{b}}(z_{\alpha/2}/h)^2$

[a]N is the number of sampling units required for a pre-specified level of reliability, $\bar{Y}$ is an initial estimate of the mean count per sampling unit, and $z_{\alpha/2}$ is the $100(1-\alpha/2)$th percentile of the standard normal distribution. For aggregated data, an estimate of either the negative binomial aggregation parameter ($\hat{k}$) or the Taylor's power law parameters ($\hat{a}, \hat{b}$), are required. The formulae omit the finite population correction.

Table 10.2 shows each of equations 10.10–10.12 rearranged so as to provide an expression for N, the sample size. Sample size calculation requires choice of a definition of reliability, and initial estimates of the mean, $\bar{Y}$, and aggregation as characterized by the negative binomial $\hat{k}$ (since the variance is specified by $\bar{Y}$ and $\hat{k}$). The negative binomial $\hat{k}$ usually varies with the mean, $\bar{Y}$, so it is unlikely that a single value of $\hat{k}$ will be appropriate for sample size calculation over a wide range of $\bar{Y}$ values. One way to circumvent this difficulty is to characterize aggregation in terms of Taylor's power law instead of the negative binomial distribution.

10.4.3 Taylor's power law

Taylor's power law describes a relationship between the observed variance and mean for count data, in which the variance of the counts is proportional to a power of the mean (section 9.5.5). The variance of the counts is estimated by $s_Y^2 = a\bar{Y}^b$. Estimated values of the coefficients must be obtained from a previous data analysis. For a sample comprising a total of N sampling units, the estimated standard error of the sample mean is $\sqrt{\hat{a}\bar{Y}^{\hat{b}}/N}$. If reliability is defined by the coefficient of variation of the sample mean, CV:

$$CV = \frac{\sqrt{\hat{a}\bar{Y}^{\hat{b}}/N}}{\bar{Y}} \quad (10.13)$$

If reliability is defined by setting the half the length of the required confidence interval of the sample mean equal to a fixed proportion of the mean, H:

$$H\bar{Y} = z_{\alpha/2}\sqrt{\frac{\hat{a}\bar{Y}^{\hat{b}}}{N}} \quad (10.14)$$

If reliability is defined by setting the half the length of the required confidence interval of the sample mean equal to a fixed positive number h:

$$h = z_{\alpha/2}\sqrt{\frac{a\bar{Y}^{\hat{b}}}{N}} \quad (10.15)$$

Table 10.2 shows each of equations 10.13–10.15 rearranged so as to provide an expression for N, the sample size. Sample size calculation requires choice of a definition of reliability, a previously calculated Taylor's power law relationship for the pathosystem in question and an initial estimate of the mean $\bar{y}$. The sample size formulae arising from rearrangement of equations 10.13 and 10.14 have the quirk that N decreases with increasing $\bar{Y}$ for $\hat{b} < 2$ but increases with increasing $\bar{Y}$ for $\hat{b} > 2$ (Buntin, 1994). In practice, values $\hat{b} > 2$ are uncommon, but when they do occur, caution should be exercised in adopting them as the basis for sample size calculations.

10.4.4 Sample size calculations

Fig. 10.2 shows how required sample size N varies with $\bar{Y}$ when Taylor's power law provides a description of variability (Boivin et al., 1990), using $CV = 0.2$ as a definition of the required level of reliability. For comparison, the variation of N with $\bar{Y}$ when the Poisson distribution provides a description of variability is shown, again using $CV = 0.2$. In both cases, N decreases with increasing $\bar{Y}$, but N is higher for Taylor's power law (aggregated data) than for the Poisson distribution (random data) as a description of variability, except at very low density.

Example 10.2. The arrays labeled Field 10.2.1 and Field 10.2.2 represent counts of lesions per plant in

	Frequency (number of plants)		
Number of lesions per plant	*Sample 10.2.1 (N = 20)*	*Sample 10.2.2a (N = 20)*	*Sample 10.2.2b (N = 40)*
0	0	1	2
1	1	0	4
2	1	1	2
3	3	2	3
4	2	4	1
5	1	2	2
6	5	3	7
7	1	2	7
8	4	0	2
9	1	1	1
10	1	1	3
11	0	0	1
12	0	0	0
13	0	1	1
14	0	1	1
15	0	0	1
16	0	1	0
17	0	0	2

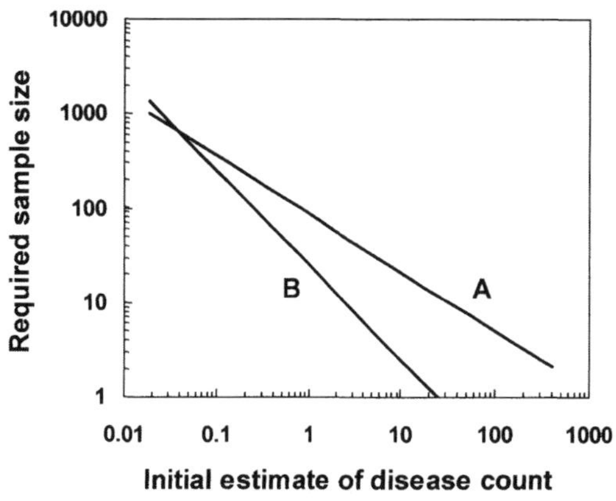

FIG. 10.2. Simple random sampling for count data. Required sample size (N) varies with the initial estimate of mean count per sampling unit $\bar{y}$, using $CV = 0.2$ as a definition of the required level of reliability. (**A**) based on Taylor's power law with $\hat{a} = 3.46$, $\hat{b} = 1.38$; (**B**) based on the Poisson distribution (see Table 10.2).

FIELD 10.2.1

0	5	1	3	5	1	4	2	8	4	5	7	4	7	2	3	7	3	6	6	3	8	4	5	8	4	5	5	5	5
5	6	6	1	1	5	8	4	4	2	4	3	9	4	1	1	4	2	5	5	3	4	10	6	2	4	6	8	3	5
3	5	5	12	9	6	8	5	5	3	4	11	5	8	8	5	9	5	7	5	4	9	7	3	6	4	4	7	6	6
1	7	7	6	5	7	8	5	4	7	5	9	8	9	6	6	7	7	3	8	4	6	2	8	9	6	6	9	4	4
4	3	4	2	3	12	3	6	6	6	7	8	10	5	4	6	7	6	9	3	7	5	5	1	6	2	4	5	7	5
12	3	4	4	5	4	2	7	6	3	7	5	5	5	7	6	3	7	5	8	7	2	6	4	8	5	4	8	4	7
2	5	2	10	2	7	7	2	6	4	6	5	4	6	6	6	5	2	6	6	8	4	6	5	7	4	9	10	8	7
9	3	6	7	5	10	5	7	8	6	6	3	6	3	5	5	7	3	4	4	3	8	7	6	7	1	8	4	11	6
6	6	9	3	3	2	5	5	5	8	3	4	7	8	7	8	3	10	7	5	7	3	7	7	7	2	7	4	3	6
6	5	11	3	8	1	3	2	3	3	5	9	8	4	10	5	8	5	5	6	5	9	5	5	9	9	3	8	6	3
6	0	7	3	7	7	4	10	5	2	9	8	10	6	3	7	5	2	9	5	5	6	9	6	6	7	4	7	5	7
10	2	2	8	10	9	3	3	5	2	6	5	10	5	3	7	10	4	8	5	9	2	8	3	3	4	1	8	8	6
4	4	10	8	4	8	8	5	9	4	2	3	10	5	6	9	6	10	6	11	5	4	7	7	9	6	11	4	3	1
6	7	3	10	4	5	3	1	2	7	6	8	2	4	2	7	5	7	5	3	3	1	3	4	4	7	8	5	4	6
3	5	9	8	4	5	5	5	9	7	6	9	9	2	4	5	6	6	5	3	7	1	5	3	9	7	2	7	6	4
6	7	3	6	1	7	7	10	3	9	3	3	3	9	4	3	9	4	6	6	4	1	10	4	4	8	10	7	5	8
7	4	10	2	4	10	5	3	5	3	6	9	5	3	3	4	9	7	7	5	2	4	6	6	4	9	3	7	8	6
3	7	6	4	6	10	8	6	10	8	6	1	6	9	7	3	5	4	3	4	7	3	6	7	4	9	9	5	7	6
9	4	0	8	8	11	5	6	8	7	6	10	5	6	6	7	6	4	5	6	1	2	3	3	7	4	5	3	7	3
10	2	8	4	7	7	3	4	6	8	6	7	7	5	4	7	2	8	8	4	1	3	5	0	7	7	1	3	5	9
5	4	6	7	3	3	4	5	5	1	6	2	5	6	8	2	7	11	5	5	7	3	3	8	14	6	3	1	5	8
5	5	4	4	6	5	3	5	7	7	9	6	4	8	7	7	11	5	4	3	4	5	8	3	1	6	9	3	6	5
7	4	10	8	3	5	9	4	4	2	6	4	5	5	5	7	4	7	7	10	5	5	3	3	5	4	3	11	6	9
6	6	3	2	7	4	4	5	7	6	5	8	5	7	3	6	5	7	4	5	4	10	6	3	6	4	9	2	9	4
3	2	5	2	3	5	3	7	4	7	2	4	3	2	8	10	9	5	9	3	3	6	7	5	2	2	5	6	4	2
6	9	7	6	11	7	4	8	3	8	5	5	3	6	3	9	4	6	7	2	6	12	4	4	4	4	7	8	11	4
2	11	3	4	4	5	2	2	5	5	6	4	9	7	2	5	5	5	6	10	6	6	6	9	3	13	6	2	1	6
4	5	10	1	5	5	2	9	4	3	8	6	5	4	6	6	9	10	3	5	7	6	3	8	3	4	9	5	4	4
4	5	6	6	4	6	5	5	2	4	4	7	6	8	5	9	7	9	5	10	3	2	7	3	4	7	2	7	7	6
5	5	5	5	8	9	6	6	5	2	6	5	4	8	3	6	10	4	4	4	3	5	4	7	4	10	11	4	8	11

FIELD 10.2.2

2	1	4	3	4	8	7	1	12	5	3	5	5	7	10	8	12	8	1	5	6	12	8	4	2	10	1	6	1	2
5	4	5	6	6	7	6	5	7	17	2	3	3	4	15	0	7	1	2	4	3	11	8	5	6	16	1	4	7	1
13	0	9	5	6	3	4	16	3	9	14	7	3	4	4	1	3	5	7	11	9	2	2	4	11	0	8	7	6	2
5	6	4	6	7	0	5	6	2	5	5	1	6	6	1	4	1	2	1	8	9	0	2	8	6	2	6	8	3	3
4	11	3	14	10	7	5	7	1	4	8	7	2	7	4	1	6	8	1	5	5	11	2	2	2	3	10	7	5	5
3	3	2	4	8		17	9	4	7	6	5	4	9	8	7	1	4	6	2	2	6	3	6	10	8	6	10	5	6
4	12	8	6	5	0	9	5	4	3	0	1	2	10	5	5	5	8	7	5	5	2	6	8	15	0	5	3	19	8
4	12	4	14	8	1	3	8	9	1	17	7	5	8	17	2	0	21	6	6	12	2	3	2	1	7	4	11	7	7
8	1	10	2	9	9	1	5	4	7	3	3	8	9	7	2	8	5	4	8	2	6	1	7	6	10	1	1	1	5
7	7	7	8	2	9	7	0	8	1	8	4	4	7	9	11	2	5	8	8	5	0	2	2	2	4	5	7	3	5
8	16	7	2	2	12	2	7	1	5	0	2	2	5	6	7	3	2	7	4	1	5	4	6	4	5	10	10	4	2
23	5	1	2	7	11	8	15	4	3	4	4	5	4	1	6	2	12	4	8	5	5	2	7	4	10	4	8	11	1
1	10	2	0	3	5	2	2	0	4	2	2	12	5	2	1	3	1	4	17	4	1	4	5	9	6	5	8	6	2
5	5	3	2	4	14	1	1	3	10	0	3	2	12	19	7	5	6	5	12	0	2	8	17	10	12	2	1	7	5
4	5	5	10	0	8	4	2	8	6	3	0	19	12	2	4	3	2	5	10	5	3	1	4	0	5	5	1	1	0
1	1	2	3	4	1	3	16	10	2	8	3	3	9	5	10	4	1	6	4	2	2	2	3	4	7	5	5	7	9
6	7	7	3	9	5	14	2	18	4	0	4	5	4	2	8	3	21	3	1	7	6	7	4	2	0	7	4	2	6
9	7	4	10	1	6	8	0	5	5	3	7	3	9	2	2	7	4	7	8	10	2	9	8	13	10	4	14	3	3
0	9	4	6	5	6	5	1	1	8	7	1	15	6	8	13	13	13	4	3	4	3	10	13	2	2	8	2	1	12
9	3	5	3	9	10	11	5	9	7	8	5	3	6	8	6	7	4	14	3	22	6	3	9	5	4	1	1	8	12
5	1	3	6	3	3	0	6	12	5	8	3	12	7	2	2	6	4	1	6	9	1	13	4	10	1	14	3	3	16
5	1	2	8	4	6	2	3	6	6	0	10	10	4	2	3	4	3	5	0	5	5	3	7	8	0	20	5	8	5
3	1	9	6	9	2	5	12	7	4	7	2	4	4	2	9	4	7	1	3	5	8	8	0	4	5	13	3	5	1
6	2	1	9	8	2	9	4	9	7	2	11	3	11	14	2	15	2	1	2	10	3	10	6	2	4	4	10	6	5
3	4	3	17	5	3	15	7	3	9	28	5	5	8	1	5	6	6	1	3	5	3	4	6	13	9	4	8	5	4

(Continued)

FIELD **10.2.2** Continued.

2	13	1	7	10	4	8	11	4	10	6	8	1	5	7	0	7	2	4	8	10	10	5	10	3	6	6	17	7	6
3	8	3	3	9	2	8	15	11	6	3	5	4	7	4	2	5	1	2	1	4	3	2	6	1	0	7	1	4	2
1	8	8	6	8	15	10	3	6	9	3	5	5	11	6	7	3	0	9	5	0	5	2	7	11	4	15	3	11	9
1	13	4	7	8	4	7	0	4	5	1	5	6	11	4	2	9	6	2	2	12	4	12	9	1	14	4	2	5	2
6	11	2	7	3	5	7	1	3	6	6	7	12	9	12	7	6	2	6	6	12	7	3	4	4	1	7	5	3	11

disease-infected crops. Initially, a random sample of 20 plants was recorded from each field (Samples 10.2.1 and 10.2.2a, respectively). Later, a second sample (Sample 10.2.2b) was recorded from Field 10.2.2. Interested readers can repeat the following calculations, as an exercise, using their own samples drawn from Fields 10.2.1 and 10.2.2.

The following summary statistics can be calculated (section 9.5.1):

$$\text{Sample 10.2.1:} \quad \bar{Y} = 5.65, \quad s_Y^2 = 6.13,$$
$$\text{Sample 10.2.2a:} \quad \bar{Y} = 6.40, \quad s_Y^2 = 17.09.$$

For Sample 10.2.1, the estimated standard error of $\bar{Y}$ is $\sqrt{6.13/20} = 0.55$ and the coefficient of variation of the sample mean, $CV \approx 10\%$. For Sample 10.2.2a, the estimated standard error of $\bar{Y}$ is $\sqrt{17.09/20} = 0.92$ and $CV \approx 14\%$. Although the estimates of mean lesion number per plant are similar for the two fields, the two means have been estimated with different levels of reliability.

To understand why this has happened, look again at the arrays representing the two fields. In Field 10.2.1, relatively few plants have lesion numbers very far from the mean. In Field 10.2.2, there are more plants with few lesions, and more with large numbers of lesions. This is an indication that the pattern of lesion number per plant in Field 10.2.2 is more aggregated (at the spatial scale at which the data are shown, i.e., the plant scale) than the pattern in Field 10.2.1. This affects the amount of sampling required to achieve a particular specified level of reliability. To see how this works, consider the two samples of 20 plants (Samples 10.2.1 and 10.2.2a), to be pilot samples that allow us to obtain some information that can be used in subsequent sampling. Since, for Field 10.2.1, the variance of the sample (6.13) is close to the mean (5.65), we could reasonably make the assumption that the Poisson distribution was an appropriate description of the frequency distribution of lesions per plant. For Field 10.2.2, where the variance of the sample (17.09) is larger than the mean (6.40), we could instead assume a negative binomial distribution for the frequency distribution of lesions per plant, with $\hat{k} = \bar{Y}^2/(s_Y^2 - \bar{Y}) \approx 4$.

Suppose $CV = 0.1$ (or 10%) is set as the required level of reliability for the sample mean. The formulae for sample size calculations are given in Table 10.2. For Field 10.2.1, the required number of sampling units (plants, in this case) is $N = 1/(5.65 \times 0.1^2) \approx 18$. As expected, this is quite close to the value $N = 20$ used for the pilot sample, which provided a value of $CV \approx 0.1$.

For Field 10.2.2, the required number of sampling units is $N = ((1/6.4) + (1/4))(1/0.1^2) \approx 40$. Recall that aggregation increases with *decreasing* $\hat{k}$ so that if the value $\hat{k} = 4$ were too large (relative to the true value), this calculated sample size would be too small. Conversely, if $\hat{k} = 4$ were too small, the calculated sample size would be too large.

Here, we see that even for a quite a moderate level of aggregation (as characterized by the negative binomial $\hat{k}$), the required number of sampling units is approximately double than that for a field with the same mean number of lesions per plant but a random pattern, if the same level of reliability ($CV = 0.1$) is to be achieved. A random sample of $N = 40$ plants (Sample 10.2.2b) was recorded from Field 10.2.2. For Sample 10.2.2b, the summary statistics are $\bar{Y} = 6.60$, $s_Y^2 = 19.37$. The estimated standard error of $\bar{Y}$ is $\sqrt{19.37/40} = 0.70$ and the sample provides the required level of reliability, with $CV \approx 0.1$.

10.4.5 Exact Poisson confidence intervals

The simplest approximate Poisson confidence interval, $\bar{Y} \pm z_{\alpha/2}\sqrt{\bar{Y}/N}$, as used in sample size formulae (section 10.4.1), has the same potential deficiencies as the approximate binomial confidence interval (section 10.3.1). For example, if a single lesion is observed in a sample of $N = 10$ plants (a plant is a sampling unit), the approximate 95% confidence interval for the mean ($\bar{Y} = 0.1$ lesions per plant) extends from -0.096 to 0.296. This is calculated using $z_{\alpha/2} = 1.96$; if the corresponding value from the *t*-distribution with 9 *df* (=2.26) were used instead, the resulting interval would be -0.126 to 0.326. This confidence interval is unsatisfactory, since the true mean count cannot be negative. Garwood (1936) derives "exact" Poisson confidence intervals analogous to the Clopper–Pearson binomial confidence intervals. Tables are provided by Fisher and Yates (1963) and Diem and Lentner (1970).

The tables refer to the *total* number of counts in the *N* sampling units. In this example, the total number of lesions on $N = 10$ plants is equal to one. The Clopper–Pearson-type 95% confidence interval is given as 0.025 to 5.57. This can be converted to an interval around the mean by dividing these limits by 10 (in this case). Thus, the Clopper–Pearson-type 95% confidence interval for the mean ($\bar{Y} = 0.1$ lesions per plant) extends from 0.0025 to 0.557. The lower limit is positive, as required, and the upper limit is considerably larger than that given by the approximate normal formula (even if the *t*-distribution is used to calculate the interval). Note that the interval is asymmetric about the mean. As in the case of the binomial distribution, Clopper–Pearson-type intervals are conservative. An alternative (less conservative) interpretation of "exact" Poisson confidence intervals is provided by Crow and Gardner (1959). For larger samples, and where the mean count is not close to zero, the approximate confidence interval based on the normal distribution is usually adequate.

Now consider a sample of N plants, in which the total number of lesion counts is zero (i.e., no disease is observed). As for disease incidence, a one-sided $100(1 - \alpha)\%$ interval can be calculated, for which the lower limit is zero and the upper limit is denoted p_U. In this case the upper limit is determined such that $\Pr(Y = 0) = \alpha = e^{-Np_U}$. This leads directly (without the need to use a Taylor series expansion) to $p_U = -\ln(\alpha)/N$, the formula on which the Rule of Three and various other approximations for p_U are based (section 10.3.3). A one-sided 95% Clopper-Pearson-type confidence interval (i.e., with $\alpha = 0.05$) for the mean number of lesions per sampling unit in a sample, size N, in which no disease is observed thus extends from 0 to $p_U \approx 3/N$. For example, no lesions are recorded in a sample of $N = 10$ plants (so that the mean is $\bar{Y} = 0$ lesions per plant). The Clopper–Pearson-type 95% confidence interval for the mean extends from 0 to $p_U \approx 3/N = 0.333$. Note that the Clopper–Pearson-type 95% confidence interval for the total number of lesions in the 10 sampling units (no lesions having been observed) extends from 0 to $Np_U = -\ln(\alpha) = 2.996 \approx 3$.

Tables for "exact" negative binomial confidence intervals are not readily available. The calculations required to obtain an interval based on tail areas of the negative binomial distribution are outlined by Gozé et al. (2003). Alternatively, Krebs (1999) provides a method (not based in any obvious way on the calculation of tail areas) using rather complicated transformations of estimated parameters.

For a sample, size N, in which no disease is observed, a one-sided $100(1 - \alpha)\%$ interval can be calculated, for which the lower limit is zero and the upper limit is denoted p_U, in this case such that $\Pr(Y = 0) = \alpha = (1 + \mu/k)^{-k}$ with $\mu = Np_U$. This leads to $p_U = k(\alpha^{-1/k} - 1)/N$. Of course, when no disease has been observed, an estimate of k cannot be calculated from the sample data. Instead, one would have to assume a value, or use a value from a previous sample. Note that:

$$\lim_{k \to \infty} \frac{k(\alpha^{-1/k} - 1)}{N} = -\frac{\ln(\alpha)}{N},$$

so that if k is thought to be large, the Poisson formula for p_U is available as an approximation.

10.5 Cluster Sampling for Disease Incidence Data

In cluster sampling, a sampling unit comprises a group (i.e., cluster) of neighboring plants or plant parts. A disease assessment of leaves on a plant, or of plants in a quadrat, are typical cluster sampling procedures. This is often convenient from a practical point of view.

In Chapter 9, the binomial distribution was used to characterize cluster-sampling data when the disease status of one individual in a sampling unit was independent of the status of the other individuals in the same unit (section 9.4.2). For plant disease incidence data collected by cluster sampling, it is often the case that the disease status of one individual in a sampling unit depends of the status of the other individuals in the same unit, such that there is a tendency for individuals in the same unit to have the same disease status. We see this in data as extra-binomial variation, where the variance of the observations is larger than the corresponding binomial variance (section 9.4.2). In a phytopathological context, this is often referred to as aggregation. Methods of characterizing aggregation, as discussed in Chapter 9 (sections 9.4.3–9.4.7) and in Madden and Hughes (1999a), provide a basis for developing sample size formulae for disease incidence data collected by cluster sampling (taking into account the extent to which the disease status of one member of a sampling unit depends on the status of other members of the same unit).

10.5.1 The binomial distribution

Suppose that a sample comprises N quadrats (sampling units), in each of which there are n plants. In each quadrat, all n plants are assessed, and recorded individually as either "healthy" or "diseased". When the binomial distribution provides an appropriate description of the frequency distribution of number of disease plants per quadrat, the sample mean disease incidence on a proportion scale is denoted $\bar{y}$ (equation 9.1) and the variance of the proportions (the y_i values) is estimated by $s_{bin}^2 = \bar{y}(1-\bar{y})/n$ (see equation 9.4).

The estimated standard error of the sample mean is $\sqrt{\bar{y}(1-\bar{y})/nN}$. If reliability is defined by the coefficient of variation of the sample mean, CV:

$$CV = \frac{\sqrt{(\bar{y}(1-\bar{y})/nN)}}{\bar{y}} \qquad (10.16)$$

If reliability is defined by setting the half the length of the required confidence interval of the sample mean equal to a fixed proportion of the mean, H:

$$H\bar{y} = z_{\alpha/2}\sqrt{\frac{\bar{y}(1-\bar{y})}{nN}} \qquad (10.17)$$

If reliability is defined by setting the half the length of the required confidence interval of the sample mean equal to a fixed positive number h:

$$h = z_{\alpha/2}\sqrt{\frac{\bar{y}(1-\bar{y})}{nN}} \qquad (10.18)$$

Table 10.3 shows each of equations 10.16–10.18 rearranged so as to provide an expression for N, the sample size. Sample size calculation requires choice of a definition of reliability and an initial estimate of mean disease incidence, $\bar{y}$, which fully specifies the variance. The binomial distribution is the appropriate description of variability when the disease status of one individual in a sampling unit is independent of the disease status of the other members of the same unit. This will rarely be the case. More often, there is a tendency for individuals in the same sampling unit to have the same disease status. Kish's (1957) account (although not phytopathologically orientated) gives a very clear view of the consequences of ignoring this tendency in a sampling context.

10.5.2 The β-binomial distribution

The β-binomial distribution may be used to characterize aggregated disease incidence data (see section 9.4.5). Suppose a sample comprises N quadrats, in each of which there are n plants. In each quadrat, all n plants are assessed, and recorded individually as either "healthy" or "diseased". When the β-binomial distribution provides an appropriate description of the frequency distribution of number of disease plants per quadrat, the sample mean disease incidence on a proportion scale is denoted $\bar{y}$ and the estimated aggregation parameter is denoted $\hat{\rho}\ (=\hat{\theta}/(\hat{\theta}+1))$. The variance of the proportions (the y_i values) is estimated by $s^2_{\beta\text{bin}} = (\bar{y}\ (1-\bar{y})/n)(1+\hat{\rho}(n-1))$ (see equation 9.11).

The estimated standard error of the sample mean is $\sqrt{\bar{y}\ (1-\bar{y})(1+\hat{\rho}\ (n-1))\ /\ nN}$. If reliability is defined by the coefficient of variation of the sample mean, CV:

$$CV = \frac{\sqrt{\bar{y}(1-\bar{y})(1+\hat{\rho}(n-1))/nN}}{\bar{y}} \qquad (10.19)$$

If reliability is defined by setting the half the length of the required confidence interval of the sample mean equal to a fixed proportion of the mean, H:

$$H\bar{y} = z_{\alpha/2}\sqrt{\frac{\bar{y}\ (1-\bar{y})(1+\hat{\rho}(n-1))}{nN}} \qquad (10.20)$$

If reliability is defined by setting the half the length of the required confidence interval of the sample mean equal to a fixed positive number h:

$$h = z_{\alpha/2}\sqrt{\frac{\bar{y}\ (1-\bar{y})(1+\hat{\rho}(n-1)}{nN}} \qquad (10.21)$$

We can think of $1+\hat{\rho}(n-1)$ as a β-binomial heterogeneity factor. Alternatively, the index of dispersion, s^2_y/s^2_{bin}, calculated from a preliminary sample, may be used as an empirical heterogeneity factor. This is a version of the deff (section 9.4.6), as defined by Kish (1995). To a good approximation, deff $= 1 + \hat{\rho}(n-1)$ for cluster-sampling data described by the β-binomial distribution. In this case, sample size calculation requires choice of a definition of reliability, and initial estimates the mean disease incidence, $\bar{y}$, and aggregation as characterized by the β-binomial $\hat{\rho}$ or the deff. Table 10.3 shows each of equations 10.19–10.21 rearranged so as to provide an expression for N, the sample size. The equations are written in terms of the deff because, in practice, it is more likely that a value of the deff will be available than a value of $\hat{\rho}$.

The β-binomial $\hat{\rho}$ usually varies with mean disease incidence, $\bar{y}$. Use of the β-binomial $\hat{\rho}$ from a single disease assessment may therefore be misleading if applied over a wide range of mean incidence. Instead, a power law relationship between the observed and binomial variances is likely to provide a better guide to the required sample size.

TABLE 10.3. Formulae for sample size calculations for disease incidence data collected by cluater simpling, with three different definitions of reliability[a].

	Reliability defined by:		
		Half-width of confidence interval equal to:	
Description of variability	*Coefficient of variation, CV*	*Proportion of mean, H*	*Fixed positive number, h*
Random (binomial distribution)	$N=(1-\bar{y})/(n\bar{y}CV^2)$	$N=[(1-\bar{y})/n\bar{y}](z_{\alpha/2}/H)^2$	$N=[(\bar{y}(1-\bar{y}))/n](z_{\alpha/2}/h)^2$
Aggregated (β-binomial distribution or deff)[b]	$N=[(1-\bar{y})(\text{deff})]/(n\bar{y}CV^2)$	$N=[((1-\bar{y})(\text{deff}))/n\bar{y}](z_{\alpha/2}/H)^2$	$N=[(\bar{y}(1-\bar{y})(\text{deff}))/n](z_{\alpha/2}/h)^2$
Aggregated (power law)	$N=(\hat{A}y^{\hat{b}-2}(1-\bar{y})^{\hat{b}})/(n^b CV^2)$	$N=[(\hat{A}\bar{y}^{\hat{b}-2}(1-\bar{y})^{\hat{b}})/n^b](z_{\alpha/2}/H)^2$	$N=[(\hat{A}(\bar{y}[1-\bar{y}])^{\hat{b}})/n^b](z_{\alpha/2}/h)^2$

[a]N is the number of sampling units, each containing n individuals, required for a pre-specified level of reliability, $\bar{y}$ is an initial estimate of disease incidence, and $z_{\alpha/2}$ is the $100(1-\alpha/2)$th percentile of the standard normal distribution. For aggregated data, an estimate of either the beta-binomial aggregation parameter $\hat{\rho}=\hat{\theta}/(\hat{\theta}+1))$ or the power law parameters $(\hat{A},\hat{b})$, are required. The formulae omit the finite population correction.

[b]deff $= 1 + \hat{\rho}\ (n-1)$.

Such a relationship can be regarded as a β-binomial sampling curve, but one in which $\hat{p}$ varies with mean incidence. More work is required, initially, to establish such a relationship, but resulting formula for sample size determination has the advantage of being equally applicable over a wide range of disease incidence.

10.5.3 The power law

Just as Taylor's power law provides a means of summarizing variability in count data from several data sets, an analogous relationship between the observed variance of the y_i values and the corresponding binomial variance may be used to summarize variability in several disease incidence data sets (section 9.4.7). The variance of the y_i values is estimated by $s_y^2 = A(s_{\text{bin}}^2)^b$ (see equation 9.17). Estimated values of the coefficients must be obtained from a previous data analysis. If this power law relationship is calculated from samples comprising quadrats (sampling units) each containing n plants, the estimated standard error of a sample mean based on N quadrats is $\sqrt{(\hat{A}(\bar{y}(1-\bar{y}))^{\hat{b}})/n^{\hat{b}}N}$.

If reliability is defined by the coefficient of variation of the sample mean, CV:

$$CV = \frac{\sqrt{\hat{A}(\bar{y}(1-\bar{y}))^{\hat{b}}/n^{\hat{b}}N}}{\bar{y}} \quad (10.22)$$

If reliability is defined by setting the half the length of the required confidence interval of the sample mean equal to a fixed proportion of the mean, H:

$$H\bar{y} = z_{\alpha/2}\sqrt{\frac{\hat{A}(\bar{y}(1-\bar{y}))^{\hat{b}}}{n^{\hat{b}}N}} \quad (10.23)$$

If reliability is defined by setting the half the length of the required confidence interval of the sample mean equal to a fixed positive number h:

$$h = z_{\alpha/2}\sqrt{\frac{\hat{A}(\bar{y}(1-\bar{y}))^{\hat{b}}}{n^{\hat{b}}N}} \quad (10.24)$$

Table 10.3 shows each of equations 10.22–10.24 rearranged so as to provide an expression for N, the sample size. Sample size calculation requires choice of a definition of reliability, a previously calculated power law relationship for the pathosystem in question and an initial estimate of mean disease incidence, $\bar{y}$. With equations 10.22 and 10.23, caution should be exercised in adopting values $\hat{b} > 2$ as the basis for sample size calculations (although such values are, in practice, uncommon).

10.5.4 Sample size calculations

The sample size formulae in Table 10.3 show N, the required number of sampling units, as a function of the estimated mean and (where appropriate) aggregation parameters, as well as the size of the sampling unit, n. The total number of individuals to be assessed to meet the required level of reliability is nN. The assumption made in presenting the formulae with N on the left-hand side is that any parameters needed for the sample size calculation will have been estimated using sampling units size n. Thus, the issue of dependence of parameter values on the size of the sampling unit is circumvented.

Generally, the number of sampling units for a required level of reliability falls as mean disease incidence increases, and is larger when there is an aggregated pattern of disease incidence. Fig. 10.3 shows how required sample size N varies with disease incidence $\bar{y}$ when the powerlaw relationship $\log_{10}(s_y^2) = 1.15 + 1.35\log_{10}(s_{\text{bin}}^2)$ (see Fig. 9.4C) provides a description of variability, using $CV = 0.2$ as a definition of the required level of reliability. For comparison, the variation of N with $\bar{y}$ when the corresponding binomial distribution provides a description of variability is shown, again using $CV = 0.2$. In both cases, N decreases with increasing $\bar{y}$, but N is higher for the power law (aggregated data) than for the binomial distribution (random data) as a description of variability, except at very low incidence ($\bar{y} < 0.01$).

Example 10.3. The arrays labeled Field 10.3.1 and Field 10.3.2 represent counts of diseased plants per quadrat (out of a total of $n = 9$) in disease-infected crops. Initially, a random sample of 25 quadrats was recorded from each field (Samples 10.3.1 and 10.3.2a, respectively). Later, a second sample (Sample 10.3.2b) was recorded from Field 10.3.2. Interested readers can repeat the following calculations, as an exercise, using their own samples drawn from Fields 10.3.1 and 10.3.2.

Number of diseased plants	*Frequency (number of quadrats)*		
	Sample 10.3.1 (N = 25)	*Sample 10.3.2a (N = 25)*	*Sample 10.3.2b (N = 56)*
0	1	4	10
1	6	4	12
2	4	7	6
3	7	1	10
4	5	4	7
5	1	4	7
6	1	1	4
7	0	0	0
8	0	0	0
9	0	0	0

The following summary statistics can be calculated (sections 9.4.1 and 9.4.2):

Sample 10.3.1: $\bar{y} = 0.293$, $s_y^2 = 0.0266$, $s_{\text{bin}}^2 = 0.0230$
Sample 10.3.2a: $\bar{y} = 0.280$, $s_y^2 = 0.0423$, $s_{\text{bin}}^2 = 0.0224$

FIELD 10.3.1

3	3	2	3	2	5	2	2	3	3	1	2
3	1	2	3	3	3	2	2	5	3	1	2
1	4	3	2	3	5	3	4	3	3	3	4
1	2	3	3	1	3	2	5	1	2	3	3
5	2	4	2	3	2	2	2	1	4	4	1
3	1	5	2	1	2	2	5	1	5	4	4
0	0	3	1	1	2	4	3	3	4	0	3
0	3	1	3	4	0	1	4	3	3	3	2
1	2	5	2	2	1	4	1	2	3	3	2
2	3	3	3	0	3	4	2	1	3	5	4
1	1	3	3	2	3	1	4	3	1	1	1
4	2	2	3	2	5	4	3	2	4	4	3
2	1	0	2	5	5	2	2	2	5	4	1
2	2	5	2	3	2	2	2	2	3	4	3
3	3	4	1	1	2	5	4	6	1	2	2
2	2	3	5	4	3	2	2	0	5	2	5
1	5	5	4	2	5	4	3	3	1	2	4
3	3	5	1	2	4	3	6	3	2	2	1
4	1	2	3	2	3	3	5	2	1	0	2
2	3	4	5	3	1	1	0	3	3	3	1
4	2	1	3	3	2	3	4	4	1	4	1
3	3	3	5	2	3	4	1	2	6	3	3
1	3	2	3	2	3	1	2	3	1	3	3
1	2	3	3	4	4	3	2	3	2	3	2
2	3	3	2	3	3	3	4	3	3	3	2
2	4	5	3	4	2	3	3	3	2	2	3
5	5	3	0	2	5	3	0	4	3	1	5
2	3	3	2	3	4	2	4	3	5	5	1
3	2	4	4	3	5	3	2	1	1	5	2
4	1	2	3	2	1	3	4	3	1	1	4
3	2	3	4	1	1	4	4	1	4	2	2
5	1	0	5	3	2	3	4	3	3	3	1
0	4	1	0	1	6	3	1	3	4	4	4
1	1	4	3	3	1	1	4	3	3	5	3
3	4	3	6	4	6	1	4	0	3	2	2
4	2	5	6	0	2	0	3	2	5	6	2
1	3	2	3	6	1	3	1	3	2	3	4
4	2	1	3	2	4	2	1	2	4	1	2
2	1	2	5	3	3	5	0	3	2	2	1
2	4	3	2	1	1	4	3	1	2	3	3
5	4	6	3	5	3	2	4	4	1	4	6
1	2	2	6	3	2	4	4	3	1	4	3
4	4	4	2	0	2	2	4	2	4	5	1
2	2	3	2	3	6	1	2	5	2	4	3
1	4	3	2	2	1	4	3	3	3	2	6
0	6	3	3	1	2	4	3	2	4	2	2
5	2	1	3	3	4	5	4	2	2	2	3
2	4	3	2	3	2	3	1	2	3	2	2
5	3	3	2	2	1	1	4	4	4	3	2
4	3	3	4	3	4	1	4	0	1	5	2

FIELD 10.3.2

2	0	5	5	5	1	4	3	2	2	1	3
4	4	0	7	3	3	1	2	5	3	4	1
1	2	5	5	0	1	0	2	1	1	1	5
0	1	2	0	0	6	1	3	3	1	4	0
5	4	0	1	5	0	0	5	0	5	0	8
3	1	1	2	0	5	1	2	3	0	8	1
7	3	4	1	0	2	3	5	2	0	1	1
9	8	2	6	2	4	4	1	2	3	1	4
1	1	2	2	4	4	4	6	4	0	1	3
6	3	3	4	4	0	6	3	1	2	0	3
1	2	4	4	0	6	3	2	4	1	3	2
4	4	2	4	4	0	1	0	2	3	3	5
1	1	3	6	0	1	1	2	4	1	2	5
2	4	0	1	7	2	5	1	3	5	6	2
4	2	2	4	2	4	7	6	6	6	4	6
1	3	2	3	7	2	6	2	2	1	2	4
0	4	1	2	4	1	2	0	0	5	3	2
1	4	4	2	7	2	2	2	0	1	2	0
0	2	6	1	2	4	3	3	5	0	1	4
2	5	6	4	2	1	1	2	3	5	1	2
1	4	0	0	6	5	3	4	1	5	3	5
0	2	3	4	2	0	1	4	5	3	6	3
1	0	1	7	4	1	2	4	4	0	6	7
3	7	3	1	8	3	2	3	7	0	0	1
1	0	1	1	4	5	2	4	2	7	0	2
1	4	4	0	5	1	3	0	5	4	2	0
3	2	3	2	2	0	1	5	2	0	8	1
1	1	2	5	2	4	6	6	4	6	1	3
4	0	8	1	2	2	2	5	1	2	6	7
7	0	0	1	2	0	6	5	3	4	2	3
5	1	0	3	5	2	0	2	7	3	5	4
1	0	1	1	1	0	6	5	2	3	4	1
2	4	3	0	3	1	5	1	2	4	3	6
3	1	5	2	2	3	0	1	2	1	3	2
4	3	1	1	5	1	3	2	3	3	1	4
6	0	2	6	2	2	2	7	6	2	0	1
0	3	2	4	1	2	4	4	2	2	2	0
2	1	3	1	7	2	2	1	2	3	7	3
2	2	0	4	6	1	6	3	3	5	4	0
3	0	0	3	0	0	1	1	6	0	2	2
2	5	1	5	5	1	5	0	1	1	2	3
1	2	2	2	1	6	5	3	2	2	2	4
2	1	3	4	2	8	0	2	0	1	3	0
5	1	0	2	6	4	3	2	3	1	1	3
0	4	0	5	3	2	5	5	3	0	2	2
9	5	2	2	1	1	0	3	4	1	5	2
2	4	2	6	2	5	2	4	1	2	4	2
0	0	0	2	3	1	0	2	4	2	4	4
5	3	5	3	5	4	3	2	6	0	6	6
1	4	5	2	2	7	5	1	1	2	0	1

From Sample 10.3.1, the index of dispersion $D = 1.16$ and χ^2 (24 *df*) = 27.74 ($P = 0.27$). This is evidence in support of the view that in Field 10.3.1, the pattern of diseased plants at the quadrat scale is indistinguishable from random (section 9.4.3). On this basis, a sample size (number of quadrats) to achieve a reliability defined by the coefficient of variation of the sample mean of $CV = 0.1$ is given by $N = (1 - 0.293)/(9 \times 0.293 \times 0.1^2) \approx 27$ (Table 10.3). This is quite close to chosen N for Sample 10.3.1, and the value of CV for this sample is $CV = \sqrt{0.0266/25}/0.293 \approx 0.1$ (equation 10.16), as required.

It is clear from the summary statistics calculated from Sample 10.3.2a that a different approach will be required for Field 10.3.2. The index of dispersion $D = 1.89$

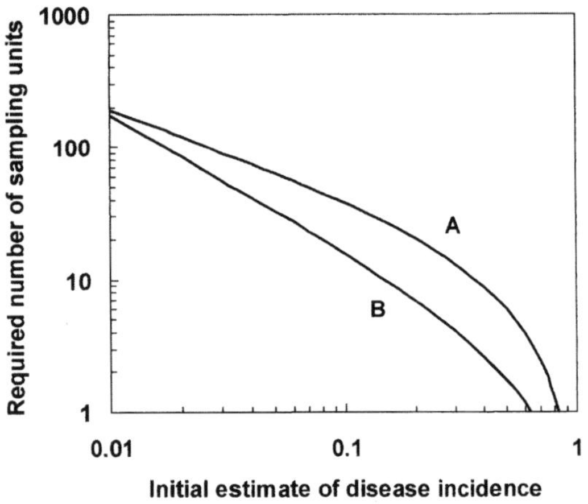

FIG. 10.3. Cluster sampling for disease incidence data. Required number of sampling units (N) varies with the initial estimate of disease incidence ($\bar{y}$), using $CV = 0.2$ as a definition of the required level of reliability. The mean number of individuals per sampling unit is $\bar{n} = 14.35$. (A) based on the power law with $\hat{A} = 10^{1.15} = 14.13$, $\hat{b} = 1.35$ (B) based on the binomial distribution (see Table 10.3).

and χ^2 (24 df) = 45.3 (P = 0.005). This is evidence in support of the view that the pattern of diseased plants at the quadrat scale is aggregated. A previously-calculated power law relationship for the pathosystem in question is available, $s_y^2 = A(s_{\text{bin}}^2)^b$, with $\hat{A} = 9.0$, $\hat{b} = 1.4$. With these values, and the initial estimate of mean disease incidence, $\bar{y} = 0.280$, a sample size (number of quadrats) to achieve a reliability defined by the coefficient of variation of the sample mean of $CV = 0.1$ is given by $N = (9.0 \times 0.280^{-0.6} \times (1-0.280)^{1.4})/(9^{1.4} \times 0.1^2) \approx 56$ (Table 10.3), about double the size of the initial sample. Sample 10.3.2b consists of a random sample of 56 quadrats recorded from Field 10.3.2. For this sample, the following summary statistics can be calculated: $\bar{y} = 0.280$, $s_y^2 = 0.0453$. Then the value of CV for this sample is $CV = \sqrt{(0.0453/56)}/0.280 \approx 0.1$, as required.

If the information from the previously calculated power law relationship was not available, an alternative would be to use the index of dispersion (or deff) calculated from Sample 10.3.2a, $s_y^2 / s_{\text{bin}}^2 = 1.89$, as an empirical heterogeneity factor. For cluster-sampling data described by the β-binomial distribution, deff $= 1 + \hat{\rho}(n-1)$, at least to a good approximation. Then, using the appropriate sample size formula from Table 10.3, $N = ((1 - 0.28) \times 1.89)/(9 \times 0.28 \times 0.01^2) \approx 54$. Thus a sample size that will meet the adopted reliability requirement is calculated.

10.5.5 Exact confidence intervals for cluster-sampling data

Wypij and Santner (1990) discuss the calculation of both Clopper–Pearson-type "exact" confidence intervals and a less conservative alternative, for β-binomial data. Wypij and Santner's BBDCI3 computer program was used to calculate Clopper–Pearson-type 95% confidence intervals for some examples of cluster-sampling data given previously in Chapter 9. For comparison, the corresponding approximate normal 95% confidence intervals were calculated. The results are shown in Table 10.4.

When the binomial distribution provides an appropriate description of variability (e.g., Bald's TSWV data from Table 9.1), the interval calculated by Wypij and Santner's BBDCI3 computer program is the "exact" Clopper–Pearson interval for a sample size $n \times N$, which may be obtained directly from statistical tables. The number of sampling units inspected (N) is large enough here for use of the approximate normal confidence intervals to be considered reasonable, and the approximate normal confidence intervals are indeed similar to the "exact" intervals (Table 10.4).

When the β-binomial distribution provides an appropriate description of variability, aggregation is relatively low, the number of sampling units is large and disease incidence is not too low (TSWV data from *Example 9.3*), the Clopper–Pearson-type interval calculated using Wypij and Santner's BBDCI3 computer program and the approximate normal interval are again similar (Table 10.4). However, when the β-binomial distribution provides an appropriate description of variability but aggregation is higher (dogwood anthracnose data from Table 9.2), there are noticeable discrepancies between the Clopper–Pearson-type intervals and the approximate normal intervals, in spite of the large number of sampling units and disease incidence not too low. In passing, note that the confidence intervals for mean dogwood

TABLE 10.4. Summary statistics and "exact" Clopper–Pearson-type and approximate normal confidence intervals for various disease incidence data sets.

Source	$\bar{y}$	$\hat{\rho}$	*n*	N	*C-P-type 95% interval*	*Approximate 95% interval*
TSWV data from Table 9.1						
BHP-OHS	0.881	–	9	40	0.843, 0.913	0.848, 0.914
BP-T	0.933	–	9	40	0.902, 0.957	0.907, 0.959
EDR-OHS	0.633	–	9	40	0.581, 0.683	0.583, 0.683
EDR-T	0.739	–	9	40	0.690, 0.783	0.694, 0.784
Dogwood anthracnose data from Table 9.2						
1990	0.153	0.343	10	168	0.126, 0.198	0.118, 0.188
1991	0.299	0.499	10	161	0.259, 0.365	0.247, 0.351
TSWV data from Example 9.3						
	0.181	0.050	9	160	0.158, 0.206	0.157, 0.205

anthracnose disease incidence for 1990 and 1991 do not overlap. This is consistent with Zarnoch et al. (1995), who showed that there was a difference between the means for the two years (while noting that this finding was subject to the caveat that the data for the two years were not independent).

Chen and Tipping (2002) suggest an alternative procedure for obtaining "exact" confidence intervals for aggregated cluster-sampling data, using the index of dispersion as an empirical heterogeneity factor. From the data collected, the following quantities are calculated: the index of dispersion (or deff) $D = s_y^2 / s_{\text{bin}}^2$,the total number of diseased elements in all sampling units inspected $\sum Y$, and the total number of elements in all sampling units inspected, $n \times N$. When the index of dispersion $D = 1$, the interval calculated using Chen and Tipping's (2002) procedure is the "exact" Clopper–Pearson interval for a sample size $n \times N$, which may be obtained directly from statistical tables.

The lower endpoint of the required $100(1 - \alpha)\%$ confidence interval is then calculated from the $\alpha / 2$ point of the inverse cumulative beta distribution with parameters

$$\left(\frac{\sum Y}{D}, \frac{nN - \sum Y}{D} + 1 \right)$$

The upper endpoint of the interval is calculated from the $1 - \alpha / 2$ point of the inverse cumulative beta distribution with parameters

$$\left(\frac{\sum Y}{D} + 1, \frac{nN - \sum Y}{D} \right).$$

The calculations are facilitated by mathematical software such as MATHCAD. Table 10.5 shows intervals calculated using Chen and Tipping's (2002) procedure for the dogwood anthracnose data from Table 9.2 (Zarnoch et al., 1995) and for the TSWV data from *Example 9.3* (Cochran, 1936). The calculated intervals are close to those calculated using Wypij and Santner's (1990) procedure (Table 9.4). Possible advantages of Chen and Tipping's (2002) procedure are that it does not depend on assumption of a particular distributional model and that it might be considered easier to calculate.

TABLE 10.5. "Exact" confidence intervals for various disease incidence data sets, obtained using Chen and Tipping's (2002) method.

Source	*D*	$\sum Y$	*nN*	*Chen and Tipping 95% interval*
Dogwood anthracnose data from Table 9.2				
1990	3.85	269	1680	0.127, 0.198
1991	5.60	500	1610	0.258, 0.368
TSWV data from Example 9.3				
	1.42	261	1440	0.158, 0.206

Chen and Tipping's (2002) procedure also provides a convenient way of finding the upper limit of a one-sided Clopper–Pearson-type $100(1 - \alpha)\%$ confidence interval for mean disease incidence when all the sampling units are empty (i.e., $\bar{y} = 0$). In this case, $\sum Y = 0$ and the upper limit of the interval, denoted p_U, is calculated from the $1 - \alpha$ point of the inverse cumulative beta distribution with parameters $(1, nN / D)$. If $D = 1$ (i.e., for binomial data), this is identical to $p_U = 1 - \alpha^{1/nN}$. If $D > 1$ (i.e., for aggregated data), this is identical to $p_U = 1 - \alpha^{D/nN}$. For instance, with $\alpha = 0.05$, $n = 10$ and $N = 5$, $p_U = 0.06$ for the binomial case ($D = 1$) and $p_U = 0.17$ for aggregated data with $D = 3$.

Another approach for the $\bar{y} = 0$ situation is to use the fact that the negative binomial is a good approximation to the β-binomial distribution at small p (the usual situation when all sampling units are empty), with $\mu = np$ and $k = p/\theta$ (see section 9.6). Madden et al. (1996) showed that the upper limit of a one-sided Clopper–Pearson-type $100(1 - \alpha)\%$ confidence interval for mean disease incidence is approximated by $p_U = -\theta \ln(\alpha) / [N \ln(1 + n\theta)]$, with $\theta = (D - 1) / (n - D)$. For example, when $D = 3$, $p_U = 0.13$ is obtained, a somewhat smaller value than obtained using Chen and Tipping's (2002) approach.

10.6 Regression Analysis of Disease Incidence Data

Disease incidence data may be collected by cluster sampling in a number of distinct areas, with the objective of making a comparison between areas. For normally distributed variables, analysis of variance (ANOVA) provides a basis for the analysis of data where comparison of sample means is the objective (see, for example, Gilligan, 1986). This approach presents some difficulties for the analysis of disease incidence data, because data in the form of proportions may fail to satisfy some of the assumptions of ANOVA (Garrett et al., 2004). The difficulties are more formidable when the proportions are based on different sample sizes (but see Cochran, 1943; Reimer, 1959). One way to face these difficulties is to adopt a generalized linear model (GLM) (McCullagh and Nelder, 1989) approach to analysis. A GLM approach accounts for important properties of non-normal data by means of an appropriate transformation (link function) to achieve a linear model, together with an explicit specification of an appropriate distribution for the response variable.

An important aspect of a GLM approach is the direct link it provides with descriptive methods for disease incidence data collected by cluster sampling, based on the binomial and β-binomial distributions. For this reason, we concentrate here on the use of logistic regression and β-binomial regression as methods for comparing disease incidence data from cluster sampling. There are, however, a number of other methods in use for the analysis of proportions. Anderson (1988), Schabenberger and Pierce

(2002) and Collett (2003) present clear reviews of the available methodology from a statistical point of view. Readers should note that GLMs were introduced in Chapter 3 and used a little in Chapter 4.

10.6.1 Logistic regression

For binomial proportions, *logistic regression* (the linear logistic model) is often an appropriate starting point, in which case the logit link function and the binomial distribution are usually specified. Collett (2003) gives a full and clear account of the statistical background, very much abbreviated in what follows.

Suppose that disease status is recorded as a binary response, so that plants are assessed as either "healthy" or "diseased", in a 2^2 factorial arrangement of plots. A logistic regression model containing terms for each of the factors and for the two-factor interaction is:

$$\text{logit}(p_{ij}) = C_0 + A_i + B_j + (AB)_{ij} \qquad (10.25)$$

in which p_{ij} is the probability that a plant exposed to the ith level of factor A and the jth level of factor B ($i = 1, 2$; $j = 1, 2$) is diseased. The term C_0 is a constant. In equation 10.25, the linear systematic component of the model, given in this case by $\eta_{ij} = C_0 + A_i + B_j + (AB)_{ij}$, is referred to as the *linear predictor* and its estimated value is denoted $\hat{\eta}$. The logit link function relates the linear component to the value of p. Whereas p is restricted to values between zero and one, logit(p) may lie anywhere between $\pm\infty$. Once $\hat{\eta}_{ij}$ has been estimated, the estimated probability that a plant is diseased can be obtained from $\hat{p}_{ij} = \exp[\hat{\eta}_{ij}]/(1 + \exp[\hat{\eta}_{ij}])$.

In practice, a sequence of models can be fitted. For 2^2 factorial arrangements, as above, this sequence (starting with the simplest model) is as follows:

$$\eta = C_0;$$
$$\eta_i = C_0 + A_i;$$
$$\eta_j = C_0 + B_j;$$
$$\eta_{ij} = C_0 + A_i + B_j;$$
$$\eta_{ij} = C_0 + A_i + B_j + (AB)_{ij}.$$

We need to be able to assess how good (or bad) is the fit of each of these models to a set of data, in order to help decide on the appropriate terms to include in the linear predictor. Typically, the models are fitted by maximum likelihood. To compare models, a quantity known as the *deviance* is defined as minus twice the logarithm of the ratio of the maximized likelihoods of a particular model in question and a model with as many unknown parameters as there are observations (the "full" model). Thus the deviance is large, indicating a poor fit, when the maximized likelihood of the model in question is small relative to that of the full model. On the other hand, the deviance is small, indicating a good fit, when the maximized likelihood of the model in question is relatively close to that of the full model.

Example 10.4. Consider the data shown in Fig. 9.1 (taken from Bald, 1937). The 2^2 factorial arrangement involved two tomato cultivars (Burwood Prize (BP), A_1; Early Dwarf Red (EDR), A_2) and two irrigation methods (overhead sprays (OHS), B_1; trenches (T), B_2). The binary response variable is TSWV-infection of tomato plants. Plots comprised 40 quadrats, each of nine plants. The observed frequencies are given in Table 10.6.

A sequence of logistic regression models is now fitted to these data. The required calculations can be carried out using most widely-available general purpose statistical software packages. The analysis as described here was carried out using EGRET for Windows (Cytel Software Corporation, Cambridge, MA).

Model	*Terms fitted in model*	*Deviance*	*df*
10.4*a*	Constant term	294.600	159
10.4*b*	Constant term, cultivar	181.374	158
10.4*c*	Constant term, irrigation method	280.606	158
10.4*d*	Constant term, cultivar, irrigation method	166.238	157
10.4*e*	Constant term, cultivar, irrigation method, cultivar × irrigation method	166.013	156

The table shows the deviance after fitting each of a sequence of logistic regression models with terms included as indicated. When a logistic regression model provides a satisfactory fit, the deviance has an approximate χ^2

TABLE 10.6. TSWV data from Bald (1937). The disease assessments were made on 12 December 1928.

Number of plants diseased (/9)	*Cultivar–irrigation method*			
	BP–OHS	*BP–T*	*EDR–OHS*	*EDR–T*
0	0	0	0	0
1	0	0	0	0
2	0	0	0	0
3	0	0	2	0
4	0	0	6	3
5	0	0	13	3
6	3	0	6	12
7	7	6	7	12
8	20	12	6	7
9	10	22	0	3

distribution, with degrees of freedom equal to the number of observations minus the number of parameters to be estimated in the particular model of current interest. Formally, this can be tested by comparing the deviance with the tabulated χ^2 distribution at the appropriate *df* (informally, the deviance of a model that is a good fit will be approximately equal to its *df*). We look first at model 10.4*e*, for which the χ^2 test gives $P = 0.28$. If the deviance after fitting this model were large (conventionally, large enough to result in a significance probability $P < 0.05$), this would be an indication that the binomial distribution provides an inadequate description of within-plot variability for the TSWV data in question. We may, instead, be dealing with extra-binomial variation, referred to as aggregation in the present context. The repercussions of this are taken up in section 10.6.2. In this case, the significance test indicates that the binomial distribution provides an acceptable description of within-plot variability. Of course, this is not surprising given what we already know about these data from Table 9.1.

It still remains to decide on the appropriate model. Comparisons of nested models are made using the *likelihood ratio statistic* (LRS). The LRS is simply a difference between two deviances. For two nested models, the deviance for the model with a larger number of terms is subtracted from the deviance for the model with a smaller number of terms: this is the LRS. The degrees of freedom for the model with a larger number of terms is subtracted from the degrees of freedom for the model with a smaller number of terms: this is the *df* for the calculated LRS. Since the deviance for each model has an approximate χ^2 distribution, the difference between the two deviances will also be approximately χ^2 (Collett, 2003).

Example 10.4 continued. For the TSWV data shown in Table 10.6, likelihood ratio statistics are tabulated below. Such tables are sometimes referred to as *analysis of deviance* tables. In the table below, the upper stratum is obtained, identical in form, whether the analysis is based on the full data set of 160 observations or just on the overall proportions from the four plots. This part of the table refers to between-plot variability, the plots being the experimental units to which the treatments were applied. In this experimental design, the cultivar × irrigation method term in the upper stratum of the table is equivalent to the residual. The lower stratum is obtained from an analysis of all 160 observations of number of infected plants out of nine. This part of the table refers to within-plot variability, the quadrats of nine plants being the sampling units.

Source of variation	*Models compared*	*LRS*	*df*	*P*
Upper stratum (between plots)				
Cultivar	10.4*a*, 10.4*b*	113.226	1	<0.001
Cultivar (adjusted for irrigation method)	10.4*c*, 10.4*d*	114.368	1	<0.001
Irrigation method	10.4*a*, 10.4*c*	13.994	1	<0.001
Irrigation method (adjusted for cultivar)	10.4*b*, 10.4*d*	15.136	1	<0.001
Cultivar × irrigation method (adjusted for cultivar and irrigation method)	10.4*d*, 10.4*e*	0.225	1	
Lower stratum (within plots)		166.01	156	0.28

The cultivar effect may be tested either taking into account the effect of irrigation method (i.e., the "adjusted" comparison of model 10.4*c* with 10.4*d*) or ignoring it (i.e., the "unadjusted" comparison of model 10.4*a* with 10.4*b*). Similarly, the irrigation method effect can be tested taking into account the effect of cultivar or ignoring it. Having noted this, we will present only the adjusted comparisons in similar analyses that follow. Significance is tested by comparison of the LRS with the tabulated χ^2 distribution at the appropriate *df*. Parameter estimates (on a logit scale) and standard errors after fitting model 10.4*d* are given here.

Model term	*Estimate*[a]	*Standard error*	*P*[b]
C_0	2.039	0.140	<0.001
Cultivar	−1.510	0.152	<0.001
Irrigation method	0.534	0.138	<0.001

[a]EGRET sets the estimates corresponding to the first level of each factor to zero and provides an estimate of the difference in response between the two levels of each factor and the standard error of that difference.

[b]EGRET tests the significance of estimated parameters by reference to the standard normal distribution. These tests provide useful guidance on the importance of parameters, but should not be regarded as providing exact *P* values.

Disease incidence is higher in cultivar Burwood Prize than in cultivar Early Dwarf Red ($P < 0.001$), and higher when irrigation is by trenches compared with overhead sprays ($P < 0.001$). The estimated linear predictors are then:

$$\begin{aligned} \text{BP–OHS}: \hat{\eta}_{11} &= 2.039, \\ \text{BP–T}: \hat{\eta}_{12} &= 2.039 + 0.534 = 2.573, \\ \text{EDR–OHS}: \hat{\eta}_{21} &= 2.039 - 1.510 = 0.529, \\ \text{EDR–T}: \hat{\eta}_{22} &= 2.039 - 1.510 + 0.534 = 1.063. \end{aligned}$$

Fitted probabilities can then be calculated from $\hat{p}_{ij} = \exp[\hat{\eta}_{ij}]/(1+\exp[\hat{\eta}_{ij}])$ as follows: BP–OHS, 0.885; BP–T, 0.929; EDR–OHS, 0.629; EDR–T, 0.743. Comparison of these fitted values with the calculated proportions shown in Table 9.1 shows good agreement.

10.6.2 β-Binomial regression

Zarnoch et al. (1995) analyzed data on dogwood anthracnose infection (caused by *Discula destructiva*) in the south-eastern United States. The 1991–1992 data show the total number of dead trees per plot, and the number dead where there was evidence that dogwood anthracnose was the likely cause, in North Carolina (39 plots) and in Virginia (20 plots) (Table 10.7). Straight away, we note that the sampling units (plots) are of variable size; that is to say, the binomial denominators (the *n*s in the table) are not constant and so expected frequencies cannot be generated to test goodness-of-fit using χ^2. However, separate C(α) goodness-of-fit tests of the binomial distribution against the specific alternative of the β-binomial (section 9.4.7) for the North Carolina and Virginia data give a significance probability $P < 0.001$ in each case. While we therefore realize from the outset that an analysis of the data in Table 10.7 will have deficiencies if it is based on the binomial distribution as a description of within-plot variability of dogwood anthracnose infection, we will proceed step-by-step through the analysis and describe (and mitigate) those deficiencies along the way.

Example 10.5. For the dogwood anthracnose data in Table 10.7, all the trees assessed were dead, and the binary response variable is "dead and infected with dogwood anthracnose" or "dead but not infected with dogwood anthracnose". The single explanatory variable is "state", a factor with two levels (North Carolina and Virginia). Of interest here is whether the proportion of overall dogwood mortality due to dogwood anthracnose differs between the two states.

The starting point is the same as in *Example 10.4*. A sequence of logistic regression models is fitted to the data:

$$\text{Model 10.5}a: \ \eta = C_0$$
$$\text{Model 10.5}b: \ \eta_i = C_0 + \text{state}_i.$$

The deviances on fitting logistic regression models to the dogwood anthracnose data in Table 10.7 (from Zarnoch et al., 1995) are then tabulated, as follows.

Model	*Terms fitted in model*	*Deviance*	*df*
10.5*a*	Constant term	282.37	58
10.5*b*	Constant term, state effect	278.12	57

On the basis of Models 10.5*a* and 10.5*b*, the effect of state is tested by calculating the LRS (the difference between the deviances) which is compared with the tabulated χ^2 distribution at the appropriate *df*. The comparison of Model 10.5*a* with Model 10.5*b* gives LRS = 4.25, 1 *df*, $P = 0.04$, indicative of a difference between North Carolina and Virginia in proportion of dogwood mortality due to dogwood anthracnose infection. The maximum likelihood estimates based on Model 10.5*b* are: North Carolina, $\text{logit}(\hat{p}) = -0.50$, SE = 0.11; Virginia, $\text{logit}(\hat{p}) = -0.16$, SE = 0.12. The difference between the two estimates (on a logit scale) is 0.34, with a standard error of 0.16 ($P = 0.04$). Conventionally, the evidence would lead us to infer a difference in between the two states. The problem with this analysis is that after fitting Model 10.5*b*, the deviance is significantly large, at 278.1 with 57 *df* ($P < 0.001$). This may be an indication that the binomial distribution provides an inadequate description of within-plot variability of dogwood anthracnose infection.

Since the standard errors of the parameter estimates from Model 10.5*b* (quoted above) are based on the assumption that the within-plot variability in dogwood anthracnose infection follows a binomial distribution, the detection of extra-binomial variation would indicate that these standard errors were too small. Thus, in such circumstances, the use of the binomial distribution to describe variability tends to exaggerate artificially the significance of comparisons. However, before embarking on an analysis of extra-binomial variation, other possible reasons for a large deviance should be investigated. These include the omission of one or more important explanatory variables from the model, and data that include particularly unusual or influential observations. Collett (2003, Chapter 5) gives details of model checking procedures. In the case of the logistic regression

TABLE 10.7. Data from the North Carolina–Virginia dogwood anthracnose survey, 1991–1992 (Zarnoch et al., 1995)[a].

North Carolina		*North Carolina*		*Virginia*	
Y	*n*	*Y*	*n*	*Y*	*n*
2	3	0	7	4	9
0	2	3	9	18	20
0	5	12	13	0	12
5	7	6	20	2	7
8	14	2	10	11	20
3	3	4	14	4	10
0	1	0	8	13	18
8	8	0	4	8	14
1	1	0	3	2	14
10	10	6	6	14	20
7	7	10	15	3	15
0	15	2	7	1	5
1	11	13	20	5	12
0	7	0	8	7	20
4	10	0	6	2	9
3	20	7	15	7	16
0	9	0	5	12	20
0	9	8	14	7	15
5	9	1	7	1	10
0	4			7	12

[a] *n* is the number of dead dogwood trees on a plot, *Y* is the number of dead dogwood trees infected by dogwood anthracnose.

analysis based on the data in Table 10.7, there are no additional explanatory variables available for inclusion. Delta-beta statistics provide a means of ascertaining the influence of each individual observation on the fit of a model (for details, see Collett, 2003). A delta-beta is an approximation of the amount that an estimated regression parameter would change if a given observation was omitted from the regression fit. Observation no. 41 (Virginia, $Y = 18$, $n = 20$) has a relatively large (negative) delta-beta (Fig. 10.4), but there is no evidence in Zarnoch et al. (1995) to support the omission of this observation from the data. We therefore proceed with an analysis that takes account of extra-binomial variation. β-Binomial regression (Williams, 1975; Crowder, 1978) allows us to fit regression models in a similar fashion to logistic regression, but instead of the assumption that the within-plot variability in dogwood anthracnose infection follows a binomial distribution, a β-binomial distribution (section 9.4.5) is assumed.

Example 10.5 continued. The sequence of regression models fitted to the data can now be extended by inclusion of the β-binomial aggregation parameter θ (section 9.4.5):

$$\text{Model 10.5}c: \quad \eta = C_0, \qquad \theta = A_0.$$
$$\text{Model 10.5}d: \quad \eta_i = C_0 + \text{state}_i, \qquad \theta = A_0.$$
$$\text{Model 10.5}e: \quad \eta_i = C_0, \qquad \theta_i = A_0 + \text{state}_i.$$
$$\text{Model 10.5}f: \quad \eta_i = C_0 + \text{state}_i, \qquad \theta_i = A_0 + \text{state}_i.$$

The term A_0 is a constant. The deviances on fitting β-binomial regression models to the data on dogwood anthracnose data in Table 10.7 (from Zarnoch et al., 1995) are then tabulated, as follows.

	Terms fitted in model			
Model	*Factor(s)*	*Aggregation parameter(s)*	*Deviance*	*df*
10.5*c*	Constant term	Common	163.56	57
10.5*d*	Constant term, state effect	Common	162.49	56
10.5*e*	Constant term	Separate	152.80	56
10.5f	Constant term, state effect	Separate	152.10	55

First, we investigate the outcome of including the β-binomial aggregation parameter. The comparison of Model 10.5*b* with Model 10.5*d* gives LRS = 115.63, 1 *df*, $P < 0.001$. Thus, the inclusion of a single aggregation parameter, common to both states, brings about a significant reduction in deviance. Note that the LRS for inclusion of a single θ into a previous model that had no aggregation parameter(s) is not tested as a value of χ^2, but by treating $\sqrt{(\text{LRS})}$ as a standardized normal deviate. Thus, in this case, $\sqrt{(115.63)} = 10.75$ is tested as a standardized normal deviate.

The comparison of Model 10.5*d* with Model 10.5*f* gives LRS = 10.39, 1 *df*, $P = 0.001$, indicating a further reduction in deviance is achieved if a separate aggregation parameter is included for each state. We note that the deviance after fitting Model 10.5*f* is still significantly large, indicating that there is variability in these data that is not characterized by the β-binomial aggregation parameter(s).

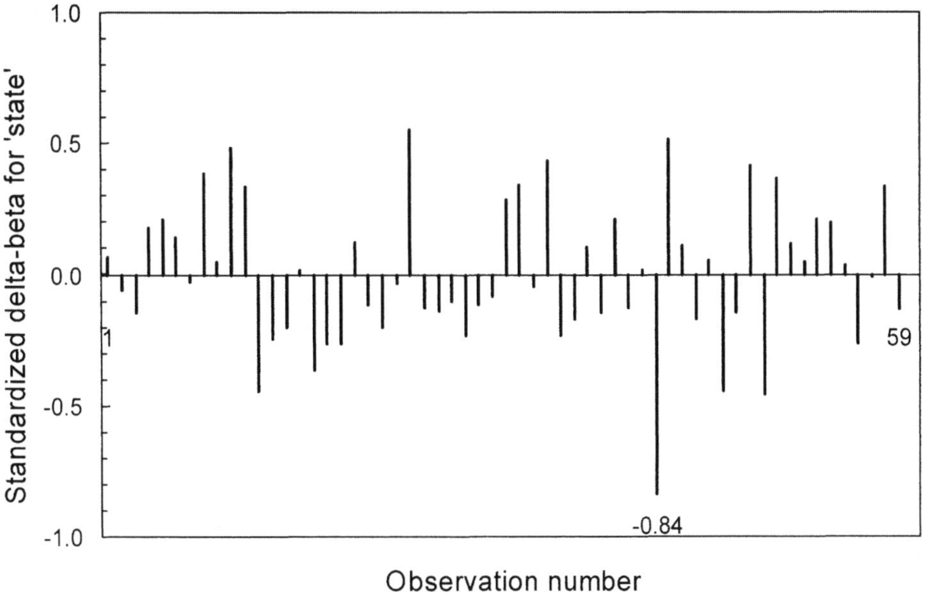

FIG. 10.4. Post-fit regression diagnostics for Model 10.5*b* fitted to the data in Table 10.7. The largest delta–beta (=−0.84) is for observation 41 (Virginia, $Y = 18$, $n = 20$).

We are now in a position to test the state effect, having taken extra-binomial variation into account by means of the β-binomial model. The comparison of Model 10.5*e* with Model 10.5*f* gives LRS = 0.70, 1 *df*, $P = 0.40$. On this basis, no significant reduction in deviance is achieved by the inclusion of state as a factor in the model, if a separate aggregation parameter is included for each state. We note that for the dogwood anthracnose data under discussion here, the same result is obtained if a single aggregation parameter, common to both states, is included. In this case, the relevant comparison is of Model 10.5*c* with Model 10.5*d* (LRS = 1.07, 1 *df*, $P = 0.30$). Parameter estimates (on a logit scale) and standard errors after fitting Model 10.5*d* are as follows.

Model term	*Estimate*[a]	*Standard error*	P^b
Between-states stratum			
C_0	−0.635	0.233	0.006
State	0.369	0.356	0.300
Within-states stratum			
A_0	0.546	0.141	

[a]EGRET sets the estimates corresponding to the first level of each factor to zero and provides an estimate of the difference in response between the two levels of each factor and the standard error of that difference.

[b]EGRET tests the significance of estimated parameters by reference to the standard normal distribution. These tests provide useful guidance on the importance of parameters, but should not be regarded as providing exact *P* values.

On the basis of Model 10.5*d*, there is no difference between North Carolina and Virginia in proportion of dogwood mortality due to dogwood anthracnose infection ($P = 0.30$). This conclusion differs from that reached on the basis on Model 10.5*b*, which took no account of extra-binomial variation. Because the dogwood anthracnose infection data in Table 10.7 are in fact significantly aggregated, the standard errors of parameter estimates from Model 5*b* are artificially small, leading to an erroneous attribution of statistical significance. The maximum likelihood estimates based on Model 10.5*d* are: North Carolina, $\text{logit}(\hat{p}) = -0.64$, SE = 0.23; Virginia, $\text{logit}(\hat{p}) = -0.27$, SE = 0.27. The difference between the two estimates (on a logit scale) is 0.37, with a standard error of 0.36 ($P = 0.30$). The most important effect of taking into account extra-binomial variation is not on the parameter estimates themselves but on their standard errors, which are inflated relative to those obtained using logistic regression.

Hughes and Madden (1998) give another example of β-binomial regression, applied to the formulation of management recommendations for black spruce (*Picea mariana*) seed orchards in Ontario (Bruhn et al. 1996). As an alternative, Hughes and Samita (1998) and Hughes et al. (1998) approach the analysis of extra-binomial variation in disease incidence data by application of logistic normal binomial (LNB) regression. To illustrate the difference between the two approaches, we first refer back to equation 10.25, for a logistic regression model containing terms for each of two factors, *A* and *B*, and for the two-factor interaction: $\text{logit}(p_{ij}) = C_0 + A_i + B_j + (AB)_{ij}$, in which p_{ij} is the (constant) probability that a plant exposed to the *i*th level of factor *A* and the *j*th level of factor *B* is diseased. Aggregation can be modeled by relaxing the assumption that the response probability for the *i*th level of factor *A* and the *j*th level of factor *B* is a constant. Then, different assumptions about the form of the variability in the response probabilities lead to different models. For example, if it is assumed that the actual response probabilities vary about a mean value and that the variability can be described by a beta distribution, this leads to the β-binomial model (see section 9.4.5). The LNB model is based, instead, on the assumption that the *logit*-transformed response probabilities vary about a mean value, and that the variability can be described by a normal distribution. That is to say, if variability in the response probabilities is described by a logistic normal distribution (Aitchison and Shen, 1980), this leads to the LNB model. Both approaches are useful for the analysis of disease incidence data. One advantage of the LNB regression approach is that it can easily be extended to the analysis of multiple scales in a hierarchy, where data are available, for example, for plots, quadrats within plots, plants within quadrats and leaves within plants. For more details, readers are referred to Madden et al. (2002) and Piepho (1999).

10.6.3 Logistic regression with deff-transformed data

Rao and Scott (1992) suggested an alternative approach to the analysis of extra-binomial variation in cluster-sampling data. The method involves a simple transformation of the observed data, using the index of dispersion (which they refer to as the deff; section 9.4.2). The data must be classified by a single factor. Essentially, the transformation involves calculation, for each sampling unit, of $Y^* = Y/\text{deff}_i$, $n^* = n/\text{deff}_i$ (with *i* denoting the level of the single factor in question), following which the transformed variables Y^*, n^* are used in logistic regression. In passing, note that the quantity *n*/deff is a version the *effective sample size* as used in the sampling literature (Kish, 1995). For aggregated data, deff > 1. Dividing *n* by the deff results in an effective sample size < *n*. This reflects the reduction in the amount of information obtained from a group of *n* individual plants or plant parts in a sampling unit, when proximity (i.e., membership of the same sampling unit) results in a tendency for individuals to have the same disease status.

For the dogwood anthracnose data in Table 10.7 (from Zarnoch et al., 1995), the deffs are equal to 4.458 and 3.239 for the North Carolina and Virginia data respectively. Using the statistical software GENSTAT, a sequence of logistic regression models is fitted to the deff-transformed data:

$$\text{Model } 10.5g: \ \eta = C_0$$
$$\text{Model } 10.5h: \ \eta_i = C_0 + \text{state}_i.$$

The deviances are then tabulated, as follows.

Model	*Terms fitted in model*	*Deviance*	*df*
10.5*g*	Constant term	69.29	58
10.5*h*	Constant term, state effect	68.17	57

The comparison of Model 10.5*g* with Model 10.5*h* gives LRS = 1.12, 1 *df*, *P* = 0.29. The maximum likelihood estimates based on Model 10.5*h* are: North Carolina, logit($\hat{p}$) = –0.495 , SE = 0.234; Virginia, logit($\hat{p}$) = –0.159, SE = 0.217. The difference between the two estimates (on a logit scale) is 0.337, with an estimated standard error of 0.319 (*P* = 0.29). Conventionally, the evidence would not lead us to infer any difference in between the two states in proportion of dogwood mortality due to dogwood anthracnose infection. After fitting Model 10.5*h*, the deviance is 68.17 with 57 *df* (*P* = 0.15). Thus the deff transformation of the data appears to have dealt with the problem of extra-binomial variation in a satisfactory manner. Compared with the previous logistic regression analysis of the dogwood anthracnose data (section 10.6.2, *Example 10.5*), the parameter estimates are similar but the standard errors are inflated. As with β-binomial regression analysis, the outcome is that an erroneous attribution of statistical significance relating to the difference between states is avoided. Possible advantages of the deff-transformation approach are that it does not depend on assumption of a particular distributional model and that it can easily be carried out with general-purpose statistical software. On the other hand, the β-binomial approach—as well as the generalized linear mixed modeling approach which involves the logistic-normal-binomial distribution (not covered here)—provides a direct link between the use of statistical probability distributions in data analysis and in the description of spatial patterns of disease, and is applicable to experimental data classified by more than a single factor.

10.6.4 Fitting statistical probability distributions

In sections 9.4.2 and 9.4.5, the binomial and β-binomial distributions were fitted to data as a way of describing frequency distributions of disease incidence. In case it is not already obvious, we now state explicitly that the methodology described in sections 10.5.1 and 10.5.2 achieves exactly the same outcome, albeit using a different parameterization. Thus we have seen, in section 10.5.1, how the use of logistic regression resulted in binomial parameter estimates for each of the four frequency distributions of TSWV disease incidence shown in Fig. 9.1.

In section 10.5.2, the use of β-binomial regression resulted in β-binomial parameter estimates for the dogwood anthracnose data from each of two states (because the *n*s differed among sampling units, no frequency distribution can be drawn). Had we approached the analysis with the objective of simply describing the data from each state, a summary such as that shown in Table 10.8 could be prepared.

While Model 10.5*d* was adequate for a comparison between the two states based on β-binomial regression, it is Model 10.5*f* that provides the analysis analogous to that shown in Table 10.8. The analysis for Model 10.5*f* is shown in Table 10.9, from which the following results are obtained. For North Carolina, logit($\hat{p}$) = –0.568, $\hat{\theta} = 1.037$; for Virginia, logit($\hat{p}$) = $-0.568 + 0.278$, $\hat{\theta} = 1.037 - 0.852$. Then:

TABLE 10.8. Output after analyses of the dogwood anthracnose data in Table 10.7 (from Zarnoch et al., 1995) using the BBD computer program (Madden and Hughes, 1994). Only an abbreviated summary is given here. Various test statistics calculated by the software are not shown.

β-*Binomial parameter estimates*	*North Carolina*	*Virginia*
$\hat{p}$	0.362	0.428
SE ($\hat{p}$)	0.058	0.052
$\hat{\theta}$	1.037	0.185
SE ($\hat{\theta}$)	0.332	0.081

TABLE 10.9. Output (an abbreviated summary) after analyses of the dogwood anthracnose data in Table 10.7, based on Model 10.5*f*.

Model term	*Estimate*[a]	*Standard error*	*P*[b]
Between-states stratum			
C_0	−0.568	0.253	0.025
State	0.278	0.331	0.402
Within-states stratum			
A_0	1.037	0.337	
State	−0.852	0.347	0.014

[a]EGRET sets the estimates corresponding to the first level of each factor to zero and provides an estimate of the difference in response between the two levels of each factor and the standard error of that difference.

[b]EGRET tests the significance of estimated parameters by reference to the standard normal distribution. These tests provide useful guidance on the importance of parameters, but should not be regarded as providing exact *P* values.

$$\text{for North Carolina, } \hat{p} = \frac{\exp(-0.568)}{1+\exp(-0.568)} = 0.362,$$
$$\hat{\theta} = 1.037,$$
$$\text{for Virginia, } \hat{p} = \frac{\exp(-0.568+0.278)}{1+\exp(-0.568+0.278)} = 0.428,$$
$$\hat{\theta} = 1.037 - 0.852 = 0.185.$$

Thus Tables 10.8 and 10.9 represent alternative parameterizations of the same analysis. This makes clear the point that descriptions of aggregated disease incidence data based on statistical probability distributions provide the basis for making comparisons when more than one data set is collected.

10.7 Regression Analysis of Count Data

In much the same way that the binomial and β-binomial distributions provide a basis both for analyzing spatial patterns and for statistical modeling of disease incidence data, the Poisson and negative binomial distributions can be applied to disease data in the form of counts. Since relatively few disease data are collected in the form of counts we provide only an abbreviated summary here. A full description of the methodology from a statistical viewpoint is provided by Cameron and Trivedi (1998).

10.7.1 Poisson and negative binomial regression

Count data present similar difficulties to those noted for incidence data (section 10.6) when comparison of sample means is the objective. Poisson regression provides a way of circumventing problems arising from the non-normality of count data, which limits the applicability of ANOVA. In this case, the natural logarithm (ln) is the link function used to achieve a linear model and the Poisson distribution is specified for the response variable. However, if Poisson regression analysis is carried out on aggregated count data, the standard errors of the estimated linear predictors, and of differences between them, are likely to be too small and so may lead to unwarranted claims of statistical significance. Negative binomial regression is one method of analyzing aggregated count data in a way that takes account of extra-Poisson variation.

Example 10.6. For this example, we return to the data from the North Carolina–Virginia dogwood anthracnose survey, 1991–1992 (Zarnoch et al., 1995) shown in Table 10.7. In the previous *Example 10.5*, the analysis related to the proportion of dogwood mortality that was due to anthracnose infection. In the present example, the analysis relates to the *number* of tree deaths due to anthracnose infection. That is, the analysis is now based just on Y, the number of dead dogwood trees infected by dogwood anthracnose. Because there is, in fact, an upper limit to tree mortality in a plot (since no more than n trees can die), the previously described analysis of the data as proportions is more appropriate. In order to justify the use of an analysis based on the number of dead trees, we would need to have values of Y that were low relative to the corresponding values of n. The main purpose of this example is to illustrate some methods for analyzing count data.

As before, the single explanatory variable is "state", a factor with two levels (North Carolina and Virginia). The analysis described below was carried out with *SAS* statistical software. Table 10.10 shows the results of three regression models fitted to the data for number of dead dogwood trees infected by dogwood anthracnose (Y). Model 10.6*a* is based on a Poisson regression analysis. The difference between states (on a ln scale) is 0.645, with SE = 0.124 (P = 0.001), but the deviance is significantly large (257.99 with 57 *df*), indicative of aggregation (extra-Poisson variation).

Model 10.6*b* is based on a negative binomial regression analysis, incorporating a single aggregation parameter common to both states. The difference between states (on a ln scale) is 0.645 as before, but now with SE = 0.314 (P = 0.045). The deviance is now a satisfactory 68.21 with 57 *df*. At this stage we could conclude that there is evidence at about the 5% significance level for a difference between states in mean dogwood mortality per plot due to dogwood anthracnose. Probably, however, we would want to check the results of the negative binomial regression analysis incorporating a separate aggregation parameter for each state, if only to see whether a firmer attribution of statistical significance (relative to the conventionally adopted 5% significance level) could be made, one way or the other.

Model 10.6*c* is based on a negative binomial regression analysis, incorporating a separate aggregation parameter for each state. The difference between states (on a ln scale) is 0.645 as before, but now with SE = 0.293 (P = 0.032). The deviance is now 64.00 with 57 *df*. Thus we

TABLE 10.10. Parameter estimates for three regression models fitted to the data for number of dead dogwood trees infected by dogwood anthracnose (Table 9.2).

		Parameter estimates			
Model	*State*	*Linear predictor*	*SE*	$\hat{k}$	*SE*
10.6a	North Carolina	1.212	0.087	–	–
	Virginia	1.856	0.088	–	–
10.6b	North Carolina	1.212	0.189	{0.909	0.238}[a]
	Virginia	1.856	0.251		
10.6c	North Carolina	1.212	0.230	0.566	0.183
	Virginia	1.856	0.181	2.000	0.863

[a]In Model 10.6*b* there is a single estimate of k of the negative binomial distribution, common to both States.

conclude that there is evidence for a difference between states in mean dogwood mortality per plot due to dogwood anthracnose. Recall that the analysis in *Example 10.5* led us to conclude that there was no difference between states with respect to the *proportion* of dogwood mortality attributable to dogwood anthracnose.

From Table 10.10, we can see that Model 10.6*c* has provided an improved description of aggregation. The common aggregation parameter fitted in Model 10.6*b* was rather too large for the North Carolina data (i.e., the data were more aggregated than as characterized by Model 10.6*b*). In contrast, common aggregation parameter fitted in Model 10.6*b* was rather too small for the Virginia data (i.e., the data were less aggregated than as characterized by Model 10.6*b*). These differences in aggregation between states are reflected Model 10.6*c* in the increase in the standard error of the linear predictor for North Carolina (relative to Model 10.6*b*) and a decrease in the standard error of the linear predictor for Virginia (again relative to Model 10.6*b*).

10.8 Group Testing with Incidence Data

Disease incidence may be estimated by means of a simple random sample (section 10.3). Suppose that the assessment of infection is made by use of a laboratory-based assay. If the units of plant material collected by sampling are tested individually, an unbiased estimate of incidence is the proportion of units infected, $\bar{y} = \Sigma Y_i / N$, in which N is the total number of units, and $Y_i = 0$ for a unit assessed as uninfected or $Y_i = 1$ for a unit assessed as infected ($i = 1, 2, \ldots, N$) (for the sake of simplicity, we assume here that the assay provides assessments that can be regarded as error free). The proportion of units infected, $\bar{y}$, is taken as an estimate of the probability that an individual unit is infected.

Alternatively, the individual units of plant material may be combined into groups, and the groups tested. Testing groups is an intuitively appealing option when few individual units are thought to be infected, and when the cost of testing is non-negligible. Less obvious, perhaps, is that a good *statistical* case for testing groups can be made (see, for example, Loyer, 1983). Using the same notation as in section 9.7.1, the probability that an individual unit is infected is denoted p_{low} (the subscript "low" indicates reference to the individual unit scale). If groups of units from the sample are formed at random in the laboratory prior to testing, the probability that a group of n units has no infected units is given by the zero term of the binomial distribution, $\Pr(Y = 0) = (1 - p_{\text{low}})^n$. The probability that a group of n units contains at least one infected unit is then $p_{\text{high}} = 1 - \Pr(Y = 0) = 1 - (1 - p_{\text{low}})^n$, in which the subscript "high" indicates reference to the group scale. On rearranging this, $p_{\text{low}} = 1 - (1 - p_{\text{high}})^{1/n}$ is obtained. An estimate of p_{high} is $\hat{p}_{\text{high}} = \Sigma Y_i / N$ (N is now the total number of groups (i.e., the number of tests), and $Y_i = 0$ for a group that tests negative for infection or $Y_i = 1$ for a group that tests positive ($i = 1, 2, \ldots, N$)). The maximum likelihood estimate of p_{low} is:

$$\tilde{p}_{\text{low}} = 1 - (1 - \hat{p}_{\text{high}})^{1/n} \qquad (10.26)$$

in which the "tilde" indicates that the estimate is obtained from data at a different scale (so in this case, an estimate at the unit scale is obtained from data at the group scale). An approximate variance can be formulated using the method outlined by Colquhoun (1971):

$$\tilde{v}_{\text{low}} \approx \frac{\hat{p}_{\text{high}}(1 - \hat{p}_{\text{high}})^{(2-n)/n}}{n^2} \qquad (10.27)$$

Or, writing as a function of $\tilde{p}_{\text{low}}$:

$$\tilde{v}_{\text{low}} \approx \frac{(1 - \tilde{p}_{\text{low}})^2((1 - \tilde{p}_{\text{low}})^{-n} - 1)}{n^2} \qquad (10.28)$$

An estimated standard error for $\tilde{p}_{\text{low}}$ is then given by $SE(\tilde{p}_{\text{low}}) = \sqrt{\tilde{v}_{\text{low}} / N}$.

Use of equation 10.26 to estimate a proportion is often referred to as *group testing*. The statistical methodology of group testing has a long association with the study of disease in human, animal and plant populations. Dorfman's (1943) early work was concerned with developing a method of classifying individuals in a population as either infected or not when testing every individual was deemed inefficient because infection was rare (the term *Dorfman screening* is sometimes used as a synonym for group testing). Here, we are concerned with group testing for estimation, and in particular with the estimation of the proportion of individual units infected. Particular examples of the application of group testing where infection in plant populations has been studied include Moran et al. (1983, 1985), Rodoni et al. (1994) and Block et al. (1999). An important phytopathologically related area for the application of group testing is in studies of infection in populations of vectors of plant diseases (e.g., Smith and Lea, 1946; Swallow, 1985; Romanow et al., 1986; Tebbs and Bilder, 2004). Use of group testing to estimate incidence (proportion of individuals infected) requires consideration of the properties of the estimator $\tilde{p}_{\text{low}}$ and of the choice of group size n. On both counts, Swallow (1985) is essential reading for plant pathologists contemplating the application of group testing methodology.

10.8.1 The estimator $\tilde{p}_{\text{low}}$

As an estimator of p, the true but unknown proportion of individuals infected, $\tilde{p}_{\text{low}}$ is biased when $n > 1$. Swallow (1985) gives the following relationships:

$$\text{bias}(\tilde{p}_{\text{low}}) = E(\tilde{p}_{\text{low}}) - p \qquad (10.29)$$

$$\mathrm{Var}(\tilde{p}_{\mathrm{low}}) = E(\tilde{p}_{\mathrm{low}} - E(\tilde{p}_{\mathrm{low}}))^2 \quad (10.30)$$

$$\mathrm{MSE}(\tilde{p}_{\mathrm{low}}) = E(\tilde{p}_{\mathrm{low}} - p)^2 = \mathrm{Var}(\tilde{p}_{\mathrm{low}}) + (\mathrm{bias}(\tilde{p}_{\mathrm{low}}))^2 \quad (10.31)$$

based on standard statistical theory, in which $E(\bullet)$ denotes expected value and $\mathrm{Var}(\bullet)$ denotes variance. The bias is the difference between the expected (i.e., average) value of the estimator $\tilde{p}_{\mathrm{low}}$ and the true value p, and the mean squared error (MSE) is the average squared deviation of $\tilde{p}_{\mathrm{low}}$ from p. For an unbiased estimator, the expected value is equal to the true value p and the mean squared error is equal to the variance. To make use of equations 10.28–10.30, formulae for the expected value and the variance of the estimator are needed. These are given by Swallow (1985):

$$E(\tilde{p}_{\mathrm{low}}) = 1 - \sum_{i=0}^{N} \left(\frac{i}{N}\right)^{1/n} \binom{N}{i} [(1-p)^n]^i [1-(1-p)^n]^{N-i} \quad (10.32)$$

$$\mathrm{Var}(\tilde{p}_{\mathrm{low}}) = \sum_{i=0}^{N} \left(\frac{i}{N}\right)^{2/n} \binom{N}{i} \times [(1-p)^n]^i [1-(1-p)^n]^{N-i} - [1 - E(\tilde{p}_{\mathrm{low}})]^2 \quad (10.33)$$

Now, for specified values of N (number of groups) and n (number of units per group), bias $(\tilde{p}_{\mathrm{low}})$ and MSE $(\tilde{p}_{\mathrm{low}})$ can be calculated as functions of p. Swallow (1985) discusses a scenario where the number of groups remains constant and the group size is increased, and concludes that choosing too large a group size should be avoided. For example, Figs. 10.5 and 10.6 are based on $N = 150$ groups, with group sizes $n = 5, 10, 25$ and 50. Fig. 10.5 shows a sequence of graphs of $E(\tilde{p}_{\mathrm{low}})$ against p, based on this scenario. When all units are tested individually ($n = 1$), $E(\tilde{p}_{\mathrm{low}}) = p$ and the estimator $\tilde{p}_{\mathrm{low}}$ is unbiased. As n increases, the extent to which the estimator is upwardly biased increases and the range of p for which the bias may be regarded as negligible decreases. Fig. 10.6 shows a sequence of graphs of MSE $(\tilde{p}_{\mathrm{low}})$ against p, based on the same scenario. Note that there is a range of p for which MSE $(\tilde{p}_{\mathrm{low}})$ is below the MSE of the unbiased estimator (which is equal to the variance), but that this range decreases as n increases. Burrows (1987) discussed an alternative estimator with superior bias and MSE properties, but this has not been widely adopted.

10.8.2 Choice of group size

From section 10.8.1, we have already seen that use of large groups increases the extent to which the maximum likelihood estimate $\tilde{p}_{\mathrm{low}}$ is biased. Large group size is also undesirable because—as Block et al. (1999) found—it

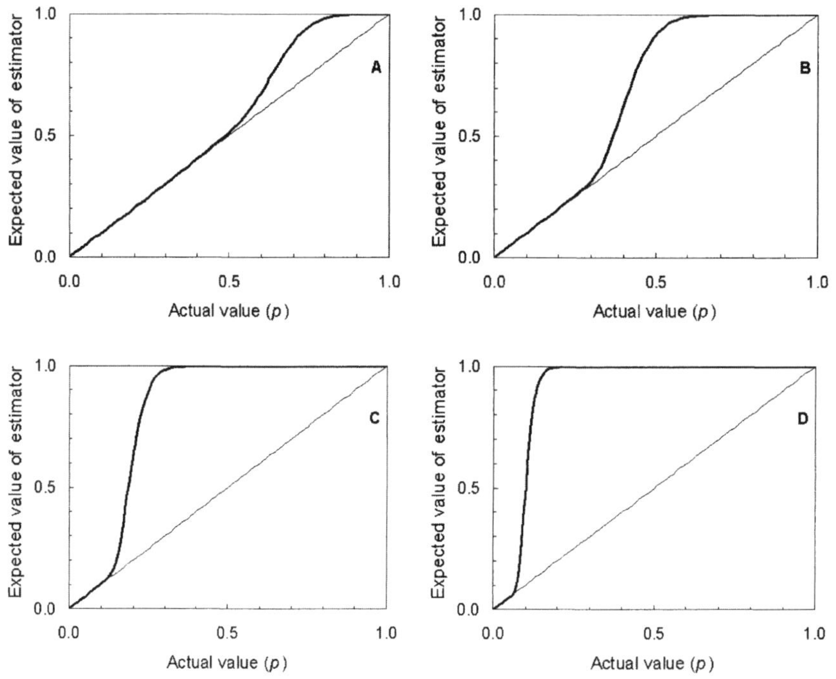

FIG. 10.5. Group testing with disease incidence data. The figure shows a sequence of graphs of the expected value of the estimator, $E(\tilde{p}_{\mathrm{low}})$, against p, the true but unknown proportion of individuals infected. The number of groups is $N = 150$ in each case. For an unbiased estimator, $E(\tilde{p}_{\mathrm{low}}) = p$ (represented by a thin straight line between (0, 0) and (1, 1) shown on the graphs). This is the case when all units are tested individually ($n = 1$). When group size $n > 1$, $\tilde{p}_{\mathrm{low}}$ is a biased estimator of p. As n increases, the extent to which the estimator is upwardly biased increases and the range of p for which the bias may be regarded as negligible decreases. (**A**) $n = 5$; (**B**) $n = 10$; (**C**) $n = 25$; (**D**) $n = 50$.

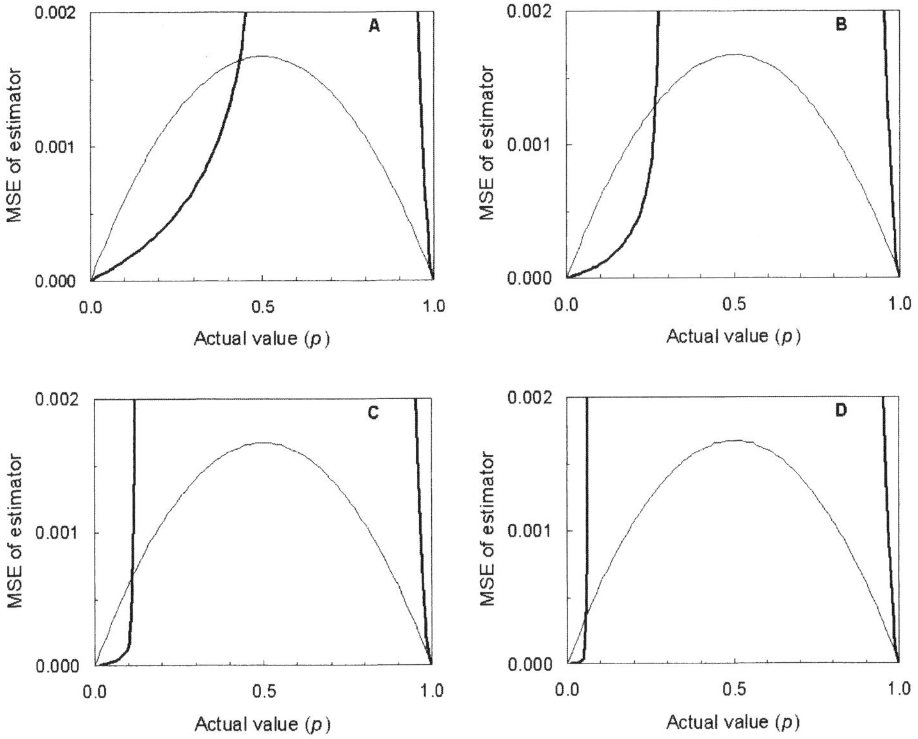

FIG. 10.6. Group testing with disease incidence data. The figure shows a sequence of graphs of the mean squared error of the estimator $\tilde{p}_{low}$, MSE ($\tilde{p}_{low}$), against p, the true but unknown proportion of individuals infected. The number of groups is $N = 150$ in each case. For an unbiased estimator, the MSE of the estimator is equal to the variance (represented by a parabola with a peak at $p = 0.5$ shown on the graphs). This is the case when all units are tested individually ($n = 1$). When group size $n > 1$, $\tilde{p}_{low}$ is a biased estimator of p. There is a range of p for which MSE ($\tilde{p}_{low}$) is below the MSE of the unbiased estimator. This range decreases as group size n increases. (**A**) $n = 5$; (**B**) $n = 10$; (**C**) $n = 25$; (**D**) $n = 50$.

increases the chance that all groups test positive. In such cases, there is little useful that can be concluded about the incidence of infection of individual units (Bhattacharayya et al., 1979). In addition to statistical considerations, the choice of group size must also take into account the characteristics of the assay used to detect infection. Specifically, the assay must be capable of discriminating between a positive group and a negative group when only one out of n units in the positive group is infected. This can be demonstrated by an experimental dilution series made up from plant material of known infection status (e.g., Hughes and Gottwald, 1998). A further issue concerning the assay used to detect infection is the possibility of misclassification errors (Chen and Swallow, 1990). If the only positive unit in a large group is not detected, the group is erroneously declared negative. The impact on $\tilde{p}_{low}$ may be large. These non-statistical factors also weigh against the use of large groups.

It is desirable to adopt a group size that results in some groups testing negative and some testing positive. Various guidelines exist. Chiang and Reeves (1962) gave:

$$n = \frac{\log(1/2)}{\log(1 - p_{low})}$$

based on an equal chance of a negative or a positive test, and translated this formula into a table of "standard" group sizes. Thompson (1962) suggested choosing n on the basis of minimizing the asymptotic variance of $\tilde{p}_{low}$, but the disadvantages of this procedure are apparent from Swallow's (1985) discussion. Swallow (1985, Table 1) adopts instead the criterion of minimizing MSE ($\tilde{p}_{low}$) and gives the optimum values of n for a range of values of N and p_{low}. Swallow (1987) discussed group size based on minimizing the cost per unit information. In this case, optimum values of n for given values of N and p_{low} were smaller than when based on minimizing MSE ($\tilde{p}_{low}$). All these guidelines for choice of group size depend on being able to provide an initial estimate of the unknown incidence of infection in individual units. If a likely range can be specified with reasonable confidence, Swallow (1985) suggests that group size n is adopted on the basis of the upper value of this range. This has the desirable outcome that group size is conservatively small.

It is *not* a prerequisite for group testing that all groups are the same size. Studies by Moran et al. (1985) and Rodoni et al. (1994) both used a sequential series of group sizes in estimating the incidence of virus infection in carnations. If so little is known in advance that even specifying a likely range for p presents difficulties, this procedure has particular advantages. Chen and Swallow (1990) give an estimator for p for use when groups are of unequal size. Hepworth (1996) discusses in some detail the statistical aspects of the procedure adopted by Rodoni et al. (1994).

10.8.3 Sample size calculations

Equipped with an initial estimate of the incidence of infection in individual units and an appropriate choice of group size, the number of groups that must be tested (N) to estimate $\tilde{p}_{\text{low}}$ with a pre-specified level of reliability may be calculated using the methods outlined in section 10.2.6 and the mean and variance formulae given in section 10.8.1. In equations 10.34–10.36 below, the initial estimate of the incidence of infection in individual units is given in terms of $\phi = (1 - \tilde{p}_{\text{low}})$ and n is the chosen group size. If reliability is defined by the coefficient of variation of the sample mean, CV:

$$N = \frac{\phi^2(\phi^{-n} - 1)}{(1-\phi)^2 n^2 CV^2} \tag{10.34}$$

If reliability is defined by setting the half the length of the required confidence equal to a fixed proportion of the mean, H:

$$N = \frac{\phi^2(\phi^{-n} - 1)}{(1-\phi)^2 n^2}\left(\frac{z_{\alpha/2}}{H}\right)^2 \tag{10.35}$$

If reliability is defined by setting the half the length of the required confidence interval of the sample mean equal to a fixed positive number h:

$$N = \frac{\phi^2(\phi^{-n} - 1)}{n^2}\left(\frac{z_{\alpha/2}}{h}\right)^2 \tag{10.36}$$

as in Worlund and Taylor (1983).

For example, if $\tilde{p}_{\text{low}} = 0.05$, $n = 25$, $CV = 0.1$, equation 10.34 gives $N = 150$ as the number of groups to be tested (for a total of $Nn = 3750$ individuals). If, instead, we put $n = 1$, then $N = 1900$ (all other things being equal). We note in passing that the sample size $N = 1900$ can also be obtained from the formula $N = (1-\bar{y})/\bar{y}CV^2$ (Table 10.1) when $\bar{y} = 0.05, CV = 0.1$. In this example, group testing requires about double the number of units, but only about eight per cent of the number of tests, compared with testing individual units.

10.8.4 Exact confidence intervals

Moran et al. (1983) investigated virus infection in potato crops using group testing. For one certified seed crop of cv. Sebago, groups comprised $n = 100$ leaflets, and five out of ten groups tested positive for *Potato virus X* (PVX). Thus $\hat{p}_{\text{high}} = 0.5$, and, from equation 10.26, $\tilde{p}_{\text{low}} = 6.91 \times 10^{-3}$. Moran et al. (1983) gave the "exact" 95% confidence interval for this $\tilde{p}_{\text{low}}$ as extending from 2.07×10^{-3} to 16.63×10^{-3}.

Chiang and Reeves (1962) and Worlund and Taylor (1983) both give details of the required (iterative) calculations for "exact" confidence intervals for a binomial proportion estimated by group testing (with equal group sizes). In addition, the former article provides a series of figures from which 95% and 99% confidence intervals for $\tilde{p}_{\text{low}}$ can be read off, for various combinations of $\hat{p}_{\text{high}}$ and n, and the latter provides a table of 90% confidence intervals for $\tilde{p}_{\text{low}}$, for various combinations of N and n. Reading off Fig. 5 in Chiang and Reeves (1962) gives an "exact" 95% confidence interval for $\tilde{p}_{\text{low}}$ in the specific example above of 2.1×10^{-3} to 16.7×10^{-3}, in good agreement with Moran et al. (1983).

The easiest way to calculate an "exact" confidence interval for a value of $\tilde{p}_{\text{low}}$ calculated from equation 10.26 is first to obtain the Clopper–Pearson interval, as in section 10.3.3, for $\hat{p}_{\text{high}}$. The upper and lower limits of this interval are then substituted into equation 10.26 in turn, and the resulting values are the limits of the corresponding interval for $\tilde{p}_{\text{low}}$. Using MINITAB, the 95% Clopper–Pearson interval for $\hat{p}_{\text{high}}$ in the specific example above extends from 0.187086 to 0.812914. Substituting these limits in turn into equation 10.26, the limits of the corresponding interval for $\tilde{p}_{\text{low}}$ extend from 2.07×10^{-3} to 16.62×10^{-3}.

When unequal group sizes are used, as in Rodoni et al. (1994), the above calculations for "exact" confidence intervals are no longer appropriate. In this case, Hepworth (1996) gives details of the required procedure.

10.8.5 Group testing using generalized linear models

Farrington (1992) puts group testing for estimation into a GLM framework. The starting point is $p_{\text{high}} = 1-(1 - p_{\text{low}})^n$ (section 10.8.1), from which $\text{CLL}(p_{\text{high}}) = \ln(n) + \text{CLL}(p_{\text{low}})$, where CLL(•) denotes the complementary log-log transformation (section 9.7.1). Consider a data set comprising M assessments of infection. For the jth assessment ($j = 1, \ldots, M$), the group size is n_j, and there are N_j groups, of which Y_j test positive for infection. Table 10.11 shows such a data set, comprising all the data (17 assessments) on PVX infection of cv. Sebago certified seed potato crops given in Tables 4 and 5 of Moran et al. (1983). The data as presented here, for illustration, pose two problems. First, there are two different group sizes (different group sizes were used in each of two seasons for which data are reported). Second, the proportion of groups testing positive varies widely between assessments, raising the prospect that an analysis of virus infection over assessments may be influenced by extra-binomial variation.

TABLE 10.11. Group-testing data from 17 assessments of PVX infection for cv. Sebago certified seed potato crops, extracted from Tables 4 and 5 of Moran et al. (1983).

Number of assessments	*Number of groups*	*Group size*	*Number (proportion) of groups testing positive*
3	20	50	0 (0.00)
7	10	100	0 (0.00)
1	20	50	1 (0.05)
1	20	50	2 (0.10)
1	10	100	1 (0.10)
1	20	50	4 (0.25)
1	20	50	9 (0.45)
2	10	100	5 (0.50)

Both problems are resolved using the approach outlined by Farrington (1992).

The data in Table 10.11 were fitted to a GLM with binomial errors and the complementary log-log link function, with the coefficient of $\ln(n)$ set equal to one (referred to as an "offset" in GLM terminology) and with no explanatory variables. As in Farrington (1992), the analysis was carried out using the statistical software GLIM. The estimated value of CLL (p_{high}) is -6.386, with an estimated standard error of 0.1927. These results can be transformed back to an incidence scale as follows:

$$\tilde{p}_{\text{low}} = 1 - \exp(-\exp(-6.386)) = 1.684 \times 10^{-3}$$

and after using standard methodology to obtain the approximate variance of a function (see, e.g., Colquhoun, 1971):

$$\text{SE}(\tilde{p}_{\text{low}}) \approx \exp(-6.386 - \exp(-6.386))0.1927$$
$$= 0.324 \times 10^{-3}$$

However, the resulting deviance of 69.7 (16 *df*) is significantly large, indicating extra-binomial variation. This is readily accounted for by means of so-called quasi-likelihood methods (Williams, 1982; Bennett, 1989), as described by Farrington (1992). The estimated value of CLL (p_{high}) is then -6.400, with an estimated standard error of 0.4248. The resulting deviance of 14.5 (16 *df*) is satisfactory. The results, transformed back to an incidence scale, are then as follows:

$$\tilde{p}_{\text{low}} = 1 - \exp(-\exp(-6.400)) = 1.660 \times 10^{-3}$$

$$\text{SE}(\tilde{p}_{\text{low}}) \approx \exp(-6.400 - \exp(-6.400))0.4248$$
$$= 0.705 \times 10^{-3}$$

The estimate $\tilde{p}_{\text{low}}$ is little changed, but its estimated standard error is inflated from the artificially small value obtained from the initial analysis. An approximate 95% confidence interval for $\tilde{p}_{\text{low}}$ then extends from 0.028×10^{-3} to 3.041×10^{-3}.

In a group-testing scenario, the quasi-likelihood methodology adopted here provides a more straightforward way of analyzing aggregated disease incidence data than distribution-based methodology. The quasi-likelihood methodology introduced by Williams (1982) can also be used as an alternative to the distribution-based methodology for the regression analysis of aggregated disease incidence data collected by cluster sampling, discussed in section 10.6.2. Hughes and Madden (1995) give a phytopathological example of the application of Williams' methods to such data.

10.9 Binomial Sampling for Count Data

Recall that in section 10.4 the estimated mean number of infections per sampling unit ($\bar{Y}$) was obtained by counting all the infections in each sampling unit inspected. The idea behind *binomial sampling* is that a relationship between the estimated mean number of infections per sampling unit and the proportion of sampling units with more than a *tally threshold* of T infections can be used to avoid the need for a full count of the number of infections. Instead, only the proportion of sampling units with more than T infections need be obtained by sampling. Most applications of binomial sampling have been in economic entomology (e.g., Jones, 1994b) and weed science (e.g., Gold et al., 1996). Furthermore, binomial sampling for insect pests and for weeds is often used in sequential sampling plans, and where the objective is classification rather than estimation. The statistical properties of estimates obtained by binomial sampling have been discussed by Binns et al. (2000). Here we restrict our attention to the estimation of mean number of infections per sampling unit from a sample of prespecified size. A special case of binomial sampling is when $T = 0$ and sampling units are simply recorded as either uninfected or infected. However, this is rarely optimal in practice.

10.9.1 Binomial sampling based on probability distributions

For a Poisson distribution with mean μ, the proportion of sampling units with more than T infections is given by:

$$p_T = 1 - \sum_{i=0}^{T} \frac{\mu^i e^{-\mu}}{i!} \qquad (10.37)$$

for $i = 0, 1, 2, \ldots, T$. If $T = 0$, equation 10.37 reduces to $p_T = p_{\text{high}} = 1 - e^{-\mu}$ (equation 9.33). In this particular case we can rearrange equation 10.37 to $\mu = -\ln(1 - p_T)$. Then, on obtaining an estimate of the proportion of sampling units with more than T infections ($\hat{p}_T$), the mean number of infections per sampling unit is estimated by $\bar{Y} = -\ln(1 - \hat{p}_T)$. Unfortunately, there is no general rearrangement of equation 10.37 leading to $\tilde{Y} = f(\hat{p}_T)$ for $T > 0$. Normally, then, a binomial sampling estimate of mean number of infections per sampling unit is obtained by numerical solution of:

$$\hat{p}_T = 1 - \sum_{i=0}^{T} \frac{\tilde{Y}^i e^{-\tilde{Y}}}{i!} \quad (10.38)$$

for $\tilde{Y}$, for a given T and an estimate $\hat{p}_T$ obtained by sampling. The question then becomes one of an appropriate choice of tally threshold T. The lower is the value of T, the easier is the process of data collection by sampling. However, this is not the only criterion for selecting a value of T, as we shall see.

Example 10.2 continued. The task here is to estimate the mean number of lesions per plant in Field 10.2.1 (section 10.4.4) by binomial sampling. As before, $CV = 0.1$ is set as the required level of reliability for the sample mean, and a preliminary sample estimate of the mean of 5.65 lesions per plant is available.

An approximate binomial sampling variance for the counts of lesions per plant is given by:

$$s_{\tilde{Y}}^2 \approx \frac{\hat{p}_T(1-\hat{p}_T)}{(d\hat{p}_T / d\tilde{Y})^2}$$

(Binns et al., 2000) in which $\hat{p}_T$ is formulated as in equation 10.38 and $d\hat{p}_T / d\tilde{Y} = \tilde{Y}^T e^{-\tilde{Y}} / T!$ Then the required sample size (number of sampling units to be inspected) N is given by:

$$N = \frac{s_{\tilde{Y}}^2}{\tilde{Y}^2 CV^2}$$

(Binns et al., 2000). N can be calculated for a range of values of T, using the required level of reliability for the sample mean ($CV = 0.1$) and a mean of 5.65 lesions per plant (from the preliminary sample). The following results are obtained.

T	0	1	2	3	4	5	6	7
N	888	182	73	43	32	28	28	32

Clearly the choice of tally threshold T has an important influence on the amount of sampling required to meet the required level of reliability. On the basis of these results, a value of $T = 5$ is selected, so $N = 28$. This is larger than the value $N = 18$ (see section 10.4.4) required when all lesions are to be counted, but it may require less effort to record plants only as having either ≤ 5 or >5 lesions, during sampling.

In a random sample of $N = 28$ plants drawn from Field 10.2.1, 15 plants had five or fewer lesions. Numerical solution of equation 10.38 for $\tilde{Y}$, with $T = 5$ and $\hat{p}_T = 15/28$, gives $\tilde{Y} = 5.89$. The realized value of CV for this sample is given by:

$$CV = \frac{\sqrt{s_{\tilde{Y}}^2 / N}}{\tilde{Y}} = 0.098$$

so the mean number of lesions per plant, as estimated by binomial sampling, has the required level of reliability.

For aggregated count data, it is possible—at least in principle—to use a similar approach, based on the negative binomial distribution rather than the Poisson (e.g. Jones, 1994b). By analogy with equation 10.37, we can write:

$$p_T = 1 - \sum_{i=0}^{T} \left(\frac{\Gamma(k+i)}{\Gamma(k)\Gamma(i+1)} \right) \left(\frac{\mu}{k} \right)^i \left(1 + \frac{\mu}{k} \right)^{-(k+i)}$$

If $T = 0$, this reduces to $p_T = p_{\text{high}} = 1 - (1 + (\mu / k))^{-k}$ (equation 9.35). In this particular case the equation can be rearranged to $\mu = k((1 - p_T)^{-1/k} - 1)$. Then (supplied with a previously-determined estimate of the negative binomial aggregation parameter k), on obtaining an estimate of the proportion of sampling units with more than T infections ($\hat{p}_T$), the mean number of infections per sampling unit is estimated by $\tilde{Y} = \hat{k}((1 - \hat{p}_T)^{-1/\hat{k}} - 1)$. As in the case of the Poisson-based analysis, there is no general rearrangement leading to $\tilde{Y} = f(\hat{p}_T)$ for $T > 0$. Note that the estimate $\tilde{Y}$ (whether obtained analytically for $T = 0$ or numerically for any T) depends on the adopted value for the negative binomial aggregation parameter. This means that the adopted value of the negative binomial aggregation parameter affects the estimate $\tilde{Y}$ in terms of both bias and precision. When all infections are counted, the estimated mean $\bar{Y}$ (equation 9.20) is unbiased, and the adopted value of the negative binomial aggregation parameter affects only the precision of this estimate.

10.9.2 Binomial sampling based on empirical models

As an alternative to the use of relationships based on statistical probability distributions, empirical models have been used to characterize relationships between the mean number of infections per sampling unit and the proportion of sampling units with more than a tally threshold of T infections. A number of such relationships have been described in the phytopathological literature, where it is normally the case that T = 0 is adopted. These are therefore relationships between mean number of infections per sampling unit ($\bar{Y}$) and disease incidence at the scale of the sampling unit ($\hat{p}_{\text{high}}$). In what follows, a, b, and c represent arbitrary constants, used to characterize these relationships.

Dillard and Seem (1990) used a logarithmic relationship of the form:

$$\ln(\bar{Y}) = a + b\hat{p}_{\text{high}}$$

to characterize a relationship between the number of uredinia per leaf and the percentage of diseased leaves in their study of common maize rust (caused by *Puccinia sorghi*) on sweet corn. Polynomial relationships of the form:

$$\bar{Y}^{1/c} = a + b\hat{p}_{\text{high}}$$

have also been investigated. In studies by Seem and Gilpatrick (1980) and Silva-Acuña et al. (1999), $c = 2$ was adopted. In Beresford and Royle (1991), $c = 3$ was adopted. Such relationships provide a basis for prediction of mean number of infections per sampling unit from sample data for disease incidence at the scale of the sampling unit. As far as applications in binomial sampling are concerned, the most widely studied relationship between the mean number of infections per sampling unit and the proportion of sampling units with more than a tally threshold of T infections is:

$$\ln(\bar{Y}) = \ln(a) + b\text{CLL}(\hat{p}_T) \qquad (10.39)$$

In economic entomology, this is often referred to as the Kono–Sugino (or K–S) equation, after the early work of Kono and Sugino (1958) (published in Japanese). Another early application of equation 10.39 is given by Gerrard and Chiang (1970). This is now the model of choice for applications of binomial sampling in the management of insect pests (e.g., Binns et al., 2000). For the particular case of $T = 0$, equation 10.39 can be obtained by rearrangement of the two-parameter generalization of the Poisson model discussed in section 9.7.2.

Once a K–S equation has been established as a suitable model for a particular pathosystem, data for $\hat{p}_T$ may be collected by sampling, and used to predict $\tilde{Y}$. There has been a considerable amount of discussion in the entomological literature relating to the calculation of an appropriate variance for this estimate. Binns and Nyrop (1992) provide a useful summary. Important components of the variance estimate for $\tilde{Y}$ are the sampling variance of the estimate $\hat{p}_T$ and the variances of the estimated parameters of the fitted model. In this context, Snedecor and Cochran (1989, section 9.9) provide a statistical discussion of the use of regression models for prediction.

Organisms being what they are, however, there would be some variation about the model even in a perfect world, resulting in a biological variance that has nothing to do with sampling error. Schaalje et al. (1991) argue that this biological variance needs to be estimated, and the variance estimate for $\tilde{Y}$ reduced accordingly. A particularly clear account of this rather complicated issue is given by Jones (1994b). Hepworth and MacFarlane (1992) discuss the implications of non-random sampling for the variance estimate for $\tilde{Y}$. Current practice for applications of the K–S equation for binomial sampling economic entomology is outlined by Binns et al. (2000), who note that "*it has not yet been fully resolved.*" Plant pathologists have yet to make much use of the K–S equation, but for those interested in its application, the extensive documentation in the entomological literature is a valuable resource.

10.10 Estimation of Disease Severity

If we are prepared to assume that mean disease severity ($\bar{y}_{\text{sev}}$) is approximately normally distributed, sample size calculations for estimation of $\bar{y}_{\text{sev}}$ by simple random sampling may be based on the formulae given by Karandinos (1976) (section 10.2.6). N is the number of sampling units (plants or plant parts) to be inspected to meet a pre-specified level of reliability for this estimate. If reliability is defined in terms of coefficient of variation of the sample mean, CV:

$$N = \frac{s^2}{\bar{y}_{\text{sev}}^2 CV^2} \qquad (10.40)$$

If reliability is defined by setting the half the length of the required confidence interval of the sample mean equal to a fixed proportion of the mean, H:

$$N = \frac{s^2}{\bar{y}_{\text{sev}}^2}\left(\frac{z_{\alpha/2}}{H}\right)^2 \qquad (10.41)$$

If reliability is defined by setting the half the length of the required confidence interval of the sample mean equal to a fixed positive number h:

$$N = s^2\left(\frac{z_{\alpha/2}}{h}\right)^2 \qquad (10.42)$$

Thus, before the sample size for the required level of reliability can be calculated, a preliminary estimate for s^2 is required, and also (where relative reliability is specified) a preliminary estimate for $\bar{y}_{\text{sev}}$.

As noted by James and Shih (1973), a method of predicting severity from data for disease incidence could be of practical importance because it would simplify the process of obtaining disease severity assessments. While most plant pathologists would agree with this, little seems to have been done as yet to develop the application of relationships between disease severity and disease incidence for sampling purposes. Various relationships between severity and incidence have been formulated (section 9.8; McRoberts et al., 2003), including, in equation 9.38, an analogue of the K–S equation (equation 10.39). However, important issues for prediction of severity from sample data for incidence remain to be addressed. In particular, the development of binomial sampling in the context of economic entomology has shown the importance of careful consideration both of tally thresholds other than zero, and of the formulation of the variance of the predicted mean. Paul et al. (2005) have recently addressed some aspects of prediction of disease severity from samples of incidence based on methodology originally developed in economic entomology.

10.11 Inverse Sampling for Disease Incidence

Suppose we are to estimate disease incidence by simple random sampling (section 10.3), and that our view ahead of sampling is that disease incidence is low. The required level of reliability for the sample estimate of incidence is that the half-length of the 95% confidence interval is equal to a fixed proportion of the mean, $H = 0.5$. From Table 10.1, initial estimates of disease incidence of 0.01, 0.02 and 0.05 lead, respectively, to calculated sample sizes of 1522, 753 and 292. Two points emerge. First, the calculated sample sizes are large. Second, a small change in the initial estimate of disease incidence makes a big difference to the calculated sample size. With an initial estimate that is only slightly too small, we may do much more sampling than is really needed. With an initial estimate that is only slightly too large, we may do far too little sampling to meet the required level of reliability. In such circumstances, we may consider *inverse sampling*. In inverse sampling, the sample size is not fixed ahead of sampling. Instead, sampling continues until a predetermined number of "positives" has been recorded. In plant pathology, it is likely that the positive/negative status of sampling units would be determined on the same basis as usual for incidence, i.e. the presence/absence of disease. However, if required, the positive/negative status of sampling units could be determined by reference to a tally threshold, as is more often the case in economic entomology. The following discussion is restricted to inverse sampling when individual plants, or plant parts, constitute the sampling units. For a discussion of inverse sampling in a cluster sampling scenario, see Madden et al. (1996).

10.11.1 How many positives?

The statistical basis for inverse sampling was provided by Haldane (1945). A useful, agriculturally-orientated account is given by Sampford (1962). Bennett (1981) discusses applications in epidemiology. An unbiased estimate of incidence is:

$$p' = \frac{n' - 1}{N - 1} \qquad (10.43)$$

in which n' is the number of positives (i.e., identified as infected) and N is the total size of the sample (positives plus negatives). The estimate often used in practice, $p' = n'/N$, is biased (the bias is only large when n' and N are small). An unbiased estimate of the variance of p', $s^2(p')$is:

$$s^2(p') = \frac{p'(1-p')}{N-2} \qquad (10.44)$$

(ignoring the finite population correction). Progress then depends on finding an appropriate approximation for the estimated standard error, based on the variance estimate given in equation 10.44. From Lui (1995):

$$\mathrm{SE}(p') \approx \sqrt{\frac{(p')^2(1-p')}{n'}} \qquad (10.45)$$

Equation 10.45 may be used to formulate equations for the number of positives (n') needed in the sample to meet the required level of reliability of the estimate p', under the different definitions of reliability given in section 10.2.6. If reliability is defined by the coefficient of variation of the sample mean, CV:

$$n' = \frac{(1-p')}{CV^2} \qquad (10.46)$$

If reliability is defined by setting the half the length of the required confidence equal to a fixed proportion of the mean, H:

$$n' = (1-p')\left(\frac{z_{\alpha/2}}{H}\right)^2 \qquad (10.47)$$

If reliability is defined by setting the half the length of the required confidence interval of the sample mean equal to a fixed positive number h:

$$n' = p'^2(1-p')\left(\frac{z_{\alpha/2}}{h}\right)^2 \qquad (10.48)$$

Suppose that, as above, the required level of reliability for the sample estimate of incidence is that the half-length of the 95% confidence interval is equal to a fixed proportion of the mean. Here, H is specified as 0.5 and, for a 95% confidence interval, $z_{\alpha/2} = 1.96$. Now, provided with an initial estimate of incidence, we can use equation 10.47 to calculate the number of positives (n')needed in the sample to meet the required level of reliability (as defined by H) of the estimate p'. For example:

Initial p'	0.01	0.02	0.05	0.10	0.15	0.20
Required n'	16	16	15	14	14	13

The values of n' obtained from equation 10.47 are rounded up to the next integer above. Note that n' is not very sensitive to the initial p' in this case, because p' is low and so in equation 10.47, the term $1 - p' \approx 1$. As an example, suppose that the initial estimate of incidence was taken to be $p' = 0.02$. We would then sample until $n' = 16$ diseased individuals had been recorded. If this number were reached when $N = 300$ individuals had been inspected, we would then have $p' = (16-1)/(300-1) = 0.05$ (equation 10.43), somewhat larger than the initial estimate. Then, from equation 10.45, $SE(p') = 0.0122$, and the realized value of H is equal to 0.48, close to the required value of 0.50.

Sampford (1962), with an extra approximation, obtains an alternative formula for the standard error:

$$SE(p') \approx \frac{p'}{\sqrt{n'-1}} \qquad (10.49)$$

If we start with $H p' = z_{\alpha/2}\, SE(p')$ and use equation 10.49, after some rearrangement:

$$n' = 1 + \left(\frac{z_{\alpha/2}}{H}\right)^2 \qquad (10.50)$$

In this case, the required n' does not depend on any assumption about p', other than that it is small. For $H = 0.5$ and $z_{\alpha/2} = 1.96$, $n' = 17$. The price paid for the extra assumption required to obtain equation 10.49 is some extra (unnecessary) sampling.

Realized values of H obtained after use of equation 10.47 to calculate n' (for various realized values of p', and with $z_{\alpha/2} = 1.96$) are as follows:

n'	*Realized* p'	*Realized H*
16	0.01	0.488
16	0.02	0.485
15	0.05	0.493
14	0.10	0.497
14	0.15	0.483
13	0.20	0.486

Note that results given by Lui (1995) based on equation 10.47 relate to the *whole* length of the confidence interval, not the half-length as here. If, instead, equation 10.50 is used to calculate n' (again with $z_{\alpha/2} = 1.96$), realized values of H are as follows:

n'	*Realized* p'	Realized H
17	0.01	0.473
17	0.02	0.470
17	0.05	0.463
17	0.10	0.451
17	0.15	0.438
17	0.20	0.425

The extra sampling carried out as the price of not providing an initial value of p' means that a confidence interval rather narrower than actually required is obtained, especially with larger values of p'. However, for values $p' \leq 0.05$, the difference between use of equation 10.47 and of equation 10.50—in terms of the amount of sampling required and the realized values of H—is not large.

10.11.2 Exact confidence intervals

In section 10.3.3, the example of a simple random sample in which one diseased plant is observed in a sample of $N = 10$ plants was considered. In that case, the Clopper–Pearson 95% interval extended from 0.0025 to 0.445. Consider now the situation in which nine healthy plants are observed and then one diseased plant, at which point sampling is halted. George and Elston (1993) discuss the special case of Clopper–Pearson-type confidence intervals based on the first occurrence of an event. From George and Elston (1993, Table 1), the 95% confidence interval for the point estimate $1/N$ extends from 0.0025 to 0.336.

Lui (1995) generalizes the approach of George and Elston (1993) to cover calculation of Clopper–Pearson-type confidence intervals when sampling continues until the required number of positives in the sample (n') is obtained, for $n' \geq 1$. These intervals apply to the (biased) maximum likelihood estimate of incidence, $p' = n'/N$. For example, for $p' = 0.01$, $n' = 16$, the 95% Clopper–Pearson-type confidence interval extends from 0.0057 to 0.015. For $p' = 0.1$, $n' = 14$, the 95% interval extends from 0.056 to 0.155.

In early work by Finney (1947, 1949), it was pointed out that tables used to obtain "exact" (Clopper–Pearson) binomial confidence intervals (Diem and Lentner, 1970; Fisher and Yates, 1963) could also be used to obtain Clopper–Pearson-type confidence intervals with inverse sampling. If inverse sampling results in n' positives in a sample of total size N, the lower limit of a two-sided $100(1-\alpha)\%$ confidence interval for incidence is the same as the lower limit of the $100(1-\alpha)\%$ binomial confidence interval for a simple random sample with n' positives in a total of N. The upper limit of a two-sided $100(1-\alpha)\%$ confidence interval for incidence is the same as the upper limit of the $100(1-\alpha)\%$ binomial confidence interval for a simple random sample with $n' - 1$ positives in a total of $N - 1$.

10.11.3 The geometric series

Finney (1949) points out that since inverse sampling data are usually collected sequentially, the data can also provide evidence on whether the condition of independence of successive observations is fulfilled. If successive observations are independent of one another, infected plants should occur at random intervals throughout the sample. Non-independence, such as long sequences of diseased plants, would increase the frequency of both longer and shorter intervals between infected plants. If a statistical significance test were indicative of deviations from randomness of intervals between infected plants, this in turn would indicate that the standard error (equation 10.45 or 10.49) and approximate confidence intervals based on it were not applicable.

Cochran (1936) considered tests of this kind in connection with the analysis of data such as those illustrated in Fig. 9.11. Consider the actual plot rows to be joined end to end in one continuous "row". It is the number of "runs" of diseased plants (N_r) that is counted. The geometric series test can be conceptualized as follows. Proceed along the row until a diseased plant is reached: if each plant has an equal and independent probability of becoming infected (denoted p), the probability that the next $Y_r - 1$ plants are diseased, and the following one healthy (a run of Y_r diseased plants), follows a geometric series. The distribution of the number of runs of Y_r diseased plants (i.e., runs of length Y_r) is given by:

$$\Pr(Y_r) = p^{Y_r - 1}(1 - p)$$

in which $Y_r = 1, 2, \ldots$, without limit. The observed proportion of diseased plants ($\bar{y}$) provides an estimate of p, and the resulting probabilities are multiplied by N_r to give expected frequencies. The observed and expected frequencies of runs of length Y_r for the first-count data in Fig. 9.11, given by Cochran (1936, page 59), are as follows.

Length of run (no. of diseased plants, Y_r)	*Observed frequency (O)*	*Expected geometric series frequency (E)*
1	164	169.5
2	33	30.7
3	9	5.6
4	1	1.0
5	0	0.2

Cochran (1936) gives the following test of independence. One estimate of p is the observed proportion of diseased plants ($\bar{y} = 261/1440 = 0.181$). The quantity:

$$\frac{\left(\sum Y - N_r\right)}{\sum Y},$$

in which N_r (=207) is the observed number of runs and $\sum Y$ (=261) is the number of diseased plants in these runs, is also an estimate of p and is greatest when there are many long runs (Cochran, 1936). In this case,

$$\frac{\left(\sum Y - N_r\right)}{\sum Y} = \frac{54}{261} = 0.207.$$

Comparison of

$$\frac{\left(\sum Y - N_r\right)}{\sum Y} - \bar{y} = 0.026$$

with its estimated standard error $\sqrt{[\bar{y}(1 - \bar{y})^2]/N_r} = 0.024$ does not provide evidence to suggest that the occurrence of diseased plants along rows is anything other than at random. Further examples of the application of this analysis to plant disease data are given in Bald (1937). Cochran (1936) also discusses versions of the analysis appropriate when runs of both healthy and diseased plants are being counted, and when there is neighbor infection (i.e., when disease spreads from an infected plant to its immediate neighbors with a specified probability).

Tests based on the geometric series can be affected by the assumption that a plot is a single long row, because we consider some plants to be adjacent that are actually not so (Cochran, 1936). If there is higher infection at the edges of the plot, artificially long runs of diseased plants may be created by the joining of plot rows. In such circumstances, Cochran (1936) suggests data from the edges of the plot be kept aside when making the test. Clearly this also needs to be borne in mind when inverse sampling.

10.12 Sequential Estimation of Disease

The statistical basis for *sequential estimation* is discussed by Anscombe (1953), but the application of sequential methods in the study of biological populations is based largely on two articles by Kuno (1969, 1972). As with inverse sampling, the main motivation for sequential sampling is that the need to provide an initial estimate of (in the present context) disease intensity is avoided. Instead, a "stop" condition is specified such that sampling is halted when an estimate of the required reliability has been obtained. In general, the stop condition is specified by noting that mean intensity per unit can be written as the accumulated total intensity divided by the accumulated number of units assessed. Thus mean intensity can be replaced in the sample size formulae given in Tables 10.1–10.3 by an expression for accumulated total intensity divided by the accumulated number of units assessed. If the resulting formulae can be rearranged so that accumulated total intensity is on the left-hand side, we will have formulae that indicate the accumulated total intensity required for an estimate of mean intensity with a pre-specified level of reliability. Where sample size calculations depend on having estimates of additional parameters (such as, for example, the aggregation parameters of statistical probability distributions), these must still be supplied ahead of sampling.

We denote the accumulated total intensity required for an estimate of mean intensity with a pre-specified level of reliability as T_N. In most cases, the formulae for sequential estimation obtained from the sample size formulae given in Tables 10.1–10.3 are of the form $T_N = f(N)$.

The practical implication of this is that knowing when to stop sampling requires more than just a simple count of sampling units inspected. Instead, a cumulative record of disease intensity is kept during sampling, on a chart that shows the *stop line* as a graphical plot of T_N against N. Sampling is halted when the cumulative record of disease intensity intersects the stop line. Contrast this with inverse sampling, where the total number of units observed does not determine when to halt sampling—only the number of positive units is involved. When discrete data such as disease incidence are collected, sequential estimation necessarily involves an approximation. The stop condition is represented by a continuous variable, so it will not be met exactly by the cumulative record of a discrete variable. Instead, sampling will be halted when the cumulative record of disease has overshot the stop condition.

In deriving formulae for sequential estimation of disease from the formulae in Tables 10.1–10.3, we omit the finite population correction. This omission requires some consideration. In the case of fixed sample size formulae, there is an opportunity for revision, before sampling, if the required sample size is too large relative to the size of the population being sampled (section 10.2.5). In the case of sequential estimation, the sample size is not specified before sampling, so at the outset it is not known whether or not the finite population correction is needed. If a small population is being sampled using sequential estimation methodology, the best policy is to build the finite population correction into the function $T_N = f(N)$, representing the stop line, from the outset. Kuno (1969) discusses this issue and gives some examples of stop lines formulated using the finite population correction. The effect of incorporating the finite population correction is that the resulting stop line will result in an earlier cessation of sampling than the corresponding stop line for a large population.

10.12.1 Sequential estimation of disease incidence from simple random sampling

Kuno (1969) and Jones (1994b) discuss the sequential estimation of proportions. The sample size formulae in Table 10.1 are all of the form $N = f(\bar{y}, \text{reliability})$, in which N is the number of sampling units (individual plants or plant parts in the case of simple random sampling) and $\bar{y}$ is disease incidence expressed as a proportion. Here, we substitute using $T_N / N = \bar{y}$ and then solve for T_N. The resulting stop line equations are given in Table 10.12. When sampling from large populations, the stop line equations for reliability defined by the coefficient of variation of the sample mean, CV, or by setting the half the length of the required confidence equal to a fixed proportion of the mean, H, increase towards upper asymptotes of $T_N = 1/CV^2$ and $T_N = (z_{\alpha/2}/H)^2$, respectively, as N increases (so that $1/N \approx 0$).

If the finite population correction is included, the stop line equation for reliability defined by the coefficient of variation of the sample mean, CV, is:

$$T_N = \frac{1 - \dfrac{N}{M}}{CV^2 + \dfrac{1}{N} - \dfrac{1}{M}}$$

in which M is the population size. The corresponding stop line equation for reliability defined by setting the half the length of the required confidence equal to a fixed proportion of the mean, H, is:

$$T_N = \frac{1 - \dfrac{N}{M}}{\dfrac{H^2}{z_{\alpha/2}^2} + \dfrac{1}{N} - \dfrac{1}{M}}$$

Both these formulae specify stop lines that reach a maximum value of T_N at some intermediate value of N, following which T_N decreases with further increases in N.

If reliability is defined by setting the half the length of the required confidence interval of the sample mean equal to a fixed positive number, h, the formula given in Table 10.3 specifies two stop lines. On a graphical plot of T_N against N ($N \geq 0$), these two lines intersect to form a closed loop (Jones, 1994b). The lines intersect at

TABLE 10.12. Stop line equations for sequential simple random sampling for disease incidence data, with three different definitions of reliability[a].

	Reliability defined by:		
		Half-width of confidence interval equal to:	
Description of variability	*Coefficient of variation, CV*	*Proportion of mean, H*	*Fixed positive number, h*
Random (binomial distribution)	$T_N = 1/(CV^2 + (1/N))$	$T_N = 1/[(H/z_{\alpha/2})^2 + (1/N)]$	$T_N = (N/2)\left(1 \pm \left(\sqrt{z_{\alpha/2}^2 - 4h^2N}/z_{\alpha/2}\right)\right)$

[a]The cumulative number of diseased plants or plant parts, T_N, is shown as a function of the number of sampling units (individual plants or plant parts), N. The term $z_{\alpha/2}$ is the $100(1 - \alpha/2)^{\text{th}}$ percentile of the standard normal distribution. The formulae omit the finite population correction.

$N = 0$ and at the maximum sample size, $N = (z_{\alpha/2}/2h)^2$. If the finite population correction is included, the two stop line equations again intersect to form a closed loop, with the maximum sample size reduced to $N = 1/[(4h^2/z^2_{\alpha/2}) + (1/M)]$.

In practice, once the appropriate stop condition has been satisfied, estimated disease incidence $\bar{y}$ and its standard error are obtained using the formulae given in section 10.3, taking N as the sample size. While recognizing that this may produce a biased estimate, the extent of this bias is usually small enough not to be of practical significance.

10.12.2 Sequential estimation for count data

The sample size formulae in Table 10.2 are adapted for sequential estimation by substituting $T_N/N = \bar{Y}$ and then solving for T_N. The resulting stop line equations are given in Table 10.13. Formulae for sequential estimation of count data are discussed by Kuno (1969), Green (1970), Binns (1975), and Hutchison (1994). By far the majority of applications of sequential sampling for count data are in economic entomology (Hutchison, 1994), but Strandberg (1973) gives a phytopathological example for estimation for cabbage black rot (caused by *Xanthomonas campestris* pv. *campestris*), using a stop line based on the negative binomial distribution.

Example 10.2 continued. The task, as before, is to estimate the mean number of lesions per plant in Fields 10.2.1 and 10.2.2 (section 10.4.4), but this time using sequential sampling. $CV = 0.1$ is again set as the required level of reliability for the sample mean. For Field 10.2.1 we use our previous information that the Poisson distribution is an appropriate description of variability. Then, from Table 10.13, $T_N = 1/0.1^2 = 100$. This stop line is shown in Fig. 10.7. In this case, since T_N is constant, we are, in effect, inverse sampling for counts, with the stop condition specified as an accumulated count of 100 lesions. If we were to take into consideration the finite population correction, the stop line equation would be:

$$T_N = \frac{1}{CV^2}\left(1 - \frac{N}{M}\right)$$

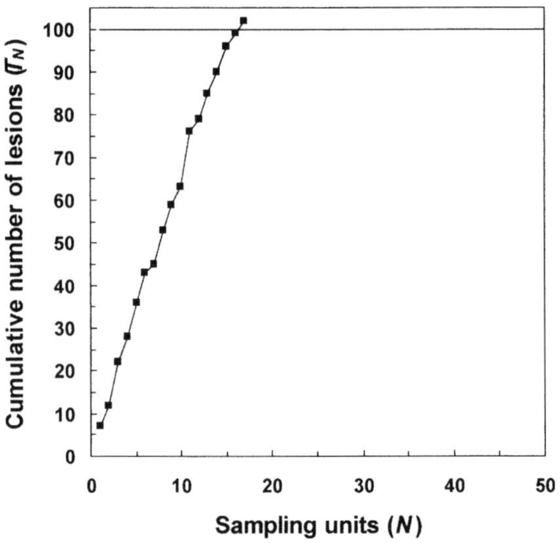

FIG. 10.7. Sequential estimation for count data. $CV = 0.1$ is set as the required level of reliability for the sample mean. For Field 10.2.1, the Poisson distribution is an appropriate description of variability. Then, from Table 10.13, the stop line is $T_N = 1/0.1^2 = 100$ (shown as a thin horizontal line on the graph). A random sample of plants is inspected, and the number of lesions recorded, plant by plant, until the line for the cumulative number of lesions (shown as a line joining the sample data points on the graph) intersects the stop line. Following this procedure for Field 10.2.1, sampling was halted when 102 lesions had been observed after inspection of 17 plants.

TABLE 10.13. Stop line equations for sequential simple random sampling for count data, with three different definitions of reliability[a].

	Reliability defined by:		
		Half-width of confidence interval equal to:	
Description of variability	*Coefficient of variation, CV*	*Proportion of mean, H*	*Fixed positive number, h*
Random (Poisson distribution)	$T_N = 1/CV^2$	$T_N = (z_{\alpha/2}/H)^2$	$T_N = (hN/z_{\alpha/2})^2$
Aggregated (Negative binomial distribution)[b]	$T_N = 1/[CV^2 - (1/\hat{k}N)]$	$T_N = 1/[(H/z_{\alpha/2})^2 - (1/\hat{k}N)]$	$T_N = (N/2)\left(-\hat{k} \pm \left(\sqrt{z^2_{\alpha/2}\hat{k}^2 + 4h^2\hat{k}N}/z_{\alpha/2}\right)\right)$
Aggregated (Taylor's power law)	$T_N = (CV^2/\hat{a})^{1/(\hat{b}-2)} N^{(\hat{b}-1)/(\hat{b}-2)}$	$T_N = (H^2/z^2_{\alpha/2}\hat{a})^{1/(\hat{b}-2)} N^{(\hat{b}-1)/(\hat{b}-2)}$	$T_N = (h^2/z^2_{\alpha/2}\hat{a})^{1/\hat{b}} N^{(\hat{b}+1)/\hat{b}}$

[a]The cumulative number of lesions (for example), T_N, is shown as a function of the number of sampling units (plants or leaves, for example), N. The term $z_{\alpha/2}$ is the $100(1-\alpha/2)$th percentile of the standard normal distribution. For aggregated data, an estimate of either the negative binomial aggregation parameter ($\hat{k}$) or the Taylor's power law parameters ($\hat{a}$, $\hat{b}$), are required. The formulae omit the finite population correction.

[b]Only the positive values of N that yield a positive value of the corresponding T_N are of interest in the present context.

On a graphical plot of T_N against N, this is a straight line that intersects the T_N axis at $T_N = CV^{-2}$ and the N axis at $N = M$.

A random sample of plants is inspected, and the number of lesions recorded, plant by plant, until the line for the cumulative number of lesions intersects the stop line. Sampling is then halted. Following this procedure for Field 10.2.1, sampling was halted when 102 lesions had been observed after inspection of 17 plants. The individual plant counts were 7, 5, 10, 6, 8, 7, 2, 8, 6, 4, 13, 3, 6, 5, 6, 3, and 3. The corresponding cumulative counts are shown in Fig. 10.7. The estimate $\bar{Y} = T_N/N$ is biased, but is nevertheless the one usually used in practice (Kuno, 1972). Thus, we have $\bar{Y} = 102/17 = 6.00$. The variance is $s_Y^2 = 7.75$, so the realized reliability level is $CV = 0.11$.

For Field 10.2.2 we use our previous information that a negative binomial distribution with $\tilde{k} = 4$ is an appropriate description of variability. Then, from Table 10.13, for $N > 1/(CV^2\hat{k}) = 25$, $T_N = 1/(0.1^2 - (1/4N))$. This stop line is shown in Fig. 10.8. If we were to take into consideration the finite population correction, the stop line equation would be:

$$T_N = \frac{\left(1 - \frac{N}{M}\right)}{CV^2 - \frac{1}{N\hat{k}}\left(1 - \frac{N}{M}\right)} \quad \text{for } N > \frac{1}{CV^2\hat{k} + M^{-1}}.$$

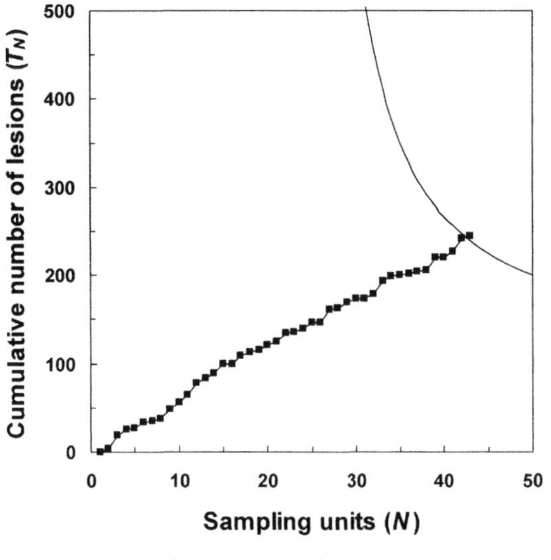

FIG. 10.8. Sequential estimation for count data. $CV = 0.1$ is set as the required level of reliability for the sample mean. For Field 10.2.2, a negative binomial distribution with $\hat{k} = 4$ is an appropriate description of variability. Then, from Table 10.13, the stop line is $T_N = 1/(0.1^2 - (4N)^{-1})$ (shown as a smooth curve in the upper right-hand part of the graph). A random sample of plants is inspected, and the number of lesions recorded, plant by plant, until the line for the cumulative number of lesions (shown as a line joining the sample data points on the graph) intersects the stop line. Following this procedure for Field 10.2.2, sampling was halted when 245 lesions had been observed after inspection of 43 plants.

A random sample of plants is inspected, and the number of lesions recorded, plant by plant. When the line for the cumulative number of lesions intersects the stop line, sampling is then halted. For Field 10.2.2, sampling was halted when 245 lesions had been observed after inspection of 43 plants. In this case, the individual plant counts were 0, 4, 15, 7, 1, 7, 1, 2, 11, 8, 9, 13, 5, 6, 10, 1, 9, 4, 3, 5, 4, 9, 2, 4, 7, 0, 14, 1, 8, 3, 1, 5, 15, 5, 1, 2, 2, 2, 15, 0, 6, 15, and 3. The corresponding cumulative counts are shown in Fig. 10.8. For this sample, $\bar{Y} = 5.70, s_Y^2 = 21.36$, and the realized reliability level is therefore $CV = 0.12$.

Binns and Nyrop (1992) provide a useful discussion of sequential sampling for estimation of counts. The adopted definition of reliability specifies an average long-run requirement rather than a criterion that will be met each and every time a sample is taken. If a sequential sampling plan is used many times, the estimates obtained generate a distribution that has a mean and variance of its own. These statistics characterize the overall reliability of the plan, rather than the reliability of an individual estimate. Evaluation techniques based on pre-existing data sets and on simulated data can be used to establish this overall reliability as part of the process of developing a sequential sampling plan (Hutchison, 1994).

10.12.3 Sequential estimation of disease incidence from cluster sampling

Formulae for estimation of disease incidence from sequential cluster-sampling data were developed by Madden et al. (1996). The sample size formulae in Table 10.3 are adapted for sequential estimation by substituting $T_N/nN = \bar{y}$ and then, where possible, solving for T_N. The resulting stop line equations are given in Table 10.14. If reliability is defined by setting the half the length of the required confidence interval of the sample mean equal to a fixed positive number h, all the stop lines are of the "closed loop" type (section 10.12.1). In some cases, specifically where a definition of relative reliability is adopted and the power law (section 9.4.7) is the appropriate description of variability, there is no simple formula for T_N as a function of N. In this situation, a stop line can be determined numerically (Madden et al., 1996). Again, the long-run performance of a sequential sampling plan is of interest, and can be investigated using pre-existing data sets and simulated data (Madden et al., 1996). Turechek and Madden (1999b) and Turechek et al. (2001) have developed sequential estimation methods for diseases of strawberry. Detailed accounts of the spatial analyses underlying the sampling plans and of the evaluation methods used establish the long-run characteristics of the plans are given.

Example 10.3 continued. Here, we will estimate disease incidence in Fields 10.3.1 and 10.3.2 (section 10.5.4) using sequential sampling. As before, $CV = 0.1$ is set as

TABLE 10.14. Stop line equations for sequential cluster sampling for disease incidence data, with three different definitions of reliability[a].

	Reliability defined by:		
		Half-width of confidence interval equal to:	
Description of variability	*Coefficient of variation, CV*	*Proportion of mean, H*	*Fixed positive number, h*
Random (binomial distribution)	$T_N = 1/(CV^2 + (1/nN))$	$T_N = 1/[(H/z_{\alpha/2})^2 + (1/nN)]$	$T_N = (nN/2)\left(1 \pm \left(\sqrt{z^2_{\alpha/2} - 4h^2nN}/z_{\alpha/2}\right)\right)$
Aggregated (β-binomial distribution or deff)[b]	$T_N = 1/[(CV^2/\text{deff}) + (1/nN)]$	$T_N = 1/[(H^2/(z^2_{\alpha/2}\text{deff})) + (1/nN)]$	$T_N = (nN/2)\left(1 \pm \left(\sqrt{z^2_{\alpha/2}\text{deff} - 4h^2nN}/(z_{\alpha/2}\sqrt{\text{deff}})\right)\right)$
Aggregated (power law)[c]	$\gamma_N = (CV^2/\hat{A})n^{3\hat{b}-2}N^{2\hat{b}-1}$	$\gamma_N = (H^2/z^2_{\alpha/2}\hat{A})n^{3\hat{b}-2}N^{2\hat{b}-1}$	$T_N = (nN/2)\left(1 \pm \sqrt{1 - 4n(h^2N/z^2_{\alpha/2}\hat{A})^{1/\hat{b}}}\right)$

[a]The cumulative number of diseased plants or plant parts, T_N, is shown as a function of the number of sampling units, N (each containing n individuals (plants or plant parts)). The term $z_{\alpha/2}$ is the $100(1 - \alpha/2)$th percentile of the standard normal distribution. For aggregated data, an estimate of either the β-binomial aggregation parameter ($\hat{\rho} = \hat{\theta}/(\hat{\theta}+1)$) or the power law parameters ($\hat{A}$, $\hat{b}$), are required. The formulae omit the finite population correction.

[b]$\text{deff} = 1 + \hat{\rho}(n-1)$.

[c]There is no simple formula for T_N as a function of N with the binary power law for the situations when reliability is defined in terms of CV or H. The left-hand side of the formula is written $\gamma_N = f(T_N) = T_N^{\hat{b}-2}(nN - T_N)^{\hat{b}}$ and the stop line is then determined numerically (see Madden et al., 1996).

the required level of reliability for the sample mean. For Field 10.3.1 we use our previous information that the binomial distribution is an appropriate description of variability. From Table 10.14, the stop line is described by $T_N = 1/(0.1^2 + (9N)^{-1})$ (Fig. 10.9). This stop line sets a low limit when N is low. For example, when N = 1, T_N = 8.25, which may be exceeded if the first quadrat inspected had all nine plants diseased. Halting sampling after inspection of a very low number of sampling units is usually regarded as undesirable, so inspection of a specified minimum number of sampling units (five, for example) is often part of sequential sampling plans based on stop lines that, like this one, have low values of T_N at low N.

A sequential random sample of quadrats (each containing n = 9 plants) is inspected, and the cumulative number of diseased plants recorded, quadrat by quadrat, until the stop line is intersected, whereupon sampling is halted. Following this procedure for Field 10.3.1, sampling was halted when 68 diseased plants had been observed after inspection of 23 quadrats. The individual quadrat values were 3, 1, 5, 4, 0, 5, 5, 1, 2, 2, 4, 1, 5, 3, 3, 2, 4, 5, 2, 3, 1, 5 and 2. The corresponding cumulative values are shown in Fig. 10.9. The estimated mean and variance (based on equations 9.1 and 9.2 respectively) are $\bar{y} = 0.329$ and $s_y^2 = 0.032$, so the realized reliability level is CV = 0.11. The estimates are biased (because they are based on a sequential sample), but in practice this is usually ignored.

For Field 10.3.2, we have a previously calculated power law relationship, $s_y^2 = A(s_{\text{bin}}^2)^b$ with $\hat{A}$ =9.0, $\hat{b}$ = 1.4 on

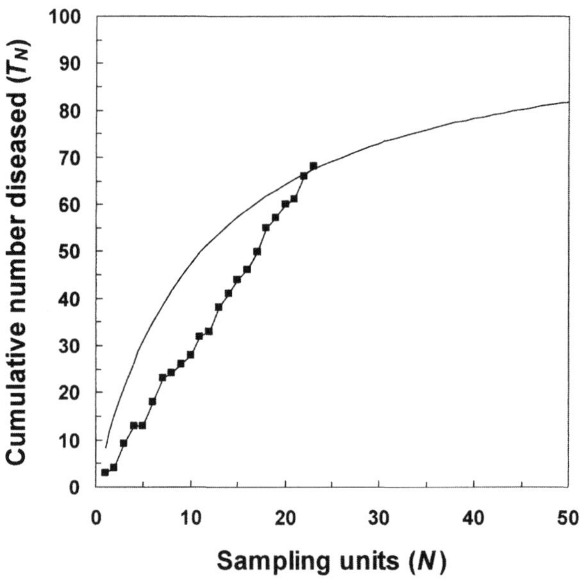

FIG. 10.9. Sequential estimation of disease incidence from cluster sampling. CV = 0.1 is set as the required level of reliability for the sample mean. For Field 10.3.1, the binomial distribution is an appropriate description of variability. Then, from Table 10.14, the stop line is $T_N = 1/(0.1^2 + (9N)^{-1})$ (shown as a smooth curve on the graph). A sequential random sample of quadrats (each containing n = 9 plants) is inspected, and the number of diseased plants is recorded until the line for the cumulative number diseased (shown as a line joining the sample data points on the graph) intersects the stop line. Following this procedure for Field 10.3.1, sampling was halted when 68 diseased plants had been observed after inspection of 23 quadrats.

which to base a stop line for sequential sampling. Since no stop line equation of the form $T_N = f(N)$ can be written down, numerical methods must be used. For each of an appropriate range of values of N, the corresponding T_N value is the one that minimizes the difference between $\gamma_N = (CV^2/\hat{A})n^{3\hat{b}-2}N^{2\hat{b}-1}$ and $\gamma_N = f(T_N) = T_N^{\hat{b}-2}(nN - T_N)^{\hat{b}}$ (Table 10.14), for given values of $\hat{A}$, $\hat{b}$, CV^2, and n (Fig. 10.10). T_N has low values at low values of N, so it would be appropriate to set a minimum of five sampling units to be inspected.

Again a sequential random sample of quadrats (each containing n = 9 plants) is inspected. Sampling continues until the cumulative number of diseased plants, recorded quadrat by quadrat, intersects the stop line, whereupon sampling is halted. Following this procedure for Field 10.3.2, sampling was halted when 145 diseased plants had been observed after inspection of 53 quadrats. In this case, the 53 individual quadrat values were recorded as follows: 4, 2, 2, 6, 0, 3, 4, 2, 6, 2, 4, 6, 2, 0, 2, 2, 2, 2, 2, 2, 0, 1, 3, 3, 6, 1, 4, 3, 7, 1, 2, 0, 5, 1, 2, 5, 0, 1, 2, 3, 2, 5, 1, 3, 2, 2, 2, 4, 5, 4, 0, 5, and 5. The corresponding cumulative values are shown in Fig. 10.10. The estimated mean and variance (based on equations 9.1 and 9.2 respectively) are $\bar{y} = 0.304$ and $s_y^2 = 0.041$, so the realized reliability level in this case is $CV = 0.09$. Again, bias in the calculated estimates is ignored (because it is usually small).

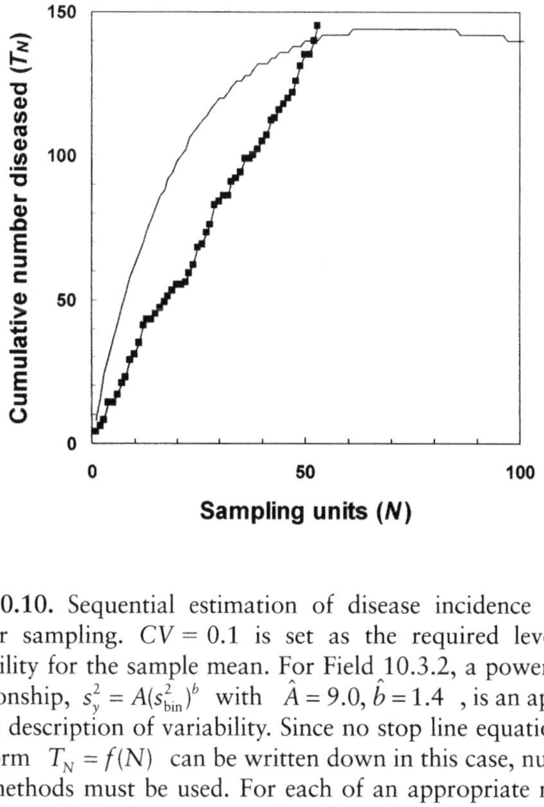

FIG. 10.10. Sequential estimation of disease incidence from cluster sampling. $CV = 0.1$ is set as the required level of reliability for the sample mean. For Field 10.3.2, a power law relationship, $s_y^2 = A(s_{\text{bin}}^2)^b$ with $\hat{A} = 9.0, \hat{b} = 1.4$, is an appropriate description of variability. Since no stop line equation of the form $T_N = f(N)$ can be written down in this case, numerical methods must be used. For each of an appropriate range of values of N, the corresponding T_N value is the one that minimizes the difference between $\gamma_N = (CV^2/\hat{A})n^{3\hat{b}-2}N^{2\hat{b}-1}$ and $\gamma_N = f(T_N) = T_N^{\hat{b}-2}(nN - T_N)^{\hat{b}}$ (Table 10.14), for given values of $\hat{A}$, $\hat{b}$, CV^2 and n (the resulting stop line is shown as a curve on the graph). A sequential random sample of quadrats (each containing $n = 9$ plants) is inspected, and the number of diseased plants is recorded until the line for the cumulative number diseased (shown as a line joining the sample data points on the graph) intersects the stop line. Following this procedure for Field 10.3.2, sampling was halted when 145 diseased plants had been observed after inspection of 53 quadrats.

10.13 Conclusions

Sampling consumes resources, so it is important that the epidemiological data obtained by sampling meet our requirements in terms of reliability while making efficient use of the resources available. To achieve these ends, sampling needs careful planning, a process that draws on the statistical descriptions of disease data discussed in Chapter 9. A knowledge of the distribution of disease, or an equivalent description of variability, is the starting point for the development of formulae for calculating sample sizes needed to achieve an estimate of disease intensity with a desired level of reliability. This desired level of reliability will not necessarily be achieved in any individual sample—it is an average level of reliability to which we aspire in the long run.

Disease data in the form of proportions are important in clinical epidemiology (Lui, 2004) and the medical literature contains a substantial amount of sampling methodology of direct relevance to plant pathologists collecting disease incidence data. Data in the form of counts are collected by economic entomologists and weed scientists. Entomologists, in particular, have developed a large body of sampling methodology (Kuno, 1991a, b; Pedigo and Buntin, 1994) that is directly applicable in plant pathology when count data are collected. The economic entomology literature is also of interest in that it provides more general guidelines for the development sampling methods in crop protection. Such guidelines can contribute to new developments in sampling in plant pathology. In this context, given the importance of disease severity as a measure of the impact of plant pathogens on their hosts, further research on sampling methods for the efficient collection of reliable severity data is a priority.

Detection methodology has developed rapidly in the era of molecular biology, but near-perfect (or even perfect) detection of disease at the level of the sampling unit is not by itself enough to deliver a high level of reliability at the population level. The estimation of disease in plant populations depends on the deployment of detection methodology in a sampling plan devised to deliver estimates of known reliability. Careful thought is needed to make the best use of new developments in detection methodology. For example, improved pathogen detection methods may make possible the use of large group sizes in group testing. This may be attractive in terms of resource use but is usually undesirable for the purpose of estimation, from a statistical point of view.

Hutchison (1994) writes, "*In general terms, sequential sampling is advantageous because it provides an estimate of population intensity with predetermined acceptable levels of precision at minimum cost.*" Developments in sampling methodology such as group testing, binomial sampling, inverse sampling, and sequential sampling can all be seen as being motivated by the need to make efficient use of resources devoted to sampling. Plant pathologists have yet to make extensive use of these and other developments in sampling methodology, but Turechek and Madden (1999b) and Turechek et al. (2001) provide examples that show how detailed attention to the statistical description of variability of plant disease can lead directly to efficient sampling procedures for application in disease management.

10.14 Suggested Readings

Collett, D. 1991. Overdispersion. In: *Modelling Binary Data*. Chapman & Hall, London (chapter 6).

Lui, K.-J. 2004. Population proportion or prevalence. In: *Statistical Estimation of Epidemiological Risk*. John Wiley & Sons, New York (chapter 1).

Madden, L. V., and Hughes, G. 1999. Sampling for plant disease incidence. *Phytopathology* 89: 1088–1103.

Perry, J. 1994. Sampling and applied statistics for pests and diseases. *Aspects Appl. Biol.* 37: 1–14.

Snedecor, G. W., and Cochran, W. G. 1989. Sample surveys. In: *Statistical Methods* (8th edition). Iowa State University Press, Ames (chapter 21).

11

Decision Making in the Practice of Plant Disease Management

It was that rarity, an easy decision.

Conrad Metcalf
("Gun, with Occasional Music" by Jonathan Lethem)

11.1 Decision Making in Disease Management

The long and close association between agriculture and the study of plant disease epidemics has meant that disease management has been a prevailing theme in plant disease epidemiology. The landmark of Vanderplank's (1963) book *Plant Diseases: Epidemics and Control* is an example of the way in which epidemiologists have sought to show how the basic principles of suppression of plant disease are derived from the knowledge of epidemiology. Crops grown in agriculture are exposed to the risk of attack by pathogens, the outcome of which may be economic loss resulting from reductions in the quantity and quality of crop yield (see Chapter 12). Here, we are interested in quantifying the level of risk to which a crop is exposed as a basis for deciding whether or not intervention aimed at disease suppression is justified. Of course, the details of this process differ from pathogen to pathogen, from crop to crop, and from place to place (e.g., Hardwick, 1998). However, the main theme of this chapter is the identification of generic aspects of the decision making process as practised in plant disease management. Throughout this chapter, the emphasis is on short-term crop protection problems involving a binary (yes/no) decision in relation to a single pathogen and a single host.

In the practice of disease management, the challenges faced by decision makers are many and various. In this chapter, therefore, we cannot hope to be comprehensive. Instead, we outline two approaches to evidence-based decision making that have application in disease management. First, sampling as a basis for decision making is discussed. The purpose of sampling is to gather information on disease intensity. This information is the evidence on which a decision about the need for some appropriate control action can be based. Sampling a crop for the purpose of deciding whether or not treatment is required entails classification of observed disease intensity in relation to some pre-specified critical intensity (threshold). This is not the same as sampling a crop for the purpose of estimating disease intensity (as described in Chapter 10). The theory of sampling for the purpose of classification has been developed most extensively in the area of statistical quality control (e.g., Guenther, 1977; Wetherill, 1977; Hald, 1981). Developments in the context of crop protection have been pioneered by M.R. Binns and J. P. Nyrop and their colleagues (e.g., Nyrop and Binns, 1991; Binns and Nyrop, 1992; Binns, 1994; Binns et al., 2000). In this chapter, we concentrate our discussion on methodology for sampling disease incidence data, because sampling for counts is already so widely discussed in the economic entomology literature.

Binns et al. (2000) draw a distinction between crop protection that is based on curative action and crop protection based on preventative action. Without necessarily wishing to adhere rigidly to this dichotomy, it is nevertheless clear that in some cases (mainly the former), sample data are the most important components of the information on which decision making is based. In others (mainly the latter), data relating to the host and the environment often play a more important role, and the evidence on which a decision is made about the need for some appropriate control action is therefore likely to be more wide ranging. Thus, the second part of this chapter is devoted to a discussion of the prediction of plant disease, using the conceptual basis of diagnostic decision making already widely applied in evidence-based medicine (e.g., Friedland, 1998). The concept of classification is again important, although the requirements for quantification of the level of risk to which a crop is exposed now go beyond the collection of sample data on disease intensity. Note that in this chapter, the calculation of risk exposure is based on empirical models, rather than the analytical models of disease progress of the type discussed in Chapters 4 and 5 (although, in principle, analytical models could also be used for this purpose).

11.2 Acceptance Sampling Preliminaries

Acceptance sampling is the generic term used to describe sampling when the objective is classification rather than estimation. Much of the terminology used in connection with acceptance sampling is borrowed from the lexicon of statistical quality control (e.g., Grant and Leavenworth, 1996; Montgomery, 1997).

11.2.1 Probability and likelihood

In Chapter 10, we faced the problem of estimating population parameters (e.g., p, μ). The solution involved drawing a sample from the population in question. Based on the sample data, a parameter estimate and a confidence interval could be calculated. For a 95% confidence interval, for example, this enables us to make a statement to the effect that there is a 95% probability that the calculated interval contains the true value of the population parameter. That is to say, given the sample data, we have made a probability-based statement about the population parameter of interest.

We now face a different problem. In the generic terminology, we wish to classify a *lot* as either acceptable or unacceptable, according to some predetermined standard defined in terms of one or more population parameters. For crop protection purposes, we can think of a lot as, for example, a field for which a decision on control action for a crop is to be made. The predetermined standard is then a parameter that applies not just to a particular field, but generally to fields in which the same type of crop is being cultivated under similar conditions. The solution to this problem involves drawing a sample from the lot. In this case, the parameter is given (in the form of the predetermined standard), and we wish to make a probability-based statement about the sample data. Specifically, we are interested in the probability that the sample data have been drawn from a lot that conforms to the standard. If this probability is sufficiently high, the lot may be deemed acceptable (i.e., the crop in the field does not require control action). On the other hand, if this probability is low, the lot may be deemed unacceptable (i.e., the crop in the field requires control action). The probability, for any given value of the parameter, of obtaining the actual data observed, is usually referred to as a likelihood in the statistical literature, following R. A. Fisher's terminology (e.g., Box, 1978). However, in accounts of applications of acceptance sampling, a notational distinction between likelihoods and probabilities is not usually drawn, and this guides our own use of terminology here.

11.2.2 Thresholds

Zadoks and Schein (1979) defined disease management as "the total of all actions, intentional or not, that serve to regulate disease levels so that they remain below the economic threshold level...". The idea behind the *economic threshold* (Stern, 1973) is, in essence, to facilitate identification of the circumstances in which it is economically advantageous to adopt certain crop protection measures. In its most widely used form, the economic threshold is a discrete choice threshold: the only options open to the grower are to apply crop protection measures or to withhold them.

Although the economic threshold is a familiar concept to most practitioners of disease management, we will follow Binns et al. (2000) in not being specific about the precise basis of the threshold that is used to guide decision making. Thus we will refer to a critical incidence, p_{CRIT}, as a generic threshold, recognizing that economic advantage is only one of a number of different factors that may motivate a decision maker.

11.2.3 The operating characteristic curve

In the acceptance sampling literature, variables assessed on a presence/absence basis, such as disease incidence for an individual plant, are referred to as *attributes*. In acceptance sampling for attributes, an *operating characteristic* (OC) *curve* is a graphical plot of the *probability of acceptance*, Pr(A), against the *proportion of defectives*, p, in a lot. In an ideal world, the OC curve would be:

$$\Pr(A) = \begin{cases} 1 & (p \leq p_{CRIT}) \\ 0 & (p > p_{CRIT}) \end{cases}$$

A version of this OC curve is illustrated in Fig. 11.1. It represents a situation in which all lots at or below the adopted decision threshold would be accepted, while all lots above the threshold would be rejected. Unfortunately, we could only hope to achieve such an OC curve if every individual plant were assessed for disease, and no errors were made in the individual assessments.

More realistically, suppose that we have decided on the appropriate critical incidence to use as a decision threshold and now wish to use it to guide our decision

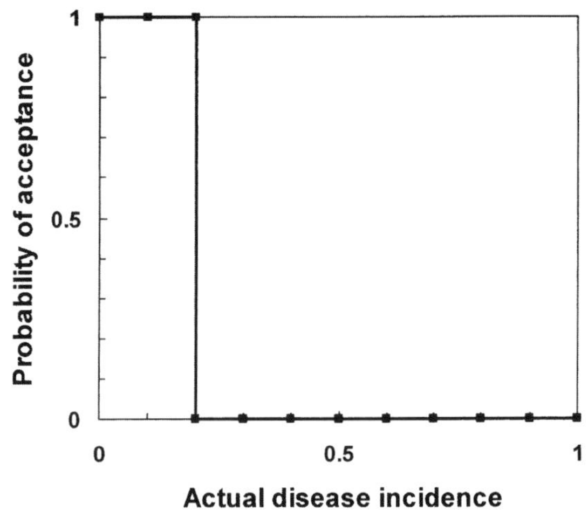

FIG. 11.1. The ideal operating characteristic curve for a sampling plan for which the adopted decision threshold is set at an actual disease incidence (p) equal to 0.2. Below this threshold, the probability of acceptance (Pr(A)) is equal to 1, above this threshold, Pr(A) is equal to 0.

making. A simple random sample of N plants is assessed for disease, and disease incidence on a proportion scale, $\bar{y}$, calculated. The decision to be made is a binary one, either *not* apply control measures (if $\bar{y}$ is at or below the adopted threshold), or to proceed with the application (if $\bar{y}$ is above the adopted threshold). While, in an ideal world, we would be certain that a value of $\bar{y}$ (based on the sample) at or below the adopted threshold meant that $p \leq p_{CRIT}$, we must recognize that there is in fact a chance that $p > p_{CRIT}$ (in which case the lot would be wrongly accepted on the basis of the sample). Similarly, we would like to be certain that a value of $\bar{y}$ above the adopted threshold meant that $p > p_{CRIT}$, but must recognize that there is a chance that $p \leq p_{CRIT}$ (in which case the lot would be wrongly rejected on the basis of the sample). No sampling plan provides a guarantee that such misclassification errors will not occur.

We note that there is an analogy here with hypothesis-testing procedures as used in experimental plant pathology. In that context, we formulate an appropriate null hypothesis about a population parameter of interest, and a corresponding alternative hypothesis. To test the null hypothesis, data are collected (using, for example, a random sampling procedure) to be able to calculate an appropriate test statistic. However, based on the sample data, there exists a chance that a null hypothesis that is really true might be rejected (a Type I error) and a chance that a null hypothesis that is really false might not be rejected in favor of the alternative (a Type II error).

For our purposes here, the OC curve shows the average probability of a decision, based on sampling, *not* to apply control measures, as a function of the true disease incidence. In disease management decision making, the OC curve can be applied in combination with the decision threshold, denoted here p_{CRIT}. Up to and including the adopted decision threshold, the decision not to apply control measures is a correct one. Above the threshold, however, the decision not to apply control measures would be incorrect. Qualitatively, a typical OC curve changes in a smooth fashion, with $\Pr(A)$ near 1 when p is much smaller than p_{CRIT}, $\Pr(A)$ near 0.5 when p is near p_{CRIT}, and $\Pr(A)$ near 0 when p is much larger than p_{CRIT} (Binns and Nyrop, 1992). Quantitatively, OC curves can be derived from the cumulative distribution functions of appropriate statistical probability distributions. Comparison of such an OC curve with Fig. 11.1 then serves as a qualitative indication of the extent to which a sampling plan departs from the ideal. When acceptance sampling for attributes (such as disease incidence) using simple random sampling, the binomial and the hypergeometric distributions are applicable.

11.2.4 The binomial distribution

For the most part, we restrict our attention here to decisions on the acceptance or rejection of a lot based on one sample drawn from that lot. In the acceptance sampling literature, this is referred to as *single sampling*. The size of this single sample and the acceptance criterion together comprise a *sampling plan*. In acceptance sampling, the acceptance criterion is usually given as an acceptance number, C, so that a sampling plan is characterized as (N, C). A lot is then acceptable if there are C or fewer defectives in a sample size N (N and C are both integers). If all we want to do is describe a sampling plan, we can simply take $C/N = p_{CRIT}$.

As noted in Chapter 10, the binomial distribution is the appropriate statistical model when sampling with replacement. In practice, however, it is often used in sampling without replacement, when the population size is sufficiently large (relative to the size of the sample) for the issue of replacement of sampling units not to be of major concern. The same is true in acceptance sampling. Suppose a sample comprising N individual elements (plants, for example) is randomly drawn from a population size M in which X plants were "diseased" and $M-X$ were "healthy". Sampling with replacement means that M and X do not change as the sample is drawn, plant-by-plant, so we can write $p = X/M$ for the true disease incidence in the population. The binomial probability of Y, the number of plants diseased (out of a total of N), is then given by:

$$\Pr(Y) = \binom{N}{Y} p^{Y}(1-p)^{N-Y} \quad (Y = 0, 1, \ldots, N). \qquad (11.1)$$

Jones (1994a) discusses a decision guideline for fungicidal control of eyespot disease of wheat (at the time, the causal pathogen was referred to as *Pseudocercosporella herpotrichoides*). Treatment was thought likely to be worthwhile if $\geq$20% of tillers were affected at growth stage 30–31 (Jones, 1994a, Table 11). Accordingly, the advice was that a sample of tillers was to be collected at the appropriate growth stage and a decision on treatment then made on the basis of the percentage of tillers affected by eyespot disease in the sample, in relation to the specified threshold criterion. In practice, decision making was based on a two-stage cluster sampling procedure (section 9.4.10), used to collect a sample of 50 tillers (Goulds and Polley, 1990). Here, the example just serves to motivate a general discussion of disease control decision making based on sampling. What follows is not intended to relate to eyespot disease management.

We consider, in the first instance, a random sample of $N = 10$ tillers. We will assume here that disease assessments of individual tillers are carried out without error, and that the plant population is sufficiently large that calculations can be based on the binomial distribution. In this example, with sample size of $N = 10$ and an acceptance number $C = 2$, a crop will be left untreated if there are two or fewer affected tillers in the sample, and will be treated if there are more than two affected tillers in the sample. This is therefore a (10,2) sampling plan. The probability of acceptance, $\Pr(A)$, is the probability that the observed number of infected tillers, Y, is less than or equal to C. The OC curve for the plan is then:

$$\Pr(A) = \Pr(Y \le C) = \sum_{Y=0}^{C} \Pr(Y) \qquad (11.2)$$

where $\Pr(Y)$ is based on the binomial distribution (equation 11.1). This is illustrated in Fig. 11.2.

Fig. 11.2A shows the binomial OC curve for a (10,2) sampling plan. Fig. 11.2B–D exemplify the calculations on which this OC curve is based. Fig. 11.2B shows the binomial distribution (equation 11.1) with $N = 10$ and $p = 0.1$. For this distribution, $\Pr(Y = 0) = 0.35$, $\Pr(Y = 1) = 0.39$ and $\Pr(Y = 2) = 0.19$. Therefore, $\Pr(Y \le 2) = 0.35 + 0.39 + 0.19 = 0.93$, and this is the probability of acceptance, $\Pr(A)$, when $p = 0.1$ in Fig. 11.2A (point marked B). At this point, $p < p_{CRIT}$ and $\Pr(A)$ is fairly close to one. Fig. 11.2C shows the binomial distribution (equation 11.1) with $N = 10$ and $p = 0.2$. For this distribution, $\Pr(Y = 0) = 0.11$, $\Pr(Y = 1) = 0.27$ and $\Pr(Y = 2) = 0.30$. Therefore, $\Pr(Y \le 2) = 0.11 + 0.27 + 0.30 = 0.68$, and this is the probability of acceptance, $\Pr(A)$, when $p = 0.2$ in Fig. 11.2A (point marked C). At this point, $p = p_{CRIT}$ and $\Pr(A)$ is fairly close to 0.5. Fig. 11.2D shows the binomial distribution (equation 11.1) with $N = 10$ and $p = 0.4$. For this distribution, $\Pr(Y = 0) = 0.01$, $\Pr(Y = 1) = 0.04$ and $\Pr(Y = 2) = 0.12$. Therefore, $\Pr(Y \le 2) = 0.01 + 0.04 + 0.12 = 0.17$, and this is the probability of acceptance, $\Pr(A)$, when $p = 0.4$ in Fig. 11.2A (point marked D). At this point, $p > p_{CRIT}$ and $\Pr(A)$ is fairly close to zero.

It can be seen (Fig. 11.2A) that the OC curve defined by equation 11.2 characterizes $\Pr(A)$ as a decreasing function of p. In this particular example, with a binomial (10,2) acceptance sampling plan, it is approximately the case that $\Pr(A) \ge 0.93$ when $p \le 0.1$, and $\Pr(A) \le 0.05$ when $p \ge 0.5$. Thus there is a high probability of lot acceptance at low values of p and a low probability of lot acceptance at high values of p, which is as we require it. We can be confident that we are making the right decision to accept a lot (i.e., not apply control measures) when the sample estimate of p, $\bar{y}$, is less than or equal to 0.1 (i.e., the sample of 10 tillers has zero or one diseased), and that we are making the

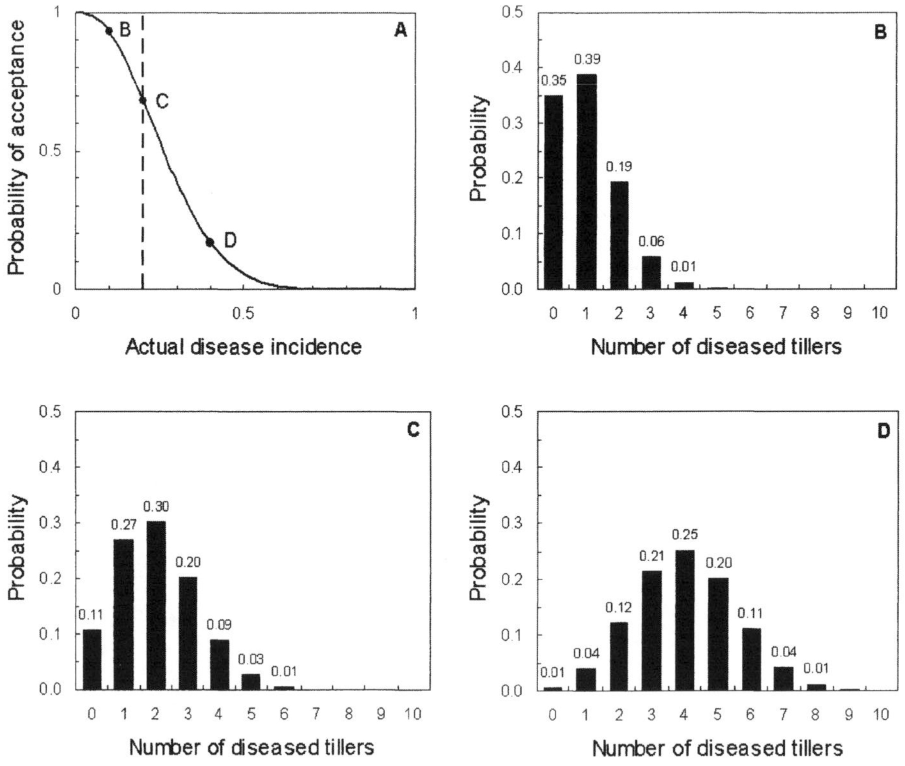

FIG. 11.2. Derivation of an operating characteristic curve, illustrated for a (10,2) sampling plan based on the binomial distribution. (**A**) The OC curve for a (10,2) sampling plan. The probability of acceptance, $\Pr(A)$, is the probability that the observed number of infected tillers (out of 10 in the sample), Y, is less than or equal to the acceptance number $C = 2$. (**B**) The binomial distribution (equation 11.1) with $N = 10$ and $p = 0.1$. For this distribution, $\Pr(Y = 0) = 0.35$, $\Pr(Y = 1) = 0.39$ and $\Pr(Y = 2) = 0.19$. Therefore, $\Pr(Y \le 2) = 0.35 + 0.39 + 0.19 = 0.93$, and this is the probability of acceptance, $\Pr(A)$, when $p = 0.1$ in Fig. 11.2A (point marked B). (**C**) The binomial distribution (equation 11.1) with $N = 10$ and $p = 0.2$. For this distribution, $\Pr(Y = 0) = 0.11$, $\Pr(Y = 1) = 0.27$ and $\Pr(Y = 2) = 0.30$. Therefore, $\Pr(Y \le 2) = 0.11 + 0.27 + 0.30 = 0.68$, and this is the probability of acceptance, $\Pr(A)$, when $p = 0.2$ in Fig. 11.2A (point marked C). (**D**) The binomial distribution (equation 11.1) with $N = 10$ and $p = 0.4$. For this distribution, $\Pr(Y = 0) = 0.01$, $\Pr(Y = 1) = 0.04$ and $\Pr(Y = 2) = 0.12$. Therefore, $\Pr(Y \le 2) = 0.01 + 0.04 + 0.12 = 0.17$, and this is the probability of acceptance, $\Pr(A)$, when $p = 0.4$ in Fig. 11.2A (point marked D).

right decision to reject a lot (i.e., apply control measures) when $\bar{y}$ is greater than or equal to 0.5 (i.e., the sample of 10 tillers has five or more diseased). For sample values $0.1 < \bar{y} < 0.5$, we cannot be as confident that we are making the right decision. The decision rule is that a crop will be left untreated if there are two or fewer affected tillers in the sample, and will be treated if there are more than two affected tillers in the sample. However, the probabilities of erroneously accepting a lot (i.e., not applying control measures where they were actually required) or of erroneously rejecting a lot (i.e., applying control measures where they were actually not required) are higher in this intermediate range of $\bar{y}$ (where the sample of 10 tillers has two, three or four diseased).

An OC curve based on a cumulative distribution function (e.g., equation 11.2) characterizes the average performance of a sampling plan. An OC curve simulation can be used to provide a graphical plot of Pr(A) against p that illustrates the likely extent of variability around the average. Using MATHCAD software, a simple small-scale simulation of the (10,2) sampling plan discussed above was carried out (Fig. 11.3). The OC curve for the (10,2) sampling plan calculated using equation 11.2 characterizes the average performance of sampling carried out according to this plan.

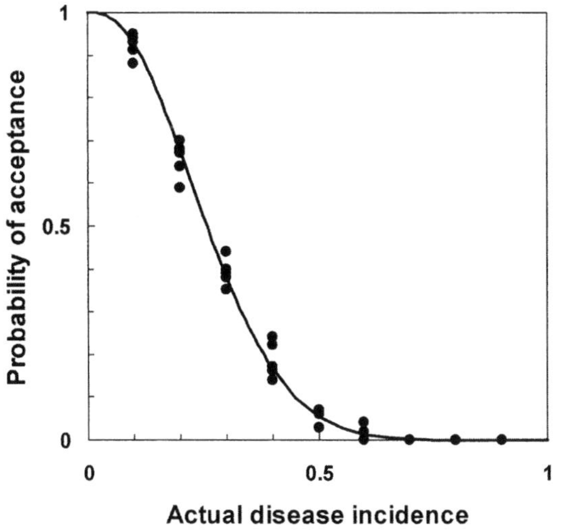

FIG. 11.3. Simulation of a (10,2) sampling plan. A sample comprised 10 numbers drawn from a binomial distribution (equation 11.1) with $N = 1$ and a specified value of p. Thus, the numbers drawn were either 0 or 1. For each specified value of p, 100 such samples were drawn and the proportion of samples for which the sum of the numbers drawn was ≤2 calculated, as an estimate of Pr(A). This procedure was repeated five times, so that at each specified value of p there were five values of Pr(A), each based on 100 samples of 10 random numbers. The specified values of p were 0.1, 0.2, 0.3, 0.4, 0.5, 0.6, 0.7, 0.8, and 0.9. The results show the correspondence between the OC curve for the (10, 2) sampling plan calculated using equation 11.2 (the solid line) and the results of the simulation based on the same plan (shown as data points).

The results of the simulation of sampling based on the same plan illustrate typical performance of the plan in application.

Note that although the OC curve for the (10,2) sampling plan has Pr(A) near 1 when p is much smaller than p_{CRIT} and Pr(A) near 0 when p is much larger than p_{CRIT}, the proximity of Pr(A) to 0.5 when $p = p_{CRIT}$ (=C/N) could be questioned. For binomial OC curves, Pr(A) is not necessarily equal to 0.5 when $p = p_{CRIT}$ (=C/N) because C and N are both integers and the binomial is a discrete probability distribution. It is possible to design acceptance sampling plans based on discrete distributions that meet the criterion that Pr(A) = 0.5 when $p = p_{CRIT}$, at least to a reasonable approximation (sections 11.3.2 and 11.3.3). However, such plans will not, in general, have the property $C/N = p_{CRIT}$.

11.2.5 The hypergeometric distribution

The binomial distribution is often appropriate as a basis for acceptance sampling plans for disease incidence. However, if we are in a situation that involves sampling from small populations (e.g., Hughes and Gottwald, 2001), we have recourse to the hypergeometric distribution. As noted in Chapter 10, the hypergeometric distribution is the exact statistical model when sampling without replacement. The hypergeometric probability of Y, the number of plants, or plant parts, diseased (out of a total of N), is given by:

$$\Pr(Y) = \frac{\binom{X}{Y}\binom{M-X}{N-Y}}{\binom{M}{N}} \quad (Y = 0, 1, \ldots, N), \qquad (11.3)$$

where the notation is as in section 11.2.4. If desired, Pr(Y) can be written as a function of p by substituting Mp (rounded to an integer value if necessary) for X (the actual number diseased in the population) in the expression on the right-hand side. As before, the probability of acceptance, Pr(A), is the probability that the observed number of infected plants or plant parts, Y, is less than or equal to the acceptance number, C. The OC curve is given by equation 11.2:

$$\Pr(Y \le C) = \sum_{Y=0}^{C} \Pr(Y)$$

but now with Pr(Y) based on the hypergeometric distribution (equation 11.3). If the sample size, N, is less than about 10% of the population size, M, the difference between the hypergeometric- and binomial-based OC curves for a given (N,C) plan is small.

Fig. 11.4 shows OC curves for three hypergeometric sampling plans with the property $C/N = p_{CRIT} = 0.2$: (10,2) with population size $M = 50$, (20,4) with

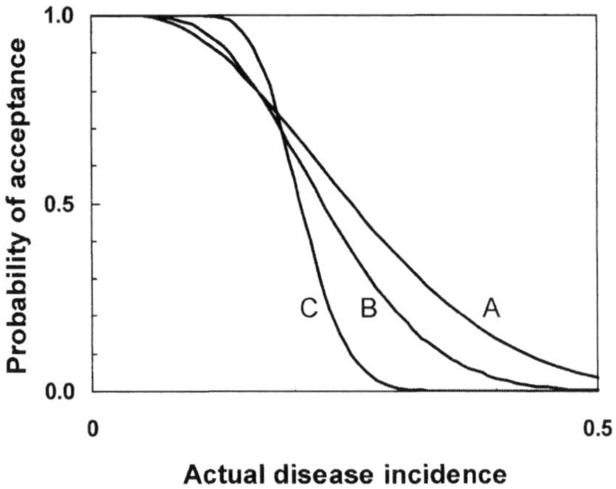

FIG. 11.4. OC curves for three hypergeometric sampling plans with the property $C/N = p_{CRIT} = 0.2$. (A), (10,2) with population size $M = 50$; (B), (20,4) with $M = 100$; (C), (100,20) with $M = 500$.

$M = 100$, and (100, 20) with $M = 500$. For all three plans, the sample size (N) is the same proportion of the population size (=0.2). The important point to note is that sampling plans that have sample sizes set as a constant proportion of a varying population size do *not* have the same OC curve. If the objective were to sample from a series of small populations in such a way that the same (average) rate of misclassification errors occurred as population size varied, keeping sample size as a constant proportion of population size would not be an appropriate method of achieving this. The calculation of sample sizes for sampling plans that fulfill this objective is discussed in section 11.3.

11.2.6 Inspection errors in simple random sampling

Misclassification errors occur when control measures are withheld from a crop that really required them, or when they are applied to crops that really did not require them. Inspection errors occur when an individual plant or plant part is wrongly assessed. Plants that are healthy may be assessed as diseased, and plants that are diseased may be assessed as healthy (section 10.3.2). Here, however, we consider only the latter type of inspection error. The analysis is based on the hypergeometric distribution (section 11.2.5), but now there is a proportion of diseased individuals, denoted ζ, erroneously recorded as healthy. As before, Y is the number of plants, or plant parts, diseased out of a total of N in the sample. Now, however, the number of plants assessed as diseased in the sample (Z) is not the same as Y (unless $\zeta = 0$). The probability distribution of Z is:

$$\Pr(Z) = \sum_{Y=Z}^{N} \left[\frac{\binom{X}{Y}\binom{M-X}{N-Y}}{\binom{M}{N}} \right] \left[\binom{Y}{Z} (1-\zeta)^{Z} \zeta^{Y-Z} \right] \qquad (11.4)$$

($Z = 0, \ldots, N$) (Johnson et al., 1980). Although quite tedious to work with using a calculator, equation 11.4 is easy to utilize with MATHCAD and similar mathematics computer programs. The probability of acceptance, $\Pr(A)$, is the probability that the observed number of plants assessed as diseased, Z (including those actually diseased and those erroneously assessed as diseased), is less than or equal to the acceptance number, C. The OC curve is given by:

$$\Pr(Z \le C) = \sum_{Z=0}^{C} \Pr(Z) \qquad (11.5)$$

with $\Pr(Z)$ based on equation 11.4.

Fig. 11.5 shows OC curves for two (50,2) sampling plans, both with $M = 100$, one based on equation 11.3 (i.e., the hypergeometric distribution with no inspection errors) and the other based on equation 11.4 (i.e., a "faulty inspection" distribution incorporating a proportion, $\zeta = 0.05$, of diseased individuals erroneously recorded as healthy). The OC curve based on equation 11.4 is shifted to the right, compared with the OC curve based on equation 11.3. This is discussed further in section 11.3.3. Johnson et al. (1991) give a comprehensive account of the analysis of inspection errors in attributes such as disease incidence (including more details of statistical probability distributions), and Suich (1990) discusses the impact of inspection errors on the OC curve. Hughes et al. (2002b) provide a phytopathological example.

11.3 Designing a Sampling Plan with a Specified OC Curve

Up to now, we have used OC curves to describe sampling plans, given the sample size and acceptance number

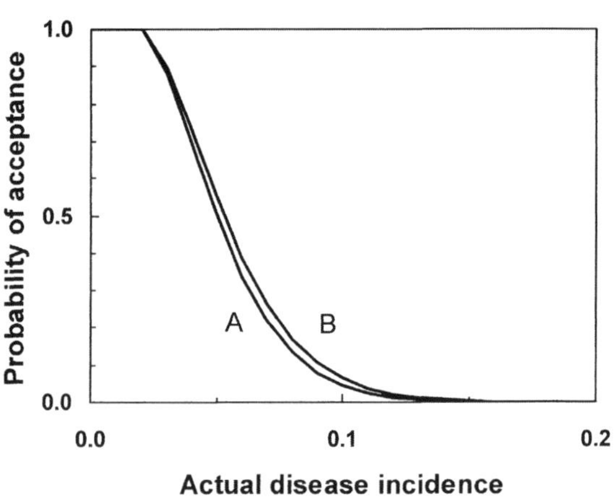

FIG. 11.5. OC curves for two (50,2) sampling plans, both with $M = 100$. (A), based on equation 11.3 (i.e., the hypergeometric distribution with no inspection errors); (B), based on equation 11.4 (i.e., a "faulty inspection" distribution incorporating a proportion, $\zeta = 0.05$, of diseased individuals erroneously recorded as healthy).

(or critical incidence) of the plan in question (and if necessary, the population size). Of greater interest, possibly, is the problem of designing an acceptance sampling plan with a specified OC curve, such that the sample size and the acceptance number can be calculated and used in practical decision making (Hamaker, 1949). In this section we consider sampling plans that have an acceptance number $C \geq 1$. Sampling plans with $C = 0$ are a special case, and are discussed separately in section 11.4.

11.3.1 Plans based on the producer's and consumer's risks

We have already used the term risk colloquially, to refer to the chance of a crop suffering damage or loss as the result of pathogen attack. As used here, a *risk* is synonymous with a probability, carrying the connotation that the probability in question is that of an undesirable outcome. The use of the term risk in this way has the advantage of being consistent with usage both in the acceptance sampling literature (e.g., Montgomery, 1997) and in epidemiology (e.g., Selvin, 1996; Lui, 2004). However, we recognize that our adopted usage carries a risk that followers of, for example, Kammen and Hassenzah (1999) may be somewhat disorientated.

A single point—a value of p_{CRIT} and the corresponding value of Pr(A)—is not sufficient to specify an OC curve, but two such points will do. The specification of an OC curve based on two points is an approach often used in industrial quality control. We think this approach has relatively little application in crop protection decision making, but nevertheless include a brief outline for reference. Any two points can be used for the specification, but typically, the points of view of a notional producer and a notional consumer are adopted. The idea is that the producer is concerned only that the acceptable lots (those with a [low] true proportion defective of p_1 or less) are accepted with high probability, denoted here $1 - \alpha$ (thus α is a [low] rate of misclassification of acceptable lots as unacceptable, referred to as the *producer's risk*). On the other hand, the consumer is concerned only that unacceptable lots (those with a [high] true proportion defective of p_2 or more) are accepted with low probability, denoted here β (thus β is a [low] rate of misclassification of unacceptable lots as acceptable, referred to as the *consumer's risk*). Using the binomial OC curve as an example, the two risk points can be specified:

$$1-\alpha \leq \sum_{Y=0}^{C} \binom{N}{Y} p_1^Y (1-p_1)^{N-Y}$$

$$\beta \geq \sum_{Y=0}^{C} \binom{N}{Y} p_2^Y (1-p_2)^{N-Y}.$$

Because the binomial is a discrete probability distribution, the specification must be given in terms of inequalities, otherwise there may be no corresponding OC curve. The task then is to find an N and a C that satisfy these inequalities, for given α, β, p_1, and p_2 (Hald, 1967, Guenther, 1973). However, this is not the most straightforward way to apply acceptance sampling methodology in disease management decision making when using a fixed sample size. In plant health decision making for field crops, (implicit) reference is usually made to a single risk point on the OC curve, where Pr(A) = 0.5. For plant health decision making in a quarantine context, where an exporter (producer) and an importer (consumer) are involved, acceptance sampling plans based on two agreed risk points may find application. Even here, however, the usual tendency is to adopt one point of view or the other (e.g., Baker et al., 1990; Harte et al., 1992; Cannon, 1998).

11.3.2 Plans based on the indifference quality level

In the terminology of acceptance sampling, the *indifference quality level* (IQL) is the point on the OC curve at which Pr(A) = 0.5. Specification of a value of p_{CRIT} for which the corresponding value of Pr(A) = 0.5 is analogous to the specification of a decision threshold in disease management (see section 11.2.3), so this approach is likely to be more widely applicable than the two-point method outlined in section 11.3.1. However (again using the binomial OC curve as an example), the specification:

$$\sum_{Y=0}^{C} \binom{N}{Y} p_{\text{CRIT}}^Y (1-p_{\text{CRIT}})^{N-Y} \approx 0.5$$

is not enough, by itself, to enable us to find values of N and C for an acceptance sampling plan with a particular OC curve. Additionally, the relative slope of the OC curve at the IQL can be specified (Hamaker, 1950; Chakraborty, 1990). While this would be enough to enable values of N and C to be calculated for an acceptance sampling plan with a particular OC curve, specification of a value for the relative slope of the OC curve at the IQL is not a task for which many plant pathologists are equipped.

Here, we will adopt the IQL approach to specify a value of p_{CRIT} for which the corresponding value of Pr(A) = 0.5, but not aim to define values of N and C for an acceptance sampling plan with a particular OC curve. Instead, we will specify a range of values of C and calculate the corresponding range of values of N. This approach yields a family of OC curves, for all of which Pr(A) ≈ 0.5 at the adopted decision threshold, while the (negative) slope of the OC curve at the decision threshold increases with sample size. We apply here some useful approximation formulae, details of which can be found in Hald (1981). Remember that the sample size

N and acceptance number C must be integers, so solutions to equations may require rounding. In sample size calculations, non-integer values of N are usually rounded up to the nearest whole number above.

For a binomial OC curve:

$$p_{\text{CRIT}} \approx \frac{C+(2/3)}{N+(1/3)}.$$

In practice, this is usually further approximated to:

$$p_{\text{CRIT}} \approx \frac{C+(2/3)}{N} \quad (11.6)$$

from which, given values of p_{CRIT} and C, values of N can be found from:

$$N = \frac{C+(2/3)}{p_{\text{CRIT}}} \quad (11.7)$$

For example, the sample sizes for binomial sampling plans with $p_{\text{CRIT}} = 0.05$ and $C = 1, 2, 3, 4$, and 5 are calculated as $N = 34, 54, 74, 94$, and 114, respectively. The OC curves for the binomial sampling plans (34,1), (74,3), and (114,5) are shown in Fig. 11.6. The design criterion that $\Pr(A) = 0.5$ at the adopted decision threshold of $p_{\text{CRIT}} = 0.05$ is observed, at least to a reasonable approximation. It can be seen from Fig. 11.6 that as the sample size increases, the (negative) slope of the OC curve at the adopted decision threshold increases. Thus, as sample size increases, the OC curve for the sampling plan resembles more closely the ideal OC curve (Fig. 11.1). A sampling plan that produces relatively few misclassification errors is characterized by an OC curve with a steeper (negative) slope at the adopted decision threshold. In acceptance sampling, a reduction in misclassification errors is thought of as an increase in accuracy. Thus increasing sample size has the effect of increasing accuracy, which is a natural result.

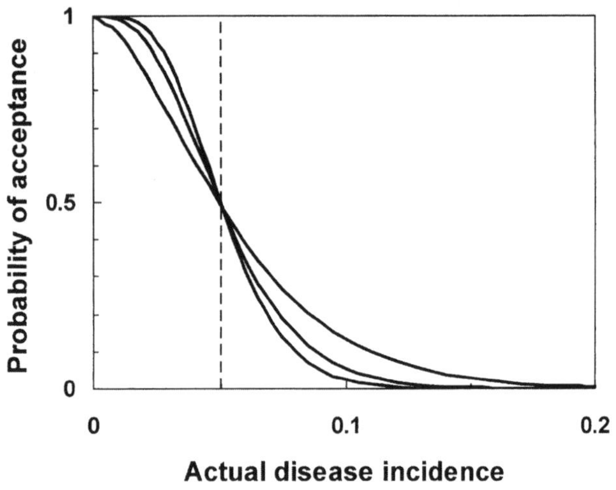

FIG. 11.6. The OC curves for the binomial sampling plans (34,1), (74,3), and (114,5) all meet the design criterion that $\Pr(A) = 0.5$ at the adopted decision threshold of $p_{\text{CRIT}} = 0.05$, at least to a reasonable approximation. As the sample size increases, the (negative) slope of the OC curve at the adopted decision threshold increases. Thus, as sample size increases, the OC curve for the sampling plan resembles more closely the ideal OC curve (Fig. 11.1).

For a hypergeometric OC curve:

$$p_{\text{CRIT}} \approx \frac{C+(2/3)-(N/3M)}{N+(1/3)-(2N/3M)}.$$

In practice, this is usually further approximated to:

$$p_{\text{CRIT}} \approx \frac{C+(2/3)}{N} + \frac{1}{3M} \quad (11.8)$$

from which, given values of M, p_{CRIT} and C, N can be found from:

$$N = \frac{M(3C+2)}{3Mp_{\text{CRIT}}+1} \quad (11.9)$$

For example, the sample sizes for hypergeometric sampling plans with $M = 100$, $p_{\text{CRIT}} = 0.05$ and $C = 1, 2, 3$, and 4 are calculated as $N = 32, 50, 69$, and 88, respectively. The sample sizes for hypergeometric sampling plans with $M = 200$, $p_{\text{CRIT}} = 0.05$, and $C = 1, 2, 3, 4$, and 5 are calculated as $N = 33, 52, 71, 91$, and 110, respectively. Thus, the calculated hypergeometric sample sizes (for finite population size) are somewhat smaller than those calculated for an indefinitely large population size, based on the binomial distribution, as one would expect. Note that equations 11.7 and 11.9 indicate that lower values of p_{CRIT} lead to higher values of sample size, N, other things being equal.

A question of particular interest is how to sample from different size lots yet keep the same OC curve (so that $\Pr(A) = 0.5$ at the adopted decision threshold, and the rate of misclassification errors does not change). Hamaker (1959) provides a simple approximate solution, based on equations 11.6 and 11.8. The notation is expanded here, so that a binomial sampling plan is characterized by its sample size and acceptance number as (N_b, C_b) and a hypergeometric sampling plan is similarly characterized by (N_h, C_h). If we equate the right hand sides of equations 11.6 and 11.8, and solve for N_h, we obtain, for given N_b, C_b, and M:

$$N_h = \frac{N_b\left(C_h+(2/3)\right)}{C_b+(2/3)+(N_b/3M)} \quad (11.10)$$

If C_h is chosen such that $N_h > MN_b/(M+N_b)$ (it is the smallest such N_h that is of interest), we can use equation 11.10 to calculate sample sizes for different size lots while maintaining (to a reasonable approximation) the same OC curve. For example, the hypergeometric sampling plans (32,1) for $M = 100$, (52,2) for $M = 200$, and (53,2) for $M = 300$, 400, and 500 all have OC curves close to the OC curve for a binomial (54,2) sampling plan (with $\Pr(A) \approx 0.5$ at the adopted decision threshold of $p_{CRIT} = 0.05$) (Fig. 11.7). In the case of the hypergeometric (32,1) sampling plan for $M = 100$, $N_h = 32$ is slightly smaller than $MN_b/(M+N_b) \approx 35$, but visual inspection shows the OC curve to be acceptable.

11.3.3 Finding a sampling plan by iteration

An interactive spreadsheet-based approach to the formulation of sampling plans with specified OC curves is presented by Ng (2003). Hughes and Gottwald (2001) used an iterative searching routine (using MATHCAD software) to calculate hypergeometric sample sizes. For IQL sampling plans, this involves finding values of N such that:

$$\left| \sum_{Y=0}^{C} \frac{\binom{X}{Y}\binom{M-X}{N-Y}}{\binom{M}{N}} - 0.5 \right| < \text{TOL} \qquad (11.11)$$

for specified values of M and p_{CRIT}, with $X = Mp_{CRIT}$ (taking the integer part). The acceptance number is specified as $C = Np_{CRIT}$ (again taking the integer part), which has the effect of restricting solutions to those where the incidence of infection in the sample is less than or equal to the incidence of infection adopted as the decision threshold, p_{CRIT}. TOL is some suitably small tolerance level for deviations of realized values of $\Pr(A)$ from the required value of 0.5. To apply this method to calculate binomial sample sizes, we find values of N such that:

$$\left| \sum_{Y=0}^{C} \binom{N}{Y} p^Y (1-p)^{N-Y} - 0.5 \right| < \text{TOL} \qquad (11.12)$$

in which p is disease incidence in the population.

The sample sizes calculated from equations 11.11 and 11.12 on setting $p_{CRIT} = 0.05$ and TOL $= 0.01$, so that all the resulting sampling plans have $0.49 \le \Pr(A) \le 0.51$ at the adopted decision threshold, are shown in Table 11.1. Sample sizes for population sizes up to and including 500 were calculated from the hypergeometric formula (equation 11.11), while sample sizes for the indefinitely large population were calculated from the binomial formula (equation 11.12). The values of N (in the body of the table) are all close to the values calculated by the approximate methods given in section 11.3.2 (right after equation 11.9). Each column in Table 11.1 yields a family of (N,C) sampling plans that all maintain the same OC curve as population size changes, at least to a reasonable approximation. A family of sampling plans based on larger sample sizes (i.e., use of columns for higher acceptance numbers) will have an OC curve with a steeper (negative) slope at the adopted decision threshold than a family based on smaller sample sizes (i.e., use of columns for lower acceptance numbers). Note that maintaining the same OC curve as population size increases does not require the sample size to increase in proportion to population size (see Montgomery, 1997).

The value of the method outlined here is that it can be used when the cumulative distribution function on which the OC curve is based is known, but an approximate formula for p_{CRIT} as a function of N is not available. For example, we may need to calculate the appropriate sample size when it is known that there is a

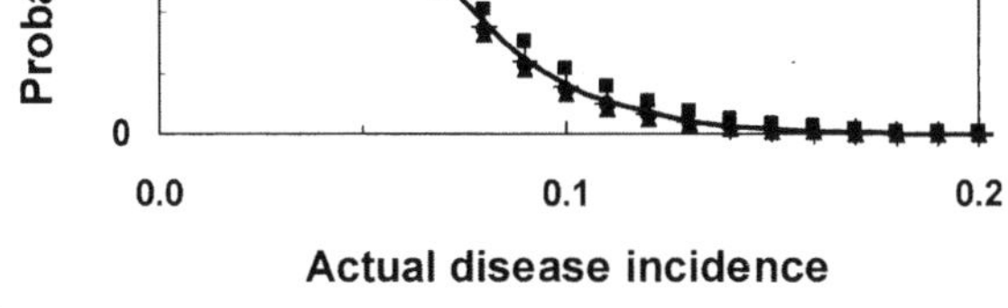

FIG. 11.7. The hypergeometric sampling plans (32,1) for $M = 100$ (■), (52,2) for $M = 200$ (▲) and (53,2) for $M = 300$ (◆), 400 (●) and 500 (+) all have OC curves close to the OC curve for a binomial (54,2) sampling plan (shown as a solid line), with $\Pr(A) \approx 0.5$ at the adopted decision threshold of $p_{CRIT} = 0.05$.

TABLE 11.1. Sample sizes, N (in the body of the table), calculated from either equation 11.11 or 11.12, with $p_{CRIT} = 0.05$ and TOL $= 0.01$, for a range of sampling plans characterized by population size (M) and acceptance number (C)[a].

Population size, M	*Acceptance number, C*				
	1	*2*	*3*	*4*	*5*
100	31	50	69	87	–
200	32	51	71	90	109
300	32	52	71	91	110
400	33	52	72	91	111
500	33	52	72	92	111
∞	33	53	73	93	112

[a]Sample sizes for population sizes up to and including 500 were calculated from the hypergeometric formula (equation 11.11), while sample sizes for the indefinitely large population were calculated from the binomial formula (equation 11.12).

proportion of diseased plants erroneously recorded as healthy. In this case, we find values of N such that:

$$\left|\sum_{Z=0}^{C} \Pr(Z) - 0.5\right| < \text{TOL} \qquad (11.13)$$

with $\Pr(Z)$ given by equation 11.4, for specified values of M, p_{CRIT} and ζ (the proportion of diseased plants wrongly assessed as healthy). The acceptance number is specified as $C = Np_{CRIT}$, and $X = Mp_{CRIT}$ (taking the integer parts of the right-hand sides of each expression). Setting $M = 200$, $p_{CRIT} = 0.05$, $\zeta = 0.05$ and TOL $= 0.01$, yields the following sampling plans: (34,1), (54,2), (74,3), (95,4), and (115,5). Thus with faulty inspection, larger sample sizes are required than for the corresponding sampling plans for error-free disease assessment, shown in Table 11.1.

It is of interest to consider these sample sizes for faulty inspections further, along lines suggested by Winkler (1985) and Pham-Gia and Turkkan (1992). The proportion of plants assessed as diseased (including both correct and incorrect assessments), denoted $\bar{y}$, is related to the true proportion diseased, p, as follows:

$$\bar{y} = p(1-\zeta) + (1-p)\delta$$

in which δ denotes the proportion of healthy plants wrongly assessed as diseased, assumed equal to zero in calculations here. (Estimation of disease incidence with inspection errors was previously discussed in section 10.3.2.) With an estimate $\bar{y}$ based on a sample size N, $\delta < \bar{y} < 1-\zeta$, the maximum likelihood estimate of p is:

$$p^* = \frac{\bar{y} - \delta}{1 - \zeta - \delta}.$$

The sample size that would be required if disease assessment were error-free is denoted N^*, and the relationship between N^* and N, the required sample size taking account of inspection errors, is:

$$N^* = \frac{Np^*(1-p^*)}{(p^* + k_1)(1 - p^* + k_2)} \qquad (11.14)$$

in which $k_1 = \delta/(1-\zeta-\delta)$, $k_2 = \zeta/(1-\zeta-\delta)$ and $N > N^*$. Now, N^* can be thought of as an effective sample size (Cussens, 1996), taking into account the possibility of faulty inspections in disease assessment. In effect, we can calculate what our sample size N is "worth" in terms of the size of a corresponding sample of error-free assessments. We find, in this case, that with $M = 200$, $p_{CRIT} = 0.05$, and $\zeta = 0.05$, the sample sizes N of the (N,C) sampling plans (34,1), (54,2), (74,3), (95,4), and (115,5) have N^* values of 32, 51, 70, 90 and 109, respectively. These are (to a good approximation) the sample sizes of the (assumed error-free) hypergeometric sampling plans for $M = 200$, $p_{CRIT} = 0.05$, given in Table 11.1.

If we specify a hypergeometric sampling plan for incidence of diseased plants, but disease assessment is not error-free, the OC curve of the actual plan in practice will be shifted to the right of that for the plan as specified. Suppose, for example, that a (50,2) plan is specified for $M = 100$, so that $p_{CRIT} = 0.05$ when $\Pr(A) = 0.5$ (Table 11.1). However, if there is a probability $\zeta = 0.05$ that a diseased plant will be erroneously assessed as healthy, $p_{CRIT} = 0.05$ when $\Pr(A) = 0.55$ (Fig. 11.4). The probability of acceptance has inadvertently been increased, with the outcome that more crops above the decision threshold will (wrongly) not be treated. The specified OC curve can be retrieved if the sample size calculated taking account of inspection errors is substituted for the calculated hypergeometric sample size. In the case of the example given in Fig. 11.4, the iterative solution of equation 11.13 for N, with $M = 100$, $p_{CRIT} = 0.05$, $C = Np_{CRIT}$ (taking the integer part), and $\zeta = 0.05$, is $N = 53$. We then find from equation 11.14 that these 53 observations are worth 50 error-free observations. If the OC curve for a (53,2) sampling plan is calculated using equation 11.5 with $M = 100$, $p_{CRIT} = 0.05$, $C = Np_{CRIT}$ (taking the integer part), and $\zeta = 0.05$, it is indistinguishable from the original hypergeometric (50,2) plan specified for $M = 100$, so that $p_{CRIT} = 0.05$ when $\Pr(A) = 0.5$, as shown in Fig. 11.5.

11.4 Zero Acceptance Number Sampling Plans

Sometimes a low level of disease intensity carries a disproportionately high risk. Examples include some diseases of horticultural crops, where even a small amount of cosmetic damage can result in a large loss in value (e.g., de Jong, 1995), some diseases that are known to spread rapidly from initially low levels (e.g., Seem et al., 1985), and some diseases that are so destructive as to warrant strict regulatory control (e.g., Clayton and Slack, 1988). In such cases, control action is triggered by the detection of disease at any level in the sample. In acceptance sampling terms, the sampling plan has an acceptance number $C = 0$. Squeglia (1994) discusses zero acceptance number acceptance sampling plans in an industrial quality control context, and Venette et al. (2002) include some discussion of the topic, from an entomological point of view, in their review. The main issue in designing sampling plans with $C = 0$ is not what happens when a defective item is detected in the sample. Rather, it is what happens if the sample contains no defectives. If we think of individual plants in a crop being sampled for disease incidence under a zero acceptance number sampling plan, then finding a single diseased plant triggers whatever is the appropriate control action.

However, if disease incidence in the sample is zero, this does not necessarily mean that the crop is disease free.

11.4.1 The operating characteristic curve

The ideal, but unachievable, OC curve for a zero acceptance number sampling plan is characterized by:

$$\Pr(A) = \begin{cases} 1 & (p = 0) \\ 0 & (p > 0) \end{cases}$$

It represents a situation in which all lots with zero defectives would be accepted, while all lots with any defectives at all would be rejected. Being realistic, we recognize that there is a chance that we may draw a sample such that $\bar{y} = 0$ (leading to acceptance of the lot) even when $p > 0$, in which case the acceptance would be erroneous. Note that with a $C = 0$ acceptance sampling plan, in the absence of inspection errors (specifically, healthy plants wrongly assessed as diseased) we cannot draw a sample that will lead us erroneously to reject a lot.

The OC curve for a $C = 0$ acceptance sampling plan based on the binomial distribution is provided by equation 11.2 with $C = 0$ and $\Pr(Y)$ from equation 11.1:

$$\Pr(A) = \Pr(Y = 0) = (1 - p)^N \qquad (11.15)$$

Note, in passing, that this equation was used previously (section 10.3.3) as the basis for calculating a one-sided confidence interval for disease incidence for the situation in which no diseased plants are observed in a sample of size N. Fig. 11.8 shows OC curves calculated from equation 11.15 with three different sample sizes, $N = 400$, 800, and 4800. This is essentially the same as Fig. 1a in Clayton and Slack (1988), where an extra curve (for $N = 1200$) is shown. Because, in zero acceptance

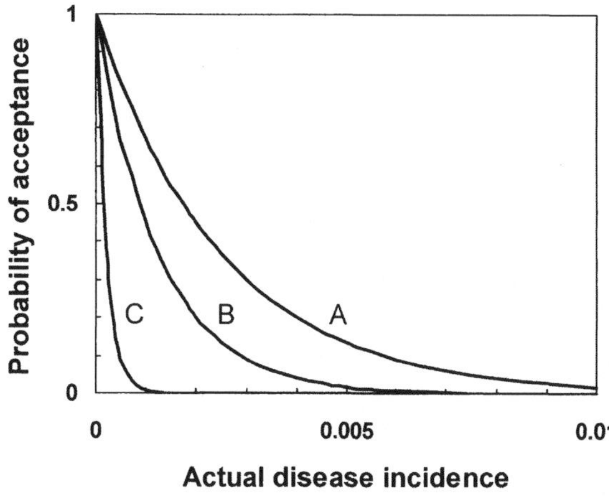

FIG. 11.8. OC curves for zero acceptance number ($C = 0$) sampling plans calculated from equation 11.15 with three different sample sizes. A: $N = 400$; B: $N = 800$; C: $N = 4800$.

number sampling, any acceptance of a lot with $p > 0$ is erroneous, Clayton and Slack (1988) refer to $\Pr(A)$ as the "probability of erroneous acceptance" (*PEA*). Clayton and Slack (1988) use the Poisson approximation for the zero term of the binomial distribution, $\exp(-Np) \approx (1 - p)^N$ (reasonable for small p), and plot $PEA = \exp(-Np)$ as a function of p. When the acceptance number $C = 0$, the OC curve has a different shape compared with OC curves for $C > 0$ (section 11.3). When $C = 0$, there is no point of inflexion in the OC curve. $\Pr(A)$ decreases monotonically with increasing p. As the sample size, N, increases, the OC curve represented by equation 11.15 increasingly resembles the ideal ($C = 0$) OC curve.

If population size is such that a $C = 0$ acceptance sampling plan should be based on the hypergeometric distribution, the OC curve is provided by equation 11.2 with $C = 0$ and $\Pr(Y)$ from equation 11.3:

$$\Pr(A) = \Pr(Y = 0) = \frac{(M - X)!\,(M - N)!}{M!\,(M - X - N)!} \qquad (11.16)$$

in which $X = Mp$ (rounded to an integer value if necessary). The OC curve for a $C = 0$ acceptance sampling plan when inspection errors occur (such that a proportion of defective units are erroneously assessed as acceptable) is provided by equation 11.5 with $C = 0$ (see, for example, Hughes et al., 2002b).

11.4.2 Sample size calculations

As we have noted, we cannot, on the basis of a sample in which no infected plants are found, say that there are no infected plants in the crop. Instead, what we would like to be able to do when no infected plants are found in a sample is make a statement that there is only a (small) specified probability that actual disease incidence exceeds a (low) specified level of incidence. Once we have decided on the specifications, we need to be able to calculate the sample size, N, that will enable us to make this statement when sampling is undertaken and no infected plants are found.

Equation 11.15 can be rearranged as:

$$N = \frac{\log(\Pr(A))}{\log(1 - p)} \qquad (11.17)$$

in which $\Pr(A)$ and p now represent particular specified values and N is the corresponding sample size required to be able to make the statement that there is only a small probability $\Pr(A)$ that actual disease incidence exceeds the low specified level p. A phytopathological example is discussed by Clayton and Slack (1988), in relation to the problem of sampling for bacterial ring rot of potato (caused by *Clavibacter michiganensis* subsp. *sepedonicus*). It is assumed here that disease assessments of individual tubers are carried out without error, that the population of tubers is sufficiently large so that calculations can be based on the binomial distribution and that the tubers inspected constitute a simple random sample of the population. It is desired that if the actual incidence of bacterial ring rot is 0.001, the probability that it will wrongly be concluded, on the basis of sampling, that a population of tubers is disease-free is at most 0.01. Then, using equation 11.17, $N = \log(0.01)/\log(1 - 0.001) = 4603$. The approximation $\Pr(A) = \exp(-Np)$, which is slightly conservative (i.e., results in a slightly larger sample size), yields $N = -[\ln(\Pr(A))/p]$; $N = -[\ln(0.01)/0.001] = 4605$, as given by Clayton and Slack (1988).

Because equation 11.16 is difficult to manipulate, an approximation formula is usually used when a sample size is required for a hypergeometric $C = 0$ acceptance sampling plan. Yamamura and Sugimoto (1995) give:

$$N \approx \left(M - \frac{Mp-1}{2}\right)(1 - \Pr(A)^{1/Mp}) \qquad (11.18)$$

which is slightly conservative compared to the exact result, while Kuno (1991b) gives:

$$N \approx M(1 - \Pr(A)^{1/Mp}) \qquad (11.19)$$

which is slightly more conservative yet. In either case, to calculate a required sample size, N, we set a specified low probability that actual disease incidence exceeds a specified low level. With specified values $\Pr(A) = 0.01$ and $p = 0.001$, sample sizes calculated from both equation 11.18 and equation 11.19 for a range of population sizes, M, are given in Table 11.2. The sample size for an indefinitely large population (again with specified values $\Pr(A) = 0.01$ and $p = 0.001$) is based on equation 11.17. Note that equations 11.17, 11.18, and 11.19 all imply that specification of smaller values of $\Pr(A)$ and p leads to larger values of N, the required sample size. In Table 11.2, the sample sizes based on equation 11.18 are the same as those given in Clayton and Slack (1988, Table 1).

TABLE 11.2. Sample sizes for some $C = 0$ acceptance sampling plans with $\Pr(A) = 0.01$ and $p = 0.001$.

Population size, M	*Sample size, N (equation 11.18)*	*Sample size, N (equation 11.19)*
5000	3009	3010
10,000	3689	3691
25,000	4204	4206
100,000	4499	4501
500,000	4582	4585
1,000,000	4593	4595
∞	4603[a]	

[a]Based on equation 11.17.

All the sample sizes in Table 11.2 produce the same OC curve (at least to a reasonable approximation), with $\Pr(A) = 1$ when $p = 0$, then decreasing monotonically towards the horizontal axis with increasing p, passing through the point $\Pr(A) = 0.01$, $p = 0.001$. It can be seen from Table 11.2 that although sample size increases with population size, the required sample size is not a constant proportion of the population size. As with sampling plans where $C \geq 1$ (section 11.2.5), it is important to bear in mind that calculating sample sizes as a constant proportion of a varying population size does *not* maintain the same OC curve.

Sample size for a zero acceptance number sampling plan when inspection errors occur (such that a proportion of defective units are erroneously assessed as acceptable) is calculated in the same way as for a $C \geq 1$ sampling plan with inspection errors (section 11.3.3). A larger sample size is required, to compensate for the effect of inspection errors, but this can be calculated from equation 11.13 with $C = 0$ (see, for example, Hughes et al., 2002b). Equation 11.14 allows calculation of the effective size (i.e., the corresponding number of error-free assessments) of a sample, taking into account the effect of inspection errors.

11.4.3 The mailroom problem

Mullin (1990), like Clayton and Slack (1988), discusses the problem of sampling for bacterial ring rot of potato. Mullin (1990) takes an extreme view (which, it turns out, is not very helpful) of the zero tolerance scenario. Consider a lot comprising $M = 200{,}000$ potato tubers in which just one tuber is infected with ring rot ($X = 1$). What is the probability that the infected tuber will be among a sample of $N = 500$ tubers, drawn at random, without replacement, from the lot? Intuitively, this probability is simply the sampling fraction $N/M = 500/200{,}000$ or 0.0025. Working from the hypergeometric distribution (equation 11.3) with $M = 200{,}000$, $N = 500$, $X = 1$ and $Y = 0$, we have:

$$\Pr(Y=0) = \frac{\binom{1}{0}\binom{200{,}000-1}{500-0}}{\binom{200{,}000}{500}}$$

or using equation 11.16:

$$\Pr(Y=0) = \frac{(200{,}000-1)!\,(200{,}000-500)!}{200{,}000!\,(200{,}000-1-500)!}.$$

Either way, we obtain $\Pr(Y = 0) = 0.9975$. Since this is the probability that the sample does *not* contain the infected tuber, the probability that the sample contains the infected tuber is equal to $1 - 0.9975 = 0.0025$.

Further examples of the calculation of $\Pr(Y = 0)$ for the hypergeometric distribution with $X = 1$ are given by McSorley and Littell (1993, Table 1).

Now consider the problem a little more generally. We wish to test the null hypothesis that one or more tubers in the lot is infected, the alternative being that none are infected. We assume that no inspection errors occur. What sample size is required so that the probability that we will wrongly reject the null hypothesis is less than or equal to some (small) specified level, denoted α? In passing, note that we cannot wrongly accept the null hypothesis in the situation as described. Now, to calculate the required sample size, we must consider the worst case scenario, that there is indeed only $X = 1$ infected tuber in the lot. Then, given a lot size M, we need a sample size N such that:

$$\Pr(Y = 0) = \frac{\binom{1}{0}\binom{M-1}{N}}{\binom{M}{N}} \leq \alpha$$

It turns out (see Wright, 1990) that this reduces to:

$$\frac{N}{M} \geq 1 - \alpha.$$

In practical terms this means that if $\alpha = 0.01$ is specified, in order that we may state with $100(1-\alpha) = 99\%$ confidence that the null hypothesis is rejected and the alternative accepted (no infected tubers having been found in the sample), we must sample 99% of the lot. This kind of sampling problem achieved some prominence when it became necessary to inspect distribution facilities of the US Postal Service for anthrax contamination, as discussed by Levy et al. (2002). The hypothesis testing scenario is unusual. Because of the undesirable consequences of not detecting infected tubers (or anthrax spores), we take the conservative null position that there are some infected tubers, and place the burden on the data to lead us to the conclusion that this is unlikely to be the case.

The large sample sizes required for high confidence or, alternatively, the low confidence obtained from small sample sizes, led Mullin (1990) to note, rather pessimistically, "the futility of relying on sampling inspection methods for quarantine disease control." As we have already noted, no sampling plan provides a guarantee against misclassification errors. However, using acceptance sampling methods such as those discussed by Clayton and Slack (1988) and McSorley and Littell (1993), it is possible to formulate a plan that enables us to make a statement that there is only a specified small probability that actual disease incidence exceeds a specified low level of incidence, no infected units having been found in the sample (see section 11.4.2).

11.5 Sequential Sampling for Classification

Suppose we have a decision rule such that (ideally) crops with true disease incidence p greater than a critical incidence p_{CRIT} will be treated (i.e., the generic terminology, these are unacceptable lots), and those with a true disease incidence of p or less will not be treated (these are acceptable lots). From what we know about the general shape of OC curves for $C \neq 0$ (e.g., Fig. 11.6), we might expect that if the true disease incidence is either substantially below or substantially above the adopted critical incidence, a relatively small sample would be sufficient to classify a crop accurately (i.e., with only a small chance of misclassification). The advantage of sequential classification methods is that they allow us to avoid the unnecessary sampling effort that may arise if a fixed sample size plan is applied to a crop where true disease incidence is either substantially below or substantially above the adopted critical incidence (Madden and Hughes, 1999a).

The widest use of acceptance sampling methods in the context of crop protection is, without doubt, in economic entomology. The methodology has been refined to facilitate the classification of pest population density in relation to decisions about the use of control measures using sequential binomial sampling for count data. This methodology is not without interest for plant pathologists (see, for example, Boivin et al., 1990), but we do not pursue it here. Instead, we refer interested readers to Binns et al. (2000) for a comprehensive account of the topic. Here, we describe sequential sampling applied to the classification of disease incidence data.

As with sequential estimation methods (section 10.12), sequential classification methods do not require specification of sample size ahead of sampling. Instead, sampling continues until a pre-specified stop condition has been met. In crop protection, the two methods of specifying stop conditions most widely used are based on Iwao (1975) (see, for example, Vincelli and Lorbeer, 1987; Gaunt and Cole, 1992; Copes et al., 2001), and on the sequential probability ratio test (SPRT) devised by Wald (1947). Binns (1994) gives a comparative account of the two methods. Here, we restrict our attention to the SPRT-based method, which is the basis for most of the modern applications of sequential classification both in crop protection and also in the wider context of statistical quality control.

11.5.1 Sequential classification with simple random sampling data

In section 11.2.3, we briefly drew attention to the analogy between sampling for classification and hypothesis testing. Consider classification based on a fixed sample size. In the case of a decision rule such that crops with true disease incidence p greater than a critical incidence

p_{CRIT} will be treated, and those with a true disease incidence of p or less will not be treated, for instance, we observe the appropriate number of sampling units and then use the data to test the null hypothesis H_0: $p \leq p_{CRIT}$ against the alternative hypothesis H_1: $p > p_{CRIT}$. When $p_{CRIT} \neq 0$, the test holds the possibility of two kinds of erroneous decisions: a Type I error (wrongly rejecting H_0 when it is true) and a Type II error (wrongly accepting H_0 when it is false).

Now, with sequential sampling for classification, a choice between *three* possible decisions is required. After each sampling unit has been observed, we can either: (i) cease sampling and accept the null hypothesis; (ii) cease sampling and accept the alternative hypothesis; (iii) observe another sampling unit. To put this in hypothesis-testing terms, two new constants, p_1 and p_2 are invoked, such that $p_2 > p_1$ and $(p_1 + p_2)/2 = p_{CRIT}$. Then the null hypothesis is H_0: $p = p_1$ and the alternative is H_1: $p = p_2$. Acceptance of the null hypothesis is interpreted as $p \leq p_{CRIT}$, while acceptance of the alternative is interpreted as $p > p_{CRIT}$. In the present context, sampling units are observed sequentially, and the cumulative total of those assessed as infected is recorded. The formulae for the stop conditions (i.e., the criteria for cessation of sampling) can be written in terms of p_1 and p_2 and two further constants, denoted α and β, that affect the characteristics of the OC curve (and so the rates of Type I and Type II errors, as discussed in more detail below). The OC curve of the sampling plan can be described in terms of p_1, p_2, α, and β. Additionally, in sequential sampling for classification, it is of interest to know the *average sample number* (ASN) of the sampling plan (because N is not a fixed quantity). In the present context, the ASN curve is a graphical plot of the average number of sampling units required to reach a decision (i.e., to reach one of the specified stop conditions) for any true value of p. The ASN curve can also be described in terms of p_1, p_2, α, and β.

OC and ASN curves based on Wald's (1947) SPRT are approximations of the curves corresponding to sequential sampling plans as usually implemented in crop protection. Wald's (1947) SPRT is based on an assumption that sampling stops when one of the specified stop conditions is reached *with no overshoot*. Since simple random sampling for disease incidence yields discrete data (the number of plants or plant parts diseased), and the stop conditions are specified in terms of numbers that are not necessarily integers (as discussed in the following paragraph), we cannot expect this assumption to be satisfied by a practical sequential sampling plan for disease incidence. The same is true, of course, for any sequential sampling plan that involves collection of discrete data. Another source of approximation arises because Wald's (1947) SPRT was derived under the assumption that a sampling plan specifies neither an arbitrary minimum nor an arbitrary maximum number of sampling units to be observed. Most practical sequential sampling plans used in crop protection do in fact specify such limits. Realized OC and ASN curves and error rates for particular practical sequential sampling plans can be calculated by simulation (Binns et al., 2000). Here, we continue with the SPRT-based versions of OC and ASN curves, in order to illustrate some generic characteristics of sequential sampling plans.

For disease incidence data, the cumulative total number of sampling units assessed as infected is denoted T_N. Sampling continues if, after observation of the previous sampling unit, $I_{low} + SN < T_N < I_{high} + SN$ (see Binns, 1994, for a derivation) in which I_{low}, I_{high}, and S are functions of p_1, p_2, α, and β (Table 11.3, binomial distribution). N is the number of sampling units assessed. Alternatively, the expressions on either side of the inequality can be thought of as formulae for two parallel straight lines on a graphical plot of T_N against N. These *stop lines* represent the criteria for cessation of sampling (Fig. 11.9). If the lower line ($T_N = I_{low} + SN$) is intersected by a line representing the observed sample data, the appropriate decision is to cease sampling and accept the null hypothesis. If the upper line ($T_N = I_{high} + SN$) is intersected by a line representing the observed sample data, the appropriate decision is to cease sampling and accept the alternative hypothesis. Otherwise, sampling continues.

Suppose that sampling commences and each sampling unit assessed in sequence is uninfected. Sampling would cease at the value of N defined by the intersection of $T_N = 0$ and $T_N = I_{low} + SN$, that is, when $N \geq -I_{low}/S$.

TABLE 11.3. Calculation of the sequential probability ratio test stop criteria for sequential classification with simple random sampling data, for both the binomial and the normal distribution[a].

Distribution	S	I_{low}	I_{high}
Binomial	$\dfrac{(\ln(q_1/q_2))}{(\ln(p_2q_1/p_1q_2))}$	$\dfrac{\ln(\beta/(1-\alpha))}{\ln(p_2q_1/p_1q_2)}$	$\dfrac{\ln((1-\beta)/\alpha)}{\ln(p_2q_1/p_1q_2)}$
Normal	$\dfrac{(\mu_1+\mu_2)}{2}$	$\dfrac{s_Y^2\ln(\beta/(1-\alpha))}{(\mu_2-\mu_1)}$	$\dfrac{s_Y^2\ln((1-\beta)/\alpha)}{(\mu_2-\mu_1)}$

[a]The cumulative total of sampling units assessed as infected is denoted T_N. Sampling continues if, after observation of the previous sampling unit, $I_{low} + SN < T_N < I_{high} + SN$ in which I_{low}, I_{high}, and S are functions (as given in the table) of the appropriate constants that characterize the OC curve (as described in the text) and N is the number of sampling units assessed. $q_1 = 1 - p_1$, $q_2 = 1 - p_2$.

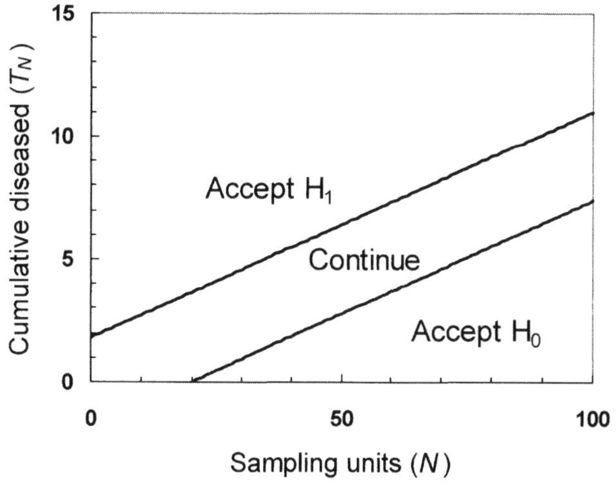

FIG. 11.9. Sequential classification for simple random sampling data. For disease incidence data, the cumulative total number of sampling units assessed as infected is denoted T_N. N is the cumulative total number of sampling units assessed. Sampling units are observed sequentially, and the cumulative total of those assessed as infected is recorded. The criteria for cessation of sampling are characterized by two parallel straight lines (the stop lines). If the lower stop line $(T_N = I_{low} + SN)$ is intersected by the line drawn for the observed data, the appropriate decision is to cease sampling and accept the null hypothesis (interpreted as $p \leq p_{CRIT}$). If the upper stop line $T_N = I_{high} + SN$ is intersected by the line drawn for the observed data, the appropriate decision is to cease sampling and accept the alternative hypothesis (interpreted as $p > p_{CRIT}$). Otherwise, sampling continues. I_{low}, I_{high}, and S are functions of p_1, p_2, α, and β (Table 11.3, binomial distribution). The stop lines shown here correspond to an SPRT-based sequential sampling plan for disease incidence data, collected by simple random sampling, with $p_1 = 0.05$, $p_2 = 0.15$, $\alpha = 0.1$, and $\beta = 0.1$.

If each sampling unit assessed in sequence is infected, sampling would cease at the value of N defined by the intersection of $T_N = N$ and $(T_N = I_{high} + SN)$, that is, when $N \geq I_{high}/(1 - S)$. Thus, for an SPRT-based sequential sampling plan for disease incidence data collected by simple random sampling, a decision to cease sampling cannot be reached while N is less than the minimum of $-I_{low}/S$ and $I_{high}/(1 - S)$ (Guenther, 1977).

The corresponding OC and ASN curves may be calculated from the formulae given for the binomial distribution in Table 11.4. A dummy variable (denoted here j) is given values at intervals over a range between about -4 and $+4$ (note that different formulae are applicable when $j \neq 0$ and when $j = 0$). Use of more values (i.e., shorter intervals) leads to smoother realizations of the OC and ASN curves. For a sequential sampling plan that involves collection of disease incidence data by simple random sampling, the formulae for the binomial distribution in Table 11.4 can be used to plot the OC curve (a graphical plot of OC(j) [the probability of acceptance] against $p(j)$) and the ASN curve (a graphical plot of ASN(j) against $p(j)$). The stop lines shown in Fig. 11.9 correspond to an SPRT-based sequential sampling plan for disease incidence data, collected by simple random sampling, with $p_1 = 0.05$, $p_2 = 0.15$, $\alpha = 0.1$, and $\beta = 0.1$. The corresponding OC curve and ASN curve for this plan are shown in Fig. 11.10. The OC curve (Fig. 11.10A) shows that the probability of acceptance is high when $p < p_1$ and low when $p > p_2$. For values of p between p_1 and p_2, the OC curve declines with increasing p. The probability of acceptance has a value ≈ 0.5 when $p = p_{CRIT} = (p_1 + p_2)/2 = 0.1$. The ASN curve (Fig. 11.10B) shows that the average number of sampling units required for a decision to be reached is low when $p < p_1$ and when $p > p_2$. Between p_1 and p_2 the average number of sampling units required for a decision to be reached is higher. The maximum of the ASN curve (at 40 sampling units, for this example) occurs close to $p = p_{CRIT} = 0.1$. The ASN curve shows the value of sequential classification: on average, relatively little sampling effort is required when p is substantially below or above the critical incidence adopted as a threshold value.

A sampling plan for sequential classification can be fine-tuned by altering the specified values of α and β. For example, the stop lines shown in Fig. 11.11 correspond to an SPRT-based sequential sampling plan for disease

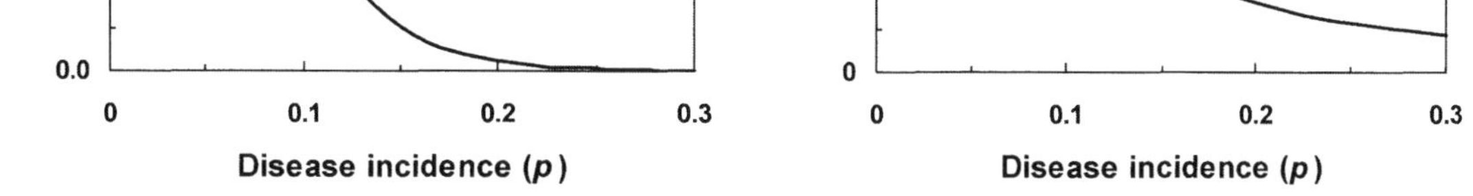

FIG. 11.10. Sequential classification for simple random sampling data. (A) Operating characteristic (OC), and (B) average sample number (ASN) curves for the sampling plan illustrated in Fig. 11.9.

TABLE 11.4. Sequential probability ratio test formulae for disease incidence (p), operating characteristic (OC), and average sample number (ASN) based on the binomial and normal distributions[a].

Distribution	*Disease incidence*	*OC curve*	*ASN curve*
Binomial	$p(j)=\frac{(1-(q_2/q_1)^j)}{((p_2/p_1)^j-(q_2/q_1)^j)}$	$OC(j)=\frac{((1-\beta)/\alpha)^j-1}{[((1-\beta)/\alpha)^j-(\beta/(1-\alpha))^j]}$	$ASN(j)=\frac{\ln(\beta/(1-\alpha))OC(j)+\ln((1-\beta)/\alpha)(1-OC(j))}{p(j)\ln(p_2q_1/p_1q_2)+\ln(q_2/q_1)}$
	$p(0)=\frac{(\ln(q_1/q_2))}{(\ln(p_2q_1/p_1q_2))}$	$OC(0)=\frac{\ln((1-\beta)/\alpha)}{\ln((1-\beta)/\alpha)-\ln(\beta/(1-\alpha))}$	$ASN(0)=\frac{-\ln((1-\beta)/\alpha)\ln(\beta/(1-\alpha))}{\ln(p_2/p_1)\ln(q_1/q_2)}$
Normal	$\mu(j)=\frac{(\mu_2+\mu_1-j(\mu_2-\mu_1))}{2}$	$OC(j)=\frac{[((1-\beta)/\alpha)^j-1]}{[((1-\beta)/\alpha)^j-(\beta/(1-\alpha))^j]}$	$ASN(j)=\frac{\ln(\beta/(1-\alpha))OC(j)+\ln((1-\beta)/\alpha)(1-OC(j))}{[[(\mu_2+\mu_1-2\mu(j))(\mu_1-\mu_2)]/2s_Y^2]}$
	$\mu(0)=\frac{(\mu_2+\mu_1)}{2}$	$OC(0)=\frac{\ln((1-\beta)/\alpha)}{\ln((1-\beta)/\alpha)-\ln(\beta/(1-\alpha))}$	$ASN(0)=\frac{-\ln((1-\beta)/\alpha)\ln(\beta/(1-\alpha))}{(\mu_2-\mu_1)^2/s_Y^2}$

[a] j is a dummy variable taking values at intervals over a range between about −4 and +4 (note that different formulae are applicable when $j \neq 0$ and when $j = 0$). For a sequential sampling plan that involves collection of data by simple random sampling, the formulae can be used to plot the OC curve (a graphical plot of OC(j) against $p(j)$) and the ASN curve (a graphical plot of ASN(j) against $p(j)$). The various constants required to characterize the curves are described in the text. $q_1 = 1 - p_1$, $q_2 = 1 - p_2$.

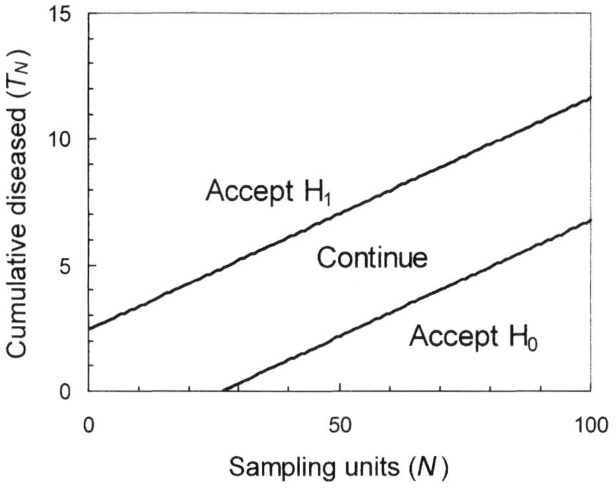

FIG. 11.11. Sequential classification for simple random sampling data. The stop lines shown here correspond to an SPRT-based sequential sampling plan for disease incidence data, collected by simple random sampling, with $p_1 = 0.05$, $p_2 = 0.15$, $\alpha = 0.05$, and $\beta = 0.05$.

incidence data, collected by simple random sampling, with $p_1 = 0.05$, $p_2 = 0.15$, $\alpha = 0.05$ and $\beta = 0.05$. By comparison with Fig. 11.9, it can be seen that the effect of decreasing α and β is to widen the gap between the two stop lines (p_1 and p_2 being unchanged). The corresponding OC curve and ASN curve for the plan characterized by the stop lines shown in Fig. 11.11 are shown in Fig. 11.12. Both curves are qualitatively similar to those shown in Fig. 11.10, but differ in the quantitative detail. For values of p between p_1 and p_2, the OC curve in Fig. 11.12A declines more steeply with increasing p than the OC curve in Fig. 11.10A, and the ASN curve in Fig. 11.12B is higher than the ASN curve in Fig. 11.10B. For the example in Fig. 11.12B, the maximum of the ASN curve is at 71 sampling units. The overall effect of decreasing the specified values of α and β is to produce a sampling plan with, on average, decreased rates of misclassification errors that requires, on average, a larger sample size.

The constants α and β do not have to have the same value; their values can be altered independently. The effects of decreasing α while β remains unchanged are mainly (but not exclusively) to increase I_{high} (and so raise the upper stop line), and to raise the part of the OC curve where $p < p_{CRIT}$ (and so reduce the rate of Type I errors). The effects of decreasing β while α remains unchanged are mainly (but not exclusively) to decrease (i.e., make more negative) I_{low} (and so lower the lower stop line), and to lower the part of the OC curve where $p > p_{CRIT}$ (and so reduce the rate of Type II errors). Decreasing either α or β (or both) raises the maximum of the ASN curve.

11.5.2 Sequential classification with cluster sampling data

Disease incidence data are often collected by cluster sampling (section 10.5). For sequential classification in a cluster sampling scenario, the cumulative total of sampling units assessed as infected is, as before, denoted T_N. Sampling continues if, after observation of the previous sampling unit, $I_{low} + SnN < T_N < I_{high} + SnN$ in which N, as before, is the number of sampling units and n is the number of elements per sampling unit (Madden and Hughes, 1999a). The simple random sampling scenario (section 11.5.1) is, in effect, cluster sampling with a cluster size of $n = 1$. The expressions on either side of the inequality can be thought of as formulae for two parallel straight lines (slope Sn) on a graphical plot of T_N against N (not nN). When the binomial distribution is an appropriate description of variability, I_{low}, I_{high}, and S are functions of p_1, p_2, α, and β as given (for the binomial distribution) in Table 11.3.

More often, disease incidence data collected by cluster sampling are aggregated, and the β-binomial distribution (section 9.4.5) and the power law (section 9.4.7) usually provide a better description of variability than the binomial distribution. β-Binomial-based formulae for SPRT stop lines are not available (Binns et al., 2000).

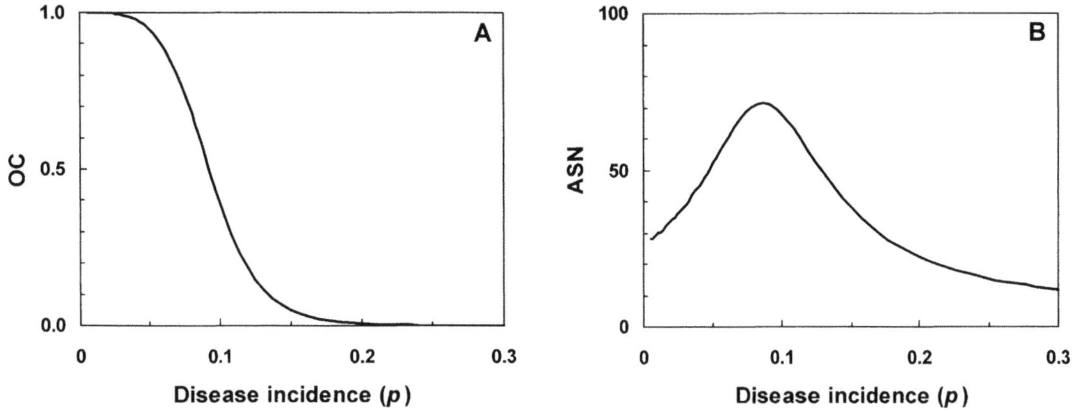

FIG. 11.12. Sequential classification for simple random sampling data. (A) Operating characteristic (OC), and (B) average sample number (ASN) curves for the sampling plan illustrated in Fig. 11.11.

Faced with this deficiency, Hoffman et al. (1996) and Nault and Kennedy (1996) used the binomial-based SPRT stop lines (for given values of p_1, p_2, α, and β) and investigated the effect of aggregation using simulated β-binomial cluster sampling data. More aggregated data tended to generate less steep OC curves (i.e., more misclassification errors) and lower ASN curves (i.e., smaller sample sizes). This might be regarded as rather unsatisfactory: it happens because the binomial-based stop lines yield a "continue sampling" region on the graphical plot of T_N against N that is too narrow for use with data that are, in fact, aggregated.

Instead, Madden and Hughes (1999a) developed SPRT stop lines for aggregated disease incidence data by assuming that mean disease incidence is (at least to a reasonable approximation) normally distributed. The formulae for SPRT stop lines based on the normal distribution (Table 11.3) can then be applied (with $\mu_1 = np_1$ and $\mu_2 = np_2$), provided an appropriate variance can be specified. Madden and Hughes (1999a) used a variance formula based on the power law (section 9.4.7):

$$s_Y^2 = A_Y(np(1-p))^b \qquad (11.20)$$

in which s_Y^2 is an estimate of the variance of the number of infected elements per sampling unit and A_Y is related to A in equation 9.17 by $A_Y = An^{2-2b}$ (and b is the same as b in equation 9.17). The approximate β-binomial stop lines, based on the SPRT for the normal distribution (Table 11.3) and the power law variance formula (equation 11.20 with $p = p_{\text{CRIT}}$) are then given by $T_N = I_{\text{low}} + SN$ and $T_N = I_{\text{high}} + SN$, in which:

$$S = \frac{\mu_1 + \mu_2}{2} = \frac{np_1 + np_2}{2} = np_{\text{CRIT}}$$

$$I_{\text{low}} = \frac{An^{1-b}(p_{\text{CRIT}}(1-p_{\text{CRIT}}))^b}{p_2 - p_1}\ln\left(\frac{\beta}{1-\alpha}\right)$$

$$I_{\text{high}} = \frac{An^{1-b}(p_{\text{CRIT}}(1-p_{\text{CRIT}}))^b}{p_2 - p_1}\ln\left(\frac{1-\beta}{\alpha}\right)$$

Note that if we wish to retain the same format as for the stop line formulae based on the binomial distribution with cluster sampling data (above), with $T_N = I_{\text{low}} + SnN$ and $T_N = I_{\text{high}} + SnN$, we write instead

$$Sn = \frac{\mu_1 + \mu_2}{2} = \frac{np_1 + np_2}{2} = np_{\text{CRIT}}, \quad S = p_{\text{CRIT}}.$$

I_{low} and I_{high} are unaffected here. An example is shown in Fig. 11.13. Either way, for aggregated cluster sampling data (characterized by $A > 1$, $b \geq 1$), these stop lines yield a "continue sampling" region on the graphical plot of T_N against N that is wider than that provided by stop lines for random cluster sampling data (characterized by $A = 1$, $b = 1$) with the same values of p_1, p_2, α, and β.

The effect of decreasing the specified values of α and β is to produce a sampling plan with, on average, decreased rates of misclassification errors that requires, on average, a larger sample size. This can be seen by reference to OC and ASN curves for the sampling plans in question. OC and ASN curves may be calculated from the formulae given for the normal distribution in Table 11.4, with s_Y^2 specified as in equation 11.20. However, the analytical formulae for SPRT-based OC curves and ASN curves (Table 11.4) are not generally used for particular applications in crop protection (Binns et al., 2000). Instead, simulation is usually used to obtain these curves, based on the assumed (discrete) distribution of disease.

11.5.3 The need for simulation

One reason for the use of simulation methods to obtain OC and ASN curves is that for data that are discrete, formulae based on the normal distribution are approximations. Even when formulae for SPRT-based OC curves and ASN curves are based directly on an appropriate (discrete) statistical probability distribution, certain assumptions on which the derivations of the formulae are based can be difficult to justify in practice (see, e.g., Fowler and Lynch, 1987). It is also the case that in applications of SPRT-based methods in crop protection, decisions are usually based on a comparison of discrete data with continuous stop lines, and arbitrary stop conditions are often specified in addition to the SPRT-based stop lines (to prevent sampling either being stopped at a very small sample size, or being continued at very large sample sizes without prospect of stopping). The analytical formulae for SPRT-based stop lines, OC curves, and ASN curves given in Tables 11.3 and 11.4 therefore represent approximations of the characteristics of practical sampling plans. Simulation methods are usually used to derive and, where necessary, adjust practical sampling plans for sequential classification. Binns et al. (2000, Chapter 5) provide a step-by-step outline of an algorithm by which such simulation methods could be implemented.

Turechek et al. (2001) give details of development and implementation of a sequential sampling plan for classification of incidence of Phomopsis leaf blight of strawberry, including evaluation by means of simulation. The raw material for such a simulation comprises an appropriate statistical description for the disease data that will be collected by sampling (see Chapter 9) including, where necessary, adopted values of aggregation parameters. For disease incidence data, this will usually involve the binomial (section 9.4.2) or the β-binomial (section 9.4.5) distribution, or the empirical power law that relates observed and binomial variances (section 9.4.7). Together with the four constants p_1, p_2, α, and β (section 11.5.1), the adopted statistical description for the disease data allows construction of initial SPRT-based stop lines

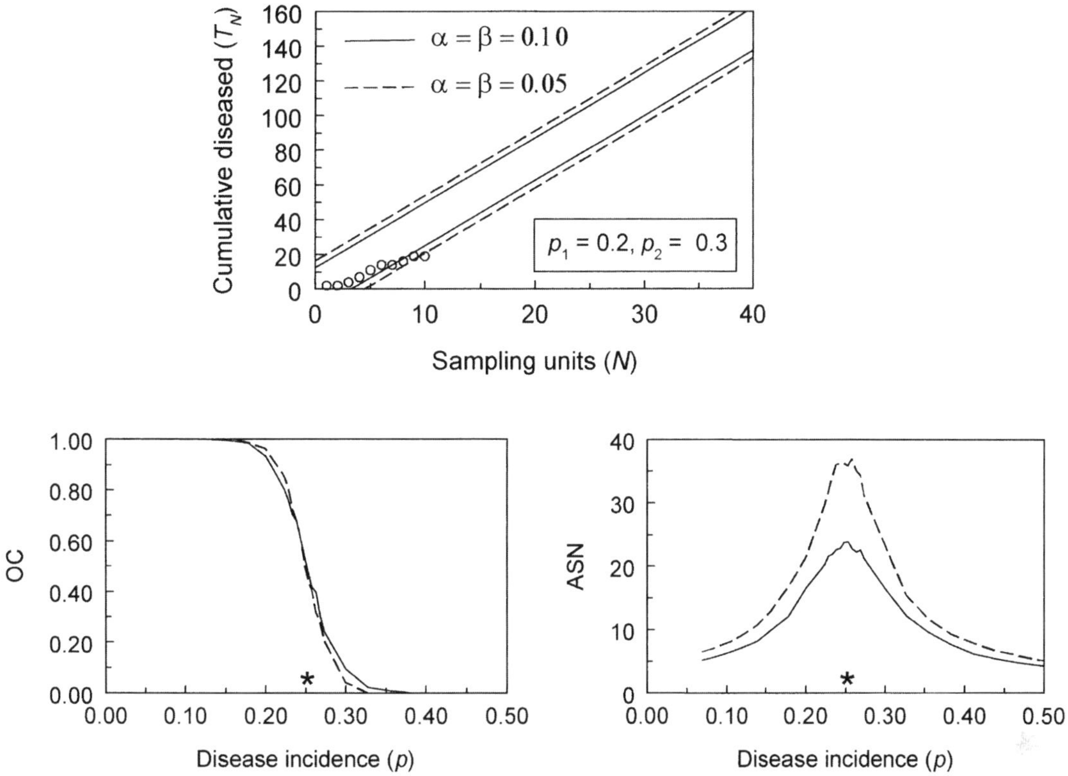

FIG. 11.13. Sequential classification for cluster sampling data. The example here is based on previously determined power law parameters for Phomopsis leaf blight of strawberry (caused by *Phomopsis obscurans*): $A_Y = 2.48$ ($A = 6.57$) and $b = 1.18$, with $n = 15$ leaflets per sampling unit (Turechek et al., 2001). The stop lines were calculated with $p_1 = 0.2$ and $p_2 = 0.3$ (so that $p_{CRIT} = 0.25$). From these values, $S = 15 \times 0.25 = 3.75$, and the intercepts I_{low} and I_{high} are ±16.5 for $\alpha = \beta = 0.05$ and ±12.5 for $\alpha = \beta = 0.10$. One example data set is illustrated, where the first 10 observations obtained by sampling were 2, 0, 2, 3, 4, 3, 0, 2, 3, and 0 diseased leaflets per sampling unit. The corresponding cumulative values are shown in the upper part of the figure (as open circles), along with both pairs of stop lines. The cumulative total number of diseased leaflets exits the "continue sampling" region at $N = 7$ for stop lines based on $\alpha = \beta = 0.10$, leading to acceptance of the null hypothesis (i.e., $p \leq 0.25$). With the example data set, a decision would have been reached and sampling halted after the first seven sampling units had been observed, in sequence, in this case. However, for stop lines based on $\alpha = \beta = 0.05$, the cumulative total number of diseased leaflets does not exit the "continue sampling" region until 10 sampling units had been observed, in sequence (again the appropriate decision is acceptance of the null hypothesis). The OC and ASN curves (lower left and lower right, respectively) were determined based on simulations from the β-binomial distribution with θ predicted by the binary power law parameters, and a minimum N of 3 was imposed. As expected, the ASN curve was higher overall, and considerably higher near p_{CRIT}, for $\alpha = \beta = 0.05$ than for $\alpha = \beta = 0.10$. At values of p below p_{CRIT}, the OC curve was higher for $\alpha = \beta = 0.05$ than for $\alpha = \beta = 0.10$; above p_{CRIT}, the converse was found. Thus, the (average) error rates were lower, as expected, when using $\alpha = \beta = 0.05$ than when using $\alpha = \beta = 0.10$ in deriving the stop lines, with the largest differences around p_1 and p_2. In such cases, a judgement would have to be made on whether the magnitude of the reduction in error rates justified the cost of the extra sampling required.

(see Table 11.3). The simulation procedure itself involves specification of an appropriate range of (in the present context) true mean disease incidence. Random data sets are then generated, using the adopted statistical description (e.g., β-binomial), such that a large number of replicate data sets are obtained at each specified point in the range of (simulated) true mean disease incidence. Each random data set generated is a simulated sequential sample from a distribution with known characteristics (mean and, where necessary, aggregation parameters). Thus the approximation introduced by using the normal distribution to describe mean disease incidence is avoided.

Taking into account any arbitrary stop conditions additional to the specified stop lines, each simulated sequential sample is analyzed. The number of sampling units assessed in order to classify a sample, and the nature of the classification are recorded. At each specified point in the range of true mean disease incidence, the average number of sampling units assessed in order to classify a sample provides a point on the realized ASN curve. The number of times that disease incidence is classified as less than or equal to p_{CRIT}, expressed as a proportion of the total number of replications, provides a point on the realized OC curve.

The characteristics of the realized OC and ASN curves of the sampling plan can now be checked and, if necessary, adjusted to obtain a sampling plan that, on average, provides an acceptable combination of error rates and sample size. If the realized ASN curve is regarded as too high, α and/or β could be increased. This has the effect of narrowing the "continue sampling" region on the graphical plot of T_N against N so that, on average, fewer sampling units are required for classification (at the cost of an increase in misclassification errors). If the realized OC curve is regarded as allowing too many misclassification errors, α and/or β could be decreased. This has the effect of widening the "continue sampling" region on the graphical plot of T_N against N so that, on average, fewer misclassification errors are made (at the cost of more sampling units being required for classification). This is illustrated in Fig. 11.13.

Simulation provides a means of evaluating sampling plans for sequential classification based on Wald's (1947) SPRT. This is an important aspect of the use of SPRT-based sequential methods in crop protection, since it overcomes discrepancies between theory and practice in sampling. We should not, however, lose sight of the fact that the simulation methodology itself requires assumptions, important among which are the statistical description adopted for the disease data to be sampled and the choice of critical incidence, p_{CRIT}. None of the preceding discussion of sequential classification methodology tells us whether an adopted value of p_{CRIT} is in fact the appropriate one for a particular crop protection decision making problem. Nyrop et al. (1998) discuss this issue with reference to OC and ASN curves.

11.6 Risk Algorithms as a Basis for Decision Making

"Disease forecasting", "disease prediction", and the development of "disease warning systems" are activities familiar to plant disease epidemiologists (see, e.g., Zadoks, 1984; Madden and Ellis, 1988; Campbell and Madden, 1990 [Chapter 15]; Hardwick 1998). It is clearly of interest, having identified the factors that lead to epidemics, to seek to use this information to provide a basis for the management of plant disease. However, in this section we do not review or evaluate any particular disease warning systems, the details of which tend to be pathosystem specific. Instead, we describe a generic approach to the decision making process as practised in plant disease management and introduce aspects of decision theory applicable to the problem.

11.6.1 Risk factors

So far, discussion of the decision making process has taken only a single *risk factor*—disease intensity (assessed in some appropriate fashion)—into account when classifying a crop as requiring protection measures, or not. However, the level of risk to which a crop is exposed may be influenced by many factors, some beyond the control of those involved in crop production, and some that are integral components of crop production systems. Meteorological factors are the most obvious example of the former (e.g., D. A. Johnson et al., 1994), while Fry (1982) lists choice of crop and cultivar, sowing date and method, rate of fertilizer application, pesticide use, tillage type and frequency, irrigation method and frequency, harvesting method, and crop storage as examples of the latter.

The first step in the formulation of disease management strategy for any particular cropping system is to identify the most important risk factors among those on the long list of possible candidates. This is facilitated by basic epidemiological studies of pathogen life cycles, and an understanding of the way in which weather and cropping factors affect the level of initial inoculum and the rate at which progress through the life cycle proceeds. To be able to identify risk factors, we need information both on the candidate risk factors and on the definitive status of the crops in which they are studied. In the widely used terminology in such analyses, a crop that definitively required treatment is referred to as a *case*, one that definitively did not is referred to as a *control*. Note that the classification into cases and controls must be made independent of the risk factors that might be used as a basis for decision making. Typically, during the development stages of management strategies in crop protection, this is a retrospective classification of untreated crops based on yield or an end-of-season disease assessment.

Example 11.1. Potato late blight. D. A. Johnson et al. (1996, 1998) analyzed relationships between weather factors and epidemics of potato late blight (caused by *Phytophthora infestans*) in south-central Washington State. Their objective was to develop a forecasting model to aid in disease management at the regional scale. Using retrospective data from the study area over the period 1970–1994, D. A. Johnson et al. (1996, model 2) identified important risk factors for potato late blight in south-central Washington as follows: whether or not the previous year was an outbreak year; the number of days on which rain fell during April and May; and the total precipitation during May when the daily minimum temperature was $\leq$5°C. Any year in the study period in which late blight was confirmed in any field in the study area was classified as an outbreak year (these are the cases), otherwise a year was classified as a non-outbreak year (these are the controls). The procedures outlined for distinguishing between outbreak years and non-outbreak years are assumed to have been free of error.

Example 11.2. Sclerotinia stem rot of oil seed rape. Researchers at the Swedish University of Agricultural Sciences (Yuen et al., 1996; Twengström et al., 1998) analyzed relationships between weather, crop, and disease factors and epidemics of Sclerotinia stem rot of spring sown oil seed rape (caused by *Sclerotinia sclerotiorum*) in Sweden. Their objective was to develop a forecasting model to predict the need for fungicide applications, working at the field scale. Stepwise logistic regression procedures were used to identify risk factors. Twengström et al. (1998) identified six important risk factors for Sclerotinia stem rot of spring sown oil seed rape in Sweden, as follows: number of years since the previous oilseed rape crop; percentage disease incidence in the previous host crop; plant population density; rainfall in the previous two weeks; weather forecast; and regional risk for apothecium development.

Recall that the economic threshold (section 11.2.2) may be used to identify circumstances in which it is economically advantageous to adopt crop protection measures. The economic threshold is a discrete choice threshold: the only options are to apply crop protection measures or to withhold them. However, the choice between these two options must be made before it is known for sure whether or not a crop will sustain economic loss resulting from reductions in the quantity and quality of crop yield. In fact, the threshold that is actually of interest is the *economic injury level* (Stern et al., 1959; Pedigo et al., 1986). In the present context, we can think of this as the lowest level of risk to which a crop is exposed that will result in economic damage. Unfortunately, by the time that economic damage can be measured directly, it is beyond prevention. The economic threshold is the level of risk exposure at which crop protection measures should be applied, in order to prevent the economic injury level from being reached. The economic threshold may be used as a basis for deciding whether or not crop protection measures are required, at a time when it is still possible to keep damage below the economic injury level. In the Sclerotinia stem rot study, data from about 800 fields, untreated with fungicide, were collected over a period of 10 years and used in the analysis. Retrospectively, the fields were divided into two groups: those with >25% diseased plants (these are the cases) and those with ≤25% diseased plants (these are the controls), 25% diseased plants having been identified as the economic injury level. Because a sampling procedure was used to assess % diseased plants, there is uncertainty attached to the classification of cases and controls. For this illustration we will, however, assume that cases and controls have been classified definitively and correctly.

11.6.2 Risk algorithms

Once the important risk factors have been identified, we need a way of combining them so that we can use data on risk factors to make a prediction of whether or not crop protection measures are required. We use the term *risk algorithm* to refer to a calculation that combines data on identified risk factors in order to make an assessment of the need for crop protection measures. When such a calculation is based on statistical procedures, the term *statistical prediction rule* (Swets et al., 2000) is synonymous. When the calculation is carried out by means of dedicated computer software, the term *decision support system* is sometimes used. Risk-based disease warning methods such as BLITECAST (for potato late blight) and Mills tables (for apple scab) are examples of risk algorithms developed in plant disease epidemiology.

A risk algorithm characterizes the relationship between a binary response variable (i.e., the requirement for crop protection measures or otherwise) and one or more risk factors (i.e., the explanatory variables). Typically, a risk algorithm is formulated from a data set where the requirement for crop protection measures, or otherwise, has been assessed retrospectively (and so definitively) for a number of crops for which various risk factors thought likely to be related to disease intensity and/or crop yield reduction have also been measured. Formulation of a risk algorithm need not necessarily be based on a statistical analysis of the data, although it is increasingly the case that statistical analysis is used. Statistical methods used in the construction of risk algorithms include logistic regression analysis and discriminant function analysis.

The disease management decision-making problem can now be outlined as follows. We wish to classify crops as requiring protection measures, or otherwise. Because the objective is to use the crop protection measures to prevent disease from becoming a problem (e.g., to prevent the economic injury level being reached), we cannot measure this requirement directly. Instead, we predict the requirement, using data relating to the important risk factors, combined via a risk algorithm. The risk algorithm generates an *indicator variable* (or, for brevity, simply an indicator) that provides a basis for a yes-or-no decision on the need for crop protection measures. Many decision-making problems in science, industry, and society can be characterized in this way (Swets et al., 2000).

Example 11.1 continued. Potato late blight. For the years of the study period 1970–1994, Table 11.5 shows values of the identified potato late blight risk factors: D. A. Johnson et al. (1996) used two different statistical methods, discriminant function analysis and logistic regression analysis, to combine these data. Press and Wilson (1978) provide a statistically orientated discussion of these methods. One advantage of logistic regression is that it is appropriate when there are risk factors that are discrete variables (i.e., not normally distributed) (Johnson et al., 1998). We reproduce both the analyses here, using MINITAB statistical software (Table 11.5).

A diagnostic test (or, for brevity, a test) comprises an indicator variable (or a set of indicator variables combined in some appropriate way) and a classification rule. For example, two tests are illustrated in Table 11.5. The first test is based on the linear discriminant functions for the non-outbreak years (DF_1) and outbreak years (DF_2).

TABLE 11.5. Risk factors for potato late blight as identified by D. A. Johnson et al. (1996): whether or not the previous year was an outbreak year (PYO, yes = 1, no = 0), the number of days rain during April and May (NDR), and the total precipitation (mm) during May when the daily minimum temperature was ≤5°C (TPM)[a].

Year	*PYO* 0 = No 1 = Yes	*NDR (days)*	*TPM (mm)*	DF_1[b]	DF_2[c]	$DF_2 > DF_1$ 0 = No 1 = Yes	*logit*($\hat{p}$)[d]	$\hat{p}$[e]	$\hat{p} > 0.5$ 0 = No 1 = Yes	*Outbreak Year* 0 = No 1 = Yes
1970	0	8	5.84	2.78	−0.38	0	−3.05	0.05	0	0
1971	0	9	6.86	3.69	1.04	0	−2.51	0.08	0	0
1972	0	9	47.29	5.79	−0.28	0	−6.13	0.00	0	0
1973	0	6	8.89	1.21	−3.41	0	−4.58	0.01	0	0
1974	0	16	7.37	9.76	11.26	1	1.84	0.86	1	1
1975	1	10	6.08	6.57	8.72	1	1.74	0.85	1	1
1976	1	12	3.30	8.15	11.74	1	3.25	0.96	1	1
1977	1	10	11.44	6.84	8.55	1	1.26	0.78	1	1
1978	1	11	14.99	7.89	9.89	1	1.57	0.83	1	0
1979	0	8	4.06	2.68	−0.32	0	−2.89	0.05	0	0
1980	0	13	37.84	8.75	5.87	0	−2.77	0.06	0	0
1981	0	8	18.54	3.44	−0.80	0	−4.18	0.02	0	0
1982	0	15	7.87	8.92	9.78	1	1.17	0.76	1	1
1983	1	9	11.86	6.00	7.07	1	0.60	0.65	1	1
1984	1	17	9.15	12.77	18.85	1	5.87	1.00	1	1
1985	1	5	11.43	2.53	1.24	0	−1.88	0.13	0	0
1986	0	8	8.13	2.90	−0.46	0	−3.25	0.04	0	0
1987	0	5	4.83	0.14	−4.73	0	−4.84	0.01	0	0
1988	0	15	11.93	9.13	9.65	1	0.81	0.69	1	0
1989	0	12	13.70	6.64	5.20	0	−1.24	0.22	0	0
1990	0	12	19.04	6.91	5.03	0	−1.71	0.15	0	1
1991	1	11	13.71	7.83	9.93	1	1.69	0.84	1	1
1992	1	12	0.51	8.00	11.83	1	3.50	0.97	1	1
1993	1	20	18.29	15.83	22.94	1	6.94	1.00	1	1
1994	1	16	25.15	12.73	16.87	1	3.81	0.98	1	1

[a]Linear discriminant functions were calculated for the non-outbreak years (DF_1) and outbreak years (DF_2). A logistic regression analysis was calculated with the binary outcome variable *Outbreak Year* (Yes or No) and the risk factors PYO, NDR, and TPM. The estimated probability that a year is an outbreak year is $\hat{p}$.

[b]$DF_1 = -4.428 + 2.052\,\text{PYO} + 0.863\,\text{NDR} + 0.052\,\text{TPM}$.

[c]$DF_2 = -11.884 + 6.191\text{PYO} + 1.462\,\text{NDR} - 0.033\,\text{TPM}$.

[d]$\text{logit}(\hat{p}) = -7.554 + 3.556\,\text{PYO} + 0.62\,\text{NDR} - 0.0895\,\text{TPM}$.

[e]$$\hat{p} = \frac{[\exp(-7.554 + 3.556\,\text{PYO} + 0.629\,\text{NDR} - 0.0895\,\text{TPM})]}{[1 + \exp(-7.554 + 3.556\,\text{PYO} + 0.629\,\text{NDR} - 0.0895\,\text{TPM})]}.$$

Using these discriminant functions and the values of the risk factors for the individual years, each year was classified either as an outbreak year (if $DF_2 > DF_1$) or a non-outbreak year (if $DF_2 \le DF_1$). The second test is based on a logistic regression analysis. The binary outcome variable was *Outbreak Year* (Yes or No) and the risk factors were those given in the Table 11.5. Using the calculated logistic regression equation and the values of the risk factors for the individual years, each year was classified either as an outbreak year (if $\hat{p} > 0.5$) or a non-outbreak year (if $\hat{p} \le 0.5$). Both tests correctly identified 11/12 outbreak years (cases) and 11/13 non-outbreak years (controls). Results of this kind are often shown in a two-way table, as follows.

	True status	
Predicted status[a]	*Outbreak year (cases)*	*Non-outbreak year (controls)*
Outbreak year $\hat{p} > 0.5$	11	2
Non-outbreak year $\hat{p} \le 0.5$	1	11
Total	12	13

[a]Using logistic regression notation.

In this example, whichever of the two methods of formulating a diagnostic test is used, most outbreak years are correctly classified (these are true positives) and so are most non-outbreak years (these are true negatives).

However, a small number of misclassification errors arise. Outbreak years wrongly classified as non-outbreak years are false negatives, while non-outbreak years wrongly classified as outbreak years are false positives. The true positive proportion (TPP) characterizes the proportion of outbreak years correctly classified. The true negative proportion (TNP) characterizes the proportion of non-outbreak years correctly classified. The false negative proportion is FNP = 1 − TPP, and the false positive proportion is FPP = 1 − TNP.

In this particular example, it turns out that discriminant function analysis and logistic regression analysis result in the same classification of years as either outbreak or non-outbreak, but such complete agreement is not inevitable. For the potato late blight data set, both methods of formulating a diagnostic test discussed above give TPP = 11/12 = 0.92 (FNP = 1/12 = 0.08) and TNP = 11/13 = 0.85 (FPP = 2/13 = 0.15). TPP is often referred to as the *sensitivity* and TNP as the *specificity* of a test. In the context of evaluation of a diagnostic test, sensitivity and specificity represent two kinds of accuracy, respectively, for cases and controls. Sensitivity (the number of true positives as a proportion of the total number of cases) and specificity (the number of true negatives as a proportion of the total number of controls) are independent of the proportions of cases and controls in a data set and can therefore be viewed as properties of a diagnostic test based on a risk algorithm. However, overall accuracy, calculated as the overall proportion (or percentage) of correct decisions, is *not* an inherent property of the test. Except when sensitivity and specificity are equal, overall accuracy depends on the proportions of cases and controls in a data set: overall accuracy = TPP × proportion of cases + TNP × proportion of controls. For the potato late blight data in *Example 11.1*, overall accuracy = ((11/12) × (12/25)) + ((11/13) × (13/25)) = (22/25) (=88%).

Example 11.3. Eyespot disease of wheat. Jones (1994a) reported the results of trials carried out between 1983 and 1989 to investigate the efficacy and cost effectiveness of fungicidal control of eyespot disease of wheat. In an analysis of the properties of the indicator of requirement for fungicide application, data from crops at 58 sites were presented. The case and control subgroups were identified on the basis of the increase in yield resulting from prochloraz treatment at GS 30-31. For cases, this increase was ≥0.2 t/ha; for controls, this increase was <0.2 t/ha. The data set comprised 41 cases and 17 controls. In this example, this indicator variable is eyespot incidence at GS 30-31, with the threshold set so that the decision was to treat when ≥20% of tillers were affected, and not to tr eat when <20% of tillers were affected. Of the 41 cases, 28 had ≥20% tillers affected, so sensitivity is 28/41 = 0.68. Of the 17 controls, 7 had <20% of tillers affected, so specificity is 7/17 = 0.41. Overall accuracy, calculated from the overall proportion of correct decisions, was (28 + 7)/(41 + 17) or 60%. However, the assessment of accuracy from the proportion of correct decisions is problematic because in this case, the test is more accurate for cases than for controls (i.e., sensitivity > specificity), and there are more cases than controls in the data set. For comparison, consider the following hypothetical example. Of 58 crops, 19 are classified as cases and 39 as controls. Of the 19 cases, 13 are correctly identified by the test, so sensitivity is 13/19 = 0.68. Of the 39 controls, 16 were correctly identified by the test, so specificity is 16/39 = 0.41. This hypothetical data set thus comprises the same number of crops as the data set of Jones (1994a), and the sensitivity and specificity are the same as in that example. However, in this case, the overall accuracy is only (13 + 16)/(19 + 39) or 50% (compared with 60% above), because the proportion of cases in the hypothetical data set is lower than in the data set of Jones (1994a).

Example 11.2 continued. Sclerotinia stem rot of oil seed rape. Risk algorithms for Sclerotinia stem rot were based on logistic regression analysis (Yuen et al., 1996; Twengström et al., 1998). The binary outcome variable was "% Diseased Plants" (≤25% or >25%) and the risk factors were number of years since the previous oilseed rape crop (>6, 3–5, 1–2), percentage disease incidence in the previous host crop (0, 1–10, 11–30, 31–100), plant population density (low, normal, high), rainfall (mm) in the previous 2 weeks (<10, 10–30, >30), weather forecast (high, variable, or low pressure) and regional risk for apothecium development (0–5, 6–10, 11–100 apothecia per 100 sclerotia). The fitted logistic regression equation could be used to estimate the probability of exceeding the economic injury level (25% diseased plants). However, to enhance ease of use of the risk algorithm, the estimated regression coefficients were multiplied by six and then rounded to multiples of five (Twengström et al., 1998), to provide a simplified risk points scale (Table 11.6).

The risk points accruing to each individual crop may be accumulated to provide a risk score, and frequency distributions of risk scores plotted separately for crops with >25% diseased plants (cases) and crops with ≤25% diseased plants (controls) (Fig. 11.14). There is overlap between the two distributions in the range 25–60 risk points. This means that as an indicator variable for the need for fungicide application for Sclerotinia stem rot, the risk score does not provide perfect discrimination between the cases (crops retrospectively identified as requiring fungicide application) and the controls (crops retrospectively identified as

TABLE 11.6. Risk factors for Sclerotinia stem rot of oil seed rape as identified by Twengström et al. (1998).

Risk factor	*Levels*	*Risk points*
Number of years since previous oil seed rape crop	>6 years	0
	3–6 years	5
	1–2 years	10
Disease incidence in previous host crop	None	0
	Low (1–10%)	5
	Moderate (11–30%)	10
	High (31–100%)	15
Plant population density	Low	0
	Normal	5
	High	10
Rainfall in previous two weeks (mm)	<10 mm	0
	10–30 mm	5
	>30 mm	10
Weather forecast	High pressure	0
	Variable	10
	Low pressure	15
Regional risk for apothecium development (apothecia per 100 sclerotia)	0–5 apothecia	0
	6–10 apothecia	10
	11–100 apothecia	15

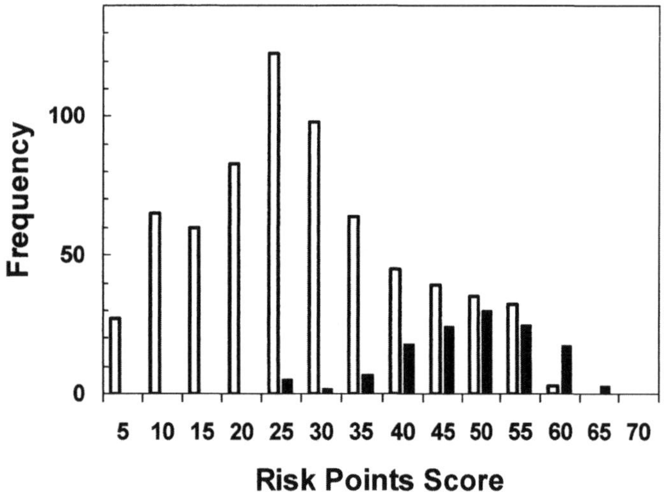

FIG. 11.14. Sclerotinia stem rot of oil seed rape. Frequency distributions of risk point scores plotted separately for crops with >25% diseased plants (cases, ■) and crops with ≤25% diseased plants (controls, □) (Twengström et al., 1998).

not requiring fungicide application). Two questions arise.

- When there is not perfect (or near-perfect) discrimination between cases and controls (as will generally be the case), how can the usefulness of an indicator be evaluated?
- When there is overlap between the frequency distributions of the indicator variable for cases and controls, how can we decide what value of the indicator variable should be adopted as a critical threshold that, if exceeded, will invoke the application of crop protection measures?

11.6.3. The receiver operating characteristic curve

Following the logistic regression analysis of the potato late blight data in Table 11.5, the natural classification rule (as adopted by D. A. Johnson et al., 1996) is to assign a year to the outbreak group if $\hat{p} > 0.5$ and to the non-outbreak group if $\hat{p} \leq 0.5$. In this particular case, this rule provides simultaneously high values of sensitivity and specificity, and there is little reason to consider alternatives. More generally, however, other classification rules may also be considered. D. A. Johnson et al. (1998) point out that if priority were given to the correct identification of cases, a critical threshold lower than $\hat{p} = 0.5$ could be adopted. This would have the effect of increasing sensitivity, but at the expense of decreasing specificity.

For the Sclerotinia stem rot data shown in Fig. 11.14, the choice of a threshold value on the risk points scale is not so straightforward. The extent of the overlap between frequency distributions for cases and controls means that there is no unequivocal "best" choice of threshold for distinguishing between crops with >25% diseased plants and crops with ≤25% diseased plants. For example, the outcomes of choosing threshold risk scores of 35, 40 and 50 points are shown in Tables 11.7 and 11.8. Table 11.7 shows the numbers of true positive, false negative, false positive, and true negative predictions achieved with each of these different choices of threshold risk score. Table 11.8 summarizes these data in terms of sensitivity and specificity. What can be seen is that if a relatively low threshold risk score is adopted, this yields a test with higher sensitivity (TPP is increased, FNP decreased) and a lower specificity (TNP is decreased, FPP increased). Conversely, adopting a relatively high threshold risk score yields a test with higher specificity (TNP is increased, FPP decreased) and a lower sensitivity (TPP is decreased, FNP increased). Note that there is no way of altering the threshold such that both FNP and FPP are simultaneously decreased.

Suppose now that instead of restricting the choice of threshold to one, two, or three values of the indicator score, we allow the threshold indicator score to vary over the whole range of possible indicator scores. A graphical plot of TPP (sensitivity) against FPP (1–specificity), known as a *receiver operating characteristic (ROC) curve*, then provides a useful summary of the characteristics of the indicator in question. ROC curves are widely used in clinical chemistry for the evaluation of diagnostic tests (e.g., Metz, 1978a; Zweig and Campbell, 1993; Swets et al., 2000). An overview of

TABLE 11.7. Sclerotinia stem rot of oil seed rape: the numbers of true positive (TP), false negative (FN), false positive (FP) and true negative (TN) predictions achieved with each of three different choices of threshold risk score.

A: Threshold = 35 risk points	*True status*	
Predicted status	*>25% diseased plants (cases)*	*≤25% diseased plants (controls)*
>25% diseased plants	117 (TP)	154 (FP)
≤25% diseased plants	14 (FN)	520 (TN)
Total	131	674
B: Threshold = 40 risk points	*True status*	
Predicted status	*>25% diseased plants (cases)*	*≤25% diseased plants (controls)*
>25% diseased plants	99 (TP)	109 (FP)
≤25% diseased plants	32 (FN)	565 (TN)
Total	131	674
C: Threshold = 50 risk points	*True status*	
Predicted status	*>25% diseased plants (cases)*	*≤25% diseased plants (controls)*
>25% diseased plants	45 (TP)	35 (FP)
≤25% diseased plants	86 (FN)	639 (TN)
Total	131	674

ROC curves in an ecological context is provided by Murtaugh (1996). Phytopathological applications of ROC curves have been pioneered by Jonathan Yuen and colleagues (e.g., Yuen et al., 1996; Twengström et al., 1998; Yuen, 2002). Madden (2006) provides another application of ROC curves and decision theory for disease management.

It is instructive, at the outset, to consider the ROC curves for two special cases: one for a hypothetical indicator that provides a test with perfect discrimination between cases and controls, the other for a hypothetical indicator that provides no discrimination at all. Fig. 11.15A shows frequency distributions of indicator scores (in arbitrary units) for cases and controls where there is no overlap between the distributions of scores for cases and controls. In this case, it is possible to choose a threshold indicator score above which fall all the cases and below which fall all the controls. Note that is normally the case that the indicator score (denoted I) is calibrated so as to be positively correlated with the perceived risk. In Fig. 11.15A, if the threshold is placed on the vertical axis (i.e., the adopted threshold indicator score $I_{CRIT} = 0$), everything is declared positive (i.e., all the indicator scores are above the threshold). This classification is correct for all the cases, so

Table 11.8. Sclerotinia stem rot of oil seed rape, summarizing the data presented in Table 11.7 for three different choices of threshold risk score[a].

Threshold risk score	*TPP (Sensitivity)*	*TNP (Specificity)*	*FNP*	*FPP*
35	117/131 = 0.89	520/674 = 0.77	0.11	0.23
40	99/131 = 0.76	565/674 = 0.84	0.24	0.16
50	45/131 = 0.34	639/674 = 0.95	0.66	0.05

[a]The true positive proportion (TPP, sensitivity) expresses the number of true positives as a proportion of the total number of cases (i.e., the proportion of outbreak years correctly classified). The true negative proportion (TNP, specificity) expresses the number of true negatives as a proportion of the total number of controls (i.e., the proportion of non-outbreak years correctly classified). The false negative proportion is FNP = 1 − TPP, and the false positive proportion is FPP = 1 − TNP.

TPP = 1. However, it is wrong for all the controls, so FPP = 1. Thus, when $I_{CRIT} = 0$, the corresponding point on the ROC curve (Fig. 11.15B) is (1,1), in the extreme top right-hand corner of the plot. As the threshold indicator score is increased over the range $I_{CRIT} = 1, 2, 3, 4$, FPP decreases (because more controls are correctly classified) but, for the hypothetical indicator in this example, TPP remains unchanged. Thus, the ROC curve follows a horizontal line from the point (1,1) towards the point (0, 1), in the extreme top left-hand corner of the plot. When $I_{CRIT} = 5$, FPP = 0 (so all the controls are correctly classified) and still TPP = 1 (so all the cases are correctly classified) and the adoption of this value of the indicator score as the threshold provides a test with perfect discrimination. On the ROC curve (Fig. 11.15B) this is the point (0,1) where, simultaneously, sensitivity and specificity are both equal to one. As the threshold indicator score is increased further, over the range $I_{CRIT} = 6, 7, 8, 9$, FPP now remains unchanged (its lower bound is zero) while TPP decreases (because more cases are wrongly classified). Thus, the ROC curve follows a vertical line from the point (0,1) down towards the point (0,0), in the extreme bottom left-hand corner of the plot. When $I_{CRIT} = 10$, everything is declared negative (i.e., all the indicator scores are at or below the threshold). This classification is wrong for all the cases, so TPP = 0. However, it is correct for all the controls, so FPP = 0 (TNP = 1). Thus, when $I_{CRIT} = 10$ (in this example), the corresponding point on the ROC curve (Fig. 11.15B) is at the origin, (0,0).

Fig. 11.15C shows frequency distributions of indicator scores (in arbitrary units) for cases and controls for a hypothetical indicator that provides no discrimination. The frequency distributions of indicator scores for cases and controls are identical, so whatever threshold indicator score is chosen, TPP = FPP. As a result, the corresponding ROC curve (Fig. 11.15D) follows a straight line (the "no

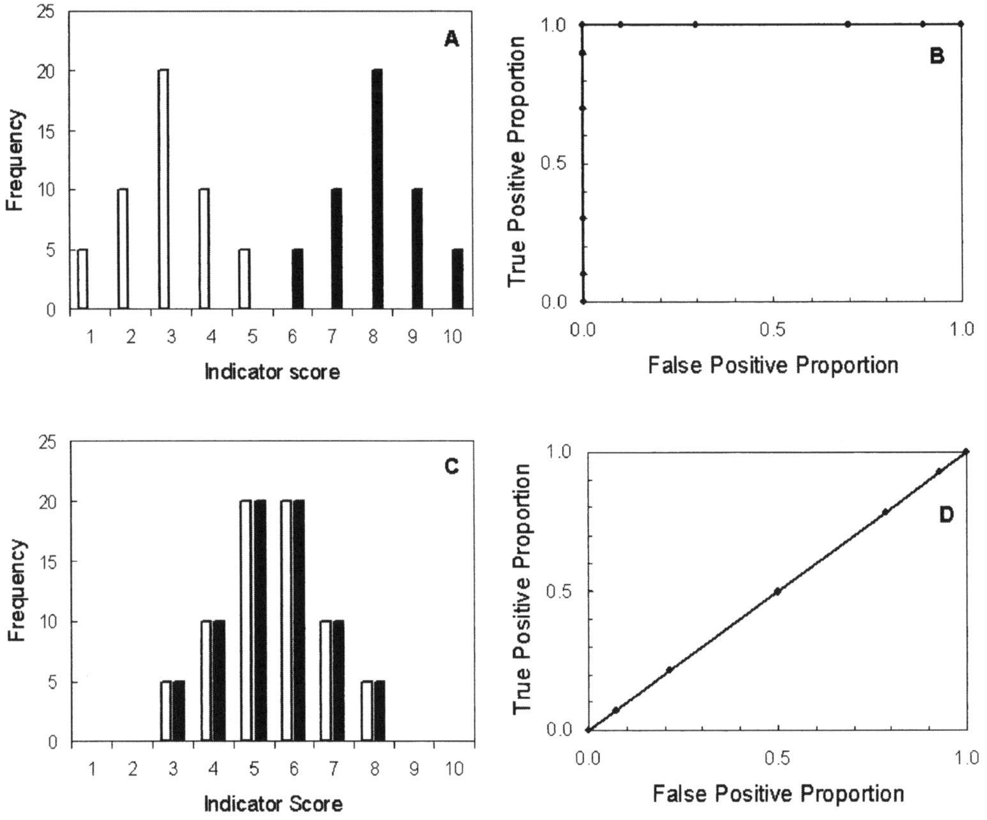

FIG. 11.15. Receiver operating characteristic (ROC) curves. (**A**) There is no overlap between the frequency distributions of indicator scores (in arbitrary units) for cases and controls. (**B**) The corresponding ROC curve follows a horizontal line from the point (1,1) towards the point (0,1), in the extreme top left-hand corner of the plot, then follows a vertical line from the point (0,1) down towards the point (0,0), in the extreme bottom left-hand corner of the plot. (**C**) There is complete overlap between the frequency distributions of indicator scores (in arbitrary units) for cases and controls. (**D**) The corresponding ROC curve follows a straight line (the "no discrimination" line) along the diagonal between the points (1,1) and (0,0).

discrimination" line) along the diagonal between the points (1,1) (low threshold, TPP = FPP = 1) and (0,0) (high threshold, TPP = FPP = 0). In passing, note that an ROC curve that falls below this no-discrimination line suggests that the indicator in question is giving consistently wrong results. Such an indicator has some discriminatory capability insofar as useful predictions could be made by inverting the results obtained by its application.

Fig. 11.14 shows frequency distributions of risk point scores for cases and controls for the Sclerotinia stem rot data of Twengström et al. (1998) discussed in *Example 11.2*. There is partial overlap between the two distributions, so we should expect to see an ROC curve that falls somewhere between the ROC curve corresponding to Fig. 11.15C (complete overlap) and the ROC curve corresponding to Fig. 11.15A (no overlap)—and preferably rather closer to the latter. This is indeed the case. Fig. 11.16 shows the ROC curve derived from Fig. 11.14, with a curvature away from the no discrimination line towards the top left-hand corner of the plot.

The ROC curve shown in Fig. 11.16 is typical of a useful indicator. Thus, plotting an ROC curve provides a way of answering the first of the two questions posed

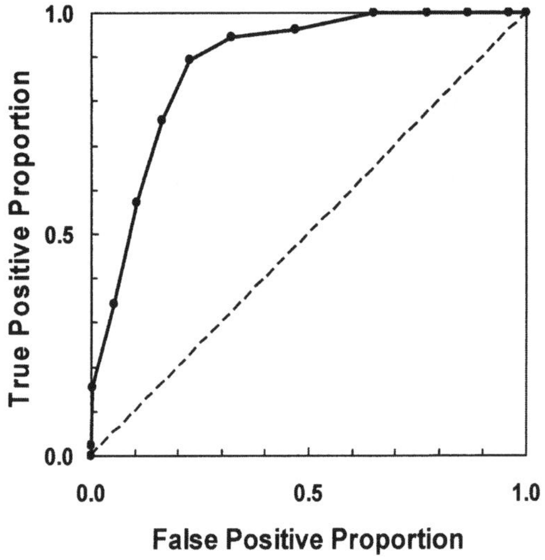

FIG. 11.16. Sclerotinia stem rot of oil seed rape. The ROC curve (solid line) corresponding to the frequency distributions of risk point scores shown in Fig. 11.14, with a curvature away from the no discrimination line towards the top left-hand corner of the plot. Points on the curve represent thresholds corresponding to different accumulations of risk point scores from 5 points (in the top right hand corner of the plot) to 70 points (in the bottom left hand corner) in increments of 5 points. The dotted line is the "no discrimination" line.

at the end of section 11.6.2, relating to evaluation of an indicator. Sometimes the area under the ROC curve is calculated, as a summary (e.g., Hanley and McNeil, 1982). The area under the ROC curve for a perfect indicator (Fig. 11.15B) is equal to one, while the area under the ROC curve for an indicator with no discriminatory capability is equal to 0.5 (Fig. 11.15D). For a useful indicator, the area under the ROC curve will be somewhere between these two extremes. In the case of Fig. 11.16, the area under the ROC curve is equal to 0.847. While values closer to the upper limit are generally preferable, the area under the ROC curve, as a summary statistic, does not convey any detail relating to the trade-off between sensitivity and specificity that is involved in choosing an appropriate threshold. This brings us to the second of the two questions posed at the end of section 11.6.2.

An indicator that has an ROC curve with a curvature towards to the top left-hand corner of the plot has desirable sensitivity and specificity characteristics, in that relatively high values of both can be achieved simultaneously with an appropriate choice of threshold indicator score. For a given choice of threshold, Youden's index J = sensitivity + specificity − 1 (=TPP − FPP) (Youden, 1950), characterizes the overall non-error rate of the corresponding test. J is equal to one for a test that enables perfect discrimination and is equal to zero for a test that provides no discrimination. For example, for the ROC curve in Fig. 11.16, a threshold of 35 risk points provides the test with the maximum value of Youden's index for the indicator in question ($J = 0.66$). Youden's index may be useful as long as the decision maker regards the two types of erroneous decision, false negatives (crops that really need treatment wrongly classified as not needing treatment) and false positives (crops that really do not need treatment wrongly classified as needing treatment) as equally undesirable. More generally, however, the trade-off between sensitivity and specificity involved in the choice of an appropriate threshold depends on the view taken by the decision maker of the relative costs the two possible erroneous decisions, as in the following qualitative account.

Choosing a threshold that falls on the ROC curve towards the top right-hand corner of the plot would mean that a high relative cost is attributed to false negative decisions, so a low threshold indicator score is set in order to reduce these. A threshold on the ROC curve at the extreme top right-hand corner of the plot (TPP = FPP = 1; or FNP = TNP = 0) means that false negative decisions would not be tolerated at all, and so crops would always be treated, whatever the circumstances (in which case forecasting is not required).

Choosing a threshold that falls on the ROC curve towards the bottom left-hand corner of the plot would mean that a high relative cost is attributed to false positive decisions, so a high threshold indicator score is set in order to reduce these. A threshold on the ROC curve at the extreme bottom left-hand corner of the plot (TPP = FPP = 0) means that false positive decisions would not be tolerated at all, and so a crop would never be treated under any circumstances (in which case forecasting is not required).

The ROC curve for an indicator provides both an overview of its usefulness in discriminating between cases and controls, and a means by which a decision maker can evaluate the consequences, in terms of the trade-off between sensitivity and specificity, of adopting a particular threshold indicator score. For a more quantitative treatment of this trade-off (albeit not in a phytopathological application), see Hughes and Madden (2003).

11.6.4 Sensitivity and specificity as conditional probabilities

Sensitivity (TPP) and specificity (TNP) can be written in the form of conditional probabilities, as follows.

- TPP is an estimate of the conditional probability $\Pr(I > I_{\text{CRIT}} \mid D+)$

(read as "the probability of an indicator score above the adopted threshold value (denoted $I > I_{\text{CRIT}}$), given that the true status of the crop was a case (i.e., requiring treatment, denoted $D+$)").

- TNP is an estimate of the conditional probability $\Pr(I \le I_{\text{CRIT}} \mid D-)$

(read as "the probability of an indicator score at or below the adopted threshold value (denoted $I \le I_{\text{CRIT}}$), given that the true status of the crop was a control (i.e., not requiring treatment, denoted $D-$)"). Similarly:

- FNP is an estimate of the conditional probability $\Pr(I \le I_{\text{CRIT}} \mid D+)$
- FPP is an estimate of the conditional probability $\Pr(I > I_{\text{CRIT}} \mid D-)$.

11.6.5 Likelihood ratios

Sensitivity and specificity are characteristics of a diagnostic test. An alternative formulation of these characteristics is provided by the calculation of two *likelihood ratios*, as follows (Go, 1998). The likelihood ratio of an indicator score above the adopted threshold value is defined as:

$$\text{LR}(+) = \frac{\Pr(I > I_{\text{CRIT}} \mid D+)}{\Pr(I > I_{CRIT} \mid D-)} = \frac{\text{sensitivity}}{1-\text{specificity}} = \frac{\text{TPP}}{1-\text{TNP}} = \frac{\text{TPP}}{\text{FPP}} \quad (11.21)$$

LR(+) is a shorthand notation for $\mathrm{LR}\,(I > I_{CRIT})$. The likelihood ratio of an indicator score below the adopted threshold value is:

$$\mathrm{LR}(-) = \frac{\Pr\left(I \le I_{\mathrm{CRIT}} \mid D+\right)}{\Pr\left(I \le I_{\mathrm{CRIT}} \mid D-\right)} = \frac{1-\text{sensitivity}}{\text{specificity}}$$
$$= \frac{1-\mathrm{TPP}}{\mathrm{TNP}} = \frac{\mathrm{FNP}}{\mathrm{TNP}} \quad (11.22)$$

LR(−) is a shorthand notation for $\mathrm{LR}\,(I \le I_{\mathrm{CRIT}})$. Usage of the term "likelihood ratio" here follows the standard terminology for the analysis of diagnostic tests (e.g., Biggerstaff, 2000).

Biggerstaff (2000) gives a geometrical interpretation of the two likelihood ratios and their relationship to the sensitivity and specificity of a diagnostic test. Fig. 11.17 shows a likelihood ratios graph based on the test described in Table 11.7A and Table 11.8, with sensitivity (TPP) equal to 0.89 and specificity (TNP) equal to 0.77 (*Example 11.2*). The false positive proportion is FPP = 1 − TNP = 0.23, and the false negative proportion is FNP = 1 − TPP = 0.11. The likelihood ratio of a positive prediction, LR(+), is estimated by TPP/FPP. In this case, LR(+) = 0.89/0.23 = 3.87. The likelihood ratio of a negative prediction, LR(−), is estimated by (1 − TPP) / (1 − FPP) = FNP/TNP. In this case, LR(−) = 0.11/0.77 = 0.14. For a (hypothetical) perfect test (i.e., one with TPP = TNP = 1), the likelihood ratios are LR(+) = +∞ and LR(−) = 0 (Go, 1998). For a (hypothetical) test with no discriminatory power, both LR(+) and LR(−) are equal to one (Go, 1998). LR(+) > 1 and LR(−) < 1 are the minimum requirements of a useful test. Ideally, we would like to have LR(+) as large as possible and simultaneously have LR(−) as small as possible (Biggerstaff, 2000). For the potato late blight indicator described in *Example 11.1*, the test with TPP = 0.92 and TNP = 0.85 has the corresponding likelihood ratios LR(+) = 6.13 and LR(−) = 0.09. For the eyespot indicator described in *Example 11.3*, the test with TPP = 0.68 and TNP = 0.41 has the corresponding likelihood ratios LR(+) = 1.15 and LR(−) = 0.78.

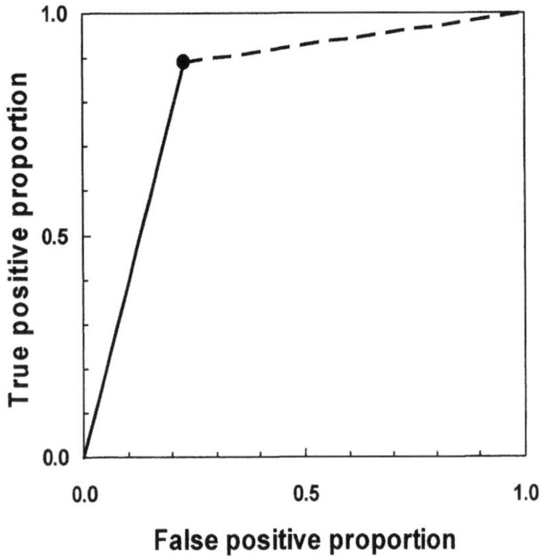

FIG. 11.17. Sclerotinia stem rot of oil seed rape. A likelihood ratios graph based on the test described in Table 11.7A and Table 11.8, with true positive proportion (TPP, sensitivity) equal to 0.89 and true negative proportion (TNP, specificity) equal to 0.77 (●) (see *Example 11.2*). The likelihood ratio of a positive prediction, LR(+), is estimated by TPP/FPP (the slope of the solid line). In this case, LR(+) = 3.87. The likelihood ratio of a negative prediction, LR(−), is estimated by (1 − TPP) / (1 − FPP) = FNP/TNP (the slope of the dashed line). In this case, LR(−) = 0.14.

11.7 Predicting the Need for Treatment

11.7.1 Bayes' theorem

The discussion so far has not been concerned directly with the problem of predicting whether or not a crop requires treatment. When sensitivity and specificity are written as conditional probabilities (section 11.6.4), it can be seen that the conditionality is such that the probability of the test result is given, conditional on the true disease status (case or control) of the crop in question. When we are forecasting plant disease, we do not know the true disease status of the crop in question. Assuming we have developed an appropriate test, and applied it, what we know is the test result ($I \le I_{\mathrm{CRIT}}$ or $I > I_{\mathrm{CRIT}}$) and what we wish to do is calculate the probability of a crop's requirement for treatment, conditional on this test result. In order to accomplish this, we need to know the sensitivity and specificity of the test (written as conditional probabilities, section 11.6.4) and the unconditional (pre-test) probability of a crop's requirement for treatment, denoted Pr(D+), based in some appropriate way on our previous experience. *Bayes' theorem* is the means by which these quantities are combined to calculate the conditional probability $\Pr(D+ \mid I > I_{\mathrm{CRIT}})$ (read as "the probability that a crop requires treatment, given a positive test result [i.e., an indicator score above the adopted threshold value]") (Yuen, 2002, 2003; Yuen and Hughes 2002). The calculation is:

$$\Pr\left(D+ \mid I > I_{\mathrm{CRIT}}\right) = \frac{\Pr\left(I > I_{\mathrm{CRIT}} \mid D+\right)\cdot\Pr\left(D+\right)}{\Pr\left(I > I_{\mathrm{CRIT}} \mid D+\right)\cdot\Pr\left(D+\right) + \Pr\left(I > I_{\mathrm{CRIT}} \mid D-\right)\cdot\Pr\left(D-\right)} \quad (11.23)$$

in which $\Pr(I > I_{\mathrm{CRIT}} \mid D+)$ is the sensitivity, $\Pr(I > I_{\mathrm{CRIT}} \mid D-)$ is 1 − specificity, and Pr(D−) is the unconditional (pre-test) probability that a crop does *not* require treatment, equal to 1 − Pr(D+). In Bayesian terminology, Pr(D+) is the *prior probability* of the requirement for treatment and $\Pr(D+ \mid I > I_{CRIT})$ is the corresponding *posterior probability* of the requirement for treatment given that there has been a prediction of this requirement. In essence, Bayes' theorem facilitates the updating of a prior probability, using evidence from risk factors, to a posterior probability.

Example 11.1 continued. Potato late blight. Outbreaks of potato late blight were identified in 12 of 25 years in commercial potato fields in south-central Washington between 1970 and 1994 (D. A. Johnson et al., 1996). About half the years in the study period were outbreak years. We therefore take $\Pr(D+) = 0.5$ to be a reasonable estimate of the prior probability of an outbreak. We have previously (section 11.6.2) calculated the sensitivity (=0.92) and specificity (=0.85) for a test based on the data in Table 11.5 (giving details of risk factors associated with outbreaks), so using equation 11.23, the posterior probability is given by:

$$\Pr(D+ \mid I > I_{CRIT}) = \frac{0.92 \times 0.5}{0.92 \times 0.5 + (1 - 0.85) \times (1 - 0.5)} = 0.86.$$

Before using the predictor, the evidence means that outbreaks are about as likely to occur as not to occur. However, if the test results in a prediction of an outbreak year, use of equation 11.23 shows that the probability of an outbreak is increased to 0.86. The conditional probability $\Pr(D+ \mid I > I_{CRIT})$ is often referred to as the *positive predictive value* (*PPV*). We can also use Bayes' theorem to calculate the conditional probability, $\Pr(D- \mid I \le I_{CRIT})$ the *negative predictive value* (NPV), as follows:

$$\Pr(D- \mid I \le I_{CRIT}) = \frac{\Pr(I \le I_{CRIT} \mid D-)\cdot \Pr(D-)}{\Pr(I \le I_{CRIT} \mid D-)\cdot \Pr(D-) + \Pr(I \le I_{CRIT} \mid D+)\cdot \Pr(D+)} \quad (11.24)$$

in which $\Pr(I \le I_{CRIT} \mid D-)$ is the specificity and $\Pr(I \le I_{CRIT} \mid D+)$ is 1 − sensitivity. NPV is the posterior probability of (in this context) a non-outbreak year, given that there has been a prediction of an non-outbreak year. In this example, we have:

$$\Pr(D- \mid I \le I_{CRIT}) = \frac{0.85 \times 0.5}{0.85 \times 0.5 + (1 - 0.92) \times (1 - 0.5)} = 0.91$$

Before using the predictor, it appears that outbreaks are about as likely to occur as not to occur but, if the test results in a prediction of an non-outbreak year, use of equation 11.24 shows that the probability of an non-outbreak year is increased to 0.91. Note that PPV and NPV are not inherent properties of a test, since they depend on $\Pr(D+)$. Rather, they relate to application of the test in a particular context.

Equations 11.23 and 11.24 can be written is a simpler format. To enable this we first define the *odds* of an event as:

$$\text{odds}(\text{event}) = \frac{\Pr(\text{event})}{1 - \Pr(\text{event})}$$

i.e.,

$$\Pr(\text{event}) = \frac{\text{odds}(\text{event})}{1 + \text{odds}(\text{event})}$$

Now, after some rearrangement of equation 11.23:

$$\text{odds}(D+ \mid I > I_{CRIT}) = \text{odds}(D+)\cdot \text{LR}(+) \quad (11.25)$$

and after some rearrangement of equation 11.24:

$$\text{odds}(D- \mid I \le I_{CRIT}) = \text{odds}(D-)/LR(-) \quad (11.26)$$

Using the scenario of *Example 11.1*, the re-statement of Bayes' theorem given in equation 11.25 is used to move from the *prior odds* of an outbreak year to the *posterior odds* of an outbreak year, given a prediction of an outbreak year. Using the data from *Example 11.1* above:

$$\text{odds}(D+ \mid I > I_{CRIT}) = \frac{0.5}{1-0.5} \times \frac{0.92}{1-0.85} = 6.13$$

and then, if required,

$$\Pr(D+ \mid I > I_{CRIT}) = \frac{6.13}{1+6.13} = 0.86, \text{ as before.}$$

Similarly, equation 11.26 is used to move from the prior odds of a non-outbreak year to the posterior odds of a non-outbreak year, given a prediction of a non-outbreak year. Using the data from *Example 11.1* above:

$$\text{odds}(D- \mid I \le I_{CRIT}) = \frac{0.5}{1 - 0.5} \times \left(\frac{1 - 0.92}{0.85}\right)^{-1} = 10.63$$

and then, if required,

$$\Pr(D- \mid I \le I_{CRIT}) = \frac{10.63}{1 + 10.63} = 0.91, \text{ as before.}$$

From equation 11.25, we can see that as long as LR(+) > 1, the effect of a test that provides a prediction of need for treatment is to increase the posterior odds of need for treatment, relative to the prior odds. From equation 11.26, we can see that as long as LR(−) < 1, the effect of a test that provides a prediction that treatment will not be required is to increase the posterior odds that treatment will not be required, relative to the prior odds. It is now clear why LR(+) > 1 and LR(–) < 1 are the minimum requirements of a useful indicator (section 11.6.5).

Other potentially useful statements of Bayes' theorem as applied in the context of disease management decision making are available. The posterior odds of requirement for treatment following a prediction that treatment will not be required is:

$$\text{odds}\left(D+ \mid I \le I_{\text{CRIT}}\right) = \text{odds}\left(D+\right) \cdot \text{LR}\left(-\right) \qquad (11.27)$$

When converted to a probability, this is equal to 1 – NPV. The posterior odds that treatment will not be required following a prediction that treatment will be required is:

$$\text{odds}\left(D- \mid I > I_{\text{CRIT}}\right) = \frac{\text{odds }(D-)}{\text{LR}(+)} \qquad (11.28)$$

which is equal to 1 – PPV when converted to a probability.

Example 12.2 continued. Sclerotinia stem rot of oil seed rape. A 20-year average for the frequency of need for control measures for Sclerotinia stem rot of oil seed rape in Uppland, east-central Sweden, is 16% (Yuen and Hughes, 2002). We, therefore, take Pr($D+$) = 0.16 as the prior probability of need for fungicide application (so the prior odds is odds($D+$) = 0.19). Different threshold values of the risk score give different values of sensitivity and specificity (Tables 11.7 and 11.8). These data can now be combined, via Bayes' theorem, to characterize predictive values.

Table 11.9 shows values of LR(+) [= sensitivity/(1 – specificity)] and the corresponding positive predictive values for risk score thresholds of 35, 40, and 50 points. A threshold of 50 points provides the largest value of LR(+). Adopting this threshold (from among those tabulated) leads to the largest posterior probability of need for fungicide application following a forecast of this need (PPV). Note that even following a prediction of the need for fungicide application, this probability (= 0.56) is not much greater than 0.5. This kind of result is discussed further in section 11.7.2. Table 11.10 shows values of LR(–) [=(1 – sensitivity)/specificity] and the corresponding negative predictive values for risk score thresholds of 35, 40, and 50 points. In this case, a threshold of 35 points provides the smallest value of LR(–). Adopting this threshold (from among those tabulated) leads to the largest posterior probability that fungicide application is not required following a forecast that application is not needed (NPV).

None of the thresholds discussed is superior overall. A threshold of 50 points is the best for confirming the need for fungicide application but the worst for confirming that fungicide application is not needed. A threshold of 35 points is the worst for confirming the need for fungicide application but the best for confirming that fungicide application is not needed. A threshold of 40 points is intermediate between the other two from both points of view. In the case of Sclerotinia stem rot of oil seed rape, values of the likelihood ratios are such that useful forecasts can be made. However, decision makers must prioritize their requirements, in terms of seeking an increase in the odds of need for treatment following a positive prediction, or an increase in the odds of no need for treatment following a negative prediction, or otherwise opting for a compromise between these two positions.

Table 11.9. Using data from Tables 11.7 and 11.8 (Sclerotinia stem rot of oil seed rape), values of the likelihood ratio of an indicator score above the threshold value LR(+) and the corresponding positive predictive (PPV) values for risk score thresholds of 35, 40, and 50 points.

	Need for treatment			
Threshold risk score	LR(+)	*Prior odds*	*Posterior odds*	*PPV*
35	3.87	0.19	0.74	0.42
40	4.75	0.19	0.90	0.48
50	6.80	0.19	1.30	0.56

Table 11.10. Using data from Tables 11.7 and 11.8 (Sclerotinia stem rot of oil seed rape), values of the likelihood ratio of an indicator score below the threshold value LR(–) and the corresponding negative predictive values (NPV) for risk score thresholds of 35, 40, and 50 points.

	No need for treatment			
Threshold risk score	LR(–)	*Prior odds*	*Posterior odds*	*NPV*
35	0.14	5.25	36.75	0.97
40	0.29	5.25	18.38	0.95
50	0.69	5.25	7.56	0.88

11.7.2 Predicting unusual events is problematic

The advantages of being able to predict correctly the occurrence of particularly rare events, or the non-occurrence of particularly common ones, hardly need spelling out. However, such events—e.g., diseases that occur very frequently or very infrequently—pose a problem from the point of view of forecasting. For example, suppose that a review of the available evidence in relation to the occurrence of a disease leads to the conclusion that the prior probability of need for fungicide application (or another appropriate control measure) is Pr($D+$) = 0.05

(odds $(D+) = 0.05 / (1 - 0.05) \approx 0.053$). Suppose also that we have a test (risk algorithm) with both high sensitivity (=0.9) and specificity (=0.9), so $LR(+) = 0.9/(1-0.9) = 9$. After a prediction of need for fungicide application, the posterior probability of this need (PPV) is $\Pr(D+ \mid I > I_{CRIT}) = 0.32$ (using Bayes' theorem). The posterior probability of need for a fungicide application exceeds the prior probability by about a factor of six (0.32 compared with 0.05) following a prediction of this need. However, the end result is that about two-thirds of the predictions of a requirement for fungicide application will be for crops that do not actually require it, since $\Pr(D- \mid I > I_{CRIT}) = 1 - \Pr(D+ \mid I > I_{CRIT}) = 0.68$. It can be seen that when the prior probability is low, most predictions of a requirement for fungicide application will be for crops that do not actually require it, unless the test used is one with a very low FPP. This is often the scenario in screening programs related to medical decision making. In that case, when disease incidence is low, FPP must be small, otherwise most positives will be false positives, causing an unnecessary burden on both the health care system and on individual patients (Metz, 1978a).

A similar problem arises in trying to predict the non-occurrence of a frequently occurring disease. In this case, non-occurrence is the unusual event and we would like to be able predict when fungicide application (or another appropriate control measure) is *not* required. A test with high sensitivity and specificity will provide an increase in posterior probability that fungicide is not required (NPV) over a low prior probability, following a prediction that fungicide application is not required. However, this posterior probability may still be such that most of the predictions that fungicide application is not required will be for crops that actually do require the application.

The point is that it is difficult to make useful forecasts where rare events are concerned. For a prior probability that is very low, or very high, it is difficult to provide posterior probabilities that will change a decision maker's view (based on this prior probability), even if a test with good sensitivity and specificity characteristics is available. In contrast, a test with lower sensitivity and specificity can provide useful predictions in cases when prior probabilities are neither very low nor very high. For one example, using a test with sensitivity = 0.75 and specificity = 0.75, $\Pr(D+) = 0.33$ is increased to $\Pr(D+|I > I_{CRIT}) = 0.60$ following a positive prediction. For another, $\Pr(D-) = 0.4$ is increased to $\Pr(D-|I \le I_{CRIT}) \approx 0.6$ following a negative prediction based on a test with sensitivity = 0.7 and specificity = 0.7. These are the kinds of probability revision (expressed in terms of the posterior probability compared with the prior probability) that may influence a decision maker's initial view. This analysis has important implications for the deployment of resources in solving disease management decision making problems (Yuen, 2003). Predictions are likely to be most useful in cases where the prior probabilities are neither very low nor very high. Thus, some formal consideration of prior probabilities is an important component of the development process for useful predictors of plant disease.

11.8 Conclusions

In economic entomology, sampling—particularly sequential sampling—has become the cornerstone of crop protection decision making. Binns (1994) is unequivocal: "For decision making, sequential sampling is often the only method to be seriously considered because it allows sampling to cease as soon as it becomes obvious which decision should be made." The methods of acceptance sampling developed for use in control of insect pests have found little direct application in the management of plant disease. This may be because lesion counts—the phytopathological equivalent of the data most frequently collected in economic entomology—are only infrequently collected in disease management. Notwithstanding the paucity of direct applications, *Sampling and Monitoring in Crop Protection* by Binns et al. (2000) is nothing less than required reading for anyone who wants to understand the theoretical basis for applications of acceptance sampling in the context of crop protection. The basis of these applications is Wald's (1947) sequential probability ratio test. This is applicable directly with disease incidence data collected by simple random sampling and by means of a normal approximation with disease incidence data collected by cluster sampling (Madden and Hughes, 1999a; Turechek et al., 2001).

Fixed sample size applications of acceptance sampling are most likely to be found in regulatory plant pathology, or when the determination of disease status of each observation is done after the sampling has ended (e.g., when using ELISA in a laboratory to determine disease incidence of a sample of leaves collected in the field). Zero acceptance number sampling plans, in particular, have been used in quarantine situations. In this context, the objective is to be able to make a statement that there is only a small specified probability that actual disease incidence exceeds a low specified level of incidence, when no infected plants are found in a sample. Sample sizes can be calculated to meet particular specifications. Note that the smaller is the specified probability that actual disease incidence exceeds a low specified level of incidence, and the lower is that level of incidence, the larger will be the required sample size.

It is more common in plant pathology than in economic entomology to base crop protection decision making on evidence beyond that provided by a sampling the harmful organism population directly (or indirectly, via its effects on the host). Prominent among the reasons for this is the fact that visual inspections of crops for disease do not generally detect latent (or pre-symptomatic) infections, which may comprise the majority of infections early in an epidemic (see Chapter 5) when control measures are likely to be most efficacious. Sampling may still have a role, but a wider range of risk factors may also be considered. A risk algorithm is any calculation that uses observations of identified risk factors from the host crop, the pathogen population and the environment to make an assessment of the need for crop protection measures. Usually data on the various risk factors are combined using an appropriate statistical method, such as logistic regression, but this is not a requirement. If an estimate of the pathogen population, or of disease intensity, is just one of a number of factors used in the risk algorithm, this estimate is obtained using sampling methods for estimation (Chapter 10), not the acceptance sampling methods outlined in this chapter. Acceptance sampling methods are appropriate when decision making is based only on the outcome of sampling the population in question, without taking other risk factors into account. The accuracy of risk algorithms (i.e., their sensitivity and specificity) can be assessed by means of receiver operating characteristic curves. Using Bayes' theorem, sensitivity and specificity are combined with information on the prior probability of, for example, the need for disease control measures, to calculate the corresponding posterior probability of the need for control measures, given the evidence related to risk factors. There is inevitably an element of uncertainty built into this process, because we want to act in order to prevent a loss (which we may express in terms of crop yield or revenue) that we think may occur in the future if we do not act. Thus, in crop protection, we need to make decisions based on predictions of their consequences.

The Bayesian analysis can be taken further to account explicitly for the costs and benefits of different decisions relating to disease management. Carlson (1970), for example, gave an economic analysis of decision making in relation to peach brown rot (caused by *Monilinia fructicola*) in California. Carlson (1970) concluded that decision theory procedures appear to be applicable when disease control costs are high relative to the product price, when the intensity of damage is highly variable from year to year and when epidemics can be predicted with some reliability. Of course, the cost of control relative to the product price may be influenced by, for example, regulatory policies or consumer demand. Shtienberg (2000) mentions the imposition of strict regulations on levels of pesticide residues in marketable products and consumer willingness to pay premiums for minimally treated produce as mechanisms that may lead to wider implementation of decision theory procedures as the basis for disease management.

Shtienberg (2000) suggests that the potential value of a crop, the destructiveness of the pathogen and the cost of disease management relative to the likely loss in revenue resulting from pathogen attack are foremost in influencing growers' perceptions concerning disease management. When the value of the crop is low, the cost of spraying relative to the likely loss in revenue resulting from pathogen attack is high, especially if the pathogen in question is only moderately destructive. Growers in this position may be motivated to avoid false positive decisions, i.e., erroneous decisions to spray a crop. When the value of a crop is high, the cost of spraying relative to the likely loss in revenue resulting from pathogen attack is low, especially if the pathogen in question is highly destructive. In such circumstances, growers may seek to avoid false negative decisions by spraying frequently, whether or not it is actually necessary. Shtienberg (2000) cites his own experience with growers of high value crops in Israel, and that of MacKenzie (1981) with potato growers in the USA, as examples of this scenario. As Shtienberg (2000) points out, putative aids to decision making that do not take the growers' perceptions of the need to apply or withhold crop protection measures are unlikely to be implemented. In effect, then, predictors require some form of operational validation in addition to the statistical validation that most are subjected to during the development process. Altman and Royston (2000) provide a useful discussion of this issue, in a clinical context, and Hughes and Madden (2003) discuss in a different context how costs of decisions (correct or incorrect) can be formally linked to ROC curves for choosing action thresholds.

As discussed here, disease management confronts us with a problem in diagnostic decision making. Because the scale at which the vast majority of plant diseases are treated is that of the population (crop) level rather than the individual (plant) level, that is the scale at which we must make our diagnosis. Thus we can think of risk algorithms and sampling plans that classify crops as requiring treatment or otherwise as population-scale diagnostics. We assume that, where diagnosis at the individual plant scale forms a component of population-scale diagnosis, the plant-scale diagnosis is effectively error-free. With immunological and molecular diagnostic techniques, this may be reasonable. For plant-scale diagnosis based on an assessment of visual symptoms, it is almost certainly not.

New and more powerful diagnostic technology will continue to become available for disease assessment at the plant scale. To be able to take maximum advantage of this technology in disease management decision making, we need to deploy new diagnostics in risk algorithms and sampling plans for population-scale diagnosis. Research in the deployment of new diagnostic technology has recently lagged behind research in the development of the technology. Ultimately, however, the continuing requirement for improved decision making in the practice of disease management means that plant-scale and population-scale diagnosis must be seen as complementary aspects crop protection.

11.9 Suggested Readings

Binns, M., Nyrop, J. P., and van der Werf, W. 2000. Sequential sampling for classification. In: *Sampling and Monitoring for Crop Protection Decision Making*. CAB International, London (Chapter 5).

Clayton, M. K., Slack, S. A. 1988. Sample size determination in zero tolerance circumstances and the implications of stepwise sampling: bacterial ring rot as a special case. *Am. Potato J.* 65:711–723.

Go, A. S. 1998. Refining probability: an introduction to the use of diagnostic tests. In: *Evidence-Based Medicine: A Framework for Clinical Practice* (D. J. Friedland, editor). Lange Medical Books/McGraw-Hill/Appleton & Lange, New York, pp. 11–33.

Madden, L.V., Hughes, G. 1999. Sampling for plant disease incidence. *Phytopathology* 89:1088–1103.

Yuen, J. 2003. Bayesian approaches to plant disease forecasting (Online). *Plant Health Progress* (DOI:10.1094/PHP-2003-1113-06-RV).

12

Epidemics and Crop Yield

Relating the yield of crops to varying levels of plant disease has remained a complex task both in theory and in practice.

B. M. Cooke

12.1 Introduction

Botanical epidemics are studied, in part, because of the many costs attributed to plant diseases in agriculture and forestry, including the costs of yield reduction (James and Teng, 1979). Many researchers, including those who have never attempted to quantify the relationship between disease intensity and yield at the population level, find a strong motivation for their work from the published accounts of the economic (and social) impact of plant diseases. The major pandemics of history—ergot in Medieval Europe, coffee rust in Sri Lanka, late blight in Ireland (and elsewhere), chestnut blight and southern corn (maize) leaf blight in the U.S., and African cassava mosaic in Africa, to name just a few—serve as a reminder of the worst-case consequences of plant disease epidemics for society as a whole (Schumann, 1991; Chapter 2 in Campbell and Madden, 1990; Horsfall and Cowling, 1978a). However, every year in every part of the world, epidemics occur that affect crop yield, or would affect yield if (often costly) controls were not being utilized. It is always hoped that increasing our understanding of epidemics will lead to a reduction in the effects of diseases on yield or to the development of more efficient and effective controls, or both.

Despite the economic importance of plant diseases, the relationship between epidemic development and resulting yield was not studied rigorously for many diseases until recent decades. Nearly 90 years ago, Lyman (1918) lamented that we had few accurate measures of crop loss. James and Teng (1979) believed this lack of crop loss information retarded the progress of plant protection as much as any other single factor. In their poetic language, Horsfall and Cowling (1978b) summarized their view of the situation:

> "Although it can see the dying elms along the streets, society in general is only vaguely aware of how much damage it suffers from the depredations of other plant pathogens. That is because we have not told society about the losses, and that is because we do not know, and that is because we have not researched it very well."

We would argue that plant pathologists *have* researched it well in the last 30+ years or so, and have a much better understanding of how plant diseases affect yield. Vanderplank (1963) clearly identified the importance of the disease:yield relationship in evaluating the outcome of epidemics, and showed results of some early empirical studies. The 1970s and 1980s were especially productive for this field of research, with many pioneering contributions by W. Clive James, Paul Teng, and others. Their work led to the development of several concepts and principles for relating epidemics to crop yield, and formulation of empirical models of the underlying processes. We review several of these concepts and principles in this chapter.

The term *crop loss assessment* is often used for the study of the relationship between attack by harmful organisms and the resulting yield (or yield loss) of crops. The disease:yield relationship falls within the study of crop loss assessment, as do the effects of infestations of insect pests and weeds on crop yields. As will become apparent in the following sections, the relationship between epidemic development and crop yield is complicated, and can depend on more variables than those typically measured when just quantifying disease increase in time and spread in space. Thus, in agreement with Lyman, we still do not adequately know what the precise losses are to various diseases. Even when a precise and repeatable functional relationship has been established between disease intensity (in the broad sense) and yield, losses for a region or country may still be uncertain. This is because, due to lack of funds or labor, we do not routinely obtain precise estimates of disease incidence or severity by sampling in commercial fields on a large scale and on a regular basis. Chapters 10 and 11 of this book cover many important issues related to sampling for plant disease intensity.

There are multiple uses for information on crop yield in relation to disease intensity. For instance, in addition to determining if a (potential) disease control method reduces disease intensity or the rate of disease increase, one can determine if the control has an effect on yield.

In an extreme case, it is possible that a control reduces disease severity but does not change crop yield very much. Related to this, one can attempt to make optimal disease management decisions based on the cost of the controls and the benefit obtained from the controls (reduced yield loss and, thus higher income) (Funt et al., 1990; Gould, 1989; Main, 1977). One can also use this information to make predictions of yield for a given location or region, which is of benefit to crop producers and the individuals and groups who purchase the commodities. The relationship between yield and disease can furthermore be used to determine the economic importance of a given disease in a given region. Through risk assessment, one can use the information to assess the threat of a non-indigenous pathogen to a region or country (NRC, 2002). Some excellent reviews and books give a great deal more information (Cooke, 1998; James, 1974; James and Teng, 1979; James et al. 1991; Nutter, 2002b; Savary et al., 2006; Teng, 1985; Chapter 8 in Zadoks and Schein, 1979).

The production of biomass by a plant, and the allocation of a portion of this biomass into the seed or other harvested plant component, are physiological processes that are influenced, at a minimum, by the environment, by diseases and other stress-causing agents (e.g., insects), and by competing plants (Johnson, 1987, 1992). In fact, a considerable understanding of crop yield production has been obtained from mechanistic modeling using physiological principles (Batchelor et al., 1993; Boote et al., 1983; Johnson and Teng, 1990). It is possible to link physiological models to population dynamic models for diseases, which generally requires the use of computer simulation and a large number of component equations, but such modeling is not necessarily needed (or desirable in some cases). Just as the complex physiological processes involved over the time span from spore penetration to the production of new spores are reduced to a single parameter representing the inverse of mean latent period (see Chapter 5), summary parameters can be used to relate the physiological processes of yield production and loss to disease (or variables that are related, directly or indirectly, to disease intensity).

We demonstrate the approach of using summary physiological parameters to model crop loss later in the chapter. But we first discuss general concepts of crop losses and the principles of data collection, followed by a presentation of the more empirical approaches to modeling that have been very popular over the last few decades. In the spirit of model parsimony, we show that a considerable understanding of crop losses can be obtained based on relatively simple models of yield in relation to epidemic and crop variables. In particular, we demonstrate how spatial heterogeneity of disease affects observed disease:yield relationships, and how temporal dynamics of disease increase can be linked to crop yield production.

12.2 Definitions and Concepts

12.2.1 Yield

Before discussing relations between yield and disease, it is helpful to present some definitions and discuss some concepts about crop loss. We base the presentation on the useful texts by Chiarappa (1971) and Zadoks and Schein (1979). Of course, as is typical in plant pathology, other authors may use the terms differently. A *crop* is a unit of plants grown to provide food, fiber, stimulants, or other products. *Yield* is the measurable produce of a crop. The produce may be seed, fruit, leaves, roots, or tubers, depending on the crop species.

It is helpful sometimes to consider different potential levels of yield for a crop. Fig.12.1 is based on Fig. 8.2 in Zadoks and Schein (1979). *Primitive yield* is the yield of landraces of crop plants in primitive agriculture (i.e., without the use of modern crop protection practices). For grain crops, primitive yields are around 1 MT/ha of dry weight. In contrast, *theoretical yield* is the yield obtained under ideal conditions, generally in small plots (e.g., experimental crops) with high levels of pesticides, fertilizers, other chemicals, and with the adoption of crop husbandry practices that could not be justified economically in full-scale modern commercial agriculture. Theoretical yield, which is sometimes called potential yield, could be considered the genetic maximum possible yield for a crop species, although this may be unknown. For grain crops, theoretical yield may be in the range of 10–20 MT/ha of dry weight. It is expected that our knowledge of theoretical yield of crops will continue to increase over time as we learn more about plant growth and development.

Between primitive and theoretical yields are attainable and actual yields. *Attainable yield* is the yield obtained when a full-scale commercial crop is grown using the full range of modern agricultural technology. Generally, this requires that a crop is grown with optimum use of pesticides, fertilizers, and so on. Attainable yield may or may

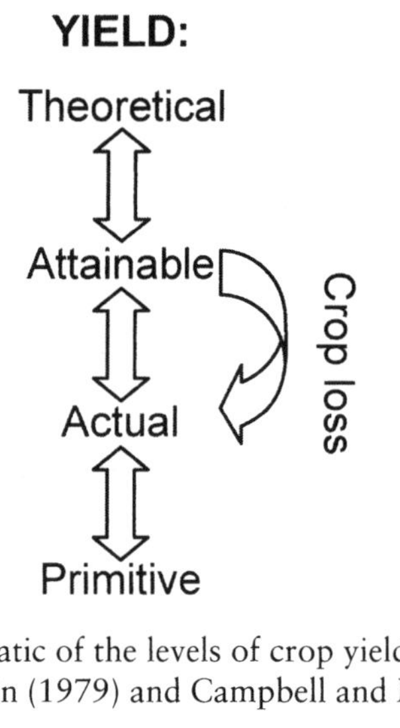

FIG. 12.1. Schematic of the levels of crop yield. Adapted from Zadoks and Schein (1979) and Campbell and Madden (1990).

not be affordable, depending on whether or not the costs of production outweigh than the income obtained from the crop. *Actual yield* is the crop yield obtained by a given set of crop husbandry practices. Actual yield generally is less than or equal to attainable yield. The difference between attainable yield and primitive yield is a loss that is avoidable, in the sense that (at least in principle) there are measures—including crop protection measures—available that could be used to raise yield from the primitive to the attainable level. In practice it will rarely be justifiable in economic terms to use the full range of modern agricultural technology. Instead, selected measures will be used to achieve an actual yield that is less than the attainable yield. In so doing, the loss corresponding to the difference between actual yield and primitive yield has been avoided. In general, the difference between attainable and actual yield can be then considered crop loss (or yield loss) (Fig. 12.1).

This conceptualization is broader in scope than the consideration of yield in relation to disease. Obviously, many factors influence yield, and reduce yield from that attainable, including a wide spectrum of harmful organisms and other environmental stress factors (e.g., lack of moisture). We are mostly concerned about the yield of crops in relation to disease epidemics in this book. Thus, as an operational definition, we consider *crop loss* to be the difference between actual yield and yield obtained in the absence of disease. The latter yield value is analogous to (but not the same as) attainable yield in the more general setting. The actual yield may be for a plant, plot, commercial field, region, or larger area (e.g., country). Depending on the particular scenario of interest, "absence of disease" may be in reference to the absence of a single disease or of a number of diseases. We use crop loss and yield loss as synonyms in this text.

There are variations on the meaning of crop loss. One common definition is: reduction in either quantity and/or quality of yield brought about by crop damage (Campbell and Madden, 1990). Crop damage here implies injuries from biotic or abiotic agents that, collectively, result in a measurable loss of yield. The general definition of the previous paragraph for crop loss does not depend, however, on use of terms such as "injury" and "damage" to represent plant disease effects on yield. Readers can refer to Main (1977) for a nice early presentation on the concepts of "injury", "damage", "stress", and "loss".

It is important to note that crop loss (or yield loss) does not imply that the harvested portion of the crop is directly reduced by a given disease during a given epidemic. Some diseases actually reduce or eliminate the harvested produce of a crop *after* the harvested portion is produced. Examples would be post-harvest diseases or other diseases where fruit become infected after being produced. Most plant diseases limit the formation of harvested produce rather than removing the previously formed produce. Some have argued that "crop loss" or "yield loss" are not appropriate terms for the consequences of most plant diseases (Cook, 1985). However, with the definition used here, crop loss applies to both the removal of previously formed produce (due to rotting, for instance) and for the unrealized gain in harvestable produce over time in a crop.

12.2.2 Impacts of disease on crops

For some purposes, it is convenient to convert yield and yield-loss values into economic terms, and to quantify the effects of diseases in terms of costs. In fact, one can explicitly quantify and study economic loss as well as biological loss (Nutter, 2002b). The reduction in quantity of yield is clearly one important cost of crop disease (James and Teng, 1979). Reduction in quality of yield is an additional cost. Blemishes on fruit and toxins in grain are two examples of quality reductions.

There are several other costs attributed to plant diseases that are outlined here. Cost of control is of major importance. This includes short-term costs, such as the cost of pesticides (e.g., fungicides for fungal diseases and insecticides for the vectors of some viral diseases), as well as the cost of applying the pesticide (including time and labor). Other short-term costs include screening of seed and plant material (transplants) for the presence of some pathogens. There can be extra costs of harvesting and grading of produce (e.g., sorting fruit) when a disease is present. Sometimes, a crop has to be replanted due to pre- or post-emergence damping off, and sometimes a crop cannot be planted due to the presence of a pathogen. Although crop rotation has many long-term benefits, there may be a short-term cost if the most desirable crop cannot be planted in a given year of the rotation. Longer term costs include breeding for resistance and developing new pesticides, and the development and testing of other controls. One potentially large cost is in trade disruptions. If the pathogen causing disease is of quarantine significance, its presence in an area can result in the (total) loss of export markets (Madden and Wheelis, 2003), even if the yield in the field is little affected by the disease. For pathogens that produce certain toxins, there is also a cost in terms of human and animal health. There can also be a cost in terms of higher food prices or unavailable foods in a region.

Some of these costs are borne directly by the grower. Others costs are paid by industry, or society as a whole. In fact, a grower could benefit from higher food prices, although society would have to bear this cost. These costs due to plant disease can be partitioned in various ways to better understand the impact of disease. Using the term "loss" for what we generally refer to as cost in this section, Zadoks and Schein (1979) developed an interesting classification system. Direct losses are experienced on the farm (i.e., directly in the agricultural sector). In contrast, indirect losses are generally experienced off the farm, by society as a whole. The direct losses can be divided into primary and secondary ones. Direct primary losses involve reductions in quantity and quality of yield as well as immediate control costs. In contrast, direct secondary losses are long term, incurred beyond the current year or growing season. These can be due to weakening of perennials and build-up of inoculum in the soil, both of which can reduce yield in future years.

It is extremely difficult to estimate the total actual cost of plant diseases to a region or a country, although Pimentel and colleagues have made valiant attempts over the years to quantify the costs due to all pests, including plant pathogens (Pimentel et al., 2000; Pimentel, 2002). Using data from the U.S. Bureau of Census and other sources, they estimate a total cost of plant disease to the U.S. alone to be over $30 billion per year. This value includes more than just the reduction in quantity of yield. Large scale estimates of costs like this require substantial extrapolation and are based on numerous assumptions about the agricultural and nonagricultural business sectors. In fact, the Pimentel estimate is larger than that obtained by James et al. (1991), although the later authors were more specific in the types of costs considered. However, even as a rough approximation, the analyses make clear the large economic importance of plant diseases.

It is useful for readers to have an appreciation for the many costs incurred due to plant diseases. However, as is common in botanical epidemiology, we focus in this chapter on reductions in quantity of yield in relation to disease development. This is the area where we can most directly show, either empirically or, in some cases, mechanistically, how the spatio-temporal dynamics of disease impacts the production of yield by crops. Some aspects of costs in disease management are covered in Chapter 11.

12.3 Data and Relationships

12.3.1 Graphs of yield and disease

Fig. 12.2 shows examples of the relationship between crop yield and some representations of the disease epidemic. It is sometimes convenient to refer to the graph of yield (or yield loss) versus disease as a "damage curve". The examples were chosen from a wide collection of

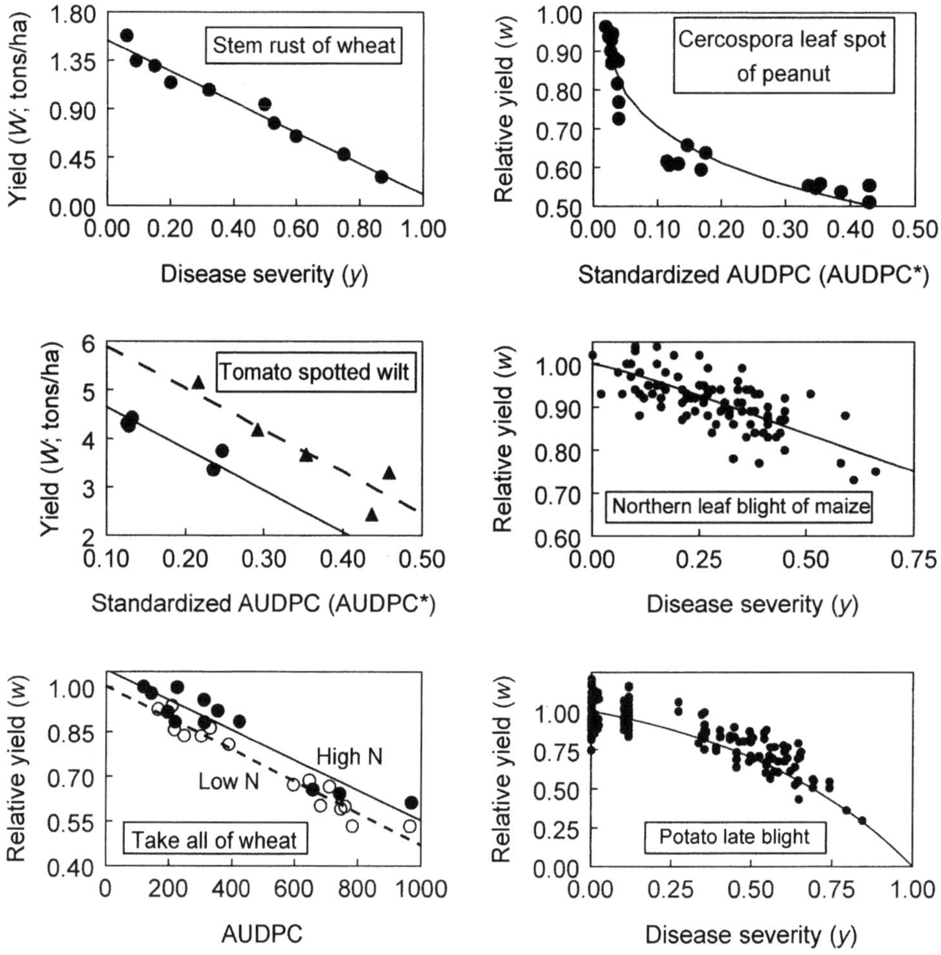

FIG. 12.2. Examples of crop yield [in relative (w) or absolute (W) units] in relation to disease intensity or area under the disease progress curve (AUDPC). Data were obtained from Vanderplank (1963) for stem rust of wheat, Madden et al. (1981a) for leaf spot of peanut, Culbreath et al. (1997) for tomato spotted wilt (with separate lines for different cultivars), Pataky et al. (1998) for northern corn (maize) leaf blight, Schoeny et al. (2001) for take all of wheat, and Ferrandino (1989) for potato late blight.

pathosystems to demonstrate the typical relationships observed. The most common way of showing the impact of the epidemic on yield is to plot yield (or yield loss) versus disease intensity (y or Y) at one time during the epidemic. One could also show disease-free intensity instead of Y. Sometimes the yield (or yield loss) is presented in "yield units" (e.g., kg/ha); alternatively a proportion or percentage of a disease-free control yield can be calculated. The single time may be at the end of the growing season, or a time that is "known" or thought to be critical for yield production of a crop (see below). Yield can also be related to disease intensity at multiple times, often summarized by the area under the disease progress curve (AUDPC; equation 4.42). Time to reach a certain intensity of disease or the rate of disease increase can also be used as representations of the epidemics.

Consider the graph in Fig. 12.2 for stem rust of wheat, obtained from a field study in 1937 and described in Greaney et al. (1941). The data were also used by Vanderplank (1963; p. 148) to show possible impacts on yield of reductions in initial disease intensity and rate of disease increase. The data show only minor variation around a straight line. However, as other graphs in Fig. 12.2 show, there is not necessarily a straight-line relation, and there can be very large variation around a bestfitting line (curve). As will be seen later, sometimes there is no significant relationship between measured disease intensity (as a proportion) at any time and yield (or loss), even though disease is affecting yield.

The practical goal of many crop loss studies is to characterize the relationships shown in Fig. 12.2, and to compare the relationships for different cultivars, treatments, locations, and so on. To do so requires measurements of yield and disease at one or more times in several experimental or sampling units and the fitting of models to the obtained data. Next we review some of the approaches for obtaining the data.

12.3.2 Obtaining data from a range of epidemics

Ideally, a data set should have a wide range of yield and disease values if one wants to characterize the yield (or loss) response to epidemic development. The principles and methods of measuring disease intensity and crop properties (such as leaf area index), as presented in Chapter 2, apply here, as do the approaches for characterizing disease development over time, as presented in Chapter 4 (and later chapters). The required data can be obtained in either of two basic ways: from conventional field (or greenhouse) experiments; or from surveys of fields with naturally occurring epidemics.

For conventional field experiments, one imposes treatments that will, in principle, result in different epidemics (e.g., differences in y_0, r_*, and K; see Chapter 4), and possibly different crop yields (if the disease in question has a yield effect). There are different types of treatments that can be used for this purpose. The first is through inoculations. One can inoculate plots with different inoculum densities, or at different number of times during the season. A single inoculation can be done either early or late. Use of inoculation treatments is most desirable when only low disease intensity would occur without inoculations. Typically, this works best when overwintering inoculum is low, or there are few infected alternate hosts in the area. An additional advantage of inoculations is that one can impose a spatial pattern to the initial infections (see Chapter 9) to relate disease intensity and pattern simultaneously to resulting yield (see section 12.7).

The second type of treatment is the use of protectants, such as fungicides for fungi and insecticides and stylet oils for insect-transmitted viruses. The use of biocontrol agents would also fit in this setting. By varying the number of pesticide applications, their timing, type of chemical, or level of active ingredient, a wide range of epidemics may be achieved. This approach works best when disease intensity would normally be high without the use of protectants, such as when there is high inoculum density in the soil or many infected alternate hosts in the region. It is possible, of course, to combine the inoculation and protection methods. Whenever possible, one should test the fungicides in the absence of disease to make sure the chemicals do not affect yield directly.

There are other ways of obtaining different epidemics in different plots. One could use different genotypes of the host, preferably isolines or near-isolines, which vary in susceptibility. One could also manipulate the environment, for instance, by using irrigation as a treatment to influence disease increase (for those diseases affected by surface wetness and/or atmospheric moisture). Or, one could use location as a treatment factor; that is, having plots in different locations. Of course, the environment may (will) vary among locations, and there may be interactions of location and genotype with disease in terms of yield response. One should be cautious in the interpretation of crop loss results based on the use of cultivars or locations to obtain different values of disease and yield. For instance, differences in yield between two near-isolines (that happen to differ in disease intensity) may be due to genetic traits independent of disease. Likewise, yield differences obtained at different locations, or with different imposed environmental conditions, may be due more to the location or environment effects than the disease effect.

Inoculation and protection are the two most useful ways of obtaining different epidemics in planned experiments. In fact, the different cultivars (genotypes) and locations (or environments in general) are utilized specifically to determine how they influence crop losses, rather than as ways of simply obtaining different epidemics and yield. In section 12.5, we show one way to quantify the effects of factors such as cultivar on disease-yield relationships. As suggested by James (1974) and Madden (1983), it is highly desirable to obtain data over a few years, at more than one location, and with a range of commonly used cultivars in order to understand and characterize crop losses. Of course, it is often not possible, because of the lack of funds or labor, for example, to obtain such comprehensive data.

In contrast to conventional experiments, by definition, one does not impose any treatments with surveys of naturally-occurring epidemics. Instead, one determines disease and yield data for fields selected by an accepted survey-sampling procedure (such as simple random sampling; see Chapter 10). This is discussed more in the next section in the context experimental and sampling units.

12.3.3 Experimental and sampling units

12.3.3.1 Planned experiments. Epidemic and yield data can be obtained from different types of experimental or sampling units. In a planned experiment, say, with different inoculation treatments, the typical experimental unit is a relatively small plot. There will be a number of replicate plots for each treatment. Depending on the pathosystem, the plot may be one or a few rows wide and a few meters long. For trees, or even other plant species, the experimental unit may be a single plant. Unless the plots are very small, disease intensity (and other variables) are observed/measured, either visually or with electronic approaches (Nutter, 2002b), in several samples; the principles and methods in Chapter 10 are applicable here. Yield can be determined for each sample (and generally expressed as a mass or volume per unit area), permitting a quantification of disease and yield at the sampling-unit scale. However, it is very common to determine a single (total or mean) yield measurement for the plot (but expressed per unit area). This necessitates the calculation of mean disease intensity (or other variable) from the samples. Then, the relationship between mean yield and mean disease intensity (or other epidemic property) is quantified. The implication of using means is addressed in section 12.7.

As an alternative to averaging the values within plots, one can directly measure disease and yield for individual plants within the plots, and utilize the individual values in later analysis. Even when the average disease intensity is very low (or very high) for a plot, there may still be a wide range of individual observations (see Chapter 9), and possibly a corresponding wide range of values for yield/plant. This approach, sometimes called the "single plant" method, has the advantage of giving a wide range of values of the response (yield) and predictor variables (e.g., disease intensity at the end of the season), which aids in the identification of the functional relationship (if one exists). However, if the single plants are located relatively close together, use of standard least squares model fitting (see Chapter 3) would be problematical, because this form of data analysis requires independent values (at least with the usual standard statistical assumptions). Moreover, to account more fully for both sampling variability (variation in yield between samples within plots) and experimental variability (variation between plots treated in the same way), any statistical model would require two error terms, one for sampling error and one for experimental error. This can be done using linear or nonlinear mixed models (Littell et al., 1996, 2006; Schabenberger and Pierce, 2002). This complexity is, unfortunately, often ignored in practice.

An alternative to the conventional plot is the microplot (Francl et al., 1987; Navas-Cortés, et al., 2000), especially common for soil-borne pathogens. As the name implies, this is a very small plot, often representing a single plant; there may also be barriers in the soil separating the plants. The microplot is different from the "single plant" sampling unit because the microplot *is* the experimental unit, not a sample within plots. (Of course, the distinction between a small plot and a microplot can be arbitrary). Microplots allow fine control over the conditions experienced by the individual plant, which may be very useful for understanding how disease and other variables affect the yield response. However, there may be high variability among microplots receiving the same treatment, and this type of experimental unit could be too artificial, if one is attempting to relate disease to yield in commercial fields.

The issue of number of replications of either plots or microplots deserves some consideration. The typical approach is to use three or four replications, as is commonly done for other field experiments. However, Teng and Oshima (1983) argue in favor of more treatments (to give more distinct epidemics) and fewer replications (when the total number of experimental units is fixed). For instance, assume that a researcher can afford 20 plots. A typical experiment might consist of five treatments (a control and four different fungicide regimes), with four replications each. This may be quite adequate to determine if the treatment has a significant effect on disease or yield. However, in crop loss experiments, the treatment is often not of direct interest, but is used to get different levels of disease and yield. A wider range of epidemics and yields may be obtainable with 10 treatments and two replications each, or even 20 treatments with one (!) replication. Obviously, the precision of any measurement goes down as the number of replications goes down. But, this approach might provide a fuller representation of the entire disease:yield relationship (although the uncertainty of each datum may be high). We feel that giving up all replication may be a bit extreme, but using fewer replications than in standard crop experimentation may be of benefit if this means that more treatments can be imposed (potentially giving more distinct epidemics).

12.3.3.2 Surveys. When one obtains data from a survey, rather than from a planned experiment, fields (orchards, etc.) are selected as samples or sampling units, based on the objectives of the investigator, and a number of sub-samples typically are obtained for disease (and other predictor variables) within each sample. The fields may be commercial ones or may even be found on university and government research facilities. Composite yield and disease values for each sample (e.g., field) are typically obtained, although yields for each sub-sample could also be obtained. The sub-samples here correspond to the samples within treatment plots in conventional experiments (with imposed treatments) or within single fields in observational studies (see Chapters 10 and 11). Often, mean disease and yield are used for each field (i.e., each sample) in the analysis.

The approach based on survey samples sometimes is known as the synoptic approach, especially when multiple variables are measured or observed for each sample (Stynes, 1980). This approach can result in a wide range of yield and disease values, which is desirable for quantifying relationships, and may reveal relationships not determined in conventional experiments, because the "right" treatments were not evaluated in the planned situation. Moreover, because the data are collected in "natural" agricultural conditions, results have the potential to be more "realistic" than found for small experimental plots (e.g., at a research center).

Since there is no control over the conditions in which the data are obtained in a survey (because no treatments are imposed), results could be misleading. For instance, there may be a negative relationship between yield and final disease intensity, but also a positive relationship between both yield and disease intensity and soil moisture content. Thus, without a wide range of realized values, it is possible that a positive relation between yield and disease intensity will be observed (because in this example, yield increases more with increasing moisture than does disease intensity). Specifically, the correlations of the observed or measured variables can be influenced by other (unmeasured) variables. For this reason, results from surveys are probably best used to *suggest* relations rather than to characterize relations. Moreover, it is seldom affordable or practical to obtain measurements of disease (or other variables) at multiple times during an epidemic in several sampled fields; thus, an investigator would not be able to evaluate AUDPC or other season-long summaries of epidemics as predictors of yield (or loss). Nevertheless, the survey approach may be the only practical way of obtaining data for some pathosystems.

12.3.4 Yield per unit area

So far, we have discussed yield and yield measurements in the general sense, without any details. A little elaboration is warranted at this point. Readers may have noticed by now that this chapter is a little different from most of the rest of the book in one respect. That is, the measurements or observations used elsewhere mostly dealt with disease intensity. Direct measurements of the crop (such as LAI; Chapter 2), if taken at all, were used to provide a basis for the calculation of disease intensity. Now, we also use measurements of crop yield in addition to disease intensity, and the portion of the crop measured as yield typically is separate from the portion of the crop infected. Of course, some diseases directly affect the harvested portion (e.g., seed diseases and fruit rots), but many of the most costly diseases do not infect the harvested portion of the crop.

For the most part, we link or relate a single yield value to each epidemic. However, crop yield can be partitioned into components, and sometimes the components are directly determined. This is demonstrated with yield of wheat. Yield can be represented as, for example, 0.5 kg/m^2 or 5 MT/ha. This total is actually the product of the number of seeds produced per unit area and the average weight per seed. A given disease could reduce total yield by reducing the number of seeds, or the individual seed weight, or both. These *yield components* can be further subdivided. That is, total yield can be represented as the product of number of plants/m^2, number of spikes/plant, number of spikelets/spike, number of seeds/spikelet and individual seed weight (Slafer, 2003). These components are often negatively correlated with each other; thus, a disease that reduces the number of plants per unit area by, say, 20%, does not automatically reduce the total yield by this percentage.

Starting in section 12.4, we primarily deal with the quantification of a single (total) crop yield value associated with each epidemic. Analysis of the components of yield requires some additional multivariate and univariate statistical methods not discussed here. Interested readers should refer to Madden (1983), Hau (1990), and Stynes (1980).

12.3.5 Expert opinion

Obtaining data from planned experiments or surveys under a wide range of environments and cropping systems is expensive and time consuming, and it is unlikely that sufficient information is available for every disease of concern. As an alternative to experiments and surveys, one can obtain estimates of crop losses from experts (such as extension specialists) for a particular crop in a given region. The expert's opinion would reflect years of experience with a given crop and disease (or diseases), and be based on a combination of previous experiments, surveys and interactions with crop producers, crop advisors and consultants, and others. This approach has been formalized for some crops, such as soybeans in the U.S. (e.g., Wrather et al., 2001, 2003).

In this book, we are more concerned with characterizing crop losses using direct observations of disease and yield, so we do not deal further with the use of expert knowledge. However, it must be acknowledged that use of expert opinion will likely remain important in obtaining estimates of losses and costs of diseases for regions and countries (e.g., Pimentel et al., 2000). One area that deserves more research is an evaluation of the methodology for eliciting quantitative information from experts, for use use in either estimation of losses or in decision making. Methods considered for eliciting disease incidence information from experts (Hughes and Madden, 2002) are relevant here.

12.4 Modeling Yield in Relation to Disease

12.4.1 Notation and general concepts

We assume that there is a relation, at least potentially, between crop yield and one or more representations of an epidemic. We primarily consider in this section the situation where the relevant epidemic components consist of disease intensity at one or more times during a growing season, or summary statistics based on these disease intensities. In section 12.6, we broaden this pool of predictor variables by viewing crop yield in a more physiological manner.

As elsewhere in this book, we use Y for disease intensity in absolute units (e.g., area of diseased leaves) and y for intensity in relative units (e.g., proportion of leaf area diseased). Likewise, H is used for disease-free intensity (e.g., healthy leaf area) and h for disease-free intensity in relative units. Thus, $Y + H = M$, $y = Y/M$, $h = H/M$, and $y + h = 1$, where for convenience here, M is the total host crop "size". The variables could be averages of several observations (samples or sub-samples) from a plot or field, or individual values. For simplicity, we do not use bar notation for averages here, since it is unnecessary for the current presentation of the concepts.

The yield of a crop per unit area under a given set of conditions (in absolute units) is represented by W. For example, W could be 5 metric tons per hectare (5 MT/ha), 500 g/m^2, 250 fruits/m^2, and so on. It is a density measurement in typical usage—if one records yield per plant, it is useful to convert the value to yield per unit area for subsequent analyses. W_0 is used to represent yield in the absence of the disease of interest, with the same units as W. Relative yield (w) is then given by: $w = W/W_0$. Loss in yield (L) is given by: $L = W_0 - W$. Loss in yield relative to the disease-free yield is given by:

$$\ell = \frac{(W_0 - W)}{W_0} = 1 - w$$

Relative yield (or loss) and disease as a proportion are unitless (dimensionless) variables. However, it is important to note that there is a fundamental difference between w (or ℓ) and y. The variable y is determined from a *single* plant individual (e.g., proportion of leaf, root, or plant diseased) or from a sample of a *single* population (e.g., proportion of diseased plants in a plot or field, or average proportion of plants diseased in a plot or field), since both Y and H are definable for a single individual or population. For instance, a plant may have 100 cm^2 of diseased leaf area and 100 cm^2 of disease-free area, giving $y = 100/(100 + 100) = 0.5$. In contrast, w (or ℓ) is determined from two *different* individuals or samples from two different populations, one with a given epidemic and one without disease. A plant population (e.g., field) can be characterized by a yield, W, but loss in absolute units (L), as well as relative yield and relative loss (w and ℓ), can only be defined in comparison to another population (e.g., field) without the disease epidemic.

A difficulty with use of L, w, and ℓ, is that sometimes (often, in fact) there are no disease-free plots in a planned and replicated experiment. For practical matters, mean yield from plots with very low Y can be used as an approximation for W_0. Sometimes, modeling (see below) provides another approach to dealing with this problem (by allowing the prediction of W_0 based on the observed data). With surveys, the problem is more challenging, even if fields with no disease are observed. This is because the fields with no disease might differ from those with disease in terms of other agronomic attributes, some or all of which may not be known to the investigator. For instance, the yield may be unusually high (or low) for reasons that have nothing to do with the level of disease intensity.

Despite the difficulties mentioned here, it is common to express loss, relative yield, or relative loss, in relation to disease intensity (either y or Y) (see Fig. 12.2). This may be because of the appeal of thinking of the impact of plant disease in relative terms, or difference from a "standard", rather than in terms of a given level of yield for a given epidemic situation. In recent years, however, there has been increasing use of W (instead of L, w, or ℓ) as the response variable in crop loss studies. This is partly because of the relevance of yield in absolute units in physiological modeling of crops (see section 12.6).

In a broad sense, we can consider yield, W, to be a function of the disease epidemic. This can be written generally as:

$$W = g\ (\text{epidemic variables; parameters}),$$

where $g(\bullet)$ is a linear or nonlinear function (see Chapter 3). For ease of expression, equations of this type can be referred to as crop loss or yield loss models, in addition to yield models, even when the response variable is yield. How the epidemic is summarized—that is, which epidemic variables are utilized in $g(\bullet)$ and in what form—serves as a useful classification system for crop loss models. We use a hybrid model classification, based on Teng (1985) and Madden and Nutter (1995).

12.4.2 Single point models

12.4.2.1 Linear models. The simplest, and perhaps most practical, empirical description of crop loss is obtained by use of a single measurement of disease intensity during the epidemic and a linear model. One version of the equation is written as:

$$W = \beta_0 - \beta_1 y \qquad (12.1)$$

in which β_0 and β_1 are parameters. In other words, $g(\bullet)$ is specifically expressed as:

$$g(y; \text{parameters}) = \beta_0 - \beta_1 y.$$

Disease intensity in absolute units (Y) could also be used (if the data were available) in the function $g(\bullet)$. The values of y (or Y) are usually based on visual measurements of symptoms, but electronic measurements (i.e., percentage reflectance of solar radiation) can also be used (e.g., Newton et al. 2004). One could also express loss (L) as a function of y, or relative yield (w) or loss (ℓ) as a function of y, using the interrelationships presented above. Equation 12.1 is known as a *single-point model* because y from a single time in the epidemic is used. Fig 12.2A is an example of a situation where a single point model is appropriate. When needed for clarity, a subscript can be added to y to indicate the time during the epidemic. The time might be arbitrary, such as the end of the season, or the time can be chosen more formally based on statistical analyses. For some crops, it has been found that a certain time period, corresponding to key growth (phenological) stages (see Chapter 2), is especially critical in yield formation, and that disease severity at this time is often (highly) correlated with yield (James, 1974; Teng, 1985; Madden and Nutter, 1995). For grain crops, the critical time period is around Feeke's growth stages 10.5. Single point models can be called critical point models for this reason.

In equation 12.1, the intercept term, β_0, can be thought of as expected yield when $y = 0$ (under a given set of crop production and environmental conditions). Thus, the intercept is a yield property of the cultivar in a given environment when the disease of interest is not affecting the crop. In this case, one could use W_0 instead of β_0 in the model; that is, $W = W_0 - \beta_1 y$. As shown below, this interpretation of β_0 depends on equation 12.1 being appropriate over the *entire* range of possible y values (0–1), not just for the observed range of y values. Thus, we prefer here not to use W_0 for the intercept term. The slope parameter β_1 in equation 12.1 represents the change in yield with change in disease intensity at the particular (single, critical) time of interest during the epidemic. If crop-loss results are being compared across cultivars, β_1 represents, in a sense, the tolerance of a given cultivar to a given disease. For instance, if two cultivars have the same W_0, then the one with the smaller β_1 (closer to 0) will have higher yield at a given level of disease intensity. This can be seen graphically from lines "a" and "b" in Fig. 12.3, where line "a" represents the more tolerant cultivar. In contrast, line "c" can represent a lower yielding cultivar (lower β_0)

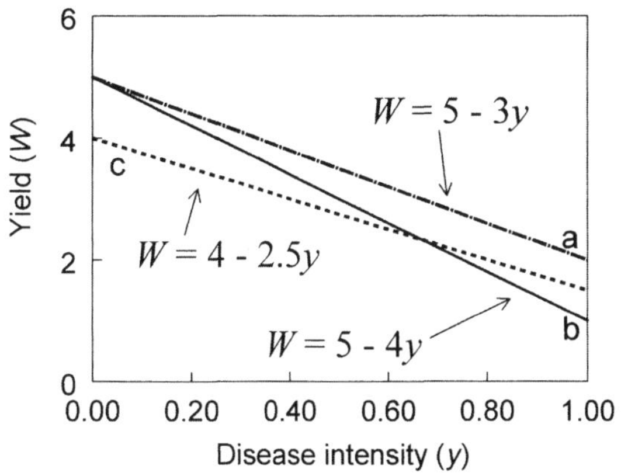

FIG. 12.3. Theoretical yield:disease relationships based on linear equation 12.1 with different slopes and/or intercepts.

compared with cultivars "a" and "b", but which is also more tolerant than the others to disease (lower β_1). There is a value of y at which cultivar "c" has a higher yield than cultivar "b" in this hypothetical situation.

The above discussion was in terms of cultivar comparisons. However, one can take the same approach to compare the yield response over two or more years, or, in general, two or more environments. Differences in β_0 reflect differences in effects of environment on yield in the absence of disease. Differences in β_1 reflect the differential effects of environments on yield response to disease.

An example of a single-point model is the one developed by Bissonnette et al. (1994) for oat yield in relation to crown rust (caused by *Puccinia coronata*).

$$\hat{W} = 4.0 - 5.7 \cdot y$$

where yield is expressed as metric tons per hectare (MT/ha) and y is proportion of disease at oat growth stage 75. The R^2 value was 0.8. In the absence of disease, expected yield is 4 MT/ha. For each increase in y by 0.01 (i.e., 1%), there is a decrease in yield of 0.057 MT. If equation 12.1 is taken as being valid for all values of y (so that $\beta_0 = W_0$), one could write the equation for loss, L, as: $\hat{L} = 4.0 - \hat{W} = 5.7 \cdot y$. The relative yield equation would be given by: $\hat{w} = (4.0 - 5.7 \cdot y)/4.0 = 1 - 1.425 \cdot y$, and the relative loss equation would be given by $\hat{\ell} = 1.425 \cdot y$.

In practice, there is no real reason to expect a straight-line relationship between W and y (see Fig. 12.2 for examples). In fact, there are some theoretical reasons for a nonlinear relationship, as will become apparent below. Various authors have used linear and nonlinear models to characterize yield (or loss) in relation to intensity. For linear models (i.e., linear in the parameters), the simplest approach to capture curvature in the W:y line is to use a transformation of y, y^*, as the predictor variable. The single-point model can then be written as:

$$W = \beta_0 - \beta_1 y^* \qquad (12.2)$$

Useful transformations include the square root and log (Romig and Calpouzos, 1970; Large and Doling, 1962; Teng et al., 1979). In equation 12.2, β_1 is the change in yield with change in y^*, not with change in y. The "intercept" β_0 represents expected yield when y^* is equal to 0, not when y is equal to 0. For instance, consider the situation when $y^* = \ln(y)$. Then $y^* = 0$ when $y = 1$; thus, β_0 represents expected yield when $y = 1$. Following with this example, with disease intensity, y, represented as a proportion, β_0 is the expected yield when y is at its highest possible value (i.e., 1), rather than its lowest (i.e.,.0). For β_0 to represent yield of disease-free plants in equation 12.2, y^* must equal 0 when y is equal to 0. With y as a proportion (the standard representation in this book), the transformation $y^* = \ln(y + 1)$ meets this criterion.

12.4.2.2 Nonlinear models. As discussed in Chapter 3, many biological phenomena are best described by nonlinear models. These kinds of models are especially useful when there are thresholds, when the response variable approaches an asymptote, and when it is desired that the same model describes different shapes for the curves of W versus y. Several nonlinear models have been proposed to describe crop loss data (Madden and Nutter, 1995; Madden, 1983), and we show one utilized by Hughes (1988) as an example. Yield can be written as:

$$W = W_0\left(\frac{1-y}{\left((1-\Theta)+\Theta(1-y)\right)^2}\right) \qquad (12.3)$$

in which W_0 and Θ ($\Theta \leq 1/2$) are parameters. As above, W_0 is the expected yield when $y = 0$. Θ is an indication of the rate of decline in W with increase in y and the shape of the curve; however, for this model, there need not be a straight-line decline. Fig. 12.4A demonstrates theoretical values of W for three choices of Θ in equation 12.3. When Θ is positive ($0 < \Theta \leq 0.5$; curve II), there is initially little decline in W with increasing y, and the rate of decline in W increases as y increases. Mathematically, the second derivative (d^2W/dy^2) is negative when Θ is between 0 and 0.5. When Θ is negative ($\Theta < 0$; curve I), there is initially a rapid decline in W with increasing y, and the rate of decline decreases as y increases; d^2W/dy^2 is positive when $\Theta < 0$. When $\Theta = 0$, there is a straight-line relation; that is, equation 12.3 reduces to $W = W_0(1 - y) = W_0 - W_0y$. For $\Theta = 0$, $d^2W/dy^2 = 0$.

Other nonlinear models can be used when there is evidence of a threshold (i.e., a value of y below which no loss occurs, denoted here a) or a lower nonzero limit to yield [i.e., a maximum possible loss (less than W_0) that is reached or approached at the highest possible y; denoted here W_{min})] (Madden, 1983; Madden and Nutter, 1995). These concepts are displayed in Fig. 12.5. Curve 1 is the simple situation of no disease threshold ($a = 0$) and $W_{min} = 0$. Curve 2 is similar to curve 1, except that $W_{min} = 1$ (i.e., no lower value of W than W_{min}). Curve 3 also is based on $W_{min} = 1$, but with a disease threshold of $a = 0.2$ (20% severity).

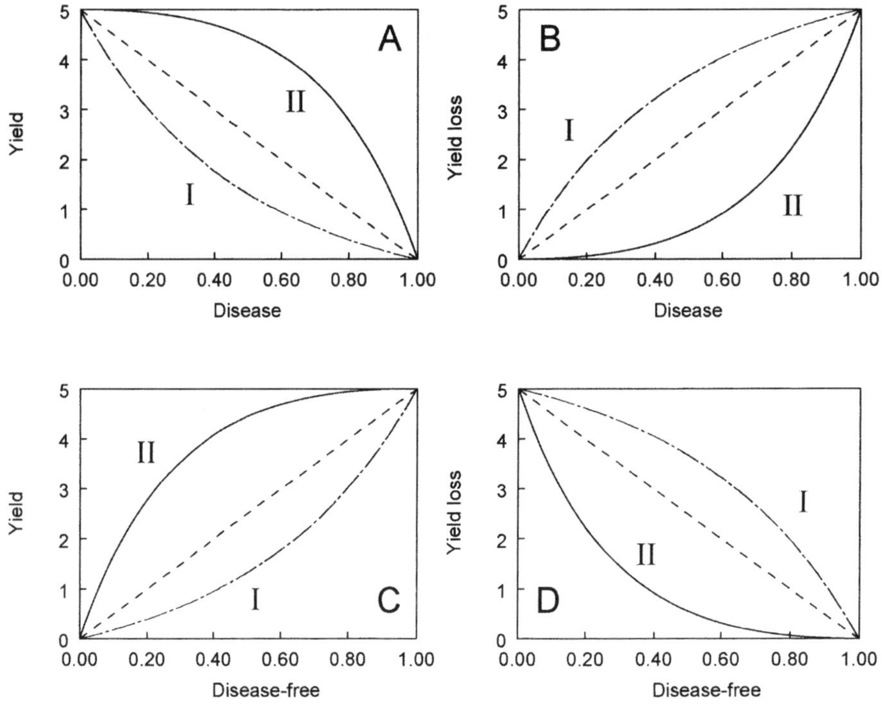

FIG. 12.4. Yield or loss versus disease intensity (y) or disease-free intensity ($h = 1 - y$) for situations with a disease-free yield (W_0) of 5. (A) Yield (W) versus y. (B) Yield loss ($L = W_0 - W$) versus y. (C) W versus h. (D) L versus h. Lines were generated using equation 12.3 with $\Theta = 0$ (the straight line), 0.5 (curve II), and -0.75 (curve I). When $\Theta = 0$, equation 12.3 is identical to equation 12.1, with $\beta_0 = W_0$, and $\beta_1 = W_0$.

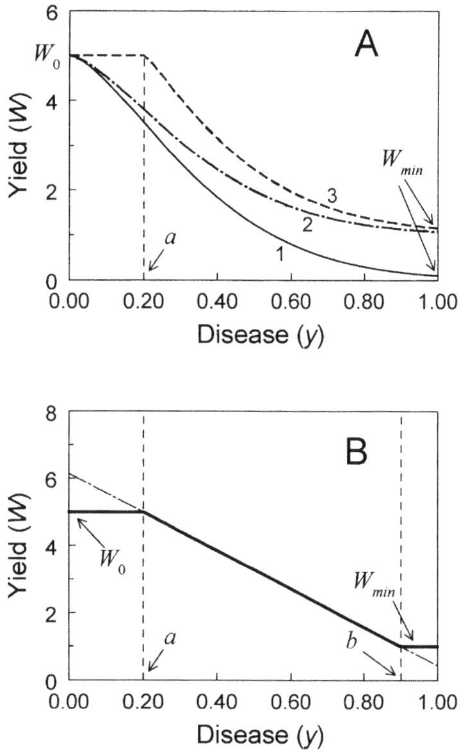

FIG. 12.5. Theoretical relationships between yield and disease intensity, based on use of equation 12.5 (**A**) and 12.6a (**B**). In all cases, $W_0 = 5$. (**A**) Parameters for the three curves were: $W_{min} = 0$, $a = 0$, $\delta = 0.4$, and $c = 1.5$ (curve 1); $W_{min} = 1$, $a = 0$, $\delta = 0.4$, and $c = 1.5$ (curve 2); and $W_{min} = 1$, $a = 0.2$, $\delta = 0.3$, and $c = 1.2$ (curve 3). (**B**) $W_{min} = 1$, $a = 0.2$, $b = 0.9$ for the thick solid line. The thin broken line in B indicates extrapolation of equation 12.6a at $y < a$ and $y > b$.

We show here a framework for a general crop loss model incorporating these (and other) ideas. A disease:yield relationship can be written as:

$$W = W_{min} + (W_0 - W_{min}) \cdot S(y) \tag{12.4}$$

in which $S(y)$ is a function of disease intensity (or, more broadly, the predictor variable) that ranges from 0 to 1 over the full range of possible y values. $S(y)$ has been called a switching (switch-off or switch-on) function (Schabenberger and Pierce, 2002) because it switches (either slowly or quickly) between 0 and 1 as y goes from its minimum to maximum (or maximum to minimum). Note that the threshold a, and generally other parameters, are part of $S(y)$. The nonlinear function in parentheses after W_0 in equation 12.3 is a switch-off function (for certain Θ values); thus, equation 12.3 is a special case of the general equation 12.4, with $W_{min} = 0$ and no disease threshold (i.e., $a = 0$). A useful choice for $S(y)$ is one minus the cumulative distribution function for a unimodal continuous random variable. Although not based on any biological mechanisms, such a choice ensures that $S(y)$ is limited by 0 and 1. Using the Weibull cumulative distribution function (Madden et al., 1981a), one can write the switch-off expression as:

$$S(y) = \exp\left(-\left(\frac{y-a}{\delta}\right)^c\right), \quad \text{if } y \geq a$$
$$S(y) = 1 \qquad \text{if } y < a$$

Then, equation 12.4 is written as:

$$W = W_{min} + (W_0 - W_{min}) \cdot \exp\left(-\left(\frac{y-a}{\delta}\right)^c\right), \quad \text{if } y \geq a$$
$$W = W_0, \qquad \text{if } y < a \tag{12.5}$$

There are five parameters, W_0 and W_{min} (both for yield), a (threshold of disease intensity on horizontal axis), δ (a measure of steepness of the W-y curve), and c (a unitless measure of curve shape). The concept of a shape parameter was discussed at some length in Chapter 4. All the curves in Fig. 12.5A were obtained with equation 12.5.

Other distribution functions certainly can easily be chosen for $S(y)$. Schabenberger and Pierce (2002) and Gregorie and Schabenberger (1996) discuss switching functions at some length for modeling other biological phenomena. It should be noted that even the simple linear model of equation 12.1 is a form of the general equation 12.4, if constraints are placed on the β_1 parameter. This can be seen by using W_0 for β_0 as described above, and re-writing equation 12.1 as:

$$W = W_0 \cdot \left[1 - \frac{\beta_1}{W_0} y\right].$$

This is a version of equation 12.4, with $S(y) = 1 - (\beta_1/W_0)y$, no disease threshold ($a = 0$), and a minimum W of 0 ($W_{min} = 0$) occurring at $y = 1$. Clearly, $S(0)$ here is equal to 1. However, for $S(1)$ to be equal to 0 (i.e., for the switch-off function to be 0 at $y = 1$), β_1 must equal W_0. It is rare to find the slope and intercept of equation 12.1 to be equal when fitted to crop loss data. This is because W_{min} is generally not 0 ($W_{min} > 0$), or the minimum yield does not occur at $y = 1$, or there is a threshold ($a > 0$). Thus, investigators may not explicitly consider these complicating issues. In fact, the actual W_{min} may occur only at a value of y beyond the largest observed y, and W_0 may occur at a value of y smaller than the smallest observed y. In these cases, there may be little evidence of a relationship other than a straight (or curved) line over the observed range of y values.

To better understand the use of equation 12.4 for yield–disease relationships, we demonstrate the model with a more general switch-off function, which is still a straight line. If the cumulative uniform distribution function (Johnson et al., 1995) is utilized for $S(y)$, one can write the switch-off function as:

$$S(y) = \frac{b}{b-a} - \frac{1}{b-a}y, \quad \text{if } a \le y \le b$$
$$S(y) = 1, \quad \text{if } y < a$$
$$S(y) = 0, \quad \text{if } y > b$$

in which a is the threshold y below which no yield reduction occurs (as already defined), and b is an upper value of y, above which no further decrease in yield occurs. The model for W can then be written as:

$$W = W_{min} + (W_0 - W_{min}) \cdot \left(\frac{b}{b-a} - \frac{1}{b-a}y \right), \quad \text{if } a \le y \le b$$
$$W = W_0, \quad \text{if } y < a$$
$$W = W_{min}, \quad \text{if } y > b \tag{12.6a}$$

The thick solid line in Fig. 12.5B is an example of the yield-disease relation based on equation 12.6a, showing the maximum and minimum yields, and the values of y where these occur. Unlike the situation with the Weibull function used for $S(y)$, there is an abrupt transition from W to W_{min}. With both the Weibull and uniform functions, however, the transitions from W_0 to W are abrupt. For values of y between a and b, one can rewrite equation 12.6a as:

$$W = \left[W_{min} + (W_0 - W_{min}) \cdot \left(\frac{b}{b-a} \right) \right] - \left(\frac{W_0 - W_{min}}{b-a} \right) y. \tag{12.6b}$$

which is, of course, a straight line function with slope given by $(W_0 - W_{min})/(b-a)$. Note that equations 12.6a and 12.6b are nonlinear in the parameters, as discussed in Chapter 3. Although a portion of the model can be considered linear (i.e., for $a < y < b$), the fact that a and b are unknown makes this model nonlinear. Readers should return to section 3.5 for more information.

For the case with a minimum yield of 0 occurring at $y = 1$ ($W_{min} = 0$; $b = 1$) and maximum yield occurring at $y = 0$ ($a = 0$), the slope is W_0, as indicated above when the simple linear equation 12.1 is appropriate for the entire range of disease intensities and there is a zero minimum yield. In the more general sense, the slope depends on the difference between the upper and lower possible yields ($W_0 - W_{min}$), and the range in disease intensities where yield varies with disease ($b-a$).

A little algebra will show that equation 12.6b predicts yield values of W_0 when $y = a$ and W_{min} when $y = b$, as required based on equation 12.6a. However, using equation 12.6b (with no constraints) would be misleading at disease intensities outside of this disease range ($y < a$, or $y > b$). This could happen in practice if all the observations of y are between the true values of a and b, and equation 12.1 appears appropriate. Then, the intercept would actually not be W_0, but be given by:

$$\beta_0 = W_{min} + (W_0 - W_{min}) \cdot \left(\frac{b}{b-a} \right).$$

This can be seen by the extrapolated (thin) line extending to the W axis in Fig. 12.5B. A similar type of error would be made in determining the minimum yield. If all observations are between a and b, there is no way (from a single data set) to know the values of a or b, and corresponding values of W_{min} and W_0, without making additional assumptions.

It is useful to consider the oat rust equation equation of Bissonette et al. (1994). With $y = 1$ in the fitted equation, one obtains $\hat{W} = -1.7$, a meaningless value. If there is no lower disease threshold ($a = 0$), then $\hat{W} = 4$. This means that the slope (5.7) is equal to $(4 - W_{min})/b$. If one can assume that $W_{min} = 0$, then $4/b = 5.7$ (the experimentally derived slope). Rearrangement shows that b is then estimated as 0.7. If these assumptions hold for this example, all disease intensities above 0.70 result in no oat yield.

The point of this discussion here is to make it clear that interpretation of model parameters depends on the appropriateness of the model over the range of disease intensities where interpretation is desired. Typically, there are observations of disease intensity near 0 (Fig. 12.2), which means that estimates of a are obtainable. When there are low values of observed disease intensity and no evidence of a leveling off of W, then it is very reasonable to assume that $a = 0$ and that the intercept term is W_0. Most of the data sets plotted in Fig. 12.2 are consistent with $a \approx 0$. However, there are clear examples of thresholds in other systems (Madden et al., 1981a) coupled with curvature of the W:y relation above the disease threshold. A bigger challenge is to determine minimum yield, since observed disease intensities may not be large enough to determine this. In situations where there is a nonlinear relationship between W and y, it may be more useful to use versions of $S(y)$ with smooth transitions to W_{min} because then there is no b parameter to estimate.

Other choices for $S(y)$ can be utilized, but we do not discuss this aspect further. More complicated functions may be needed for more complicated $W : y$ relationships. For instance, as discussed by Teng (1985), low levels of y may actually promote plant growth and yield production per unit area, so that there is an actual increase in W with increase in y (when y is small).

12.4.2.3 Model fitting. The single-point models are generally fitted to data using either linear or nonlinear least squares methodology (Neter et al., 1983; Schabenberger and Pierce, 2002), although maximum likelihood could also be used. The methodology of Chapter 3 is relevant for this empirical modeling. The approach for a linear model is demonstrated in section 12.5.

12.4.3 Some considerations regarding the response and predictor variables in single-point (and other) models

Before considering alternatives to single-point models, it is useful to consider consequences of alternative expressions for the yield and disease variables used in the models. These concepts apply to the models that are presented in later sections. We consider only the situation for a linear switch-off function, with $a = 0$ and $b = 1$ in equation 12.6b, so that there is a straight-line relation between W and y over the full range of disease values. This means we can use equation 12.1 for the yield-disease relation, with W_0 for the intercept (β_0) and ($W_0 - W_{min}$) for the slope (β_1). Two examples are shown in Fig. 12.6A, which differ in the disease-free yield (β_0) but not in the rate of decline of yield with increasing disease (β_1). The two lines could, for instance, represent two different cultivars.

As mentioned above, it is fairly common to relate relative yield to disease intensity. This is done by dividing W by the disease-free yield. As usual, we use lower-case w for relative yield (i.e., $w = W/W_0$). To obtain w in practice (from experiments or surveys), one would divide each W by the mean yield for observations (plots, commercial fields, etc.) with no disease, or by predicting W_0 from a regression of W on y, and then dividing by this intercept value (assuming no disease threshold). By dividing each W by W_0, one is forcing the relative yield at $y = 0$ to be 1. Conversion from W to w has implication in interpretation in the yield-disease model. For instance, using W_0 for β_0 in equation 12.1, the model for relative yield is:

$$w = \frac{W_0 - \beta_1 y}{W_0} = 1 - \frac{\beta_1}{W_0} y \qquad (12.7)$$

In equation 12.7, the slope of relative yield versus disease is an inverse function of the disease-free yield, and the intercept is independent of disease-free yield. This is demonstrated in Fig. 12.6B, which shows the corresponding relative yield versions of the absolute yield relationships shown in Fig. 12.6A. When the slopes for two groups (cultivars, years, locations, etc.) are the same in terms of absolute yield, the slopes must be different in terms of relative yield, if the value of W_0 used in obtaining w is not the same for all groups. In particular, the larger the disease-free yield, the smaller the slope (closer to 0) for relative yield versus disease intensity.

Figures 12.6A and 12.6B showed how equal slopes on one scale can become different slopes on another scale. The converse can also happen. Fig. 12.6C demonstrates a situation where the slopes (and intercepts) are not the same for the model of W in relation to y for two groups. As shown in Fig. 12.6D, conversion to relative yield now results in equal slopes for the two groups (because of the chosen values of intercepts and slopes in Fig. 12.6C). Thus, as seen in Fig. 12.6, comparison of groups and interpretation of parameter estimates will be affected by the choice of W or w for the response variable. This is not an issue if the same W_0 is used for all groups in

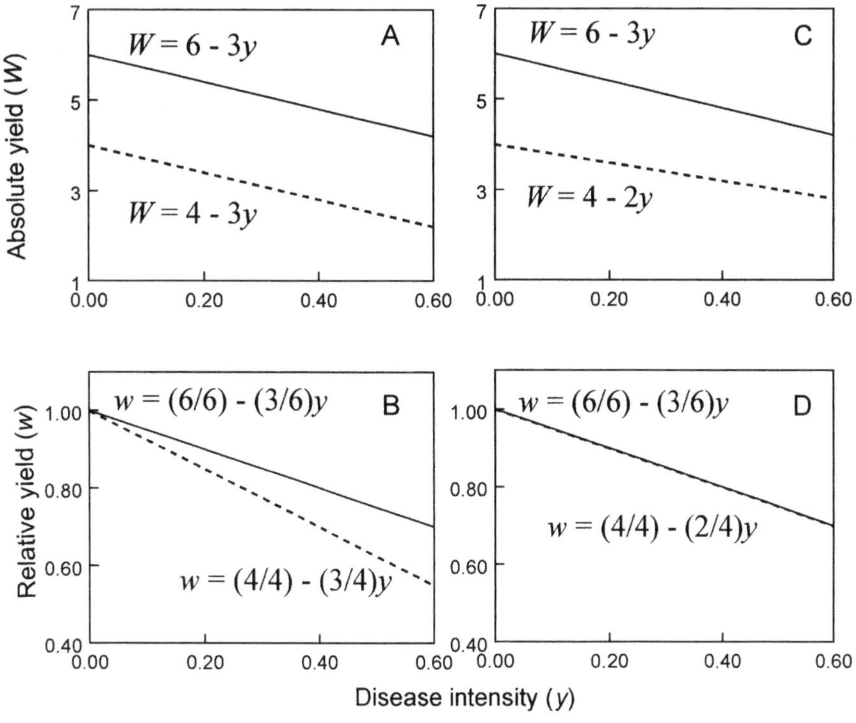

FIG. 12.6. Absolute versus relative yield based on use of linear equation 12.1 (with parameter values given in the frames). (A) Two different disease-free yields. (B) Relative yields corresponding to the absolute yield in (A). (C) Different disease-free yields and the rate of decline of yield with increasing disease. (D) Relative yields corresponding to the absolute yields in (C).

obtaining w values. In this case, the differences of slopes for $W:y$ models are maintained with $w:y$ models. However, simply using a single W_0 across all groups when the data suggest different maximum yields for the different groups could be misleading, since the intercept would then not necessarily correspond to the disease-free yield (for some of the groups).

Researchers do not always present crop loss results in terms of yield (relative or absolute) versus disease intensity, although we mostly express the relationship in this manner for internal consistency in the book. It is helpful for the reader to see some alternative representations, in order to avoid confusion in interpretation of published graphs. We demonstrate this for equations 12.1 and 12.3 (with two different values of the Θ parameter) by returning to the four panels of Fig. 12.4. We already considered Fig. 12.4A for our typical presentation of W versus y. This decreasing relation between W and y is reversed when yield loss ($L = W_0 - W$) is plotted versus y (Fig. 12.4B). This is obvious, of course. What may not be as obvious is that the shapes of the curves (other than the straight lines) are also "reversed" when L is used rather than W.

For reasons that should become clear in section 12.6, it is becoming more common to quantify crop losses in terms of disease-free crop leaf area ("disease-free intensity", H or h; see also Chapter 5) instead of disease intensity. Figs. 12.4C and 12.4D show the relationship between W and h, and L and h, respectively, all based on the original curves in Fig. 12.4A. For all the curves in Fig. 12.4, we leave it to the reader to work through the algebra, starting with equation 12.3 (as an example), to determine the expressions for L in relation to y, and for both W and L in relation to h. All of the relations shown in Fig. 12.4 for yield or loss in absolute units can also be expressed for relative yield (w) and relative loss (l). Careful consideration of the abscissa and ordinate in crop loss curves can thus avoid confusion in presentation and comparison of results.

12.4.4 Multiple-point models

Single-point models could be criticized because they are based on only a small portion of a disease progress curve. For instance, consider the three generated disease progress curves in Fig. 12.7. One might hypothesize that the epidemic depicted by curve "a" has the lowest yield since y is higher than the other epidemics early in the season. Or, if $t = 90$ (for instance) is a critical time for yield production, one could hypothesize that the epidemic depicted by curve "c" has the lowest yield since y is higher than the other epidemics at the end of the season. However, if a single-point loss model was developed based on y at $t = 75$ (y_{75}), then all three epidemics would have the same estimated/predicted yield because y_{75} is identical for all three epidemics. Of course, if 75 days post planting *is* the critical time for yield formation (and

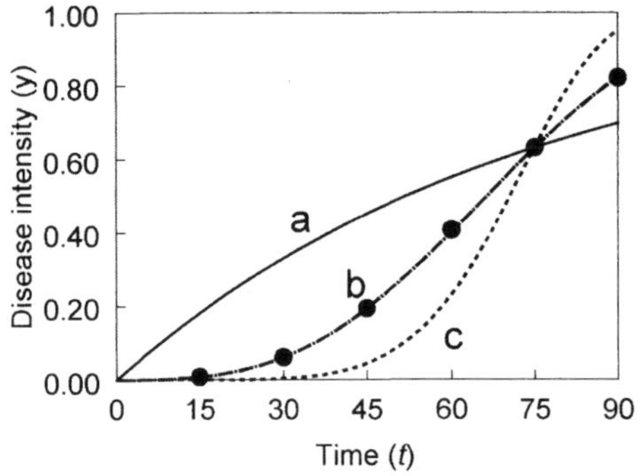

FIG. 12.7. Three theoretical disease progress curves. Points are specific values of y at selected times for curve b.

the only time of relevance, hypothetically), then the model based on y_{75} *is* the appropriate one, and yield would truly be the same for all three cases.

In general, it is not expected that a very narrow time window will be satisfactory for relating yield to the disease epidemic. Thus, a more general framework for relating yield to disease is needed. The remaining subsections of section 12.4 deal with this requirement.

Multiple-point models relate yield (or loss) to disease intensity at two or more times during an epidemic. The multiple-point models typically used are linear, but this is not a requirement. An example is:

$$W = \beta_0 - \beta_1 y_1 - \beta_2 y_2 - \beta_3 y_3 \cdots \quad (12.8)$$

in which the subscripts on the y values are labels to indicate different times during the epidemic (not necessarily three consecutive assessment times), or different growth stages, and the βs are parameters. For instance, the y values could correspond to the points (•) on curve "b" in Fig. 12.7. The concept behind a multiple point model goes back at least to Kirby and Archer (1927) who showed that yield estimates of wheat could be improved by using more than one assessment of stem rust. More formal consideration of models of this type can be dated to the important papers by Burleigh et al. (1972) and James et al. (1972).

The multiple-point model can also be formulated in terms of loss (L) or as relative yield (w) or relative loss (ℓ). Moreover, disease intensity can be in relative or absolute units, but the former is more common. The minus signs in equation 12.8 will become plus signs in the model versions with L or ℓ as the response variable, because loss typically increases with increasing disease intensities. Another variation of the model is to use *change* in disease intensity between times instead of actual measurements of disease at individual times. For instance, one could use Δy_1 and Δy_2 as predictor variables in the model, where Δy_1 is the change in y between times 1 and 2, and Δy_2 is the change in y between times 2 and 3. The parameters then are usually considered to be weights given to each disease value, or change in disease value between times, in influencing yield.

An example multiple-point model, shown with estimated parameters, is one developed for relative loss of peanut yield in Egypt in relation to Cercopora leaf spot severity (Madden et al., 1981b):

$$\hat{\ell} = 0.06 + 4.21\Delta y_1 - 0.24\Delta y_2 + 1.33\Delta y_3$$

in which the three predictor variables are changes in disease severity over 14-day periods. The coefficient of determination was 0.78. In this example, the estimated β_2 parameter is negative (−0.24), which implies that, all other things being equal, that the higher the incremental increase in disease between the two dates, the lower the relative loss. Estimates of parameters with the opposite sign of that expected are not uncommon with multiple-point models that are based on y or Δy (e.g., Burleigh et al., 1972; Teng et al., 1979). Although it is possible for yield to increase somewhat with increasing disease severity (at low mean severities), a much more likely explanation for unexpected parameter values is *bias* of the estimates caused by correlation of the y values.

As shown in Chapter 4, y at a given time is correlated with y at the previous time—this is a consequence of the cumulative population-dynamic process that generates typical disease progress curves. When the predictor variables in any regression model are correlated, the estimated parameter for a variable (e.g., β_2), and its standard error, depend on what other predictor variables are in the model (Neter et al., 1983). Thus, each disease predictor variable is not providing as much information as might be implied by just looking at the fitted equation.

In practice, disease data are obtained at multiple times for several epidemics and models are fitted by using least squares (multiple) regression (or possibly maximum likelihood) analysis (see Chapter 3; Madden, 1983). Because of the high correlation of ys across times, it is likely that some of the predictor variables will not be significant contributors (in a statistical sense) to the multiple regression equation, even if they are all significantly related when considered individually (and biologically relevant). In other words, a subset of the y variables will probably be able to provide as much information for yield estimation as all the y variables do when taken together. The subset of the y values can be chosen using formal or informal stepwise regression methods (Neter et al., 1983), as discussed at some length by Madden (1983). Appraisal of the residual plots can suggest transformations of the predictor or response variables. Modeling is empirical in nature for multiple point models (see Chapter 3) and sometimes very good fits to the data are obtained. However, both the estimated parameters and the significant variables can differ among studies, or even among years in a single study, making it difficult to draw general conclusions, especially since the direct interpretation of the β values is hindered by the correlation of the predictors. Plus, it is difficult to obtain measurements of disease intensity at the exact same time in different studies, making it challenging to compare results across studies. For these reasons, multiple point models are not used as often as one might initially think. In fact, it is quite possible for a single-point model to provide as good a fit (in terms of R^2) as a multiple-point model (e.g., Madden et al., 1981b).

Nevertheless, there is a valuable concept underlying multiple-point models: the entire epidemic potentially influences crop yield, and the importance of a given intensity of disease (in relative or absolute units) depends (at least in part) on the time during the season when that given disease intensity occurs. Fortunately, there are alternatives to multiple point models that are consistent with this concept.

12.4.5 Integral models

The area under the disease progress curve (AUDPC) is commonly used as an overall synthesis of polycyclic and monocyclic diseases. This area can be written as:

$$\text{AUDPC} = \int_{t_0}^{t_F} y_t \cdot dt \qquad (12.9a)$$

in which t_0 and t_F are the times of first and last disease assessments, respectively. As shown in Chapter 4 (section 4.7.2.6), the epidemic parameters $r*$ and y_0, together with the time duration of the epidemic (i.e., $t_F - t_0$), can be used to determine AUDPC. In general, increasing values of $r*$, y_0, and $t_F - t_0$ lead to increasing areas. Even when a particular disease progress model is not fitted to the epidemic data, it is easy to approximate AUDPC of equation 12.9a using equation 4.42. In fact, the most common way of obtaining AUDPC is with equation 4.42 rather than based on a fitted growth model to the disease progress curve. It is also possible to define an area under the curve for disease intensity in absolute units (Y), or for disease-free intensity (h or H); one simply substitutes the variable of interest for y in equation 12.9a (or equation 4.42). One can standardize for duration of the epidemic by dividing AUDPC by $t_F - t_0$ (to obtain AUDPC*).

Vanderplank (1963) may have been the first to suggest that AUDPC is related to crop yield. Many investigators have since found this variable to be quite useful in predicting yield or loss (as reviewed in, for example, Teng, 1985; Madden, 1983; Nutter, 2002b). Some of the graphs in Fig. 12.2 show yield in relation to AUDPC. A simple linear model is written as:

$$W = \beta_0 - \beta_1 \text{AUDPC} \qquad (12.10)$$

That is, y in equation 12.1 is replaced by AUDPC. As with single-point models, transformations of the predictor variable can be used (e.g., $\sqrt{\text{AUDPC}}$), if plots of W versus AUDPC, or the residual plot from a regression analysis (see Chapter 3), indicate that a transformation is needed. Moreover, AUPDC can be used in nonlinear models (e.g., equations 12.3 and 12.5) instead of y at a single time (Madden and Nutter, 1995). In general, the models in section 12.4 can be use to relate yield (or loss) to AUDPC. We call models such as equation 12.10 and others that use AUDPC as *integral models*, or simply AUDPC models.

A recent example of an integral model, shown with estimated parameters, is one developed for the effects of Stagonosporpora blotch on wheat yield (MT/ha) in Australia (Bhathal et al., 2003):

$$\widehat{W} = 3.6 - 4.2 \times 10^{-3}\text{AUDPC}$$

which had an R^2 of 0.83. We show the coefficient of AUDPC based on measurements of y on a proportion scale (but note that the authors used y on a percentage scale).

By using AUDPC, the entire epidemic (from t_0 to t_F) is utilized in predicting W. Because of the cumulative aspect of the area, high (or even moderate) values of y early in the epidemic result in a higher area than high values of y only late in the epidemic. For instance, the AUDPC values for the three curves in Fig. 12.7 are 38, 25, and 21, for curves a, b, and c, respectively. If equation 12.10 is appropriate, then the lowest yield would, indeed, correspond to epidemic "a".

In one sense, an integral model uses the same information to predict W (or w or L or ℓ) as does a multiple-point model. However, the integral model does this with (considerably) fewer parameters. In a statistical sense, an integral model is more parsimonious than a multiple-point one. Instead of having a parameter for the y value at each time point, there is one parameter (β_1 in equation 12.10) for all disease intensities. This is especially advantageous when one does not have a large number of data points to be used in model development. Of course, it is still quite possible that all y values should not be given equal weight in determining an integral variable. For instance, it is possible (for some diseases) that very late season severity has no (or little) effect on yield. One can calculate a new area defined as:

$$A^* = \int_{t_0}^{t_F} f(y_t) \cdot dt \qquad (12.9b)$$

where $f(y_t)$ is some function of y (Shaw and Royle, 1987; Shtienberg et al., 1990). A simple example for the function is $f(y_t) = \omega_t \cdot y_t$, in which ω_t is a weight for the given time. Unlike the situation with multiple-point models, the weights are chosen prior to modeling W in relation to A^*. Two possibilities would be $\omega_t = t$ and $\omega_t = 1/t$, to account for increasing and decreasing importance over time, respectively, of y in determining yield. From an empirical modeling perspective, one could calculate A^* with different weight functions, and determine which one provides the best predictions of W. This is analogous to evaluating y and transformations of y in single-point loss models.

As indicated above, AUDPC has been found to be very useful for characterizing the effects of plant diseases on crop yield in several studies. By generalizing the integral variable from AUDPC to A^*, one clearly increases the flexibility of linear and nonlinear crop loss models. However, the modeling approach is still empirical, and does not necessarily account for known physiological aspects of yield production by crops (see section 12.6). Nevertheless, it is important to re-emphasize the important general principle of using models with an integral-type variable to characterize the cumulative effects of disease on yield. Several later sections of the chapter will make this clear.

12.4.6 Other predictor variables in empirical models

So far we have focused on the relationship between disease intensity at one or more times, or its integration (summation) over time (e.g., AUDPC), and the resulting crop yield (or loss). Other terms, however, can be used to summarize an epidemic. One could use an estimate of the rate parameter (r_*) for an appropriate population growth model or the secondary infection rate (i.e., transmission) parameter of a more complicated compartmental model (e.g., equation 5.30), as well as time until a certain level of disease intensity is reached (t_y'). An example of the latter is the time until first symptoms are observed—the basic premise is that the longer it takes to reach the specified disease value, the higher the yield. When the model contains only one predictor variable, it may still be called a single-point model (see section 12.4.2), even if the predictor is not an estimate of disease intensity obtained at a single time. However, we mostly use this model label when the predictor variable is disease intensity.

An example model is the one developed by Bejarano-Alcázar et al. (1997) for Verticillium wilt of cotton. Number of bolls/plant, relative to the disease-free control, can be predicted with:

$$\widehat{w} = -5 + 1.4 \cdot X_1 - 1.1 \cdot X_2$$

where X_1 is the log of the time (measured as physiological time or growing degree days) to first symptoms and X_2 is the log of the rate of increase of disease intensity. The R^2 was 0.68. As expected, the parameter for time to disease was positive and the parameter for rate of increase was negative. Because the predictor variables involved logs, the estimated intercept term represents predicted relative yield when the logs of the time and rate terms are 0, well outside the range of actual predictor values. Thus, the intercept has no biological meaning here.

As demonstrated in Chapter 4, the time to reach any value of y is directly related to y_0 (initial disease intensity) and r_*. Of course, y at any time is also a function of these same parameters. Moreover, AUDPC is determined directly from y_0, r_*, and the time between the first and the final disease assessments ($t_F - t_0$). So, AUDPC captures many of the features of epidemics that are of relevance in determining yield, a likely reason behind its popularity in empirical models of crop losses.

With some pathosystems, it may still be beneficial to use some of these epidemic components individually in crop loss models. This could be the case when there is a precise (and possibly linear) relation between W and $r*$, for example, but there is a less precise and nonlinear relation between W and AUDPC. This essentially involves use of multiple regression equations, such as equation 12.8, with the different epidemic components substituted for the y variables on the right-hand side.

Multiple linear regression models also provide a way of incorporating other variables (unrelated–at least directly–to disease) of possible significance in determining yield. For instance, fertilizer concentrations and summary weather variables (e.g., rainfall) could be incorporated into models. A model could be written as:

$$W = \beta_0 - \beta_1 y + \alpha_1 X_1 + \alpha_2 X_2 + \alpha_3 X_3 + \cdots \qquad (12.11)$$

in which the Xs refer to variables other than disease intensity or AUDPC, and the αs are parameters. This approach is of benefit for data collected in surveys of commercial fields, because many factors (soil nutrition, weather) cannot be controlled or, because of lack of randomization, cannot be discounted in terms of characterizing the impact of plant diseases on yield (or loss). In addition to use of classical linear regression-type models, more elaborate univariate and multivariate models can be used, based on methods such as: path analysis (vanBruggen and Arneson, 1986), correspondence analysis (Savary and Zadoks, 1992b; Savary et al., 1995), principal components (Hau and Kranz, 1990; Shane, 1987), or discriminant analysis (Francl et al., 1987). We do not pursue this subject here, but refer readers to these cited articles as well as more general articles on analysis by Madden (1983), Madden and Nutter (1995), and Hau and Kranz (1990).

12.5 An Example Analysis

We demonstrate the fitting of empirical models to crop loss data here, based on some of the results reported in Navas-Cortés et al. (2000) for the effects of Fusarium wilt on chickpea yield. For each of three planting dates, two different cultivars ("P-2245" or "PV-61") were grown in microplots infested with three different densities of either race 0 or 5 of *Fusarium oxysporum* f.sp. *ciceris*. We consider the data for one of the growing seasons (1987/88) studied. The modeling performed by Navas-Cortés et al. (2000) mostly involved relative yield, but we consider the actual yield (g/microplot row) here for demonstration purposes. Relative yield was determined based on disease-free yields, which varied with planting date.

Disease intensity was estimated throughout the epidemics, and the Richards model (equation 4.23a) was fitted to the disease progress data. The rate of disease increase was summarized with the weighted mean rate of increase (ϖ; equation 4.50), which adjusts for different shapes of the curves. Other variables were also calculated, including AUDPC and the time to initial symptoms. The means across four replications are shown in Fig. 12.8 for the one data set considered here. There was,

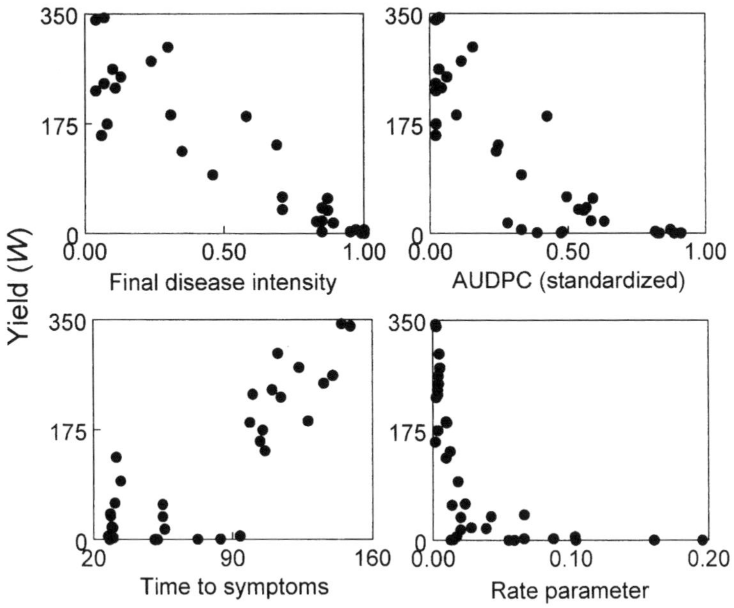

FIG. 12.8. Relationship between yield of chickpeas (g per microplot row) and final intensity of Fusarium wilt, caused by *Fusarium oxysporum* f. sp. *ciceris*, area under the disease progress curve (AUDPC), time to first symptoms, and the mean weighted rate parameter of the Richards disease progress model (see equation 4.50). Observations are from one of the datasets of Navas-Cortés et al. (2000). Note: the original paper dealt with relative yield, but we are showing absolute yield here.

as expected, a decrease in *W* with increasing final disease intensity (y_F), AUDPC, and the estimated rate parameter. Only the $W : y_F$ relationship appeared to be linear. There was also a positive relationship between *W* and time to symptoms; this is expected based on the other graphs, since there is a known negative relationship between the rate parameter of a disease progress model and time to reach a certain disease value (see Fig. 4.10). We show the analysis only for the relationship $W : y_F$. The data are shown in Fig. 12.9 in the form of a SAS program. More details on the use of SAS are found in sections 4.6 and 4.7 of Chapter 4, and also in Chapter 7. The analyses described here can be done by many statistical programs.

Values of yield extended from about 300 g/microplot row at $y \approx 0$ down to 0 at $y \approx 1$, so there was no evidence of a nonzero minimum yield (i.e., we can take $W_{min} = 0$ and $b = 1$, see equation 12.6a). Moreover, there was no evidence of *W* leveling off at a maximum value for some level of disease intensity above 0; that is, there was no (low) level of disease intensity below which there was no apparent yield response to disease intensity (as seen in Fig. 12.5), and, thus, no disease threshold for a yield response. This corresponds to $a = 0$ in equation 12.6a. So, we first fitted—using ordinary least squares regression analysis (see Chapter 3)—the statistical version of linear equation 12.1 (same as equation 12.6b with $a = 0$, $W_{min} = 0$, and $b = 1$) to the data. This is:

$$W_i = \beta_0 - \beta_1 y_{F,i} + \varepsilon_i \tag{12.12}$$

in which the subscript *i* refers to the individual observations. Without a threshold of disease, β_0 is the same as W_0 in equation 12.4 (the maximum disease). The data are graphed in Fig. 12.10, with different symbols for the cultivars and pathogen races. Estimated parameters (with standard errors in parentheses) were: $\hat{\beta}_0 = 277.8\,(14.3)$ and $\hat{\beta}_1 = 280.9\,(20.3)$. The R^2 value was 0.85. The fitted line is shown in Fig. 12.10A. (Note that the SAS program in Fig. 12.9 does not fit this simple model, but the more complicated one discussed below. Readers should refer to Fig. 4.14 for information on use of the REG procedure of SAS). For linear models, regression procedures assume that there are additions

```
data a;
        input race cult y_F W @@;
        datalines;
5       2245       0.97 5.7     0     2245     0.10    261.4
5       2245       1.00 0.8     0     2245     0.13    249.1
5       2245       0.99 0.6     0     2245     0.31    188.7
5       2245       1.00 0.0     0     2245     0.08    174.3
5       2245       1.00 0.0     0     2245     0.06    156.2
5       2245       1.00 0.0     0     2245     0.85    19.9
5       2245       0.95 2.6     0     2245     0.71    37.7
5       2245       1.00 0.0     0     2245     0.83    18.8
5       2245       1.00 0.2     0     61       0.24    274.0
5       61         0.69 141.0   0     61       0.04    339.7
5       61         0.58 186.5   0     61       0.07    343.3
5       61         0.30 296.3   0     61       0.11    231.5
5       61         0.89 16.5    0     61       0.04    226.8
5       61         0.87 36.4    0     61       0.07    238.6
5       61         0.87 55.6    0     61       0.71    57.8
5       61         0.85 2.5     0     61       0.46    92.9
5       61         0.85 40.5    0     61       0.35    130.6
5       61         1.00 5.5
;
title 'Navas-Cortes et al. Take all of wheat; 1987/88';
title2 'Yield (W) versus final disease intensity (y_F)';

proc mixed data=a;
        class race cult;
        model W = cult race y_F cult*y_F race*y_F / solution ;
        estimate 'int 2245_r5' int 1 cult 0 1 race 0 1 ;
        estimate 'int 61__r5' int 1 cult 1 0 race 0 1 ;
        estimate 'int 2245_r0' int 1 cult 0 1 race 1 0 ;
        estimate 'int 61_r0' int 1 cult 1 0 race 1 0 ;
        estimate 'slope 2245_r5'  y_F 1 cult*y_F 0 1  race*y_F 0 1 ;
        estimate 'slope 61_r5'  y_F 1 cult*y_F 1 0  race*y_F 0 1 ;
        estimate 'slope 2245_r0'  y_F 1 cult*y_F 0 1  race*y_F 1 0 ;
        estimate 'slope 61_r0'  y_F 1 cult*y_F 1 0  race*y_F 1 0 ;
run;
```

FIG. 12.9. SAS program for reading in the data for yield of chickpeas and final intensity of Fusarium wilt, and for fitting a linear model to the data, with intensity as a continuous variable and pathogen race and host crop cultivar as class (factor) variables (see equation 12.13). Data from Navas-Cortés et al. (2000).

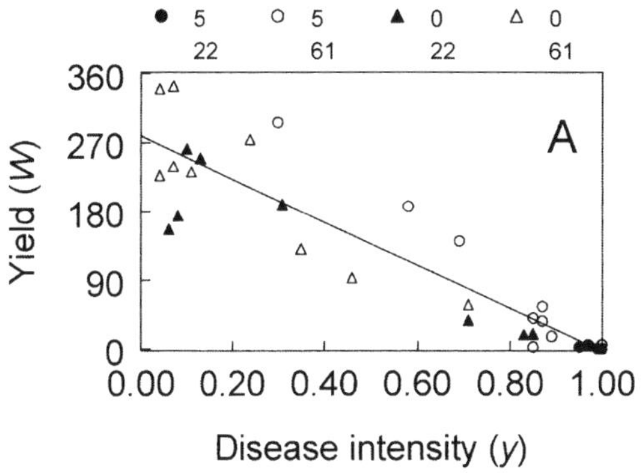

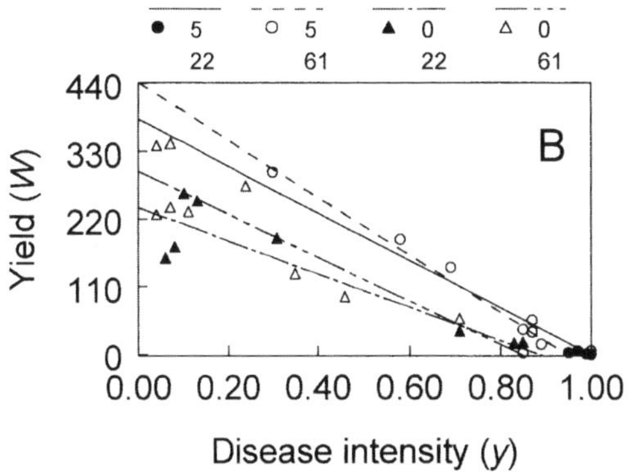

FIG. 12.10. Relationship between yield of chickpeas (g per microplot row) and final intensity of Fusarium wilt, caused by *Fusarium oxysporum* f. sp. *ciceris*. Symbols represent combination of pathogen race (0 or 5) and cultivar ("61" for "PV-61" and "22" for "P-2245"). (**A**) Fit of equation 12.1 to the data. (**B**) Fit of equation 12.13 to the data, which includes additive effects of race and cultivar on the intercept and the slope.

between terms, but we have written the linear models for yield with a subtraction between the intercept term and the disease term so that β_1 is a positive number when there is a decline in W with increase in y. Thus, the output from a regression program lists the slope as −280.9 rather than as a positive value. To be consistent with the model presentations in this chapter, one simply changes the sign.

As discussed previously, when equation 12.6b is appropriate, the slope is $W_0 - W_{min}$, or just W_0 when $W_{min} = 0$. For the model fitted to the example data, the slope and intercept were both around 280, consistent with the theory. A plot of the residuals (not shown here) indicated that a linear relationship was appropriate. However, there was some evidence of residual variability being dependent on W (or y_F), with higher variation at large W (small y_F) and low variation at small W (large y_F). We re-fit the model using weighted least squares, with weights equal to $(W + 1)^{-1/2}$ (for a strictly empirical analysis). Estimated parameters (and standard errors) were: $\hat{\beta}_0 = 262.6$ (15.3) and $\hat{\beta}_1 = -265.8$ (16.4). The R^2 value was 0.886.

The analysis so far ignored the effects of cultivar and pathogen race. As the next step, one could conduct a separate analysis for each combination of cultivar and race, and then compare the estimated parameters with t-tests, but we do not pursue this here. For one thing, for some combinations of cultivar and race (e.g., "P-2245" and race 5), there was a very limited range of W and y_F values, making it difficult to obtain reasonable estimates of the slope and intercept from the small number of data points available. Instead, we take a more formal overall approach, and conduct a covariance analysis, with y_F as the continuous variable, and cultivar (C) and race ($\mathfrak{R}$) as class or factor variables. This approach was taken in sections 4.7.1 and 4.7.2 for comparing disease progress curves. It is very important that readers understand those sections, because we do not repeat the details or background here.

In pseudo-equation form, we can write an equation for yield as:

$$\begin{aligned}\text{YIELD} = {} & \text{CONSTANT} + \text{CULTIVAR} + \text{RACE} \\ & + \text{DISEASE} + (\text{CULTIVAR} \times \text{DISEASE}) \\ & + (\text{RACE} \times \text{DISEASE}) + \text{ERROR}\end{aligned}$$

The intercept (in this case, maximum yield in the absence of disease) is given by CONSTANT + CULTIVAR + RACE, and the slope is given by DISEASE + (CULTIVAR × DISEASE) + (RACE × DISEASE), where the factor variables take on different values depending on the cultivar and race. As a covariance model, we expand equation 12.12 to write:

$$W_{ijk} = \beta_0 + C_j + \mathfrak{R}_k - \beta_1 y_{F,ijk} + \beta_{1j} y_{F,ijk} + \beta_{1k} y_{F,ijk} + \varepsilon_{ijk} \qquad (12.13)$$

in which: β_0 is a constant, C_j is the effect of the jth cultivar (j = 61, 2245) and $\mathfrak{R}_k$ is the effect of the kth race (k = 0, 5) on the intercept (W at $y_F = 0$); β_1 is (as before) a slope for y_F; and β_{1j} is the effect of the jth cultivar and β_{1k} is the effect of the kth race on the slope. The subscript i refers to the specific individual observation (different planting dates and inoculum densities, which are not being explicitly modeled here). We do not include the interactions of cultivar and race on the intercept or slope, since we felt there was not sufficient number of observations to estimate these terms precisely. Plus, this would have returned us to the problem of having a very limited range of observations for some combinations of cultivar, race, and disease intensity.

The MIXED procedure of SAS (Fig. 12.9) fits equation 12.13 to the data, and a portion of the output is shown in Fig. 12.11. The F-tests of fixed effects indicate that both factors affected the intercept (e.g., F = 6.12 for cultivar) and slope (e.g., F = 5.93 for interaction of disease and cultivar). Thus, the height of the lines and their slopes were affected by the experimental factors. The estimated parameters are shown as the solution for fixed effects. The estimated intercept and slope for

Solution for Fixed Effects

Effect	race	cult	Estimate	Standard Error	DF	t Value	Pr > |t|
Intercept			385.29	50.2986	29	7.66	<.0001
cult		61	59.1997	23.9387	29	2.47	0.0195
cult		2245	0	.	.	.	.
race	0		-145.85	47.6390	29	-3.06	0.0047
race	5		0	.	.	.	.
y_F			-386.18	53.0687	29	-7.28	<.0001
y_F*cult		61	-82.8731	34.0188	29	-2.44	0.0212
y_F*cult		2245	0	.	.	.	.
y_F*race	0		117.04	58.4111	29	2.00	0.0545
y_F*race	5		0	.	.	.	.

Type 3 Tests of Fixed Effects

Effect	Num DF	Den DF	F Value	Pr > F
cult	1	29	6.12	0.0195
race	1	29	9.37	0.0047
y_F	1	29	135.02	<.0001
y_F*cult	1	29	5.93	0.0212
y_F*race	1	29	4.01	0.0545

Estimates

Label	Estimate	Standard Error	DF	t Value	Pr > |t|
int 2245_r5	385.29	50.2986	29	7.66	<.0001
int 61__r5	444.49	46.1847	29	9.62	<.0001
int 2245_r0	239.44	19.1136	29	12.53	<.0001
int 61_r0	298.64	15.9270	29	18.75	<.0001
slope 2245_r5	-386.18	53.0687	29	-7.28	<.0001
slope 61_r5	-469.06	57.4753	29	-8.16	<.0001
slope 2245_r0	-269.15	34.7108	29	-7.75	<.0001
slope 61_r0	-352.02	35.7857	29	-9.84	<.0001

FIG. 12.11. Output of the SAS program in Fig. 12.9 for fitting equation 12.13 to the dataset of Navas-Cortés et al. (2000).

each combination of cultivar and race are determined from the displayed estimated parameters (*solution*). For instance, the estimated intercept for "PV-61" and race 0 is $\hat{\beta}_0 + \hat{C}_{61} + \hat{\Re}_0 = 385.3 + 59.2 - 145.8 = 298.7$ The program listing shows how to estimate all intercepts and slopes. As discussed in section 4.7, the parameterization utilized by SAS is to specify the last level of a factor as 0. Thus, some of the estimate statements are redundant with the solution portion of the output, although they are shown to help the reader see the process of determining the values. For example, the intercept for "P-2245" and race 5 is just $\hat{\beta}_0 = 385.3 + 0 + 0 = 385.3$ (which is displayed directly as the Intercept in Fig. 12.11). Note that the displayed slopes are negative, because of the assumed addition signs between terms in linear models.

With only two cultivars, significant F-tests for the main effect and interaction with disease intensity indicate that "PV-61" and "P-2245" differ in terms of disease-free yield and the for the rate of decline in yield with increasing disease. In particular, disease-free yield is 59.2 g/microplot row higher with "PV-61" than with the other cultivar, but the decline in yield with disease is 82.9 g (per unit disease) steeper. With more than two cultivars, contrasts would be needed (see Fig. 4.16) to determine which pairs of cultivars were different. The predicted lines for the combinations of pathogen race and host cultivar are shown in Fig. 12.10. In contrast to the situation where the experimental factors were not considered, a residual plot did reveal a need to use weights. The MIXED procedure does not calculate an R^2 value. One trick (for linear models with no random-effect terms, other than the residual) is to square the correlation between observed and predicted W values (done outside of the procedure). This produces an R^2 of 0.92.

The reader is reminded that the analysis done here is based on means across replications. With the individual replicate data, one could more fully assess the effects of the experimental factors, by fitting a model that also includes random effects appropriate for the split-plot (Schabenberger and Pierce, 2002) type of design used by Navas-Cortés et al. (2000). Additional modeling could be done for AUDPC and other predictor variables, both for yield and relative yield as the response variable. Navas-Cortés et al. (2000) should be consulted for detailed analysis of relative yield.

12.6 Mechanistic Approaches to Crop Loss Assessment

Over the last decade or so, considerable effort has been made to develop models that (1) provide a more mechanistic characterization of crop losses in relation to disease, and (2) are simple enough to be useful for description and comparison of results for different diseases and host crops. We present some of these key concepts in the following sections.

12.6.1 General considerations based on crop physiology

"Yield is the final outcome of crop growth and development processes occurring throughout the growing season" (Slafer, 2003). Photosynthetically active leaf area is a major determinant of crop yield. More than a half century ago, Watson (1947) stated that yield is directly related to the leaf area index (LAI; see Chapter 2) integrated over time during a season. In a casual way, this is the basis for the empirical loss models of section 12.4. That is, for a fixed LAI for all observations, and a necrotic disease (such as a leaf spot), a field with $y = 0.5$ for severity will have half the disease-free LAI as one with $y = 0.0$; presumably, the former field will have a lower yield than the latter, depending on the plant and pathogen species, and current environment.

If the observations (fields, plots, rows, plants, etc.) have different LAI values, then there is a difficulty with the above scenario. For instance, suppose (as an extreme example) that the field with $y = 0.5$ (e.g., 50% of foliage is covered by lesions) had a total LAI of 8 and the field with $y = 0.0$ had a total LAI of 4. Then the disease-free LAI is the same in both fields, even though there is a substantial difference in disease intensity as a proportion. The most direct way to overcome this difficulty is to express disease in absolute units. With y as the proportion of LAI that is diseased, then absolute disease severity is $Y = y \cdot \text{LAI}$. Likewise, disease-free or healthy LAI is simply $H = (1 - y) \cdot \text{LAI}$. By definition, $Y + H = \text{LAI}$ for diseases of foliage when disease is determined as a proportion of leaf area. With the idealized example here, one would expect both fields to have the same yield if the only impact of the disease is to reduce disease-free LAI, and diseased LAI makes no contribution to yield.

It is possible to measure Y directly using visual, electronic, or other methods (see Chapter 2). With a measurement or estimate of LAI, one can then obtain H from LAI$-Y$. It is more common, however, to obtain H from measurements of (total) LAI and proportion diseased. In the crop loss assessment literature, H (when determined in units of leaf area density) is often known as HLAI [i.e., healthy leaf area index; $\text{HLAI} = H = (1 - y) \cdot \text{LAI}$]. Depending on the circumstances, H can be used more broadly than HLAI; for instance, number of healthy roots, area of healthy roots, number of healthy plants, and number of healthy fruits (see Chapter 5). Note, the calculation of HLAI as $(1 - y) \cdot \text{LAI}$ is applicable only when y is *not* representing defoliation. Sometimes defoliation is included as disease severity, and sometimes it is not. HLAI can be defined in either case, but its calculation depends on how disease is estimated.

Just as y is integrated over time to determine AUDPC, HLAI can be integrated over time to obtain another type of area. One simply substitutes HLAI for $f(y)$ in equation 12.9b to obtain so-called *healthy area duration* (HAD). One could then characterize crop losses by modeling yield (or yield loss) as a function of HAD instead of as a function of AUDPC (or A^* in general) (e.g., Parker et al., 2004). Only when LAI is constant for all observations (all plots or fields) is there an exact (negative) relationship between AUDPC and HAD. As demonstrated with Fig. 12.4, the direction of the relationship between W and the predictor variable is reversed when the predictor is representing disease-free intensity instead of disease intensity. Thus, one expects a positive relationship between W and HAD if the disease of interest is reducing yield.

As an aside, we point out that we used algebraic notation in equations for most of the book, with single-character letters for variables and parameters. A major exception was the use of AUDPC for area under the disease progress curve. However, in the crop loss assessment literature dealing with physiological approaches to characterize yield, it is very common to use mnemonics such as HLAI, HAD, and other multiple-character terms to be defined below, instead of single-character symbols. We have tended to partly follow this convention in this section of the book. Thus, whereas M was used for total plant population "size" ($M = Y + H$) elsewhere in this book, here we use LAI instead of M to explicitly refer to total leaf area.

Coupled with the differences in LAI (and HLAI) between fields, is the fact that total LAI (even without disease) is not constant within a field during the growing season (e.g., Béasse et al., 2000). Since "yield is formed continually from sowing to harvest" (Slafer, 2003), one must account for the dynamics of LAI to adequately relate this variable to yield. Foliar diseases obviously affect HLAI, but some diseases (both foliar and root, as well as systemic ones) may affect total LAI through the production of new HLAI over time and the loss of diseased (or healthy) LAI through disease-caused defoliation (Johnson, 1992). In Chapter 6 we dealt with some of the issues related to disease progress when total host size (e.g., represented here by LAI) is not fixed during the course of the epidemic. Below we address the topic in terms of crop losses.

Monteith (1977) greatly refined Watson's (1947) concept of yield by indicating that dry matter or biomass production of a crop is a direct function of the radiation interception by the crop canopy integrated over time. The key parameter of this function is the efficiency (rate) at which the radiation is converted into biomass. In the Monteith approach, LAI only matters because it can be used to determine how much solar radiation is intercepted by the crop. In a highly significant paper, Waggoner and Berger (1987) applied Monteith's general approach to plant diseases, and presented ways to model crop loss within the radiation-interception framework. In a follow-up article, Johnson (1987) elaborated on the Waggoner and Berger approach and showed how different types of diseases (or pests, in general) can affect crop losses in terms of the radiation interception and use of the radiation, and also showed some of the limitations of the method. Several authors have since used this valuable approach to characterize the impact of diseases and pests on yield (see the many references in Bergamin Filho et al., 1997).

12.6.2 Radiation interception and yield

The growth and development of all plants, including the production of the harvested plant portion (i.e., yield) of crop plants, is powered by radiant energy originating from the sun. Not all wavelengths of solar radiation matter in driving plant growth (Lee, 1978; see also section 2.4.3 of this book). Radiation in the wavelengths from around 400 to 800 nm, known as photosynthetically active radiation (PAR), is most important. As a rule of thumb, PAR comprises about half of the total solar short-wave radiant energy incident at the Earth's surface. The amount of radiant energy incident per unit area per unit time is known as *irradiance* (or insolation) (R, $MJ.m^{-2}.day^{-1}$). Irradiance at the Earth's surface is affected by many factors, including time of year and time of day, latitude, topography (e.g., aspect), and atmospheric conditions. However, irradiance at the surface is generally between 50 and 80% of the irradiance at the top of the atmosphere, and varies little from day to day during a growing season (Lee, 1978).

In studies of crop growth (and crop loss) we are usually interested in the amount of incident radiation that is intercepted by a crop canopy. The calculation of daily radiation interception (RI, $MJ.m^{-2}.day^{-1}$) uses the fact that fractional radiation interception (Ψ, a dimensionless proportion) is a function of LAI (Monteith, 1977):

$$\Psi = 1 - \exp(-k \cdot \mathrm{LAI}) \qquad (12.14)$$

in which k is a positive parameter (a type of "extinction coefficient"), generally taking values from 0.3 to 1, depending on the canopy architecture. In canopies with flat, overlapping leaves, k is close to 1, and in canopies with erect (vertical) leaves, k is about 0.3. One can statistically estimate k if one has experimental data for the radiant energy flux density (obtained with a radiometer) both immediately above and under crop canopies over a range of LAIs. However, for crop loss studies, it is more common to use published values of k for the type of canopy being studied. Given a value of k and the current LAI, daily radiation interception is $\mathrm{RI} = \Psi \cdot R$. Since R is multiplied by a unitless value to obtain RI, RI is also known as an irradiance (radiant energy per unit area and time); however, it is quite common to refer to RI simply as radiation interception.

In meteorology, equation 12.14 is known as Beer's Law, but readers of this book should recognize it as being equivalent to the monomolecular equation for y (see equation 4.10, with LAI for time, and k for r_M), and as one minus the zero term of the Poisson distribution (equation 5.2b). An obvious property of Beer's Law is that the curve for Ψ is concave to the LAI axis. The fraction intercepted rises rapidly at first, but the magnitude of the increase in Ψ decreases with increasing LAI. Depending on the value of k, there is very little increase in Ψ at LAI values above ~4. Equation 12.14 is graphed in Fig. 12.12.

Assuming that crop biomass production, in general, or yield production, in particular, is a direct function of RI (Monteith, 1977), the curves shown in Fig. 12.12 have major implications for plant diseases. If the main effect of a disease is to reduce LAI (say, through defoliation), then a given disease would not have much of an impact on yield if LAI is normally high in the absence of disease. This is especially true if k is high. For instance, with k = 0.9 and an LAI of 7 without the disease, Ψ is 0.998, meaning that virtually all incident radiations are intercepted. If the disease reduced LAI by half to 3.5, Ψ would be 0.957, indicating that virtually all the incident radiation is still intercepted by the foliage. For this scenario, a discernable change in yield would only occur when LAI is reduced by much more than 50%. With a lower k, corresponding to a more erect foliage, the changes in LAI needed to have an effect on Ψ (and hence RI) would not have to be as large.

We use W_T to indicate the total biomass (as dry matter, DM) produced by a crop per unit area (e.g., $g\ DM.m^{-2}$, $kg\ DM.ha^{-1}$) by the end of the season. The moisture content is quite different in different crops (potato tubers versus

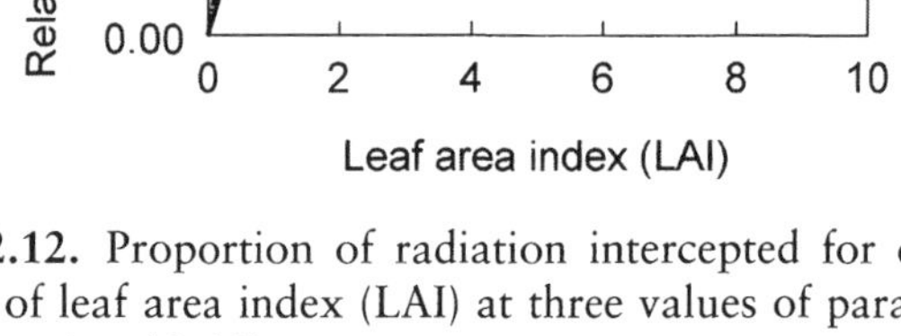

FIG. 12.12. Proportion of radiation intercepted for different values of leaf area index (LAI) at three values of parameter k (see equation 12.14).

wheat seed versus apple fruits at harvest, for instance), so conversion to dry matter is useful for comparing results from different systems. The essential concept underlying Monteith's analysis is that biomass is a function of daily radiation interception integrated (i.e., summed) over the growing season. The model can be written as:

$$W_T = \int_{t_0}^{t_F} \mathrm{RUE} \cdot \mathrm{RI} \cdot dt = \int_{t_0}^{t_F} \mathrm{RUE} \cdot [R \cdot (1 - \exp(-k \cdot \mathrm{LAI}))] \cdot dt \qquad (12.15)$$

in which RUE is a term known as the *radiation use efficiency* (g DM.MJ^{-1}). RUE may be a constant or a variable (see below), reflecting the rate at which radiation intercepted by the crop canopy is converted into crop biomass through photosynthesis. Clearly, RUE would not be the same at every day of a growing season; however, it turns out that a constant works well in equation 12.15 in predicting biomass, at least in some circumstances (notably for crops growing in good agronomic conditions). When RUE is a constant, it can be moved to the left-hand side of the integration symbol in equation 12.15. Then:

$$W_T = \mathrm{RUE} \cdot \int_{t_0}^{t_F} [R \cdot (1 - \exp(-k \cdot \mathrm{LAI}))] \cdot dt$$

The integral gives the amount of radiation intercepted by the crop canopy, accumulated over an appropriate time interval such as the growing season (units of MJ.m^{-2}), and W_T then has the appropriate units of g DM.m^{-2}. Not all crop biomass produced is converted into harvestable yield. Early in the season, photosynthate is incorporated into roots, stems, shoots, and leaves, and vegetative parts in general. Later, photosynthate is incorporated into fruit, seeds, or other harvestable products. Harvested yield biomass (W) often can be determined as a simple fraction of W_T (often called the *harvest index*).

One way to account for the impact of foliar disease on intercepted radiation is to multiply the total daily radiation interception (RI) by the proportion of disease-free leaf area; that is, one can define the daily *healthy radiation interception* (HRI) as:

$$\mathrm{HRI} = (1 - y) \cdot \mathrm{RI} = (1 - y) \cdot [R \cdot (1 - \exp(-k \cdot \mathrm{LAI}))] \qquad (12.16a)$$

where y is specifically (here) the proportion of current LAI affected by disease (not total severity, if defoliation is included for total), and R is, as before, the irradiance immediately above the crop canopy. In other words, the daily radiation interception by *all* foliage is determined (RI), and this value is then reduced as a linear function of the area of foliage that is currently disease-free. An alternative for estimation of HRI would be to modify the fractional radiation interception (Ψ) by a direct function of HLAI; that is, one could use:

$$\mathrm{HRI} = R \cdot [1 - \exp(k \cdot (1 - y) \cdot \mathrm{LAI})] \qquad (12.16b)$$

Waggoner (1990) did not believe the latter expression would properly account for the shading of lower leaves by upper ones in determining HRI. Most investigators working on crop losses in relation to plant disease simply use equation 12.16a, and we use this equation below.

Fig. 12.13 demonstrates these concepts of leaf area and intercepted radiation for fixed values of k and R, with a dynamic LAI, and logistic progress for y. For the

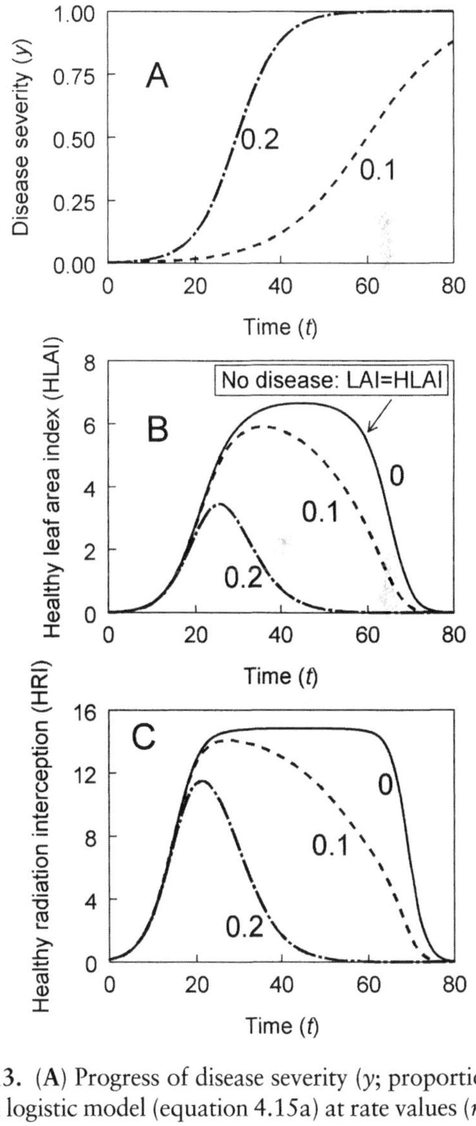

FIG. 12.13. (A) Progress of disease severity (y; proportion scale) based on logistic model (equation 4.15a) at rate values (r_L) of 0.1 and 0.2/day (indicated on the graph). (B) Healthy leaf area index [HLAI = $(1 - y) \cdot$ LAI] versus time for the two epidemics depicted in (A) and for the case without disease ("0"). LAI was generated based on a generalized logistic model consisting of a cubic function for time. In this example, disease did not affect LAI, so that y includes only diseased leaf area (so there is no disease-caused defoliation). (C) Healthy radiation intercepted [HRI; equation 12.16a, with k = 0.7 and incident radiation (R) of 15 MJ.m^{-2} day^{-1}) versus time corresponding to the curves in (B).

situation without disease, one can see that daily radiation interception (RI) stays relatively constant from about 20 to 60 days ("0" curve in Fig. 12.13C), even though LAI is changing quite a bit. This is a consequence of Beer's law when LAI is high. With disease, both LAI and RI are reduced in direct proportion to y at any time, based on the definitions of HLAI and HRI used here. The situation (and thus the curves in Fig. 12.13) can be more complicated if disease also reduces total LAI (through defoliation), which is not shown here.

The basic premise of Waggoner and Berger (1987) is that total biomass of a crop in the presence of a foliar disease can be determined by simply substituting HRI for RI in equation 12.15:

$$W_T = \int_{t_0}^{t_F} \mathrm{RUE} \cdot \mathrm{HRI} \cdot dt = \int_{t_0}^{t_F} \mathrm{RUE} \cdot (1-y)\left[R \cdot (1-\exp(-k \cdot \mathrm{LAI}))\right] \cdot dt \qquad (12.17a)$$

In other words, intercepted radiation drives photosynthesis, which in turn determines biomass production. Photosynthesis (in the model) occurs only where the disease does not occur, thus, only some of the intercepted radiation is utilized by the plant (depending on the severity of disease). With less intercepted radiation to utilize (HRI < RI, if $y > 0$), the biomass production is lower than in the absence of disease. So, one (mechanistic) way that plant diseases reduce yield is to reduce the interception of radiation available for photosynthesis, via a reduction in the area of photosynthetically functioning leaf tissue. The other way that diseases may affect yield (or plant biomass in general) is to reduce the efficiency at which HRI is converted into biomass. Many diseases (including systemic, foliar, and root diseases) affect the photosynthesic efficiency of the plant, even in parts of the plant where the pathogen is not present (see section 12.6.4). Of course, a given pathogen can affect both RI and RUE in a given host crop.

One can think of equation 12.15 as a special case of 12.17a because, when there is no disease, HRI and RI are identical. To express actual yield specifically (and not total biomass) in terms of RI and RUE, one can write:

$$W = \int_{t_0}^{t_F} \mathrm{RUE}^* \cdot \mathrm{HRI} \cdot dt = \int_{t_0}^{t_F} \mathrm{RUE}^* \cdot (1-y)[R \cdot (1-\exp(-k \cdot \mathrm{LAI}))] \cdot dt \qquad (12.17b)$$

in which RUE* is a scaled version of RUE. That is, RUE* equals RUE multiplied by a factor to reflect the fraction of total biomass that represents the harvested yield of the crop. Yield in published papers may or may not be converted to dry matter (or some other standard moisture content), and irradiance values presented may be total solar short wave radiation per unit area and time or it may be just photosynthetically active radiation (PAR) per unit area and time. Thus, readers should be careful in comparing published results among studies.

When RUE* is (or is taken to be) constant over the growing season, the term can be moved outside the integral. Equation 12.17b can be written as:

$$W = \mathrm{RUE}^* \cdot \int_{t_0}^{t_F} \mathrm{HRI} \cdot dt \qquad (12.18a)$$

The integral term was called *healthy area absorption* (HAA = $\int$ HRI · dt; $\mathrm{MJ.m^{-2}}$) by Waggoner and Berger, and this label is fairly common. Equation 12.18a can be written as:

$$W = \mathrm{RUE}^* \cdot \mathrm{HAA} \qquad (12.18b)$$

which is an equation for a straight-line relationship between W and HAA, with slope of RUE* and intercept of 0.

In principle, use of equation 12.7b, 12.18a, or 12.18b can result in a much more mechanistic characterization of crop losses than models based on disease intensity at individual times or integrated over time (AUDPC). Biomass production in general, and yield production in particular, are determined by how much radiation is intercepted by the crop and the efficiency at which the radiation is utilized. Leaf area only matters in that it affects the interception of radiation. In this context, diseases only matter in that they can:

1. Change the healthy radiation interception (HRI), which is done by lowering healthy leaf area, either by
   i. directly reducing *total* LAI (such as through defoliation), or by
   ii. "occupying" (i.e., replacing) some or all of the healthy leaf area with lesions;
2. Alter the efficiency at which the crop utilizes intercepted radiation, specifically for the production of harvestable yield.

It is possible, of course, for some diseases to affect both RUE (or RUE*) and HRI.

12.6.3 Characterizing crop losses in relation to HAA and RUE

In the last few years there has been considerable interest in using equation 12.18b for modeling crop yield (and hence loss) for various pathosystems. Bergamin Filho et al. (1997) cite many of the articles. In Fig. 12.14 there is an example taken from part of the study by Bergamin

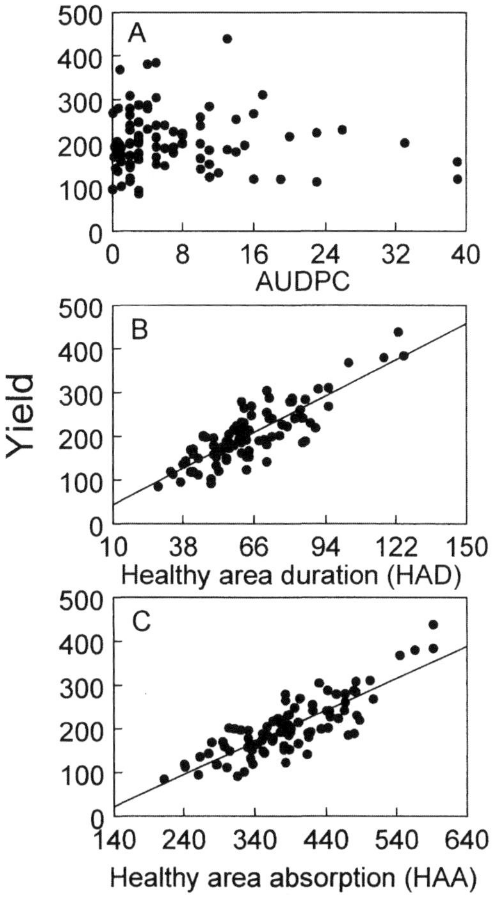

FIG. 12.14. Relationship between yield (g m^{-2}) of Phaseolus beans and: (**A**) area under the disease progress curve (AUDPC), for angular leaf spot (caused by *Phaeoisariopsis griseola*); (**B**) Healthy area duration [HAD; see equation 12.9b, with HLAI for $f(y)$]; and (**C**) Healthy area absorption [HAA; see equation 12.18b, with 12.16a for HRI]. Results from Bergamin Filho et al. (1997) for the May planting of cultivar "Carioca" in Paraná, Brazil.

Filho. Data are for angular leaf spot of *Phaseolus* beans, caused by *Phaeoisariopsis griseola*. Mean irradiance was determined to be 12.4 $MJ.m^{-2}$ day^{-1}, and leaf area at the weekly sampling times was estimated indirectly from measurements of the maximum width of the central leaflet of observed plants. For k (equation 12.14), a value of 0.7 was assumed. As shown in Fig. 12.14A, there was no discernable relationship between yield and AUDPC for this data set. Yield was expressed as weight of seeds (at 12% moisture) per unit area (g m^{-2}). This was a situation with very low mean disease severity. However, there was a clear positive relationship between yield and both HAD and HAA (Fig. 12.14B,C). Although some authors find a nonlinear relationship between yield and HAD, in this case a linear model appeared satisfactory to characterize these data. However, the relationship between yield and HAD varied among data sets for the tested bean cultivar (Bergamin Filho et al., 1997).

There was a linear relationship between yield and HAA (Fig. 12.14C). This indicates that RUE* was constant over the time period when HLAI measurements were being taken (see equations 12.18a, b), or that deviation of RUE* from a constant value was not great enough to result in any observed curvature in the relationship (given the inherent measurement and sampling variation in disease intensity, yield, and HAA). Based on linear regression analysis (equation 12.18b, but with an additional intercept term; see Chapter 3), the estimate of RUE* was 0.73 g MJ^{-1} (SE = 0.052) and the estimate of the intercept was −80 g m^{-2} (SE = 20.6). R^2 was 0.70. Note that the minimum observed value of HAA was 212 $MJ.m^{-2}$; thus, one must be very cautious in interpreting the intercept (W at HAA = 0) in a physical sense. A little algebra shows that the regression results indicate a predicted yield of 0 at −(−80)/0.73 = 109.6 $MJ.m^{-2}$, roughly half the smallest observed HAA. One would actually need to have observed HAA and yield values at and below 110 to confirm that no yield is produced at low (but nonzero) HAA. If this were found, a more general model for W in relation to HAA could be justified, possibly using equation 12.4, with S(HAA) as the switching function.

Although not shown here, the estimate of RUE* (0.73 g MJ^{-1}) was consistent for different experiments with the same cultivar. Note that the estimated RUE* is for harvested seed (not total biomass) at 12% moisture content. As shown in Bergamin Filho et al. (1997), conversion of RUE* to RUE (i.e., adjusting for moisture content and total *dry* matter production, not just the harvested portion) results in an RUE of ~1.6 g DM MJ^{-1}, typical of many crop species (Russell et al., 1989).

To see the theoretical relationships between yield and radiation interception, curves were generated based on equation 12.17a for a situation where LAI does not depend on y (although, of course, HLAI does depend directly on y). Predicted yield was plotted versus HAD (not used to generate yield) and also HAA. As shown in Fig. 12.15A, when equation 12.17a with a constant RUE* is appropriate (so that equations 12.17a and 18b are identical), a wide diversity of lines are produced when W is plotted against HAD (lines "a" to "d"), and the lines are only straight when LAI is constant (a and b). However, these lines all collapse to a single straight line when W is plotted versus HAA (Fig. 12.15). Thus, a wide range of yield responses to disease (as determined by HAD) "disappear" when yield is related directly to intercepted radiation by healthy plant tissue integrated over time (when the model is appropriate).

Curve "e" in Fig. 12.15 represents the same situation as curve "d", except that RUE* depends on the rate of disease increase (see figure legend). That is, here the value of RUE* is 0.7 when no disease is present, and decreases as the rate parameter of the disease progress model increases. This is a more complicated scenario than when RUE* takes two values (with and without disease). With a nonconstant efficiency term (across the different epidemics), or over time for each epidemic (not shown in graph), there is curvature to the W:HAA line, as expected.

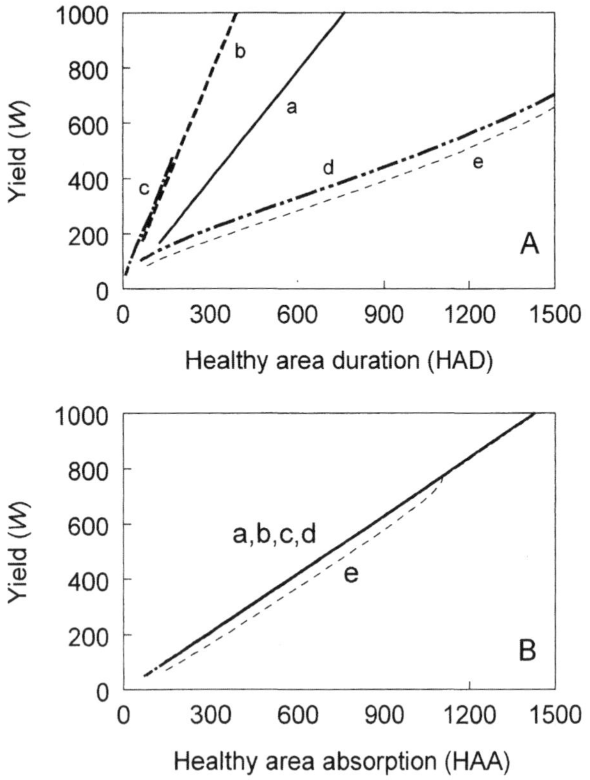

FIG. 12.15. Relationship between yield (W; g m^{-2}) and (A) healthy area duration (HAD; days), and (B) healthy area absorption (HAA; $MJ.m^{-2}$). Lines were generated using equation 12.17b, with $k = 0.9$, $R = 15$ $MJ.m^{-2}$ day^{-1}, and except for line "e", RUE* = 0.7 g MJ^{-1}. Lines correspond to a constant low ("b") or high ("a") leaf area index (LAI), or a dynamic LAI that increases to a low ("c") or high ("d") maximum, and then declines towards 0. Disease intensity (y) increased according to the logistic model (see equation 4.15a), and total LAI at a given time was not affected by y (so that there is no disease-caused defoliation). For "e", RUE* was a function of $\exp(-r_L)$; that is, the efficiency decreased with increasing rate parameter for the logistic disease progress curves. Equation 12.17b was solved numerically by the MATHCAD program.

12.6.4 Virtual lesions

The previous sections discussed a model for crop loss based on the premises that: (1) diseased leaf area (i.e., LAI−HLAI) does not contribute to yield production by a crop; (2) yield production (or biomass production in general) is driven by intercepted radiation (quantified by HRI integrated over time); and (3) the efficiency of conversion of radiation into biomass by disease-free leaf area may be affected (reduced) by the presence of the disease anywhere in the plant units (quantified by RUE). In reality, of course, there may still be some photosynthesis occurring in diseased areas, and the RUE may be a function of proximity of the healthy area relative to the diseased areas (i.e., lower RUE for the healthy parts of a leaf with some lesions, by comparison with a totally disease-free leaf).

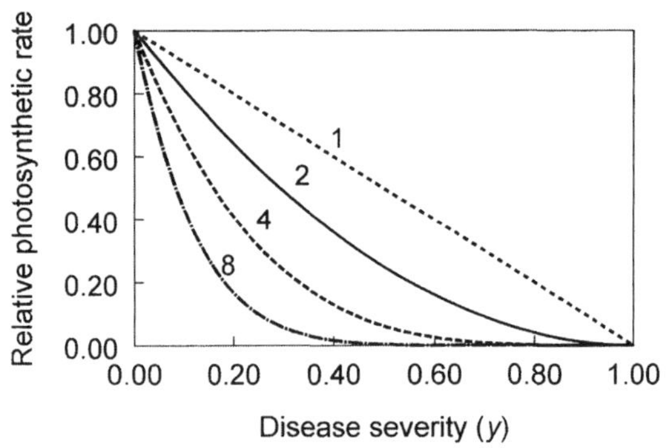

FIG. 12.16. Relative photosynthetic rate of leaf tissues (P_y/P_0) in relation to disease severity (y), based on equation 12.19. Numbers refer to exponent parameter (φ) in model.

There are alternatives to the Monteith/Waggoner/Berger approach for characterizing yield in relation to disease, and we present one approach here based on direct measurements of photosynthesis (typically under controlled-environment conditions). Bastiaans (1991) showed that a curvilinear relation is obtained when one plots photosynthetic rate of leaves versus disease severity of leaves. Fig. 12.16 is an example of the typical relations found by Bastiaans and others (e.g., Bassanezi et al., 2001; Lopes and Berger, 2001; Robert et al., 2004, 2005). Observations are obtained by electronically measuring photosynthetic rate for a large number of leaves (or other plant units) and estimating the corresponding proportion of leaf area covered by lesions. Here, the rate is standardized by dividing the photosynthetic rate for a given leaf with a given severity (P_y) by the mean rate for disease-free leaves (P_0). The curves are based on the model:

$$\frac{P_y}{P_0} = (1-y)^{\varphi} \qquad (12.19)$$

in which φ is a unitless parameter. The derivation of this model is explained in Bastiaans (1991). When $\varphi = 1$, equation 12.19 can be written as $P_y/P_0 = 1-y$, or $P_y = P_0 - P_0 y$, meaning that photosynthesis declines linearly with increasing disease severity. In other words, when $\varphi = 1$, the disease-free portion of the leaf tissue (based on symptoms) is not affected by the disease (at least in terms of photosynthesis). This is analogous to the situation discussed above when RUE is not affected by disease.

In the majority of cases studied, the standardized photosynthetic rate declines faster with increasing y than predicted by equation 12.19 with $\varphi = 1$. In particular, P_y/P_0 is smaller than $1-y$, or equivalently, P_y is smaller than $P_0 - P_0 y$. This corresponds to a value of $\varphi > 1$. For some diseases, values of φ as high as 8 or higher have been found (e.g., Bassanezi et al., 2004; Lopes and Berger, 2001). This means that the photosynthetic rate of the apparently disease-free leaf tissue is affected by the disease, and that the reduction in photosynthesis of the disease-free portion of the leaf depends on the magnitude of (estimated) disease severity. In terms of the Monteith/Waggoner/Berger approach, this corresponds to situations where diseases reduce RUE, with RUE dependent on the magnitude of disease intensity (i.e., where RUE is not constant).

Bastiaans has shown how one can view $1-(P_y/P_0)$ (with equation 12.19 for P_y/P_0) as a "virutal lesion size" or "virtual lesion area" (on a proportion scale). Virtual lesion size is greater than the observed lesion size when $\varphi > 1$. We call the virtual lesion size y_v ($y_v \geq y$). One can write this as:

$$y_v = 1-(1-y)^{\varphi} \quad (12.20)$$

The concept is that the total photosynthetic rate for a leaf can be thought of as being determined by the fraction of the total leaf that is unaffected by disease (the virtual disease-free area), and that the photosynthetic rate for this fraction is the same as what would occur if the disease was not present at all. So, instead of thinking of no photosynthesis where visible lesions are present and a reduced average level of photosynthesis per unit area in the symptomless portion of a diseased leaf, one can think of a larger-than-observed (virtual) lesion area, and a smaller-than-observed (virtual) disease-free area. Photosynthesis rate per unit area is then conceived to be zero for the entire virtual lesion area, and be unaffected (same as with a healthy leaf) by disease in the virtual disease-free area. This is a conceptual approach and is not essential to the use of equation 12.19. That is, one can interpret equation 12.19 directly as showing how relative photosynthetic rate is affected by the visibly diseased area.

The virtual lesion concept can be applied to both controlled-environment and field studies. With the supposition, discussed above, that crop yield is dependent on the rate of biomass production per unit of intercepted radiation (RUE*), which is determined, in part, by photosynthetic rate per unit of leaf area, the curves in Fig. 12.16 for $\varphi > 1$ would suggest that RUE will be affected by many diseases (see sections 12.6.2 and 12.6.3). Moreover, one would also expect that RUE is not just reduced by many diseases, but that the degree of reduction depends on disease severity. This would give a curve, rather than a straight line, when yield is plotted against HAA. Nevertheless, it is common to find that equation 12.18b is satisfactory for representing crop yield. This may be due to the "averaging" of many heterogeneous factors affecting crop yield over time and space in the epidemics, leading to equation 12.18b as a parsimonious (albeit simplified, by definition) description of crop losses.

One useful approach for incorporating the results on photosynthetic rate on crop loss is to substitute y_v for y in equation 12.17a (or simplified equation 12.18b for a fixed RUE*). In other words, $1-y$ is replaced by $(1-y)^{\varphi}$, where y is determined from visual assessments of disease severity in the field and φ is determined from separate studies (i.e., is specified before HAA is determined). Bassanezi et al. (2001) took this approach for three different pathosystems, and found that the fit of equation 12.18b to yield data when using $(1-y)^{\varphi}$ was as least as good as when using $(1-y)$.

More studies clearly are needed to determine how well this approach works for other crops in other environments. It should be pointed out that Robert et al. (2005) have recently shown in a very thorough study that virtual-lesion results can be strongly influenced by the disease-assessment approach. In particular, the relationship between photosynthetic rate and leaf rust of wheat (in controlled experiments) depended on whether disease severity was measured as just sporulating lesion area, sporulating + necrotic lesion area, or sporulating + chlorotic + necrotic lesion area (total diseased area), with the later resulting in φ values close to 1 and the other measurements giving much larger values of φ. Care must be taken, therefore, in comparing results (especially the estimates of φ) among studies with the same pathogen and host, if different investigators are measuring different components of total visible disease severity.

12.6.5 Type I and Type II curves

The HLAI/HRI/RUE approach to crop loss characterization definitely leads to some valuable insights in the relationships between crop yield and disease intensity. As mentioned above, for crops with high LAI and a disease that impacts yield primarily (only?) through reductions in LAI, increasing disease intensity will only have an appreciable affect on yield when intensity is high. This is because small-to-moderate changes in LAI will have little impact on fractional radiation interception when LAI is high (see Fig. 12.12). Only at high intensities will the reduction in LAI be large enough for HRI to be markedly affected. In principle, this may result in the so-called Type II curve shown in Fig. 12.4 when yield is plotted versus disease intensity (or AUDPC). Here, disease intensity includes both lesions (or other symptoms) and defoliation.

As discussed in section 12.6.4, diseases often affect photosynthetic rate and the resulting RUE of the apparently disease-free leaf tissue. For diseases that impact yield primarily through reductions in RUE, the impact of disease on yield is apparent even at small intensity. In principle, this may result in the so-called Type I curve shown in Fig. 12.4 when yield is plotted versus disease intensity (or AUDPC). Of course, it can be expected that diseases will affect both HRI and RUE, resulting in intermediate types of curves.

The labels (Type I and Type II) for the shapes of crop loss curves are originally due to Mumford and Norton (1987). Hughes (1988), Ferrandino (1989), and Johnson (1987) explore crop loss relationships in the context of these two fundamental relations in Fig. 12.4. Although the two curve types can be motivated by HLAI/HRI/RUE concepts, the curves can also be hypothesized based on ecological arguments (Hughes, 1988; Hughes et al., 1989). If there is compensation for the removal of LAI by diseases, such as by the increased RUE in plants (or plant parts) not infected, a Type II curve may result. In contrast, if there is substantial intraspecific competition within the pathogen population (so that the impact of individual infections decreases as the number of infections increases), a Type I curve may result. Some ecological factors, such as spatial patterns of disease (see Chapter 9), can also alter the shape of these curve types, as explained below in section 12.7.

Despite the arguments made here about the relationship between crop-loss curve type and underlying mechanism driving crop losses, one should be very cautious in (over-) interpreting observed relationships between yield and disease intensity (or AUDPC, or other predictor variables derived from epidemics). First of all, a disease may reduce yield by lowering RUE and have absolutely no affect on LAI (and hence HRI) (see discussion in Johnson, 1987). In this situation, there is *no* range in disease intensities (as characterized in terms of visible symptoms such as lesions), and hence no meaning to a plot of W versus y (or Y or HLAI). Furthermore, the linkage between HLAI/HRI/RUE and the Type I/II curves works better for short-term and recognizable damage to crops (such as mechanical defoliation) than for cumulative processes operating over the length of a growing season. This can be seen with a hypothetical example depicted in Fig. 12.17. The true relationship between W and disease intensity at a particular time (y_t), or AUDPC, is a straight line in each of the frames of Fig. 12.17; the labels indicate the true predictor variable for each curve. Consider the graph in Fig. 12.17C. The line marked "50"corresponds to the yield model:

$$W = \beta_0 - \beta_1 y_{50}$$

For the lines (curves) marked "30", "70", and "AU" in Fig. 12.17C, the model was of the same form, but the predictor variable was y_{30}, y_{70}, and AUDPC, respectively (although yield was plotted versus y_{50}). In all situations, the y values on the horizontal axis were generated using the logistic model with different rates (i.e., the different values of the predictor variable were obtained from epidemics with the same y_0 and K values, but different r_L values).

In Fig. 12.17C, W predicted by the four models is plotted versus y_{50}. A straight line is found, as expected, only for the case where W truly is a linear function of y_{50}. If the true relationship was a straight line between W and y_{30}, an approximate Type II curve would result when W is plotted versus y_{50} (see "30" in Fig. 12.17C). If the true relationship was a straight line between W and y_{70}, an approximate Type I curve would result when W is plotted versus y_{50} (see "70" in Fig. 12.17C). A more complicated curve would result when the true relationship was a straight line between W and AUDPC. The other frames of Fig. 12.17 show the same kind of result when the chosen variable for the horizontal axis does not correspond to the actual predictor variable generating the response values. The converse can also be

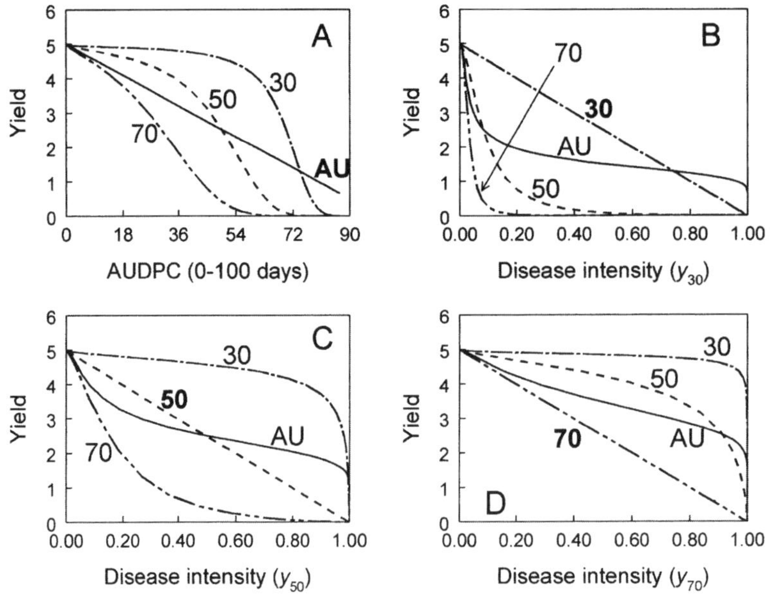

FIG. 12.17. Theoretical relationships between yield (W) and either disease intensity at a particular time (y_t) or area under the disease progress curve (AUDPC). The true relationship is a linear (straight-line) relation between W and: disease intensity at one of the three labeled times (y at t = 30, 50, or 70) or AUDPC. In each frame, the straight line is obtained when W is plotted versus the *correct* (or true) predictor variable. The other curves correspond to plotting W versus an *"incorrect"* predictor. In all cases, epidemics were generated with a logistic model (equation 4.15a) with different r_L values (with fixed y_0 and K = 1).

expected to be true: if the true relationship was a Type I (or II) curve at a particular time, an approximate straightline relationship may be found when W is plotted versus a predictor variable that did not generate the yield values (i.e., y at a different time).

The problem is that an investigator would not know the true predictor variable ahead of time, and may plot W versus y at different times, or versus AUDPC, and develop empirical models for the relations. This could result in incorrect interpretation of the underlying process [e.g., giving a mechanistic interpretation to a type I curve when, in reality, the true relationship was a straight line (but for a different time)]. As shown here, the shape of the curve can vary considerably over the course of the epidemic, even if the underlying process is not changing. Moreover, the investigator may not have measurements of disease intensity at the "correct" time(s).

The reason for the results in Fig. 12.17 is the nonlinear nature of disease progress curves. If one plotted y_{30} versus y_{70} for different epidemics (corresponding to different r_L values of the logistic model, all with the same y_0 and K), for example, one would obtain a curve and not a straight line. The reader can easily do this with equation 4.15a at a fixed value of y_0 (e.g., 0.005). For r_L from 0.01 up to 0.4/day, one can obtain two columns of y data for plotting, one for $t = 30$ and one for $t = 70$. As can be seen in the different frames of Fig. 12.17, a wide range of curve shapes for yield versus disease can be obtained even when the true relationship was a straight line (i.e., generated by model that was linear in the parameters and variables). Of course, epidemics are never exactly logistic, there can be (considerable) heterogeneity in disease intensity and yield, and we do not expect W to be a simple straight-line function of y or AUDPC, based on the arguments made in the previous sections. Thus, there is typically considerable variation in W in relation to y (or to AUDPC), and unambiguous functional relations may not be found. The point of the graphs in Fig. 12.17 is to show that considering only temporal slices of epidemics may result in misleading interpretations, because of the complex dynamics of disease progress over time. As we pointed out in the previous sections, plots of W versus AUDPC can be misleading if the true relation involves HAD (and LAI is not fixed). Nevertheless, if one's goal is to simply predict or describe crop loss, enabling the comparison of treatments (cultivars, control methods, and so on), then highly precise results may (but not necessarily will) be obtained using disease intensity at particular times.

12.6.6 Time of infection

In the above presentation, the crop could be considered analogous to a factory, with sunlight providing the energy used to drive the factory. Mathematically, we could consider the crop "size" (e.g., LAI, HLAI) as a continuous random variable, converted to HRI with a nonlinear function (using Beer's law), which when multiplied by a scaling factor (RUE) and integrated over time, resulted in a prediction of yield (equation 12.17b). Here, an alternative approach is presented that may be appropriate for systemic diseases of plants, such as those caused by viruses, phytoplasmas, spiroplasmas, and some soil-borne pathogens. For these types of diseases, time of infection is often a good determinant of final yield (or yield loss). For instance, with controlled inoculations of plants, the impact of infection on yield often declines with increasing time during the season when plants are inoculated (Fargette and Vié, 1995; Madden and Nutter, 1995; Madden et al., 2000a). Plants infected early in the season may have a very low yield (perhaps 0), but plants infected late may have the same (or nearly the same) yield as plants not infected at all. Theoretical curves for w and ℓ versus time of infection are shown in Fig. 12.18, based on results in Madden et al. (2000a).

In terms of the HLAI/HRI/RUE approach, one can think of RUE as being a function of time of infection—the later

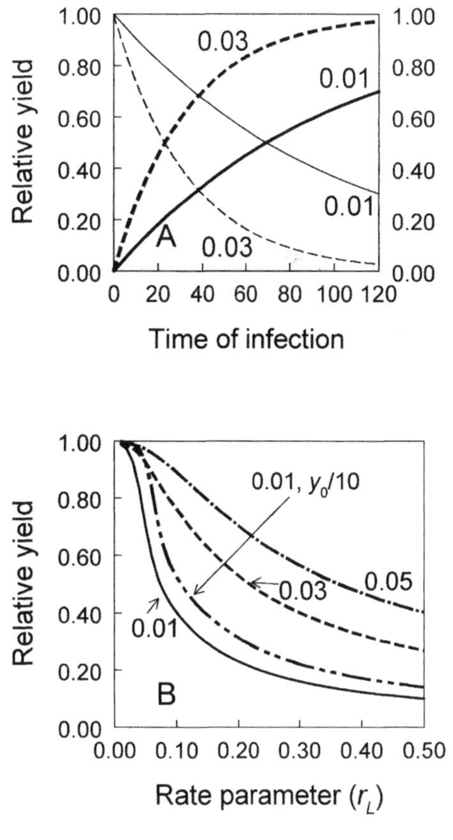

FIG. 12.18. (A) Relative yield ($w = W/W_0$; left-hand axis and thin curves) and relative loss (ℓ; right-hand axis and thick curves) versus time of infection (t). Based on equation 12.22c, with $w_{min} = 0$ and two values of γ displayed on the curves. (B) relative yield for a population based on equation 12.23, with logistic disease increase and fixed y_0 (see equation 4.15a), with three values of γ indicated on the curves (0.01, 0.03, and 0.05). The case with $\gamma = 0.01$ and a 90% reduction in y_0 ("0.01, y_0/10") is also shown.

the infection, the less that RUE is affected by disease. It is straightforward to model crop yield in terms of time of infection when investigators control the infection time through inoculations, so that all plants become infected at the same time in experimental units (e.g., plots). Simple exponential functions, such as used in Fig. 12.18, are effective for this. However, plants do not all become infected at the same time in a field during a natural epidemic (see Chapters 4–5). Indeed, the time of infection of individual plants is seldom known; rather, the total number or proportion of plants diseased at selected times is often recorded. Readers of earlier chapters know how to characterize epidemics with fairly simple (e.g., the logistic) or more complicated (e.g., the coupled differential equation of Chapter 5) models. To complete the model for crop loss of a population, we require the proportion of *new* infections over a small time interval at each time during the epidemic, given by $(dy/dt) \cdot dt$.

Madden et al. (2000a) and Madden and Nutter (1995) showed how to link models for disease progress and yield-response to time of infection to characterize crop loss for systemic diseases of the type discussed here. For model development, we can start with a version of equation 12.4 for yield of individual plants (or groups of plants all infected at the same time), but expressed in terms of time of infection rather than in terms of y.

$$W = W_{\min} + (W_0 - W_{\min}) \cdot S(t) \qquad (12.21)$$

Here, the switching function, $S(t)$, ranges from 0 at time 0 (i.e., when plants are infected at $t = 0$) to 1 at large t (when plants become infected late in the epidemic). In other words, $S(t)$ is now a switch-on function since W increases as time of infection increases. For convenience, we use the cumulative exponential distribution function, $S(t) = 1 - \exp(-\gamma t)$, where γ is a parameter with units of 1/time. By dealing with disease incidence, we are not (too) concerned with changing host size during the epidemic, so it is easier to deal with proportions diseased and fractional yield ($w = W/W_0$) and yield loss ($\ell = 1 - w$). Dividing equation 12.21 by W_0, and then subtracting from 1 (to obtain relative loss), one obtains (after some algebra) the expression:

$$\ell(t) = (1 - w_{\min}) \cdot (1 - S(t)) \qquad (12.22a)$$

in which $w_{\min}$ is $W_{\min}/W_0$ (relative minimum yield). We added an explicit t to the loss function to make it clear that it is defined in this section in terms of time and not disease intensity. For the chosen expression for $S(t)$, this can be written as:

$$\ell(t) = (1 - w_{\min}) \cdot \exp(-\gamma t) \qquad (12.22b)$$

This was called the "zeta" function [$\zeta(t)$] by Madden et al. (2000a), with α used for $1 - w_{\min}$, the maximum possible loss (on a relative scale). Relative yield for infection at time t is simply written as:

$$w(t) = 1 - (1 - w_{\min}) \cdot \exp(-\gamma t) \qquad (12.22c)$$

The values of $w_{\min}$ and γ indicate how the crop tolerates infection. Loss is high (or yield is low; see Fig. 12.18) when γ is low (near 0), so that infections late in an epidemic have (nearly) the same effect as infections early. A small $w_{\min}$ indicates that loss is high (or yield is low) when plants are infected early.

To determine the relative yield for a crop with an epidemic in which plants become infected at different times (starting at $t = 0$), one multiplies equation 12.22a by the number of new infections over short time spans at each time [$(dy/dt)dt$ on a proportion scale]. Then, integrating over time and subtracting the result from 1 gives:

$$w_C = 1 - \left[\int_0^{t_F} \left(\frac{dy}{dt} \right) \cdot \ell(t) \cdot dt + \vartheta_0 \right] \qquad (12.23)$$

in which ϑ_0 is a constant that specifies the loss occurring from the initial infection event [at $t = 0$; ϑ_0 equals $1 - w_{\min}$ times initial y (y_0)]. If there is no increase in y over time ($dy/dt = 0$), then $w_C = 1 - \vartheta_0$. Note that we place a C subscript on w here to indicate the relative yield for the entire crop, when plants are (potentially) infected at different times. Instead of writing dy/dt as a function of y, it is more useful in this context to write the rate as a function of t. For instance, dy/dt for the simple logistic model is obtained by taking the derivative of equation 4.15b with respect to t; this is written as:

$$\frac{dy}{dt} = \frac{r_L \left(\dfrac{1 - y_0}{y_0} \right) \exp(-r_L t)}{\left(1 + \left(\dfrac{1 - y_0}{y_0} \right) \exp(-r_L t) \right)^2}$$

Interestingly, there may not be an analytical solution to equation 12.23 for many simple disease progress models. Fortunately, with computer programs like MATHCAD or MATHEMATICA, it is straight-forward to solve this equation numerically. Fargette and Vié (1995) used a version of this approach in a more complicated simulation model to explore the impact of sanitation and resistance on epidemics of cassava mosaic virus in the multiple-year cassava cropping cycle.

The lower frame of Fig. 12.18 shows the crop yield (on a relative scale) obtained from epidemics with logistic disease progress, in which each epidemic started with $y_0 = 0.005$ (0.5% infected at time 0). One can see the impact of different rates of disease increase and three different "tolerance" values (γ) on final yield. At very low r_L, meaning that only a small proportion of plants become infected, relative yield is high for all values of γ. At larger epidemic rates, however, the effects of γ become quite obvious, with large reductions in yield when γ is close to 0. Among other things, one can use graphs such as this one to develop strategies for control—is it more effective in a given system, for instance, to reduce rate of disease increase or to select (or breed for) cultivars that are more tolerant of disease (measured here by γ)?

The curve marked "0.01, $y_0/10$" is based on the same condition as the one marked "0.01", except there is a 10-fold reduction in initial incidence for all epidemics. One can see the impact of controlling the initial disease level for polycyclic diseases on crop yield is minor compared with changing the rate of increase or γ. This is consistent with the results in Chapter 4 on the relative importance of r_L and y_0 for compound interest diseases.

It is quite possible to use other functions for $S(t)$ (and hence ℓ) and dy/dt in equation 12.23, and to expand the model to accommodate the more complicated linked differential equation models for disease dynamics (see Chapter 5). In such cases, results almost always will be based on numerical solutions rather than analytical ones. Fitting equation 12.23 to field data is challenging because the model must be numerically solved (in the general case) for any set of parameters. Madden et al. (2000a) developed a protocol for iteratively estimating parameters and found that the model provided reasonable fits to observed crop loss data for soybean mosaic virus.

12.6.7 Discussion

Crop growth and development, and the production of biomass that is harvested as yield, are complicated phenomena. We have shown that some basic considerations of crop physiology can lead to an interpretable representation of crop yield in relation to plant disease epidemics. In particular, the HLAI/HRI/RUE model provides a useful foundation for determining the effects of plant disease on crop yield. With this approach, diseases can be seen to affect intercepted radiation by the healthy leaf area [i.e., through a reduction in LAI and/or increase in proportion of the leaf area covered by lesions], or the utilization of the intercepted radiation [through alteration of physiological processes involved in photosynthesis and the transport of photosynthates (RUE)]. Some diseases will likely affect both HRI (a nonlinear function of LAI and disease intensity; equation 12.17b) and RUE. This approach to modeling is especially useful when total LAI changes considerably during the epidemics and when disease affects the total LAI (e.g., through defoliation), and not just the proportion diseased.

The integration of disease intensity over time, in the form of AUDPC, is a popular and useful variable when characterizing crop losses. The HLAI/HRI/RUE approach builds on the AUDPC model by integrating RUE $\times$ HRI over time. When RUE* is constant, the predictor variable HAA can be constructed for use with a crop loss model (see equation 12.18b) with a parameter equal to RUE*. Although this conceptualization is useful for many types of plant diseases, there also are specialized approaches of value for some types of diseases. In particular, for systemic diseases, the time of plant infection may be the major determinant of yield, partly because this determines the RUE for the plant from this point on.

In standard use of the RUE:HRI approach in plant pathology, it is typically assumed that RUE (or RUE*) is constant, so that the variable HAA can be constructed and used in a linear model for yield (with slope of RUE*). Clearly, the efficiency at which radiant energy is converted into biomass and ultimately into the harvested portion of the crop is not constant throughout an entire season, and only a portion of the total growing season may be relevant for yield formation (Johnson, 1987, 1992). Moreover, all portions of the plant, or even all portions of the disease-free leaf area do not contribute equally to formation of yield (Batchelor et al., 1993; Béasse et al., 2000). For example, the top (flag) leaf of a wheat plant is especially important in the yield of the crop (James and Teng, 1979). The environment certainly varies over time and over space in a field (e.g., because of fertility and soil-moisture gradients), likely affecting yield. At best, one can think of RUE* as roughly constant, at least for the purpose of analysis. This is no different from considering a fixed r_L for logistic growth or fixed βH_0 in the coupled differential-equation model for epidemics. These population-dynamic values are not exactly unvarying, but use of constants often works well for a parsimonious description of epidemics. Likewise, a constant RUE* does seem to work well in some systems. However, there is empirical evidence that a constant RUE* is not always appropriate (e.g., Bergamin Filho et al., 1997). Moreover, published results on relative photosynthetic rate (and virtual lesion sizes) would suggest that diseases not only affect RUE* (assuming that photosynthetic rate determines, at least in part, radiation use efficiency), but that the magnitude of RUE* may depend on the magnitude of y (or Y)—in other words, that RUE* is a variable.

There are various mechanistic models for crop growth and yield production (e.g., Batchelor et al., 1993; Johnson and Teng, 1990; Termorshuizen and Rouse, 1993). These are usually of the simulation type, with numerous variables and parameters, typically constructed based on a reductionist philosophy of breaking down the system into many components and subcomponents. These models can be quite useful when one is interested in the underlying processes controlling yield. We do not discuss these here because we are much more concerned about the population dynamics of diseases. One very important concept has emerged from the detailed modeling and analysis of crop physiology that is of relevance. In particular, Boote et al. (1983) have concluded that pests in general, and pathogens in particular, affect crop growth and development, and yield formation, in seven possible ways. Pests and pathogens act as:

1. Tissue consumers;
2. Leaf senescence accelerators;
3. Stand reducers;
4. Light stealers;
5. Photosynthetic rate reducers;
6. Assimilate sappers; and
7. Turgor reducers.

Of course, a given pathogen may cause detrimental effects through several of these actions. The value of this classification is that physiological models can be used to account for these processes within crop simulators. Interestingly, as pointed out by Johnson (1987), the first four actions primarily affect LAI (or HLAI), and the last three primarily affect RUE. Thus, the approach in the previous sections is fully consistent with a more reductionist (and detailed) consideration of crop loss.

Johnson (1992) further extended the linkage between the HLAI/HRI/RUE and the more detailed physiological considerations by developing and testing of a multiple-pest computer simulator based directly on the principles of radiation interception and use for potatoes. To quote from Johnson's article:

> "A principal advantage of the RI/RUE approach is that physiologic detail can be minimized while at the same time maintaining mechanistic integrity. This advantage also allows for closer ties to theoretical approaches [population dynamics of disease] and for more effort to be directed towards epidemiologic consideration of integrating and understanding disease and insect effects on crop growth."

12.7 Spatial Heterogeneity

12.7.1 General concepts

So far, we have virtually ignored the effects of spatial heterogeneity of disease on crop loss in this chapter. As mentioned in Chapters 9 and 10, plant disease intensity is usually heterogeneous in fields or other areas of interest, and this spatial heterogeneity can have a large influence on the precision of estimated parameters (such as means) and on predictions of responses. We do not readdress the precision issue here, other than to remind the reader that the more variation of the random variable (disease or yield, for instance), the less precise is the estimate of the mean (or other population parameter). In this section we discuss another important effect of heterogeneity of disease on crop loss results.

A diversity of crop loss relations, in the form of yield versus a measure of the epidemic (such as y at a single time or AUDPC), is shown in Fig. 12.2. The individual data points in graphs of this type can consist of the yield and disease (or disease free) value for individual plants; however, usually some averaging is first performed. The average yield and disease severity per row or per plot, or some other group of individual plants, is usually used to determine the relation, if any, between yield and disease intensity. Averaging could even be carried out over larger scales, such as would be done to determine the mean yield per state or country. Even when the data points correspond to individual plants, some (implicit) averaging or totaling likely occurs. For instance, the yield and disease severity of an individual grape vine actually consist of the total weight of harvested berries from all shoots of that vine, and the average disease severity across shoots for that vine, respectively.

At first it might seem that the only effect of averaging is to reduce the variation around the line (curve) for the yield-disease relationship. As we show in the next section, however, the averaging may actually change the shape of the curve.

12.7.2 Models

For ease of presentation, we assume that W (the response variable) is affected by disease intensity at a single time (y; the predictor variable), or that W can be predicted by just one predictor variable, y. All the points discussed here apply to any predictor variable, including AUDPC, r_L, time to symptoms, HAD, and HAA. In general notation, we can write $W = g(y)$ for the functional relationship between W and y. For reasons that will become soon apparent, we expand the model notation of section 12.4 and write this as:

$$W(y) = g(y) \qquad (12.24)$$

to make it clear that W is a function of y. As an example, we assume that $g(y)$ is given by the right-hand side of equation 12.3, that is,

$$g(y) = W_0 \left(\frac{1-y}{\left((1-\Theta)+\Theta(1-y)\right)^2} \right) \qquad (12.25)$$

Any expression could be used for $g(y)$ that characterizes the change in W with change in y. In terms of the general formulation of equation 12.4, the choice here of $g(y)$ means that $W_{min} = 0$, $b = 1$, $a = 0$, and the switching function is:

$$S(y) = [(1-y)\cdot((1-\Theta)+\Theta(1-y))^{-2}].$$

Suppose now that this expression for $g(y)$ is the fundamental one at the lowest hierarchical scale (e.g., that of the individual plant) of interest, but that we want to express the relationship at a higher scale (e.g., that of the plot or field). We add an i subscript to indicate specific values of disease and yield (over which averaging occurs), and write equation 12.24 as:

$$W(y_i) = g(y_i)$$

More specifically for this demonstration, we assume that y and W take on only a finite set of values (i.e., that the variables are discrete), and i is the label for the different values (not the different sampling units). Mean values of y and W are given by:

$$\bar{y} = \sum_i y_i \cdot \Pr(y_i) \qquad (12.26a)$$

$$\bar{W}(y) = \sum_i W(y_i) \cdot \Pr(y_i) \qquad (12.26b)$$

where $\Pr(y_i)$ is the probability of the specific y_i. With data in the form of a sample of y_i values from a population of interest, these probabilities are estimated by the fraction of y_i values with the particular (discrete) value of disease intensity i (for each value of i). These equations for means are equivalent to equation 9.1, but with the data in categories or classes. Chapter 9 includes details of how to calculate these probabilities for some common statistical distributions used in plant disease epidemiology. If equation 12.25 is appropriate, one might naturally then want to see how actual mean yield $[\bar{W}(y)]$ is related to the yield that would be predicted by using $\bar{y}$ in equation 12.24 $[g(\bar{y})]$. One might initially think that $\bar{W}(y) = g(\bar{y})$. However, this would not necessarily be correct, as shown by Hughes (1988) and Hughes et al. (1989).

Consider the situation where $W_0 = 5$ (in appropriate units, such as g/m^2) and $\Theta = 0.5$ in equation 12.25. Suppose we have the following six pairs of observations $[y, W(y)]$: (0.2, 4.94) (0.2, 4.94), (0.4, 4.69) (0.4, 4.69), (0.6, 4.08), (0.6, 4.08), where the yield values come directly from equation 12.25. Since there are three unique y values, in equal fraction, $\Pr(y_i) = 1/3$ for all classes. The calculated averages, based on equations 12.26a, b, are:

$$\bar{y} = (1/3) \cdot 0.2 + (1/3) \cdot 0.4 + (1/3) \cdot 0.6 = 0.4$$

and

$$\bar{W}(y) = (1/3) \cdot 4.94 + (1/3) \cdot 4.69 + (1/3) \cdot 4.08 = 4.57.$$

However, the predicted W based on use of $\bar{y}$ in equation 12.25, $g(\bar{y})$, is 4.687. To distinguish this value from the actual mean of the yields, we write $W(\bar{y}) = g(\bar{y}) = 4.687$. The values 4.57 and 4.687 are similar, but not identical. The difference between the results of the two calculations of W arises because the chosen crop loss function is not a straight-line function (it is nonlinear in the variables and parameters; see Chapter 3). In mathematical terms (noting that the mean is an expectation), a function of an expectation of a variable $[W(\bar{y})]$ is not equal to the expectation of the function of a variable [i.e., $\bar{W}(y)$], unless the function in question is linear. The relationship between these ways of considering yield $[\bar{W}(y)$ versus $W(\bar{y})]$ can be understood by considering a mathematical approximation for a function, as shown in the next section.

12.7.3 An approximation (but a good one)

A function of a variable $[g(y)]$ can be written as a Taylor series expansion of the function around a particular value (μ). The expansion is written as:

$$g(y) = g(\mu) + (y - \mu) \cdot g'(\mu) + [(y - \mu)^2 / 2] \cdot g''(\mu) + \cdots \qquad (12.27)$$

where g' and g'' represent the first and second derivatives of the $g(\bullet)$ function, evaluated at $y = \mu$ [e.g., $g'(\mu) = dg(\mu)/dy$]. Sometimes μ is called the solution locus. First derivatives were used frequently throughout the book, such as in the form of dy/dt in Chapter 4 and dy/ds in Chapter 8. The second derivative represents the change in the first derivative with unit increase in the predictor variable (y in this case). We take μ to be the mean of the population of y values. The series increases to indefinitely high order (powers), but typical usage involves using only the first three terms, as shown here. Thus, a Taylor series expansion is an approximation for a function, and the accuracy of the approximation depends on the form of the function and the value μ being used. Taylor series expansions are of great value in many mathematical and statistical applications. Ferrandino (1989) utilized this approach in crop loss modeling. Allen (1988) used this methodology for modeling biological responses to a variable environment.

One can use the rules of expectations (Stuart and Ord, 1994) and take expectations of both sides of equation 12.27:

$$E[g(y)] \cong g(\mu) + E[(y - \mu)] \cdot g'(\mu) + E\left\{\frac{(y - \mu)^2}{2}\right\} \cdot g''(\mu) \qquad (12.28a)$$

We note that the expectation of a constant (such as μ) is just the constant, and the expectation of a function of a constant (not a variable) is just the function of the constant. It can be shown that $E[(y - \mu)] = 0$ when μ is the mean of a variable; that is, the average difference between a random variable and its mean is always 0. Furthermore, $E[(y - \mu)^2]$ is, by definition, the variance of y; that is, $E[(y - \mu)^2] = \sigma^2$. Thus, one can write equation 12.28a as:

$$E[g(y)] \cong g(\mu) + \frac{\sigma^2 \cdot g''(\mu)}{2} \qquad (12.28b)$$

In practice, the parameters μ and σ^2 are unknown and one uses their estimates $\bar{y}$ and s^2, respectively, in the equation. The mean of the $g(y)$ values, $\overline{g}(y)$, used above, is equivalent to $E[g(y)]$. So, we can now write:

$$\overline{g}(y) \cong g(\overline{y}) + \frac{s^2 \cdot g''(\overline{y})}{2} \tag{12.28c}$$

This can be put in terms of crop yield by noting from section 12.7.2 that $W(\bullet) = g(\bullet)$. Thus, equation 12.26b for mean or expected yield can be *approximated* as:

$$\overline{W}(y) \cong W(\overline{y}) + \frac{s^2 \cdot W''(\overline{y})}{2} \tag{12.29}$$

in which $W''(\overline{y})$ is the second derivative. Based on this equation, mean or expected yield (say, for a crop row, plot, field, etc.) equals the predicted (or estimated) yield at the mean value of y ($\overline{y}$), plus the product of the variance of y (not the variance of W) divided by 2 and the second derivative of the crop loss function.

If $W(y)$ is given by equation 12.25, then one finds that the second derivative is:

$$W''(y) = W_0\left(-2\Theta\frac{2+\Theta y - 3\Theta}{(\Theta y - 1)^4}\right)$$

Of course, every different function for W will have a different second derivative. For the six y values in the example, $s^2 = 0.032$ (which the reader can confirm on their own). For reasons we do not go into here, it may be more appropriate to use the maximum likelihood estimate of the variance rather than the moment (or least squares) estimate shown here. The maximum likelihood estimate is given by $(N - 1)/N$ times the moment estimate; obviously, this will only have a noticeable effect at small N. The maximum likelihood estimate is $(5/6)\cdot 0.032 = 0.0267$. The estimated second derivative at $y = \overline{y}$ (when $\Theta = 0.5$), $W''(\overline{y})$, is equal to -8.545. Using equation 12.29, we estimate the mean yield as:

$$\overline{W}(y) = 4.687 + \frac{0.0267\cdot(-8.545)}{2} = 4.57$$

which is the same as the directly calculated mean yield.

In the example, the mean yield is less than that predicted by use of $\overline{y}$. Mathematically, this is because the second derivative is negative in this example. The magnitude of the difference between $\overline{W}(y)$ (the correct value in this context) and $W(\overline{y})$ depends on the size of the second derivative and the variance of the disease values. The larger the variance, the larger the discrepancy between (true) mean yield and that predicted based just on $\overline{y}$. If all values of y are the same, the variance is 0, and $\overline{W}(y) = W(\overline{y})$. The second way that $\overline{W}(y) = W(\overline{y})$ is when the second derivative is 0; this only happens for a straight-line function (such as $W = b_0 - b_1 y$). So, only with an underlying straightline relationship is there no effect of heterogeneity on the realized W:y line.

Some of the effects of spatial heterogeneity are demonstrated in Fig. 12.19, which is based on the use of equation 12.25 for yield, with $\Theta = 0.5$ (Type II curve) or -0.5 (Type I curve). For demonstration purposes, we assume that disease incidence is used as the measure of intensity. Based on the material in Chapter 9, the reader should note that s^2 can be defined in terms of estimated mean y. In the figure, three values for s^2 were used:

- 0 (i.e., no variation; y is fixed at all locations—the original situation),
- $\overline{y}(1-\overline{y})/n$ [i.e., random situation ("ran."; y is represented by a binomial distribution, with $n = 10$ individuals per sampling unit; and
- $(\overline{y}(1-\overline{y})/n)\cdot(n/2)$ [i.e., one-half the maximum possible variance at all $\overline{y}$ ("1/2")]. In this last case, the index of dispersion (D) is equal to $n/2$ (=5 here).

As can be seen in Fig. 12.19, increasing heterogeneity causes increasing deviation from the original relationship, whether the curve is convex or concave to the disease axis. As shown by Ferrandino (1989), the effects of increasing heterogeneity is to "straighten the line" (when s^2 is a constant proportion of the random case). Ferrandino gave clear examples of the effect with different data sets.

These mathematical results are consistent with ecological or crop growth considerations. If a Type II curve is obtained because of compensation effects of the plants, then aggregation of disease will decrease the magnitude of compensation, and hence reduce yield. This is because plant damage occurs in patches when there is spatial heterogeneity, so there may be fewer (or no) neighboring plants to compensate for a diseased individual. If a Type I curve is obtained because of intraspecific competition of the pathogen, then aggregation of disease will increase

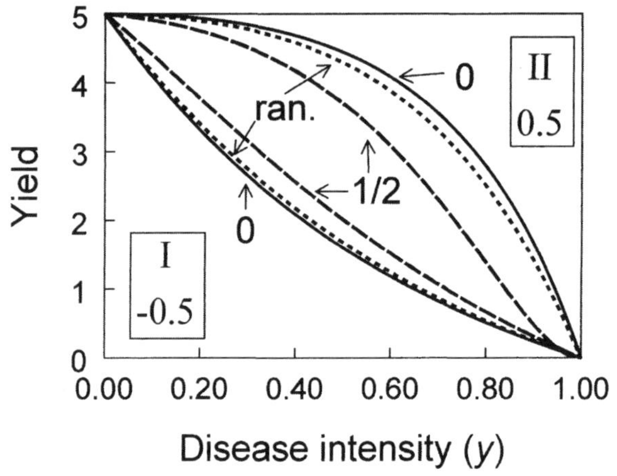

FIG. 12.19. Yield (W) in relation to disease intensity, based on equation 12.29, with $W(\overline{y})$ given by equation 12.25. Labels refer to magnitude of the estimated variance of disease intensity, s^2 ["0" for a zero variance; "ran." for binomial (random) given by equation 9.4); and 1/2 of maximum possible variance of disease incidence with $n = 10$ plants per sampling unit]. Numbers in boxes with refer to Θ values of the yield model and the category of yield-disease curve (type I or II).

the magnitude of the competition (since more pathogen or disease "individuals" will be found on the same plants or plant units (e.g., leaves). Thus, yield at a given mean disease intensity will be increased by comparison with the homogeneous disease situation.

To conclude this section, we note that the problem can be addressed from the opposite perspective. So far, we have emphasized how averaging can change observed relationships compared with the underlying relationship. Most data sets analyzed are based on some averaging (or summing). Thus, if there is any curvature to the observed relationship between yield and disease, it is likely that the underlying relationship at the individual level (e.g. plant) exhibits even stronger curvature.

12.8 Discussion and Conclusions

James and Teng (1979) believe that the lack of information on crop losses has hindered the progress of plant protection as much as any other single factor. The lack of information has two causes: (1) lack of precise estimates of disease intensity, which is really a problem with inadequate resources applied to determining disease intensity in regions, states, or provinces, and countries; and (2) inadequate understanding of the relationship between disease intensity and yield. This chapter dealt with issues related to the second item.

Although plant disease epidemics usually result in a reduction in yield compared to the no-disease situation, characterizing the relationship between the epidemic and crop yield (or loss), in either absolute or relative units, is challenging. For descriptive and comparative purposes (such as comparing treatments, cultivars, years, and so on), a range of empirical linear or nonlinear models have been used to great advantage over the last several decades. Models based on AUDPC (or a function of integrated disease or disease-free intensity over time) are especially appealing since they summarize the entire epidemic effect on yield with a small number of parameters.

As many researchers have observed, the relationship between epidemic characteristics such as AUDPC and yield is often quite variable, making it difficult to develop a general understanding of crop loss using purely empirical models. Among other things, spatial heterogeneity of disease intensity and the selected times during epidemics at which disease is assessed can influence results, as well as the form that yield is represented (absolute or relative units). When LAI is not constant, it may be beneficial to relate yield to integral variables that directly use LAI, such as HAD. In many circumstances, it may be even more beneficial to relate yield to healthy radiation interception integrated over time. With a constant efficiency of radiation use (RUE), a summary predictor variable, HAA, can be constructed to predict yield. Results of work on relative photosynthetic rate (Bastiaans, 1991) of diseased leaves suggest, however, that a constant RUE cannot be assumed.

For systemic diseases, such as those caused by many viruses, explicitly coupling time of infection to disease dynamics may be of value in understanding crop loss processes. Of course, simulation modeling can also be undertaken if sufficient information is available about the crop and disease in question.

It was shown in previous chapters that strategies for control can be evaluated using models of epidemics. This approach can certainly be applied to crop loss assessment. For instance, suppose that relative yield is represented (consistently) by the very simple model: $w = 1 - \beta_1 y_{75}$, where y_{75} is disease intensity at $t = 75$. If one wanted to use a management strategy that ensured (assuming the model is correct) a relative yield no lower than, say, $\breve{w} = 0.9$, one can rearrange the yield equation to solve for disease: $y_{75} = (1-\breve{w})/\beta_1$. If $\beta_1 = 1$, then the highest acceptable intensity of disease at time 75 is 0.1 (10%). This value of disease can also be placed in terms of disease dynamics. If y increases logistically, one can substitute equation 4.15a for y_{75} (with $t = 75$), and solve for initial disease intensity and the rate parameter. A little algebra results in the following equation:

$$\ln\left(\frac{1-\breve{w}}{\beta_1 - 1 + \breve{w}}\right) = \ln\left(\frac{y_0}{1-y_0}\right) + 75 r_L$$

Combinations of y_0 and r_L can be found that give a relative yield of $\breve{w}$. Fig. 12.20 shows this for several indicated values of $\breve{w}$ (0.95, 0.90, 0.80, 0.60, and 0.40) when $\beta_1 = 1$. For example, if $y_0 = 0.01$, a relative yield of 0.90 is obtained with $r_L = 0.032$/day. As shown for polycyclic diseases in earlier chapters, high initial disease intensity is only tolerable if r_L is very low. At moderate or higher r_L, y_0 must be very low to achieve the goal of a relative yield no lower than $\breve{w}$. As the selected yield value decreases, higher y_0 and r_L are acceptable. This approach can be adapted for other disease progress models and generalized for other crop loss models, including those based on integral predictor variables

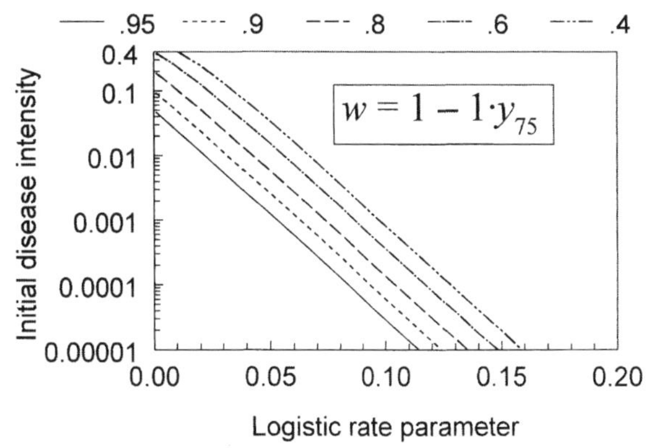

FIG. 12.20. Combinations of initial disease intensity (y_0) and logistic rate parameter (r_L) that result in the indicated predictions of relative yield (0.95, 0.9, 0.8, 0.6, and 0.4) based on the indicated yield model (in box).

(HAD and HAA). In some cases, it may be necessary to iteratively determine the epidemic parameters that give the desired yield value(s).

A considerable amount of research still lies ahead for those who are concerned about crop loss assessment. Among other things, additional work is needed on: exploring linkages between virtual-lesion results and the HLAI/HRI/RUE models for yield; a more thorough evaluation of the linkages between time-of-infection and disease dynamics; and an evaluation of the value and usage of crop loss models for Bayesian decision making in the framework presented in Chapter 11. Although not dealt with here in any detail at all, assessment and modeling of economic losses is far from adequate in general. As shown by Paveley et al. (2001), it should be possible to use HLAI/HRI/RUE models to develop optimal control decisions (in an economic sense).

One further topic not discussed in this chapter, other than in very general terms, is the situation with multiple diseases and other pests. Clearly, plant populations are affected by multiple diseases at the same time, and there certainly can be interactions of diseases in terms of crop yield. Various authors have dealt with this subject in some detail (e.g., Johnson, 1992; Francl et al., 1987; Robert et al., 2004; Savary and Zadoks, 1992a,b; Teng, 1987). From a strictly empirical approach, linear models such as equation 12.11 can be utilized. At the other extreme, mechanistic simulation models offer the most potential for developing an understanding of the physiological processes involved (Johnson, 1992), although it is unlikely that sufficient data bases of physiological results will be obtained for most crops and diseases to allow this type of modeling as a general approach for characterizing crop losses. It is noteworthy, however, that the virtual lesion and HLAI/HRI/RUE models allow direct extension to multiple diseases, although the specific approach for handling multiple diseases is not unambiguous (Johnson, 1987, 1992; Lopes and Berger, 2001; Paveley et al., 2001).

12.9 Suggested Readings

Cooke, B. M. 1998. Disease assessment and yield loss. In: *The Epidemiology of Plant Diseases* (D. G. Jones, editor). Kluwer, Dordrecht, pp. 42–72.

Francl, L. J., and Neher, D. A. 1997. *Exercises in Plant Disease Epidemiology*. APS Press, St Paul, MN (chapters 28–31).

James, W. C., Teng, P. S., and Nutter, F. W., Jr. 1991. Estimated Losses of Crops From Plant Pathogens. in: *CRC Handbook of Pest Management in Agriculture* (D. Pimentel, editor). CRC Press, Boca Raton, FL, pp. 15–51.

Johnson, K. B. 1992. Evaluation of a mechanistic model that describes potato crop losses caused by multiple pests. *Phytopathology* 82: 363–369.

Madden, L. V., G. Hughes, and M. E. Irwin. 2000. Coupling disease-progress-curve and time-of-infection functions for predicting yield loss of crops. *Phytopathology* 90: 788–800.

Teng, P. S., editor 1987. *Crop Loss Assessment and Pest Management*. APS Press, St Paul, MN.

References

Abramowitz, M., and Stegun, I. A. 1965. *Handbook of Mathematical Functions*, Dover Publications Inc., New York.

Adler, F. R. 1992. The effects of averaging on the basic reproductive ratio. *Math. Biosci.* 111: 89–98.

Agresti, A. 1990. *Categorical Data Analysis*, John Wiley and Sons, New York.

Agrios, G. N. 2005. *Plant Pathology,* 5th Edn. Academic Press, New York.

Aitchison, J., and Shen, S. M. 1980. Logistic-normal distributions: Some properties and uses. *Biometrika* 67: 261–272.

Allen, J. C. 1988. Averaging functions in a variable environment: A second-order approximation method. *Environ. Entomol.* 17: 621–625.

Almeida, A. M. R., Bergamin Filho, A., and Amorim, L. 1994. Disease progress of soybean bud blight and spatial pattern of diseased plants. *Z. Pflanzenkr. Pflanzenschutz* 101: 386–392.

Altman, D. G., and Royston, P. 2000. What do we mean by validating a prognostic model? *Statist. Med.* 19: 453–473.

Amorim, L., Bergamin Filho, A., and Hau, B. 1993. Analysis of progress curves of sugarcane smut on different cultivars using functions of double sigmoid pattern. *Phytopathology* 83: 933–936.

Anderson, D. 1988. Some models for overdispersed binomial data. *Aus. J. Statist.* 30: 125–148.

Anderson, R. M., and May, R. M. 1991. *Infectious Diseases of Humans: Dynamics and Control*, Oxford University Press, Oxford.

Anderson, R. M., Gordon, D. M., Crawley, M. J., and Hassell, M. P. 1982. Variability in the abundance of animal and plant species. *Nature* 296: 245–248.

Andrade-Piedra, J. L., Forbes, G. A., Shtienberg, D., Grünwald, N. J., Chacón, M. G., Taipe, M. V., Hijmans, R. J., and Fry, W. E. 2005. Qualification of a plant disease simulation model: Performance of the LATEBLIGHT model across a broad range of environments. Phytopathology 95: 1412–1422.

Anscombe, F. J. 1950. Sampling theory of the negative binomial and logarithmic series distributions. *Biometrika* 37: 358–382.

Anscombe, F. J. 1953. Sequential estimation. *J. R. Statist. Soc.* B15: 1–21.

Arneson, P. A. 2001. Plant disease epidemiology. *The Plant Health Instructor*. DOI: 10.1094/PHI-A-2001-0524-01 [http://www.apsnet.org/education/AdvancedPlantPath/Topics/Epidemiology/Epidemiology.htm].

Aronson, D. G., and Weinberger, H. F. 1978. Multidimensional nonlinear diffusion arising in population genetics. *Adv. Math.* 30: 33–76.

Asea, G., Bigirwa, G., Adipala, E., Owera, S. A. P., Pratt, R. C., and Lipps, P. E. 2002. Effect of *Cercospora zeae-maydis* infested residue on progress and spread of grey leaf spot of maize in central Uganda. *Ann. Appl. Biol.* 140: 177–185.

Assefa, H., van den Bosch, F., and Zadoks, J. C. 1995. Focus expansion of bean rust in cultivar mixtures. *Plant Pathol.* 44: 503–509.

Aylor, D. E. 1986. A framework for examining inter-regional aerial transport of fungal spores. *Agric. For. Meteorol.* 38: 263–288.

Aylor, D. E. 1999. Biophysical scaling and the passive dispersal of fungus spores: Relationship to integrated pest management strategies. *Agric. For. Meteorol.* 97: 275–292.

Aylor, D. E. 2003. Spread of plant disease on a continental scale: Role of aerial dispersal of pathogens. *Ecology* 84: 1989–1997.

Aylor, D. E., and Ferrandino, F. J. 1989. Temporal and spatial development of bean rust epidemics initiated from an inoculated line source. *Phytopathology* 79: 146–151.

Aylor, D. E., and Ferrandino, F. J. 1990. Initial spread of bean rust close to an inoculated bean leaf. *Phytopathology* 80: 1469–1476.

Bailey, D. J., and Gilligan, C. A. 1999. Dynamics of primary and secondary infection in take-all epidemics. *Phytopathology* 89: 84–91.

Baker, R. 1978. Inoculum potential. In: *Plant Disease; An Advanced Treatise. Vol. II: How Disease Develops in Populations* (J. G. Horsfall and E. B. Cowling, editors). Academic Press, New York, pp. 137–158.

Baker, R. T., Cowley, J. M., Harte, D. S., and Frampton, E. R. 1990. Development of a maximum pest limit for fruit flies (*Diptera: Tephritidae*) in produce imported into New Zealand. *J. Econ. Entomol.* 83: 13–17.

Bald, J. G. 1937. Investigations on 'spotted wilt' of tomatoes III: Infection in field plots. Bulletin 106. Melbourne, Australia: Council for Scientific and Industrial Research.

Barnhart, H. X., Haber, M., and Song, J. 2002. Overall concordance correlation coefficient for evaluating agreement among multiple observers. *Biometrics* 58: 1020–1027.

Barreto, M. L. 1993. The dot map as an epidemiological tool: A case study of *Schistosoma mansoni* infection in an urban setting. *Int. J. Epidemiol.* 22: 731–741.

Bartlett, M. S. 1964. Spectral analysis of two-dimensional point processes. *Biometrika* 44: 299–311.

Bassanezi, R. B., Amorim, L., Bergamin Filho, A., Hau, B., and Berger, R. D. 2001. Accounting for photosynthetic efficiency of bean leaves with rust, angular leaf spot and anthracnose to assess crop damage. *Plant Pathol.* 50: 443–452.

Bassanezi, R. B., Bergamin Filho A., Amorim, L., Gimenes-Fernandes, N., Gottwald, T. R., and Bové J. M. 2003. Spatial and temporal analyses of citrus sudden death as a tool to generate hypotheses concerning its etiology. *Phytopathology* 93: 502–512.

Bastiaans, L. 1991. Ratio between virtual and visual lesion size as a measure to describe reduction in leaf photosynthesis of rice due to leaf blast. *Phytopathology* 81: 611–615.

Basu, P. K., Lin, C. S., and Binns, M. R., 1977. A comparison of sampling methods for surveying alfalfa foliage diseases. *Can. J. Plant Sci.* 57: 1091–1097.

Batchelor, W. D., Jones, J. W., Boote, K. J., and Pinnschmidt, H. O. 1993. Extending the use of crop models to study pest damage. *Trans. ASAE* 36: 551–558.

Bates, D. M., and Watts, D. G. 1988. *Nonlinear Regression Analysis and Its Applications.* John Wiley and Sons, New York.

Béasse, C. B., Ney, B., and Tivoli, B. 2000. A simple model of pea (*Pisum sativum*) growth affected by *Mycosphaerella pinodes. Plant Pathol.* 49: 187–200.

Bejarano-Alcázar, Blanco-López, M. A., Melero-Vara, J. M., and Jiménez-Díaz, R. M. 1997. The influence of verticillium wilt epidemics on cotton yield in southern Spain. *Plant Pathol.* 46: 168–178.

Bennett, B. M. 1981. On the use of the negative binomial in epidemiology. *Biom. J.* 23: 69–72.

Bennett, R. J. 1979. *Spatial Time Series,* Pion, Ltd., London.

Bennett, S. 1989. An extension of Williams' method for overdispersion models. *GLIM Newslett.* 17: 12–18.

Benson, D. M. 1994. Inoculum. In: *Epidemiology and Management of Root Diseases* (C. L. Campbell and D. M. Benson, editors). Springer, Berlin, pp. 1–33.

Beresford, R. M., and Royle, D. J. 1991. The assessment of infectious disease for brown rust (*Puccinia hordei*) of barley. *Plant Pathol.* 40: 374–381.

Bergamin Filho, A., and Amorim, L. 1996. *Doençcas de Planta Tropicais: Epidemiologia e Controle Econômico*, Editora Agronômica Ceres Lda, São Paulo, Brasil.

Bergamin Filho, A., Carneiro, S. M. T. P. G., Godoy, C. V., Amorim, L., Berger, R. D., and Hau, B. 1997. Angular leaf spot of Phaseolus beans: Relationships between disease, healthy leaf area, and yield. *Phytopathology* 87: 506–515.

Berger, R. D. 1981. Comparison of the Gompertz and logistic equations to describe plant disease progress. *Phytopathology* 71: 716–719.

Berger, R. D., and Luke, H. H. 1979. Spatial and temporal spread of oat crown rust. *Phytopathology* 69: 1199–1201.

Bhathal, J. S., Loughman, R., and Speijers, J. 2003. Yield reduction in wheat in relation to leaf disease from yellow (tan) spot and spetoria nodorum blotch. *Eur. J. Plant Pathol.* 109: 435–443.

Bhattacharayya, K. B., Karandinos, M. G., and DeFoliart, G. R. 1979. Point estimates and confidence intervals for infection rates using pooled organisms in epidemiologic studies. *Am. J. Epidemiol.* 109: 124–131.

Biggerstaff, B. J. 2000. Comparing diagnostic tests: A simple graphic using likelihood ratios. *Statist. Med.* 19: 649–663.

Binns, M. 1975. Sequential estimation of the mean of a negative binomial distribution. *Biometrika* 62: 433–440.

Binns, M. R. 1994. Sequential sampling for classifying pest status. In: *Handbook of Sampling Methods for Arthropods in Agriculture* (L. P. Pedigo and G. D. Buntin, editors). CRC Press Inc., Boca Raton, FL, pp. 137–174.

Binns, M. R., and Nyrop, J. P. 1992. Sampling insect populations for the purpose of IPM decision making. *Ann. Rev. Entomol.* 37: 427–453.

Binns, M. R., Nyrop, J. P., and van der Werf, W. 2000. *Sampling and Monitoring in Crop Protection: The Theoretical Basis for Developing Practical Decision Guides*, CABI Publishers, Oxon, U. K.

Bissonnette, S. M., D'Arcy, C. J., and Pedersen, W. L. 1994. Yield loss in two spring oat cultivars due to *Puccinia coronata* f. sp. *avenae* in the presence or absence of barley yellow dwarf virus. *Phytopathology* 84: 363–371.

Birch, C. P. D. 1999. A new generalized logistic sigmoid growth equation compared with the Richards growth equation. *Ann. Botany* 83: 713–719.

Blaker, H. 2000. Confidence curves and improved exact confidence intervals for discrete distributions. *Can. J. Statist.* 28: 783–798.

Bliss, C. I. 1958. The analysis of insect counts as negative binomial distributions. *Proc. 10th Int. Congress Entomol.* 2: 1015–1032.

Block, C. C., Hill, J. H., and McGee, D. C. 1999. Relationship between late-season severity of Stewart's bacterial wilt and seed infection in maize. *Plant Dis.* 83: 527–530.

Boivin, G., Kushalappa, A. C., and Brodeur, L. 1990. Spatial dispersion and binomial sequential sampling plan for *Cercospora carotae* on carrots. *Can. J. Plant Pathol.* 12: 209–212.

Boote, K. J. 1982. Growth stages of peanut (*Arachis hypogaea* L.). *Peanut Sci.* 9: 35–39.

Boote, K. J., Jones, J. W., Mishoe, J. W., and Berger, R. D. 1983. Coupling pests to crop growth simulators to predict yield reductions. *Phytopathology* 73: 1581–1587.

Bourke, P. M. A. 1964. Emergence of potato blight, 1843–46. *Nature* 203: 805–808.

Box, G. E. P. 1979. Robustness in the strategy of scientific model building. In: *Robustness in Statistics* (R. L. Launer and G. N. Wilkinson, editors). Academic Press, New York, pp. 201–236.

Box, J. F. 1978. *R. A. Fisher, the Life of a Scientist*, John Wiley and Sons, Inc., New York.

Bozdogan, H. 2000. Akaike's information criterion and recent developments in information complexity. *J. Pmath. Psychol.* 44: 62–91.

Brain, P., and Cousens, R. 1989. An equation to describe dose responses where there is stimulation of growth at low doses. *Weed Res.* 29: 93–96.

Bramson, M. 1983. Convergence of solutions of the Kolmogorov equation to traveling waves. Memoirs Am. Math. Soc. 44, no. 285.

Brassett, P. R., and Gilligan, C. A. 1988. A model for primary and secondary infection in botanical epidemics. *Z. Pflanzenkr. Pflanzenschutz* 95: 352–360.

Bringmann, W. G., Luck, H. E., Miller, R., and Early, C. E., editors. 1997. *A Pictorial History of Psychology*, Quintessence Publishing Co, Inc., Chicago.

British Mycological Society. 1947. The measurement of potato blight. *Trans. Br. Mycol. Soc.* 31: 140–141.

Brooks, F. E. 2002. Brown root rot disease in American Samoa's Tropical rain forests. *Pacific Sci.* 56: 377–387.

Brown, H., and Prescott, R. 2001. *Applied Mixed Models in Medicine*, John Wiley and Sons, Chichester, UK.

Browning, J. A., and Frey, K. J. 1969. Multiline cultivars as a means of disease control. *Annu. Rev. Phytopathol.* 7: 355–382.

Bruhn, J. A., and Fry, W. E. 1981. Analysis of potato late blight epidemiology by simulation modeling. *Phytopathology* 71: 612–616.

Bruhn, J. N., Mihail, J. D., and Meyer, T. R. 1996. Using spatial and temporal patterns of *Armillaria* root disease to formulate management recommendations for Ontario's black spruce (*Picea mariana*) seed orchards. *Can. J. Forest Res.* 26: 298–305.

Brunner, E., Domhof, S., and Langer, F. 2002. *Nonparametric Analysis of Longitudinal Data in Factorial Experiments*, John Wiley and Sons, New York.

Buckland, S. T., Burnham, K. P., and Augustin, N. H. 1997. Model selection: An integral part of inference. *Biometrics* 53: 503–618.

Buntin, G. D. 1994. Developing a primary sampling program. In: *Handbook of Sampling Methods for Arthropods in Agriculture* (L. P. Pedigo and G. D. Buntin, editors). CRC Press, Boca Raton, FL, pp. 99–115.

Burleigh, J. R., Eversmeyer, M. G., and Roelfs, A. E. 1972. Linear equations for predicting wheat leaf rust. *Phytopathology* 62: 947–953.

Burrough, P. A., 1987. Spatial aspects of ecological data. In: *Data Analysis in Community and Landscape Ecology* (R. H. G. Jongman, C. J. F. ter Braak, and O. F. R. van Tongerman, editors). PUDOC, Wageningen, pp. 213–251.

Burrows, P. M. 1987. Improved estimation of pathogen transmission rates by group testing. *Phytopathology* 77: 363–365.

Butt, D. J., and Royle, D. J. 1990. Multiple regression analysis in the epidemiology of plant diseases. In: *Epidemics of Plant Diseases: Mathematical Analysis and Modeling* (J. Kranz, editor), 2nd Edn. Springer, Berlin, pp. 143–180.

Cameron, A. C., and Trivedi, P. K. 1998. *Regression Analysis of Count Data*, Cambridge University Press, Cambridge.

Campbell, C. L. 1998. Disease progress in time: Modeling and data analysis. In: *Plant Disease Epidemiology* (D. G. Jones, editor). Kluwer, London, pp. 181–206.

Campbell, C. L., and Madden, L. V. 1990. *Introduction to Plant Disease Epidemiology*, John Wiley and Sons, New York.

Campbell, C. L., and Benson, D. M., editors. 1994. *Epidemiology and Management of Root Diseases*, Springer, Berlin.

Campbell, C. L., and Neher, D. A. 1994. Estimating disease severity and incidence. In: *Epidemiology and Management of Root Diseases* (C. L. Campbell and D. M. Benson, editors). Springer, Berlin, pp. 117–147.

Campbell, C. L., and Noe, J. P. 1985. The spatial analysis of soilborne pathogens and root diseases. *Annu. Rev. Phytopathol.* 23: 129–148.

Campbell, C. L., and Van Der Gaag, D. J. 1993. Temporal and spatial dynamics of microsclerotia of *Macrophomina phaseolina* in three fields in North Carolina over four to five years. *Phytopathology* 83: 1424–1440.

Cannon, R. M. 1998. Sampling to comply with a maximum pest limit. *Biometrics* 54: 847–858.

Carlson, G. A. 1970. A decision theoretic approach to crop disease prediction and control. *Am. J. Agric. Econ.* 52: 216–223.

Carrasco, J. L., and Jover, L. 2003. Estimating the generalized concordance correlation coefficient through variance components. *Biometrics* 59: 849–858.

Chakraborty, T. K. 1990. The determination of indifference quality level single sampling attribute plans with given relative slope. *Sankhyā: Ind. J. Statist.* 52B: 238–245.

Chan, M.-S., and Jeger, M. J. 1994. An analytical model of plant virus disease dynamics with rouging and replanting. *J. Appl. Ecol.* 31: 413–427.

Charest, J., Dewdney, M., Paulitz, T., Philion, V., and Carisse, O. 2002. Spatial distribution of *Venturia inaequalis* airborne ascospores in orchards. *Phytopathology* 92: 769–779.

Chellemi, D. O., Rohrbach, K. G., Yost, R. S., and Sonoda, R. M. 1988. Analysis of the spatial pattern of plant pathogens and diseased plants using geostatistics. *Phytopathology* 78: 221–226.

Chen, C., and Tipping, R. W. 2002. Confidence interval of a proportion with over-dispersion. *Biom. J.* 44: 877–886.

Chen, C. L., and Swallow, W. H. 1990. Using group testing to estimate a proportion, and to test the binomial model. *Biometrics* 46: 1035–1046.

Chiang, C. L., and Reeves, W. C. 1962. Statistical estimation of virus infection rates in mosquito vector populations. *Am. J. Hygiene* 75: 377–391.

Chiarappa, L., editor. 1971. Crop loss assessment methods. Supplement 3. FAO Manual on the Evaluation and Prevention of Losses by Pests, Diseases, and Weeds. Commonwealth Agric. Bureaux, Farnham Royal (UK).

Chtioui, Y., Francl, L. J., and Panigrahi, S. 1999. Moisture prediction from simple micrometeorological data. *Phytopathology* 89: 668–672.

Chuang, T. Y., and Jeger, M. J. 1987. Relationship between incidence and severity of banana leaf spot in Taiwan. *Phytopathology* 77: 1537–1541.

Clark, P. J., and Evans, F. C. 1954. Distance to nearest neighbor as a measure of spatial relationships in populations. *Ecology* 35: 445–453.

Clayton, M. K., and Slack, S. A. 1988. Sample size determination in zero tolerance circumstances and the implications of stepwise sampling: Bacterial ring rot as a special case. *Am. Potato J.* 65: 711–723.

Cliff, A. D., and Ord, J. K. 1981. *Spatial Processes: Models and Applications.* Pion, London.

Clopper, C. J., and Pearson, E. S. 1934. The use of confidence or fiducial limits illustrated in the case of the binomial. *Biometrika* 26: 404–413.

Coakley, S. M., Line, R. F., and McDaniel, L. R. 1988. Predicting stripe rust severity on winter wheat using an improved method for analyzing meteorological and rust data. *Phytopathology* 78: 543–550.

Cobb, N. A. 1892. Contribution to an economic knowledge of the Australian rusts (Uredineae). *Agric. Gazt. (NSW)* 3: 60.

Cochran, W. G. 1936. The statistical analysis of field counts of diseased plants. *Suppl. J. R. Statist. Soc.* 3: 49–67.

Cochran, W. G. 1943. Analysis of variance for percentages based on unequal numbers. *J. Am. Statist. Assoc.* 38: 287–301.

Cochran, W. G. 1977. *Sampling Techniques*, 3rd Edn. John Wiley and Sons, Inc., New York.

Cohen, J. 1960. A coefficient of agreement for nominal scale. *Edu. Psychol. Measure.* 20: 37–46.

Cohen, J. 1968. Weighted kappa: Nominal scale agreement with provision for scaled disagreement or partial credit. *Psychol. Bull.* 70: 213–220.

Cole, D. L. 1981. Diseases of groundnut (*Arachis hypogaea* L.) I. Fungicde spray effects on *Cercospora arachidicola* and *Phoma arachidicola* leaf infection, kernel yield and pod rots. *Zimbabwe J. Agric. Res.* 19: 101–110.

Collett, D. 1991. *Modelling Binary Data*, Chapman and Hall, London.

Collett, D. 2003. *Modelling Binary Data*, 2nd Edn. Chapman and Hall, London.

Colquhoun, D. 1971. *Lectures on Biostatistics*, Clarendon Press, Oxford.

Cook, R. J. 1985. Editorial: Use of the term "crop loss". *Plant Dis.* 69: 95.

Cooke, B. M. 1998. Disease assessment and yield loss. In: *The Epidemiology of Plant Diseases* (D. G. Jones, editor). Kluwer, Dordrecht, pp. 42–72.

Cooke, B. M. 2006. Disease assessment and yield loss. Pages 43–80 In: *The Epidemiology of Plant Diseases*, 2nd Edn. (B. M. Cooke, D. G. Jones, and B. Kayle, editors). Springer, The Netherlands.

Cooke, B. M., Jones, D. G., and Kayle, B. 2006. *The Epidemiology of Plant Diseases*, 2nd Edn. Springer, The Netherlands.

Copes, W. E., Scherm, H., and Ware, G. O. 2001. Sequential sampling to assess the incidence of infection by *Monilinia vaccinii-corymbosi* in mechanically harvested rabbiteye blueberry fruit. *Phytopathology* 91: 348–353.

Cousens, R. 1985. A simple model relating yield loss to weed density. *Ann. Appl. Biol.* 107: 239–252.

Cowger, C., Wallace, L. D., and Mundt, C. C. 2005. Velocity of spread of wheat stripe rust epidemics. *Phytopathology* 95: 972–982.

Cowling, E. B., and Horsfall, J. G. 1978. Prologue: How disease develops in populations. In: *Plant Disease: An Advanced Treatise. Vol. II: How Disease Develops in Populations* (J. G. Horsfall and E. B. Cowling, editors). Academic Press, New York, pp. 1–15.

Cox, K. D., and Scherm, H. 2001. Gradients of primary and secondary infection by *Monilinia vaccinii-corymbosi* from point sources of ascospores and conidia. *Plant Dis.* 85: 955–959.

Cressie, N. A. C. 1993. *Statistics for Spatial Data (Revised Edition)*, Wiley, New York.

Crow, E. L., and Gardner, R. S. 1959. Confidence intervals for the expectation of a Poisson variable. *Biometrika* 46: 441–453.

Crowder, M. J. 1978. Beta-binomial anova for proportions. *Appl. Statist.* 27: 34–37.

Culbreath, A. K., Todd, J. W., Gorbet, D. W., Shokes, F. M., and Pappu, H. R. 1997. Field response of new peanut cultivar UF 91108 to tomato spotted wilt virus. *Plant Dis.* 81: 1410–1415.

Cussens, J. 1996. Effective sample size in a dichotomous process with noise. *Commun. Statist.—Theor. Meth.* 25: 1233–1246.

Daamen, R. A. 1986. Measures of disease intensity in powdery mildew (*Erysiphe graminis*) of winter wheat. 2. Relationships and errors of estimation of pustule number, incidence and severity. *Netherlands J. Plant Pathol.* 92: 207–222.

Dale, M. R. T. 1999. *Spatial Pattern Analysis in Plant Ecology*, Cambridge University Press, Cambridge.

Dale, M. R. T., Dixon, P., Fortin, M.-J., Legendre, P., Myers, D. E., and Rosenberg, M. S. 2002. Conceptual and mathematical relationships among methods for spatial analysis. *Ecography* 25: 558–577.

Daley, D. J., and Gani, J. 1999. *Epidemic Modelling: An Introduction*. Cambridge University Press, Cambridge.

Dallot, S., Gottwald, T., Labonne, G., and Quiot, J.-B. 2003. Spatial pattern analysis of sharka disease (plum pox virus strain M) in peach orchards of southern France. *Phytopathology* 93: 1543–1552.

de Jong, P. D. 1995. Sampling for detection: Leek rust as an example. *Int. J. Pest Manag.* 41: 31–35.

de Jong, P. D., and de Bree, J. 1995. Analysis of the spatial distribution of rust-infected leek plants with the Black–White join-count statistic. *Eur. J. Plant Pathol.* 101: 133–137.

de Koeijer, K. J., and van der Werf, W. 1995. Effect of beet yellowing viruses on light interception and light use efficiency of the sugarbeet crop. *Crop Prot.* 14: 291–297.

de Wit, C. T., and Rabbinge, R. 1979. Systems analysis and dynamic simulation. *EPPO Bull.* 9: 149–153.

De Wolf, E. D., Madden, L. V., and Lipps, P. E. 2003. Risk assessment models for wheat Fusarium head blight epidemics based on within-season weather data. *Phytopathology* 93: 428–435.

Del Ponte, E. M., Shah, D. A., and Bergstrom, G. C. 2003. Spatial patterns of Fusarium head blight in New York wheat fields suggest role of airborne inoculum. Online. *Plant Health Progress*. DOI: 10.1094/PHP-2003-0418-01-RS.

Diekmann, O. 1978. Thresholds and traveling waves for the geographical spread of infections. *J. Math. Biol.* 6: 109–130.

Diekmann, O. 1979. Run for your life. A note on the asymptotic speed of propagation of an epidemic. *J. Diff. Equ.* 33: 58–73.

Diekmann, O., and Heesterbeek, J. A. P. 2000. *Mathematical Epidemiology of Infectious Diseases: Model Building, Analysis, and Interpretation*, John Wiley and Son, Ltd., Chichester.

Diekmann, O., Heesterbeek, J. A. P., and Metz, J. A. J. 1990. On the definition and the computation of the basic reproduction ratio R_0 in models for infectious diseases in heterogeneous populations. *J. Math. Biol.* 28: 365–382.

Diekmann, O., Heesterbeek, H., and Metz, J. A. J. 1995. The legacy of Kermack and McKendrick. In: *Epidemic*

Models: Their Structure and Relation to Data (D. Mollison, editor). Cambridge University Press, Cambridge, pp. 95–115.

Diem, K., and Lentner, C. 1970. *Documenta Geigy Scientific Tables*, 7th Edn. J. R. Geigy SA, Basle, Switzerland.

Diggle, P. J. 1983. *Statistical Analysis of Spatial Point Patterns*, Academic Press, London.

Diggle, P. J. 2003. *Statistical Analysis of Spatial Point Patterns*, 2nd Edn. Arnold, London.

Diggle, P. J., and Chetwynd, A. G. 1991. Second-order analysis of spatial clustering for inhomogeneous populations. *Biometrics* 47: 1155–1163.

Dillard, H. R., and Seem, R. C. 1990. Incidence-severity relationships for common maize rust on sweet corn. *Phytopathology* 80: 842–846.

Dittrich, V., Ernst, G. H., and Ruesch, O. 1990. Resistance mechanisms in sweet-potato whitefly (*Homoptera: Aleyrodidae*) populations from Sudan, Turkey, Guatemala and Nicaragua. *J. Econ. Entomol.* 83: 1665–1670.

Dixon, P. M. 2002. Ripley's *K* function. In: *Encyclopedia of Environmetrics* (A. H. El-Shaarawi and W. W. Piegorsch, editors). Vol. 3, John Wiley and Sons, Ltd, Chichester, pp. 1796–1803.

Donner, A. 1986. A review of inference procedures for the intraclass correlation coefficient in the one-way random effects model. *Int. Statist. Rev.* 54: 67–82.

Dorfman, R. 1943. The detection of defective members of large populations. *Ann. Math. Statist.* 14: 436–440.

Draper, N. R., and Smith, H. 1998. *Applied Regression Analysis*, John Wiley and Sons, New York.

Duhamel de Monceau, H. L. 1728. Explication physique d'une maladie qui fait périr plusieurs plantes dans le Gastinois, et particulièrement le safran. *Mém. Acad. Sci. Math. Phys.* (Paris): 100–112.

Dungan, J. L., Perry, J. N., Dale, M. R. T., Legendre, P., Citron-Pousty, S., Fortin, M.-J., Jakomulska, A., Miriti, M., and Rosenberg, M. S. 2002. A balanced view of scale in spatial statistical analysis. *Ecography* 25: 626–640.

Durr, P., and Pfeiffer, D. 2002. Preface. *Prevent. Vet. Med.* 56: 1–3.

Durrett, R., and Levin, S. 1993. Stochastic spatial models: A user's guide to ecological applications. *Philos. Trans. R. Soc. Lond. B* 343: 329–350.

Edelstein-Keshet, L. 1988. *Mathematical Models in Biology*. Birhauser, McGraw-Hill, Philadelphia.

Edminster, T. W. 1978. Concepts for using modeling as a research tool. Technical Manual 520. U.S. Department of Agriculture, Beltsville, MD, USA.

El-Allaf, S. M., Madden, L. V., and Lipps, P. E. 2003. Incidence-severity relationships for Fusarium head blight on wheat. *Phytopathology* 93: S23.

Elliott P., Wakefield, J. C., Best, N. G., and Briggs, D. J., editors. 2000. *Spatial Epidemiology: Methods and Applications*, Oxford University Press, Oxford.

Emge, R. G., and Shrum, R D. 1976. Epiphytology *of Puccinia striiformis* at five selected locations in Oregon during 1968 and 1969. *Phytopathology* 66: 1406–1412.

English, J. T., and Mitchell, D. J. 1994. Host roots. In: *Epidemiology and Management of Root Diseases* (C. L. Campbell and D. M. Benson, editors). Springer, Berlin, pp. 34–64.

Evans, M., Hastings, N., and Peacock, B. 2000. *Statistical Distributions*, 3rd Edn. John Wiley and Sons, New York.

Everitt, B. S. 1999. *Dictionary of Statistics*, Cambridge University Press, U. K.

Falloon, R. E., Viljanen-Rollinson, S. L. H., Coles, G. D., and Poff, J. D. 1995. *New Zealand J. Crop Horticul. Sci.* 23: 31–37.

Fargette, D, and Vié, K. 1995. Simulation of the effects of host resistance, reversion, and cutting selection on incidence of African cassava mosaic virus and yield losses in cassava. *Phytopathology* 85: 370–375.

Farr, D. F., Bills, G. F., Chamuris, G. P., and Rossman, A. Y. 1989. *Fungi on Plants and Plant Products in the United States*, APS Press, St Paul, MN.

Farrington, C. P. 1992. Estimating prevalence by group testing using generalized linear models. *Statist. Med.* 11: 1591–1597.

Fauquet, C., and Fargette, D. 1990. African cassava mosaic-virus—etiology, epidemiology, and control. *Plant Dis.* 74: 404–411.

Ferrandino, F. J. 1989. A distribution-free method for estimating the effect of aggregated plant damage on crop yield. *Phytopathology* 79: 1229–1232.

Ferrandino, F. J. 1993. Dispersive epidemic waves: I. Focus expansion within a linear planting. *Phytopathology* 83: 795–802.

Ferrandino, F. J. 1996. Length scale of disease spread: Fact or artifact of experimental geometry. *Phytopathology* 86: 806–811.

Ferrandino, F. J. 1998. Past nonrandomness and aggregation to spatial correlation: 2DCORR, a new approach for discrete data. *Phytopathology* 88: 84–91.

Ferrandino, F. J. 2005. The explicit dependence of quadrat variance on the ratio of clump size to quadrat size. *Phytopathology* 95: 452–462.

Filajdic, N., and Sutton, T. B. 1992. Chemical control of Alternaria blotch of apples caused by *Alternaria mali. Plant Dis.* 76: 126–130.

Filipe, J. A. N., and Gibson, G. J. 1998. Studying and approximating spatio-temporal models for epidemic spread and control. *Philos. Trans. R. Soc. Lond. B* 353: 2153–2162.

Filipe, J. A. N, and Maule, M. M. 2003. Analytical methods for predicting the behaviour of population models with general spatial interactions. *Math. Biosci.* 183: 15–35.

Filipe, J. A. N., and Maule, M. M. 2004. Effects of dispersal mechanisms on spatio-temporal development of epidemics. *J. Theor. Biol.* 226: 125–141.

Finney, D. J. 1947. Errors of estimation in inverse sampling. *Nature*, 160: 196–197.

Finney, D. J. 1949. On a method of estimating frequencies. *Biometrika*, 36: 233–234.

Fisher, R. A. 1925. *Statistical Methods for Research Workers*, Oliver and Boyd, Edinburgh.

Fisher, R. A. 1937. The wave of advance of advantageous genes. *Ann. Eugenetics* 7: 355–369.

Fisher, R. A., and Yates, F. 1963. *Statistical Tables for Biological, Agricultural and Medical Research*, 6th Edn. Longman, London.

Fitt, B. D. L., and McCartney. 1986. Spore dispersal in relation to epidemic models. In: *Plant Disease Epidemiology, Vol. 1: Population Dynamics and Management* (K. J. Leonard and W. E. Fry, editors). Macmillan Publication, New York, pp. 311–345.

Fitt, B. D. L., Gregory, P. H., Todd, A. D., McCartney, H. A., and Macdonald, O. C. 1987. Spore dispersal and plant disease epidemics: A comparison between two empirical models. *J. Phytopathol.* 118: 227–242.

Forbes, G. A., and Jeger, M. J. 1987. Factors affecting the estimation of disease intensity in simulated plant structures. *Z. PflKrankh. PflSchultz.* 94: 113–120.

Forbes, G. A., and Korva, J. T. 1994. The effect of using a Horsfall-Barratt scale on precision and accuracy of visual estimation of potato leaf blight severity in the field. *Plant Pathol.* 43: 675–682.

Forster, M., and Sober, E. 1994. How to tell when simpler, more unified, or less ad hoc theories will provide more accurate predictions. *Br. J. Philos. Sci.* 45: 1–35.

Fortin, M.-J., Dale, M. R. T., and ver Hoef, J. 2002. Spatial analysis in ecology. In: *Encyclopedia of Environmetrics* (A. H. El-Shaarawi and W. W. Piergorsch, editors). Vol. 4, John Wiley and Sons Ltd, Chichester, pp. 2051–2058.

Fowler, G. W., and Lynch, A. M. 1987. Sampling plans in insect pest management based on Wald's sequential probability ratio test. *Environ. Entomol.* 16: 345–354.

Fox, R. T. V. 1998. Plant disease diagnosis. In: *The Epidemiology of Plant Diseases* (D. G. Jones, editor). Kluwer Publication, Dordrecht, pp. 14–41.

Fracker, S. B. 1936. Progressive intensification of uncontrolled plant-disease outbreaks. *J. Econ. Entomol.* 29: 923–940.

Frampton, V. L., Linn, M. B., and Hansing, E. E. 1942. The spread of virus diseases of the yellow type under field conditions. *Phytopathology* 32: 799–808.

Francl, L. J. 2001. The disease triangle: A plant pathological paradigm revisited. *The Plant Health Instructor*. DOI: 10.1094/PHI-T-2001-0517-01 [http://www.apsnet.org/education/search/abstracts/triangle.asp].

Francl, L. J. 2004. Squeezing the turnip with artificial neural nets. *Phytopathology* 94: 1007–1012.

Francl, L. J., Madden, L. V., Rowe, R. C., and Riedel, R. M. 1987. Potato yield loss prediction and discrimination using preplant population densities of *Verticillium dahliae* and *Pratylenchus penetrans*. *Phytopathology* 77: 579–584.

Francl, L. J., and Neher, D. A. 1997. *Exercises in Plant Disease Epidemiology*, APS Press, St. Paul, MN.

Frantzen, J., and van den Bosch, F. 2000. Spread of organisms: Can traveling and dispersive waves be distinguished? *Basic Appl. Ecol.* 1: 83–91.

Friedland, D. J. (editor) 1998. *Evidence-Based Medicine: A Framework for Clinical Practice*, Lange Medical Books, New York.

Fry, W. E. 1978. Quantification of general resistance of potato cultivars and fungicide effects for integrated control of potato late blight. *Phytopathology* 68: 1650–1655.

Fry, W. E. 1982. *Principles of Plant Disease Management*. Academic Press, New York.

Funt, R. C., Ellis, M. A., and Madden, L. V. 1990. Economic analysis of protectant and disease-forecast-based curative fungicide spray programs for control of apple scab and grape black rot in Ohio. *Plant Dis.* 74: 638–642.

Garrett, K. A., Madden, L. V., Hughes, G., and Pfender, W. F. 2004. New applications of statistical tools in plant pathology. *Phytopathology* 94: 999–1003.

Garson, G. I., and Moser, E. B. 1995. Aggregation and the Pearson Chi-Square statistic for homogeneous proportions and distributions in ecology. *Ecology* 76: 2258–2269.

Garwood, F. 1936. Fiducial limits for the Poisson distribution. *Biometrika* 28: 437.

Gäumann, E. 1946. *Pflanzliche Infektionslehre*, Birkhäuser, Basel.

Gäumann, E. 1950. *Principles of Plant Infection* (translated by W. B. Brierly), Lockwood and Son, London.

Gaunt, R. E. 1987. Measurement of disease and pathogens. In: *Crop Loss Assessment and Pest Management* (P. S. Teng, editor). APS Press, St. Paul, MN, pp. 6–18.

Gaunt, R. E., and Cole, M. J. 1992. Sequential sampling for wheat stripe rust management. *Crop Protection* 11: 138–140.

George, V. T., and Elston, R. C. 1993. Confidence limits based on the first occurrence of an event. Statist. Med. 12: 685–690.

Gerrard, D. J., and Chiang, H. C. 1970. Density estimation of corn rootworm egg populations based upon frequency of occurrence. *Ecology* 51: 237–245.

Gibbons, J. D. 1976. *Nonparametric Methods for Quantitative Analysis*, Holt, Rinehart, and Winston. New York.

Gibson, G. 1997a. Markov Chain Monte Carlo methods for fitting spatiotemporal stochastic models in plant epidemiology. *Appl. Statist.* 46: 215–233.

Gibson, G. J. 1997b. Investigating mechanisms of spatiotemporal epidemic spread using stochastic models. *Phytopathology* 87: 139–146.

Gibson, G. J., and Austin, E. J. 1996. Fitting and testing spatio-temporal stochastic models with applications in plant epidemiology. *Plant Pathol.* 45: 172–184.

Gibson, G. J., Gilligan, C. A., and Kleczkowski, A. 1999. Predicting variability in biological control of a plant-pathogen system using stochastic models. *Proc. R. Soc. Lond. B* 266: 1743–1753.

Gilles, T., Evans, N., Fitt, B. D. L., and Jeger, M. J. 2000. Epidemiology in relation to methods for forecasting light leaf spot (*Pyrenopeziza brassicae*) severity on winter oilseed rape (*Brassica napus*) in the UK. *Eur. J. Plant Pathol.* 106: 593–605.

Gilles, T., Fitt, B. D. L., McCartney, H. A., Papastamati, K., and Steed, J. M. 2001. The roles of ascospores and conidia of *Pyrenopeziza brassicae* in light leaf spot epidemics on winter oilseed rape (*Brassica napus*) in the UK. *Ann. Appl. Biol.* 138: 141–152.

Gilligan, C. A. 1985. Construction of temporal models: III. Disease progress of soil-borne pathogens. In: *Advances in Plant Pathology. Vol. 3: Mathematical Modelling of Crop Disease*, Academic Press, London, pp. 67–106.

Gilligan, C. A. 1986. Use and misuse of the analysis of variance in plant pathology. *Adv. Plant Pathol.* 5: 225–261.

Gilligan, C. A. 1988. Analysis of the spatial pattern of soilborne pathogens. In: *Experimental Techniques in Plant Disease Epidemiology* (J. Kranz and J. Rotem, editors). Springer, Berlin, pp. 85–98.

Gilligan, C. A. 1990. Comparison of disease progress curves. *New Phytol.* 115: 223–242.

Gilligan, C. A. 1994. The dynamics of infection by the take all-fungus on seminal roots of wheat—sensitivity analysis of a stochastic simulation model. *New Phytol.* 128: 539–553.

Gilligan, C. A. 1995. Modelling soil-borne pathogens with special emphasis on spatial aspects of disease: Reaction-diffusion models. *Can. J. Plant Pathol.* 17: 96–108.

Gilligan, C. A. 2002. An epidemiological framework for disease management. Adv. Bot. Res. 38: 1–64.

Gilligan, C. A., and Kleczkowski, A. 1997. Population dynamics of botanical epidemics involving primary and secondary infection. *Philos. Trans. R. Soc. Lond. B* 352: 591–608.

Gilligan, C. A., Brassett, P. R., and Campbell, A. 1994. Computer simulation of early infection of cereal roots by the take-all fungus—a detailed mechanistic simulator. *New Phytol.* 128: 515–537.

Gilligan, C. A., Gubbins, S., and Simons, S. A. 1997. Analysis and fitting of an SIR model with host response to infection load for a plant disease. *Philos. Trans. R. Soc. Lond. B* 352: 353–364.

Gitaitis, R. D., Dowler, C. C., and Chalfant, R. B. 1998. Epidemiology of tomato spotted wilt in pepper and tomato in southern Georgia. *Plant Dis.* 82: 752–756.

Go, A. S. 1998. Refining probability: An introduction to the use of diagnostic tests. In: *Evidence-Based Medicine: A Framework for Clinical Practice* (D. J. Friedland, Editor). Lange Medical Books/McGraw-Hill, New York, pp. 11–33.

Godoy, C. V., Carneiro, S. M. T. P. G., Iamauti, M. T., Dalla Pria, M., Amorim, L., Berger, R. D., and Bergamin Filho, A. 1997. Diagrammatic scales for bean diseases: Development and validation. *Z. Pflanzenkrankheiten Pflanzehschutz [J. Plant Dis. Prot.]* 104: 336–345.

Goel, N. S., and Norman, J. M. 1990. *Instrumentation for Studying Vegetation Canopies for Remote Sensing in Optical and Thermal Infrared Regions*, Harwood Academic Publication, GmbH, London.

Gold, H. J. 1989. Decision analytic modeling for plant disease control. Pages 84–122 In: *Plant Disease Epidemiology, Vol. 2: Genetics, Resistance, and Management* (K. J. Leonard and W. E. Fry, editors). McGraw-Hill, New York, pp. 84–122.

Gold, H. J., Bay, J., and Wilkerson, G. G. 1996. Scouting for weeds, based on the negative binomial distribution. *Weed Sci.* 44: 504–510.

Gottwald, T. R., and Graham, J. H. 1990. Spatial pattern analysis of epidemics of citrus bacterial spot in Florida citrus nurseries. *Phytopathology* 80: 181–190.

Gottwald, T. R., and Timmer, L. W. 1995. The efficacy of windbreaks in reducing the spread of citrus canker caused by *Xanthomonas campestris* pv. *citri*. *Trop. Agric (Trinidad)* 72: 194–201.

Gottwald, T. R., Ritchie, S. M., and Campbell, C. L. 1992. LCOR2—spatial correlation analysis software for the personal computer. *Plant Dis.* 76: 213–215.

Gottwald, T. R., Cambra, M., Moreno, P., Camarasa, E., and Piquer, J. 1996. Spatial and temporal analyses of citrus tristeza in eastern Spain. *Phytopathology* 86: 45–55.

Gottwald, T. R., Graham, J. H., and Riley, T. D. 1997. The influence of spray adjuvants on exacerbation of citrus bacterial spot. *Plant Dis.* 81: 1305–1310.

Gottwald, T. R., Garnsey, S. M., and Borbón, J. 1998. Increase and patterns of spread of citrus tristeza virus infections in Costa Rica and the Dominican Republic in the presence of the brown citrus aphid, *Toxoptera citricida*. *Phytopathology* 88: 621–636.

Gottwald, T. R., Gibson, G. J., Garnsey, S. M., and Irey, M. 1999. Examination of the effect of aphid vector population composition on the spatial dynamics of citrus tristeza virus spread via stochastic modelling. *Phytopathology* 89: 603–608.

Gottwald, T. R., Sun, X., Riley, T., Graham, J. H., Ferrandino, F., and Taylor, E. L. 2002. Geo-referenced spatiotemporal analysis of the urban citrus canker epidemic in Florida. *Phytopathology* 92: 361–377.

Goudriaan, J., and van Laar, H. H. 1994. Modeling potential crop growth processes. Kluwer Academic Publication, London.

Gould, H. J. 1989. Decision analytic modeling for plant disease control. In: *Plant Disease Epidemiology, Vol. 2: Genetics, Resistance, and Management*, McGraw-Hill, New York, pp. 84–122.

Goulds, A., and Polley, R. W. 1990. Assessment of eyespot and other stem base diseases of winter wheat and winter barley. *Mycol. Res.* 94: 819–822.

Gozé, E., Nibouche, S., and Deguine, J.-P. 2003. Spatial and Probability distribution of *Helicoverpa armigera* (Hübner) (Lepidoptera: Nocutidae) in cotton: Systematic sampling, exact confidence intervals and sequential test. *Environ. Entomol.* 32: 1203–1210.

Grant, E. L., and Leavenworth, R. S. 1996. *Statistical Quality Control*, 7th Edn. The McGraw-Hill Companies, Inc., New York.

Greaney, F. J., Woodward, J. C, and Whiteside, A. G. O. 1941. The effect of stem rust on the yield, quality, chemical composition, and milling and baking qualities of Marquis wheat. *Sci. Agric.* 22: 40–60.

Green, D. E., II, Burpee, L. L., and Stevenson, K. L. 1998. Canopy reflectance as a measure of disease in tall fescue. *Crop. Sci.* 38: 1603–1613.

Green, R. H. 1970. On fixed precision level sequential sampling. *Res. Popul. Ecol.* 12: 249–251.

Gregoire, T. G., and Schabenberger, O. 1996. Nonlinear mixed-effects modeling of cumulative bole volume with spatially correlated within-tree data. *J. Agric. Biol. Environ. Statist.* 1: 107–119.

Gregory, P. H. 1948. The multiple-infection transformation. *Ann. Appl. Biol.* 35: 412–417.

Gregory, P. H. 1961. *The Microbiology of the Atmosphere*, John Wiley and Sons, New York.

Gregory, P. H. 1968. Interpreting plant disease dispersal gradients. *Annu. Rev. Phytopathol.* 6: 189–212.

Gregory, P. H. 1973. *The Microbiology of the Atmosphere*, 2nd Edn. Leonard Hill, London.

Gregory, P. H., and Read, D. R. 1949. The spatial distribution of insect-borne plant-virus diseases. *Ann. Appl. Biol.* 36: 475–482.

Groth, J. V., Ozmon, E. A., and Busch, R. H. 1999. Repeatability and relationship of incidence and severity measures of scab of wheat caused by *Fusarium graminearum* in inoculated nurseries. *Plant Dis.* 83: 1033–1038.

Gubbins, S., and Gilligan, C. A. 1997. Persistence of host-parasite interactions in disturbed environments. *J. Theor. Biol.* 188: 241–258.

Gubbins, S., Gilligan, C. A., and Kleczkomski, A. 2000. Population dynamics of plant-parasite interactions: Thresholds for invasion. *Theor. Popul. Biol.* 57: 219–233.

Guenther, W. C. 1973. A sample size formula for the hypergeometric. *J. Qual. Technol.* 5: 167–170.

Guenther, W. C. 1977. Sampling Inspection in Statistical Quality Control. Charles Griffin and Company Ltd, London and High Wycombe.

Gumpertz, M. L. 1997. Testing binary response variables for spatial autocorrelation. In: *Exercises in Plant Disease Epidemiology* (L. J. Francl and D. A. Neher, editors). APS Press, St Paul, MN, pp. 78–84.

Gumpertz, M. L., and Ristaino, J. B. 1997. Spatial autocorrelation: Methods for continuous variables. In: *Exercises in Plant Disease Epidemiology* (L. J. Francl and D. A. Neher, editors). APS Press, St Paul, MN, pp. 72–77.

Gumpertz, M. L., Larkin, R. P., and Ristaino, J. B. 1997. Geostatistical applications in epidemiology. In: *Exercises in Plant Disease Epidemiology* (L. J. Francl and D. A. Neher, editors). APS Press, St Paul, MN, pp. 94–99.

Haase, P. 1995. Spatial pattern-analysis in ecology based on Ripley K-function—introduction and methods of edge correction. *J. Vegetation Sci.* 6: 575–582.

Hald, A. 1967. The determination of single sampling attribute plans with given producer's and consumer's risk. *Technometrics* 9: 401–415.

Hald, A. 1981. *Statistical Theory of Sampling Inspection by Attributes*, Academic Press, London.

Haldane, J. B. S. 1945. On a method of estimating frequencies. *Biometrika* 33: 222–225.

Hamaker, H. C. 1949. Lot inspection by sampling. *Philips Tech. Rev.* 11: 176–182.

Hamaker, H. C. 1950. The theory of sampling inspection plans. *Philips Tech. Rev.* 11: 260–270.

Hamaker, H. C. 1959. Adjusting single sampling plans for finite lot size. *Appl. Statist.* 8: 210–214.

Hanley, J. A., and McNeil, B. J. 1982 The meaning and use of the area under a receiver operating characteristic (ROC) curve. *Radiology* 143: 29–36.

Hardwick, N. V. 1998. Disease forecasting. In: *The Epidemiology of Plant Diseases* (D. G. Jones, editor). Kluwer Publishers, Dordrecht, pp. 207–230.

Harte, D. S., Baker, R. T., and Cowley, J. M. 1992. Relationship between preentry sample size for quarantine security and variability of estimates of fruit fly (*Diptera: Tephritidae*) disinfestation treatment efficacy. *J. Econ. Entomol.* 85: 1560–1565.

Hartman, J. R. Parisi, L., and Bautrais. 1999. Effect of leaf wetness duration, temperature, and conidial inoculum dose on apple scab infections. *Plant Dis.* 83: 531–534.

Harveson, R. M., and Rush, C. M. 2002. The influence of irrigation frequency and cultivar blends on the severity of multiple root diseases in sugar beets. *Plant Dis.* 86: 901–908.

Hau, B. 1988. Modelling epidemics of polycyclic foliar diseases and development of simulators. In: *Experimental Techniques in Plant Disease Epidemiology* (J. Kranz, editor). Springer, Berlin, pp. 267–277.

Hau, B., and Kranz, J. 1990. Mathematics and statistics for analysis in epidemiology. In: *Epidemics of Plant Diseases: Mathematical Analysis and Modeling*, 2nd Edn. (J. Kranz, editor). Springer, Berlin, pp. 12–52.

Hau, B., Amorim, L., and Bergamin Filho, A. 1993. Mathematical functions to describe disease progress curves of double sigmoid pattern. *Phytopathology* 83: 928–932.

Hau, F. C., Campbell, C. L., and Beute, M. K. 1982. Inoculum distribution and sampling methods for *Cylindrocladium crotalariae* in peanuts field. *Plant Dis.* 66: 568–571.

Hebert, T. T. 1982. The rationale for the Horsfall-Barratt plant disease assessment scale. *Phytopathology* 72: 1269.

Heesterbeek, J. A. P., and Zadoks, J. C. 1987. Modelling pandemics of quarantine pests and diseases: Problems and perspectives. *Crop Protect.* 6: 211–221.

Hengveld, R. 1989. *Dynamics of Biological Invasions*, Chapman and Hall, London.

Hepworth, G. 1996. Exact confidence intervals for proportions estimated by group testing. *Biometrics* 52: 1134–1146.

Hepworth, G., and MacFarlane, J. R. 1992. Variance of the estimated population density from a presence-absence sample. *J. Econ. Entomol.* 85: 2240–2245.

Hoffman, M. P., Nyrop, J. P., Kirkwyland, J. J., Riggs, D. M., Gilrein, D. O., and Moyer, D. D. 1996. Sequential sampling plan for scheduling control of lepidopteran pests of fresh market sweet corn. *J. Econ. Entomol.* 89: 386–395.

Holdenrieder, O., Pautasso, M., Weisberg, P. J., and Lonsdale, D. 2004. Tree diseases and landscape processes: The challenge of landscape pathology. *Trends Ecol. Evol.* 19: 446–452.

Holgate, P. 1989. Approximate moments of the adès distribution. *Biom. J.* 31: 875–883.

Holliday, P. 1989. *A Dictionary of Plant Pathology*, Cambridge University Press, Cambridge, UK.

Holt, J., and Chancellor, T. C. B. 1996. Simulation modeling of thespread of ricetungro virus desease: The potential for management by rouging. *J. Appl. Ecol.* 33: 927–936.

Holt, J., Jeger, M. J., Thresh, J. M., and Otim-Nape, G. W. 1997. An epidemiological model incorporating vector population dynamics applied to African cassava mosaic virus disease. *J. Appl. Ecol.* 34: 793–806.

Holt, J., Colvin, J., and Muniyappa, V. 1999. Identifying control strategies for tomato leaf curl virus disease using an epidemiological model. *J. Appl. Ecol.* 36: 625–633.

Horne, J. K., and Schneider, D. C. 1995. Spatial variance in ecology. *Oikos* 74: 18–26.

Horsfall, J. G., and Cowling, E B. 1978a. Some epidemics man has known. In: *Plant Disease: An Advanced Treatise. Vol. II: How Disease Develops in Populations* (J. G. Horsfall and E. B. Cowling, editors). Academic Press, New York, pp. 17–32.

Horsfall, J. G., and Cowling, E. B. 1978b. Pathometry: The measurement of plant diseases. In: *Plant Disease: An Advanced Treatise. Vol. II: How Disease Develops in Populations* (J. G. Horsfall and E. B. Cowling, editors). Academic Press, New York, pp. 119–136.

Horsfall, J. G., and Dimond, A. E., editors. 1960. *Plant Pathology: An Advanced Treatise, Vol. 3: The Diseased Population.* Academic Press, New York.

Huber, L., Madden, L. V., and Fitt, B. D. L. 1998. Rain-splash and spore dispersal: A physical perspective. In: *Plant Disease Epidemiology* (D. G. Jones, editor). Kluwer, London, pp. 348–370.

Hudelson, B. D., Clayton, M. K., Smith, K. P., and Upper, C. D. 1993. Modeling of superimposed spatial pattern of bacterial brown spot. *Phytopathology* 83: 430–438.

Hudelson, B. D., Clayton, M. K., Smith, K. P., and Upper, C. D. 1997. Detection and description of spatial patterns of snap beans using cyclic samples. *Phytopathology* 87: 33–41.

Hughes, G. 1988. Spatial heterogeneity in crop loss assessment models. *Phytopathology* 78: 883–884.

Hughes, G., and Gottwald, T. R. 1998. Survey methods for assessment of citrus tristeza virus incidence. *Phytopathology* 88: 715–725.

Hughes, G., and Gottwald, T. R. 1999. Survey methods for assessment of citrus tristeza virus incidence when *Toxoptera citricida* is the predominant vector. *Phytopathology* 89: 487–494.

Hughes, G., and Gottwald, T. R. 2001. Survey methods for assessment of *Citrus tristeza virus* incidence in citrus nurseries. *Plant Dis.* 85: 910–918.

Hughes, G., and Madden, L. V. 1992. Aggregation and incidence of disease. *Plant Pathol.* 41: 657–660.

Hughes, G., and Madden, L. V. 1993. Using the beta-binomial distribution to describe aggregated patterns of disease incidence. *Phytopathology* 83: 759–763.

Hughes, G., and Madden, L. V. 1995. Some methods allowing for aggregated patterns of disease incidence in the analysis of data from designed experiments. *Plant Pathol.* 44: 927–943.

Hughes, G., and Madden, L. V. 1998. Comment: Using spatial and temporal patterns of *Armillaria* root disease to formulate management recommendations for Ontario's black spruce (*Picea mariana*) seed orchards. *Can. J. Forest Res.* 28: 154–158.

Hughes, G., and Madden, L. V. 2002. Some applications for eliciting expert knowledge of plant disease epidemics and their application in cluster sampling for disease incidence. *Crop Protect.* 21: 203–215.

Hughes, G., and Madden, L. V. 2003. Evaluating predictive models with application in regulatory policy for invasive weeds. *Agric. Syst.* 76: 755–774.

Hughes, G., and Samita, S. 1998. Analysis of patterns of pineapple mealybug wilt disease in Sri Lanka. *Plant Dis.* 82: 885–890.

Hughes, G., Monreal, A. R., and Vara, P. C. 1989. Characterizing the effect of spatially heterogeneous pest injuy and crop yields. *Crop Res. (Hort. Res.)* 29: 37–49.

Hughes, G., Madden, L. V., and Munkvold, G. P. 1996. Cluster sampling for disease incidence data. *Phytopathology* 86: 132–137.

Hughes, G., McRoberts N., Madden, L. V., and Gottwald, T. R. 1997. Relationships between disease incidence at two levels in a spatial hierarchy. *Phytopathology* 87: 542–550.

Hughes, G., Munkvold, G. P., and Samita, S. 1998. Application of the logistic-normal-binomial distribution to the analysis of Eutypa dieback disease incidence. *Int. J. Pest Manag.* 44: 35–42.

Hughes, G., Gottwald, T. R., and Levy, L. 2002a. The use of hierarchical sampling in the surveillance program for Plum pox virus incidence in the United States. *Plant Dis.* 86: 259–263.

Hughes, G., Gottwald, T. R., and Yamamura, K. 2002b. Survey methods for assessment of *Citrus tristeza virus* incidence in urban citrus populations. *Plant Dis.* 86: 367–372.

Hughes, G., McRoberts, N., and Madden, L. V. 2004. Daamen's incidence-severity relationship revisited. *Eur. J. Plant Pathol.* 110: 759–761.

Hutchison, W. D. 1994. Sequential sampling to determine population density. In: *Handbook of Sampling Methods for Arthropods in Agriculture* (L. P. Pedigo and G. D. Buntin, editors). CRC Press, Boca Raton, FL, pp. 207–243.

Ingram, W. R., and Green, S. M. 1972. Sequential sampling for bollworms on raingrown cotton in Botswana. *Cotton Grow. Rev.* 49: 265–275.

Irwin, M. E., Ruesink, W. G., Isard, S. A., and Kampmeier, G. E. 2000. Mitigating epidmeics caused by non-persistently transmitted aphid-borne viruses: The role of the pliant environment. *Virus Res.* 71: 185–211.

Ives, P. M., and Moon, R. D. 1987. Sampling theory and protocol for insects. In: *Crop Loss Assessment and Pest Management* (P. S. Teng, editor). APS Press, St Paul, MN, pp. 49–75.

Iwao, S. 1975. A new method of sequential sampling to classify populations relative to a critical density. *Res. Popul. Ecol.* 16: 281–288.

James, W. C. 1971. *A Manual of Assessment Keys for Plant Diseases*, Canada Dept. of Agric., Publ. No. 1458. APS Press, St Paul, MN.

James, W. C. 1974. Assessment of plant diseases and losses. *Annu. Rev. Phytopathol.* 12: 27–48.

James, W. C., and Shih, C. S. 1973. Relationship between incidence and severity of powdery mildew and leaf rust on winter wheat. *Phytopathology* 63: 183–187.

James, W. C., and Teng, P. S. 1979. The Quantification of production constraints associated with plant diseases. In: *Applied Biology*, Vol. IV. (T. H. Coaker, editor). Academic Press, New York, pp. 201–267.

James, W. C., Shih, C. S., Hodgson, W. A., and Callbeck, L. C. 1972. The quantitative relationship between late blight of potato and loss in tuber yield. *Phytopathology* 62: 92–96.

James, W. C., Teng, P. S., and Nutter, F. W. Jr 1991. Estimated losses of crops from plant pathogens. In: *CRC Handbook of Pest Management in Agriculture* (D. Pimentel, editor). CRC Press, Boac Raton, FL, pp. 15–51.

Jeger, M. J. 1982. The relation between total, infectious, and postinfectious diseased plant tissue. *Phytopathology* 72: 1185–1189.

Jeger, M. J. 1983. Analysing epidemics in time and space. *Plant Pathol.* 32: 5–11.

Jeger, M. J. 1986a. The potential of analytic compared with simulation approaches to modeling in plant disease epidemiology. In: *Plant Disease Epidemiology, Vol. 1: Population Dynamics and Management* (K. J. Leonard and W. E. Fry, editors). Macmillan Publication, New York, pp. 255–281.

Jeger, M. J. 1986b. Asymptotic behaviour and threshold criteria in model plant disease epidemics. *Plant Pathol.* 35: 355–361.

Jeger, M. J. 1987. The influence of root growth and inoculum density on the dynamics of root disease epidemics: Theoretical analysis. *New Phytol.* 107: 459–478.

Jeger, M. J. (editor) 1989. *Spatial Components of Plant Disease Epidemics*, Prentice Hall, Englewood Cliffs, NJ.

Jeger, M. J. 1999. Improved understanding of dispersal in crop pest and disease management: Current status and future directions. *Agric. For. Meteorol.* 97: 331–349.

Jeger, M. J. 2000. Theory and plant epidemiology. *Plant Pathol.* 49: 651–658.

Jeger, M. J. 2004. Analysis of disease progress as a basis for evaluating disease management practices. *Annu. Rev. Phytopathol.* 42: 61–82.

Jeger, M. J., and Lyda, S. D. 1986. Epidemics of Phymatotrichum root rot (*Phymatotrichum omnivorum*) in cotton: Environmental correlates of final incidence and forecasting criteria. *Ann. Appl. Biol.* 109: 523–534.

Jeger, M. J., and Thresh, J. M. 1993. Modelling reinfection of replanted cocoa by swollen shoot virus in pandemically diseased areas. *J. Appl. Ecol.* 30: 187–196.

Jeger, M. J., and van den Bosch, F. 1994a. Threshold criteria for model plant disease epidemics. I. Asymptotic results. *Phytopathology* 84: 24–27.

Jeger, M. J., and van den Bosch, F. 1994b. Threshold criteria for model plant disease epidemics. II. Persistence and endemicity. *Phytopathology* 84: 28–30.

Jeger, M. J., and Viljanen-Rollinson, S. L. H. 2001. The use of the area under the disease-progress curve (AUDPC) to assess quantitative disease resistance in crop cultivars. *Theor. Appl. Genet.* 102: 32–40.

Jeger, M. J., Jones, D. G., and Griffiths, E. 1983. Disease spread of nonspecialized fungal pathogens from inoculated point sources in intraspecific mixed stands of cereal cultivars. *Ann. Appl. Biol.* 102: 237–244.

Jeger, M. J., van den Bosch, F., Madden, L. V., and Holt, J. 1998. A model for analyzing plant-virus transmission characteristics and epidemic development. *IMA J. Math. Appl. Med. Biol.* 15: 1–18.

Jeger, M. J., Dutmer, M. Y., and van den Bosch, F. 2002. Modelling plant virus epidemics in a combined field-nursery system. *IMA J. Math. Appl. Med. Biol.* 19: 1–20.

Jeger, M. J., Holt, J., van den Bosch, F., and Madden, L. V. 2004. Epidemiology of insect-transmitted plant viruses: Modeling disease dynamics and control interventions. *Physiol. Entomol.* 29: 1–14.

Johnson, C. S. 1991. Plant mortality distribution and crop losses in flue-cured tobacco. *Plant Dis.* 75: 390–394.

Johnson, D. A., Alldredge, J. R., and Allen, J. R. 1994. Weather and downy mildew epidemics of hop in Washington state. *Phytopathology* 84: 524–527.

Johnson, D. A., Alldredge, J. R., and Vackoch, D. L. 1996. Potato late blight forecasting models for the semi-arid environment of south-central Washington. *Phytopathology* 86: 480-484.

Johnson, D. A., Alldredge, J. R., and Hamm, P. B. 1998. Expansion of potato late blight forecasting models for the Columbia Basin of Washington and Oregon. *Plant Dis.* 82: 642–645.

Johnson, K. B. 1987. Defoliation, disease, and growth: A reply. *Phytopathology* 77: 1495–1497.

Johnson, K. B. 1992. Evaluation of a mechanistic model that describes potato crop losses caused by multiple pests. *Phytopathology* 82: 363–369.

Johnson, K. B., and Teng, P. S. 1990. Coupling a disease progress model for early blight to a model of potato growth. *Phytopathology* 80: 416–425.

Johnson, N. L., and Kotz, S. 1970. *Continuous Univariate Distributions*. John Wiley amd Sons, New York.

Johnson, N. L., Kotz, S., and Sorkin, H. L. 1980. "Faulty inspection" distributions. Commun. Statist. *Theor. Meth.* A9: 917–922.

Johnson, N. L., Kotz, S., and Wu, X. 1991. *Inspection Errors for Attributes in Quality Control*, Chapman and Hall, London.

Johnson, N. L., Kotz, S., and Wu, X. 1991. *Inspection Errors for Attributes in Quality Control*, Chapman and Hall, London.

Johnson, N. L., Kotz, S., and Balakrishnan, N. 1994. *Continuous Univariate Distributions*, Vol. 1, 2nd Edn. John Wiley and Sons, New York.

Johnson, N. L., Kotz, S., and Balakrishnan, N. 1995. *Continuous Univariate Distributions*, Vol. 2. 2nd Edn. John Wiley and Sons, New York.

Johnson, N. L., Kotz, S., and Kemp, A. 1996. *Discrete Distributions*, 2nd Edn. John Wiley and Sons, New York.

Jones, D. G. (editor). 1998a. *The Epidemiology of Plant Diseases*. Kluwer Academic Plublishers, Dordrecht.

Jones, D. G. 1998b. An introduction to plant disease epidemiology. In: *Plant Disease Epidemiology* (D. G. Jones, editor). Kluwer, London, pp. 3–13.

Jones, D. R. 1994a. Evaluation of fungicides for control of eyespot disease and yield loss relationships in winter wheat. *Plant Pathol.* 43: 831–846.

Jones, V. P. 1994b. Sequential estimation and classification procedures for binomial counts. In: *Handbook of Sampling Methods for Arthropods in Agriculture* (L. P. Pedigo and G. D. Buntin, editors). CRC Press, Boca Raton, FL, pp. 175–205.

Jones, W. H. S. (Translator). 1972. *Hippocrates I. Loeb Classical Library*, Heinemann, London.

Jovanovic, B. D., and Levy, P. S. 1997. A look at the Rule of Three. *Am. Statist.* 51: 137–139.

Jovanovic, B. D., and Zalenski, R. J. 1997. Safety evaluation and confidence intervals when the number of observed events is small or zero. *Anna. Emerg. Med.* 30: 301–306.

Julius, S. 2004. Letter to the editors of Biometrics. *Biometrics* 60: 285–285.

Kadane, J. B., and Wolfson, L. J. 1998. Experiences in elicitation. *Statistician* 47: 3–19.

Kammen, D. M., and Hassenzah, D. M. 1999. *Should We Risk It? Exploring Environmental, Health, and Technological Problem Solving*, Princeton University Press, Priceton, NJ.

Kampmeijer, P. and Zadoks, J. C. 1977. EPIMUL, a simulator of foci and epidemics in mixtures, multilines, and mosaics of resistant and susceptible plants. Simulation Monographs. Pudoc, Wageningen.

Karandinos, M. G. 1976. Optimum sample size and comments on some published formulae. *Bull. Entomol. Soc. Am.* 22: 417–421.

Keeling, M. J. 1999. The effects of local spatial structure on epidemiological invasions. *Proc. R. Soc. Lond. B* 266: 859–867.

Keeling, M. J. 2000. Simple stochastic models and their power-law type behaviour. *Theor. Popul. Biol.* 58: 21–31

Keeling, M. J., Brooks, S. P., and Gilligan, C. A. 2004. Using conservation of pattern to estimate spatial parameters from a single snapshot. *P. N. A. S.* 101: 9155–9160.

Keitt, T. H., Bjornstad, O. N., Dixon, P. M., and Citron-Pousty, S. 2002. Accounting for spatial pattern when modeling organism-environment interactions. *Ecography* 25: 616–625.

Kelly, M., and Meentemeyer, R. 2002. Landscape dynamics of the spread of sudden Oak death. *Photo. Eng. Rem. Sens.* 68: 1001–1009.

Kelly, N. M., and Tuxen, K. 2003. WebGIS for monitoring "sudden oak death" in coastal California. *Comput., Environ. Urban Syst.* 27: 527–547.

Kemp, A. W. 1987. Families of discrete distributions satisfying Taylor's power law. *Biometrics* 43: 693–699.

Kenward, M. G., and Roger, J. H. 1997. Small sample inference for fixed effects from restricted maximum likelihood. *Biometrics* 53: 983–997.

Kermack, W. D., and McKendrick, A. G. 1927. A contribution to the mathematical theory of epidemics. *Proc. R. Soc. Lond. Ser. A* 115: 700–721.

Keyfitz, T. W. 1968. *Introduction to the Mathematics of Populations*, Addison—Weley publishing company, Massachusetts.

Kim, K. S., Wang, T. C., and Yang, X. B. 2005. Simulation of apparent infection rate to predict severity of soybean rust using a fuzzy logic system. *Phytopathology* 95: 1122–1131.

King, T. S., and Chinchilli, V. M. 2001. A generalized concordance coefficient for continuous and categorical data. *Statist. Med.* 20: 2131–2147.

Kirby, R. S., and Archer, W. A. 1927. Disease of cereal and forage crops in the United States in 1926. *Plant Dis. Rep. Suppl.* 53: 110–208.

Kish, L. 1957. Confidence limits for clustered samples. *Am. Sociol. Rev.* 22: 154–165.

Kish, L. 1995. *Survey Sampling*, John Wiley and Sons, Inc., New York.

Kleczkowski, A., Bailey, D. J., and Gilligan, C. A. 1996. Dynamically generated variability in plant-pathogen systems with biological control. *Proc. R. Soc. Lond. B* 263: 777–783.

Kocks, C. G., and Ruissen, M. A. 1996. Measuring field resistance of cabbage cultivars to black rot. *Euphytica* 91: 45–53.

Kolmogorov, A., Petrovsky, I., and Piscounov, N. 1937. Etude de l'équation de la diffusion avec croissance de la auantité de matiere et son application a un probleme biologique. Moscou Universtet. *Bull. Math.* 1: 1–25.

Kolnaar, R. W., and van den Bosch, F. 2001. Effect of temperature on epidemiological parameters of *Puccinia lagenophorae*. *Plant Pathol.* 50: 363–370.

Kono, T., and Sugino, T. 1958. On the estimation of the density of rice stems infested by the rice stem borer. *Jpn J. Appl. Entomol. Zool.* 2: 184–188.

Kot, M., Lewis, M. A., and van den Driessche, P. 1996. Dispersal data and the spread of invading organisms. *Ecology* 77: 2027–2042.

Kowalchuk, R. K., Keselman, H. J., Algina, J., and Wolfinger, R. D. 2004. The analysis of repeated measurements with mixed-model adjusted F tests. *Edu. Psychol. Measure.* 64: 224–242.

Kranz, J. (editor). 1974a. *Epidemics of Plant Diseases: Mathematical Analysis and Modeling*, Springer, Berlin.

Kranz, J. 1974b. Comparison of epidemics. *Annu. Rev. Phytopathol.* 12: 355–374.

Kranz, J. 1980. Comparative epidemiology: An evaluation of scope, concepts and methods. In: *Comparative epidemiology; A Tool for Better Disease Management* (J. Palti and J. Kranz, editors). PUDOC, Wageningen, The Netherlands, pp. 18–28.

Kranz, J. 1988. Measuring plant disease. In: *Experimental Techniques in Plant Disease Epidemiology* (J. Kranz and J. Rotem, editors). Springer, Berlin, pp. 35–50.

Kranz, J. (editor). 1990a. *Epidemics of Plant Diseases: Mathematical Analysis and Modeling*, 2nd Edn. Springer, Berlin.

Kranz, J. 1990b. Epidemics, their mathematical analysis and modeling: An introduction. In: *Epidemics of Plant Diseases: Mathematical Analysis and Modeling*, 2nd Edn. (J. Kranz, editor). Springer, Berlin, pp. 1–11.

Kranz, J. 1996. Epidemiologie der Pflanzenkrankheiten. Eine Einführung, Ulmer.

Kranz, J. 2003. *Comparative Epidemiology of Plant Diseases*, Springer, Berlin.

Krause, M. S., Madden, L. V., and Hoitink, H. A. J. 2001. Effect of potting mix microbial carrying capacity on biological control of Rhizoctonia damping-off of radish and Rhizoctonia crown and root rot of Poinsettia. *Phytopathology* 91: 1116–1123.

Krebs, C. J. 1989. *Ecological Methodology*, HarperCollins Publishers Inc., New York.

Krebs, C. J. 1998. *Ecological Methodology*, 2nd Edn. Addison Wesley Longman, Inc., CA.

Kühn, J. G. 1858. Die Krankheiten der Kulturgewächse; Ihr Uraschen und Ihre Verhütung. Bosselman, Berlin. Kuno, E. 1969. A new method of sequential sampling to obtain the population estimates with a fixed level of precision. *Res. Popul. Ecol.* 11: 127–136.

Kuno, E. 1969. A new method of sequential sampling to obtain the population estimates with a fixed level of precision. *Res. Popul. Ecol.* 11: 127–136.

Kuno, E. 1972. Some notes on population estimation by sequential sampling. *Res. Popul. Ecol.* 14: 58–73.

Kuno, E. 1991a. Sampling and analysis of insect populations. *Annu. Rev. Entomol.* 36: 285–304.

Kuno, E. 1991b. Verifying zero-infestation on pest control: A simple sequential test based on the succession of zero-samples. *Res. Popul. Ecol.* 33: 29–32.

Lalancette, N., Ellis, M. A., and Madden, L. V. 1988. Development of an infection efficiency model for *Plasmopara viticola* on American grape based on temperature and duration of leaf wetness. *Phytopathology* 78: 794–800.

Lalancette, N., Madden, L. V., and Ellis, M. A. 1988. A quantitiative model for describing the sporulation of *Plasmopara viticola* on grape leaves. *Phytopathology* 78: 1316–1321.

Lamari, L. *ASSESS: Image Analysis Software for Plant Disease Quantification*, APS Press, St Paul, MN.

Lambert, D. H., Villareal, R. L., and MacKenzie, D. R. 1980. A general model for gradient analysis. *Phytopathol. Z.* 98: 150–154.

Landa, B. B., Navas-Cortes, J. A., Hervas, A., Jimenex-Diaz, R. M. 2001. Influence of temperature and inoculum density of *Fusarium oxysporum* f. sp *ciceris* on suppression of fusarium wilt of chickpea by rhizosphere bacteria. *Phytopathology* 91: 807–816.

Landis, J., and Koch, G. 1977. The measurement of observer agreement for categorical data. *Biometrics* 33: 159–174.

Large, E. C. 1952. The interpretation of progress curves for potato blight and other plant diseases. *Plant Pathol.* 1: 109–117.

Large, E. C. 1954. Growth stages in cereals: Illustrations of the Feekes scale. *Plant Pathol.* 3: 128–129.

Large, E. C. 1966. Measuring plant disease. *Annu. Rev. Phytopathol.* 4: 9–28.

Large, E. C., and Doling, D. A. 1962. Effect of mildew on yield of winter wheat. *Plant Pathol.* 12: 128–130.

Larkin, R. P., Gumpertz, M. L., and Ristaino, J. B. 1995. Geostatistical analyisis of Phytophthora epidemic development in commercial bell pepper fields. *Phytophathology* 85: 191–203.

Lawless, J. F. 1982. *Statistical Models and Methods for Lifetime Data*, John Wiley and Sons, New York.

Lawson, A. B. 2001. *Statistical Methods in* Spatial Epidemiology, John Wiley, New York.

Lee, R. 1978. *Forest Microclimatology*, Columbia University Press, New York.

Leemis, L. M. 1986. Relationships among common univariate distributions. *Am. Statist.* 40: 143–146.

Legendre, P., and Legendre, L. 1998. *Numerical Ecology*, 2nd Edn. Elsevier Science B. V., Amsterdam.

Legendre, P., Dale, M. R. T., Fortin, M.-J., Gurevitch, J., Hohn, M., and Myers, D. 2002. The consequences of spatial structure for the design and analysis of ecological field surveys. *Ecography* 25: 601–615.

Leonard, K. J., and Fry, W. E., editors. 1989. *Plant Disease Epidemiology, Vol. 2: Genetics, Resistance, and Management*. McGraw-Hill, New York.

Levy, P. S., Hsia, J., Jovanovic, B., and Passaro, D. 2002. Sampling mailrooms for presence of anthrax spores: A curious property of the hypergeometric distribution under an unusual hypothesis testing scenario. *Chance* 15: 19–21.

Li-Cor. 1990. *LAI-2000 plant canopy analyzer. Insturction Manual*, LI-COR, Lincoln, NE.

Liebhold, A. M., and Gurevitch, J. 2002. Integrating the statistical analysis of spatial data in ecology. *Ecography* 25: 553–557.

Lin, C. S., Poushinsky, G., and Mauer, M. 1979. An examination of five sampling methods under random and clustered distribution using simulation. *Can. J. Plant Sci.* 59: 121–130.

Lin, L., Hedayat, A. S., Sinha, B., and Yang, M. 2002. Statistical methods in assessing agreement: Models, issues, and tools. *J. Am. Statist. Assoc.* 97: 257–270.

Lin, L. I.-K. 1989. A concordance correlation coefficient to evaluate reproducibility. *Biometrics* 45: 255–268.

Lin, L. I.-K., and Chinchilli, V. 1997. Rejoinder to the Letter to the Editor from Atkinson and Nevill. *Biometrics* 53: 777–778.

Lindow, S., and Webb, R. R. 1983. Quantification of foliar plant disease symptoms by microcomputer-digitized video image analysis. *Phytopathology* 73: 520–524.

Littell, R. C., Milliken, G. A., Stroup, W. W., and Wolfinger, R. D. 1996. *SAS System for Mixed Models*, SAS Publication, Cary, NC.

Littell, R. C., Milliken, G. A., Stroup, W. W., Wolfinger, R. D., and Schabenberger, O. 2006. *SAS System for Mixed Models*, 2nd Edn. SAS Publication, Cary, NC.

Liu,, W. M., Hethcote, H. W., and Levin, S. A. 1987. Dynamical behavior of epidemiological models with nonlinear incidence rates. *J. Math. Biol.* 25: 359–380.

Loomis, R. S., and Bennett, C. W., 1966. Competitive relationships in virus-infested sugar beet fields. *J. Am. Soc. Sugar Beet Technol.* 14: 218–231.

Lopes, D. B., and Berger, R. D. 2001. The effects of rust and anthracnose on the photosynthetic competence of diseased bean leaves. *Phytopathology* 91: 212–220.

Louis, T. A. 1981. Confidence intervals for a binomial parameter after observing no successes. *Am. Statist.* 35: 154.

Loyer, M. W. 1983. Bad probability, good statistics and group testing for binomial estimation. *Am. Statist.* 37: 57–59.

Ludwig, J. A., and Reynolds, J. F. 1988. *Statistical Ecology: A primer on Methods and Computing*, John Wiley and Sons, New York.

Lui, K.-J. 1995. Confidence limits for the population prevalence rate based on the negative binomial distribution, *Statist. Med.* 14: 1471–1477.

Lui, K.-J. 2004. *Statistical Estimation of Epidemiological Risk*. John Wiley and Sons, New York.

Lundquist, J. E. 2005. Landscape pathology—forest pathology in the era of landscape ecology. In: *Forest Pathology from Genes to Landscapes* (J. E. Lundquist and R. C. Hamelin, editors). APS Press, St Paul, MN, pp. 155–165.

Lyman, G. R. 1918. The relation of phytopathologists to plant disease survey work. *Phytopathology* 8: 219–228.

MacHardy, W. E. 1996. *Apple Scab: Biology, Epidemiology, and Management*. APS Press, St Paul, MN.

MacKenzie, D. R. 1981. Scheduling fungicide applications for potato late blight with BLITECAST. *Plant Dis.* 65: 394–399.

Madden, L. V. 1983. Measuring and modeling crop losses at the field level. *Phytopathology* 73: 1591–1596.

Madden, L. V. 1986. Statistical analysis and comparison of disease progress curves. In: *Plant Disease Epidemiology* (K. Leonard and W. E. Fry, editors). Macmillan Publishing Co., New York, pp. 55–84.

Madden L. V. 1989. Dynamic nature of within-field disease and pathogen distributions. In: *Spatial Components of Plant Disease Epidemics* (M. J. Jeger, editor). Prentice Hall, NJ, USA, pp. 96–126.

Madden, L. V. 1992. Rainfall and the dispersal of fungal spores. *Adv. Plant Pathol.* 8: 39–79.

Madden, L. V. 2006. Botanical epidemiology: Some key advances and its continuing role in disease management. *Eur. J. Plant Pathol.* 115: 3–23.

Madden, L. V., and Boudreau, M. A. 1997. Effect of strawberry density on the spread of anthracnose caused by *Colletotrichum acutatum*. *Phytopathology* 87: 828–838.

Madden, L. V., and Campbell, C. L. 1986. Description of virus disease epidemics in time and space. In: *Plant Virus Epidemics: Monitoring, Modelling, and Predicting Outbreaks* (G. D. McLean, R. G. Garrett, and W. G. Ruesink, editors). Academic Press, Sydney, pp. 273–293.

Madden, L. V., and Campbell, C. L. 1990. Nonlinear disease progress curves. In: *Epidemics of Plant Diseases: Mathematical Analysis and Modeling*, 2nd Edn. (J. Kranz, editor). Springer, Berlin, pp. 181–229.

Madden, L. V., and Ellis, M. A. 1988. How to develop plant disease forecasters. In: *Experimental Techniques in Plant Disease Epidemiology* (J. Kranz and J. Rotem, editors). Springer, Berlin, pp. 191–208.

Madden, L. V., and Hughes, G. 1994. BBD—Computer software for fitting the beta-binomial distribution to disease incidence data. *Plant Dis.* 78: 536–540.

Madden, L. V., and Hughes, G. 1995. Plant disease incidence: Distributions, heterogeneity, and temporal analysis. *Annu. Rev. Phytopathol.* 33: 529–564.

Madden, L. V., and Hughes, G. 1999a. Sampling for plant disease incidence. *Phytopathology* 89: 1088–1103.

Madden, L. V., and Hughes, G. 1999b. An effective sample size for predicting plant disease incidence in a spatial hierarchy. *Phytopathology* 89: 770–781.

Madden, L. V., and Hughes, G. 2002. Plant epidemics, models, and analysis. In: *Encyclopedia of Environmetrics*, Vol. 3 (A. El-Shaarawi and W. Piegorsch, editors). John Wiley and Sons, New York, pp. 1557–1562.

Madden, L. V., and Nutter, F. W. Jr 1995. Modeling crop losses at the field scale. *Can. J. Plant Pathol.* 17: 124–137.

Madden, L. V., and van den Bosch, F. 2002. A population-dynamics approach to assess the threat of plant pathogens as biological weapons against annual crops. *BioScience* 52: 65–74.

Madden, L. V., and Wheelis, M. 2003. The threat of plant pathogens as weapons against U. S. crops. *Annu. Rev. Phytopathol.* 41: 155–176.

Madden, L. V., Pennypacker, S. P., Antle, C. E., and Kingsolver, C. H. 1981a. A loss model for crops. *Phytopathology* 71: 685–689.

Madden, L. V., Pennypacker, S. P., and Kingsolver, C. H. 1981b. A comparison of crop loss models. *Phytopathol. Z.* [*J. Phytopathol.*] 101: 196–201.

Madden, L. V., Louie, R., Abt, J. J., and Knoke, J. K. 1982. Evaluation of tests for randomness of infected plants. *Phytopathology* 72: 195–198.

Madden, L. V., Louie, R., and Knoke, J. K. 1987a. Temporal and spatial analysis of maize dwarf mosaic epidemics. *Phytopathology* 77: 148–156.

Madden, L. V., Pirone, T. P., and Raccah, B. 1987b. Temporal analysis of two viruses increasing in the same tobacco fields. *Phytopathology* 77: 974–980.

Madden, L. V., Pirone, T. P., and Raccah, B. 1987c. Analysis of spatial patterns of virus-diseased tobacco plants. *Phytopathology* 77: 1409–1417.

Madden, L. V., Reynolds, K. M., Pirone, T. P., and Raccah, B. 1988. Modeling of tobacco virus epidemics as spatio-temporal autoregressive integrated moving-average processes. *Phytopathology* 78: 1361–1366.

Madden, L. V., Knoke, J. K., and Louie, R. 1990a. Spread of maize chlorotic dwarf virus by its leafhopper vector, *Graminella nigrifrons*. *Phytopathology* 80: 291–298.

Madden, L. V., Raccah, B., and Pirone, T. P. 1990b. Modeling plant disease increase as a function of vector numbers: Nonpersistent viruses. *Res. Popul. Ecol.* 32: 47–65.

Madden, L. V., Hughes, G., and Ellis, M. A. 1995a. Spatial heterogeneity of the incidence of grape downy mildew. *Phytopathology* 85: 269–275.

Madden, L. V., Nault, L. R., Murral, D. J., and Apelt, M. R. 1995b. Spatial pattern analysis of the incidence of aster yellows disease in lettuce. *Res. Population Ecol.* 37: 279–289.

Madden, L. V., Hughes, G., and Munkvold, G. P. 1996. Plant disease incidence: Inverse sampling, sequential sampling, and confidence intervals when observed mean incidence is zero. *Crop Protect.* 15: 621–632.

Madden, L. V., Hughes, G., and Irwin, M. E. 2000a. Coupling disease-progress-curve and time-of-infection functions for predicting yield loss of crops. *Phytopathology* 90: 788–800.

Madden, L. V., Jeger, M. J., and van den Bosch, F. 2000b. A theoretical assessment of the effects of vector-virus transmission mechanism on plant virus disease epidemics. *Phytopathology* 90: 576–594.

Madden, L. V., Turechek, W. W., and Nita, M. 2002. Evaluation of generalized linear mixed models for analyzing disease incidence data obtained in designed experiments. *Plant Dis.* 86: 316–325.

Maffia, L. A., and Berger, R. D. 1998. Models of plant disease epidemics. II: Gradients of bean rust. *J. Phytopathol.* 147: 199–206.

Main, C. E. 1977. Crop destruction—The raison d'être of plant pathology. In: *Plant Disease: An Advanced Treatise; Vol. I, How Disease is Managed* (J. G. Horsfall and E. B. Cowling, editors). Academic Press, New York, pp. 55–78.

Mak, T. K. 1988. Analysing intraclass variation for dichotomous variables. *Appl. Statist.* 37: 344–352.

Mantel, N. 1967. The detection of disease clustering and a generalized regression approach, *Cancer Res.* 27: 209–220.

Marcus, R. 1991. Deterministic and stochastic logistic-models for describing increase of plant-diseases. *Crop Protect.* 10: 155–159.

Marcus, R., Fishman, S., Talpaz, H., Salomon, R., and Bar-Joseph, M. 1984. On the spatial distribution of citrus tristeza virus disease. *Phytoparasitica* 12: 45–52.

Marshall, E. J. P. 1988. Field-scale estimates of grass weed populations in arable land. *Weed Res.* 28: 191–198.

McCartney, H. A., and Fitt, B. D. L. 1985. Construction of dispersal models. In: *Advances in Plant Pathology, Vol 3: Mathematical Modelling of Crop Disease* (C. A. Gilligan, editor). Academic Press, London, pp. 107–143.

McCartney, H. A., and Fitt, B. D. L. 1998. Dispersal of foliar fungal plant pathogens: Mechanisms, gradients, and spatial patterns. In: *Plant Disease Epidemiology* (D. G. Jones, editor). Kluwer, London, pp. 138–160.

McCullagh, P., and Nelder, J. A. 1989. *Generalized Linear Models*, 2nd Edn. Chapman and Hall, London.

McKinney, H. H. 1923. Influence of soil temperature and moisture on infection of wheat seedlings of *Helminthosporium sativum*. *J. Agric. Res.* 26: 195–218.

McLean, G. D., Garrett, R. G., and Ruesink, W. G., editors. 1986. *Plant Virus Epidemics: Monitoring, Modelling, and Predicting Outbreaks*, Academic Press, Sydney.

McRoberts, N., Hughes, G., and Madden, L. V. 1996. Incorporating spatial variability into simple disease progress models for crop pathogens. *Asp. Appl. Biol.* 46: 75–82.

McRoberts, N., Hughes, G., and Madden, L. V. 2003. The theoretical basis and practical application of relationships between different disease intensity measurements in plants. *Ann. Appl.Biol.* 142: 191–211.

McSorley, R., and Littell, R. C. 1993. Probability of detecting nematode infestations in quarantine samples. *Nematropica* 23: 177–181.

Mead, R., Curnow, R. N., and Hasted, A. M. 1993. *Statistical Methods in Agriculture and Experimental Biology*, 2nd Edn. Chapman and Hall, London.

Merrill, W. 1977. *Theory and Concepts of Plant Pathology*. Department of Plant Pathology, The Pennsylvania State University, University Park, PA, USA.

Metz, C. E. 1978a. Basic principles of ROC analysis. *Nucl. Med.* 8: 283–298.

Metz, J. A. J. 1978b. The epidemic in a closed population with all susceptibles equally vulnerable; some results for large susceptible populations and small initial infections. *Acta Biotheoretica* 27: 75–123.

Metz, J. A. J., and Diekmann, O. 1980. *The Dynamics of Physiologically Structured Populations*, Lecture Notes in Biomathematics 68. Springer, Berlin.

Mihail, J. D., and Alcorn, S. M. 1987. *Macrophomina phaseolina* spatial patterns in a cultivated soil and sampling strategies. *Phytopathology* 77: 1126–1131.

Mila, A. L., Carriquiry, A. L., and Yang, X. B. 2004. Logistic regression modeling of prevalence of soybean Sclerotinia stem rot in the North-Central region of the United States. *Phytopathology* 94: 102–110.

Millar, R. L. 1978. Phytopathological terminology: Epidemiology. *Phytopathology* 68: 682–683.

Milliken, G. A., and Johnson, D. E. 2002. *Analysis of Messy Data, Vol. III: Analysis of Covariance*, Chapmann and Hall/CRC, Boac Raton, FL.

Minogue, K. P. 1986. Disease gradients and the spread of disease. In: *Plant Disease Epidemiology: Population Dynamics and Management*, Vol. 1. (K. J. Leonard, and W. E. Fry, editors). Macmillan Publication, New York, pp. 285–310.

Minogue, K. P. 1989. Diffusion and spatial probability models for disease spread. In: *Spatial Components of Plant Disease Epidemics* (M. J. Jeger, editor). Prentice Hall, Englewood Cliff, NJ, pp. 127–143.

Minogue, K. P., and Fry, W. E. 1983a. Models for the spread of disease: Model description. *Phytopathology* 73: 1168–1173.

Minogue, K. P., and Fry, W. E. 1983b. Models for the spread of disease: Some experimental results. *Phytopathology* 73: 1173–1176.

Miyai, S., Kiritani, K., and Nakasuji, F. 1986. Models of epidemics of rice dwarf. In: *Plant Virus Epidemics; Monitoring, Modelling and Predicting Outcomes* (G. D. McLean, R. G. Garrett, and W. G. Ruesink, editors). Academic Press, Sydney, pp. 459–480.

Modjeska, J. S., and Rawlings, J. O. 1983. Spatial correlation analysis of uniformity data. *Biometrics* 39: 373–384.

Mollison, D. 1972. The rate of spatial propagation of simple epidemics. In: *Proceeding Symposium Mathematical Statistics and Probability*, Vol. 3, (D. Mollison, editor). University of California Press, Berkeley, pp. 579–614.

Mollison, D. 1977. Spatial contact models for ecological and epidemic spread. *J. R. Statist. Soc. B* 39: 283–326.

Mollison, D. 1995. The structure of epidemic models. In: *Epidemic Models: Their Structure and Relation to Data* (D. Mollison, editor). Cambridge University Press, Cambridge, pp. 17–33.

Monteith, J. L. 1977. Climate and efficiency of crop production in Britain. *Philos. Trans. R. Soc. Lond. B* 281: 277–294.

Montgomery, D. C. 1997. *Introduction to Statistical Quality Control* 3rd Edn., John Wiley and Sons, Inc., New York.

Mora-Aguilera, G., Nieto-Angel, D., Campbell, C. L., Teliz, D., and Garcia, E. 1996. Multivariate comparison of papaya ringspot epidemics. *Phytopathology* 86: 70–78.

Moran, J. R., Garrett, R. G., and Fairweather, J. V. 1983. Strategy for detecting low levels of potato viruses X and S in crops and its application to the Victorian Certified Seed Potato Scheme. *Plant Dis.* 67: 1325–1327.

Moran, J. R., Wilson, J. M., Garrett, R. G., and Smith, P. R. 1985. ELISA indexing of commercial carnations for carnation mottle virus using a urease-antibody conjugate. *Plant Pathol.* 34: 467–471.

Moran, P. A. P. 1950. Notes on continuous stochastic phenomena. *Biometrika* 37: 17–23.

Morgan, P. H., Mercer, L. P., and Flodin, N. W. 1975. General model for nutritional responses of higher organisms. *Proc. Natl Acad. Sci. USA* 72: 4327–4331.

Mullin, J. 1990. A note on determining the reliability of sample size for assessing disease in stored potato tubers. *Irish J. Agric. Res.* 29: 141–144.

Mumford, J. D., and Norton, G. A. 1987. Economics of integrated pest control. In: *Crop Loss Assessment and Pest Management* (P. S. Teng, editor). APS Press, St Paul, MN, pp. 191–200.

Mundt, C. C. 1989. Use of the modified Gregory model to describe primary disease gradients of wheat leaf rust produced from area sources of inoculum. *Phytopathology* 79: 241–246.

Mundt, C. C., and Browning, J. A. 1985. Development of crown rust epidemics in genetically diverse host populations: Effects of genotype unit area. *Phytopathology* 75: 607–610.

Munkvold, G. P., Duthie, J. A., and Marois, J. J. 1993. Spatial patterns of grapevines with Eutypa dieback in vineyards with or without perithecia. *Phytopathology* 83: 1440–1448.

Murphy, J. F., Zehnder, G. W., Schuster, D. J., Sikora, E. J., Polston, J. E., and Kloepper, J. W. 2000. Plant growth-promoting rhizobacterial mediated protection in tomato against Tomato mottle virus. *Plant Dis.* 84: 779–784.

Murray, D. J., and Ross, H. E. 1988. E. H. Weber and Fechner psychophysics. In: *G. T. Fechner and Psycology*, Passavia Universitatsverlagund-Druck GmbH, Passau, Germany, pp. 79–86.

Murray, J. D. 1991. *Mathematical Biology. Mathematics Texts Volume 19*, Springer, Berlin.

Murtaugh, P. A. 1996. The statistical evaluation of ecological indicators. *Ecol. Appl.* 6: 132–139.

Myers, J. H. 1978. Selecting a measure of dispersion. *Environ. Entomol.* 7: 619–621.

Nachman, G. 1981. A mathematical model of the functional relationship between density and spatial distribution of a population. *J. Animal Ecol.* 50: 453–460.

Nault, L. R. 1997. Arthropod transmission of plant viruses: A new synthesis. *Ann. Entomol. Soc. Am.* 90: 521–541.

Nault, B. A., and Kennedy, G. C. 1996. Sequential sampling plans for use in timing insecticide applications for control of European corn borer (*Lepidoptera: Pyralidae*) in potato. *J. Econ. Entomol.* 89: 1168–1176.

Navas-Cortés, J. A., Hau, B., and Jiménez-Díaz, R. M. 2000. Yield loss in chickpeas in relation to development of Fusarium wilt epidemics. *Phytopathology* 90: 1269–1278.

Neter, J., Wasserman, W., and Kutner, M. H. 1983. *Applied Linear Regression Models*. Richard, D. Irwin, Inc., Homewood, IL.

Neubert, M. G., and Caswell, H. 2000. Demography and dispersal: Calculation and sensitivity analysis of invasion speed for structured populations. *Ecology* 81: 1613–1628.

Newcombe, R. G. 1998. Two-sided confidence intervals for the single proportion: Comparison of seven methods. *Statist. Med.* 17: 857–872.

Newton, A. C., Hackett, C. A., Lowe, R., and Wale, S. J. 2004. Relationship between canopy reflectance and yield loss due to disease in barley. *Ann. Appl. Biol.* 145: 95–106.

Ng, S. H. K. 2003. Designing attribute acceptance sampling plans. *J. Online Math. Appl., 3*. Retrieved April 13, 2005, from http://joma.org/mathDL/4/?pa=content&sa=view Document&nodeId=428.

Ngugi, H. K., King, S. B., Holt, J., and Julian, A. M. 2001. Simultaneous temporal progress of sorghum anthracnose and leaf blight in crop mixtures with disparate patterns. *Phytopathology* 91: 720–729.

Nicholson, M., and Barry, J. 1995. Inferences from spatial surveys about the presence of an unobserved species. *Oikos* 72: 74–78.

Nilsson, H. E. 1995. Remote sensing and image analysis in plant pathology. *Annu. Rev. Phytopathol.* 33: 489–527.

Nilsson, H. E., and Johnsson, L. 1996. Hand-held radiometry of barley infected by barley stripe disease in a field experiment. *J. Plant Dis. Prot.* 103: 517–526.

Nisbet, R. M., and Gurney, W. S. C. 1982. *Modelling Fluctuating Populations*, Wiley, New York.

Nita, M., Ellis, M. A., and Madden, L. V. 2003. Reliability and accuracy of visual estimation of Phomopsis leaf blight of strawberry. *Phytopathology* 93: 995–1005.

NRC. 2002. *Predicting Invasions of Nonindigenous Plants and Plant Pests*, National Research Council. National Academy Press, Washington.

NRC. 2003. *Countering Agricultural Bioterrorism*, National Research Council. National Academy Press, Washington.

Nutter, F. W. Jr 1997. Quantifying the temporal dynamics of plant virus epidemics: A review. *Crop Prot.* 16: 603–618.

Nutter, F. W. Jr 2002a. Disease assessment. In: *Encyclopedia of Plant Pathology* (O. C. Maloy and T. D. Murray, editors). John Wiley and Sons, New York, pp. 312–323.

Nutter, F. W. Jr 2002b. Disease losses. In: *Encyclopedia of Plant Pathology* (O. C. Maloy and T. D. Murray, editors). John Wiley and Sons, New York, pp. 340–351.

Nutter, F. W. Jr, and Schultz, P. M. 1995. Improving the accuracy and perception of disease assessments: Selection of methods and use of computer-aided training programs. *Can. J. Plant Pathol.* 17: 174–184.

Nutter, F. W. Jr, Littrell, R. H., and Brenneman, T. B. 1990. Utilization of a mulitspectral radiometer to evaluate fungicide efficacy to control late leaf spot of peanut. *Phytopathology* 80: 102–108.

Nutter, F. W. Jr, Gleason, M. L., Jenco, J. H., and Christians, N. C. 1993. Assessing the accuracy, intra-rater repeatability, and inter-rater reliability of disease assessment systems. *Phytopathology* 83: 806–812.

Nutter, F. W. Jr, Esker, P. D., and Coelho Netto, R. A. 2006. Disease assessment concepts and the advancements made in improving the accuracy and precision of plant disease data. *Eur. J. Plant Pathol.* 115: 95–103.

Nyrop, J. P., and Binns, M. R. 1991. Quantitative methods for designing and analyzing sampling programs for use in pest management. In: *CRC Handbook of Pest Management in Agriculture*, 2nd Edn.,Vol. II, (D. Pimentel and A. A. Hanson, editors). CRC Press Inc., Boca Raton, FL, pp. 67–132.

Nyrop, J. P., Binns, M. R., and van der Werf, W. 1999. Sampling for IPM decision-making: Where should we invest time and resources? *Phytopathology* 89: 1104–1111.

Okubo, A. 1988. *Diffusion and Ecological Problems: Mathematical Models. Biomathematics 10*, Springer, New York.

Olkin, I., Gleser, L. J., and Derman, C. 1980. *Probability Models and Applications*, Macmillan Publishing Co. Inc., New York.

Ord, J. K. 1982. Aggregation. In: *Encyclopedia of Statistical Sciences*, Vol. 1 (S. Kotz and N. L. Johnson, editors). John Wiley and Sons, New York, pp. 30–34.

Othmer, H. G., Dunbar, S. R., and Alt, W. 1988. Models of dispersal in biological systems. *J. Math. Biol.* 26: 263–298.

Otim-Nape, G. W., Bue, A., Thresh, J. M., Baguama, Y., Ogwal, S., Ssemakula, G. N., Acola, G., Byabakama, B., Colvin, J., Cooter, R. J., and Martin, A. 2000. *The Current Pandemic of Cassava Mosaic Virus Disease in East Africa and its Control*, Natural Resources Institute, Chatham, U. K.

Palti, J., and Kranz, J. 1980. *Comparative Epidemiology: A Tool for Better Disease Management*, Pudoc, Wageningen, The Netherlands.

Palumbo, J. C., Horowitz, A. R., and Prabhaker, N. 2001. Insecticidal control and resistance management of *Bemisia tabaci*. *Crop Prot.* 20: 739–765.

Papastamati, K., van den Bosch, F., Welham, S. J., Fitt, B. D. L., Evans, N., and Steed, J. M. 2002. Modelling the daily progress of light leaf spot epidemics on winter oilseed rape (*Brassica napus*), in relation to *Pyrenopeziza brassicae* inoculum concentrations and weather factors. *Ecol. Model.* 148: 169–189.

Park, A. W., Gubbins, S., and Gilligan, C. A. 2001. Invasion and persistence of plant parasites in a spatially structured host population. *Oikos* 94: 162–174.

Parker, S. K., Nutter, F. W. Jr, and Gleason, M. L. 1997. Directional spread of Septoria leaf spot in tomato rows. *Plant Dis.* 81: 272–276.

Parker, S. R., Shaw, M. W., and Royle, D. J. 1995. The reliability of visual estimates of disease severity on cereal leaves. *Plant Pathol.* 44: 856–864.

Parker, S. R., Welham, S., Paveley, N. D., Foulkes, J., and Scott, R. K. 2004. Tolerance of septoria leaf blotch in winter wheat. *Plant Pathol.* 53: 1–10.

Pataky, J. K., and Headrick, J. M. 1988. Relationships between common rust incidence and severity on a susceptible and a partially resistant sweet corn hybrid. *Phytopathology* 78: 1155–1160.

Pataky, J. K., Raid, R. N., du Toit, L. J., and Schueneman, T. J. 1998. Disease severity and yield of sweet corn hybrids with resistance to northern leaf blight. *Plant Dis.* 82: 57–63.

Patil, G. P., and Stiteler, W. M. 1974. Concepts of aggregation and their quantification: A critical review with some new results and applications. *Res. Popul. Ecol.* 15: 238–254.

Paul, P. A., El-Allaf, S. M., Lipps, P. E., and Madden, L. V. 2005. Relationships between incidence and severity of Fusarium head blight on winter wheat in Ohio. *Phytopathology* 95: 1049–1060.

Paveley, N. D., Sylvester-Bradley, R., Scott, R. K., Craigon, J., and Day, W. 2001. Steps in predicting the relationship of yield on fungicide dose. *Phytopathology* 91: 708–716.

Pearl, R., and Reed, L. J. 1920. Rate of growth of the population of the United States since 1790 and its mathematical presentation. *Proc. Natl Acad. Sci. USA* 6: 275–288.

Pedigo, L. P., and Buntin, G. D., editors. 1994. *Handbook for Sampling Methods for Arthropods in Agriculture*, CRC Press, Boca Raton, FL.

Pedigo, L. P., Hutchins, S. H., and Higley, L. G. 1986. Economic injury levels in theory and practice. *Annu. Rev. Entomol.* 31: 341–368.

Perry, J. 1994. Sampling and applied statistics for pests and diseases. *Aspects Appl. Biol.* 37: 1–14.

Perry, J. N. 1981. Taylor's power law for dependence of variance on mean in animal populations. *Appl. Statist.* 30: 254–263.

Perry, J. N. 1998. Measures of spatial pattern for counts. *Ecology* 79: 1008–1017.

Perry, J. N., and Dixon, P. 2002. A new method for measuring spatial association in ecological count data. *Ecoscience* 9: 133–141.

Perry, J. N., and Taylor, L. R. 1985. Adès: New ecological families of species-specific frequency distributions that describe repeated spatial samples with an intrinsic power-law variance-mean property. *J. Anim. Ecol.* 54: 931–953.

Perry, J. N., and Taylor, L. R. 1988. Families of distributions for repeated samples of animal counts. *Biometrics* 44: 881–890.

Perry, J. N., Winder, L., Holland, J. M., and Alston, R. D. 1999. Red-blue plots for detecting clusters in count data. *Ecol. Lett.* 2: 106–113.

Perry, J. N., Liebhold, A. M., Rosenberg, M. S., Dungan, J., Miriti, M., Jakomulska, A., and Citron-Pousty, S. 2002. Illustrations and guidelines for selecting statistical methods for quantifying spatial pattern in ecological data. *Ecography* 25: 578–600.

Pethybridge, S. J., and Madden, L. V. 2003. Analysis of spatiotemporal dynamics of virus spread in an Australian hop garden by stochastic modeling. *Plant Dis.* 87: 56–62.

Pethybridge, S. J., and Turechek, W. W. 2003. Analysis of the association among three viruses infecting hop in Australia. *Plant Pathol.* 52: 158–167.

Pethybridge, S. J., Madden, L. V., Griggs, J., and Wilson, C. R. 2004. Species composition and abundance of aphids in Australian hop gardens and their impact on spatiotemporal patterns of *Carlavirus* epidemics. *Plant Pathol.* 53: 498–507.

Pfeiffer, D. U., and Morris, R. S. 1994. Spatial analysis techniques in veterinary epidemiology. *The Kenya Veterinarian* 18: 483–485.

Pham-Gia, T., and Turkkan, N. 1992. Information loss in a noisy dichotomous sampling process. *Commun. Statist.—Theor. Meth.* 21: 2001–2018.

Philippi, T. E. 1993. Multiple regression: Herbivory. In: *Design and Analysis of Ecological Experiments* (S. M. Scheiner and J. Gurevitch, editors). Chapman and Hall, New York, pp. 183–210.

Pielaat, A., and van den Bosch, F. 1998. A model for dispersal of plant pathogens by rain splash. *IMA (Inst. Math. Appl.) J. Math. Appl. Bio. Med.* 15: 117–134.

Pielaat, A., Madden, L. V., and Gort, G. 1998. Spores splashing under different environmental conditions: A modeling approach. *Phytopathology* 88: 1131–1140.

Pielou, E. C. 1977. *Mathematical Ecology*, John Wiley and Sons, New York.

Piepho, H.-P. 1999. Analysing disease incidence data from designed experiments by generalised linear mixed models. *Plant Pathol.* 48: 668–674.

Piepho, H.-P., Büchse, A., and Richter, C. 2004. A mixed modeling approach for randomized experiments with repeated measurements. *J. Agron. Crop Sci.* 190: 230–247.

Pimentel, D., editor. 2002. *Biological Invasions: Economic and Environmental Costs of Alien Plant, Animal, and Microbe Species*, CRC Press, Boca Raton, FL.

Pimentel, D., Lach, L., Zuniga, R., and Morrison, D. 2000. Environmental and economic costs associated with non-indigenous species in the United States. *BioScience* 50: 53–65.

Plumb, R. T., and Thresh, J. M. 1983. *Plant Virus Epidemiology: The Spread of Insect-borne Viruses*, Blackwell, Oxford, UK.

Powelson, M. L., and Rowe, R. C. 1993. Biology and management of early dying of potatoes. *Annu. Rev. Phytopathol.* 31: 111–126.

Prentice, R. L. 1986. Binary regression using an extended beta-binomial distribution, with discussion of correlation induced by covariate measurement error. *J. Am. Statist. Assoc.* 81: 321–327.

Press, S. J., and Wilson, S. 1978. Choosing between logistic regression and discriminant analysis. *J. Am. Statist. Assoc.* 73: 699–705.

Price, T. V., and Williams, B. L. 1990. Studies on the severity and incidence of paspalum leaf blight. *Aust. J. Agric. Res.* 41: 377–392.

Punja, Z., Smith, V. L., Campbell, C. L., and Jenkins, S. F. 1985. Sampling methods and extraction procedures for estimation of numbers, spatial pattern and temporal distribution of sclerotia of *Sclerotium rolfsii* in soil. *Plant Dis.* 69: 469–474.

Qu, Y., Beck, G. J., and Williams, G. W. 1990. Polya-Eggenberger distribution: Parameter estimation and hypothesis tests. *Biom. J.* 2: 229–242.

Quinn, G. P., and Keough, M. J. 2002. *Experimental Design and Data Analysis for Biologists*, Chapters 2–6. Cambridge University Press, Cambridge, UK.

Rahme, E., and Joseph, L. 1998. Prevalence estimation for a rare disease: Adjusted maximum likelihood. *The Statist.* 47: 149–158.

Rao, J. N. K., and Scott, A. J. 1992. A simple method for the analysis of clustered binary data. *Biometrics* 48: 577–585.

Rapilly, F. 1991. *L'Épidémiologie en Pathologie Végétale. Mycoses Aériennes*, INRA, Paris.

Ratkowsky, D. A. 1983. *Nonlinear Regression Modeling: A Unified Practical Approach*, Marcel Dekker, Inc., New York.

Ratkowsky, D. A. 1990. *Handbook of Nonlinear Regression Models*, Marcel Dekker, Inc., New York.

Reddingius, J. 1971. Notes on the mathematical theory of epidemics. *Acta Biotheor.* 20: 125–157.

Reich, R. M., and Lundquist, J. E. 2005. Use of spatial statistics in assessing forest diseases. In: *Forest Pathology from Genes to Landscapes* (J. E. Lundquist and R. C. Hamelin, editors). APS Press, St Paul, MN, pp. 127–143.

Reimer, C. 1959. Statistical analysis of percentages based on unequal numbers, with examples from entomological research. *Can. Entomol.* 91: 88–92.

Renshaw, E. 1991. *Modelling Biological Populations in Space and Time*, Cambridge University Press, Cambridge.

Reynolds, K. L., and Arneson, P. A. 1997. Simulation of potato late blight epidemics. In: *Exercises in Plant Disease Epidemiology* (L. J. Francl and D. A. Neher, editors). APS Press, St Paul, MN, pp. 161–165.

Reynolds, K. M., and Madden, L. V. 1988. Analysis of epidemics using spatio-temporal autocorrelation. *Phytopathology* 78: 240–246.

Reynolds, K. M., Madden, L. V., and Ellis, M. A. 1988a. Spatio-temporal analysis of epidemic development of leather rot of strawberry. *Phytopathology* 78: 246–252.

Reynolds, K. M., Madden, L. V., and Ellis, M. A. 1988b. Effect of weather variables on strawberry leather rot epidemics. *Phytopathology* 78: 822–827.

Richards, F. J. 1959. A flexible growth function for empirical use. *J. Exp. Bot.* 10: 290–300.

Richards, F. J. 1969. The quantitative analysis of growth. In: *Plant Physiology*, Vol. 5A (F. C. Steward, editor). Academic Press, New York, pp. 3–76.

Ridout, M. S., and Xu, X.-M. 2000. Relationships between several quadrat-based statistical measures used to characterize spatial aspects of data on disease incidence. *Phytopathology* 90: 568–575.

Ridout, M. S., Demétrio, C. G. G., and Firth, D. 1999. Estimating intraclass correlation for binary data. *Biometrics* 55: 137–148.

Ripley, B. D. 1976. The second-order analysis of stationary point processes. *J. Appl. Probab.* 13: 255–266.

Ripley, B. D. 1977. Modelling spatial patterns (+ discussion). *J. R. Statist. Soc.* B39: 172–212.

Ripley, B. D. 1979. Tests of 'randomness' for spatial point patterns. *J. R. Statist. Soc.* B41: 368–374.

Ristaino, J. B., and Gumpertz, M. L. 2000. New frontiers in the study of dispersal and spatial analysis of epidemics caused by species in the genus *Phytophthora*. *Annu. Rev. Phytopathol.* 38: 541–576.

Robert, C., Bancal, M.-O., Nicolas, P., Lannou, C., and Ney, B. 2004. Analysis and modeling of effects of leaf rust and *Septoria tritici* blotch on wheat growth. *J. Exp. Bot.* 55: 1079–1094.

Robert, C., Bancal, M.-O., Ney, B., and Lannou, C. 2005. Wheat leaf photosynthesis loss due to leaf rust, with respect to lesion development and leaf nitrogen status. *New Phytologist* 165: 227–241.

Rodoni, B. C., Hepworth, G., Richardson, C., and Moran, J. R. 1994. The use of a sequential batch testing procedure and ELISA to determine the incidence of five viruses in Victorian cut-flower Sim carnations. *Aust. J. Agric. Res.* 45: 223–230.

Rodriguez-De Gonzalez, P., Cho, J. J., Ullman, D., and Nelson, S. C. 1995. Effect of different initial spatial patterns of zucchini yellow mosaic virus (ZYMV) on the spatial and temporal disease spread in zucchini squash (Abstract). *Phytopathology* 85: 1122.

Rogan, W. J., and Gladen, B. 1978. Estimating prevalence from the results of a screening test. *Am. J. Epidemiol.* 107: 71–76.

Romanow, L. R., Moyer, J. W., and Kennedy, G. G. 1986. Alteration of efficiencies of acquisition and inoculation of watermelon mosaic virus 2 by plant resistance to the virus and to an aphid vector. *Phytopathology* 76: 1276–1281.

Romig, R. W., and Calpouzos, L. 1970. The relationship between stem rust and loss in yield of spring wheat. *Phytopathology* 60: 1801–1805.

Robinson, R. A. 1976. *Plant Pathosystems*, Springer, Berlin.

Ross, R. 1911. *The Prevention of Malaria*, 2nd Edn. Murray, London.

Ross, R. 1915. Some *a priori* pathometric equations. *Br. Med. J.* 1: 546–547.

Rotem, J. 1988. Techniques of controlled-condition experiments. In: *Experimental Techniques in Plant Disease Epidemiology* (J. Kranz and J. Rotem, editors). Springer, Berlin, pp. 19–31.

Roughgarden, J. 1979. *Theory of Population Genetics and Evolutionary Ecology: An Introduction*, MacMillan, New York.

Roumagnac, P., Pruvost, O., Chiroleu, F., and Hughes, G. 2004. Spatial and temporal analyses of bacterial blight of onion caused by *Xanthomonas axonopodis* pv. *allii*. *Phytopathology* 94: 138–146.

Ruesink, W. G. 1980. Introduction to sampling theory. In: *Sampling Methods in Soybean Entomology* (M. Kogan and D. C. Herzog, editors). Springer, New York, pp. 61–78.

Russell, G., Jarvis, P. G., and Monteith, J. L. 1989. Absorption of radiation by canopies and stand growth. In: *Plant Canopies: Their Growth, Form and Function* (G. Russell, B. Marshall, and P. G. Jarvis, editors). Cambridge University Press, UK, pp. 21–39.

Ryan, C. C., and Birch, R. G. 1978. Phytopathological terminology: Epiphytotic vs. epidemic. *Phytopathology* 68: 681.

Sackett, K. E., and Mundt, C. C. 2005a. Primary disease gradients of wheat stripe rust in large field plots. *Phytopathology* 95: 983–991.

Sackett, K. E., and Mundt, C. C. 2005b. The effects of dispersal gradient and pathogen life cycle components on epidemic velocity in computer simulations. *Phytopathology* 95: 992–1000.

Sampford, M. R. 1962. *An Introduction to Sampling Theory with Applications to Agriculture*, Oliver and Boyd, Edinburgh.

Samuels, M. L., and Lu T.-F. C. 1992. Sample size requirements for the back-of-the-envelope binomial confidence interval. *Am. Statist.* 46: 228–231.

Sandland, R. I., and McGilchrist, C. A. 1979. Stochastic growth curve analysis. *Biometrics* 35: 253–271.

Sanogo, S., and Yang, X. B. 2004. Overview of selected multivariate statistical methods and their use in phytopathological research. *Phytopathology* 94: 1004–1006.

Santner, T. J., and Duffy, D. E. 1989. *The Statistical Analysis of Discrete Data*, Springer, New York.

Savary, S., and Zadoks, J. C. 1992a. Analysis of crop loss in the multiple pathosystem groundnut-rust-leaf-spot. I. Six experiments. *Crop Prot.* 11: 99–109.

Savary, S., and Zadoks, J. C. 1992b. Analysis of crop loss in the multiple pathosystem groundnut-rust-leaf-spot. III. Correspondence analysis. *Crop Prot.* 11: 229–239.

Savary, S., Madden, L. V., Zadoks, J. C., and Klein-Gebbinck, H. W. 1995. Use of categorical information and correspondence analysis in plant disease epidemiology. In: *Advances in Botanical Research (Incorporating Advances in Plant Pathology*, Vol. 21 (J. A. Callow, editor). Academic Press, London, pp. 213–240.

Savary, S., Teng, P. S., Willocquet, L., and Nutter, F. W. Jr 2006. Quantification and modeling of crop losses: A review of purposes. *Annu. Rev. Phytopathol.* 44: 89–112.

Sawada, M. 1999. Rookcase: An Excel 97/2000 Visual Basic (VB) add-in for exploring global and local spatial autocorrelation. *Bull. Ecol. Soc. Am.* 80: 231–234.

Schaalje, G. B., Butts, R. A., and Lysyk, T. J. 1991. Simulation studies of binomial sampling: A new variance estimator and density predictor, with special reference to the Russian wheat aphid (*Homoptera: Aphididae*). *J. Econ. Entomol.* 84: 140–147.

Schaalje, G. B., McBride, J. B., and Fellingham, G. W. 2002. Adequacy of approximations to distributions of test statistics in complex mixed linear models. *J. Agric. Biol. Environ. Statist.* 7: 512–524.

Schabenberger, O., and Pierce, F. J. 2002. *Contemporary Statistical Models for the Plant and Soil Sciences*, CRC Press, Boca Raton, FL.

Schabenberger, O., Tharp, B. E., Kells, J. J., and Penner, D. 1999. Statistical tests for hormesis and effective dosages in herbicide dose responses. *Agron. J.* 91: 713–721.

Scherm, H. 2004. Applications of survival analysis in botanical epidemiology. *Phytopathology* 94: 1022–1026.

Schoeny, A., Jeuffroy, M.-H., and Lucas, P. 2001. Influence of take-all epidemics on winter wheat yield formation and yield loss. *Phytopathology* 91: 694–701.

Schumann, G. L. 1991. *Plant Diseases: Their Biology and Social Impact*, APS Press, St Paul, MN.

Scott, J. B., Hay, F. S., Wilson, C. R., Cotterill, P. J., and Fist, A. J. 2003. Spatiotemporal analysis of epiphytotics of downy mildew of oilseed poppy in Tasmania, Australia. *Phytopathology* 93: 752–757.

Scott, P. R., and Bainbridge, A. 1978. *Plant Disease Epidemiology*, Blackwell, Oxford, UK.

Seber, G. A. F., and Wild, C. J. 1989. *Nonlinear Regression*, John Wiley and Sons, New York.

Seem, R. C. 1984. Plant disease incidence and severity relationships. *Annu. Rev. Phytopathol.* 22: 133–150.

Seem, R. C. 1988. The measurement and analysis of the effects of crop development on epidemics. In: *Experimental Techniques in Plant Disease Epidemiology* (J. Kranz and J. Rotem, editors). Springer, Berlin, pp. 51–68.

Seem, R. C., and Gilpatrick, J. D. 1980. Incidence and severity relationships of secondary infections of powdery mildew on apple. *Phytopathology* 70: 851–854.

Seem, R. C., Magarey, P. A., McCloud, P. L., and Wachtel, M. F. 1985. A sampling procedure to detect grapevine downy mildew. *Phytopathology* 75: 1252–1257.

Segarra, J., Jeger, M. J., and van den Bosch, F. 2001. Epidemic dynamics and patterns of plant diseases. *Phytopathology* 91: 1001–1010.

Selvin, S. 1996. *Statistical Analysis of Epidemiologic Data*, 2nd Edn. Oxford University Press, New York.

Shah, D. A., and Bergstrom, G. C. 2002. A rainfall-based model for predicting the regional incidence of wheat seed infection by *Stagonospora nodorum* in New York. *Phytopathology* 92: 511–518.

Shah, D. A., and Madden, L. V. 2004. Nonparametric analysis of ordinal data in designed factorial experiments. *Phytopathology* 94: 33–43.

Shah, D. A., Bergstrom, G. C., and Ueng, P. P. 2001. Foci of *Stagonospora nodorum* blotch in winter wheat before canopy development. *Phytopathology* 91: 642–647.

Shah, D. A., Clear, R. M., Madden, L. V., and Bergstrom, G. C. 2002. Summarizing the regional incidence of seed-borne fungi with the β-binomial distribution. *Can. J. Plant Pathol.* 24: 168–175.

Shane, W. 1987. The use of principal components analysis and cluster analysis in crop loss assessment. In: *Crop Loss Assessment* (P. S. Teng and S. V. Krupa, editors). Misc. Publ. No. 7, Agricultural Exper. Station, Univ. of Minnesota, St Paul, pp. 139–149.

Shaw, M. W. 1994. Modelling stochastic processes in plant pathology. *Annu. Rev. Phytopathol.* 32: 523–544.

Shaw, M. W. 1995. Simulation of population expansion and spatial pattern when individual dispersal distributions do not decline exponentially with distance. *Proc. R. Soc. Lond. B* 259: 243–248.

Shaw, M. W., and Royle, D. J. 1987. A new method for disease-loss studies used to estimate effects of *Pyrenophora teres* and *Rhynchosporium secalis* on winter barley. *Ann. Appl. Biol.* 110: 247–262.

Sherwood, R. T., Berg, C. C., Hoover, M. R., and Zeiders, K. E. 1983. Illusions in visual assessment of Stagonospora leaf spot of orchardgrass. *Phytopathology* 73: 173–177.

Sheskin, D. J. 1997. *Handbook of Parametric and Nonparametric Statistical Procedures*, CRC Press, Boca Raton, FL.

Shigesada, N., and Kawasaki, K. 1997. *Biological Invasions: Theory and Practice. Oxford Series in Ecology and Evolution*, Oxford University Press, New York.

Shigesada, N., Kawasaki, K., and Teramoto, E. 1986. Travelling periodic waves in heterogeneous environments. *Theor. Popul. Biol.* 30: 143–160.

Shiyomi, M. 1981. Mathematical ecology of spatial pattern of biological populations. *Bull. Natl Inst. Agric. Sci.* A27: 1–29.

Shiyomi, M., and Takai, A. 1979. The spatial pattern of infected or infested plants and negative hyper-geometric series. *Jpn J. Appl. Entomol. Zool.* 23: 224–229.

Shiyomi, M., and Yoshimura, J. 2000. Measures of spatial heterogeneity for species occurrence or disease incidence with finite-counts. *Ecol. Res.* 15: 13–20.

Shoukri, M. M. 2004. *Measures of Interobserver Agreement*, Chapman and Hall/CRC, Boca Raton, FL.

Shoukri, M. M., and Pause, C. A. 1999. *Statistical Methods for Health Sciences*, 2nd Edn. CRC Press, Boca Raton, FL.

Shtienberg, D. 2000. Modelling: The basis for rational disease management. *Crop Prot.* 19: 747–752.

Shtienberg, D, Bergeron, S. N., Nicholson, A. G., Fry, W. E., and Ewing, E. E. 1990. Development and evaluation of a general model for yield loss assessment in potatoes. *Phytopathology* 80: 466–472.

Shurtleff, M. C., and Averre, C. W. III. 1997. *Glossary of Plant-Pathological Terms*, APS Press, St Paul, MN.

Silva-Acuña, R., Maffia, L. A., Zambolim, L., and Berger, R. D. 1999. Incidence-severity relationships in the pathosystem *Coffea arabica—Hemileia vastatrix*. *Plant Dis.* 83: 186–188.

Skellam, J. G. 1948. A probability distribution derived from the binomial distribution by regarding the probability of success as a varia ble between sets of trials. *J. R. Statist. Soc. B* 10: 257–261.

Skellam, J. G. 1951. Random dispersal in theoretical populations. *Biometrica* 38: 196–218.

Slafer, G. A. 2003. Genetic basis of yield as viewed from a crop physiologist's perspective. *Ann. Appl. Biol.* 142: 117–128.

Slopek, S. W. 1989. An improved method of estimating percent leaf area diseased using a 1 to 5 disease assessment scale. *Can. J. Plant Pathol.* 11: 381–387.

Smith, D. M. 1983. Maximum likelihood estimation of the parameters of the beta-binomial distribution. *Appl. Statist.* 32: 196–204.

Smith, K. M., and Lea, D. F. 1946. The transmission of plant viruses by aphids. *Parasitology* 37: 25–37.

Snedecor, G. W., and Cochran, W. G. 1989. *Statistical Methods*, 8th Edn. Iowa State University Press, Ames, IA.

Southwood, T. R. E. 1978. *Ecological Methods* 2nd Edn. Chapman and Hall, London.

Sprent, P., and Smeeton, N. C. 2001. *Applied Nonparametric Statistical Methods*, 3rd Edn. Chapman and Hall, London.

Squeglia, N. L. 1994. *Zero Acceptance Number Sampling Plans.* 4th Edn. ASQ Quality Press, Milwaukee, WI.

Sserubombwe, W. S., Thresh, J. M., Otim-Nape, G. W., and Osiru, D. O. S. 2001. Progress of cassava mosaic virus disease and whitefly vector populations in single and mixed stands of four cassava varieties grown under epidemic conditions in Uganda. *Ann. Appl. Biol.* 138: 161–170.

Stacey, A. J., Truscott, J. E., Asher, M. J. C., and Gilligan, C. A. 2004. A model for the invasion and spread of Rhizomania in the United Kingdom: Implications for disease control strategies. *Phytopathology* 94: 209–215.

Stern, V. M. 1973. Economic thresholds. *Annu. Rev. Entomol.* 18: 259–280.

Stern, V. M., Smith, R. M., van den Bosch, R., and Hagen, K. S. 1959. The integrated control concept. *Hilgardia* 29: 81–99.

Stevens, N. E. 1933. The dark ages in plant pathology in America: 1830–1870. *J. Washington Acad. Sci.* 23: 435–447.

Stevens, D. B. 1960. Cultural practices in disease control. In: *Plant Pathology: An Advanced Treatise, Vol. 3: The Diseased Population* (J. G. Horsfall and A. E. Dimond, editors). Academic Press, New York, pp. 357–429.

Strandberg, J. 1973. Spatial distribution of cabbage black rot and the estimation of diseased populations. *Phytopathology* 63: 998–1003.

Strange, R. N. 1993. *Plant Disease Control*, Chapman and Hall, London.

Stuart, A., and Ord, J. K. 1994. *Kendall's Advance Theory of Statistics, Vol. 1; Distribution Theory*, 6th Edn. Edward Arnold Publishers, London.

Stynes, B. A. 1980. Synoptic methodologies for crop loss assessment. In: *Crop Loss Assessment* (P. S. Teng and S. V. Krupa, editors). Misc. Publ. No. 7, Agricultural Exper. Station, University of Minnesota, St Paul, MN, pp. 166–175.

Suich, R. 1990. The effects of inspection errors on acceptance sampling for nonconformities. *J. Quality Technol.* 22: 314–318.

Sun, P., and Madden, L. V. 1997. Using a normal approximation to test for the binomial distribution. *Biom. J.* 39: 553–544.

Sutton, J. C., Gillespie, T. J., and James, T. D. W. 1988. Electronic monitoring and use of microprocessors in the field. In: *Experimental Techniques in Plant Disease Epidemiology* (J. Kranz and J. Rotem, editors). Springer, Berlin, pp. 99–113.

Sutula, C. L., Gillett, J. M., Morrissey, S. M., and Ramsdell, D. C. 1986. Interpreting ELISA data and establishing the positive-negative threshold. *Plant Dis.* 70: 722–726.

Swallow, W. H. 1985. Group testing for estimating infection rates and probabilities of disease transmission. *Phytopathology* 75: 882–889.

Swallow, W. H. 1987. Relative mean squared error and cost considerations in choosing groups size for group testing to estimate infection rates and probabilities of disease transmission. *Phytopathology* 77: 882–859.

Swets, J. A., Dawes, R. M., and Monahan, J. 2000. Better decisions through science. *Scientific Am.* 283: 70–75.

Swinton, J., and Gilligan, C. A. 1996. Dutch elm disease and the future of the elm in the U. K.: A quantitative analysis. *Philos. Trans. R. Soc. Lond. B* 351: 605–615.

Takai, A., and Shiyomi, M. 1980. Model and analysis of spatial pattern of infected or infested plants. *Jpn J. Appl. Entomol. Zool.* 24: 234–240.

Tanne, E., Marcus, R., Dubitzky, E., and Raccah, B. 1996. Analysis of progress and spatial pattern of corky bark in grapes. *Plant Dis.* 80: 34–38.

Tarone, R. E. 1979. Testing the goodness of fit of the binomial distribution. *Biometrika* 66: 585–590.

Taylor, L. R. 1961. Aggregation, variance and the mean. *Nature* 189: 732–735.

Taylor, L. R. 1984. Assessing and interpreting the spatial distributions of insect populations. *Annu. Rev. Entomol.* 29: 321–357.

Taylor, L. R., Woiwod, I. P., and Perry, J. N. 1979. The negative binomial as a dynamic ecological model for aggregation, and the density dependence of k. *J. Anim. Ecol.* 48: 289–304.

Taylor, L. R., Taylor, R. A. J., Woiwod, I. P., and Perry, J. N. 1983. Behavioural dynamics. *Nature* 303: 801–804.

Tebbs, J., and Bilder, C. 2004. Confidence interval procedures for the probability of disease transmission in multiple-vector-transfer designs. *J. Agric., Biol., Environ. Statist.* 9: 75–90.

Teng, P. S. 1981. Validation of computer models of plant disease epidemics: A review of philosophy and methodology. *Z. Pflanzenkr. Pflanzenschutz.* 88: 49–63.

Teng, P. S. 1985. Construciton of predictive models: II. Forecasting crop losses. In: *Advances in Plant Pathology, Vol. 3; Mathematical Modelling of Crop Disease* (C. A. Gilligan, editor). Academic Press, London, pp. 179–206.

Teng, P. S. 1987. Quantifying the relationship between disease intensity and yield loss. In: *Crop Loss Assessment and Pest Management* (P. S. Teng, editor). APS Press, St Paul, MN, pp. 105–113.

Teng, P S., and Oshima, R. J. 1983. Identification and assessment of losses. In: *Challenging Problems in Plant Health* (T. Kommedahl and P. H. Williams, editors). American Phytopathological Society, St Paul, MN, pp. 69–81.

Teng, P. S., Close, R. C., and Blackie, M. J. 1979. Comparison of models for estimating yield loss caused by leaf rust (*Puccinia hordei*) on Zephyr barley in New Zealand. *Phytopathology* 69: 1239–1244.

Termorshuizen, A. J, and Rouse, D. I. 1993. Towards a mechanistic model for the *Verticillium dahliae*-potato system. *Neth. J. Plant Pathol.* 99 (Suppl. 3): 201–218.

Thackray, D. J., Smith, L. J., Cheng, Y., Perry, J. N., and Jones, R. A. C. 2002. Effect of strain-specific hypersensitive resistance on spatial patterns of virus spread. *Ann. Appl. Biol.* 141: 45–59.

Thieme, H. R. 1977. A model for the spatial spread of an epidemic. *J. Math. Biol.* 4: 337–351.

Thieme, H. R. 1979. Asymptotic estimates of the solutions of nonlinear integral equations and asymptotic speed for the spread of populations. *J. fur Reine und Angewandte Mathematic* 306: 94–121.

Thompson, H. R. 1956. Distribution of distance to nth neighbor in a population of randomly distributed individuals. *Ecology* 37: 391–394.

Thompson, K. H. 1962. Estimation of the proportion of vectors in a natural population of insects. *Biometrics* 18: 568–578.

Thompson, W. R. 1924. La théorie mathématique de l'action des parasites entomophagues et le facteur du hasard. *Annales de la Faculté des Sciences Marseilles* 2: 69–89.

Thrall, P. H., Bever, J. D., Mihail, J. D., and Alexander, H. M. 1997. The population dynamics of annual plants and soil-borne fungal pathogens. *J. Ecol.* 83: 313–328.

Thresh, J. M. 1983. Progress curves of plant virus diseases. In: *Applied Biology*, Vol. 8. (T. H. Coaker, editor). Academic Press, New York, pp. 1–85.

Thresh, J. M. 1985. Plant virus dispersal. In: *The Movement and Dispersal of Agriculturally Important Biotic Agents* (D. R. MacKenzie, C. S. Barfield, G. G. Kennedy, and R. D. Berger, editors). Claitor's Publication, Baton Rouge, LA, pp. 51–106.

Titchener, E. B. 1928. *A Text-Book of Psychology*, The Macmillan Company, New York.

Truscott, J. E., Webb, C. R., and Gilligan, C. A. 1997. Asymptotic analysis of an epidemic model with primary and secondary infection. *Bull. Math. Biol.* 59: 1101–1123.

Turechek, W. W., and Madden, L. V. 1999a. Spatial pattern analysis of strawberry leaf blight in perennial production systems. *Phytopathology* 89: 421–433.

Turechek, W. W., and Madden, L. V. 1999b. Spatial pattern analysis and sequential sampling for the incidence of leaf spot on strawberry in Ohio. *Plant Dis.* 83: 992–1000.

Turechek, W. W., and Madden, L. V. 2000. Analysis of the association between the incidence of two spatially aggregated foliar diseases of strawberry. *Phytopathology* 90: 157–170.

Turechek, W. W., and Madden, L. V. 2001. Effect of scale on plant disease incidence and heterogeneity in a spatial hierarchy. *Ecol. Model.* 144: 75–93.

Turechek, W. W., Ellis, M. A., and Madden, L. V. 2001. Sequential sampling for incidence of Phomopsis leaf blight of strawberry. *Phytopathology* 91: 336–347.

Twengström, E., Sigvald, R., Svensson, C., and Yuen, J. 1998. Forecasting Sclerotinia stem rot in spring sown oilseed rape. *Crop Prot.* 17: 405–411.

Unger, F. 1833. *Die Exantheme der Pflanzen*, Gerold, Vienna.

Upton, G. J. G., and Fingleton, B. 1985. *Spatial Data Analysis by Example. Vol. 1: Point Pattern and Quantitative Data*, Wiley, Chichester.

vanBruggen, A. H. C., and Arneson, P. A. 1986. Path coefficient analysis of effects of *Rhizoctonia solani* on growth and development of dry beans. *Phytopathology* 76: 874–878.

van den Bosch, F., Zadoks, J. C., and Metz, J. A. J. 1988a. Focus expansion in plant disease. I: The constant rate of focus expansion. *Phytopathology* 78: 54–58.

van den Bosch, F., Zadoks, J. C., and Metz, J. A. J. 1988b. Focus expansion in plant disease. II: Realistic parameter-sparse models. *Phytopathology* 78: 59–64.

van den Bosch, F., Frinking, H. D., Metz, J. A. J., and Zadoks, J. C. 1988c. Focus expansion in plant disease. III: Two experimental examples. *Phytopathology* 78: 919–925.

van den Bosch, F., Metz, J. A. J., and Diekmann, O. 1990a. The velocity of population expansion. *J. Math. Biol.* 28: 529–565.

van den Bosch, F., Verhaar, M. A., Buiel, A. A. M., Hoogkamer, W., and Zadoks, J. C. 1990b. Focus expansion in plant disease. IV: Expansion rates in mixtures of resistant and susceptible hosts. *Phytopathology* 80: 598–602.

van den Bosch, F., Hengeveld, R., and Metz, J. A. J. 1992. Analysing the velocity of animal range expansion. *J. Biogeography* 19: 135–150.

van den Bosch, F., Metz, J. A. J., and Zadoks, J. C. 1999. Pandemics of focal plant disease, a model. *Phytopathology* 89: 495–505.

Vanderplank, J. E. 1946. A method for estimating the number of random groups of adjacent diseased plants in a homogeneous field. *Trans. R. Soc. S. Africa* 31: 269–278.

Vanderplank, J. E. 1960. Analysis of epidemics. In: *Plant Pathology: An Advanced Treatise, Vol. 3: The Diseased Population* (J. G. Horsfall and A. E. Dimond, editors). Academic Press, New York, pp. 229–289.

Vanderplank, J. E. 1963. *Plant Diseases: Epidemics and Control*, Academic Press, New York.

Vanderplank, J. E. 1968. *Disease Resistance in Plants*, Academic Press, New York.

Vanderplank, J. E. 1975. *Principles of Plant Infection*, Academic Press, London.

Vanderplank, J. E. 1984. *Disease Resistance in Plants*, 2nd Edn. Academic Press, New York.

van der Zaag, D. E. 1956., Overwintering en epidemiologie van *Phytophthora infestans*, tevens enige nieuwe bestrijdingsmogelijkheden. *Tijdschr. Plantenziekten.* 62: 89–156.

Venette, R. C., Moon, R. D., and Hutchison, W. D. 2002. Strategies and statistics of sampling for rare individuals. *Annu. Rev. Entomol.* 47: 143–174.

Verhulst, V. 1838. Notice sur la loi que la population suit dans son accroissement. *Corr. Math. et Phys. publ. par A.* Quetelet, T. X: 113–121.

Vincelli, P. C., and Lorbeer, J. W. 1987. Sequential sampling plan for timing initial fungicide application to control Botrytis leaf blight of onion. *Phytopathology* 77: 1301–1303.

Von Martius, C. F. P. 1842. Die Kartoffel-Epidemie der letzten Jahre oder die Stockfäule und Räude der Kartoffein. Verlag der Königlich Bayerischen Akamemie der Wissenschaften. Münich.

Vereijssen, J., Schneider, J. H. M., Termorshuizen, and Jeger, M. J. 2003. Comparison of two disease assessment methods for assessing Cercospora leaf spot in sugar beet. *Crop Prot.* 22: 201–209.

Waggoner, P. E. 1974. Simulation of epidemics. In: *Epidemics of Plant Diseases: Mathematical Analysis and Modeling* (J. Kranz, editor). Springer, Berlin, pp. 137–160.

Waggoner, P. E. 1978. Computer simulation of epidemics. In: *Plant Disease: An Advanced Treatise. Vol II: How Disease Develops in Populations* (J. G. Horsfall and E. B. Cowling, editors). Academic Press, New York, pp. 203–222.

Waggoner, P. E. 1981. Models of plant disease. *BioScience* 31: 325–319.

Waggoner, P. E. 1990. Assembling and using models of epidemics. In: *Epidemics of Plant Diseases: Mathematical Analysis and Modeling*, 2nd Edn. (J. Kranz, editor). Springer, Berlin, pp. 230–260.

Waggoner, P. E., and Aylor, D. E. 2000. Epidemiology: A science of patterns. *Annu. Rev. Phytopathol.* 38: 71–94.

Waggoner, P. E., and Berger, R. D. 1987. Defoliation, disease, and growth. *Phytopathology* 77: 393–398.

Waggoner, P. E., and Horsfall, J. G. 1969. EPIDEM, a simulator of plant disease written for a computer. *Conn. Agric. Exp. Stn. Bull.* 698.

Waggoner, P. E., and Rich, S. 1981. Lesion distribution, multiple infection, and the logistic increase of plant disease. *Proc. Natl Acad. Sci. USA* 78: 3292–3295.

Waggoner, P. E., Horsfall, J. G., and Lukens, R. J. 1972. EPIMAY, a simulator of southern corn leaf blight. *Conn. Agric. Exp. Stn. Bull.* 729.

Wald, A. 1947. *Sequential Analysis*, John Wiley and Sons, New York.

Ward, H. M. 1901. *Disease in Plants*, Macmillan, London.

Ward, J. S., and Ferrandino, F. J. 1999. New derivation reduces bias and increases power of Ripley's L index. *Ecol. Model* 116: 225–236.

Watson, D. J. 1947. Comparative physiological studies on the growth of field crops: I. Variation in net assimilation rate and leaf area between species and varieties, and within and between years. *Ann. Bot.* 11: 41–76.

West, J. S., Bravo, C., Oberti, R., Lemaire, D., Moshou, D., and McCartney, H. A. 2003. The potential of optical canopy measurement for targeted control of field crop diseases. *Annu. Rev. Phytopathol.* 41: 593–614.

Weinberger, H. F. 1978. Asymptotic behaviour of a model in population genetics. In: *Nonlinear Partial Differential Equations and Applications* (J. M. Chadam, editor). Springer, Berlin, pp. 47–98.

Weinberger, H. F. 1982. Long-time behaviour of a class of biological models. *SIAM J. Math. Anal.* 13: 353–396.

Wetherill, G. B. 1977. *Sampling Inspection and Quality Control*, 2nd Edn. Chapman and Hall, London.

Wheeler, T. A., Madden, L. V., Riedel, R. M., and Rowe, R. C. 1994. Distribution and yield-loss relations of *Verticillium dahliae, Pratylenchus penetrans, P. scribneri, P. crenatus,* and *Meloidogyne hapla* in commercial potato fields. *Phytopathology* 84: 843–852.

Whetzel, H. H. 1929. The terminology of phytopathology. *Proc. Second International Congress of Plant Science.* pp. 1204–1215.

White, K. A. J., and Gilligan, C. A. 1998. Spatial heterogeneity in three-species, plant-parasite-hyperparasite, systems. *Philos. Trans. R. Soc. Lond. B* 353: 543–557.

Williams, D. A. 1975. The analysis of binary responses from toxicological experiments involving reproduction and teratogenicity. *Biometrics* 31: 949–952.

Williams, D. A. 1982. Extra-binomial variation in logistic linear models. *Appl. Statist.* 31: 144–148.

Wilson, L. T., and Room, P. M. 1982. The relative efficiency and reliability of three methods for sampling arthropods in Australian cotton fields. *J. Aust. Ent. Soc.* 21: 175–181.

Wilson, L. T., and Room, P. M. 1983. Clumping patterns of fruit and arthropods in cotton, with implications for binomial sampling. *Environ. Entomol.* 12: 50–54.

Winkler, R. L. 1967. The quantification of judgement: Some methodological suggestions. *J. Am. Statist. Assoc.* 62: 1105–1120.

Winkler, R. L. 1985. Information loss in noisy and dependent processes. In: *Bayesian Statistics 2* (J. M. Bernardo, M. H. DeGroot, D. V. Lindley, and A. F. M. Smith, editors). Elsevier Science Publishers B. V., North-Holland, pp. 559–570.

Winkler, R. L., Smith, J. E., and Fryback, D. G. 2002. The role of informative priors in zero-numerator problems: Being conservative versus being candid. *Am. Statist.* 56: 1–4.

Wolfe, M. S. 1985. The current status and prospects of multiline cultivars and variety mixtures for disease resistance. *Annu. Rev. Phytopathol.* 23: 251–273.

Wolfinger, R. D. 1993. Covariance structure selection in general mixed models. *Commun. Statist. Simul. Computat.* 22: 1079–1106.

Wolfinger, R. D. 1996. Heterogeneous variance-covariance structures for repeated measures. *J. Agric. Biol. Environ. Statist.* 1: 205–230.

Wolfinger, R. D., and Lin, X. 1997. Two Taylor-series approximation methods for nonlinear mixed models. *Comput. Statist. Data Anal.* 25: 465–490.

Worlund, D. D., and Taylor, G. 1983. Estimation of disease incidence in fish populations. *Can. J. Fish Aquat. Sci.* 40: 2194–2197.

Wrather, J. A., Stienstra, W. C., and Koenning, S. R. 2001. Soybean disease loss estimates for the United States from 1996 to 1998. *Can. J. Plant Pathol.* 23: 122–131.

Wrather, J. A., Koenning, S. R., and Anderson, T. R. 2003. Effect of diseases on soybean yields in the United States and Ontario (1999 to 2002). *Plant Health Progress.* DOI: 10-1094/PHP-2003-0325-01-RV [http://www. apsnet. org].

Wright, T. 1990. When zero defectives appear in a sample: Upper bounds on confidence coefficients of upper bounds. *Am. Statist.* 44: 40–41.

Wu, B. M., van Bruggen, A. H. C., Subbarao, K. V., and Pennings, G. G. H. 2001. Spatial analysis of lettuce downy mildew using geostatistics and geographic information systems. *Phytopathology* 91: 134–142.

Wypij, D., and Santner, T. J. 1990. Interval estimation of the marginal probability of success for the beta-binomial distribution. *J. Statist. Comput. Simul.* 35: 169–185.

Xiao, C. L., Hao, J. J., and Subbarao, K. V. 1997. Spatial patterns of microsclerotia of *Verticillium dahliae* in soil and Verticillium wilt of cauliflower. *Phytopathology* 87: 325–331.

Xu, X.-M., and Madden, L. V. 2002. Incidence and density relationships of powdery mildew on apple. *Phytopathology* 92: 1005–1014.

Xu, X.-M., and Madden, L. V. 2003. Considerations for the use of SADIE statistics to quantify spatial patterns. *Ecography* 26: 821–830.

Xu, X.-M., and Madden, L. V. 2004. Use of SADIE statistics to study spatial dynamics of plant disease epidemics. *Plant Pathol.* 53: 38–49.

Xu, X.-M., and Madden, L. V. 2005. Interrelationships among SADIE indices for characterizing spatial patterns or organisms. *Phytopathology* 95: 874–883.

Xu, X.-M., and Ridout, M. S. 1998. Effects of initial epidemic conditions, sporulation rate, and spore dispersal gradient on the spatio-temporal dynamics of plant disease epidemics. *Phytopathology* 88: 1000–1012.

Xu, X.-M., and Ridout, M. S. 2000. Effects of quadrat size and shape, initial epidemic conditions, and spore dispersal gradient on spatial statistics of plant disease epidemics. *Phytopathology* 90: 738–750.

Xu, X.-M., and Ridout, M. S. 2001. Effects of prevailing wind direction on spatial statistics of plant disease epidemics. *J. Phytopathol.* 149: 155–166.

Xu, X.-M., Robinson, J. D., Berrie, A. M., and Harris, D. C. 2001. Spatio-temporal dynamics of brown rot (*Monilinia fructigena*) on apple and pear. *Plant Pathol.* 50: 569–578.

Yamamura, K., and Sugimoto, T. 1995. Estimation of the pest prevention ability of import plant quarantine in Japan. *Biometrics* 51: 482–490.

Yang, X., Madden, L. V., and Brazee, R. D. 1991. Application of the diffusion equation for modelling splash dispersal of point-source pathogens. *New Phytologist* 118: 295–301.

Yang, X.-B., and TeBeest, D. O. 1992. Dynamic pathogen distribution and the logistic increase of plant disease. *Phytopathology* 82: 380–383.

Youden, W. J. 1950. Index for rating diagnostic tests. *Cancer* 3: 32–35.

Yuen, J. 2002. Disease management decisions—making decisions that matter. In: *The BCPC Conference—Pests and Diseases 2002, Proceedings Volume 2,* The British Crop Protection Council, Farnham, UK, pp. 889–896.

Yuen, J. 2003. Bayesian approaches to plant disease forecasting. Online. *Plant Health Progress.* DOI: 10.1094/PHP-2003-1113-06-RV.

Yuen, J., Twengström, E., and Sigvald, R. 1996. Calibration and verification of risk algorithms using logistic regression. *Eur. J. Plant Pathol.* 102: 847–854.

Yuen, J. E., and Hughes, G. 2002. Bayesian analysis of plant disease prediction. *Plant Pathology* 51: 407–412.

Zadoks, J. C. 1971. Systems analysis and the dynamics of epidemics. *Phytopathology* 61: 600–610.

Zadoks, J. C. 1974. The role of epidemiology in modern Phytopathology. *Phytopathology* 64: 918–923.

Zadoks, J. C. 1981. Mr Duhamel's 1728 treatise on the violet root rot of saffron crocus: 'Physical explanation of a disease that perishes several plants in the Gastinois, and Saffron in particular'. Mededelingen van de Landbouwhogeschool (Wageningen) 81–87.

Zadoks, J. C. 1984. A quarter century of disease warning, 1958–1983. *Plant Dis.* 68: 352–355.

Zadoks, J. C. 2001. Plant disease epidemiology in the twentieth century: A picture by means of selected controversies. *Plant Dis.* 85: 808–816.

Zadoks, J. C., and Kampmeijer, P. 1977. The role of crop populations and their deployment, illustrated by means of a simulator, EPIMUL76. *Ann. N.Y. Acad. Sci.* 287: 164–190.

Zadoks, J. C., and Koster, L. M. 1976. A historical survey of botanical epidemiology: A sketch of the development of ideas in ecological Phytopathology. Mededelingen Landbouwhogeschool Wageningen (The Netherlands) 76–12.

Zadoks, J. C., and Schein, R. D. 1979. *Epidemiology and Plant Disease Management*, Oxford University Press, Oxford.

Zadoks, J. C., and van den Bosch, F. 1994. On the spread of plant disease: A theory on foci. *Annu. Rev. Phytopathol.* 32: 503–521.

Zadoks, J. C., Chang, T. T., and Konzak, C. F. 1974. A decimal code for the growth stages of cereals. *Weed Res.* 14: 415–421.

Zahareiva, T. D., de Milliano, W. A. J., and Zadoks, J. C. 1984. Powdery mildew on winter wheat in Bulgaria, 1980: Relations between disease incidence, disease severity and yield. Netherlands *J. Plant Pathol.* 90: 41–54.

Zarnoch, S. J., Anderson, R. L., and Sheffield, R. M. 1995. Using the β-binomial distribution to characterize forest health. *Can. J. For. Res.* 25: 462–469.

Zeger, S. L., and Liang, K.-Y. 1992. An overview of methods for the analysis of longitudinal data. *Statist. Med.* 11: 1825–1839.

Zeng, S. 1986. *Handbook of Plant Disease Epidemiology*, [in Chinese].

Zhang, X.-S., Holt, J., and Colvin, J. 2000a. Mathematical models of host plant infections by helper-dependent virus complexes: Why are helper viruses always avirulent? *Phytopathology* 90: 85–93.

Zhang, X.-S., Holt, J., and Colvin, J. 2000b. A general model of plant-virus disease infection incorporating vector aggregation. *Plant Pathol.* 49: 435–444.

Zucchini, W. 2000. An introduction to model selection. *J. Math. Psychol.* 44: 41–61.

Zweig, M. H., and Campbell, G. 1993. Receiver-operating characteristic (ROC) plots: A fundamental evaluation tool in clinical medicine. *Clin. Chem.* 39: 561–577.

Index

(*Page numbers in italics refer to figures.*)

A

B

C

E

I

J

K

L

M

N

O

Q

R

S